Springer Proceedings in Physics **402**

Indexed by Scopus

The series Springer Proceedings in Physics, founded in 1984, is devoted to timely reports of state-of-the-art developments in physics and related sciences. Typically based on material presented at conferences, workshops and similar scientific meetings, volumes published in this series will constitute a comprehensive up to date source of reference on a field or subfield of relevance in contemporary physics. Proposals must include the following:

- Name, place and date of the scientific meeting
- A link to the committees (local organization, international advisors etc.)
- Scientific description of the meeting
- List of invited/plenary speakers
- An estimate of the planned proceedings book parameters (number of pages/articles, requested number of bulk copies, submission deadline).

Please contact:

For Americas and Europe: Dr. Zachary Evenson; zachary.evenson@springer.com
For Asia, Australia and New Zealand: Dr. Loyola DSilva; loyola.dsilva@springer.com

Jeremy Witzens · Joyce Poon ·
Lars Zimmermann · Wolfgang Freude
Editors

The 25th European Conference on Integrated Optics

Proceedings of ECIO 2024, June 17–19, Aachen, Germany

Springer

Editors
Jeremy Witzens
RWTH Aachen University
Aachen, Nordrhein-Westfalen, Germany

Joyce Poon
Max Planck Institute of Microstructure Physics
Halle (Saale), Germany

Lars Zimmermann
Leibniz Institute for High Performance Microelectronics
Frankfurt (Oder), Brandenburg, Germany

Wolfgang Freude
Karlsruhe Institute of Technology
Karlsruhe, Baden-Württemberg, Germany

ISSN 0930-8989 ISSN 1867-4941 (electronic)
Springer Proceedings in Physics
ISBN 978-3-031-63377-5 ISBN 978-3-031-63378-2 (eBook)
https://doi.org/10.1007/978-3-031-63378-2

This Springer imprint is published by the registered company Springer Nature Switzerland AG
The registered company address is: Gewerbestrasse 11, 6330 Cham, Switzerland

Introduction to the Proceedings of ECIO 2024

Ein herzliches Willkommen to the **25th edition of the European Conference on Integrated Optics (ECIO)** to be held at RWTH Aachen University from 17th to 19th June 2024!

ECIO has rotated between European hubs of integrated photonics research since 1981. Until 2016, it was held every two years, and since then every year, to better respond to the growing interest in this field. Every year, ECIO brings together experts from Europe and around the world working on diverse integrated photonics platforms, such as silicon, III-V compound semiconductors, and polymers.

Aachen is home to several university institutes dedicated to optics and photonics, such as the Institute of Integrated Photonics (IPH) of the Faculty of Electrical Engineering, the Chair for Technology of Optical Systems (TOS), and the Chair for Laser Technology (LLT) from the Faculty of Mechanical Engineering, and the 1st Institute of Physics. The Fraunhofer Institute of Laser Technology (ILT), AMO GmbH, and the nearby Forschungszentrum Jülich provide a large-scale research infrastructure in this field. These research organizations are complemented by a rich ecosystem of photonics start-ups and larger photonics companies located in Aachen.

This year's ECIO is a collaborative effort between organizers from several German institutions: RWTH Aachen, Karlsruhe Institute of Technology (KIT), Technical University of Berlin (TUB), Leibniz Institute for Innovative Microelectronics (IHP), and Max Planck Institute for Microstructure Physics (MPI-Halle).

We received 174 contributed submissions, of which 60 have been accepted for oral presentation and 58 for poster presentation, corresponding roughly to an overall acceptance rate of 2/3, evenly split between the two categories. These presentations will build on the impetus provided by our 5 keynote and 20 invited speakers.

Some statistics are noteworthy to reflect the current interests in the photonics research community: 11 talks will be on quantum photonics, including systems based on squeezed states, single photon emitters, and ion traps, closely followed by programmable photonics with 10 talks on neuromorphic computing, non-volatile phase shifters, and polarization/mode diversity management, as well as 10 talks on hybrid laser integration. Other areas of interest include nonlinear photonics (7 talks), photonic integrated circuits for free-space communications and sensing, optoelectronic devices, microwave photonics, visible light photonics and sensing, thin-film lithium niobate (with 6 talks in each category), photon-phonon interaction (5 talks) and optical MEMS (4 talks). The remaining papers are divided between assembly and test, datacom systems, and polymer photonics. We are also pleased to see a greater participation from industry in ECIO this year, both through invited speakers and from received submissions.

We gratefully acknowledge the support of our Technical Program Committee, who nominated invited speakers and rigorously reviewed the submissions, as well as the support of our exhibitors and sponsors, without which the conference would not be possible in its current form.

Jeremy Witzens
General Chair

Joyce Poon
Co-chair

Lars Zimmermann
Technical Program Committee Chair

Wolfgang Freude

Organization

Technical Program Committee

Pavel Cheben
Daoxin Dai
Carlos Errando-Herranz
Dominic Gallagher
Sangsik Kim
David Marpaung
Delphine Marris Morini
Shayan Mookherjea
Francesco Morichetti
Pascual Muñoz
Richard Penty
Günther Roelkens
Isaac Sackey
Nicola Sambo
Pablo Sanchis Kilders
Anna-Lena Schall-Giesecke
Maziar P. Nezhad
Elizaveta Semenova
Stephan Suckow
Nathalie Vermeulen
J. Gonzalo Wangüemert-Perez
Kevin Williams
Hon Ki Tsang
Lars Zimmermann (TPC-Chair)

Sponsors and Exhibitors

LIGENTEC	**LIGENTEC** is a B2B company manufacturing Photonic Integrated Circuits (PIC) for customers in high-tech areas such as communication on ground and space, quantum technologies, LiDAR, and biosensors. LIGENTEC commercializes the all-nitride-core technology. With the low loss offering of PICs with active components, LIGENTEC enables the customers to develop their products in the Industrial Revolution 4.0. LIGENTEC is a European company strategically headquartered in close proximity to the prestigious Swiss Federal Institute of Technology (EPFL) in Lausanne, Switzerland, with offices located near X-Fab in Corbeil-Essonnes, France. Our geographical presence extends worldwide, with our cutting-edge chips finding applications across the globe. In partnership with a major European volume partner, we are well-equipped to supply large-scale markets, while ensuring efficient and effective delivery of our innovative solutions.
LioniX INTERNATIONAL	**LioniX International** is a leading global provider of silicon nitride photonic integrated circuits (PICs), modules, and MEMS—since 2001. Their expertise covers chip fabrication, electro-optical system architecture, and module assembly and packaging. From design to device, their vertically integrated approach supports their customers during the whole product development cycle.
Photon Design.	**Photon Design** was started in 1992 and now provides a wide range of innovative photonics CAD tools to 35+ countries around the world, supplying most of the world's leading photonics companies, universities, and government research labs. CAD products include tools for both passive and active (semiconductor) components and optical circuit modeling. The company has a team of the brightest people in photonics modeling, ready to help you develop innovative solutions for tomorrow's photonics applications.

SIEMENS EDA	**Siemens EDA**, represented by EDA Solutions, is a world leader in electronic hardware and software design solutions for the world's most successful electronic, semiconductor, and systems companies. The pace of innovation in electronics is constantly accelerating. To enable customers to deliver life-changing innovations faster and become market leaders, we are committed to delivering the world's most comprehensive portfolio of electronic design automation (EDA) software, hardware, and services. The company's IC Design solutions include the specialty silicon photonics solution L-Edit Photonics which is complemented by the Caliber Platform for equation-based design rules. Photonic simulation is supported through integration with our partners: Ansys, Luceda, Optiwave, and VPIphotonics.
THORLABS	**THORLABS**, a vertically integrated photonics products manufacturer, was founded in 1989 to serve the laser and electro-optics research market. As that market has spawned a multitude of technical innovations, Thorlabs has extended its core competencies in an effort to play an ever-increasing role to serve the Photonics Industry at the research end, as well as the industrial, life science, medical, and defense segments. The organization's highly integrated and diverse manufacturing assets include semiconductor fabrication of Fabry–Perot, DFB, and VCSEL lasers; fiber towers for drawing both silica and fluoride glass optical fibers; MBE/MOCVD epitaxial wafer growth reactors; extensive glass and metal fabrication facilities; advanced thin-film deposition capabilities; and optomechanical and optoelectronic shops.
YOKOGAWA	Meet the precision makers! **Yokogawa Test & Measurement** is the world's most trusted measurement partner. With over 100 years of experience, we have built our reputation on understanding the needs of scientists and engineers that require both the accuracy of a measurement system and its precision where stability and reproducibility ensure that repeated measurements under unchanged conditions show the same results. Widely used in communications R&D for their precision and sensitivity, our OSAs are used in many fields, from industrial production and LED lighting to biophotonics, healthcare, and quantum computing. Yokogawa's OSAs are also valued by scientific researchers for their superior performance compared to instruments such as spectrometers.

chilas	**Chilas** harvests the best of both Si_3N_4 and the InP worlds, resulting in external cavity tunable lasers with ultra-narrow linewidth. The lasers can be tuned over the C-band covering a wavelength range larger than 100 nm and have a linewidth of <5 kHz. Chilas lasers have a unique combination of characteristics including ultra-narrow linewidth, broad and continuous wavelength tuning, high output power, and low noise. Laser sources operating at new wavelengths are developed constantly expanding the portfolio from visible wavelengths to just beyond the L-telecom band at 1700 nm. Chilas lasers can be used in a number of applications such as telecom, quantum key distribution, and advanced optical sensing including fiber sensors and LiDAR.
AMO	**AMO**'s mission as a research-oriented company is to efficiently close the gap between university research and industrial application. For this purpose, AMO identifies those topics from basic research that seem particularly suitable for industrial implementation and demonstrates these in application-oriented technology. In joint projects and bilateral cooperation, research and development results are transferred to industry for maintenance and creation of jobs. Thus, nanotechnology is expected to provide considerable potential for application areas such as information technology, biotechnology, and environmental technology. Headed by Prof. Max Lemme, AMO operates a high level 400 m^2 cleanroom. Furthermore, AMO offers a range of services from consulting to prototype development.
Fraunhofer HHI	**Fraunhofer HHI** in Berlin is a world-leading research institution specializing in optical communication networks, photonic components, and fiber optical sensor systems. With a focus on advancing optical technology for telecommunications, sensing, and quantum communications, HHI conducts cutting-edge research and development in areas such as fiber optics, integrated photonics, terahertz spectroscopy, and high-speed data transmission. Its work encompasses the design of photonic components with high frequencies >100 GHz for advanced optical networks, including lasers, modulators, and detectors. Moreover, hybrid photonic integrated circuits are developed based on PolyBoard, SiN, and thin-film lithium niobate single-mode waveguides in combination with active elements made of InP, GaAs, or GaN. Fraunhofer HHI drives the development of next-generation photonic technologies and enables new areas of use like terabit transmitters for data centers or functionalized micro-ring resonator PICs for innovative point-of-care diagnostics.

	Optica, Advancing Optics and Photonics Worldwide, is the society dedicated to promoting the generation, application, archiving, and dissemination of knowledge in the field. Founded in 1916, it is the leading organization for scientists, engineers, business professionals, students, and others interested in the science of light. Optica's renowned publications, meetings, online resources and in-person activities fuel discoveries, shape real-life applications and accelerate scientific, technical and educational achievement.
	PHIX is a world-leading packaging and assembly foundry for photonic integrated circuits (PICs), building optoelectronic modules based on all major PIC technology platforms (such as indium phosphide, silicon photonics, silicon nitride, and planar lightwave circuits) in scalable manufacturing volumes. We specialize in chip-to-chip hybrid integration, coupling to fiber arrays, and interfacing of DC and RF electrical signals. By offering our knowledge already at the chip design stage, we ensure ease of scale-up for volume manufacturing and provide a one-stop shop for PIC assembly.

AIXEMTEC
a schunk company

Automation Specialist in Precise Assembly and Testing Machines in Photonics, Part of the Schunk Group:
AIXEMTEC is a leading company in the field of high-precision assembly and testing technologies for photonic components. As part of the Schunk Group since January 1, 2024, AIXEMTEC benefits from a strong financial and technological foundation, enabling it to offer innovative solutions in a global context. The Schunk Group, a German foundation with a broad international presence, provides AIXEMTEC with a network of 65 locations in 26 countries.
Our main product, the Fiberbeast Gen. 2, is an advanced platform for the assembly and testing of fiber optics and photonically integrated circuits (PICs). These systems are essential for the production of products in the fields of telecommunications, datacom, and advanced quantum technologies. For example, the Fiberbeast Gen. 2 enables the highest coupling efficiencies even with the smallest mode field diameters, making it particularly suitable for precise alignment and coupling in single-mode fiber optic systems.
The machine platform of AIXEMTEC is characterized by its modular architecture, allowing flexible adaptation to various assembly and testing requirements. Through the use of collaborative micromanipulators and advanced sensor systems, including computer vision and non-contact height sensors as well as power measurement heads or spectrometers, the system supports precise and repeatable assembly and testing processes. These technologies are crucial for achieving submicrometer accuracy in positioning and thus optimizing the optical performance of photonic assemblies and are deployed in AIXEMTEC's desktop machines, stand-alone production machines, and automated parallelized production lines.
Integration into the Schunk Group expands AIXEMTEC's capabilities in global services and support, enabling the company to quickly and efficiently respond to customer requirements. This strategic partnership not only ensures the long-term development of AIXEMTEC but also provides our customers with the security and support of a globally positioned technology group.

Contents

Programmable Photonics

Optical MEMS

Tuesday Poster Session

Wednesday Poster Session

Lasers and Amplifiers

Widely Tunable External Cavity Laser Across the 1634-1777 nm Spectrum with Sub-kHz Linewidth

Fathema Farjana(✉), Albert van Rees, and Dimitri Geskus

Chilas BV, Hengelosestraat 500, Enschede, The Netherlands
f.farjana@chilasbv.com

Abstract. The wavelength region of around 1700 nm has a myriad of applications including gas sensing (i.e. CH_4, C_2H_2, C_2H_4, C_3H_8), spectroscopy, LIDAR, and medical diagnostics. In this work we report a chip-integrated laser for these applications, having a wide tuning range from 1634–1777 nm, through the hybrid integration of an off-the-shelf gain chip and a low-loss wavelength-tunable mirror integrated on a Si_3N_4 chip. The laser shows an output power up to 7.7 mW and an intrinsic linewidth of 0.54 kHz.

Keywords: Hybrid integration · silicon nitride · gas sensing · tunability · linewidth

1 Introduction

Semiconductor lasers made through the hybrid integration of semiconducting optical amplifiers (SOAs) with photonic integrated chips (PICs) exhibit unique characteristics. In particular, integration of an SOA with a Si_3N_4-based feedback chip enables wide tuning and very narrow intrinsic linewidths [1]. Although such lasers are well developed for telecom applications in the 1550 nm wavelength range, extension towards the 1700 nm range [2] has started because it is of interest for several application reasons. This eye safe wavelength region is suitable for application in LIDAR systems [3]. Gases like CH_4, C_2H_2, C_2H_4, and C_3H_8 can be detected using absorption lines in this range [4]. Also, this laser is viable for ultrahigh-resolution optical coherence tomography (UHR-OCT) [5], for high absorption and penetration of 1700 nm by human tissue with lipids, while other uses include eye surgery [6].

In this study we demonstrate a hybrid integrated and widely tunable extended cavity diode laser (ECDL), functioning within the range of 1634–1777 nm. The laser consists of a gain chip coupled to a feedback circuit with two micro-ring resonators (MRRs). We demonstrate wide tuning over a range of more than 140 nm. The laser exhibits a side-mode suppression ratio (SMSR) better than 44 dB and an ultra-narrow intrinsic linewidth of 0.54 kHz.

J. Witzens et al. (Eds.): ECIO 2024, SPPHY 402, pp. 3–7, 2024.
https://doi.org/10.1007/978-3-031-63378-2_1

2 Design and Simulation

The design of the hybrid integrated tunable laser diode is illustrated in Fig. 1. The laser comprises a SOA serving as the gain medium and a Si_3N_4-based waveguide circuit forming the wavelength-tunable feedback circuit. To operate in the 1700 nm range, we use an off-the-shelf SOA (Thorlabs SAF1091). The role of the feedback circuit is twofold: firstly, it provides frequency-selective feedback to enable single-frequency operation, and secondly it extends the cavity length with minimal losses, thereby increasing the photon lifetime and reducing the laser's intrinsic linewidth. Single-frequency operation is achieved through two interconnected MRRs with slightly different free spectral range (FSR) in the cavity, leveraging the Vernier effect.

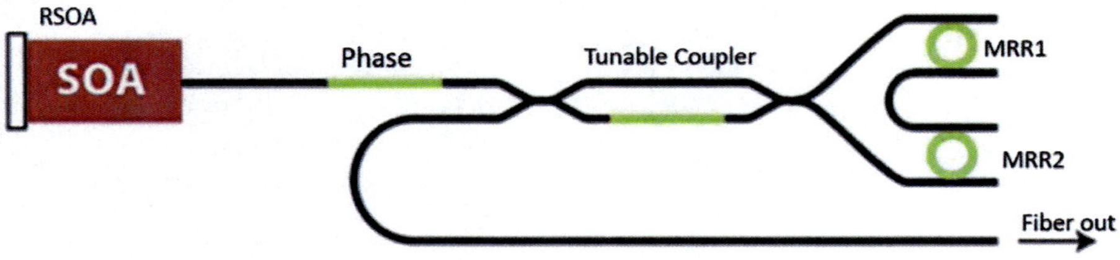

(a)

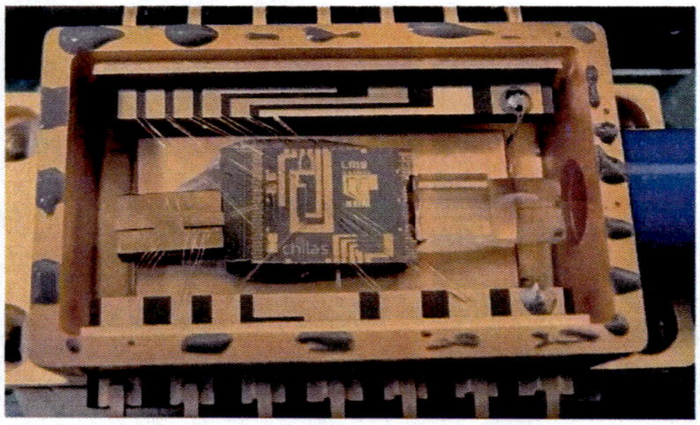

(b)

Fig. 1. (a) Schematic design of the laser with a semiconductor optical amplifiers (SOA), coupled to a feedback chip based on silicon nitride waveguides, shown with the black lines, and with heaters, as shown in green, on top of rings, phase section and a tunable coupler. (b) Photograph of the laser module in a 14-pin butterfly package. Both chips and the output fiber are all assembled together, and the heaters are wire bonded to the pins of the butterfly package.

The TriPleX™ platform of silicon nitride (Si_3N_4) waveguides embedded in a silicon oxide (SiO_2) cladding was chosen as the waveguide platform for the feedback chip, because it provides low propagation losses (<0.1 dB/cm) and has a wide transparency range extending to 2350 nm [7]. Moreover, this platform facilitates lateral tapering of the waveguides at the facet to achieve efficient coupling with the gain chip and a single-mode fiber output.

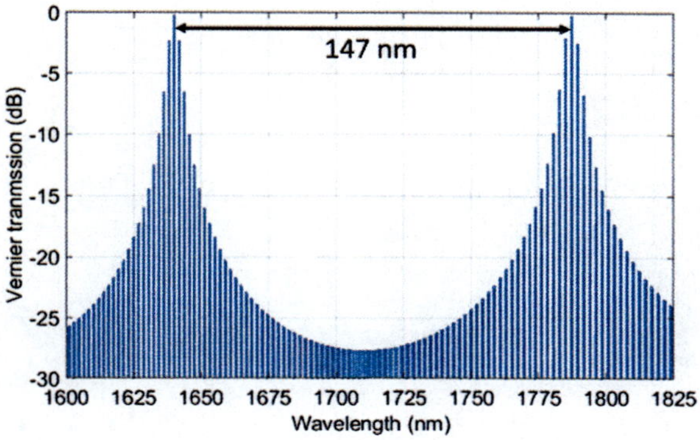

Fig. 2. The simulated Vernier transmission of the design spanning from around 1640–1787 nm (Vernier FSR is 147 nm)

As shown in Fig. 1(a) heaters are added for phase tuning by applying voltage on them which results in local refractive index change of the waveguides. The MRRs can be tuned via these heater elements to enable targeting wavelengths across the Vernier filter range. Furthermore, a heater atop the cavity waveguide, termed the cavity phase section, can be used to align cavity resonance with ring resonances for single-mode and high-power laser operation. Additionally, a tunable out-coupler controls the fraction of light coupled out from the cavity to the fiber output. The feedback circuit of the design consists of two mirroring resonators with circumferences of 930 and 943.108 μm respectively, to provide a Vernier filter range of 147 nm around the central wavelength of 1700 nm. In Fig. 2 the simulated vernier FSR is shown for the design.

The gain chip and feedback chip, together with the fiber array are placed in a compact 14-pin butterfly package as shown in Fig. 1(b).

3 Experimental Characterization

To characterize the tuning range of the laser we used an optical spectrum analyzer (Yokogawa (AQ6374)). Figure 3(a) depicts the measured spectra between 1638–1743 nm. The measured power was calibrated using an optical power meter. Tuning was achieved by varying the heater voltage on the rings. For achieving single mode operation and maximum power, the rings, phase section and tunable out-coupler was fine-tuned. The laser was operated at a pump current of 250 mA and the temperature was maintained at 21 °C for the whole measurement. From Fig. 3(a) it can be seen that the laser has a SMSR between 44–62 dB. Also, the laser has shown an output power up to 7.7 mW for higher frequencies (Fig. 3(a)).

Because the OSA could not record tuning spectra beyond 1750 nm wavelength, we used a separate spectrometer (Horiba IHR550 series with a PbS sensor and set to low resolution configuration) to record additional spectra over the full tuning range of the laser. These spectra from 1634–1777 nm are shown in Fig. 3(b).

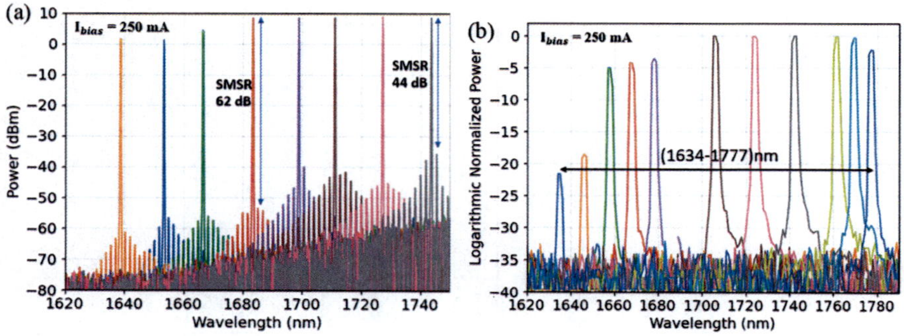

Fig. 3. (a) Superimposed tuning spectra of the laser from 1638–1743 nm. Here, the measured power was calibrated using a power meter. (b) Full tuning spectra of the laser 1634–1777 nm is recorded with a spectrometer, which is around the designed mirror FSR of 147 nm.

To characterize the frequency stability of the laser, the frequency noise of the laser was measured using a High-Finesse LWA 1k-1550. For this measurement, the pump current was again set to 250 mA. Fine-tuning of the ring resonators and phase section was done to ensure that the laser operates in a single mode. The measured power spectral density of the frequency noise, as function of Fourier frequencies, is displayed in Fig. 4. At lower Fourier frequencies, the noise is following an inverse relation with the Fourier frequency, typical for technical noise. Beyond 1 MHz frequency, the noise levels out to white noise with an average value of 169 Hz2/Hz. Multiplying this value by π gives intrinsic (Schawlow-Townes) linewidth of the laser of 0.54 kHz.

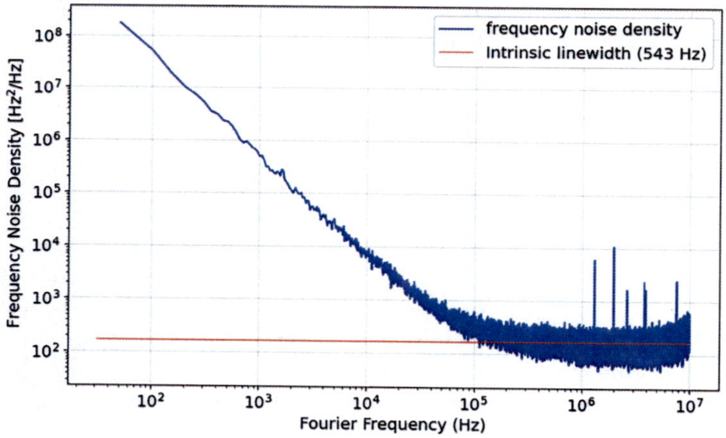

Fig. 4. Laser frequency noise power spectral density (PSD) as function of Fourier frequencies, averaged over 10 measurements. At low Fourier frequencies, the noise follows an inverse trend, which is typical for technical noise. As Fourier frequencies increase, the noise stabilizes to a white noise level of 169 Hz2/Hz, indicating an intrinsic laser linewidth of 0.54 kHz.

4 Conclusion

The extensive application range of the 1700 nm wavelength has spurred numerous research efforts aimed at developing lasers within this range. In this study, we present a compact and robust, widely tunable laser capable of covering this application-rich spectrum from 1634 nm to 1777 nm, while delivering an ultra-narrow linewidth of 543 Hz. We achieved a maximum output power of 7.7 mW with a side mode suppression ratio (SMSR) exceeding 44 dB. These results highlight the laser's potential to serve as a versatile tool for various applications such as gas sensing, LiDAR, and medical diagnostics. Additionally, its wavelength agility makes it well-suited for spectroscopy purposes.

Acknowledgements. We thank Dawson Bonneville from the integrated optical systems group University of Twente/Aluvia photonics for his support on using the spectrometer. And we thank Ronald Dekker from LioniX International for his swift calculations on the waveguide properties.

References

1. Boller, K.-J., et al.: Hybrid Integrated Semiconductor Lasers with Silicon Nitride Feedback Circuits. Photonics **7**(1), 4 (2019). Author, F.: Article title. Journal 2(5), 99–110 (2016)
2. Brian Sia, J.X., et al.: Sub-kHz linewidth, hybrid III-V/silicon wavelength-tunable laser diode operating at the application-rich 1647–1690 nm. Opt. Express **28**(17), 25215–25224 (2020)
3. Wang, K., et al.: 1645 nm coherent Doppler wind lidar with a single-frequency Er-YAG laser. Opt. Express **28**(10), 14694–14704 (2020)
4. Zheng, K., et al.: Near-infrared broadband cavity-enhanced spectroscopic multigas sensor using a 1650 nm light emitting diode. ACS Sens. **4**(7), 1899–1908 (2019)
5. Cadroas, P., et al.: All-fiber femtosecond laser providing 9 nJ, 50MHz pulses at 1650 nm for three-photon microscopy. J. Opt. **19**(6), 065506 (2017)
6. Crotti, C., et al.: Wavelength optimization in femtosecond laser corneal surgery. Invest. Ophthalmol. Visual Sci. **54**(5), 3340–3349 (2013)
7. Roeloffzen, C.G.H., et al.: Low-Loss Si3N4 TriPleX optical waveguides: technology and applications overview. IEEE J. Sel. Top. Quantum Electron. **24**, 4400321 (2018)

External Cavity 637-nm Laser with Increased RSOA-to-PIC Alignment Tolerance and a Filtered Sagnac-Loop Reflector with Single Output Waveguide

Georgios Sinatkas[1(✉)], Arijit Misra[1,2], Florian Merget[1], and Jeremy Witzens[1]

[1] Institute of Integrated Photonics, RWTH Aachen University, 52074 Aachen, Germany
gsinatkas@iph.rwth-aachen.de

[2] Cisco Optical GmbH, Nordostpark, 90411 Nuremberg, Germany

Abstract. The design of a 637-nm wavelength, photonic-integrated-circuit-based external cavity laser (PIC-based ECL) aimed at quantum technology applications is presented together with first experimental results. The PIC is designed to provide relaxed alignment tolerance for coupling to a reflective semiconductor optical amplifier (RSOA) gain chip. This is achieved by using a multi-mode edge coupler (MMEC) in place of the usually employed single-mode coupling schemes. A 1-dB-penalty misalignment tolerance of up to ± 2.4 μm can be achieved in the plane of the chip, creating a path towards reliable flip-chip integration at short wavelengths. The power coupled to the PIC is fed to a Sagnac-loop reflector, filtered by a pair of ring resonators operated in Vernier configuration for providing the required frequency selective optical feedback. The ring resonators are designed to have different loaded Q-factors and they are asymmetrically coupled to bus and drop waveguides with suitably engineered directional couplers to provide single output waveguide emission. Moreover, requirements for high output power and narrow linewidths are balanced. Finally, preliminary measurements strongly suggest lasing in the fabricated devices, with further performance optimization being currently carried out.

Keywords: Semiconductor lasers · external cavity lasers · silicon nitride · photonic integrated circuits · semiconductor optical amplifier

1 Introduction

Photonic-integrated-circuit-based external cavity lasers (PIC-based ECLs) are attracting increased attention due to their reduced footprint, narrow linewidth, and tunability, which make them ideal sources for on-chip applications. Several PIC-based ECLs have been demonstrated in the infrared regime [1–3] for use in coherent communications, sensing, and metrology. However, there is a growing interest for similar demonstrations in the visible regime [4, 5] to serve the emerging fields of bio-photonics and quantum

J. Witzens et al. (Eds.): ECIO 2024, SPPHY 402, pp. 8–16, 2024.
https://doi.org/10.1007/978-3-031-63378-2_2

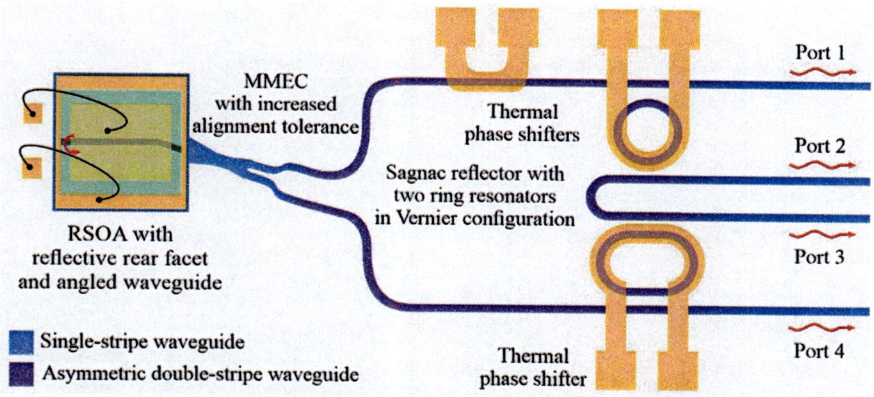

Fig. 1. Schematic representation of the ECL architecture.

technology. In this work, a PIC-based ECL at 637 nm is presented, that is intended for exciting nitrogen vacancy centers in diamond, one of the most promising platforms for quantum information processing. In the following sections, the design of the PIC is presented together with preliminary ECL measurements.

2 Design

With reference to Fig. 1, the ECL is configured as a Fabry–Pérot cavity between the broadband, highly reflective back facet of an RSOA and a frequency selective reflector on the PIC. A highly misalignment-tolerant multi-mode edge coupler (MMEC) is used for the RSOA-to-PIC coupling, designed to equally split the coupled power between two single-mode on-chip waveguides, irrespectively of the in-plane misalignment between the two chips [6]. These are looped back onto each other to form a Sagnac loop, which provides a filtered reflection by interposing a pair of thermally tuned ring resonators with slightly different free-spectral ranges (FSRs), configured in an add-drop structure. By spectrally aligning their resonances, a broader effective FSR is achieved (Vernier effect), resulting in a single resonance in the gain bandwidth of the RSOA, facilitating single-mode lasing operation. The ring resonators are asymmetrically side coupled by suitably configured directional couplers, having, thus, different loaded Q-factors [7], which allow for single-output operation and for balancing between high-output-power and narrow-linewidth requirements.

2.1 Silicon Nitride Platform

Silicon-nitride (SiN) technology is selected due to its low absorption at visible wavelengths. Two waveguide types are employed: a single-stripe (SS) and an asymmetric double-stripe (ADS) geometry, embedded in a silicon dioxide (SiO_2) cladding [Fig. 2(a), (b)]. The thickness of the bottom SiN stripe is set equal to 27 nm to achieve optimal overlap between the RSOA-beam profile and the mode profile of the MMEC in the vertical direction (Sect. 2.2). The top SiN layer and the intermediate SiO_2 have thicknesses

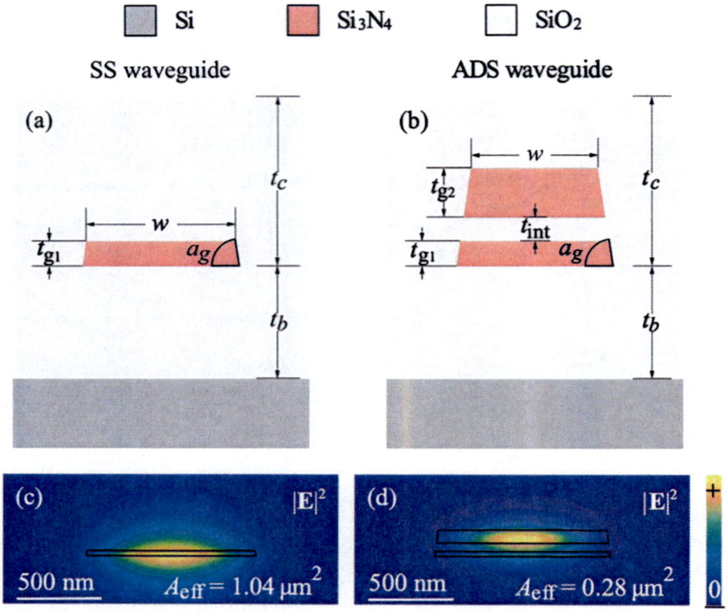

Fig. 2. (a) SS and (b) ADS waveguide with $t_{g1} = 27$ nm, $t_{int} = 50$ nm, $t_{g2} = 80$ nm, $a_g = 82°$, $t_c = 4$ μm, $t_b = 8$ μm. (c), (d) Respective TE_0 mode.

of 80 nm and 50 nm, respectively. The etching angle a_g equals 82°. The components were designed into the process of LioniX International [8].

The weak confinement of the SS waveguide is suitable for matching the large beam profile of RSOAs and optical fibers [Fig. 2(a)], while the ADS waveguide offers tighter confinement [Fig. 2(b)] and it is favorable for bends and ring resonators. The transition between the two waveguide types is achieved using vertical tapers, interposed at the two (four) output ports of the MMEC (chip) [Fig. 1].

Waveguide loss stems mainly from scattering at the side SiN waveguide walls due to roughness after etching. To mitigate this effect, the interconnect waveguide width is selected wide enough to minimize the overlap of the mode with the side walls, while maintaining single-mode operation. Based on the dispersion diagrams in Fig. 3 and considering TE operation, the single-mode waveguide width for the SS (ADS) waveguide is selected equal to 1.5 μm (0.55 μm). In the MMEC, the narrowest width is set to 2.5 μm, which supports both TE_0 and TE_1 modes as required for its operation.

2.2 Components

Alignment Tolerant Multi-mode Edge Coupler. Compared to its single-mode counterparts, the MMEC relaxes the requirements for alignment in the plane of the PIC. This is particularly useful when a complete PIC-RSOA integration is attempted, using, e.g., the flip-chip bonding technique with vertical self-alignment [9, 10]. The MMEC is designed into the SS waveguide, considering the beam characteristics of the RSOA

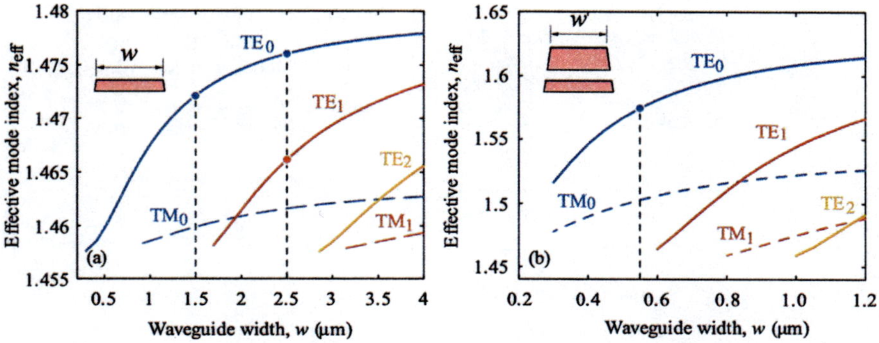

Fig. 3. Mode dispersion diagrams for the (a) SS and (b) ADS waveguide as a function of width.

(EXALOS AG). The latter has an elliptical profile with a mode field diameter (MFD) equal to 3.85 μm (0.82 μm) in the horizontal (vertical) direction [Fig. 4(a)], which is optimally matched by the TE_0 mode of an SS waveguide with a thickness of 27 nm and a width of 4.5 μm [Fig. 4(b)].

In case of misalignment [Fig. 4(c)], part of the emitted RSOA power couples to higher order modes due to the large MMEC width. Figure 4(d) shows the increase in the horizontal misalignment tolerance from ±0.8 μm to ±1.5 μm, defined at 1-dB drop from the peak coupling-efficiency, obtained if only the power coupled to the next higher order mode (TE_1) is recovered. This is achieved by incorporating a 1x2 multi-mode interference (MMI) splitter as part of the MMEC (Fig. 5), with a length that ensures that the TE_0 and TE_1 modes reach the output waveguides in quadrature to equally split the power. The narrowest L_2-long section of the MMEC is chosen to filter out modes with orders higher than TE_1. The misalignment tolerance can be further increased to ±2.4 μm for a 7-μm wide MMEC [Fig. 4(d)], with a slight IL penalty occurring at $\Delta x = 0$, but equally high peak coupling efficiency at slight lateral displacements. The calculations in Fig. 4(d) are performed using overlap integrals between the TE_0, TE_1 waveguide modes and the RSOA beam profile. The two designs of the MMEC are listed in Table 1.

Table 1. MMEC designs with different 1-dB misalignment tolerance (units: microns).

Design	w_1	w_2	L_1	L_2	L_3	L_4	L_5	s	Tolerance
#1	4.5	4.0	40	29.273	30	6	100	2	±1.5
#2	7.0	4.0	50	16.012	30	6			±2.4

Ring-Resonator-based Sagnac-Loop Reflector. This component is responsible for providing the required frequency selective feedback. It employs the ADS waveguide due to its tighter confinement, allowing for smaller bend radii and, thus, higher FSR. Setting the limit of maximum bend loss to 0.01 dB/cm, which is two orders of magnitude lower than the expected propagation loss (1 dB/cm), a minimum radius of 25 μm is

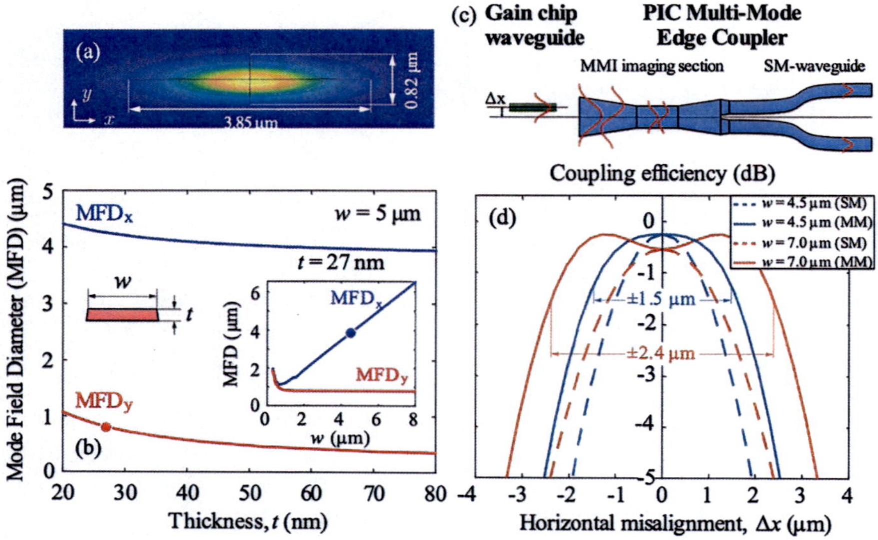

Fig. 4. (a) RSOA beam and near-field MFDs in the horizontal and vertical direction. (b) MFD for the TE_0 mode of the SS waveguide as a function of thickness and width (inset). (c) Schematic of the RSOA-to-PIC coupling for horizontal misalignment Δx. (d) Coupling efficiency between the RSOA beam and the TE_0, TE_1 modes of a 4.5-µm and 7-µm wide SS edge coupler as a function of Δx.

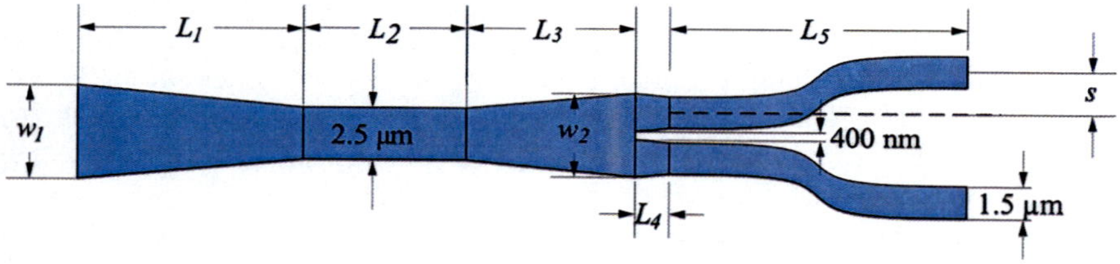

Fig. 5. Schematic of the MMI splitter with all pertinent dimensions labeled.

considered for the 550 nm wide waveguide [Fig. 6(a)], resulting in an FSR ≈ 1.42 nm and approximately four resonances in the 6-nm RSOA gain bandwidth.

To suppress the extra resonances, the Vernier effect is employed in two serially cascaded rings with slightly different radii and FSRs (1.41 nm and 1.31 nm). By tuning the ring resonances to the desired wavelength using thermal phase shifters [Fig. 6(b)], the effective FSR of the total system increases to $FSR_A \times FSR_B / |FSR_A - FSR_B| = 17$ nm, which is wide enough to allow for a single resonance in the gain spectrum of the RSOA.

In the literature, identical directional couplers are usually employed. In this work, an asymmetric design is suggested, optimally engineered to route the majority of the output power to a single output port. Such a design is expected to minimize undesired back-reflections from the waveguide ends that could destabilize the laser operation. In addition, it facilitates the integration of the PIC with fiber optics and isolators.

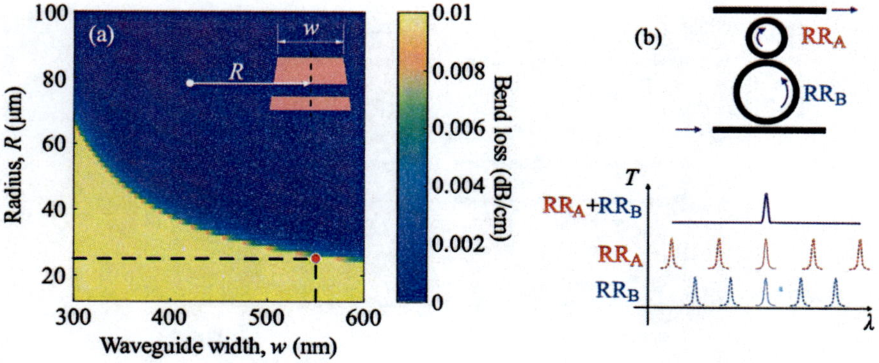

Fig. 6. (a) Bend losses for the ADS waveguide as a function of its width and bend radius. (b) Vernier effect in a pair of rings with slightly different FSRs in cascade configuration. The resulting add-drop filter features a wider FSR.

For the design, the circuit model representation of the add-drop filter is employed [11]. With reference to Fig. 7(a), Port 1 is selected as the nominal output port. The power exiting Port 4 [Fig. 7(b)] can be theoretically zeroed out if the transmission amplitude coefficients t_{B4}, t_{B3} for the respective directional couplers with gaps g_{B4}, g_{B3} satisfy the critical-coupling condition [11],

$$t_{B4} = a_B\, t_{B3}, \tag{1}$$

with a_B the round-trip amplitude attenuation for $\mathrm{RR_B}$. Considering lossless couplers, Eq. (1) can be equivalently expressed using the respective coupling coefficients as

$$\kappa_{B4} = \sqrt{1 - (1 - \kappa_{\mathrm{B3}}^2)a_B}. \tag{2}$$

Similarly, the power reaching Port 2 in Fig. 7(b) can be fully suppressed as long as

$$\kappa_{\mathrm{A2}} = \sqrt{1 - (1 - \kappa_{\mathrm{A1}}^2)a_A}\,. \tag{3}$$

Consequently, the power coupled to Ports 1 and 3 as well as the total optical feedback are rendered functions of κ_{A1} and κ_{B3}.

This still leaves the issue of minimizing the power being coupled to Port 3 [Fig. 7(a)]. This is addressed by designing $\mathrm{RR_B}$ to have a much lower loaded Q-factor (~31K) than $\mathrm{RR_A}$ (~300K). Providing $\mathrm{RR_B}$ with high waveguide coupling coefficients such that the intrinsic losses play a lesser role, on the one hand, allows for both coupled waveguides to experience close-to-critical coupling, minimizing the power lost to Port 3. Providing $\mathrm{RR_A}$ with a high Q-factor, on the other hand, allows for better filtering and lower linewidth operation.

As expected, the numerical optimization in Fig. 8(a) and (b) confirms that the power reaching Port 1 (Port 3) becomes maximum (minimum) for low κ_{A1} (high κ_{B3}) values. The on-resonance transmission to Port $N = \{1, 3\}$ is reduced to the analytical form

$$T_N = \left(\frac{a_X^2 - 1}{t_{XN} a_X^2 - \frac{1}{t_{XN}}} \right)^2, \tag{4}$$

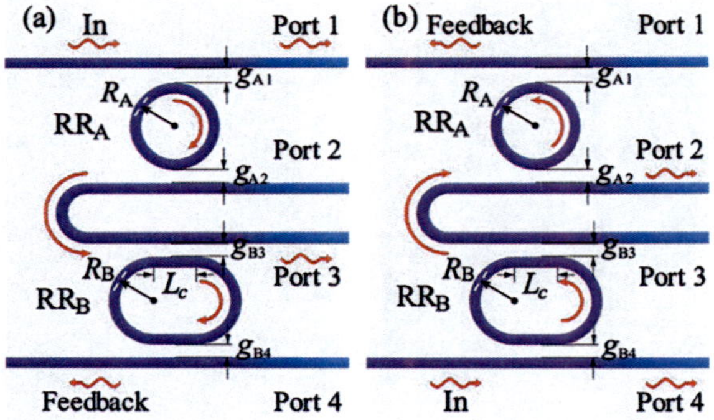

Fig. 7. Wave propagation through the ring-resonator-based Sagnac-loop reflector in the two possible directions.

with $X = \{A, B\}$ indicating the respective side-coupled resonator, after substituting Eqs. (2) and (3) in the formula for the through-port transmission of an add-drop ring-resonator filter [11]. Fixing κ_{B3} at 0.2, the coupling coefficient κ_{A1} can be varied to achieve the required balance between output power [Fig. 8(a)] and optical feedback [Fig. 8(c)] or, equivalently, ECL linewidth [Fig. 8(d)]. The latter is estimated using the formalism described in [12, 13]. After optimizing all four coupling coefficients, suitable directional-coupler geometries in Fig. 7 can be obtained through full-wave simulations.

3 Experimental Results

In Fig. 9, preliminary measurements from the hybrid RSOA-to-PIC setup [Fig. 9(g)] are presented. The PIC under test has a 7-μm wide MMEC (Table 1) and is coupled to an RSOA mounted on an alignment stage. The ROSA is biased at 78 mA with an output power of ~7 mW. RR_A is a perfect ring ($R_A = 24.97$ μm), while RR_B has a racetrack geometry, with $R_B = 23.78$ μm and $L_c = 10$ μm. The gap values equal $g_{A1} = 541$ nm, $g_{A2} = 425$ nm, $g_{B3} = 421$ nm, $g_{B4} = 415$ nm, while the calculated κ values are 0.0326,

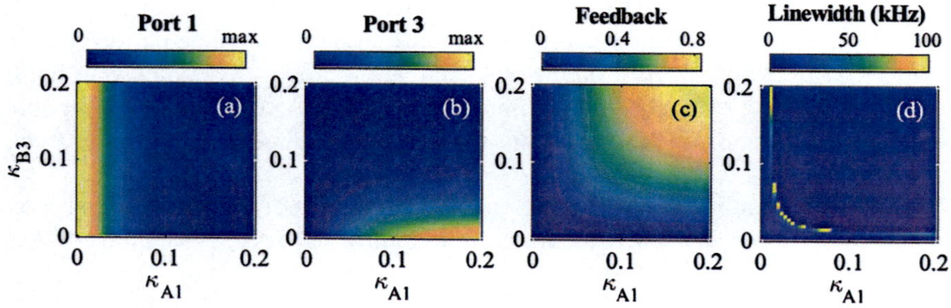

Fig. 8. Change in the power at (a) Port 1 and (b) Port 3, (c) optical feedback, and (d) theoretical ECL linewidth as a function of the coupling coefficients κ_{A1} and κ_{B3}.

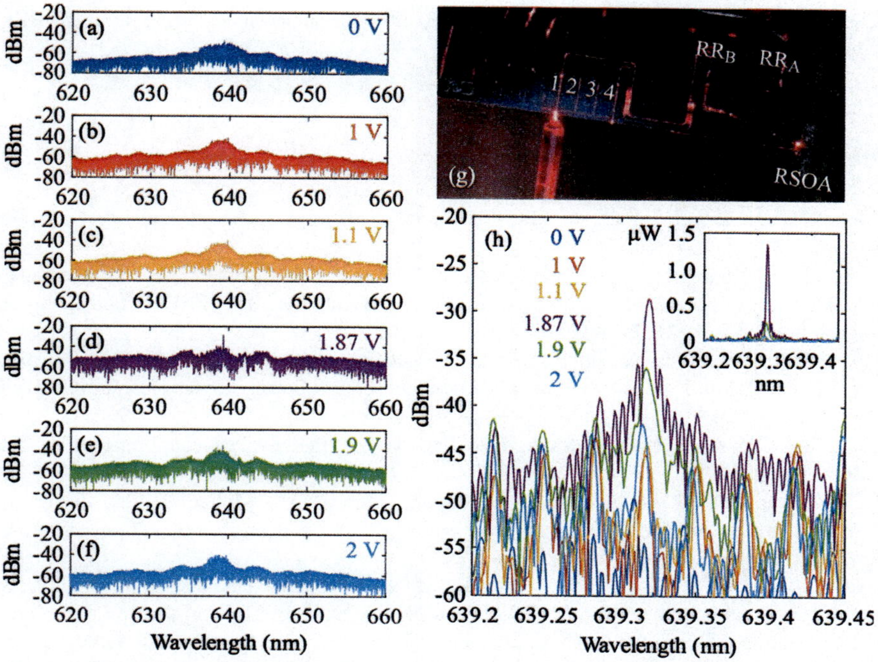

Fig. 9. (a)-(f) Spectrum evolution for six different biasing of RR_B, with RR_A remaining unbiased. At 1.87 V, the two ring resonances overlap and a distinctive peak is evidenced in the emission. (g) Snapshot of the hybrid RSOA-to-PIC coupling. (h) Close-up view of the spectra in (a)-(f). Inset: in linear scale.

0.0687, 0.21, 0.2188, respectively. By simply biasing RR_B while keeping RR_A unbiased, the resonance of the former is redshifted until both resonances overlap and a distinctive light-peak manifests itself [Fig. 9(d)]. This peak cannot be attributed to filtered amplified spontaneous emission (ASE) since the latter does not have to pass through the Sagnac loop to reach Port 1. An increase in bias shifts the RR_B resonance further, beyond RR_A, and the effect ceases. From Fig. 9(h), the peak center is measured at 639.32 nm with a linewidth of 4.93 pm, limited by the 10 pm resolution bandwidth of the optical spectrum analyzer (Thorlabs OSA201C). More accurate linewidth measurements will be performed using, e.g., delayed self-heterodyne interferometric detection [14]. The presented preliminary results strongly suggest the manifestation of lasing, with many degrees of freedom being still available for optimizing the performance.

4 Conclusions

The design of a PIC-based ECDL at 637 nm was presented, including an MMEC with relaxed alignment tolerance and an asymmetric ring-resonator-based Sagnac-loop reflector, designed for achieving single-output operation and flexible optimization between output power and linewidth. Preliminary measurements strongly indicate lasing, with a linewidth value limited by the OSA resolution.

Acknowledgments. Funded by the Federal Ministry of Education and Research (BMBF) - 13N15965.

References

1. Fan, Y., et al.: Hybrid integrated InP-Si_3N_4 diode laser with a 40-Hz intrinsic linewidth. Opt. Express **28**, 21713 (2020)
2. Tran, M.A., Huang, D., Guo, J., Komljenovic, T., Morton, P.A., Bowers, J.E.: Ring-resonator based widely-tunable narrow-linewidth Si/InP integrated lasers. IEEE J. Select. Topics Quantum Electron. **26**, 1–14 (2020)
3. Ghannam, I., Shen, B., Merget, F., Witzens, J.: Silicon nitride external cavity laser with alignment tolerant multi-mode RSOA-to-PIC interface. IEEE J. Select. Topics Quantum Electron. **28**, 1–10 (2022)
4. Franken, C.A.A., et al.: Hybrid-integrated diode laser in the visible spectral range. Opt. Lett. **46**, 4904 (2021)
5. Corato-Zanarella, M., Gil-Molina, A., Ji, X., Shin, M.C., Mohanty, A., Lipson, M.: Widely tunable and narrow-linewidth chip-scale lasers from near-ultraviolet to near-infrared wavelengths. Nat. Photon. **17**, 157–164 (2023)
6. Romero-Garcia, S., Bin Shen, Merget, F., Marzban, B., Witzens, J.: Alignment tolerant couplers for silicon photonics. IEEE J. Select. Topics Quantum Electron. **21**, 765–778 (2015)
7. Christopoulos, T., Tsilipakos, O., Sinatkas, G., Kriezis, E.E.: On the calculation of the quality factor in contemporary photonic resonant structures. Opt. Express **27**, 14505–14522 (2019)
8. Roeloffzen, C.G.H., et al.: Low-Loss Si3N4 TriPleX optical waveguides: technology and applications overview. IEEE J. Select. Topics Quantum Electron. **24**, 1–21 (2018)
9. Moscoso-Mártir, A., et al.: Hybrid silicon photonics flip-chip laser integration with vertical self-alignment. In: 2017 Conference on Lasers and Electro-Optics Pacific Rim (CLEO-PR), paper s2069. IEEE, Piscataway, New Jersey (2017)
10. Kluge, I., et al.: Flip-chip-integrated silicon nitride ECL at 640nm with relaxed alignment tolerances. In: Reed, G.T. and Knights, A.P. (eds.) Silicon Photonics XVII. p. 34. SPIE, San Francisco, United States (2022)
11. Bogaerts, W., et al.: Silicon microring resonators. Laser Photon. Rev. **6**, 47–73 (2012)
12. Patzak, E., Sugimura, A., Saito, S., Mukai, T., Olesen, H.: Semiconductor laser linewidth in optical feedback configurations. Electron. Lett. **19**, 1026 (1983)
13. Kazarinov, R., Henry, C.: The relation of line narrowing and chirp reduction resulting from the coupling of a semiconductor laser to passive resonator. IEEE J. Quantum Electron. **23**, 1401–1409 (1987)
14. Bai, Z., et al.: Narrow-linewidth laser linewidth measurement technology. Front. Phys. **9**, 768165 (2021)

Widely Tunable GaSb/Si_3N_4 Vernier Hybrid Laser Emitting Around 2.55 µm

Samu-Pekka Ojanen(✉), Nouman Zia, Jukka Viheriälä, Eero Koivusalo, Joonas Hilska, Heidi Tuorila, and Mircea Guina

Optoelectronics Research Centre, Physics Unit, Tampere University, 33720 Tampere, Finland
samu-pekka.ojanen@tuni.fi

Abstract. The development and characterization of a GaSb/Si_3N_4 tunable Vernier hybrid laser is presented. We combined a GaSb-based reflective semiconductor amplifier to provide feedback and gain, with a reflective Si_3N_4 circuit to provide wavelength tuning and filtering. The Si_3N_4 circuit comprises two thermally tunable ring resonators within a loop mirror configuration towards Vernier type operation. Operating at an emission wavelength of 2.55 µm, this laser exhibits a low threshold current of 190 mA and achieves a continuous-wave output power of 6.4 mW at 325 mA. Notably, it demonstrates a wide tuning range of 170 nm (2474–2644 nm) at 325 mA injection current, delivering a consistent output power of 5 mW within the central band (2520–2620 nm) and maintaining >1 mW even at the band edges. These characteristics provide unique advantages for possible use in a wide range of sensing applications, from spectroscopic sensing of gases to detection of biomarkers.

Keywords: hybrid lasers · GaSb · silicon nitride photonics

1 Introduction

Over recent years, photonic integration has seen a notable surge in interest due to its ability to shrink photonic systems, facilitating more cost-effective solutions and enabling a wide range of applications spanning from conventional datacom transceivers to miniaturized optical sensors. In particular, the 2–4 µm wavelength range has gained interest [1] since many industrial and environmental gases like H_2S C_2H_4, CH_4, CO, CO_2 and N_2O exhibit strong absorption there [2]. Moreover, biomarkers like lactate, urea, and glucose also exhibit absorption within this wavelength spectrum [3, 4], making it highly attractive for medical applications.

Generally speaking, hybrid integrated lasers have been successfully demonstrated at wavelengths of 2 µm [5], 2.3 µm [6], and 2.6 µm [7], leveraging the silicon-on-insulator (SOI) platform. Recently, silicon nitride-on-insulator (SNOI) circuits have emerged as a promising alternative to SOI [8, 9]. In contrast to SOI, SNOI exhibits a lower index contrast of approximately 38%, rendering it less susceptible to variations in waveguide structure width and index. This attribute is particularly advantageous in sensing applications where precise targeting of narrow wavelength regions is crucial. However,

J. Witzens et al. (Eds.): ECIO 2024, SPPHY 402, pp. 17–20, 2024.
https://doi.org/10.1007/978-3-031-63378-2_3

the majority of SNOI research has concentrated on the visible or telecommunication wavelength ranges [10, 11], with relatively limited exploration at wavelengths beyond 2 μm [9, 12].

Here, we present details of recent results concerning a new GaSb/SNOI tunable Vernier hybrid laser platform [13]. To this end, we have deployed a GaSb-based type-I quantum well (QW) heterostructure gain waveguide, which was edge-coupled to a reflective SNOI circuit comprising two thermally tunable ring resonators within a loop mirror. A schematic illustration of the laser circuit is provided in Fig. 1. The Vernier hybrid laser showcases a wide tuning range of 170 nm centered at 2.55 μm.

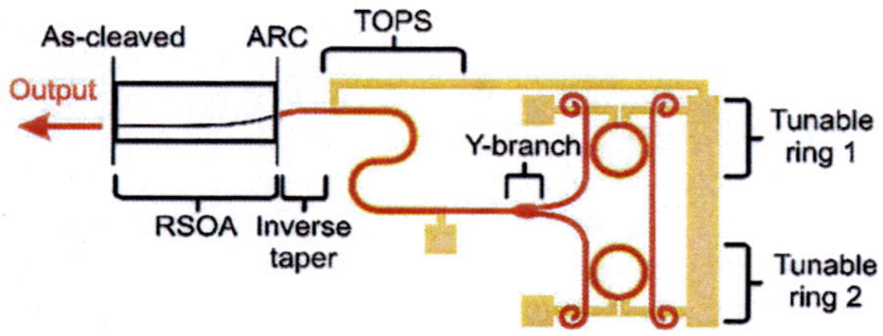

Fig. 1. Schematic of the GaSb/Si_3N_4 tunable Vernier hybrid laser circuit.

2 Measurements

The SNOI chip was manufactured utilizing open-access foundry services (LIGENTEC). The laser includes a GaSb-based reflective semiconductor optical amplifier (RSOA) with a ridge waveguide (RWG) architecture. The RWG utilizes a J-shaped geometry with a tilted (7°) front facet with an anti-reflectance coating (ARC) to minimize back reflections and maximize transmission, and a straight as-cleaved rear facet to provide ~30% reflection.

The light from the RSOA was coupled to an 800-nm-thick SNOI waveguide. An inverse taper tapering from 1.65 μm to 0.4 μm towards the facet was utilized to maximize coupling; the coupling loss was measured to be ~3 dB at 2.55 μm. The 1.65 μm SNOI waveguide width was chosen to prevent the excitation of higher order lateral modes. The waveguide propagation loss of 0.67 dB/cm was measured at 2.68 μm utilizing a Norcada distributed feedback (DFB) laser. A 700-μm-long thermo-optic phase shifter (TOPS), consisting of a resistive heater layer placed above the waveguide was utilized to allow phase-matching the RSOA with the SNOI circuit to maximize power and spectral purity. A loop mirror was formed within the SNOI circuit by utilizing a Y-branch with a measured insertion loss of 0.1 dB at 2.68 μm. Two thermally tunable ring resonators, with radii of 100 μm and 103.105 μm and a coupling gap of 650 nm, were placed inside the loop mirror to form a Vernier reflector with a simulated free spectral range (FSR) of 169 nm.

To measure the hybrid laser, the RSOA was placed on a heatsink that was stabilized to room temperature (23 °C), and the SNOI PIC was placed on a closed-loop piezo stage. The output signal was coupled to a multimode fiber connected to a photodiode, and the waveguides were actively aligned by maximizing the signal. Figure 2 a) shows the IL curve of the Vernier hybrid laser around 2.55 μm emission wavelength. The blue curve was measured by applying CW current to the RSOA and measuring the output power from 0 mA to 400 mA. The laser shows a threshold current of 190 mA and reaches a maximum CW output power of 6.4 mW at 322 mA, which is the highest output power measured from hybrid lasers at this wavelength range. The IL curve shows strong modulation with minima corresponding to maximum phase mismatch between the RSOA and the PIC, whereas the maxima correspond to maximum phase-matching. The orange curve was measured by tuning the TOPS at each injection current value to maximize the output signal. As is evident from the measurements presented, this suppresses the modulation significantly, clearly showing the importance of phase-matching.

To measure the tuning spectrum of the Vernier hybrid laser, a constant current of 325 mA was injected into the RSOA and one of the ring resonators was tuned by applying current to the heater layer above it. The spectra measured at different operation points are plotted in blue in Fig. 2. The output shows a broad tuning range of 170 nm, from 2474 nm to 2644 nm. Additionally, the output power of the laser was measured as a function of wavelength with constant injection current of 325 mA, and the result is plotted in orange in Fig. 2. The output power stays above 1 mW over the whole tuning band, which is a sufficient power level for most sensing applications.

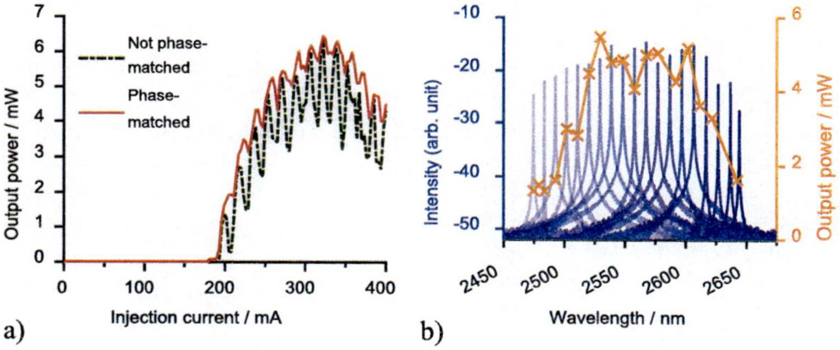

Fig. 2. a) IL curve of the Vernier hybrid laser with (orange) and without (blue) phase-matching, b) the spectrum of the laser at different ring resonator operation points (blue line, left y-axis), and phase-matched output power as a function of wavelength (orange line, right y-axis) when injecting constant 325 mA current into the RSOA.

In conclusion, we have demonstrated the first GaSb/Si_3N_4 hybrid laser in the 2.5–3.0 μm wavelength region. The Vernier hybrid laser showed a maximum CW output power of 6.4 mW and a broad tuning range of 170 nm (2474–2644 nm) with >1 mW output power over the whole tuning range. This is the highest CW output power beyond 2.1 μm [5], and the broadest tuning range beyond 1.7 μm [11], reported for PIC-based hybrid lasers. The broad tuning range of the Vernier hybrid laser makes it attractive for multispecies spectroscopy. Future research will focus on integrating the RSOA on the PIC and extending the wavelength towards and beyond 3 μm.

References

1. Tran, M.A., Huang, D., Bowers, J.E.: Tutorial on narrow linewidth tunable semiconductor lasers using Si/III-V heterogeneous integration. APL Photonics **4**, 111101 (2019)
2. Gordon, I.E., et al.: The HITRAN2016 molecular spectroscopic database. J. Quant. Spectrosc. Radiat. Transf. **203**, 3–69 (2017)
3. Amerov, A.K., Chen, J., Arnold, M.A.: Molar absorptivities of glucose and other biological molecules in aqueous solutions over the first overtone and combination regions of the near-infrared spectrum. Appl. Spectrosc. **58**, 1195–1204 (2004)
4. Lin, P.T., et al.: Label-free glucose sensing using chip-scale mid-infrared integrated photonics. Adv. Opt. Mater. **4**, 1755–1759 (2016)
5. Wang, R., Malik, A., Šimonytė, I., Vizbaras, A., Vizbaras, K., Roelkens, G.: Compact GaSb/silicon-on-insulator 20x μm widely tunable external cavity lasers. Opt. Express **24**, 28977 (2016)
6. Wang, R., et al.: Widely tunable 23 μm III-V-on-silicon Vernier lasers for broadband spectroscopic sensing. Photonics Res. **6**, 858 (2018)
7. Ojanen, S.-P., et al.: GaSb diode lasers tunable around 2.6 μ m using silicon photonics resonators or external diffractive gratings. Appl. Phys. Lett. **116**, 081105 (2020)
8. Siew, S.Y., et al.: Review of silicon photonics technology and platform development. J. Light. Technol. **39**, 4374–4389 (2021)
9. Zia, N., et al.: Widely tunable 2 μm hybrid laser using GaSb semiconductor optical amplifiers and a Si_3N_4 photonics integrated reflector. Opt. Lett. **48**, 1319 (2023)
10. Franken, C.A.A., et al.: Hybrid-integrated diode laser in the visible spectral range. Opt. Lett. **46**, 4904 (2021)
11. Guo, Y., et al.: Hybrid integrated external cavity laser with a 172-nm tuning range. APL Photonics **7**, 066101 (2022)
12. Luke, K., Okawachi, Y., Lamont, M.R.E., Gaeta, A.L., Lipson, M.: Broadband mid-infrared frequency comb generation in a Si_3N_4 microresonator. Opt. Lett. **40**, 4823 (2015)
13. Ojanen, S., et al.: Widely tunable (2.47–2.64 μm) hybrid laser based on GaSb/GaInAsSb Quantum-wells and a low-loss Si_3N_4 photonic integrated circuit. Laser Photonics Rev. **17**(7), 2201028 (2023)

Pound-Drever-Hall Laser Frequency Stabilization of Tunable 1.55 μm Monolithically Integrated Semiconductor Lasers Using an Integrated Phase Modulator

Rachel Jones(✉), Kevin Williams, and Erwin Bente

Eindhoven University of Technology, 5612 AZ Eindhoven, The Netherlands
r.a.jones@tue.nl

Abstract. Report on the advancement in the integration of the Pound-Drever-Hall (PDH) stabilization scheme on a single InP photonic integrated circuit by monolithically integrating a tunable laser and an electro-optic phase modulator (EOPM). A semi-insulating substrate allows a sufficiently reduced electrical cross-talk between the laser and the external EOPM to allow PDH stabilization. A novel implementation of the PDH lock is demonstrated, whereby an EOPM internal to the integrated lasing cavity is modulated. The frequency noise suppression of both implementations is measured and compared to those achieved with a bulk phase modulator. A 58 dB frequency noise suppression at 10–100 Hz is achieved.

Keywords: Laser frequency stabilization · monolithic integration · narrow linewidth

1 Introduction

Integrated narrow linewidth continuous wave semiconductor lasers have long been vital for sensing and metrology. Although free running lasers with linewidths <100 kHz have been monolithically integrated on III-V platforms [1, 2] their stability will significantly deteriorate over time due to environmental disturbances and electrical noise. To provide optically stable light sources over longer timescales, it is necessary to employ a frequency stabilizing feedback system. An established stabilization method is the Pound-Drever-Hall (PDH) scheme [3]. This scheme enables a single frequency laser to be locked to the resonance of an optical cavity to better than 1/100 of the full width at half maximum bandwidth of the optical resonance. Whilst the integration of the electronic circuitry on a single CMOS IC has been demonstrated [4], with regards to the optical circuit, thus far, only either the light source or the optical cavity has been successfully integrated on a single Photonic Integrated Circuit (PIC) [5–7]. Previous attempts to integrate both the laser and the electro-optical phase modulator (EOPM) on the same substrate and implement PDH had been prevented by strong electrical cross-talk between the frequency selective filters in the laser and the EOPM in the stabilization circuit [8].

J. Witzens et al. (Eds.): ECIO 2024, SPPHY 402, pp. 21–27, 2024.
https://doi.org/10.1007/978-3-031-63378-2_4

In this work, we capitalize on the development of a semi-insulating platform and improve the circuit design to frequency stabilize an integrated laser using an integrated EOPM. We further investigate an implementation of the PDH lock using an intra-cavity modulation of the laser. Finally, we compare both implementations to a system where an integrated laser is locked using a bulk fiber-coupled lithium niobate phase modulator.

1.1 The Pound-Drever-Hall Stabilization System

The PDH scheme, shown in Fig. 1, is a form of opto-electronic control system in which the frequency of a free running laser is compared to that of a stable reference. Here, we use the resonance modes of a hermetically sealed high finesse etalon (F = 3000) with a FWHM of 700 kHz and drift rate less than 1 MHz per day as our fixed reference. The light from the laser is frequency modulated, inducing sidebands equally spaced from the carrier. For the PDH scheme, a modulation depth (ϕ_{PDH}) of 1.08 is optimal. When the laser frequency deviates from the resonance mode of the cavity, the carrier will be partly reflected and interfere with each of these sidebands. The resulting beat pattern is detected on the photo-diode at the modulation frequency and subsequently isolated using a mixer and low pass filter. The result is the PDH open loop error signal, the sign and magnitude of which indicate how to correct the laser frequency to align with the cavity resonance. This becomes the input for the PID servo controller (Toptica FALC110), which will generate the control signal to feed back to the laser to align its frequency through the EOPM to that of the reference, with the aim of reducing the error signal to zero [9].

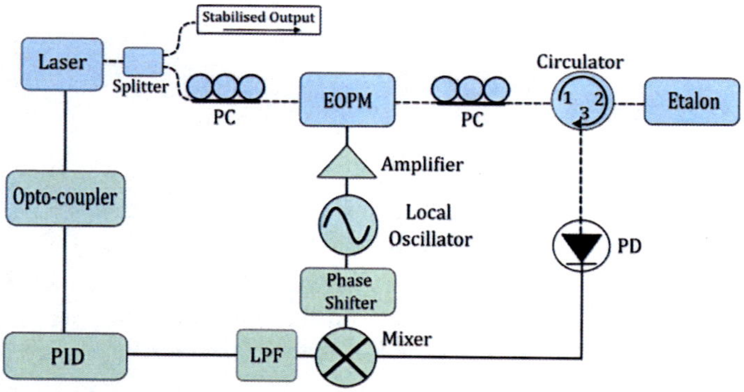

Fig. 1. Schematic of PDH frequency stabilization scheme. Shown in blue and green are the optical and electrical components, respectively.

2 Laser Frequency Stabilization Results

2.1 PDH Stabilization Using Integrated Laser and Integrated EOPM

In our first implementation of the scheme, we monolithically integrate the Distributed Bragg Reflector intra-cavity ring resonator (DBR-RR) laser [2], a tunable power splitter and an EOPM on a PIC fabricated on a semi-insulating platform. On this platform the n-doped layer is on top of a semi-insulating substrate. Every on-chip component has its own ground, thereby reducing the electrical crosstalk through the ground plane. Additionally, the tunable splitter and external EOPM are surrounded by a 'N isolation trench' which helps further isolate their ground planes from that of the laser.

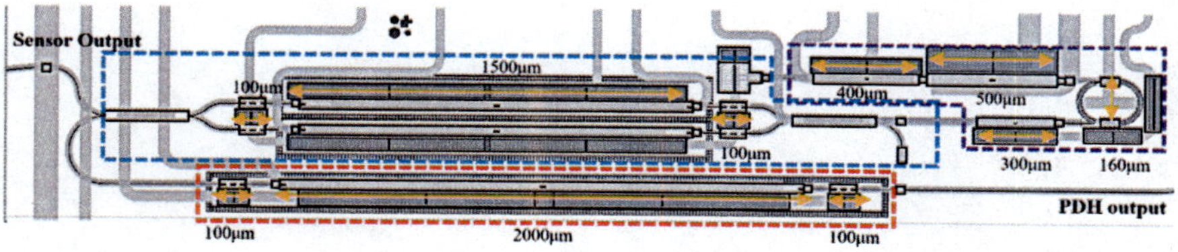

Fig. 2. Schematic of circuit layout. The purple, blue and red boxes indicate the laser, tunable splitter and the external EOPM respectively. There are two optical outputs, the sensor output port on the LHS and PDH output port on the RHS.

Figure 2 shows the layout of our PIC. The red and purple dashed box encircle the integrated EOPM and DBR intra-cavity ring laser respectively. The EOPM has a length of 2 mm and a measured 24°/V·mm, meaning that a modulation amplitude of 1.27 V is required for ϕ_{PDH}. Adjacent to the EOPM, are two shorted EOPMs of length 100 µm to prevent leakage current along the waveguide from disturbing the laser.

The PIC is mounted on an aluminum sub-mount and kept at 17 °C using a water-cooler. Electrical signals are applied using wire-bonds from a printed circuit board to the on-chip bond-pads. The laser is driven at 80 mA from a low noise current source (ILX-LDV-3260B) whilst all unused tunable elements are shorted to ground. Two lensed fibers are coupled to the angled facets of the PIC collecting the optical signals from the PDH output (RHS) and the sensor output (LHS), the latter we use for monitoring purposes. Both are recorded using a high-resolution optical spectrum analyzer (OSA, APEX 2641B) with a resolution bandwidth of 5 MHz.

Figure 3 a) shows the spectrum of light collected at the PDH output port when no modulation is applied to the EOPM. The sidebands present on this spectrum, at approximately ±94 MHz and 20 dBm below the carrier signal, are a result of interference with FM local radio signals.

Subsequently, a 60 MHz sinusoidal signal with an amplitude of 140 mV is applied to the EOPM, external to the lasing cavity, through SMA ports on the sub-mount to minimise the possibility of cross-talk occurring off the PIC. Figures 3 b) and c) show the optical spectra of light at the PDH output port and the monitoring output port respectively. In Fig. 3 b) we observe sidebands induced at multiples of the modulation frequency, ±60 MHz, ±120 MHz and ±180 MHz, from the carrier. We find a significantly smaller voltage of 140 mV is required to reach a modulation depth of 1.08. Additionally in Fig. 3 c), where light is collected at the sensor output port, we also observe sidebands at ±60 MHz and ±120 MHz from the carrier, 8.5 dB and 21.5 dB below the carrier.

These two results, namely the increased modulation efficiency and the modulated signal observed at the monitoring port, indicate electrical cross-talk still exists between the EOPM and the laser. We attribute this cross-talk to insufficient electrical isolation between active components. As the N-isolation trench cannot cross the waveguide, this

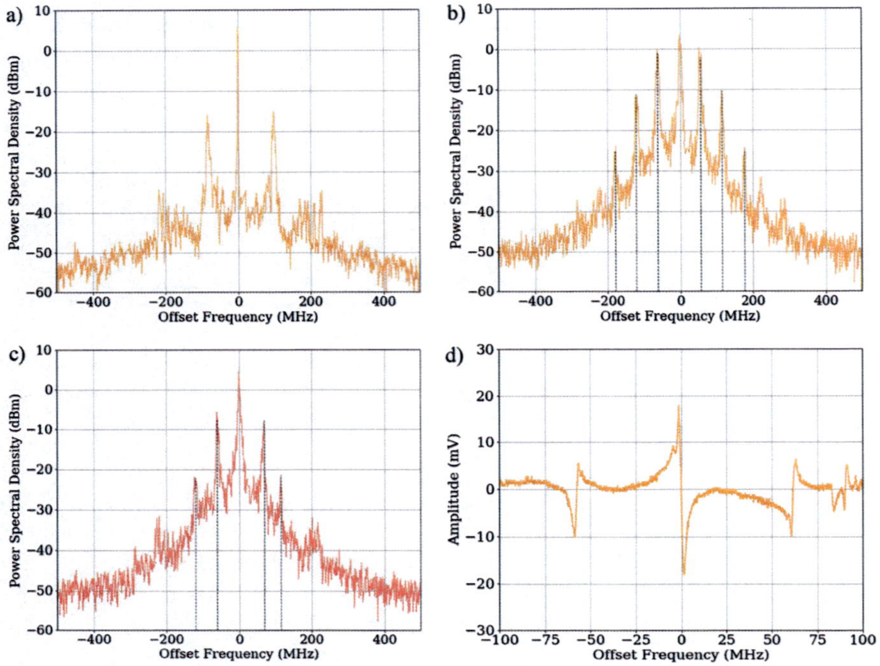

Fig. 3. **a)** and **b)** Optical spectra of light at the PDH output port when external EOPM modulation is off and on respectively, **c)** Spectrum of light at the monitoring output port when the external EOPM is modulated and **d)** PDH error signal.

means there is still a connection between the ground of the external EOPM and the rest of the circuit.

Nonetheless, evidence of crosstalk is reduced sufficiently to allow for the error signal to be obtained, see Fig. 3 d), and the laser to be stabilized which has a slope of 18.8 mV/MHz around the zero offset.

2.2 PDH Stabilization Using Integrated Laser and EOPM Internal to Lasing Cavity

In our second implementation we demonstrate, for the first time to the authors' knowledge, a PDH lock using an EOPM inside the integrated laser cavity. Unlike an external EOPM, the electrical cross talk introduced to other circuit components is expected to be far less since a lower voltage is required for the phase modulation of the laser output. The EOPM inside the laser has a length of 502 μm and forms the arcs of the intra-cavity ring filter. Figure 4 a) shows a plot of the relative power in first carrier and first three sidebands as the modulation depth is increased. We find that it takes a modulation amplitude of 5.8 mV to reach ϕ_{PDH}, over one order of magnitude less than that needed for an external EOPM. The resulting error signal when the laser is tuned across the resonance mode is shown in Fig. 4 b). The resulting slope of the error signal around zero frequency offset is 18.2 mV/MHz.

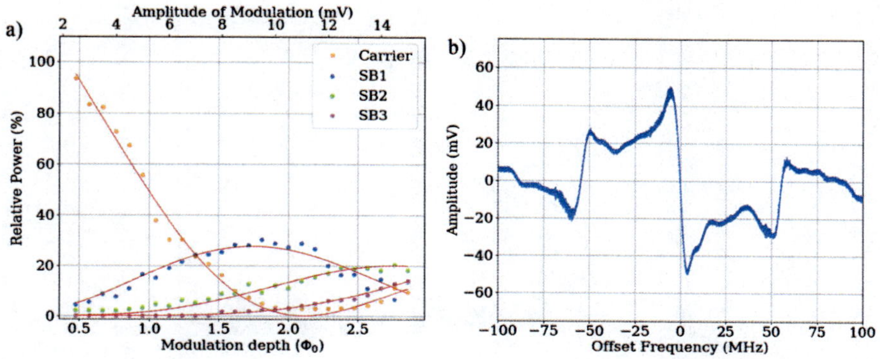

Fig. 4. **a)** Relative power in carrier and sidebands vs modulation depth and amplitude of frequency modulation (dots show measured data and red curve shows fitted squared Bessel function) and **b)** error signal versus the offset frequency from a resonance mode of the cavity using intra-cavity EOPM.

3 Comparison Between Methods of PDH Stabilization

Figure 5 shows the power spectral density of the frequency noise of an actively stabilized laser for each implementation of the PDH discussed previously and that of the laser when in open loop configuration, referred to as free running. The control loop settings are optimized for the suppression of low frequency noise. The free running laser clearly exhibits $1/f^1$ between 100 Hz–10 kHz. Comparing the frequency noise for the free running (in purple) laser to that of the locked lasers using the bulk EOPM (in green) and internal EOPM (in blue) we observe a noise suppression of 57dB between 10–100 Hz. Whilst the implementation using the integrated EOPM external to the lasing cavity (in orange) provides a suppression of only 37 dB. For each implementation the frequency noise spectra cross that of the free running laser at approximately 300 kHz, indicating the bandwidth of our lock.

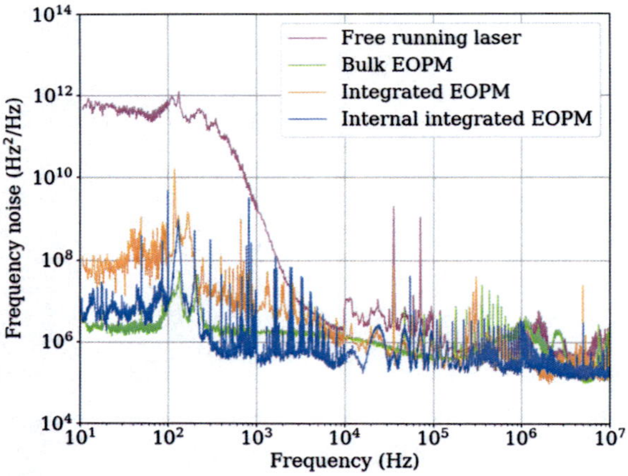

Fig. 5. Power Spectral Density of the frequency noise spectra for a free running integrated laser and the laser frequency locked using the different configurations, namely a bulk lithium niobate PM (green), an integrated EOPM external to the lasing cavity (orange) and an integrated EOPM internal to the lasing cavity (blue).

4 Conclusion

We have reported on two distinct ways in which a tunable laser and EOPM are monolithically integrated on a single InP PIC to enable a PDH stabilization lock. To the best of our knowledge, this is the first time the laser and the EOPM have been successfully integrated on the same PIC. In our second implementation we demonstrated a novel intra-cavity frequency modulation. The benefits of this latter approach are: 1) it requires a smaller amplitude of modulation (almost two orders of magnitude), significantly reducing the disturbance to the tunable elements in and outside the laser; 2) a smaller physical footprint; 3) no need for fabrication on a semi-insulating substrate. This implementation showed a frequency noise suppression comparable to that using a bulk PM. Although not useful for all applications, given the stabilized output is frequency modulated, one such example where we believe this could be useful is for a heterodyne displacement measurement [10].

References

1. Kumar, R., et al.: A 10-kHz intrinsic linewidth coupled extended-cavity DBR laser monolithically integrated on an InP platform. Opt. Lett. **47**(9), 2346–2349 (2022)
2. Andreou, S., Williams, K., Bente, E.A.J.M.: Monolithically integrated InP-based DBR lasers with an intra-cavity ring resonator. Opt. Express **27**, 26281–26294 (2019)
3. Drever, R.W.P., Hall, J.L., et al.: Laser phase and frequency stabilization using an optical resonator. Appl. Phys. B **31**, 77–105 (1983)
4. Idjadi, M.H., Aflatouni, F.: Integrated Pound−Drever−Hall laser stabilization system in silicon. Nat. Commun. **8**, 1209 (2017)
5. Spencer, D.T., et al.: Stabilization of heterogeneous silicon lasers using Pound-Drever-Hall locking to Si_3N_4 ring resonators. Opt. Express **24**(12), 13511–13517 (2016)

6. Rees, A.V., et al.: Long-term absolute frequency stabilization of a hybrid-integrated InP-Si3N4 Diode Laser. IEEE Photonics J. **15**(5), 1502408 (2023)
7. Andreou, S., Williams, K., Bente, E.A.J.M.: Frequency stabilization of an InP-based integrated diode laser deploying electro-optic tuning. IEEE Photonics Technol. Lett. **31**(24), 1983–1986 (2019)
8. Andreou, S., Williams, K., Bente, E.A.J.M.: Steady-state analysis of the effects of residual amplitude modulation of InP-based integrated phase modulators in Pound–Drever–Hall frequency stabilization. IEEE Photonics J. **11**(3), 1–14 (2019)
9. Black, E.: An introduction to Pound–Drever–Hall laser frequency stabilization. Am. J. Phys. **69**(1), 79–87 (2001)
10. Číp, O., et al.: Small displacement measurements with subatomic resolution by beat frequency measurements. Meas. Sci. Technol. **18**, 2005–2013 (2007)

High Power 1.8 W Tunable Laser Based on CMOS Compatible Power Amplifier

Neetesh Singh[1(✉)], Jan Lorenzen[1], Muharrem Kilinc[1], Kai Wang[2], Milan Sinobad[1], Henry Francis[3], Jose Carriera[3], Michael Geiselmann[3], Umit Demirbas[1], Mikhail Pergament[1], Sonia M. Garcia-Blanco[2], and Franz X. Kärtner[1,4]

[1] Center for Free-Electron Laser Science CFEL, Deutsches Elektronen-Synchrotron DESY, Hamburg, Germany
neetesh.singh@desy.de
[2] Integrated Optical Systems, MESA+ Institute for Nanotechnology, University of Twente, 7500AE Enschede, The Netherlands
[3] LIGENTEC SA, EPFL Innovation Par L, Chemin de La Dent-d'Oche 1B, 1024 Switzerland, Ecublens, Switzerland
[4] Department of Physics, Universität Hamburg, Jungiusstr. 9, 20355 Hamburg, Germany

Abstract. We demonstrate a high-power tunable laser based on a CMOS-compatible power-amplifier. We show amplified power up to 1.8 W with a tunability bandwidth of 60 nm from 1.83 μm to 1.89 μm, limited only by the seed laser. Additionally, high power amplification up to 1.95 μm is demonstrated with the potential for tunability over a larger window, 1.8 to > 2 μm.

Keywords: Lasers · Silicon Photonics · High Power Amplifier

1 Introduction

High power tunable lasers and amplifiers have various applications, from commercial to fundamental research. However, currently high-power capability falls under the realms of benchtop solid state and fiber systems owing to their large energy storage capacity, i.e. large system volume, *area* x *length*. In integrated photonics tight confinement that affords small footprint and compact circuitry is in conflict with the large volume requirement and therefore the power is usually limited to a few mW range. Thanks to the high gain offered by semiconductor gain media [1, 2], and low loss gain-ion implanted long waveguides: recently fabricated with the specialized Damascene process [3], power in the 10s of mW range can be achieved. However, due to the small gain-mode-area (area of the optical mode overlapping with the gain ions), energy storage capacity remains small and the reliance on benchtop amplifier for most of the applications continued to be an outstanding challenge, limiting most of the demonstration to the laboratory environment. For example, integrated LIDAR, coherent photonic RADAR, telecommunication, optical frequency combs and synthesizers, and photonic devices that are being investigated for space applications, all are currently deploying bench-top amplifiers to reach the required signal levels. Moreover, CMOS-compatible high-power amplifier and lasers

J. Witzens et al. (Eds.): ECIO 2024, SPPHY 402, pp. 28–31, 2024.
https://doi.org/10.1007/978-3-031-63378-2_5

can potentially be used in data centers and short-range communication systems in place of expensive and noisier semiconductor optical amplifiers (SOA).

2 Results

Large mode area technology has revolutionized the high-power amplification capabilities of fiber amplifiers. Owing to the larger mode the nonlinearity becomes negligible, the gain saturation power increases, and the energy storage capacity increases significantly. Such LMA devices on an integrated platform with mass producibility is highly desirable. Recently we demonstrated integrated CMOS-compatible LMA lasers and watt-class amplifiers [4–6]. In this follow-on study, we apply LMA amplifiers for high-power tunable laser applications and show lasing power reaching up to 1.8 W within a footprint of <4.5 mm^2. We demonstrate power levels between 1600 mW to 1800 mW for seed power levels of 80 mW to 120 mW, limited only by the seed source power. We note that the seed power used here are at the level currently achievable with integrated tunable lasers [1].

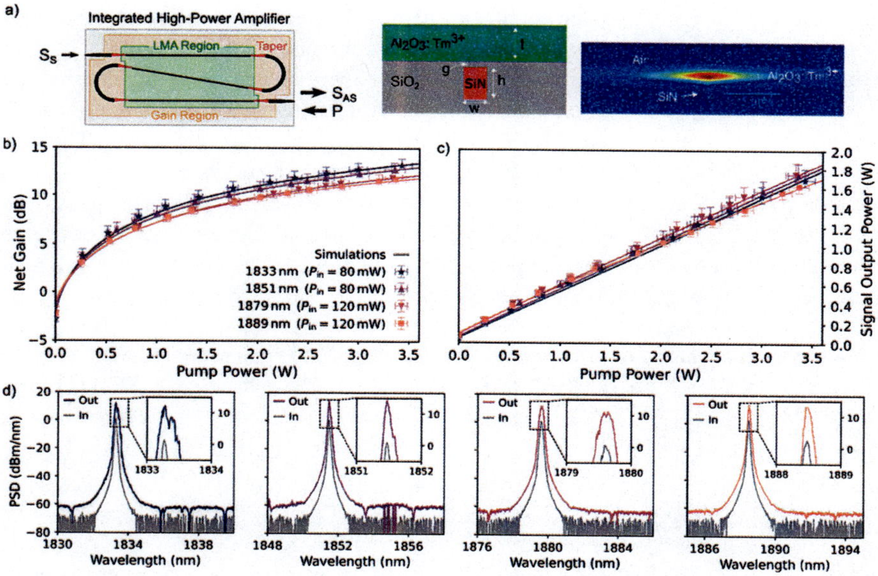

Fig. 1. a) A schematic of the device, with signal S_S launched in counter propagating configuration with the pump (P) launched from the other side. S_{AS} is the amplified signal. On the right is shown the device cross-section and the signal mode profile. b) and c) are the net gain and the amplified signal output power. d) Optical spectra taken at different wavelengths at the input and output of the amplifier. The ordinate shows relative power difference.

The tunable range demonstrated is from 1830 nm to 1890 nm limited only by the laser tuning bandwidth. The amplified power achieved is 25 times higher than the best semiconductor amplifiers (SOA) at 2 μm demonstrated to date [7] and surpasses many benchtop amplifiers [8]. Such high-power systems will not only replace many benchtop systems in the near future but will open up application landscapes with far-reaching implications. The device mainly consists of two sections: high confinement small footprint mode guiding section and high-power amplification LMA section. In the high confinement section, the mode is well confined within the silicon nitride (SiN) layer, which is required for achieving efficient pump and signal coupling as well as tight bends which facilitates a compact circuit. In the LMA region the mode area is ~ 30 μm^2 and 32 μm^2 around the signal (1.85 μm) and the pump (1.61 μm) wavelength, respectively. The length of the amplifier is ~ 6 cm. The LMA and the high mode confinement regions are connected with adiabatic tapers. The passive components of the device were fabricated in a silicon photonics foundry, *LIGENTEC SA*, which was then coated with an active alumina layer using an RF-sputtering tool at a thulium concentration of $6 \times 10^{20}/cm^3$ at the University of Twente. In this study, we used thulium gain ions due to ease in accessibility, but the device is scalable for other gain ions such as erbium and ytterbium for telecom and 1 μm window applications. The device was pumped with a pump around 1.61 μm and the seed laser was a home-built tunable Thulium: YLF solid-state laser pumped by a multimode 4 W, 792 nm diode [9]. The coupling loss was 2.4 dB/facet and 2.2 dB/facet for the signal and the pump, respectively. The measured gain and amplified signal power are shown in Fig. 1 b and c for different seed wavelengths. The amplified signal power reached between 1600 mW to 1800 mW on-chip. The net gain ranged from 11.5 dB to 13.5 dB. With lower power seed signal higher gain up to 16 dB is achievable before the onset of parasitic lasing from facet reflections. The stability of the devices was tested for 2 h without much fluctuations in power. The noise figure was measured to be around 5 dB for the seed power tested here.

In conclusion, we have demonstrated a very high-power tunable laser based on a CMOS-compatible high-power amplifier. The amplifier is based on newly developed LMA waveguides, having a high output beam quality and power levels comparable and in many cases surpassing bench-top fiber systems.

Acknowledgment. This work is supported by EU Horizon 2020 Framework Programme - Grant Agreement No.: 965124 (FEMTOCHIP), and Deutsche Forschungsgemeinschaft (SP2111) project number 403188360.

References

1. Kobayashi, N., et al.: Silicon photonic hybrid ring-filter external cavity wavelength tunable lasers. J. Light. Technol. **33**, 1241–1246 (2015)
2. Volet, N., et al.: Semiconductor optical amplifier at 2 μm wavelength on silicon. Laser Photonics Rev. **11**, 1600165 (2017)
3. Liu, Y., et al.: A photonic integrated circuit based erbium doped amplifier. Science **376**, 1309–1313 (2022)
4. Singh, N., et al.: Towards CW mode locked laser on chip – a large mode area and NLI for stretched pulse mode locking. Opt. Express **28**, 22562–22579 (2020)
5. Singh, N., et al.: Silicon photonics-based high energy passively Q-switched laser. Nat. Photonics 1–7 (2024)
6. Singh, N., et al.: Watt-class CMOS-compatible power amplifier. arXiv:2306:12940

7. Juodawlkis, P.W., et al.: High-power, low-noise 1.5-μm slab-coupled optical waveguide (SCOW) emitters: physics, devices, and applications. IEEE J. Sel. Top. Quantum Electron. **17**, 1698–1714 (2011)
8. PicoLuz – Thulium doped fiber amplifier and www.thorlabs.com. C-band EDFA. www.civillaser.com
9. Demirbas, U., et al.: Continuous-wave Tm: YLF laser with ultrabroad tuning (1772–2145 nm). Opt. Express **30**, 41219–41239 (2022)

Micro-Transfer-Printed O-Band GaAs QD-On-Si Widely Tunable Laser

Jing Zhang[1,2(✉)], Yang Liu[1,2], Ali Uzun[1,2], Evangelia Delli[1,2], Laurens Bogaert[1,2], Senbiao Qin[1,2], Konstantin Morozov[3], Sergey Mikhrin[3], Rimböck Johanna[4], Ruggero Loi[5], Peter Ossieur[6], Guy Lepage[7], Peter Verheyen[7], Dimitrios Velenis[7], Chiara Marchese[7], Joris Van Campenhout[7], Geert Morthier[1,2], and Gunther Roelkens[1,2]

[1] Photonics Research Group, INTEC, Ghent University - IMEC, Ghent, Belgium
jingzhan.Zhang@ugent.be
[2] Center for Nano- and Biophotonics, Ghent University - IMEC, Ghent, Belgium
[3] Innolume GmbH, Konrad-Adenauer-Allee 11, 44263 Dortmund, Germany
[4] EV Group E.Thallner GmbH, DI Erich Thallner Str. 1, 4782 St. Florian Am Inn, Austria
[5] X-Celeprint Ltd., Lee Maltings, Dyke Parade, Cork, Ireland
[6] IDLab, INTEC, Ghent University – IMEC, Ghent, Belgium
[7] IMEC, Kapeldreef 75, 3001 Heverlee, Belgium

Abstract. We present an O-band GaAs QD-on-Si integrated widely tunable laser realized by micro-transfer-printing a prefabricated GaAs SOA coupon on an advanced silicon photonics wafer. The resulting device exhibits single mode operation with >45 dB side-mode-suppression-ratio over a 30 nm wavelength tuning range in the O-band. The waveguide-coupled power is measured to be around 3.5 mW.

Keywords: Silicon photonics · micro-transfer printing · widely tunable laser

1 Introduction

Silicon photonics (SiPh) has experienced a rapid development in the past decades. Different platforms featuring low-loss waveguides alongside a variety of integrated functionalities, such as low-loss Si and SiN waveguides, high-speed Ge photodiodes, Ge electro-absorption modulators, high-speed doped-Si modulators etc., have been established. By utilizing these readily available building blocks in the process design kit (PDK) that are provided by the foundries, complex photonic integrated circuits (PICs) with compact sizes can be mass-produced using the mature CMOS facilities. These PICs can be used in a wide range of applications, especially in the field of data- and telecom where high-performance optical transceivers and low-power, compact switches are heavily in demand in the giant datacenters. However, the lack of high-performance and low-cost integrated light sources and optical gain elements becomes one of the most challenging hurdles in further minimizing the size and at the same time in improving

J. Witzens et al. (Eds.): ECIO 2024, SPPHY 402, pp. 32–36, 2024.
https://doi.org/10.1007/978-3-031-63378-2_6

the performance of the overall system-on-chip. Different approaches have been developed to overcome this issue including hybrid integration using flip-chip [1] or photonic wire bonding [2] and heterogeneous integration through (multiple) die-to-wafer bonding [1]. However, these methods either suffer from limited throughput (flip-chip, photonic wire bonding) or require significant modification to the established SiPh back-end process flow (die-to-wafer bonding) to accommodate the III-V processing. Direct epitaxial growth of III-V materials on SiPh wafers is another promising approach for the realization of low cost integrated light sources, especially for those employing quantum dots (QDs) active regions due to their high defects tolerance character. However, this technology is not yet mature enough as some remaining issues such as leveling between III-V and Si layers, imperfect material quality, etc. still need to be addressed [3, 4]. In this work, we present the integration of GaAs QD tunable lasers on a SiPh substrate using micro-transfer printing. The SiPh wafer has a full back-end layer stack and was fabricated in the imec 300 mm SiPh pilot line. The resulting device demonstrated single mode operation with over 45 dB side-mode-suppression- ratio (SMSR) and over 30 nm tuning range in the O-band. The maximum waveguide-coupled power was found to be around 3.5 mW.

2 Micro-transfer Printing

The concept of micro-transfer printing (MTP) was proposed by the Rogers group at Illinois university in 2004 [5]. This technology relies on the use of an elastomeric Polydimethylsiloxane (PDMS) stamp which is able to pick up one or multiple micro-scale device\material film coupons that were released from their native substrate and which can subsequently be transfer printed onto a target substrate. The pick-up and printing procedure relies on the controlling of the retrieving velocity of the stamp while an in-plane shear displacement can be applied in the printing step to ease the releasing of the device from the stamp. This technology shows significant flexibility as for the device of choice, in terms of its shape, size, thickness, and material. It also allows for the manipulation of these components in a massively parallel manner. In the past decade this approach was introduced in the field of integrated photonics where a wide range of micro-scale components, such as 2D materials, $LiNbO_3$ material film coupons, various III-V optoelectronic devices and even electronic chiplets have been successfully transfer-printed on different target substrates [1, 6]. This technology can leverage the advanced photonics platforms as only a slight modification is required to the existing SiPh back-end process flow. Particularly, minor adjustments are required to enable the local back-end layer opening for the intimate contact of the device of interest onto the exposed Si device layer. MTP also exhibits immense potential for scalable integration. Meanwhile advanced transfer-printing tools have been developed in the past years and the state-of-the-art models are able to handle 200 mm and 300 mm substrates while at the same time providing up to 500 nm @ 3σ or higher alignment accuracy, depending on the sizes of the device arrays. The transfer printing process is fast with each printing cycle only taking around 40 s [1, 6].

3 Design and Fabrication of the Tunable Laser

The silicon waveguide circuits were realized on imec's 300 mm SiPh platform, which consists of a 220 nm crystalline silicon device layer with a 160 nm thick poly-Si overlay layer and a complex back-end layer stack including 2 metal intercontact layers and exposed Al bond pads. The widely tunable laser features a classic configuration, as shown in Fig. 1. It consists of a Mach-Zehnder interferometer (MZI)-based tunable reflector, a cascaded micro-ring-based Vernier filter comprising a pair of micro-rings, an in-cavity phase section, a Sagnac-loop mirror formed by connecting the two ports of a 1 × 2 MMI and an in-cavity semiconductor optical amplifier (SOA). The micro-rings are designed with 580 nm wide rib waveguides. Slightly different radii (31.5 μm and 30 μm) are used to provide a combined vernier free spectral range (FSR) over 40 nm while the coupling gaps of the directional couplers are set to be 200 nm to have a loaded Q factor around 10K. Micro-heaters are integrated in the laser cavity to thermally tune the reflectivity of the reflector, the resonances of the micro-rings and the roundtrip phase of the cavity mode. To boost the output power, an additional amplifier is also connected to the output port of the laser (but it was not co-integrated with the laser in this work). The output of the booster is connected to an MZI-based switch with one of its output ports connected to a grating coupler and the other one connected to a high-speed Si modulator to jointly serve as an integrated transmitter.

A slightly modified back-end process flow that consists of a local back-end removal was used in this fabrication at imec, therefore the SiPh wafers are immediately ready for the transfer printing of III-V components. The size of the recess is 2.57 mm × 100 μm, enabling the printing of 2.16 mm × 45 μm device coupons. The vertical pitch of the SOA array is 280 μm. Before MTP, a thin layer of DVS-BCB is spray-coated on the Si chip, serving as a bonding agent while at the same time it helps to planarize the exposed poly-Si layer. Following, a 15-min soft-bake at 150 °C was performed to evaporate the solvent in the DSV-BCB layer. Next, the target chip and a source sample were loaded on an X-Celeprint μTP-100 lab-scale tool for the MTP process, which was carried out at room temperature. Final post-printing processes include BCB curing, n-via opening and metallization. More information about the SOA design and fabrication process flow can be found in [7, 8]. Figure 2 shows a microscope image of a representative fabricated widely tunable laser.

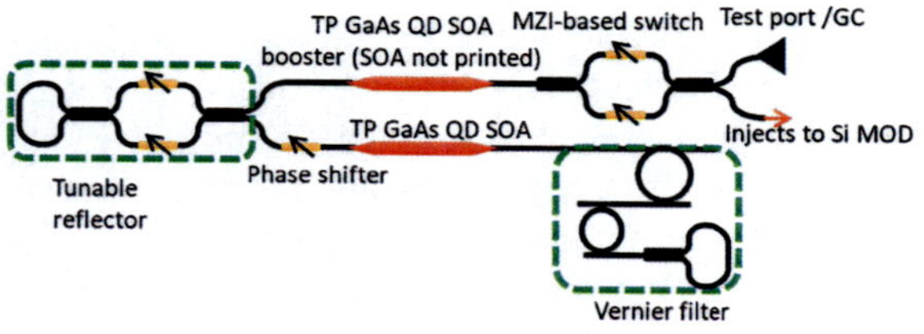

Fig. 1. Schematic layout of the GaAs QD-on-Si widely tunable laser.

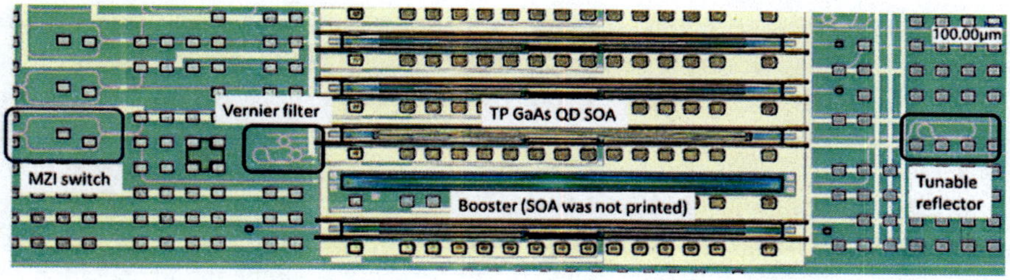

Fig. 2. Microscope image of a representative resulting device.

4 Characterization

The resulting GaAs QD-on-Si widely tunable laser was characterized on a vertical setup using a temperature controller set at 20 °C. A set of Keithley current sources and DC probes were used to drive the transfer-printed SOA and the integrated micro-heaters. The output power was collected using a standard single mode fiber through the output grating coupler indicated in Fig. 2, and is then split into two channels through a 10/90 splitter, with the 10% branch launched to a HP 81531A power meter while 90% of the light

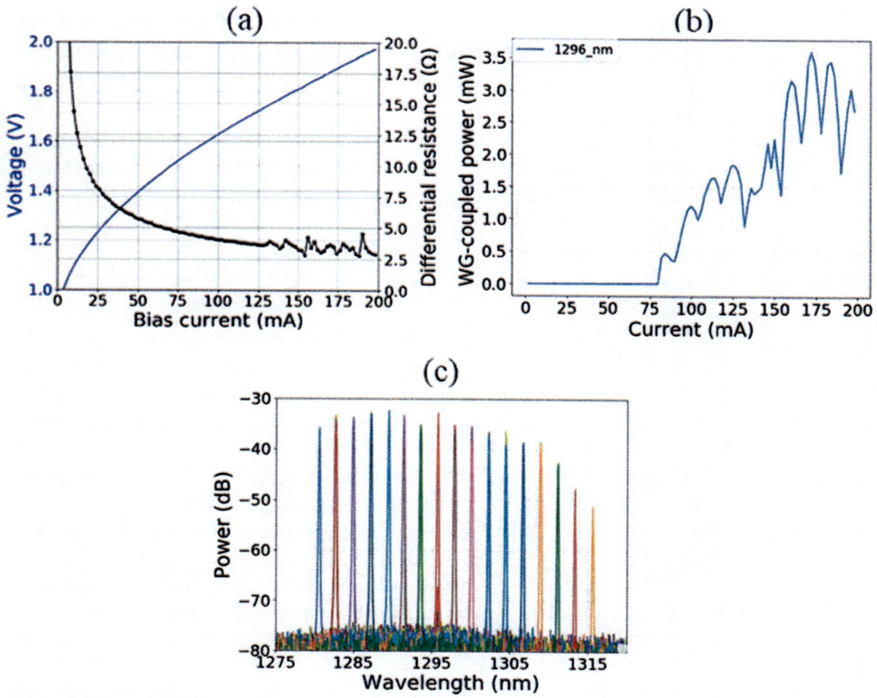

Fig. 3. Performance of the fabricated tunable laser at 20 °C. (a) Measured current-voltage curve with the calculated differential resistance, (b) Light-current curve and (c) Coarse wavelength tuning.

goes to an Agilent 8164B optical spectrum analyzer (OSA). Figure 3(a) shows the IV response and calculated differential resistance with respect to the applied bias current. A differential resistance as low as 3 Ω was obtained at a bias current of 200 mA. Figure 3(b) shows the recorded light-current curve after calibrating out the grating coupler loss. It reveals a threshold of around 80 mA and a maximum waveguide-coupled power of 3.5 mW (after calibrating out the losses from the MZI switch and the grating coupler). The periodical fluctuation of the output power corresponds to the longitudinal cavity mode hop. By tuning the micro-rings and the phase section, single mode operation can be easily achieved with over 45 dB side-mode-suppression-ratio. Figure 3(c) shows the superposition of a set of recorded output spectra obtained by tuning one of the micro-rings and the phase section while the SOA was biased at 180 mA. A wavelength tuning over 30 nm was demonstrated. Continuous wavelength tuning, however, is not demonstrated in this work, but can also be realized by simultaneously adjusting the two micro-rings and the phase section.

5 Conclusion

An O-band GaAs QD-on-Si widely tunable laser was demonstrated on a SiPh platform including a back-end layer stack. No modification was implemented to the existing SiPh back-end process flow used in the imec's 300 mm SiPh pilot line, expect for the addition of a local back-end opening. A 2.16 mm long GaAs QDs SOA coupon was used in this demonstration and the resulting device exhibits single longitudinal mode operation with over 45 dB side-mode suppression and over 30 nm wavelength tuning in the O-band.

Acknowledgements. This work was supported by the European Union's Horizon 2020 research and innovation programs under agreement No. 825453 (CALADAN), No. 101070332(PHORMIC), No. 101070560 (PUNCH) and the ESA project ProtoBIX.

References

1. Roelkens, G., et al.: Micro-Transfer Printing for Heterogeneous Si Photonic Integrated Circuits. IEEE J. Sel. Top. Quantum Electron. **29**(3: Photon. Elec. Co-Inte. and Adv. Trans. Print.), 1–14 (2023)
2. Lindenmann, N., et al.: Photonic wire bonding: a novel concept for chip-scale interconnects. Opt. Express **20**, 17667–17677 (2012)
3. Park, J.-S., Tang, M., Chen, S., Liu, H.: Heteroepitaxial growth of III-V semiconductors on silicon. Crystals **10**(12), 1163 (2020). https://doi.org/10.3390/cryst10121163
4. Shi, Y., et al.: Novel adiabatic coupler for III-V nano-ridge laser grown on a Si photonics platform. Opt. Express **27**(26), 37781–37794 (2019)
5. Sun, Y., Rogers, J.A.: Fabricating semiconductor nano/microwires and transfer printing ordered arrays of them onto plastic substrates. Nano Lett. **4**(10), 1953–1959 (2004)
6. Corbett, B., et al.: Transfer printing for silicon photonics. Semicond. Semimet. **99**, 43–70 (2018)
7. Zhang, J., et al.: Micro-transfer-printed O-band GaAs QD-on-Si DFB Laser on an advanced silicon photonics platform. In: ECIO, Netherlands, 19–21 April (2023)
8. Zhang, J., et al.: Micro-Transfer-Printed O-band GaAs QD III-V-on-Si DFB Laser. In: ECIO, Italy, p.paper W.P.13, (2022)

Widely Tunable Laser on IMOS Platform

Tasfia Kabir(✉), Yi Wang, Stefano Tondini, Yuqing Jiao, Kevin Williams, and Martijn Heck

Eindhoven University of Technology, 5612 AZ Eindhoven, The Netherlands
t.kabir@tue.nl

Abstract. We introduce a compact widely tunable laser on an InP membrane on silicon platform. The laser features an intracavity filter employing a Mach-Zehnder interferometer structure, with an optimized design for a single mode operation. Our results demonstrate a 28 mA threshold current, a maximum output power of 2 dBm on chip, effective single-mode operation with a side-mode-suppression ratio (SMSR) of 30–45 dB, and a broad tuning range of 40 nm. The tuning map of the laser was determined using a search algorithm.

Keywords: Photonic integrated circuit · semiconductor tunable laser · InP membrane

1 Introduction

Widely tunable lasers play a crucial role as optical sources in diverse applications, including dense wavelength division multiplexing [1], optical coherence tomography [2], and gas sensing [3]. In the realm of sensing or communication, attributes like energy efficiency, full integration with all necessary components as well as compactness are paramount. Photonic integrated circuit (PIC) based lasers can offer such benefits. Monolithic InP chips have demonstrated the successful integration of all components required for widely tunable lasers, including a semiconductor optical amplifier (SOA), modulators, and waveguides [4–6]. However, the limitation of a low index contrast hinders the achievement of low bend radii for the waveguides, resulting in larger device footprints and restricted scalability in the monolithic platform. The InP membrane on silicon or IMOS technology [7], as depicted in Fig. 1, emerges as a promising solution due to its high index contrast, enabling low bend radius of 5 µm. IMOS has exhibited substantial potential in realizing high-density PICs, ensuring energy efficiency and scalable manufacturing, given its use of a silicon substrate.

Previously on the IMOS platform, widely tunable Vernier ring lasers with a 25 nm tuning range and a good side-mode suppression ratio (SMSR) exceeding 30 dB was demonstrated by Pogoretskiy *et al.* [8]. An extended tuning range of 50 nm was achieved recently with an AMZI-type structure by Naoki *et al.* [9]. In this paper we introduce a Mach-Zehnder interferometer (MZI)-based intracavity widely tunable laser design on IMOS, along with corresponding measurement results. Notably, this tunable laser showcases the best performance on the IMOS platform, combining low threshold current

J. Witzens et al. (Eds.): ECIO 2024, SPPHY 402, pp. 37–42, 2024.
https://doi.org/10.1007/978-3-031-63378-2_7

and wide tunability range. The laser exhibits a maximum output power of −5 dBm on fiber, a threshold current of 28 mA. A refinement in circuit design, building upon previous demonstrations, guarantees a robust single-mode operation with a side-mode suppression ratio (SMSR) surpassing 30 dB, peaking at 45 dB. The achieved tuning range spans 40 nm, ranging from 1555 nm to 1595 nm, covering sections of the C and L bands in communication. The full wavelength tuning map is created using an automated search routine. On the IMOS platform, this laser occupies only half the area compared to analogous designs previously introduced on the InP monolithic platform [10].

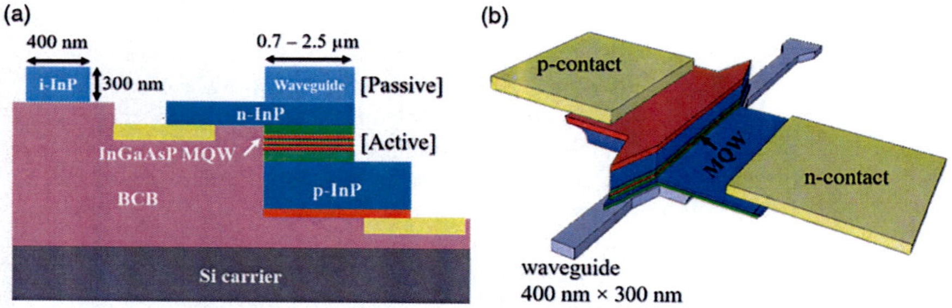

Fig. 1. a) Cross-sectional view of the IMOS twin-guide active–passive integration scheme. b) 3D artist's impression of an IMOS amplifier with adiabatic taper coupled to passive waveguide [6].

2 Design

The laser was fabricated on an in-house active-passive fabrication run of the IMOS platform. A double-sided fabrication process was used to integrate the active with the passive circuitry, as shown in Fig. 1. They are coupled vertically through an evanescent taper, forming a twin guide structure [11]. The sidewalls of the gain section are etched from two directions of the membrane to provide optical confinement which will also allow for electrical contact. The passives are made of strip waveguides that are then buried in a polyimide planarization layer, on top of which lies the metal tracks and the thermo-optic phase shifters (TOPSs). The fabricated chip is shown in Fig. 2b.

The design of the laser is shown in Fig. 2a. The laser cavity is formed by placing a semiconductor optical amplifier (SOA) section and a wavelength selective intracavity interferometric filter between broadband photonic crystal mirrors with bandwidth >200 nm [12]. The front mirror has a reflectivity of 80% and the back mirror is almost 99% reflective. Light is coupled out of the circuit using a focusing grating coupler with a measured coupling loss of 5 to 7 dB. A 500 μm long 4-quantum-well SOA with a gain peak around 1560 nm is used. The intra-cavity filter is split into four arms with a multi-mode interferometer (MMI) tree. The total footprint of the laser is 1.1 mm × 0.4 mm.

Each arm of the laser contains a TOPS of length 220 μm. These arms have a different waveguide length. This length difference determines the shape of the filter. The filter transmission spectrum is shifted by adjusting the current on the TOPS, thereby attaining a constructive interference of fields in the four arms of the filter for a specific wavelength. While adding more arms can make the filter narrower, it also makes the circuit larger and tuning the laser more complicated [13]. Simulations showed that four arms are enough if we choose the right length differences. To ensure an SMSR of over 30 dB and a broad tuning range of 40 nm, the simulation-derived arm length differences are 8.24 μm, 90.64 μm, and 107.12 μm, respectively. The resulting shape of the filter is given in Fig. 2c.

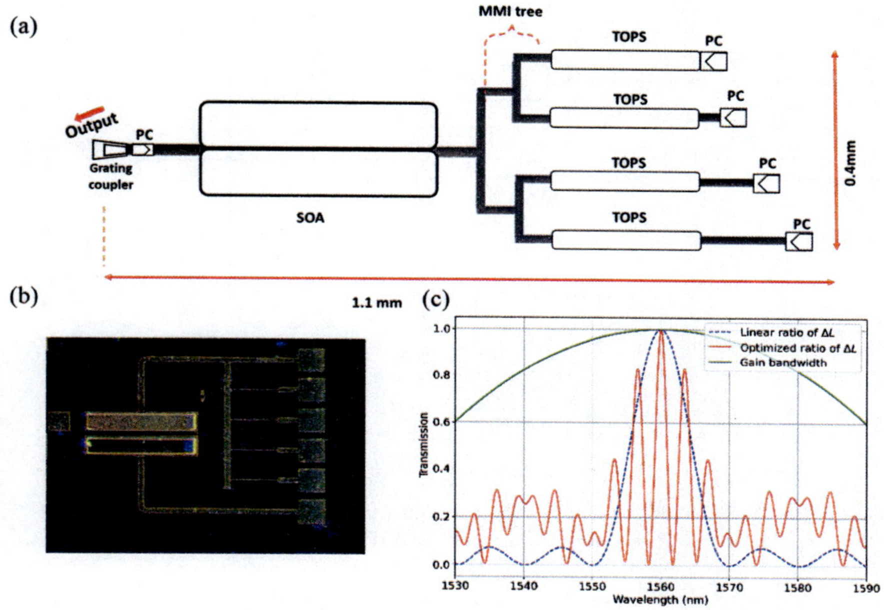

Fig. 2. a) Schematic of laser design. b) Microscope image of fabricated tunable laser. c) The transmission spectra of designed filter in red solid lines, overlapped with gain bandwidth in green. The dashed blue line depicts a linear 4 arm configuration.

3 Experimental Setup and Search Algorithm

The measurement setup is shown in Fig. 3. The DUT is placed upon a copper holder with a water cooler connection, a thermistor, and a Peltier cell. The temperature is stabilized at 10 °C. The fiber-coupled output of the laser goes to a 90–10 splitter, 90% towards a tunable band pass optical filter (EXFO-XTA50) coupled with power meter (Agilent 8614B) and 10% towards an optical spectrum analyzer (Yokogawa AQ6375) to collect the full spectral response of the DUT. Current is applied using multi-channel current sources (Keithley 2600) to tune the intracavity filter over the desired wavelength range, and the SOA current is supplied from a current source (PRO8000).

The stochastic search optimizer involved in this work is the Globalized Bounded Nelder-Mead method [14]. It was set to scatter bias current configurations to the TOPS controllers of the intracavity filter within a 4-dimensional search domain restricted to [0, 15] mA. The optimization routine is fed back by a power meter, which collects the optical signal passing through the external tunable filter. The target function is based on optical power optimization. The fine-tuning of the algorithm allowed it to converge within 200 iterations with a tolerance of 10^{-2} on the absolute error of the target function. For a single wavelength point, the search was repeated three times to limit the probability of the optimizer being trapped into local optima. This procedure has been repeated over multiple instances for different wavelengths, eventually leading to the full tuning of the laser.

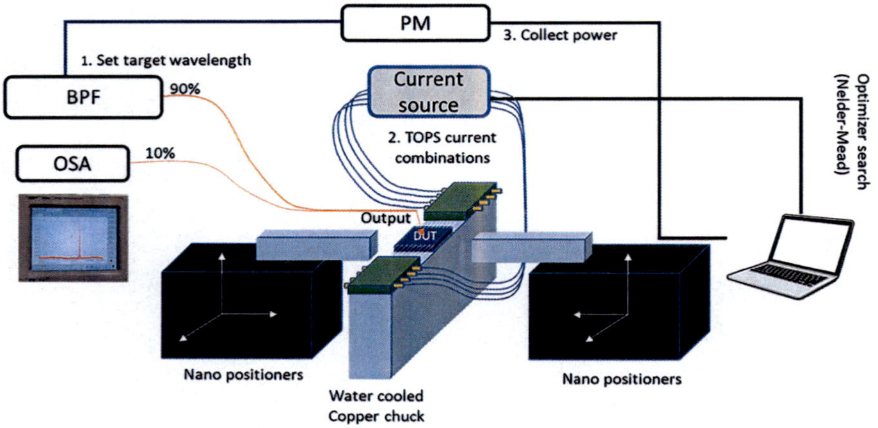

Fig. 3. Experimental setup for automated tuning measurement. Output from the laser DUT is split 90% into band-pass filter onto a power meter and 10% into optical spectrum analyzer.

4 Results and Discussion

The device's light-current-voltage (LIV) characteristics in Fig. 4a exhibit a commendable maximum output power of −5 dBm in fiber and 2 dBm on chip (at 7 dB coupling losses), with the threshold current identified at around 28 mA. This threshold current is significantly lower than previously measured lasers with similar configurations on this platform [8]. The SMSR ranges between 30 to 45 dB, as explained by Fig. 4b, showcasing the laser's single-mode operation via a high-resolution OSA. The tuning capability of the laser controlled through the Nelder Mead algorithm, achieving a broad tuning range of 40 nm. Figure 4c and the accompanying 13 spectra show that the laser could be tuned for 40 nm with an SMSR of >30 dB maintained throughout.

Figure 4d demonstrates the target wavelengths as dictated by the algorithm, and the attained peak wavelength. The points were attained at 0.375 nm step size. As can be seen, the obtained wavelengths mostly follow the target wavelengths. However, gaps can be seen, most notably in the ranges 1565–1568 nm and 1575–1583 nm. This could be caused by reflections, from additional cavities being formed between PC mirrors, grating couplers and MMIs. It is conceivable that certain design constraints, like a limited shift range in TOPM or practical challenges such as issues with chip cooling or SOA biasing, might hinder the provision of specific driving combinations for the laser to emit in certain spectral regions. This is currently being studied.

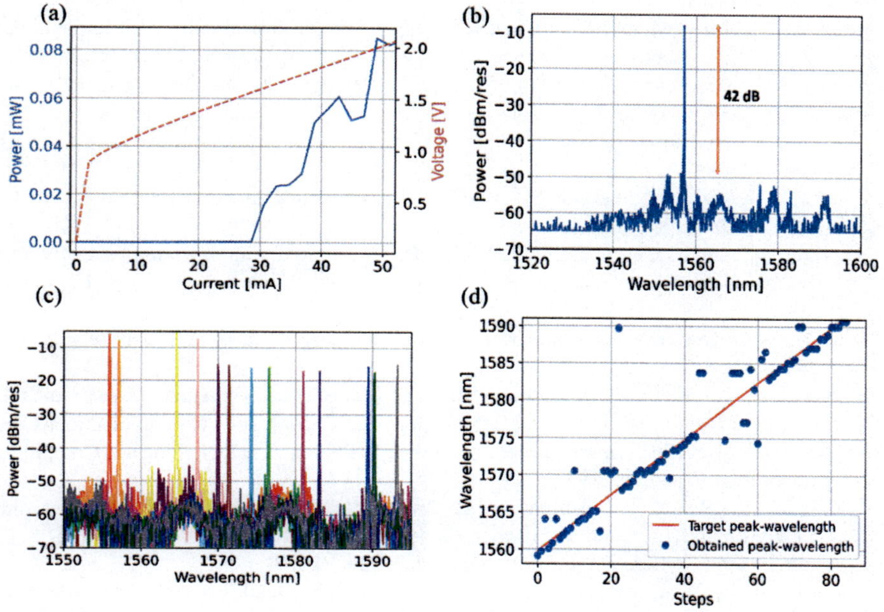

Fig. 4. a) LIV characteristics of laser. b) Single mode spectra showing 42 dB SMSR c) Selected laser spectra showing 40 nm tuning range can be accessed d) Target peak wavelength and wavelength attained with search algorithm, with SOA current set at 50 mA.

5 Conclusions

We present the best-performing demonstration of a tunable laser based on IMOS, with a low threshold current of 28 mA and a wide tuning range of 40 nm and good SMSR values of 30–42 dB. Notably, we achieved power levels of −5 dBm on fiber, which on chip considering 7 dB coupling losses is around 2 dBm. The tunability range covers significant portions of the C and L bands. The design of the circuit was optimized for a narrow shape of filter without increasing complexity, which resulted in good SMSR values. The tuning process was fully automated, facilitated by a search algorithm. The tuning range of the laser is limited by the available gain bandwidth of the SOA and can be further improved by improving the thermal performance of the SOA. This work thus expands the IMOS platform with a new and optimized components.

Acknowledgements. Synergia consortium and NWO Zwaartekracht Project "Research Centre for Integrated Nanophotonic.

References

1. Tsuzuki, K., et al.: Full C-band tunable DFB laser array copackaged with INP Mach-Zehnder modulator for DWDM optical communication systems. IEEE J. Sel. Top. Quantum Electron. **15**(3), 521–527 (2009)
2. Hazan, J., Couka, T., Pajkovic, R., Williams, K., Bente, E.: Control strategy for a monolithically integrated widely tunable laser system on InP for optical coherence tomography. In: Novel In-Plane Semiconductor Lasers XXI, vol. 12021, pp. 53–63 (2022)
3. Hänsel, A., Heck, M.J.R.: Feasibility of telecom-wavelength photonic integrated circuits for gas sensors. Sensors **18**(9), 2870 (2018)
4. Smit, M.: InP-based generic integration technology. In: Advanced Photonics for Communications (2014)
5. Latkowski, S., et al.: Novel widely tunable monolithically integrated laser source. IEEE Photonics J. **7**, 1–9 (2015)
6. Andreou, S., Williams, K.A., Bente, E.A.J.M.: Electro-optic tuning of a monolithically integrated widely tuneable InP laser with free-running and stabilized operation. J. Lightwave Technol. **38**(7), 1887–1894 (2020)
7. Jiao, Y., et al.: INP membrane integrated photonics research. Semicond. Sci. Technol. **36**, 013001 (2020)
8. Pogoretskiy, V., van der Tol, J., Smit, M., Jiao, Y.: Monolithically integrated widely tunable laser on an InP membrane circuits. In: 2019 24th OptoElectronics and Communications Conference (OECC) and 2019 International Conference on Photonics in Switching and Computing (PSC), Fukuoka, Japan, pp. 1–3 (2019)
9. Takahashi, N., Wang, Y., Nishiyama, N., Williams, K., Latkowski, S., Jiao, Y.: Compact widely-tunable laser on an InP membrane on silicon. In: 2023 Opto-Electronics and Communications Conference (OECC), Shanghai, China (2023)
10. Kabir, T., Tondini, S., Hazan, J., Heck, M.J.R.: 40-nm widely tunable laser with Mach-Zehnder interferometer based intra-cavity filter. In: 2023 IEEE Photonics Conference (IPC), Orlando, FL, USA (2023)
11. Wang, Y., et al.: Efficiency-boosted semiconductor optical amplifiers via mode-division multiplexing. Optica **10**, 1153–1160 (2023)
12. Reniers, S.F.G., Wang, Y., Williams, K.A., Van Der Tol, J.J.G.M., Jiao, Y.: Characterization of waveguide photonic crystal reflectors on indium phosphide membranes. In: IEEE J. Quantum Electron. **55**(6), 1–7, 6400107 (2019)
13. Castellan, C., Tondini, S., Mancinelli, M., Kopp, C., Pavesi, L.: Low crosstalk silicon arrayed waveguide gratings for on-chip optical multiplexing. In: Silicon Photonics: From Fundamental Research to Manufacturing (2018)
14. Tondini, S., Castellan, C., Medina, M.A., Pavesi, L.: Automatic initialization methods for photonic components on a silicon-based optical switch. Appl. Sci. **2019**, 9 (1843)

Ultrafast Tunable Photonic Integrated E-DBR Pockels Laser

Anat Siddharth[1(✉)], Simone Bianconi[1], Zheru Qiu[1], Rui N. Wang[2], Mohammad J. Bereyhi[2], Tobias J. Kippenberg[1], and Johann Riemensberger[1,3]

[1] Swiss Federal Institute of Technology Lausanne (EPFL), 1015 Lausanne, Switzerland
anat.siddharth@epfl.ch
[2] Luxtelligence SA, 1015 Lausanne, Switzerland
[3] Norwegian University of Science and Technology, 7491 Trondheim, Norway

Abstract. The hybrid integration of III-V gain media with low-loss photonic integrated circuits has advanced the performance of laser systems in terms of laser stability and coherence. The advent of lithium niobate on insulator (LNOI) photonics for self-injection locking and external cavities has allowed to increase in the frequency tuning rates of such lasers beyond the peta-hertz per second regime. Yet such lasers have up to now been difficult to operate and require complex photonic circuits and have only achieved low output powers. Here, we demonstrate a hybrid integrated frequency-agile laser based on an InP RSOA and an extended electro-optic Bragg waveguide reflector (E-DBR) manufactured with a wafer-scale lithium niobate-on-insulator process. The E-DBR laser can be tuned over 10 GHz with a tuning efficiency of more than 550 MHz/V with megahertz speed, and good linearity while featuring high optical output power of more than 10 mW and kilohertz linewidth.

Keywords: Integrated photonics · Lasers · Electro-optic material

1 Introduction

Hybrid photonic integrated lasers have revolutionized laser technology by combining the scale manufacturing of simple laser diodes with a performance, in particular frequency stability, optical linewidth [1], and tuning linearity [2], otherwise only achievable with bulk cavity lasers. The introduction of ferroelectric thin films photonic integrated circuits such as lithium niobate on insulator (LNOI) [3] has further allowed to substantially increase the tuning speeds of the hybrid integrated lasers [4] or on-chip second harmonic generation [5]. A particularly attractive architecture for hybrid integrated lasers is extended Bragg gratings first implemented in the 1990s. They do not require high-Q optical microresonators and allow for high-power operation. Recent demonstrations of an LNOI-based E-DBR laser [6] did not achieve performance that is superior to existing commercial DFB lasers. Here, we present an optimized hybrid integrated LNOI-based E-DBR laser with high output power, tuning efficiency, linearity, and low linewidth surpassing contemporary DFB lasers for applications such as frequency-modulated LiDAR and anemometry, optical communications, and metrology.

J. Witzens et al. (Eds.): ECIO 2024, SPPHY 402, pp. 43–46, 2024.
https://doi.org/10.1007/978-3-031-63378-2_8

2 Measurements and Results

Figure 1a shows the structure of the hybrid integrated Pockels E-DBR laser where a Reflective Semiconductor Optical Amplifier (RSOA) with 90% back facet reflection is edge-coupled to the LNOI photonic chip using an angled and outwards tapered LNOI waveguid Figure 1b presents a photograph of the experimental setup showing the hybrid integra-tion of the LNOI PIC and the RSOA on a copper chip holder. Electric pin probes supply the tuning voltage. The photonic chip is depicted in Fig. 1c and has a size of 10 × 5 mm featuring 35 devices. Figure 1d presents a False-colored scanning electron micro-scope (SEM) image showing the LNOI waveguide with post gratings. The LNOI wave-guide and gratings are fabricated with a diamond-like carbon hard mask [1] on com-mercial wafer substrates with 400 nm thickness and 200 nm etch depth for a waveguide ridge width of 1.0 μm. A shallow etching is favored here to increase the coupling be-tween the waveguide mode and the grating posts that have a diameter of 400 nm are placed at a distance of 400 nm from the waveguide sidewall. The posts provide a weak perturbation of the effective refractive index of the low-confined fundamental TE mode at the Bragg wavelength and constructively add up the desired optical reflection at this wavelength. The grating is designed as a third-order grating with a period of 1.28 μm and a grating back-reflection strength of 6.6 cm^{-1}. Chip-to-lensed fiber coupling is achieved with a double inverse taper structure that features less than 1.5 dB coupling loss.

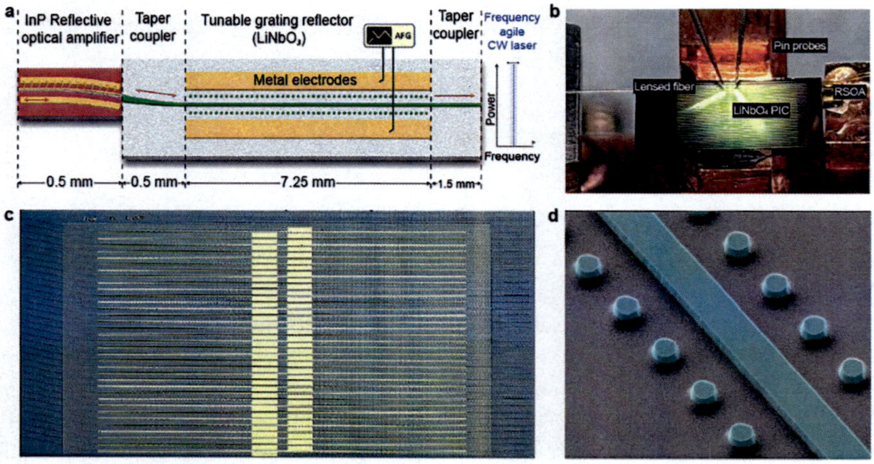

Fig. 1. Schematic showing the structure of the hybrid integrated Pockels E-DBR laser. (b) Photograph illustrating the experimental setup for operating the laser.

The optical setup for E-DBR laser characterization is depicted in Fig. 2a. Figure 2b shows the single-sided power spectral density (PSD) of the laser frequency noise. This measurement was done by heterodyne beat note spectroscopy by beating a reference cavity-locked external cavity diode laser with the E-DBR laser. The recorded data for the in-phase and quadrature components of the beatnote were processed by Welch's method

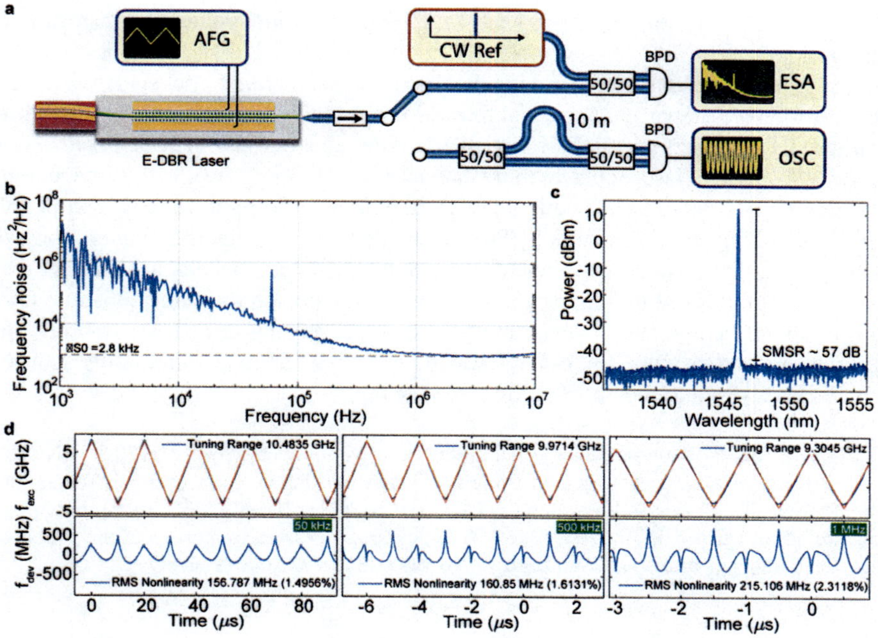

Fig. 2. (a) Optical setup for linewidth and tuning measurement of E-DBR Laser; AFG: Arbitary function generator; CW Ref: Cavity-stabilized reference Laser; ESA: Electrical Spectrum Analyzer; OSC: Oscillo-scope; BPD: Balanced photodiode. (b) Single sideband power spectral density of frequency noise of the Pockels E-DBR laser. (c) Optical spectrum of the Pockels E-DBR laser with more than 57 dB SMSR (RBW: 0.02 nm). (d) Laser frequency tuning curves measured with 10 Mach-Zehnder fiber interferometer spectrograms of the laser tuning applying 19 Vpp (peak-to-peak) triangular waveforms at 50 kHz (left), 500 kHz (middle), and 1 MHz (right) across the metal electrodes.

[8] to retrieve the single-sided phase noise power spectral density $S_{\phi\phi}$ that was converted to frequency noise S_{ff}. The laser frequency noise reaches a value (white noise floor) of 891 Hz^2/Hz at a 2 MHz offset, corresponding to the intrinsic laser linewidth of 2.8 kHz, which is superior to prior demonstrations based on a Vernier racetrack filter [5]. The noise spike at 70 kHz is attributed to a mechanical flapping mode of the photonic chip. Figure 2c shows the optical spectrum of the self-injection locked photonic integrated laser at the wavelength of 1546 nm with a side-mode suppression ratio (SMSR) of more than 57 dB and an output power of 10.3 mW. This is the highest recorded output power for a hybrid integrated external cavity laser based on an LNOI photonic integrated circuit. The frequency tuning and linearity were measured with a 10 m long fiber-based Mach-Zehnder interferometer. The frequency tuning is extracted using a Hilbert-transform-based phase extraction algorithm [9] and the extracted frequency tuning curves for triangular voltage modulation are depicted in Fig. 2d. We demonstrate the high-speed tuning of the photonic integrated Pockels E-DBR laser and show the time-frequency pattern 50 kHz, 500 kHz, and 1 MHz. The bottom panel row depicts the deviation of the measured frequency tuning pattern from a perfect triangular frequency modulation

with RMS deviations ranging from 1.5% to 2.3% without any voltage predistortion or compensation.

In summary, we have demonstrated the first hybrid integrated Pockels E-DBR laser with performance, in particular optical linewidth and tuning linearity, that is superior to existing InP DFB lasers and outperforms previous demonstrations of LNOI-based hybrid integrated lasers. Our laser achieves more than 10 mW of output power with a free running linewidth of 3 kHz and 10 GHz tuning range with only 19 V_{pp} input, which is compatible with CMOS charge pump technology [10], allowing for full miniaturization of a LiDAR engine. For further noise suppression, the photonic integrated Pockels E-DBR laser may be self-injection locked to a high-Q LNOI microresonator on the same substrate [11]. Also, hybrid packaging is a robust method to improve long-term laser stability and reduce laser frequency noise at offset frequencies below 100 kHz, eliminating acoustic instabilities inherent to the optical chip testing setup.

Acknowledgements. This publication was supported by Contract W911NF2120248 (NINJA) from the Defense Advanced Research Projects Agency (DARPA), Microsystems Technology Office (MTO). This work is also supported by HDLN (High Density Lithium Niobate), EIC Transition grant agreement no. 101113260. A.S. acknowledges the support from the European Space Technology Centre with ESA Contract No. 4000135357/21/NL/GLC/my.

References

1. Li, B., et al.: Reaching fiber-laser coherence in integrated photonics. Opt. Lett. **46**(20), 5201–5204 (2021). https://doi.org/10.1364/OL.439720
2. Lihachev, G., et al.: Low-noise frequency-agile photonic integrated lasers for coherent ranging. Nat. Commun. **13**(1), 3522 (2022). https://doi.org/10.1038/s41467-022-30911-6
3. Boes, A., et al.: Lithium niobate photonics: unlocking the electromagnetic spectrum. Science **379**(6627), eabj4396 (2023). https://doi.org/10.1126/science.abj4396
4. Snigirev, V., et al.: Ultrafast tunable lasers using lithium niobate integrated photonics. Nature **615**(7952), 411–417 (2023). https://doi.org/10.1038/s41586-023-05724-2
5. Li, M., et al.: Integrated Pockels laser. Nat.e Commun. **13**(1), 5344 (2022). https://doi.org/10.1038/s41467-022-33101-6
6. Xue, S., et al.: Integrated DBR-based Pockels Laser. In: CLEO 2023 (2023), Paper SM2J.6, Optica Publishing Group, May 2023, p. SM2J.6 (2023). https://doi.org/10.1364/CLEO_SI.2023.SM2J.6
7. Li, Z., et al.: High density lithium niobate photonic integrated circuits. Nat. Commun. **14**(1), 4856 (2023). https://doi.org/10.1038/s41467-023-40502-8
8. Welch, P.: The use of fast fourier transform for the estimation of power spectra: a method based on time averaging over short, modified periodograms. IEEE Trans. Audio Electroacoust. **15**(2), 70–73 (1967). https://doi.org/10.1109/TAU.1967.1161901
9. Ahn, T.J., Kim, D.Y.: Analysis of nonlinear frequency sweep in high-speed tunable laser sources using a self-homodyne measurement and Hilbert transformation. Appl. Opt. **46**(13), 2394 (2007)
10. Lukashchuk, A., et al: Photonic-electronic integrated circuit-based coherent LiDAR en-gine. arXiv, June 10 (2023). https://doi.org/10.48550/arXiv.2306.07990
11. Siddharth, A., et al.: Hertz-linewidth and frequency-agile photonic integrated extended-DBR lasers. arXiv, June 05 (2023). https://doi.org/10.48550/arXiv.2306.03184

External Net Gain in Monolithically Integrated Si_3N_4-Al_2O_3:Er^{3+} Spiral Waveguide Amplifiers

Dawson B. Bonneville[1,4](✉), Carlos E. Osornio Martinez[1,4], Ivo Hegeman[2,4], Quentin Coulaud[3,4], Meindert Dijkstra[1,4], and Sonia M. García-Blanco[1,4]

[1] Integrated Optical Systems, MESA+ Institute for Nanotechnology, University of Twente, 7500 AE Enschede, The Netherlands
d.b.bonneville@utwente.nl
[2] LioniX International, 7521 AN Enschede, The Netherlands
[3] Keopsys Industries, 22300 Lannion, France
[4] Universite de Rennes, CNRS, Institut Foton, UMR 6082, 35000 Rennes, France

Abstract. External net gain is demonstrated in the C-band using reactively sputtered Al_2O_3:Er^{3+} spiral waveguide amplifiers monolithically integrated with Si_3N_4. On-chip powers up to 50 mW are measured, with >30 dB internal gain at 1532 nm and external gain of ~5 dB from 1530–1560 nm for a 20 cm waveguide with an erbium concentration of 1.5 ions/cm^3.

Keywords: Aluminium oxide · Waveguide amplifier · Reactive sputtering

1 Introduction

Reactively sputtered polycrystalline Al_2O_3 has recently shown itself to be a viable host material for very high amplifier powers using rare-earth doping [1]. Additionally, it demonstrates broadband operation with low losses [2], and lends itself well as a reactively sputtered material to being monolithically integrated onto other waveguide platforms [3]. To demonstrate this towards the realization of LIDAR components and packaged amplifiers, Al_2O_3:Er^{3+} was monolithically integrated onto the LioniX TriPleX platform, with underlying passive components such as edge couplers, on-chip 1480/1550 nm WDMs and adiabatic transitions for low-loss power transfer between the two layers. This process has been shown [4] to provide promising results towards C-band amplifiers, and results here show for the first time external fiber-fiber gain on the platform.

2 Experimental Details

To measure the samples, a 127 μm pitch fiber array was used, as well as isolators for the 2–1480 nm pumps used for bidirectional pumping (500 mW output power each), a tuneable C-band laser and a fiber coupled free space u-bench with a <1500 nm pass filter

D.B. Bonneville and C.E. Osornio Martinez—These authors both contributed equally to this work.

J. Witzens et al. (Eds.): ECIO 2024, SPPHY 402, pp. 47–50, 2024.
https://doi.org/10.1007/978-3-031-63378-2_9

before the detector was used, which was an OSA, calibrated with a thermal detector. A 1.75 μm wide waveguide amplifier with a length of 20 cm was fabricated in a 750 nm thick layer with a dopant concentration of 1.5×10^{20} ions/cm^3. The $Al_2O_3{:}Er^{3+}$ layer was reactively sputtered and subsequently patterned following procedures outlined in [5], which was vertically aligned to an underlying LioniX TriPleX layer. Adiabatic transitions were used to couple between the layers, which characterized on their own demonstrate 0.3 dB losses/coupler for an 800 μm long taper.

The results for the 20 cm sample are shown in Fig. 1. Parasitic lasing was observed when pumping the chip at maximum powers, which was eliminated by using index matching gel. By using the signal enhancement (SE) method, internal on-chip gain of >30 dB was achieved at 1532 nm. Additionally, the waveguide demonstrates very low background losses, down to 0.11 dB/cm at 1630 nm shown in Fig. 1b). In Fig. 1d) the amplifier launched signal power shown has been corrected for the losses caused by the underlying passive components such as the vertical couplers, on-chip WDMs, passive waveguide losses, and fiber-chip coupling. This results in a peak power of 16.2±1.7 dBm or roughly 40–60 mW of on-chip output power.

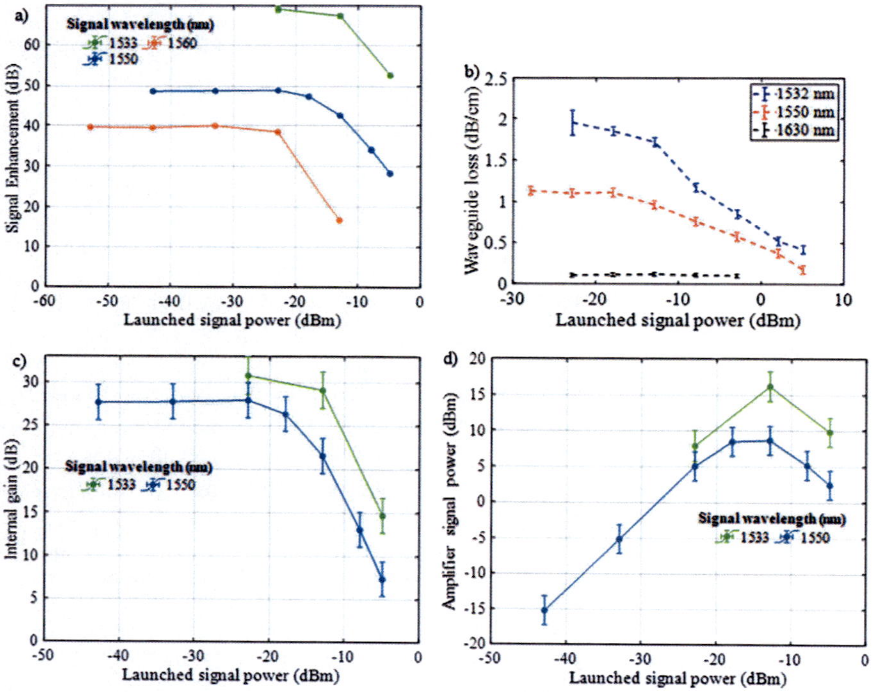

Fig. 1. Characterization of a 20 cm long spiral waveguide amplifier: a) Signal enhancement as a function of launched signal power for three signal wavelengths; b) Waveguide losses as a function of launched signal power; c) Internal net gain at 1533 nm and 1550 nm as a function of launched signal power; d) On-chip signal output power.

By comparing the output power from the fiber array using an integrating sphere, to the output power collected after propagating through the amplifier, net gain can be measured. Here, to demonstrate it using an OSA, the incident power to the amplifier was measured via calibration to the additional OSA losses. By comparing this incident power to the measured power on the OSA, net gain was calculated after measuring for a variety of wavelengths, shown in Fig. 2.

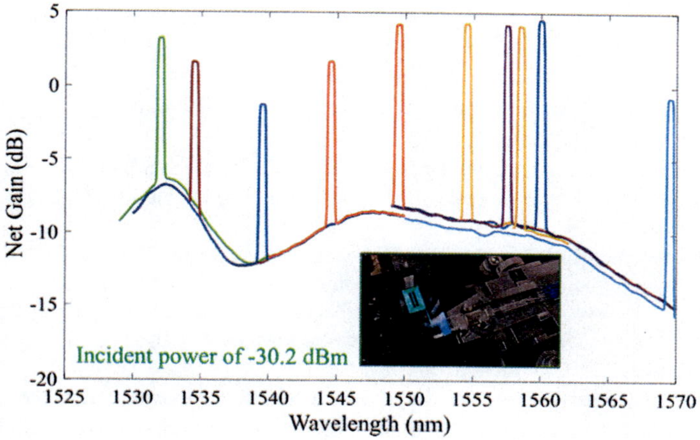

Fig. 2. Net fiber-to-fiber gain at different wavelengths. Inset: photo of the amplifier during test with fiber array.

As shown in Fig. 2, the highest off-chip gain was observed for 1560 nm, which is ~30 nm away from the peak internal gain wavelength of 1532 nm. Here a peak external net gain of 7 dB was measured for an incident power of −40 dBm. This is shown in Fig. 3 with an inset of the calibrated output power from the chip for 1560 nm across a broad power range.

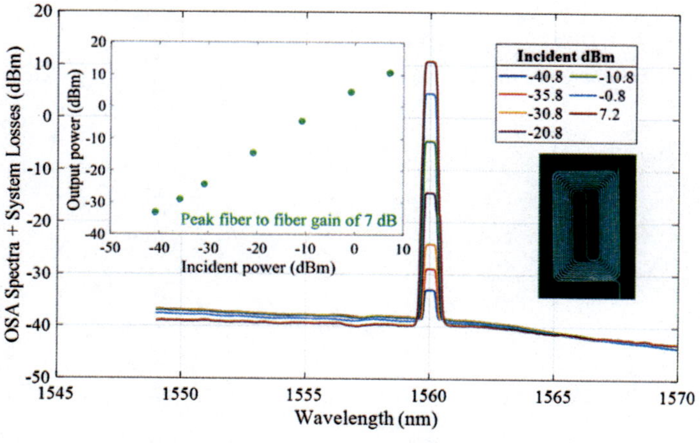

Fig. 3. Calibrated output power from the PIC amplifier for different incident signal powers at 1560 nm, showing a linear increase with a gain of 7 dB (inset).

The discrepancy between the on-chip gain observed in Fig. 1, and off-chip gain in Figs. 2 and 3 is anticipated to be a cause of additional losses caused by a spectral shift of the on-chip WDMs. On-going work towards optimization of those on-chip passive devices such as WDMs, adiabatic vertical transitions, and edge couplers has promise to improve the results show here towards even higher output powers. Additionally, further optimization of the amplifier concentration, length and cross section is a promising pathway towards improving in the future.

References

1. Singh, J.L., Sinobad, M., et al.: Watt-class CMOS-compatible power amplifier. arXiv:2306.12940 [physics.optics] (2023)
2. Hendriks, W.A.P.M., Dijkstra, M., Hegeman, I., et al.: Poly-crystalline low-loss aluminium oxide waveguides. In: Proceedings of SPIE - The International Society for Optical Engineering, vol. 11689 (2021)
3. Mu, J., Dijkstra, M., Korterik, J., et al.: High-gain waveguide amplifiers in Si3N4 technology via double-layer monolithic integration. Photonics Res. **8**(10), 1634–1641 (2020)
4. Bonneville, D.B., Osornio-Martinez, C.E., Dijkstra, M., García-Blanco, S.M.: Integration of Amplifiers in Silicon Nitride Photonic Circuits. In: Presented at OFC Conference, San Diego, CA, USA, 7 March (2023). Invited Talk
5. Emmerik, C.I., Hendriks, W.A.P.M., Stok, M.M., et al.: Relative oxidation state of the target as guide line for depositing optical quality RF reactive magnetron sputtered Al_2O_3 layers. Opt. Mater. Express **10**(6), 1451–1462 (2020)

Optoelectronic Devices

III-V Electro-Absorption Modulation and Detection Devices Integrated to 220 nm Silicon-on-Insulator

Owen Moynihan(✉), Samir Ghosh, James O'Callaghan, Brendan Roycroft, Kevin Thomas, Emanuele Pelucchi, and Brian Corbett

Tyndall National Institute, University College Cork, Cork, Ireland
owen.moynihan@tyndall.ie

Abstract. III-V indium phosphide (InP) ridge coupons capable of electro-absorption modulation and detection are integrated on a 220 nm silicon-on-insulator platform using transfer printing. The coupons employ a PIN structure design that can be actively biased to efficiently modulate and detect light with a bandwidth of >30 nm from 1550–1580 nm. Across the C-band, the grating coupled devices provide an extinction ratio (ER) of 25 dB and detect light with a responsivity of over 0.85 A/W, with an insertion loss (IL) at 0 V of 3.5 dB at 1580 nm and 6.5 dB at 1550 nm.

Keywords: Heterogeneous integration · micro-transfer printing · electro-optic modulation · photo detection · Quantum-confined-Stark-effect (QCSE)

1 Introduction

While silicon photonics (SiPh) comprising of silicon-on-insulator (SOI), is the predominant photonics integrated circuit (PIC) platform [1], there are still drawbacks when considering its limited active device selection. For example, SiPh does not have an efficient monolithic laser/amplifier due to its indirect bandgap [2], while modulation is limited to carrier dispersion effects. This results in several millimetres of device length for Mach-Zehnder interferometers or high fabrication precision and low spectral bandwidth for micro-ring resonator modulator configurations [3]. Due to these limitations, the silicon modulator performance remains inferior to InP-based modulators. Germanium (Ge) detectors involve multiple additional fabrication steps such as dopant implants and high-temperature processes. One solution to these issues is the heterogeneous integration of active devices composed of suitable materials such as III-V and lithium niobate onto SiPh. This integration can take place via flip-chip mounting [4], wafer bonding [5] or micro-transfer printing (μTP) [6].

In this work, III-V PIN waveguide coupons are fabricated and integrated on 220 nm SOI via transfer printing. The waveguiding, modulation and detection characteristics are measured through both surface normal absorption and waveguide transmission using external vertical grating coupling. The InAlGaAs multiple quantum well (MQW) structures act as both electro-absorption modulator (EAM) and photodiode (PD). While most

J. Witzens et al. (Eds.): ECIO 2024, SPPHY 402, pp. 53–57, 2024.
https://doi.org/10.1007/978-3-031-63378-2_10

PDs work far above the bandgap of the absorber material, this device intentionally works through the band edge, utilizing the quantum confined Stark effect (QCSE) to spectrally modulate and detect light by application of voltage [7]. A ridge waveguide design is implemented for a long absorption length, or transparent guiding, as needed. This is the first time a transfer printed III-V EAM device has been designed and produced for SiPh, to the best of the authors knowledge.

2 Device Design and Fabrication

The epitaxial structure was grown on an InP substrate, consisting of a thin (200 nm) n-InP region, an intrinsic region including a 10 layer MQW active region and a thick (1600 nm) p-InP region, with an InGaAs p-contact layer. The QCSE shift was measured using a 1 mm diameter mesa PIN diode test structure with annular contacts via surface normal absorption, which can be seen in Fig. 1(a). With biasing from 0 V–10 V, a band-edge shift is of >100 nm is measured (1528 nm to 1648 nm). Figure 1(b) shows the quadratic dependence of the absorption band edge energy with the applied voltage.

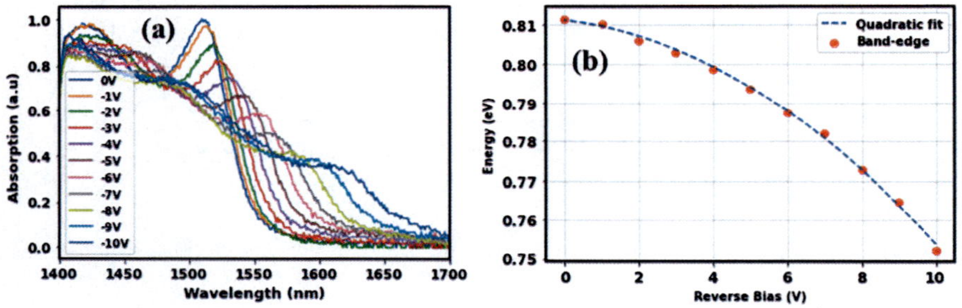

Fig. 1. (a) Surface normal absorption spectrum from 0 V to −10 V of the III-V epitaxial structure and (b) the extracted band-edge energy as a function of voltage.

An evanescently coupled ridge was designed for the III-V - SOI integration (Fig. 2(a)), whereby the coupon is printed directly on the SOI waveguide. The III-V ridge was designed to have a high mode overlap with the active region, with the simulated QW overlap being 22.5%, for a 3 μm wide ridge. Using the collinear silicon waveguide, the overlap with the loss-inducing p-doped region is 12.0% compared to 14.2% without the underlying silicon. The coupling to/from the ridge was designed using tapers to gradually move the light between the SOI waveguide and III-V ridge. Through simulations, the taper was designed to maintain a single-mode transition from the SOI to the EAM for high transmission between the waveguides, with 99% transmission for zero and 62% for 0.5 μm misalignment, respectively. The coupons, comprising of partially finished devices, were fabricated using combined electron-beam and optical lithography together with plasma etching to define the ridge, n-metal, coupon, and photoresist tethers, which were then undercut using a diluted $FeCl_3$ solution. SOI waveguides and gratings were fabricated using electron-beam lithography and plasma etching. The devices were printed directly to the 220 nm SOI over the waveguide structure, where the bond pads were then deposited. The printing accuracy was measured to be ± 0.5 μm. The device design cross-section when printed to the SOI substrate and fabricated device with lithography layers are shown in Fig. 2.

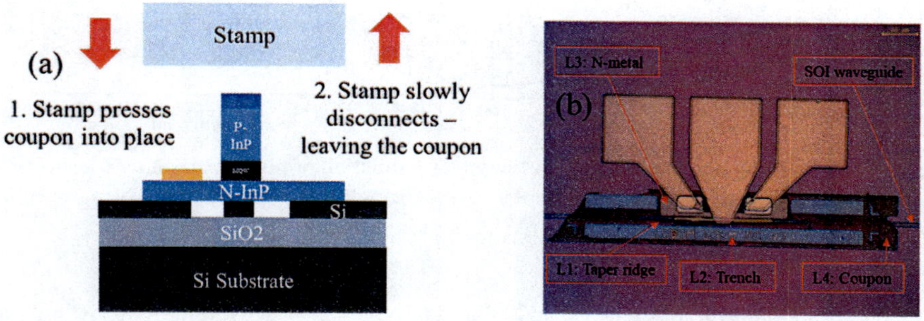

Fig. 2. (a) A schematic of the device cross-section (not to scale) and integration technique. (b) Integrated device and its components.

3 Device Characteristics

A 3 μm ridge with a device length (including in/out tapers of 100 μm) of 250 μm was measured by coupling light in and out of the circuit and measuring the transmission power and modulator photocurrent as a function of wavelength and bias. The waveguide power was extracted by transmission measurements of a reference circuit without a modulator. Figure 3 (a) shows the measured photocurrent vs. wavelength at different biases. The extracted transmission spectra of the device at different biases are shown in in Fig. 3 (b), which were used to calculate the extinction ratio (ER) and insertion loss (IL). Fabry-Perot fringes are measured in the spectra, which are due to the SOI waveguide taper between the grating and the single-mode waveguide region. For the best device (assumed zero misalignment), a 30 dB ER for a bias between 0 V to −6 V together with an IL of 6.5 dB was measured at 1550 nm. The ER remains greater than 25 dB across 1565 nm to 1580 nm with biasing from 0 V to −8 V, with an IL of 3.5 dB. The higher IL at 1550 nm is due to residual band edge absorption. The static characterization has a performance equivalent to state-of-the-art EAM modulation in monolithic InP platforms, in terms of ER and IL [8].

The detection characteristics similarly utilize the QCSE effect to shift the normally transparent material to absorb. The responsivity from 1550 nm to 1580 nm vs bias is shown in Fig. 4(a) and a comparison of the light vs dark current is shown in Fig. 4(b). As the material is mostly transparent without bias, there is very low responsivity at 0 V for 1550 nm. With increased bias, efficient absorption of light is achieved (at bias > −4 V) where a responsivity of 0.85 A/W and higher is calculated. This responsivity is calculated by measuring the light power before entering the circuit and subtracting the loss from the grating coupler (assuming two grating couplers are identical) of a reference grating (loss = 5.2 dB). As the taper is actively absorbing, the simulated loss from the taper misalignment is not included (~0.8 dB), and this is a lower bound for responsivity.

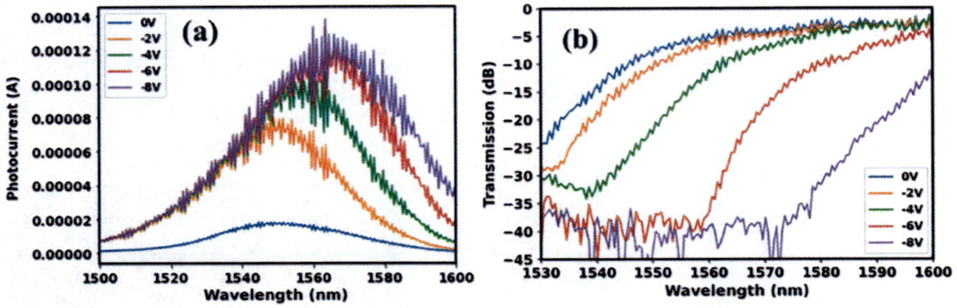

Fig. 3. (a) Received photocurrent vs. wavelength as a function of reverse bias after transmission through the circuit, including grating couplers, (b) extracted transmission vs wavelength for the modulator from 0 V to −8 V, confirming QCSE shift in the integrated device.

The detection quantum efficiency is calculated to be 87% for wavelengths 1570 nm and 1580 nm at −8 V.

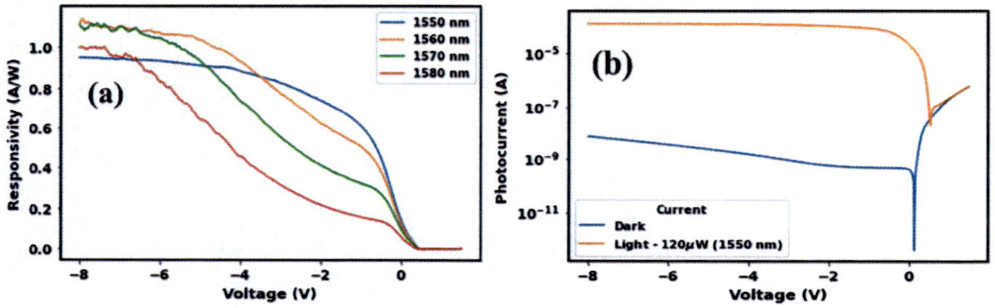

Fig. 4. (a) Responsivity vs. bias for 1550 nm to 1580 nm, (b) Comparison of light and dark current at 1550 nm.

4 Conclusion

InP EAM/PD device coupons have been successfully integrated on to a 220 nm SOI platform using transfer printing with their transmission and absorption characteristics measured. The devices are capable of functioning as a modulator and detector, displaying a responsivity of over 0.85 A/W at −4 V and an ER of >30 dB at 1550 nm, with an overall 25 dB ER over a 30 nm range. The IL is measured to be 6.5 dB and 3.5 dB for 1550 nm and 1580 nm respectively. The QCSE shift in the material was over 100 nm at −10 V and can be used to spectrally modulate transmission and absorption.

Acknowledgements. This work was supported by SFI through IPIC (12/RC/2276-P2) and the PIADs programme. The authors acknowledge the fabrication facility and services at Tyndall National Institute.

References

1. Bogaerts, W., Chrostowski, L.: Silicon photonics circuit design: methods, tools and challenges. Laser Photonics Rev. (2018). https://doi.org/10.1002/lpor.201700237
2. Kaur, P., Boes, A., Ren, G., Nguyen, T.G., Roelkens, G., Mitchell, A.: Hybrid and heterogeneous photonic integration. APL Photonics **6**(6) (2021). https://doi.org/10.1063/5.0052700
3. Reed, G.T., Mashanovich, G., Gardes, F.Y., Thomson, D.J.: Silicon optical modulators. Nat. Photonics (2010). https://doi.org/10.1038/nphoton.2010.179
4. Marinins, A., et al.: Wafer-scale hybrid integration of InP DFB lasers on Si photonics by flip-chip bonding with sub-300 nm alignment precision. IEEE J. Sel. Top. Quantum Electron. (2023). https://doi.org/10.1109/JSTQE.2022.3223641
5. Tang, Y., Peters, J.D., Bowers, J.E.: Over 67 GHz bandwidth hybrid silicon electro-absorption modulator with asymmetric segmented electrode for 13 μm transmission. Opt. Express (2012). https://doi.org/10.1364/oe.20.011529
6. Corbett, B., Loi, R., O'Callaghan, J., Roelkens, G.: Transfer printing for silicon photonics. Semiconductors Semimetals **99** (2018). https://doi.org/10.1016/bs.semsem.2018.08.001
7. Miller, D.A.B., et al.: Band-edge electroabsorption in quantum well structures: The quantum-confined stark effect. Phys. Rev. Lett. **53**(22) (1984). https://doi.org/10.1103/PhysRevLett.53.2173
8. Trajkovic, M., Blache, F., Debregeas, H., Williams, K.A., Leijtens, X.J.M.: Increasing the speed of an InP-based integration platform by introducing high speed electroabsorption modulators. IEEE J. Sel. Top. Quantum Electron. (2019). https://doi.org/10.1109/JSTQE.2019.2913727

Ge-fin Photodiodes with 3-dB Bandwidths Well Beyond 110 GHz for O-Band Receiver Subsystems

Daniel Steckler[1(✉)], Stefan Lischke[1], Anna Peczek[1], and Lars Zimmermann[1,2]

[1] IHP-Leibniz Institut für Innovative Mikroelektronik, 15236 Frankfurt (Oder), Germany
steckler@ihp-microelectronics.com

[2] Technische Universität Berlin, FG Silizium-Photonik, 10587 Berlin, Germany

Abstract. We present Ge-fin photodiode variants with 3dB-bandwidths from 68 GHz to more than 110 GHz, responsivities from 0.5 AW^{-1} to 0.8 AW^{-1} (-2 V, $\lambda = 1310$ nm) and dark currents of 150 nA at -2 V. Furthermore we evaluate the power handling capabilities of our Ge photodiodes. These devices outperform other Ge photodiodes and are on par with state-of-the-art InP-based photodiodes.

Keywords: Germanium photodiode · Germanium Fin · Silicon photonics · O-Band · Data center · IM-DD · Transceivers

1 Introduction

Terabit connectivity based on optical interconnect technology is enabling in quickly growing markets such as large scale data centers or artificial intelligence and machine learning infrastructure. The majority of the present transceiver optics operate in the O-band and use direct detection. For systems beyond 200 G/lane both IM-DD and coherent implementations are currently under discussion [1, 2]. While silicon photonic implementations are highly attractive due to scaling related cost reduction effects, increasing the bandwidth of foundry manufactured O-band photodiodes significantly has not received much attention. In this article we shall study a waveguide integrated O-band photodiode fabricated in IHP's BiCMOS pilotline.

In [3] we have for the first time demonstrated ultra-fast Ge-fin photodiodes with 3dB-bandwidths up to 265 GHz at C-Band wavelengths. As we have presented in [4] the tensile strained Ge-region of the PD also enables operation of these devices in the L-Band. In this work we focus on the behavior of these Ge-fin photodiodes at the O-band wavelength of 1310 nm.

For this study we investigated three Ge PD variants of different Ge widths. In contrast to vertical drift field photodiodes, where the i-region is adjusted by layer thicknesses e.g. during epitaxial growth, the width of laterally formed Ge-fins and by this their i-region can arbitrarily be adjusted by device layouts. Therefore, devices with various

J. Witzens et al. (Eds.): ECIO 2024, SPPHY 402, pp. 58–63, 2024.
https://doi.org/10.1007/978-3-031-63378-2_11

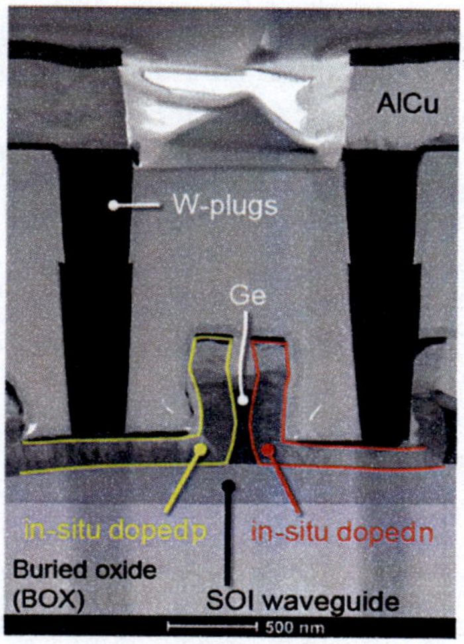

Fig. 1. Cross-sectional TEM-image of germanium photodiode with nominal width of 100 nm [1]

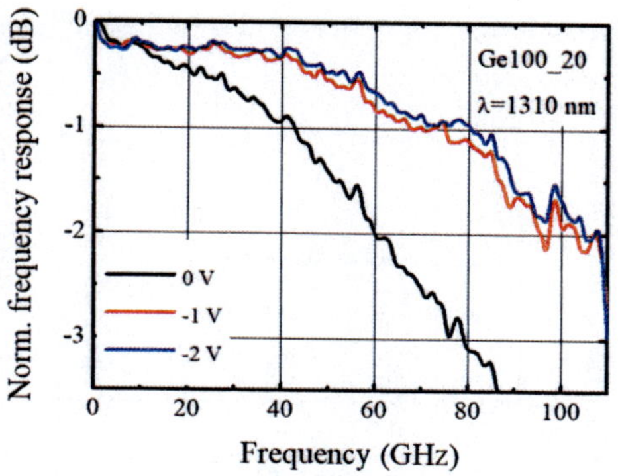

Fig. 2. Opto-electrical frequency responses of photodiode Ge100_20 at various applied biases for an optical input power of −5 dBm at photodiode

combinations of OE bandwidths and responsivities can be realized side by side [5]. Details of the fabrication can be found in [3].

Obviously, wider structures yield higher responsivities, but on the expense of OE bandwidths due to longer photo carrier drift time constants.

By preventing doping of the light-absorbing Ge-region the transport of photogenerated carriers is not governed by slow diffusion processes, such that the photodiodes' bandwidth is mainly determined by the drift velocity of the carriers and the RC-time constant of the device.

For our investigation we chose Ge-PDs with nominal Ge widths of 100 nm, 250 nm and 450 nm and a length of 20 μm. According to the naming scheme from [3, 5] the PDs are termed 'Ge100_20', 'Ge250_20' and 'Ge450_20'.

2 Measurement Results and Discussion

A cross-sectional TEM-image of our fastest photodiode variant with a Ge width of 100 nm is depicted in Fig. 1. The frequency response of these devices was measured on wafer level with a Keysight 110 GHz lightwave component analyzer at the O-band wavelength of $\lambda = 1310$ nm (Fig. 2). At biases of −1 V and −2 V the device Ge100_20 shows an excellent RF-performance with an opto-electrical 3 dB-bandwidth well beyond 110 GHz. Due to the high built-in field of the narrow device they still yield a high 3 dB-bandwidth of 80 GHz at zero bias. To the best of our knowledge these are the fastest Ge photodiodes demonstrated in the O-Band and are on par with high-speed III–V photodiodes. However the frequency response of the PD Ge100_20 is not as flat at 1310 nm as we have measured for the narrowest PDs at 1550 nm presented in [3, 5]. The investigation of this effect is still ongoing. Our current hypothesis is that at

1310 nm most of the light is absorbed on a shorter distance than at 1550 nm, such that most of the photocurrent is concentrated in a smaller volume and the resistance of the device is therefore larger at O-band wavelengths compared to measurements at C-band wavelengths. While the wider devices are mainly drift limited, the narrowest variant is approaching a 3-dB-bandwidth limitation due to its RC-time constant.

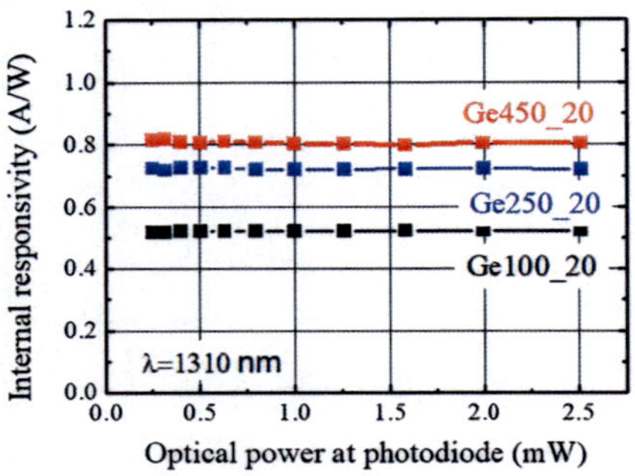

Fig. 3. Internal responsivity vs optical power at photodiode (One representative chip for each variant) at −2 V bias for $\lambda = 1310$ nm

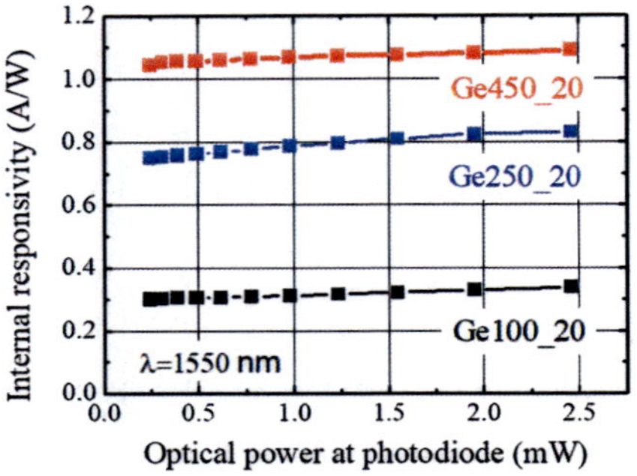

Fig. 4. Internal responsivity vs optical power at photodiode (One representative chip for each variant) at −2 V bias for $\lambda = 1550$ nm

To evaluate the behavior of our PDs at high optical powers we measured the internal responsivity of our PDs with an incident light power from 0.3 mW up to 2.5 mW at 1310 nm (Fig. 3). For the sake of comparison we also included responsivity measurements up to 2.5 mW at the C-band wavelength 1550 nm (Fig. 4). Fiber to waveguide coupling is enabled by 1D grating couplers. The internal responsivities were estimated under consideration of the respective coupling losses of about 3 dB.

At 1550 nm and optical input power of 0.3 mW the respective responsivities are 0.3 AW^{-1} (Ge100_20), 0.75 AW^{-1} (Ge250_20) and 1.05 AW^{-1} (Ge450_20). For increasing optical power a slight increase of the responsivities can be observed. We attribute this increase to self-heating: For larger optical powers, a larger photocurrent is being generated. The photocurrent leads to a temperature increase of the Ge-region due to Joule heating, which goes along with a shrinkage of the bandgap [6].

By contrast, the responsivities at 1310 nm remain constant at optical powers up to 2.5 mW, since Ge already yields a high base absorption at 1310 nm, which cannot be further increased by shrinking the bandgap. The respective responsivities at 1310 nm are 0.52 AW^{-1} (Ge100_20), 0.72 AW^{-1} (Ge250_20) and 0.81 AW^{-1} (Ge450_20).

Based on these measurements the quantum efficiency of the narrowest device Ge100_20 is higher at 1310 nm than at 1550 nm, which is most likely due to the better mode confinement in the narrow Ge-fin at shorter wavelengths. In the wider structures the quantum efficiencies are on a similar level, but the responsivity in the C-band is still higher, since more photons per joule of optical energy are transmitted at higher wavelengths.

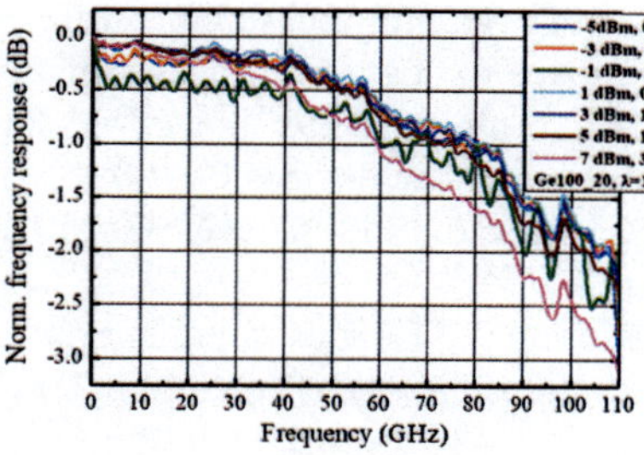

Fig. 5. Opto-electrical frequency responses of Ge100_20 for various input powers (at photodiode) from –5 dBm to 7 dBm at λ = 1310 nm

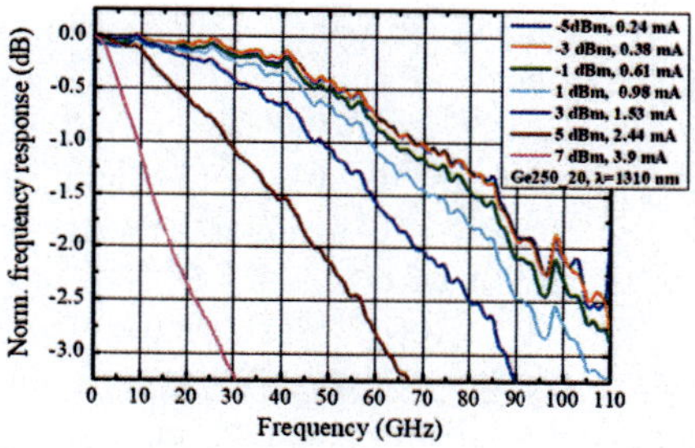

Fig. 6. Opto-electrical frequency responses of Ge250_20 for various input powers (at photodiode) from –5 dBm to 7 dBm at λ = 1310 nm

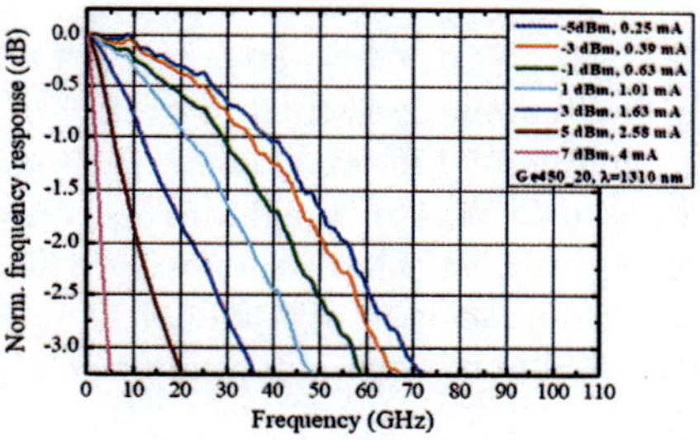

Fig. 7. Opto-electrical frequency response of Ge450_20 for various input powers (at photodiode) from –5 dBm to 7 dBm at λ = 1310 nm

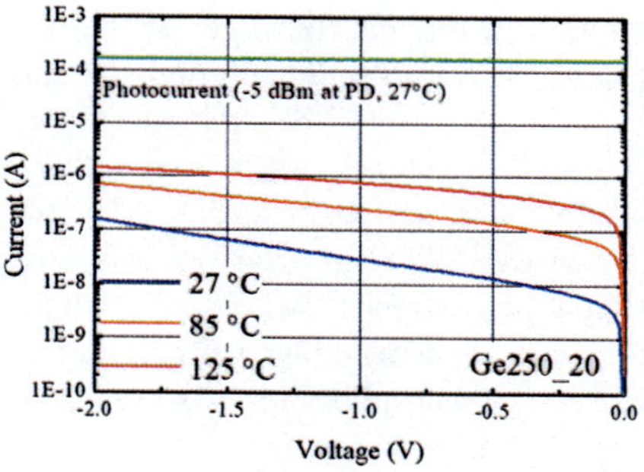

Fig. 8. Dark currents of PD variant Ge 250_20 measured at −2 V at temperatures of 27 °C, 85 °C and 125 °C and photocurrent measured at −5 dBm at λ = 1310 nm

Table 1. Benchmark of various state-of-the-art waveguide coupled photodiodes at λ = 1310 nm

	Runge [7]	Chen [8]	Boeuf [9]	Rakowski [10]	This work		
					Ge100	Ge250	Ge450
Material	InP/InGaAs	Ge	Ge	Ge	Ge	Ge	Ge
3-dB OE BW (GHz)	>67	67	>60	50	>110	>110	65
Int. Responsivity at 1310 nm (AW^{-1})	0.5	0.93	>1	0.9	0.52	0.72	0.82
Dark current (nA)	<1 (–2 V)	3.6 (–2 V)	80 (–2 V)	<40		150 (–2 V)	

Normalized frequency responses of the three Ge PD variants, measured at different optical powers are shown in Figs. 5, 6 and 7. Hereby the optical power of the incident light at the PD was swept from −5 dBm to 7 dBm to evaluate the RF power handling capabilities of the devices at 1310 nm. At low input powers the devices Ge100_20 and Ge250_20 yield 3dB-bandwidths well beyond 110 GHz and the widest device Ge450_20 reaches at 3dB-bandwidth of 68 GHz at −2 V bias.

Notably Ge100_20 yields a high 3-dB-bandwidth of 110 GHz even at high optical input powers of 7 dBm with a generated photocurrent of 3 mA. The Ge250_20 PD still achieves a 3-dB bandwidth of 110 GHz at $P_{opt} = -1$ dBm/$I_{ph} = 0.61$ mA and then degrades for higher input powers. At the widest device Ge450_20 the bandwidth-degradation already starts at $P_{opt} = -3$ dBm/$I_{ph} = 0.39$ mA photocurrent and decreases significantly for higher input powers. We suspect that the photo generated carriers generate an electric field which counteracts the applied electric field and therefore weaken the drift velocity of the carriers in the i-region. Since the electric field for a fixed applied voltage is significantly larger in the narrow devices the carrier velocity remains high despite the large generated photocurrents.

In Fig. 8 we provide dark currents for the variant Ge250_20 measured from 27 °C up to 125 °C on one wafer site. At 27 °C the dark current is about 150 nA at −2 V and increases up to 1.3 μA at −2 V for a temperature of 125 °C. Compared to other state-of-the-art PDs the dark currents are relatively high. We attribute this to the current fabrication approach, which relies on dry etching of Ge, potentially provoking defects at the Ge sidewalls. However given that the generated photocurrent is still a few orders of magnitudes higher our PDs are suitable to be used in receiver circuits even in high temperature environments.

To benchmark our results we compare our photodiodes with other state of the art Ge- and III-V photodiodes at the operational wavelength of $\lambda = 1310$ nm in Table 1.

3 Conclusion

In this work we have investigated germanium photodiode variants with 3-dB bandwidths from 68 GHz to well beyond 110 GHz with Ge widths of 450 nm, 250 nm and 100 nm at the O-band wavelength of 1310 nm. The photodiode variants yielded responsivities from 0.5 AW^{-1} to 0.8 AW^{-1} at −2 V and $\lambda = 1310$ nm. The very high-speed photodiode operation at O-band wavelengths enables silicon-photonics based receivers for future data-center interconnects with symbol rates exceeding 200 GBaud.

Acknowledgement. The authors gratefully acknowledge the support by the German Research Foundation (DFG) projects ULTRA 2 (ZI 1283-5-2) and EPIC-SENSE 2 (ZI 1283-6-2), BMBF project CBQD (16KIS1941), HorizonEurope 101092834 - CoRaLi-DAR.

References

1. Seiler, P.M., et al.: Toward coherent O-band data center interconnects. Front. Optoelectron. **14**, 414–425 (2021). https://doi.org/10.1007/s12200-021-1242-0

2. Seiler, P.M., et al.: Monolithically integrated o-band coherent ROSA featuring 2D grating couplers for self-homodyne intra data center links. IEEE Photon. J. **15**(3), 1–6, Art no. 6601306 (2023). https://doi.org/10.1109/JPHOT.2023.3272476
3. Lischke, S., et al.: Ultra-fast germanium photodiode with 3-dB bandwidth of 265 GHz. Nat. Photon. **15**, 925–931 (2021). https://doi.org/10.1038/s41566-021-00893-w
4. Steckler, D., et al.: Germanium fin photodiode with 3db-bandwidth > 110 GHz and high l-band responsivity. In: 2023 IEEE silicon photonics conference (SiPhotonics), pp. 1–2. Washington, DC, USA, (2023). https://doi.org/10.1109/SiPhotonics55903.2023.10141946
5. Peczek, A., et al.: Versatile germanium photodiodes with 3dB bandwidths from 110 GHz to 265 GHz. ECS Trans. **109**, 21 (2022). https://doi.org/10.1149/10904.0021ecst
6. Vainshtein, I.A., et al.: Applicability of the empirical Varshni relation for the temperature dependence of the width of the band gap. Phys. Solid State **41**(6), 905–908 (1999). https://doi.org/10.1134/1.1130901
7. Runge, P., et al.: Waveguide integrated InP-based photodetector for 100Gbaud applications operating at wavelengths of 1310 nm and 1550 nm. In: 2015 european conference on optical communication (ECOC), pp. 1–3. Valencia, Spain (2015). https://doi.org/10.1109/ECOC.2015.7341912
8. Chen, H., et al.: 1 V bias 67 GHz bandwidth Si-contacted germanium waveguide p-i-n photodetector for optical links at 56 Gbps and beyond. Opt. Express **24**(5), 4622–4631 (2016). https://doi.org/10.1364/OE.24.004622. PMID: 29092290
9. Boeuf, F., et al.: A silicon photonics technology for 400 Gbit/s applications. In: 2019 IEEE international electron devices meeting (IEDM), pp. 33.1.1–33.1.4. San Francisco, CA, USA (2019). https://doi.org/10.1109/IEDM19573.2019.8993627
10. Rakowski, M., et al.:45 nm CMOS—silicon photonics monolithic technology (45CLO) for next-generation, low power and high speed optical interconnects. In: 2020 Optical Fiber Communications Conference and Exhibition (OFC), pp. 1–3. San Diego, CA, USA (2020)

Integrated-SiGe Waveguide Photodetector in the 5.2–10 μm Wavelength Range Operating at Room Temperature

Thi Hao Nhi Nguyen[1(✉)], Victor Turpaud[1], Natnicha Koompai[1], Jonathan Peltier[1], Stefano Calcaterra[2], Giovanni Isella[2], Jean-René Coudevylle[1], Carlos Alonso-Ramos[1], Laurent Vivien[1], Jacopo Frigerio[2], and Delphine Marris-Morini[1]

[1] Centre de Nanosciences et de Nanotechnologies, CNRS, Université Paris-Saclay, 91120 Palaiseau, France
thi-hao-nhi.nguyen@universite-paris-saclay.fr

[2] L-NESS, Dipartimento di Fisica, Politecnico di Milano, Polo di Como, Via Anzani 42, 22100 Como, Italy

Abstract. This study presents a significant advancement in mid-infrared (MIR) photodetection, using a PIN diode incorporated into a graded-index SiGe photonic waveguide operating at room temperature. The photodetector exhibits broadband mid-infrared detection at a bias voltage of −15 V. The internal responsivity of the device is reported as 1 mA/W in the 5–10 μm infrared range.

Keywords: photodetector · silicon · germanium · mid-IR wavelength

1 Introduction

Silicon photonics in the mid-infrared spectral range holds significant promise for advancing on-chip spectroscopic systems, with various applications in environmental monitoring, medical diagnosis, and free-space communication [1]. Among various active photonic components, the photodetector is a critical device for the development of these sensing systems. Mid-IR photodetectors have been extensively developed, relying on intersubband absorption in III–V materials with multiquantum-well heterostructures [2], narrow band gap materials like HgCdTe [3], or defect-mediated absorption in Silicon-on-Insulator (SOI) [4], Germanium-on-Silicon (GOS) [5], or SiGe graded-index platforms [6]. This work demonstrates a broadband mid-IR photodetector exploiting a PIN diode embedded into a graded-index SiGe waveguide. At room temperature, the device shows high-speed performance with a 3 dB bandwidth of 32 MHz, covering mid-IR wavelengths from 5.2 μm to 10 μm.

J. Witzens et al. (Eds.): ECIO 2024, SPPHY 402, pp. 64–68, 2024.
https://doi.org/10.1007/978-3-031-63378-2_12

2 Design and Experimental Results

2.1 Design

Figure 1 illustrates the cross section of the integrated SiGe photodetector in the mid-IR spectral range, with the refractive index profile on the left and a doping profile on the right. Refractive index engineering is obtained by tuning the Ge concentration in the SiGe alloy during the epitaxial growth. To collect the photocurrent induced by light absorption in the waveguide, a PIN diode is embedded into the optical waveguide. The structure thus begins with a 3-μm-thick layer of $Si_{0.6}Ge_{0.4}$, epitaxially grown on a non-doped Si substrate, followed by a 3-μm-thick graded $Si_{1\text{-}x}Ge_x$ layer (x ranges from 40% to 100%). Subsequently, a 2-μm-thick layer of $Si_{0.3}Ge_{0.7}$ is finally grown on the top, not only to effectively confine the optical mode in the photonic waveguide and but also shift optical mode away from the upper metallic contact to minimize insertion loss. Regarding to the doping levels, the first 1-μm-thick $Si_{0.6}Ge_{0.4}$ layer from the bottom is N-doped, with a doping concentration of 1×10^{18} cm^{-3}. In contrast, the upper 300-nm-thick layer of $Si_{0.3}Ge_{0.7}$ is P-doped, with a concentration of 5×10^{18} cm^{-3}. The region sandwiched between the N and P regions is non-intentionally doped. Since the SiGe layer is 8 μm-thick, the etching depth of the waveguide is set at 7.5 μm first to confine the optical mode, but also simultaneously to be able to define the bottom ohmic contact on the N-doped region. The signal line electrode comprises a 300 nm-thick and 4 μm-wide gold layer on the top of the 6 μm-wide waveguide, while ground lines are defined on each side of the waveguide on the highly N-doped layer.

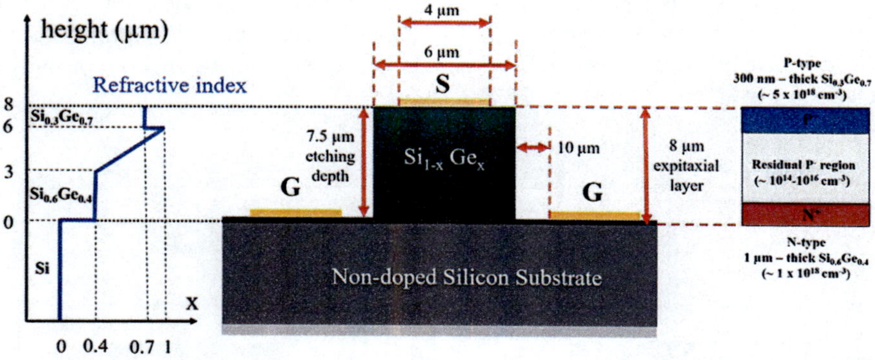

Fig. 1. Cut view of the SiGe integrated-waveguide photodetector. On the left, the refractive index profile is illustrated along the vertical direction, showing the variation in the fraction of Ge (x) in $Si_{1\text{-}x}Ge_x$ to optimize the optical confinement. On the right, the PIN doping profile in the active region of the SiGe photodetector is provided, indicating the specific parts corresponding p-type and n-type.

2.2 Experimental Results

To evaluate the sensitivity of the photodetector, a broadband QCL (wavelength from 5.2 to 11.2 μm) is employed with a pulsed rate of 100 kHz and a 5% duty cycle.

Photocurrent measurement is conducted at a reversed voltage of −15 V and at the room temperature. The photocurrent is initially amplified by a current amplifier before reaching the synchronized oscilloscope. To evaluate the performance of the photodetector, both the amplitude of the photocurrent generated by the integrated SiGe photodetector and its rising/falling times are examined. Figure 2 depicts the photocurrents collected from the integrated SiGe waveguide at the wavelength of 5.5 μm observed in the oscilloscope at zero-bias voltage and at a reverse bias voltage of −15 V, respectively. A zoom-in on the rising and falling regions are analyzed to deduce the operational bandwidth. The minimum value between rising time τ_r and falling time τ_f is extracted, giving 10.78 ns corresponding to an electro-optical bandwidth of of 32 MHz. Furthermore, from the amplitude of the electrical signal collected, both external and internal responsivity can be deduced as followed:

$$R_{ext} = \frac{I_{pp}}{P_{ext\ \text{peak-to-peak}}} = \frac{I_{pp}.duty\ cycle}{P_{ext\ avearage}} \tag{1}$$

$$R_{int} = \frac{I_{pp}}{P_{int\ \text{peak-to-peak}}} \tag{2}$$

where I_{pp} refers to the peak-to-peak amplitude of photocurrent, $P_{ext\ average}$ (resp. $P_{ext\ peak\text{-}to\text{-}peak}$) is the measured average (resp. $P_{peak\text{-}peak}$) external optical power before the input of photonic circuit, out of the chip. $P_{int\ peak\text{-}to\text{-}peak}$ is the peak-to-peak power coupled in the photonic circuit, evaluated at the entrance of the integrated SiGe photodetector by taking into account the coupling loss at the photonic facet and the propagation loss along the 2-mm-long photonic waveguide before reaching photodetector. From that, both internal and external responsivity of SiGe photodetector are deduced from 5.2 μm to 10 μm wavelength, as illustrated in Fig. 3. It can be seen that internal responsivity is flat around 1 mA/W.

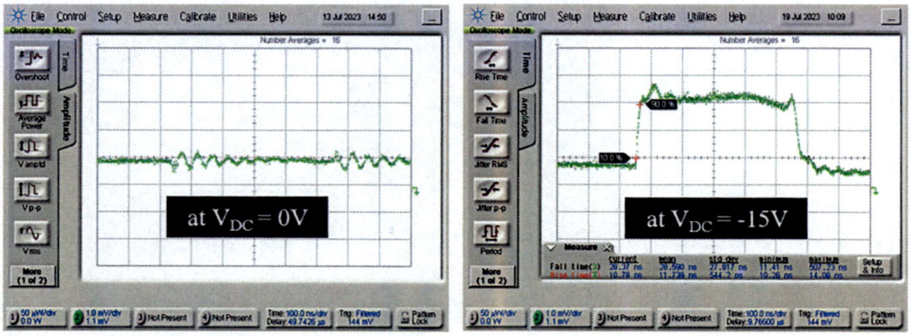

a) b)

Fig. 2. Photocurrents collected from the integrated SiGe waveguide at the wavelength of 5.5 μm are observed in the oscilloscope a) at zero-bias voltage and b) at a reverse bias voltage of −15 V. The oscilloscope interface also displayed the rising and falling times of the photocurrent, which allows to calculate the 3 dB electro-optical bandwidth of the device to be 32 MHz.

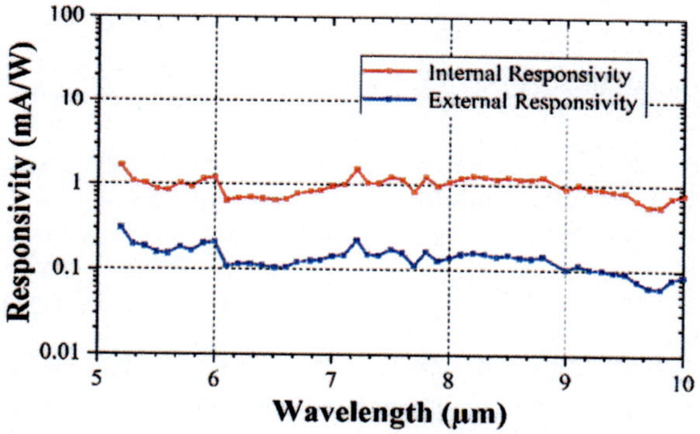

Fig. 3. The SiGe photodetector responsivity is characterized across the 5.2 μm to 10 μm wavelength range. Internal (red) and external (blue) responsivities are provided, representing the device's efficiency in converting light to electrical signals.

The physical mechanism responsible for observed photocurrents in this integrated PIN SiGe photodetector remains missing and required further investigation. One possible hypothesis could rely on sub-bandgap absorption attributed to defects or dislocations. The previous development of integrated photodetector on SOI platform has demonstrated defect-mediated sub-bandgap absorption, particularly in wavelength range between 2.2 μm to 2.4 μm [4]. Lauren et al. also observed photocurrents in a PIN junction embedded in a GOS waveguide at 3.8 μm wavelength [5]. More recently, Schottky based graded-index SiGe photodetector was demonstrated in 5–8 μm wavelength range [6]. Indeed, the epitaxial growth of SiGe on Si substrate inevitably leads to the nucleation of dislocations. Thus, it is believed that this sub-bandgap absorption mechanism could explain the observed electro-optical response.

3 Conclusion

This work focuses on the demonstration of an integrated photodetector based on a PIN diode embedded in a graded SiGe optical waveguide. This detector can operate in a wide wavelength range of the mid-IR spectrum (5.2–10 μm wavelength). The sensitivity of the device is characterized, showing an internal responsivity up to 1 mA/W in the 5.2–10 μm range. The electro-optical bandwidth of the photodetector is demonstrated at 32 MHz.

Acknowledgements. The fabrication of the device was partially performed at the Platform de MicroNanoTechnologie/C2N, which is partially funded by the "Conseil Général de l'Essonne". This work was partly supported by the French RENATECH network. Co-funded by the European Union (ERC-Electrophot, 101097569). Views and opinions expressed are however those of the author(s) only and do not necessarily reflect those of the European Union or the European Research Council. Neither the European Union nor the granting authority can be held responsible for them.

Research Funding. ANR Light-up Project (ANR-19-CE24-0002-01). Fondazione Cariplo, Grant No. 2020–4427 (project MILESTONE), ERC-Electrophot (101097569).

References

1. Mashanovich, G.Z., et al.: Group IV mid-infrared photonics. Opt. Mater. Express **8**(8), 2276–2286 (2018)
2. Hakl, M., et al.: Ultrafast quantum-well photodetectors operating at 10 μm with a flat frequency response up to 70 GHz at room temperature. ACS Photon. **8**(2), 464–471 (2021)
3. Rogalski, A.: Toward third generation HgCdTe infrared detectors. J. Alloys Compd. **371**(1–2), 53–57 (2004)
4. Grote, R.R., et al.: Extrinsic photodiodes for integrated mid-infrared silicon photonics. Optica **1**, 264–267 (2014)
5. Reid, L., Nedeljkovic, M., Cao, W., Mastronardi, L., Slavik, R., Mashanovich, G.: Photodetection at 3.8 μm using intrinsic monolithic integrated Germanium photodiodes. In: IEEE International Conference Gr. IV Photonics GFP, pp. 3–4 (2021)
6. Nguyen, T.H.N., et al.: Room temperature-integratedphotodetector between 5 μm and 8 μm wavelength. Adv. Photon. Res. **4**(2), 2200237 (2023)

Assembly and Test

Repeatability of Automated Edge Coupling for Wafer Level Testing

Anna Peczek[1(✉)], Tino Minner[2], Quan Yuan[2], Christian Mai[1], Dan Rishavy[2], and Lars Zimmermann[1,3]

[1] IHP – Leibniz-Institut für innovative Mikroelektronik, Im Technologiepark 25, 15236 Frankfurt (Oder), Germany
anna.peczek@ihp-microelectronics.com
[2] FormFactor Inc., 9100 SWG Gemini Drive, Beaverton, OR 97008, USA
[3] Technische Universität Berlin, FG Silizium-Photonik, Einsteinufer 25, 10587 Berlin, Germany

Abstract. In foundries environment the optical characterization of Photonic Integrated Circuits (PICs) for process control and device performance verification requires a high throughput test solution. So far, wafer-level characterization of PICs has been limited to the grating couplers as the optical input/output interface. We propose an alternative optical probing technique suitable for automated on-wafer measurements based on edge coupling.

Here, we evaluate a solution based on FormFactor's CM300-SiPh probe station and the Pharos Probe. A brief overview of the system is given and the repeatability of the coupling across the wafer is investigated.

Keywords: Silicon Photonics · PIC testing · edge coupling · wafer-level characterization

1 Introduction

As the field of silicon photonics applications expands, so does the demand for low loss solutions of photonic integrated circuits (PICs). The well-known grating couplers used for many years have many drawbacks such as: high polarization dependence, reduced bandwidth and relatively low coupling efficiency.

Recently, the edge coupler has become a prominent candidate for photonic foundries as a coupling interface to the chip. The low polarization dependence, low coupling loss and compatibility with advanced packaging solutions make this interface very attractive for many applications. However, testing on wafer level using edge coupling has lacked behind grating coupler based probing due to the complexity of the method and problems that need to be overcome. The calibration automation was always complex and heavily dependent on the skills of the operator. In recent years, there have been some concepts and experiments in the literature regarding edge coupling on wafer level [1–3]. To date, though, a fully automated solution has not been demonstrated. In this paper we would like to evaluate the fully automated turn key solution that has been developed by FormFactor Inc. We will provide an overview of the system and show repeatability results obtained on a 200 mm photonic wafer.

J. Witzens et al. (Eds.): ECIO 2024, SPPHY 402, pp. 71–75, 2024.
https://doi.org/10.1007/978-3-031-63378-2_13

2 System Overview

2.1 Probe Station Environment

The overall system is based on fully-automated Probe Station (FormFactor CM300xi) with an advanced eVue microscope system and 6-axis optical positioner (based on PI HexaPod H-811 and NanoCube P-611.3). The optical probe is a fiber array with 3D printed lenses and will be described more in Sect. 2.2. The automated algorithm is moving the optical probe in Y and Z directions to find the maximum signal intensity and thus the best coupling point. As shown in Fig. 1 a well-defined Gaussian beam profile can be achieved. The measurement instruments consist of Keysight tunable laser source, polarization controller and optical power meter. To provide a feedback channel for alignment purposes, the output power from the DUT is split 90 to 10 between the Keysight power meter and the probe station power meter (Newport 2103).

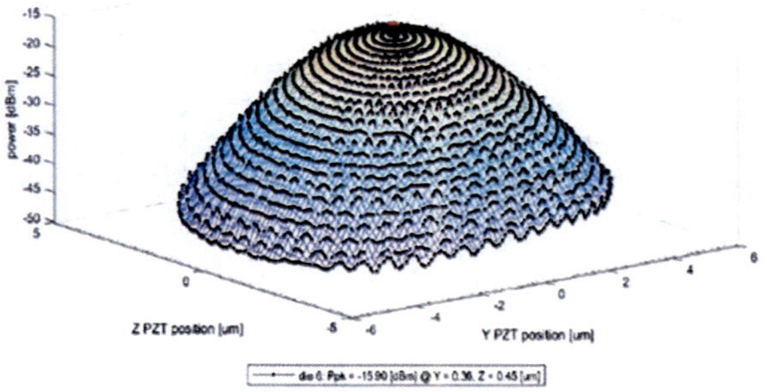

Fig. 1. Power coupling as a function of YZ NanoCube positions after alignment

2.2 Optical Probe

The optical probe, called the Pharos Probe, consists of a fiber array and 3D printed tips fabricated on the facet of the fiber array. The design of the probe must be matched with the mode field diameter (MFD) of the edge coupler on the chip. The construction of the probe allows it to be used with either deep trenches or V-grooves. However, there are certain design requirements for both types of interface. For the Pharos Probe with 40 μm diameter lenses, suitable for waveguide spot size of approximately 6 μm, the minimum trench width and depth are 100 μm and 60 μm, respectively. A specially designed fiducial marker (shown in Fig. 2a and b) is attached to the optical probe and is used to align the optical waveguide by identifying the current position of the probe. The design of the fiducial is optimized to provide good contrast for computer vision (using the eVue microscope) without overlapping with the background image. The Pharos Probe is mounted on a holder which is attached to the positioner arm. In addition, the Z-sensor (electrical capacitive sensor), shown in Fig. 2c, must be attached to the holder as well, to control the distance between the wafer surface and the optical probe. In this case, it will be a negative value while the probe is placed below the wafer surface.

Figure 2a shows a simplified view of the Pharos Probe in a trench. The width and depth of the trench fabricated on the test wafer are 160 μm and 450 μm, respectively. The depth of the trench should exceed the wafer grinding thickness to permit singulation without dicing through the trenches. For the Pharos Probe used in the experiments, the optimum working distance is 28 μm.

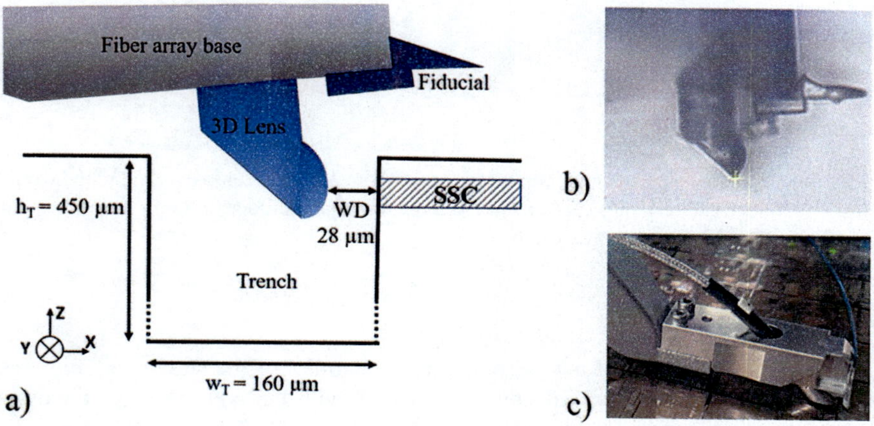

Fig. 2. Scheme (not in scale) of the Pharos Probe and the deep trench with dimensions of the current sample (a). A microscopic view of the Pharos Probe with fiducial (b) and the real photo of the Pharos Probe and Z-sensor mounted in the holder (c).

In order to use the function of the automatic alignment of the Pharos Probe to the waveguide, an automated calibration has to be performed. The procedure includes the calibration of the coordinate positioning (X, Y, Z) as well as the angular adjustment (pitch, yaw, roll). Finally, a dedicated photodiode with mirrors installed on the probe station can be used to measure the output power of the Pharos Probe and thus calibrate the setup loss, including the optical probe.

3 Automated Wafer-Level Testing with Edge Coupling

To evaluate the automation performance of the FormFactor solution, we performed fully automated optical tests on silicon waveguides fabricated on a 200 mm wafer.

3.1 Device Under the Test

The reference wafer with a test die area of 4.8 mm^2 was fabricated in 0.25 μm photonic technology in IHP's clean room. The test die consists of three 220 nm wire waveguides with different lengths and a Germanium photodiode, as shown in Fig. 3. Together there are 7 spot size converters fabricated along 2.3 mm long trench. The deep trench is designed to be coupled with a Pharos Probe with up to eight 3D tips and 250 μm pitch. The wide 160 μm trench ensures collision-free alignment of the Pharos Probe.

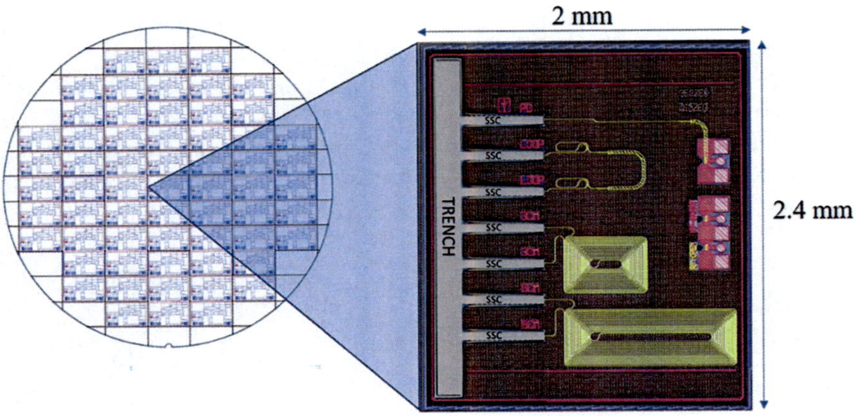

Fig. 3. 200 mm wafer map and enlarged microscope photo of the test chip overlaid with layout of the test structures. The test chip consists of, from the top, a photodiode, a 2 mm long waveguide loop, a 3 cm and a 6 cm long waveguides.

3.2 Experimental Results

To prove the functionality of the system and repeatability of the automated alignment through the edge couplers, we selected 31 chips on the wafer and measured the output power of the short waveguide loop structure. The box plot in Fig. 4 shows the coupling deviation, calculated as the difference from the die mean value of the output power divided by 2. The optical power was measured over 20 test cycles. A standard deviation of 0.2 dB was achieved.

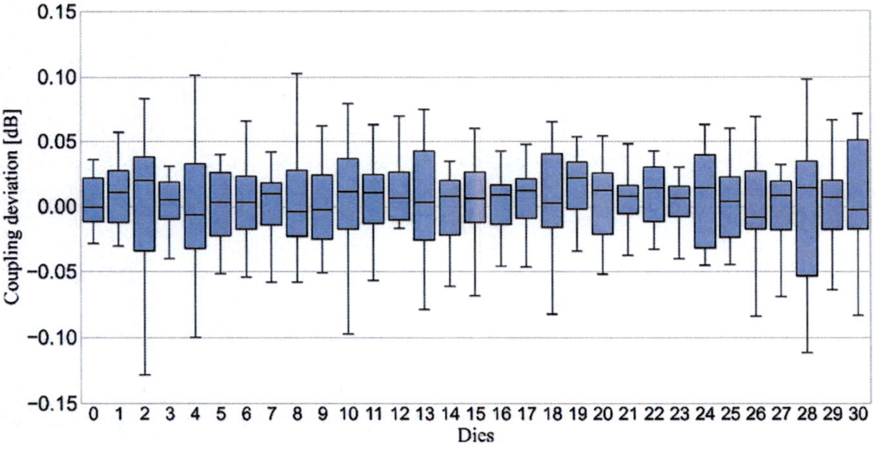

Fig. 4. Deviation of the coupling loss obtained from 20 test cycles on 31 dies on the wafer.

4 Conclusions

The fully automated measurement setup proposed by FormFactor is suitable for reliable wafer-level PIC testing with edge coupling. All building blocks of the setup, including the Pharos Probe, together with automated calibration, provide a complete operator independent measurement solution. The integrated optical fiducial enables accurate calibration and allows fast alignment of the Pharos Probe to the optical waveguide in YZ direction.

The performed repeatability tests of the alignment procedure on wafer level were based on 20 test cycles on 31 dies. We achieved a standard deviation of 0.2 dB. There is a big potential for PIC testing on wafer to substitute vertical coupling with edge coupling, especially in semiconductor foundries environment.

Acknowledgements. This work was supported in part by the German Research Foundation (DFG) through projects ULTRA 2 (ZI 1283-5-2) and EPIC-SENSE 2 (ZI 1283-6-2), by the Federal Ministry of Education and Research (BMBF) through project CBQD (16KIS1941), and by the European Commission through project HorizonEurope 101092834 - CoRaLi-DAR.

References

1. Trappen, M., et al.: 3D-printed optical probes for wafer-level testing of photonic integrated circuits. Opt. Express **28**, 37996–38007 (2020)
2. Jabon, K.M.: Edge-coupled active and passive wafer-scale measurements on 300 mm silicon photonics wafers. In: Dong, P., Kani, J., Xie, C., Casellas, R., Cole, C., Li, M., (eds.) Optical Fiber Communication Conference (OFC). OSA Technical Digest (Optica Publishing Group, 2021), paper M3A.1 (2021)
3. Huang, S.-H., et al.: Robust edge coupling probe applied in wafer-level optical testing. In: 2023 IEEE 73rd Electronic Components and Technology Conference (ECTC), pp. 1754–1759. Orlando, FL, USA (2023). https://doi.org/10.1109/ECTC51909.2023.00300

Demonstration of Micro-transfer Printing Thick Optical Components on Glass and Silicon Wafers

Saif Wakeel(✉), Padraic E. Morrissey, HowYuan Hwang, and Peter O'Brien

Photonics Packaging and System Integration Group, Tyndall National Institute, University College Cork, Cork, Ireland
saif.wakeel@tyndall.ie

Abstract. Micro-transfer printing has emerged as a promising technique for the wafer-scale assembly of photonic and electronic devices. However, conventional methods face challenges when dealing with thick components (>250 μm) due to the limitations of tether-based approaches. In this study, we present a novel tether-less micro-transfer printing approach enabled by UV-curable dicing tape, demonstrating the successful printing of thick optical components (>300 μm) on both silicon and glass wafers. The optimization of pick-up and printing process parameters for the successful printing of coupons is discussed. The shear strength of printed coupons on 2 μm InterVia coated wafers is investigated and surface roughness is analyzed. As a result, micro-optics as thick as 1000 μm were successfully printed. The shear strength of printed coupons varied from 11–19 MPa which is comparable to and higher than a few optical epoxies. The overall process of μTP has no further effect on microstructure and surface roughness, no delamination was found in SEM images.

Keywords: Micro-transfer printing · micro-optics · photonic devices · wafer-scale assembly

1 Introduction

Micro-transfer printing is a wafer-bonding approach where thousands of devices/coupons can be picked mechanically by a polydimethylsiloxane (PDMS) stamp and placed on a designated target wafer. These coupons are bonded to target wafers via InterVia or BCB [1]. So far, photonic devices such as lasers, photodiodes, modulators, ring resonators, amplifiers, micro-LED, and micro-lenses have been successfully printed [2–6]. The advantages of micro-transfer printing over other bonding processes are scalability, efficient utilization of wafers, pre and post-fabrication testing opportunities, and alignment accuracy [2]. Conventional methods of micro-transfer printing, which rely on defining tethers and undercutting for coupon release, are inadequate for thick components (>250 μm) due to the weakness of tethers, as they cannot withstand the bulk structure.

Hence, researchers have showcased the printing of flat coupons with a thickness exceeding 250 μm by employing a UV-curable dicing tape pick-up technique [6, 7].

J. Witzens et al. (Eds.): ECIO 2024, SPPHY 402, pp. 76–81, 2024.
https://doi.org/10.1007/978-3-031-63378-2_14

While micro-optical components like lenses and mirrors for photonic applications can be as thick as 1000 μm [8], there has been no prior demonstration of printing such components on silicon and glass target wafers. Consequently, this paper is dedicated to exploring the transfer printing of thick micro-optics.

2 Fabrication of Coupons and Printing Process

The challenge with the conventional micro-transfer printing approach is that thick components cannot be released due to weak tethers holding bulk coupons and the inability of the etching release process [6]. An alternative to this is a UV-curable dicing tape release in which source wafers are diced on UV-curable dicing tape and picked up by PDMS stamp. The process steps of tether-less micro-transfer printing approach and schematic of machine loaded with wafers are given in Fig. 1.

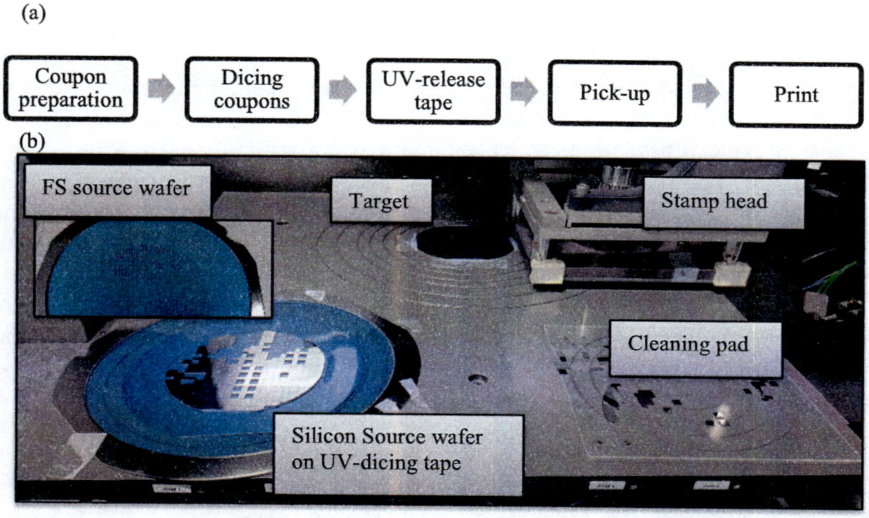

Fig. 1. (a) The process flow for tether-less micro-transfer printing and (b) machine set-up.

In general, the tether-less process starts with defining devices and placing them on a UV-curable blue tape followed by dicing and UV-curing. The source and target wafers are placed in the designated location shown in Fig. 1. Overall, the number of process steps in the tether-less method is less than conventional method as defining the tether and removing the tether is not included [2, 6]. The detailed methodology of the tether-less micro-transfer printing method is as follows. 500 μm thick silicon and fused-silica wafers, polished on both sides, underwent dicing to produce coupons of various dimensions. These included silicon coupons measuring $360 \times 1000\ \mu m^2$ and $3 \times 3\ mm^2$, as well as fused-silica coupons sized at $815 \times 2315\ \mu m^2$ and $1160 \times 6110\ \mu m^2$, referred to as pseudo-micro-lens arrays.

Also, a 1000 μm thick actual micro-lens array from Axetris micro-optics supplier was placed on UV-curable dicing tape [6]. The tape was later exposed to UV with an MA6 mask aligner as per recommended doses. The 4-inch silicon and fused-silica target wafers were spin-coated with 2 μm InterVia and soft-baked at 90 °C for 1 min. Different sizes of stamps were used depending on coupon sizes but overall, the selected stamp was in the 30–70% range of the area of coupons [6]. The overdrive distance, shear speed, and x, and y shear distances were optimized during printing trials for successful printing. The shear test of printed coupons on silicon and glass was done using the MIL-STD-883 standard and the micro-structure of printed coupons was observed using FEI Quanta 650 Scanning Electron Microscopy (SEM). The surface roughness measurement was done with the Bruker AFM machine.

3 Results and Discussion

Optimizing the pick-up and printing parameters such as overdrive distance/speed, pick-up force, and shear distance is critically important for picking up large footprint devices ($T_c > 300$ μm) due to the contact area and adhesion force dependencies [9]. A pick-up force of up to 1 g is needed to release the coupons from UV-dicing tape because of the large volume of coupons (large footprint and thickness) [6]. Out of at least 20 trials, the first successful pick-up was done with a minimum overdrive of 30 μm from the origin of the source with an overdrive speed of 15 μm/s. Optimization of shear displacement/speed is extremely important as this ensures the adhesion transfer at the interface of stamp-source-target. Increasing the shear displacement will ensure the efficient release of the stamp from the coupon during printing as pull-up or normal force required to delaminate the stamp will reduce [10]. Microscopic images of printed dummies and actual micro-lenses on silicon and glass wafers are shown in Fig. 2. All silicon and glass coupons were printed successfully without any sliding effect with a printing yield of 90%.

Fig. 2. Coupons printed on (a) silicon wafer, (b) glass wafer and (c) actual Silicon fused silica lenses printed on glass wafers.

The adhesion strength of printed coupons on target wafers can be determined by the shear force required to break the connection between coupons and wafers. The average shear strength of the fused-silica coupons printed on a glass wafer was 11 MPa whereas the shear strength of fused-silica coupons on a silicon wafer was 19 MPa. These shear strengths were comparable to optical epoxies used in various studies [11, 12] which makes InterVia a promising adhesive. Also, the shear strength of printed samples after curing was approx. 10–20 times higher than before curing (Fig. 3) due to polymerization and cross-linking [13].

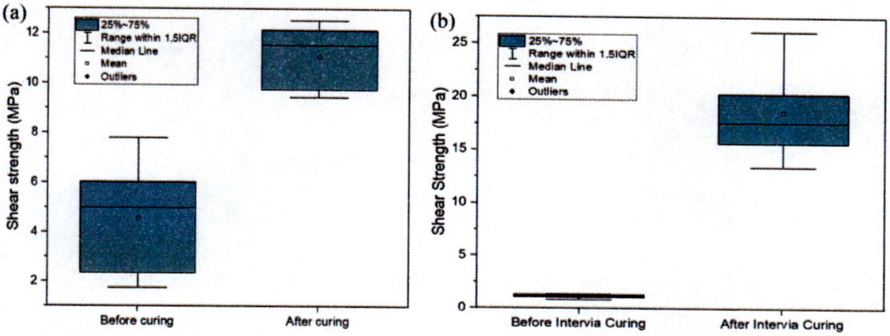

Fig. 3. Shear strength plots of pseudo-fused-silica lenses on (a) glass wafer and (b) silicon wafer.

SEM images of printed fused-silica coupons on target glass and silicon wafers shows no delamination at the interface between coupons and target (Fig. 4). It is important to avoid delamination as it can increase the chances of mechanical failure of bonded devices when subjected to mechanical or thermomechanical stress [15].

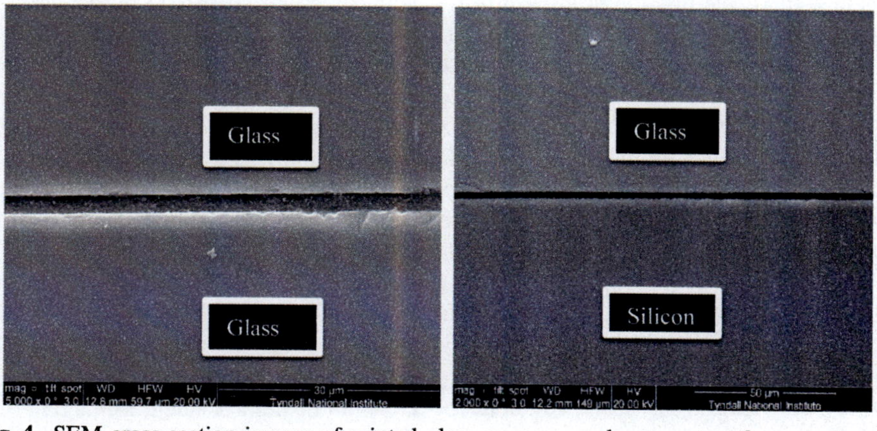

Fig. 4. SEM cross-section images of printed glass coupon on glass target wafer and glass on silicon target wafer.

During the micro-transfer printing process, the PDMS stamp makes contact with the surface of the lens, potentially altering its microstructure and surface roughness due to factors like shear distance, speed, and retraction. These changes in roughness could impact the refraction of light through the micro-lenses [14]. The root mean square (rms) surface roughness of the printed coupons was measured at 1.13 nm before printing and

1.2 nm after printing. This indicates that the printing process does not significantly affect the surface roughness or microstructure, as evidenced by AFM images in Fig. 5.

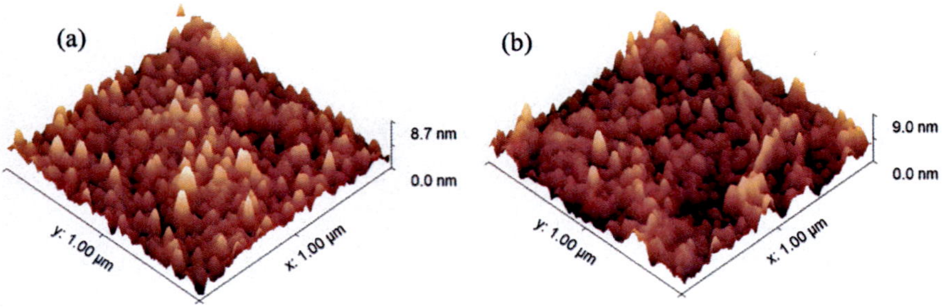

Fig. 5. AFM images of fused-silica coupon (a) before and (b) after printing showing nearly no change in the microstructure of micro-lenses before and after printing.

Outlook

This research has shown the successful printing of thick micro-optics (>300 µm) on both silicon and glass wafers utilizing UV-curable dicing tape. Through the optimization of process parameters, the study achieved successful printing, resulting in shear strength values for coupons on the target ranging from 11 to 19 MPa, a range comparable to that of optical epoxies. Notably, the microstructure and surface roughness of the coupons remained unaffected by the printing process. This is the first attempt to micro-transfer print micro-optics.

From this research, UV-curable dicing tape emerges as a viable method for lifting thick coupons, as conventional tethers lack the strength to support the bulk structure of such coupons. Tethers, typically composed of photoresist, silicon, or SiN applied via spin coating, face limitations when dealing with thicknesses exceeding 300 µm (Tc > 300 µm). The micro-transfer printing process facilitates the passive placement of thousands of micro-lens arrays onto a desired wafer, enabling scalable assembly with a single pick-up and print cycle lasting less than 40 s.

This technology holds the potential to integrate micro-optics for various applications, including enhancing the coupling efficiency in chip-to-fiber coupling for CCD and CMOS sensors by collimating and focusing light. It can also facilitate the development of 3D field cameras, improve the refraction coefficient of solar concentrators, enable beam shaping of lasers, and enhance the depth of field of integral imaging 3D displays.

Additionally, this work could pave the way for placing thousands of thick photonic and electronic integrated circuits [16] onto desired substrates to realize optical and electrical pluggable packages [17] along with bonding materials like solder balls [15].

References

1. Corbett, B., Loi, R., O'Callaghan, J., Roelkens, G.: Transfer printing for silicon photonics. In: Semiconductors and semimetals, vol. 99, pp. 43–70. Elsevier. (2018)

2. Corbett, B., Loi, R., Zhou, W., Liu, D., Ma, Z.: Transfer print techniques for heterogeneous integration of photonic components. Prog. Quantum Electron. **52**, 1–17 (2017)
3. Shaban, Z.: Micro-transfer printing of micro-structured, ultra-thin light-emitting devices. https://hdl.handle.net/10468/14474 (2023)
4. Zhang, J., et al.: III-V-on-silicon widely tunable laser realized using micro-transfer-printing. In: 45th European Conference on Optical Communication, IET (2019)
5. Haq, B.: III-V-on-Si SOAs and DFB/DBR lasers realised using micro-transfer printing. Doctoral dissertation, Ghent University (2021)
6. Wakeel, S., et al.: Micro-transfer printing of thick optical components using a tether-free UV-curable approach. J. Opt. Microsyst. **4**(1), 011003 (2024)
7. Oswalt, K., et al.: Micro transfer printing various thickness components directly from dicing tape. In: IEEE 73rd Electronic Components and Technology Conference (ECTC). IEEE (2023)
8. Gradkowski, K., Stegall, D., Mackey, D., Naughton, A., Smith, T., O'Brien, P.: Demonstration of a single-mode expanded-beam connectorized module for photonic integrated circuits. J. Lightwave Technol. **41**(11), 3470–3478 (2023)
9. Buehler, K., et al.: Micro-transfer-printing and potential process optimizations by FEA. In: 2019 20th international conference on thermal, mechanical and multi-physics simulation and experiments in microelectronics and microsystems (EuroSimE), pp. 1–8. IEEE (2019)
10. Carlson, A., et al.: Shear-enhanced adhesiveness transfer printing for use in deterministic materials assembly. Appl. Phys. Let. **98**(26) (2011)
11. Bohle. UV-Klebstoff Verifix LV 740," 1–2 (n.d.). https://www.bohle.com/en-gb/products/glass-bonding/adhesives/2474/uv-adhesive-verifix-lv-74042
12. Epoxy Technology. EPO-TEK 353ND, pp. 1–2 (2014). https://www.epotek.com/docs/en/Datasheet/353ND.pdf
13. Omnexus. Plastic and elastomers, Tensile Strength - Definition, Units, Formula & Test Methods, specialchem.com
14. RP photonics. https://www.rp-photonics.com/microlens_arrays.html. https://doi.org/10.61835/b96
15. Saif, W.: Comparison of performance of no-clean and water-soluble fluxes in fine-pitch flip-chip package/Saif Wakeel (Doctoral dissertation, Universiti Malaya) (2021)
16. Lerner, R., et al.: Integration of GaN HEMTs onto silicon CMOS by micro transfer printing. In: 28th ISPSD. IEEE (2016)
17. Hwang, H.Y., et al.: Pluggable silicon photonic MEMS switch package for data centre. In: 2018 IEEE 20th Electronics Packaging Technology Conference (EPTC), pp. 46–49. Singapore (2018). https://doi.org/10.1109/EPTC.2018.8654270

Thin-Film Lithium Niobate

Standardized TFLN Photonic Integrated Circuits Platform

Hamed Sattari(✉), Ivan Prieto, Homa Zarebidaki, Jacopo Leo, Gregory Choong, Mattia Orvietani, Fatemeh Arefi, Alberto Della Torre, Yves Petremand, Michele Palmieri, Olivier Dubochet, and Michel Despont

Swiss Center for Electronics and Microtechnology (CSEM), Neuchatel, Switzerland
hamed.sattari@csem.ch

Abstract. We present a thin-film lithium niobate (TFLN) on insulator photonic integrated circuits (PICs) platform based on a 150 mm wafer-scale manufacturing technology. The platform comprises a library of standardized building blocks, so-called process design kit (PDK). Low loss waveguides and fast electro optical modules are reliably manufactured in a well-defined foundry process addressing the demands of various fields from telecom to quantum applications.

Keywords: Thin-Film Lithium Niobate · Platform · Electro-Optics · Foundry Process · Process Design Kit

1 Introduction

Photonic integrated circuits technology has been flourished tremendously in the last decade mainly driven by the telecommunication industry. This advancement has brought the opportunity to also investigate the potential of PICs in many other domains, such as LIDAR, environment monitoring, health diagnosis and quantum. Therefore, various photonic material platforms are being developed which offer a reliable technology to address the emerging needs in these fields. Although the potential use cases of PICs are diverse, the manufacturing technology of PICs and the essential facility for this purpose are not accessible to every photonic designer. Therefore, various foundries are offering their manufacturing technologies to the PIC customers through a fabless scheme by implementing process design kits (PDK). A PDK is a library of well tested building blocks that comply with a set of design rules ensuring a reliable fabrication process. Among various PIC materials addressing the light generation, routing, manipulation and detection functionalities, lithium niobate in the form of thin-film lithium niobate (TFLN) on insulator stands out, thanks to its unique combination of optical and piezoelectric properties. However, it is worth noting that its manufacturing technology has not yet reached a level of maturity to fully meet the demands of the PIC domain.

Recent progress in bonding single crystal thin films of LiNbO3 onto silicon substrates opens new avenues to leverage LiNbO3 advantages in PICs, offering benefits in miniaturization, cost reduction, scalable manufacturing, and integration [1–3]. Consequently, in recent years, substantial progress has been achieved in demonstrating a diverse

J. Witzens et al. (Eds.): ECIO 2024, SPPHY 402, pp. 85–89, 2024.
https://doi.org/10.1007/978-3-031-63378-2_15

range of electro-optic (EO) and nonlinear optical effects in TFLN through pivotal experiments. These encompass the development of high-speed, low V_{π} EO modulators [4], Kerr and EO optical frequency combs [5], wavelength conversion [6], supercontinuum generation [7], and self-referencing [8, 9]. However, the pivotal step in guaranteeing the transition of these breakthroughs into an industrially scalable technology lies in setting up a dependable, high-quality, and high-yield fabrication process, ultimately culminating in the establishment of a foundry for TFLN PICs.

At CSEM, we have established an accessible TFLN PIC foundry, supported by a dependable PDK. Our technology enables the smooth incorporation of both passive and active elements, representing a notable advancement in integrated photonics. This milestone creates opportunities for diverse applications, including data/telecommunication, LIDAR, sensing, and quantum initiatives.

2 TFLN PIC Platform Technology Nodes

A schematical cross-sectional illustration of our TFLN PIC platform developed on 150 mm lithium niobate on insulator (LNOI) wafers are presented in Fig. 1. Our fabrication technology is based on commercially available 150 mm LNOI wafers which consist of a stack with 600 nm thick mono-crystal X-cut LiNbO3 layer on top of a 4.7 μm buried thermal oxide (BOX) layer. The main technology layers include: three waveguide layers (device layer), two metal layers, a VIA, and a deep grid for the chip release. An optical image of the chip area showing a few of the manufactured layers is presented in Fig. 1(c). The detail of the technology nodes building the core of the platform is described in the following.

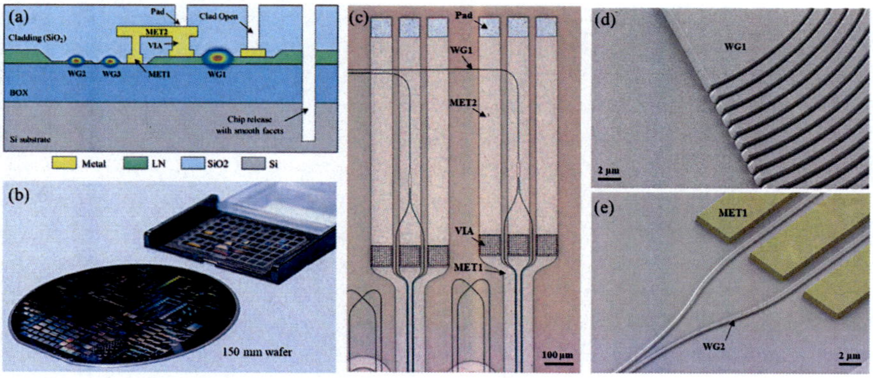

Fig. 1. (a) Schematic of the TFLN PIC platform's technology cross section, (b) 150 mm LNOI wafer with fabricated PICs, (c) Optical microscope image of the TFLN PIC after the clad opening step, (d) SEM image of a grating coupler patterned in WG1 layer, and (e) SEM image of an electro optical modulator in WG2 and MET1 layers (MET1 layer false-colored).

The device layer encompasses three waveguide types: a rib waveguide with a ridge height of 400 nm (WG1), a thin rib waveguide with a ridge height of 100 nm (WG2), and a thin strip waveguide of 200 nm height. Depending on the waveguide width, these WGs could serve to light routing from visible to infrared spectrum thanks to the wide transparency window of lithium niobate. To create the WG layer, initially the TFLN device layer of 600 nm is etched 400 nm using an optimized ion milling technology. This forms the WG1 layer. We then use the remaining 200 nm slab to manufacture the WG2 and WG3 layers using the same etching method. The passive components such as low loss waveguides, grating couplers, multimode interference couplers (MMIs), directional couplers, etc. are patterned in these steps. Note that, availability of three waveguide layers enables to devise complex passive structures such as efficient edge couplers for light coupling between the chip and the fibers. A precise alignment between the waveguide layers is essential to obtain efficient components if devised using more than one WG layer. An SEM image of a grating coupler in WG1 demonstrating smooth sidewalls and a dense pattern is shown in Fig. 1(d). The next step after the device layer patterning is to metallize the device layer (MET1). This metallization step serves mainly to the active components such as EO modulators and thermal phase shifters. Furthermore, the layer could be exploited to form electrodes for periodic poling of the TFLN layer (PPLN). Figure 1(e) demonstrates an SEM image of an EO modulator in WG2 and MET1 layers. In order to allow complex electrical circuitry and routing, we implement a two-layer metallization strategy with a vertical separation between the metal layers MET1 and MET2, and VIA connection to connect the two layers. MET1 and MET2 layers are fabricated through a process that enables excellent MET1-MET2 and MET1-WG alignment. Combination of these technology layers allows to realize highly efficient modulators with several degree of freedom in RF circuitry design for traveling-wave modulators. As the next alternating step, an additional oxide cladding layer is deposited on top of MET2 layer. The cladding protects the PICs from environmental damages and creates a symmetric index contrast in top and bottom of the waveguides for efficient spot size converters. Finally, the cladding is locally removed to create electrical access to electrodes and allows for clad-less waveguides and resonators if needed (e.g., for dispersion engineering and sensing applications). Thanks to two step cladding process, two metallization layers, and the clad opening access, various VIA opening designs and RF impedance matching circuitry could be implemented by the PIC designers. Furthermore, availability and combination of these technology steps in a monolithic platform could be of interest for the heterogenous/hybrid integration and packaging purposes. As the last step in the process, a deep reactive ion etching (DRIE) is done to release the chips from the wafer yielding to smooth chip facets that are essential for efficient optical interfacing.

3 Outlook and Conclusion

Our TFLN PIC platform enables a myriad of choices for the photonic designers not only to provide solutions for the current needs in the telecom industry but also to devise novel solutions for the other domains. Internet of things (IOT), FMCW LIDARs, integrated quantum computation, neuromorphic computing, various sensing schemes for the health and environment monitoring, etc., within both the science and technology playgrounds, could benefit from offerings of this platform. Particularly, having fast EO light manipulation, wavelength conversion, low loss waveguides, and compatibility with the heterogeneous and hybrid integration strategies, all in one PIC platform outstand the TFLN photonics among other PIC platforms. As a cornerstone of this platform, we have demonstrated low loss waveguides with a propagation loss of <0.2 dB/cm [10] and MMIs with insertion loss of <0.15 dB [11]. In the future reports we will present detailed performance of the PDK building blocks. We continue enriching the PDK by regular characterization of the building blocks and design optimization iterations.

To conclude, the unique combination of material properties and manufacturing developments in the TFLN PIC platform provides a technology which is elusive using other PIC material platforms. Our PDK development process from the design to testing steps pushes the technology to offer a wide range of reliable building blocks as schematically presented in Fig. 2. Such a PDK empowers complex PICs with small footprint and low power consumption, promising for fast large-scale PIC solutions.

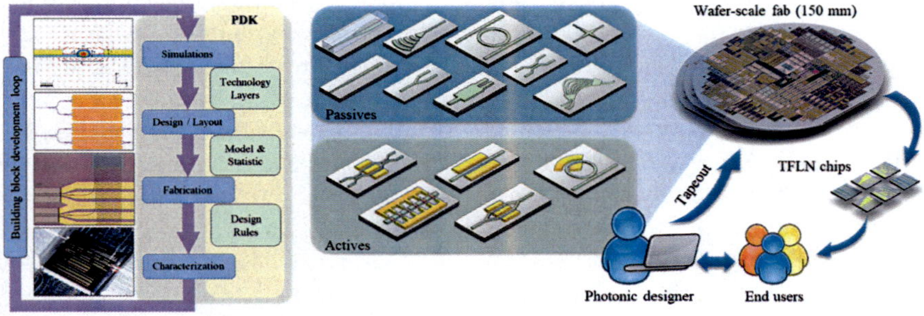

Fig. 2. Envisioned TFLN PIC building blocks offered via a reliable PDK to the end users through a fabless scheme.

Acknowledgments. This project has received funding from the European Union's Horizon 2020 (grant agreement No. 101016138), European Union under GA No. 101070506, and Swiss National Science Foundation (SNSF) (194693).

References

1. Zhu, D., et al.: Integrated photonics on thin-film lithium niobate. Adv. Opt. Photon. AOP **13**(2), 242–352 (2021). https://doi.org/10.1364/AOP.411024
2. Saravi, S., Pertsch, T., Setzpfandt, F.: Lithium Niobate on insulator: an emerging platform for integrated quantum photonics. Adv. Opt. Mater. **9**(22) (2021). https://doi.org/10.1002/adom.202100789
3. Zhang, M., Wang, C., Kharel, P., Zhu, D., Lončar, M.: Integrated lithium niobate electro-optic modulators: when performance meets scalability. Optica OPTICA **8**(5), 652–667 (2021). https://doi.org/10.1364/OPTICA.415762

4. Wang, C., et al.: Integrated lithium niobate electro-optic modulators operating at CMOS-compatible voltages. Nature **562**(7725), 101–104 (2018). https://doi.org/10.1038/s41586-018-0551-y
5. Zhang, M., et al.: Broadband electro-optic frequency comb generation in a lithium niobate microring resonator. Nature **568**(7752), 373–377 (2019). https://doi.org/10.1038/s41586-019-1008-7
6. Wang, C., et al.: Second harmonic generation in nano-structured thin-film lithium niobate waveguides. Opt. Exp. **25**(6), 6963 (2017). https://doi.org/10.1364/OE.25.006963
7. Yu, M., Desiatov, B., Okawachi, Y., Gaeta, A.L., Lončar, M.: Coherent two-octave-spanning supercontinuum generation in lithium-niobate waveguides. Opt. Let. **44**(5), 1222 (2019). https://doi.org/10.1364/OL.44.001222
8. Okawachi, Y., et al.: Chip-based self-referencing using integrated lithium niobate waveguides. Optica OPTICA **7**(6), 702–707 (2020). https://doi.org/10.1364/OPTICA.392363
9. Obrzud, E., et al.: Stable and compact RF-to-optical link using lithium niobate on insulator waveguides. APL Photon. **6**(12) (2021)
10. Leo, J., et al.: Wafer-scale fabrication of low-loss waveguides in lithium niobate on insulator (LNOI) integrated photonics platform. In: European Conference and Exhibition on Optical Communication (2022)
11. Monney, A., et al.: Statistical characterization of MMI beam splitters on thin film lithium niobate on insulator (LNOI) platform at telecom wavelength. In: The european conference on lasers and electro-optics (2023)

Compact Silicon-Rich SiN/LiNbO$_3$ Mach Zehnder and Microring Modulators

Clément Ben Braham[1,2], Ali Belarouci[3], Carlos Alonso-Ramos[2], Regis Orobtchouk[3], Laurent Vivien[2], and Yohan Désières[1](✉)

[1] Université Grenoble Alpes, CEA-LETI, Minatec, 17 rue des Martyrs, 38054 Grenoble, France
yohan.desieres@cea.fr

[2] Centre de Nanosciences et de Nanotechnologies, Université Paris-Saclay, CNRS, 91120 Palaiseau, France

[3] Université de Lyon, Institut des Nanotechnologies de Lyon, INSA Lyon, CNRS, UMR 5270, Lyon, France

Abstract. We report on the demonstration of Mach Zehnder and microring modulators based on a hybrid silicon-rich SiN_x /LiNbO$_3$ waveguide at a wavelength of 1550nm. By using low-temperature PECVD deposition, we obtained non-stoichiometric silicon nitride with a refractive index of ~2.66 which was used to fabricate racetrack ring resonators of 80μm bending radius with a 100 μm long modulating section. We measured intrinsic quality factors up to 1.1×10^5, corresponding to propagation losses of 1–1.2 dB/cm in the range 1500 nm–1600 nm. A 5 mm-long SiN_x/LiNbO$_3$ Mach Zehnder modulator with a 2.8 V.cm efficiency and a 30 dB extinction ratio has been demonstrated.

Keywords: Lithium Niobate · Silicon Nitride · Heterogeneous Integration · Modulator

1 Introduction

Lithium niobate (LN) has a large transparency window, low intrinsic losses, and presents a large variety of nonlinear effects [1]. The integration of thin-film lithium-niobate (TFLN) into silicon or silicon nitride optical platforms can be useful for a wide range of applications, from high-speed communications, quantum communications, frequency conversion, to beam steering [2]. However, the realization of partially etched TFLN waveguides, with demonstrated losses as low as 0.2 dB/m [3], remains a challenge in an industrial, CMOS-compatible foundry. This is due to the complexity of LN etching and lithium-based material introduction within a CMOS environment. Hybrid platforms based either on TFLN bonding on high-temperature stoichiometric Si_3N_4 (SiN) [4], transfer printing [5] or rib loading [6, 7] using low-temperature deposition techniques are good candidates to fully or partially overcome those drawbacks. The refractive index of existing SiN ($n \leq 2$), however, remains lower than those of LN (2.14 and 2.21) at telecom wavelengths, which brings limits to the efficient confinement of guided optical modes and to the use of compact diffractive structures. SiN/TFLN-based hybrid platforms thus

J. Witzens et al. (Eds.): ECIO 2024, SPPHY 402, pp. 90–96, 2024.
https://doi.org/10.1007/978-3-031-63378-2_16

suffer from low-compacity circuits, as bending radii of ~300 μm have to be used to maintain low losses [8]. Alternative approaches based on TiO_2 [9] or hydrogenated amorphous silicon [6] have a much stronger refractive index contrast but suffer from loss and stability issues. In this work, we use silicon-rich SiN (Si-rich SiN) with a refractive index up to ~2.65 to realize compact microring modulators and Mach Zehnder modulators (MZM), with efficiencies comparable to other rib-loaded approaches. The geometry of hybrid waveguides used in this work as well as typical optical and electrical field distributions in dephasing sections are presented in Fig. 1.

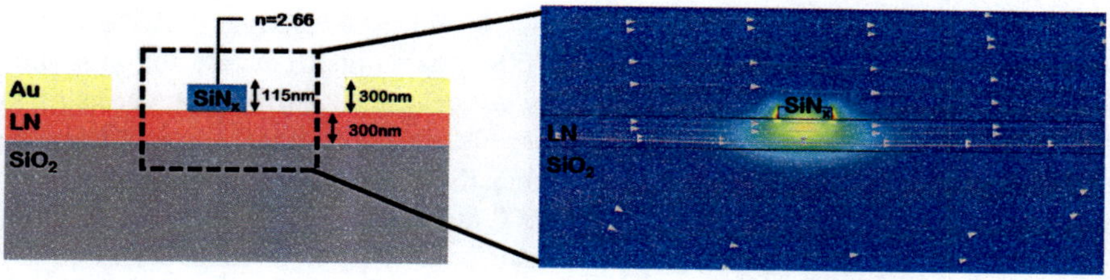

Fig. 1. Cross section of the hybrid silicon rich SiN/LN modulator (left). Optical and electrostatic fields distributions in the modulator simulated through Comsol Multiphysics (right).

2 Design and Fabrication

Racetrack microring resonators and asymmetric Mach-Zehnder modulators were designed using Lumerical, Comsol Mutiphysics and HFSS. The optimized modulation section shows a 44% overlap of the electric and optical fields in the active LN layer while maintaining metal-induced optical losses below 1 dB/cm. Due to the high refractive index (n = 2.66) of silicon rich SiN, losses remained below 0.01 dB/90° were simulated for bending radii of 80 μm. A decrease of the radius down to 40 μm leads to 0.1 dB/90° losses. These values can be compared to losses of 0.01 dB/90° for 300 μm bending radii reported in [8]. A Si-rich SiN thickness of 115nm was found to maximize the efficiency. A $V_\pi L_\pi$ of 2.7 V.cm has been simulated, taking into account our current fabrication constraints, which limit the electrode gap to a minimal value of 5 μm. The simulated efficiency is however already comparable to state-of-the-art SiN/LN modulators [5, 7]. The bending radii could be further reduced to 30 μm for 0.01 dB/90° losses by increasing the SiN layer thickness to 200 nm, but at the expense of a reduced efficiency of 4.4 V.cm. Travelling wave electrodes have also been optimized to reach 80 GHz electro-optical bandwidth.

Hybrid waveguides, racetrack microring resonators, and MZM was fabricated using the non-stoichiometric silicon-rich SiN (deposited by PECVD) on a 300 nm thick X-cut TFLN wafer. The refractive index of the SiN layer was estimated at 2.66 at a wavelength of 1550 nm was estimated by ellipsometry. The layer's thickness was measured at 116 nm. The SiN layer was patterned by e-beam lithography and ICP-RIE. A LOR10A/AZ1512HS bilayer stack was used for the lift-off process in order to fabricate gold electrodes. The bilayer stack was patterned using a Heidelberg μPG101 laser lithography tool, followed by a developing process and e-beam evaporation (MEB550-Plassys) of a 300 nm thick gold layer with a chromium adhesion layer. The bilayer stack was lifted using an ultrasonic-assisted acetone bath for the AZ1512HS resist, followed by a 2.4% TMAH-based solution bath for the LOR10A layer.

3 Experimental Results

Insertion losses around 6 dB are measured using grating couplers. Waveguide losses are extracted from a set of 36 microring resonators with variable coupling gaps and an 80 μm bending radius, as presented in Fig. 2. For a wavelength of 1550 nm, the microrings are over-coupled for gaps below 760 nm and under-coupled for gaps above 780 nm. For each ring resonator, individual resonances are used to estimate propagation losses [8]. The values are then averaged over the entire set of resonance frequencies. The extracted losses are highlighted in Fig. 2. Propagation losses are found in the range of 0.5 dB/cm to 1.5 dB/cm between 1500 and 1600 nm and reach 1.2 dB/cm at 1550 nm. The latter value corresponds to an intrinsic quality factor of 1.1×10^5 for our 80 μm-radius critically coupled resonator. PECVD SiN waveguides are known to exhibit much higher losses around a wavelength of 1520 nm (by as much as a 4 dB/cm loss gain) due to the presence of nitride-hydrogen (N-H) bonds in PECVD SiN [10]. Only a 0.2 dB/cm increase in losses around that wavelength can be observed in our hybrid waveguides. This was attributed to the partial confinement of light in LN, which is expected to have lower losses than SiN [1]. Cascaded waveguide bends were measured to assess bend losses; 0.07 dB/90° for an 80 μm radius bend and 0.11 dB/90° for a 40 μm radius bend were extracted. These values make our devices among the best net losses per ring round trip on a PECVD SiN/TFLN hybrid platform [2, 7, 8] while reducing the bend radius by a factor ~ 4.

Electrodes were deposited on an 80-μm-long bend radius racetrack microring resonator with a 100-μm-long modulation section. The measured extinction ratio was 23 dB. Intrinsic quality factors for critically coupled microrings were estimated at 20×10^3 after electrode deposition. The degradation of the quality factor is attributed to partial etching of the silicon-rich SiN by the TMAH solution during the lift-off process. Such degradation can be avoided for future samples by using an NMP-based solvent to remove the LOR10A layer. The microring resonance tunability measurement is not straightforward, as charge drift under a DC applied voltage prevents measuring resonance shifts. We therefore used a triangular signal of f = 50 kHz to sweep voltage between 0 V and 10 V. We then adjust the input laser wavelength λ_i near resonant wavelength λ_{res}. The optical output of the device is sent to a photodetector coupled to a trans-impedance amplifier with a logarithmic conversion function. The voltage V_i associated with minimal transmission for the working wavelength λ_i is then extracted. The extraction of the microring tunability is presented in Fig. 3 and estimated at 1.06 pm/V ±0.09 pm/V.

This value is in reasonable agreement with the efficiency simulated at 1.23 pm/V for the fabricated geometry. The MZM's modulation performances are measured by injecting light from a continuous laser through a polarization-maintaining fiber into the chip. An extinction ratio of 30 dB is reached at 0V. MZM electrodes are polarized between 0 to 10 V with a 50-kHz-sawtooth signal. The response of a 5-mm-long MZM driven in

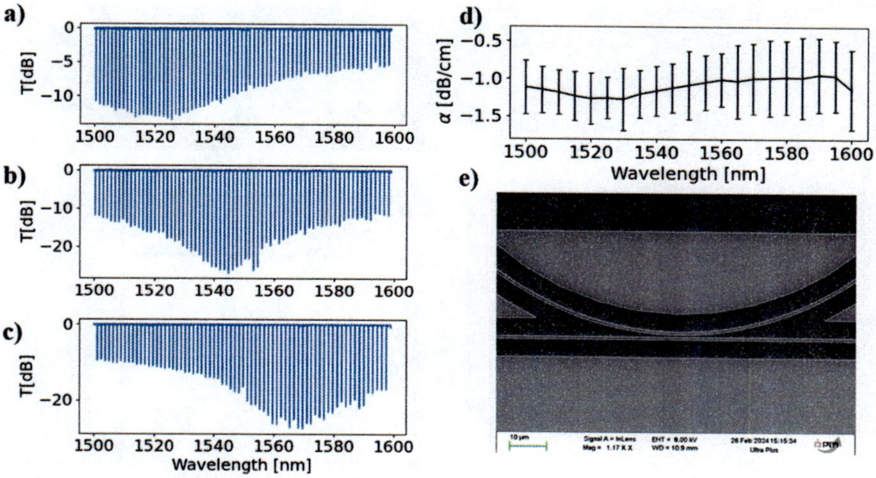

Fig. 2. Normalized microring transmission signals for a gap of 610 nm (a), 740 nm (b) and 790 nm (c). Measured propagation losses using our set of 36 microrings (d). SEM image of the coupling section of a microring (e)

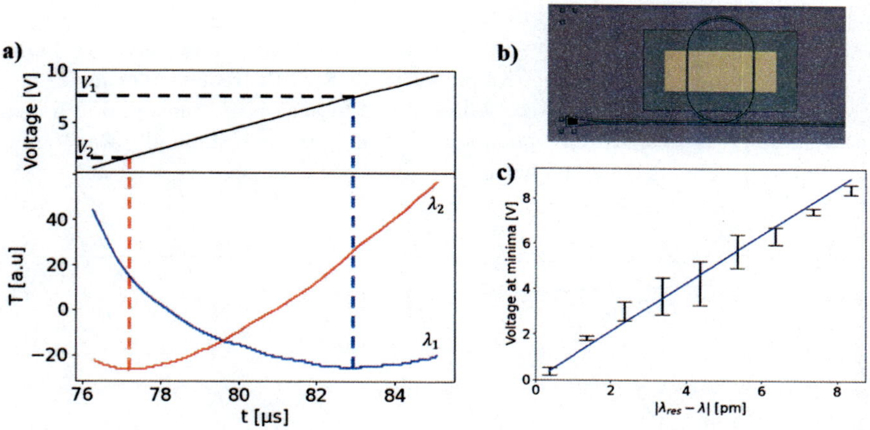

Fig. 3. Measured optical transmitted signal during voltage sweep for wavelengths $\lambda_1 = 1548.633$ nm and $\lambda_2 = 1548.639$ nm (a). Optical microscope image of micoring resonator (b). Extracted tunability of a microring resonator (c)

the push-pull configuration is presented in Fig. 4. A $V_\pi L_\pi$ of ~2.8 V.cm was inferred from the transmission response. This value is in good agreement with our simulated efficiency of 2.7 V.cm.

The partial etching of the Si-rich SiN during the lift-off process increased the MZM's insertion losses from 18 dB to 34 dB, preventing direct measurement of the electro-optical (EO) bandwidth. Electrical measurements of our traveling wave electrode were performed using a Vectorial Network Analyser (VNA) and are presented in Fig. 5a. An

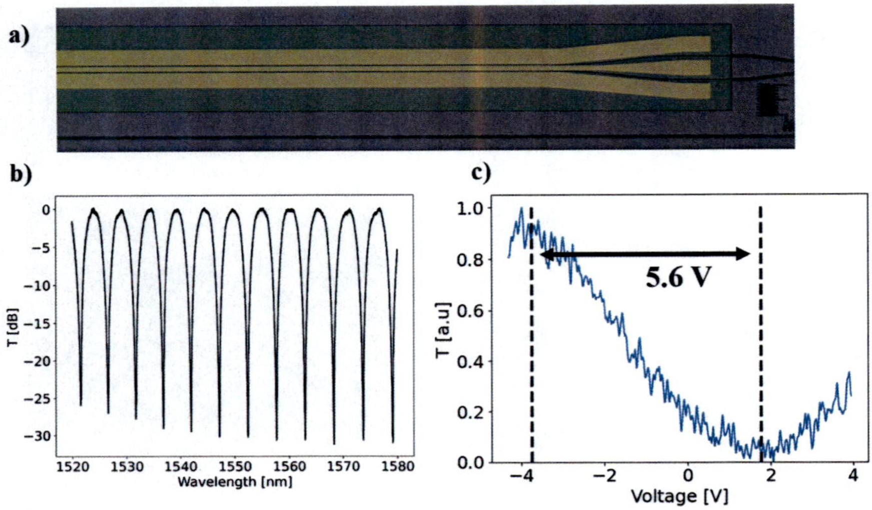

Fig. 4. Optical microscope image of the proposed MZM (a). Normalized transmission spectrum of a 5 mm-long MZM (b). Modulated transmission of the 5mm-long MZM around 1550 nm in the push-pull configuration as a function of the applied voltage (c).

EO bandwidth around 67 GHz (Fig. 5b) estimated using a well-known analytical model [11] is shown in Fig. 5b. The analytical model of the MZM's frequency response is fed with the measured impedance, effective index, and losses of our transmission line and the simulated group index of the optical mode. This estimated bandwidth was found to be limited by a 12% mismatch between the indices of the optical and RF modes.

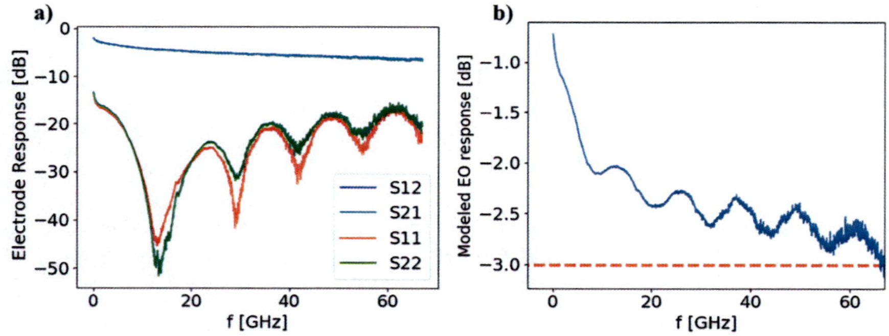

Fig. 5. Measured electrical response of the travelling wave electrodes of a 5-mm-long MZM (a). Modeled electro-optical (EO) response of a 5-mm-long MZM, deduced from electrode's electrical response and the simulated group index of the optical waveguide in the dephasing section (b)

4 Conclusion

We demonstrated hybrid Silicon-rich SiN/TFLN MZM and microring modulators with 80-μm-long bending radii exhibiting respectively $V_{\pi}L_{\pi}$ around 2.8 V.cm and peak shift of about 1 pm/V. Thanks to the high refractive index of the Silicon-rich SiN used in this study (n ~ 2.65), our devices are more compact than their PECVD counterparts [4, 7, 8] while exhibiting similar modulation efficiencies. We believe these Si-rich SiN loaded waveguides can help bridge the performance gap between current hybrid SiN/TFLN platforms and TFLN-etched platforms.

Acknowledgment. This work is part of the IPCEI Microelectronics and Connectivity and was supported by the French Public Authorities within the frame of France 2030. This work was partially supported by the French RENATECH network, implemented at the Upstream Technological Platform LTM in Grenoble (PTA/LTM).

This work has also been partially funded by the European Union (ERC-CRYPTONIT, under the Grant Agreement 101097804). Views and opinions expressed are however those of the author(s) only and do not necessarily reflect those of the European Union or the European Research Council. Neither the European Union nor the granting authority can be held responsible for them.

The authors would like to acknowledge Philippe Grosse and Nicholas Dunoyer for their supports during electro-optical measurements.

The authors would like to acknowledge Samson Edmond as well for his help during sample processing.

References

1. Weis, R.S., Gaylord, T.K.: Lithium niobate: summary of physical properties and crystal structure. Appl. Phys. A **37**, 191–203 (1985). https://doi.org/10.1007/BF00614817
2. Zhu, D., et al.: Integrated photonics on thin-film lithium niobate. Adv. Opt. Photon. **13**, 242–352 (2021)
3. Shams-Ansari, A., et al.: Reduced material loss in thin-film lithium niobate waveguides. APL Photon. 2022 **7**(8), 081301 (2022)
4. Churaev, M., Wang, R.N., Riedhauser, A., et al.: A heterogeneously integrated lithium niobate-on-silicon nitride photonic platform. Nat. Commun. **14**, 3499 (2023). https://doi.org/10.1038/s41467-023-39047-7
5. Vanackere, T., et al.: Heterogeneous integration of a high-speed lithium niobate modulator on silicon nitride using micro-transfer printing. APL Photon. **8**(8), 086102 (2023). (Author, F.: Contribution title. In: 9th International Proceedings on Proceedings, pp. 1–2. Publisher, Location (2010))
6. Wang, Y., Chen, Z., Cai, L., Jiang, Y., Zhu, H., Hui, H.: Amorphous silicon-lithium niobate thin film strip-loaded waveguides. Opt. Mater. Express **7**, 4018–4028 (2017)
7. Ahmed, A.N.R., Nelan, S., Shi, S., Yao, P., Mercante, A., Prather, D.W.: Subvolt electro-optical modulator on thin-film lithium niobate and silicon nitride hybrid platform. Opt. Lett. **45**, 1112–1115 (2020)
8. Ahmed, A.N.R., Nelan, S., Shi, S., Yao, P., Mercante, A., Prather, D.W.: Tunable hybrid silicon nitride and thin-film lithium niobate electro-optic microresonator. Opt. Lett. **44**, 618–621 (2019)
9. Zhu, X., et al.: Phase modulation using a titanium dioxide strip on lithium niobate. Opt. Mater. Express **12**, 3296–3302 (2022)

10. Kempf, E., et al.: SiN half-etch horizontal slot waveguides for integrated photonics: numerical modeling, fabrication, and characterization of passive components. Opt. Express **30**, 4202–4214 (2022)
11. Modulators, G.G.: Semiconductor devices for high-speed optoelectronics, pp. 356–439. Cambridge University Press (2009)

Foundry Fabricated Thin-Film Lithium Niobate Electro-Optic Modulators for Blue Light

Tianyi Liu[1,2(✉)], Vahid Ansari[1], Engjell Bebeti[1], John N. Straguzzi[1], Alperen Govdeli[1,2], Wesley D. Sacher[1], and Joyce K. S. Poon[1,2]

[1] Max Planck Institute of Microstructure Physics, Weinberg 2, 06120 Halle (Saale), Germany
tianyi.liu@mpi-halle.mpg.de

[2] Department of Electrical and Computer Engineering, University of Toronto, 10 King's College Road, Toronto, ON M5S 3G4, Canada

Abstract. We demonstrate a thin-film lithium niobate electro-optic modulator operating at a blue wavelength of $\lambda = 445$ nm. The modulator was fabricated in a generic lithium niobate photonic foundry process on a 150-mm diameter thin-film lithium niobate wafer. The Mach-Zender modulator has 1.9-mm long phase shifters and exhibits a low half-wave voltage of 1.80 V, a high extinction ratio >15 dB, and an estimated device insertion loss of 3 dB.

Keywords: Visible photonics · Thin-film lithium niobate · Electro-optic modulators

1 Introduction

Integrated photonics aims to realize versatile, scalable, and power-efficient optical functionalities on microchips. Emerging applications beyond fiber optic communications, such as laser scanning displays [1], underwater signaling [2], quantum control [3] and wireless optical communication [4], requires integrated photonic circuits operating in the visible spectral range ($\lambda = 380$–750 nm). Currently, the most popular choice of waveguide core materials for visible wavelengths are silicon nitride (SiN) and aluminum oxide (Al_2O_3) [5]. These materials lack free carriers and second-order susceptibility ($\chi^{(2)}$), which poses a challenge for integrating high-speed modulation functionality into the existing photonic platforms.

Lithium niobate distinguishes itself as a promising material for high-speed chip-scale modulators in the visible spectrum, due to its large Pockels coefficient and wide transparency window which extends to $\lambda = 350$ nm. In particular, thin-film lithium niobate (TFLN) platforms leverage high-confinement waveguides, enabling small footprint routing, dense integration of photonic components, and better modulation performances over their bulk counterparts. A concern with short-wavelength TFLN modulators has been the photorefractive effect [6] and higher waveguide loss due to sidewall roughness scattering [7].

J. Witzens et al. (Eds.): ECIO 2024, SPPHY 402, pp. 97–101, 2024.
https://doi.org/10.1007/978-3-031-63378-2_17

TFLN electro-optic (EO) modulators have been reported in short visible wavelengths like green [8, 9] and blue [10, 11] previously. In this work, we demonstrate the feasibility of visible spectrum TFLN modulators in a generic, wafer-scale, lithium niobate photonic foundry process at HyperLight Corporation. Our blue-light modulator has a 1.9-mm-long phase shifter section, and it exhibits a low half-wave voltage (V_π) of 1.80 V, a high extinction ratio (ER) over 15 dB, and an on-chip insertion loss of ~3 dB. So far, we have verified EO modulation up to ~130 MHz using this modulator at $\lambda = 488$ nm. Ongoing experiments are continuing, and the results will be presented at the conference.

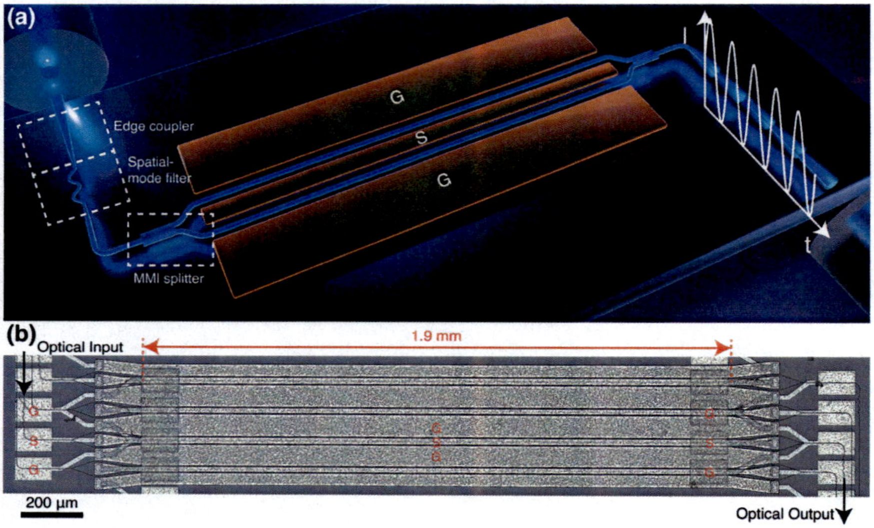

Fig. 1. (a) Schematic illustration of the integrated thin-film lithium niobate Mach-Zender modulator, including edge couplers, spatial-mode filters, multimode interference (MMI) splitters, and the push-pull modulation section. (b) Stitched optical microscope image of four Mach-Zender modulator devices at a 100 μm pitch. The modulation section length is 1.9 mm.

2 Design and Characterization

Figure 1(a) illustrates the schematic of our modulator. The Mach-Zehnder interferometer (MZI) modulator operates in a push-pull configuration. The MZI includes adiabatic taper edge couplers, S-bend waveguide spatial mode filters, 1x2 multimode interference (MMI) splitters, and a 1.9-mm-long modulation section.

The modulator devices are fabricated by HyperLight Corporation in a multi-project wafer run. A microscope image of the densely arranged modulator array is shown in Fig. 1(b). Adjacent modulators share a common ground electrode. The pitch between modulators is 100 μm, and the four modulators occupy a total footprint of 1.6 mm^2.

We characterize the modulator at wavelength $\lambda = 445$ nm with a multi-wavelength Coherent OBIS Galaxy laser source, using optical input power (at the chip facet) of 0.96

mW. The polarization is calibrated to transverse electric mode for efficient EO modulation. Input laser light is edge-coupled into the TFLN chip with cleaved single-mode fibers (Nufern S405-XP). Transmitted light is collected using another fiber and detected with a silicon photoreceiver (New Focus 2032, high gain setting). An oscilloscope (Keysight DSOX4054A) is used for collecting the photoreceiver signal. The input electric modulation signal is generated with an arbitrary function generator (Tektronix AFG31252) and sent to the ground-signal-ground metal contacts via three tungsten probes (MPI Corporation). To measure the device at a higher modulation frequency of 130 MHz, we used a different wavelength $\lambda = 488$ nm from the same laser source to input a higher optical power of 20.7 mW into the device. The modulated light is measured with a Thorlabs PDA015A2 photodetector, which has a higher detection bandwidth.

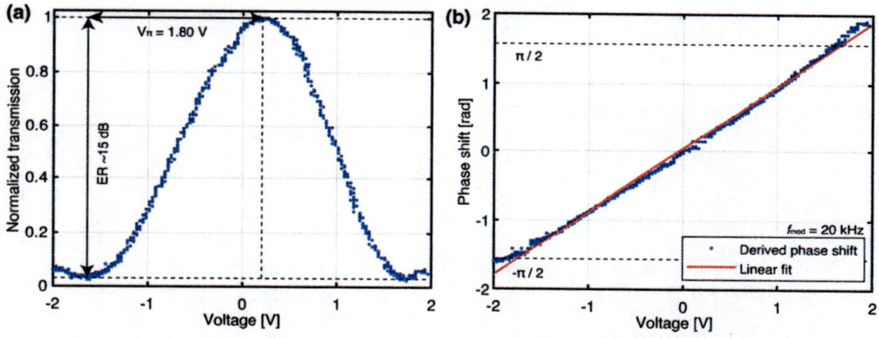

Fig. 2. (a) Normalized modulator transmission over applied voltage, measured at a modulation frequency of 20 kHz, wavelength $\lambda = 445$ nm, optical power of 0.96 mW. (b) Derived phase change per one modulator branch over applied voltage. The linear fit of the result is shown in the red line.

3 Results

The on-chip modulator insertion loss is estimated to be ~3 dB, after removing the fiber-to-chip coupling loss. The total insertion loss of the device was 26 dB, dominated by the fiber-to-chip coupling loss of 11.5 dB per facet. The high fiber-to-chip insertion loss is caused by the mode mismatch between the edge coupler on the chip and the fiber mode.

DC performance characterization is performed at 20 kHz frequency with a triangle waveform. Figure 2(a) plots the normalized optical transmission over applied voltage. The result shows a V_π of 1.80 V, corresponding to voltage-length product of 0.34 V-cm, and the ER is found to be over 15 dB. Typical DC voltage-length products of TFLN modulators at 1550 nm and 784 nm are around 2 V-cm and 0.8 V-cm respectively [12, 13]. Since the voltage-length product is proportional to the wavelength, our results, which are at a wavelength of 455 nm, are comparable with those values. The phase shift in a single modulator branch is extracted from the transmission data and plotted in Fig. 2(b). Fitting results gives a phase-shifting efficiency of $\Delta\varphi/\Delta V = 0.29\pi\ V^{-1}$ and shows good linearity.

We measured the ER as a function of the electrical modulation frequency, f_{mod}, and peak-to-peak voltage, as shown in Fig. 3(a). The ER starts to decrease at high frequencies when $f_{mod} > 30$ kHz. This effect is mainly due to the photoreceiver's limited frequency response, and it becomes more significant with higher voltage as the modulator enters the overdrive regime. The ER also decreases at low frequencies when $f_{mod} < 10$ kHz, which can be attributed to the photorefractive effect in lithium niobate.

We measured the modulation response at higher frequencies at a wavelength of $\lambda = 488$ nm. The photodetector signal is averaged using the built-in function of the oscilloscope, due to the photodetector's noise floor. Figure 3(b) shows the normalized modulator transmissions when driven at a frequency of 130 MHz. The transmission pattern appears to be overdriven under an applied peak-to-peak voltage of 2.5 V. We are currently limited by the frequency response of our measurement setup. The modulator EO bandwidth is not fully characterized yet, and we are continuing experiments to investigate the bandwidth limits of the EO modulation.

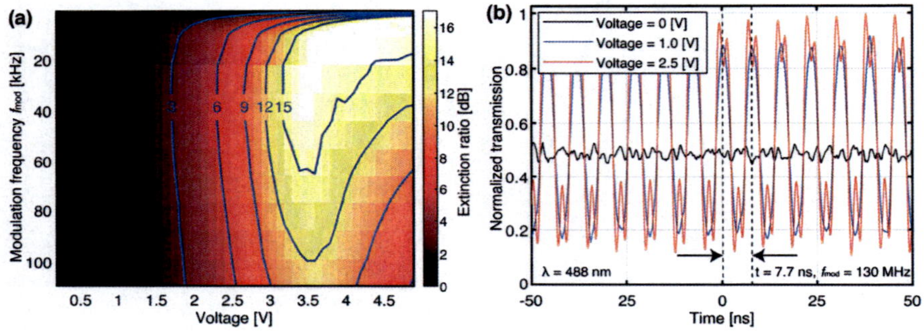

Fig. 3. (a) Measured ER as a function of frequency and peak-to-peak voltage. Blue lines show the contours of the ER. (b) Normalized transmission at the modulation frequency of 130 MHz, measured with optical wavelength $\lambda = 488$ nm and optical power of 20.7 mW.

4 Conclusion

In summary, we have demonstrated a foundry-fabricated thin-film lithium niobate EO modulator at blue wavelengths. The modulator features a voltage-length product of 0.34 V-cm and ER over 15 dB. Further EO performance characterization at GHz-frequencies is ongoing and will be discussed during the conference.

Acknowledgements. The modulator devices were designed and measured at the Max Planck Institute of Microstructure Physics and fabricated by HyperLight Corporation. We thank Dr. Lingyan He for her technical inputs and helpful discussions. We thank Dr. Andrei Stalmashonak for assistance with the experimental setup.

References

1. Raval, M., Yaacobi, A., Watts, M.R.: Integrated visible light phased array system for autostereoscopic image projection. Opt. Lett. **43**, 3678 (2018)
2. Wu, T.-C., et al.: Blue laser diode enables underwater communication at 12.4 Gbps. Sci. Rep. **7**, 40480 (2017)
3. Niffenegger, R.J., et al.: Integrated multi-wavelength control of an ion qubit. Nature **586**, 538–542 (2020)
4. Haas, H., Sarbazi, E., Marshoud, H., Fakidis, J.: Chapter 11 - Visible-light communications and light fidelity. In: Willner, A.E. (ed.) Optical fiber telecommunications VII, pp. 443–493. Academic Press (2020)
5. Poon, J.K.S., et al.: Silicon photonics for the visible and near-infrared spectrum. Adv. Opt. Photon. **16**, 1 (2024)
6. Zhang, M., et al.: Integrated lithium niobate electro-optic modulators: when performance meets scalability. Optica **8**, 652 (2021)
7. Desiatov, B., et al.: Ultra-low-loss integrated visible photonics using thin-film lithium niobate. Optica **6**, 380–384 (2019)
8. Li, C., et al.: High modulation efficiency and large bandwidth thin-film lithium niobate modulator for visible light. Opt. Express **30**, 36394 (2022)
9. Renaud, D., et al.: Sub-1 Volt and high-bandwidth visible to near-infrared electro-optic modulators. Nat. Commun. **14**, 1496 (2023)
10. Xue, S., et al.: Full-spectrum visible electro-optic modulator. Optica **10**, 125 (2023)
11. Celik, O.T., et al.: Visible electro-optic modulator array in thin-film lithium Niobate for atomic qubit control. In: Frontiers in Optics + Laser Science 2023 (FiO, LS) FM5B.2. Optica Publishing Group, Tacoma, Washington (2023)
12. Wang, C., et al.: Integrated lithium niobate electro-optic modulators operating at CMOS-compatible voltages. Nature **562**(7725), 101–104 (2018)
13. Valdez, F., Mere, V., Mookherjea, S.: 100 GHz bandwidth, 1 volt integrated electro-optic Mach-Zehnder modulator at near-IR wavelengths. Optica **10**, 578–584 (2023)

Polymer Photonics

Efficient Modelling of 3D-Printed Freeform Waveguides by a Dedicated Beam-Propagation Method (BPM) Based on Transformation Optics

Sina Foroutan-Barenji, Jonas Krimmer, Wolfgang Freude, and Christian Koos(✉)

Institute of Photonics and Quantum Electronics (IPQ), Karlsruhe Institute of Technology (KIT), Engesserstr. 5, 76131 Karlsruhe, Germany
{sina.barenji,christian.koos}@kit.edu
https://www.ipq.kit.edu/index.php

Abstract. An efficient numerical algorithm is developed to model wave propagation in 3D-printed freeform waveguides, so-called photonic wire bonds (PWBs). The method utilizes transformation-optics (TO) in combination with curvilinear meshing to transform the original 3D-freeform PWB into a straight waveguide with anisotropic material properties. This waveguide is then modeled using an anisotropic full-vectorial beam propagation method (BPM). We demonstrate the viability of our TO-BPM algorithm and benchmark its performance against a finite-difference-time-domain (FDTD) method. Our approach shows a 600-fold reduction of the computation time in combination with a mean percentage error of 3.2% in power transmission simulation relative to the FDTD.

Keywords: Photonic integrated circuit · photonic wire bond · transformation optics · beam propagation method

1 Introduction

Photonic integrated circuits (PICs) are key components for a wide variety of applications that range from high-speed communications [1], artificial intelligence [2], ultra-fast signal processing [3], optical metrology [4] to life sciences and biophotonics [5]. Technologically, PICs mainly rely on planar configurations made through established microfabrication methods involving layer deposition and 2D patterning via high-resolution electron-beam or deep-UV lithography. Over the previous years, these fabrication methods have been augmented by multi-photon lithography, enabling in-situ fabrication of 3D freeform structures and significantly enhancing the functionality of planar PICs [6–8]. A particularly interesting approach relies on 3D-printed freeform waveguides, so-called photonic wire bonds (PWBs), that connect optical chips, e.g., within high-performance photonic multi-chip modules [6,7]. A key aspect of the PWB concept is the

J. Witzens et al. (Eds.): ECIO 2024, SPPHY 402, pp. 105–110, 2024.
https://doi.org/10.1007/978-3-031-63378-2_18

ability to compensate for unavoidable positioning inaccuracies of the underlying optical chips by adapting the PWB trajectory and shape to the actual positions of the coupling facets. However, these positions are acquired only during the fabrication process itself and need to be translated into a low-loss PWB design within a few seconds prior to printing. This is currently done by computationally efficient heuristic PWB design algorithms that simply choose a smooth trajectory to connect the optical interfaces of interest. However, while this technique is fast and leads to acceptable results, it cannot predict the loss of the PWB. On the other hand, first-principle Maxwell's equations solvers based on the finite-difference time-domain (FDTD) have been explored for PWB modeling [9], but they still are far too slow even if adapted to the specific case of moderately bent 3D freeform waveguides by using transformation optics (TO). An efficient Maxwell-based PWB modeling and design tool is hence still lacking.

In this paper, we present an efficient Maxwell-based algorithm specifically geared towards modeling wave propagation in 3D-printed PWBs. The method relies on a fully vectorial beam-propagation technique (BPM) and combines it with TO. More specifically, we use the TO in combination with curvilinear meshing to transform the original 3D freeform PWB into a straight waveguide with anisotropic material properties, which is then modeled by BPM. We demonstrate the viability of our TO-BPM algorithm and benchmark its performance against an FDTD method. For technically relevant PWB structures exhibiting a moderate degree of bending, our approach shows a 600-fold reduction of the computation time in combination with a mean percentage error of 3.2% in power transmission simulation relative to the FDTD reference values. To the best of our knowledge, our work represents the first implementation of a fully vectorial BPM that can handle 3D freeform waveguides that have non-plane trajectories. We believe that this technique can be key to highly efficient real-time routing and design of PWB structures in high-throughput optical assembly.

2 Methods

The BPM has been employed for the efficient simulation of wave propagation both in isotropic and anisotropic media, but the method, assuming paraxial propagation, faces limitations in handling freeform bent structures. Different BPM algorithms have been explored to accelerate electromagnetic simulations of integrated optical waveguides. Xu *et al.* [10] reported a full-vectorial BPM method applicable to planar waveguides with anisotropic material. Lu *et al.* [11] proposed a scheme based on a scalar BPM with a curvilinear coordinate system to model bent slab waveguides. Kawamura *et al.* [12] utilized full-vectorial BPM in combination with coordinate transformation to model tapered planar waveguide.

As depicted in Fig. 1(a), PWBs establish optical connections between different PIC chips with vastly dissimilar emission directions. Consequently, the simulation domain for optimizing the PWB trajectory can be large. The TO technique [9] significantly shrinks the simulation region for a 3D freeform waveguide in (x, y, z)-space by transforming the curved waveguide into a straightened waveguide in a virtual (ξ, η, ζ)-space, see simulation regions marked by red boxes in Fig. 1(b). The transformed media in virtual space are generally anisotropic [9], even if the structure in the original space is made from isotropic material with permittivity ε and permeability μ. As demonstrated in [9] for planar waveguides, this method results in anisotropic, but diagonal permittivity and permeability tensors. To apply this method to non-planar waveguides, the numerical solver must be able to consider off-diagonal elements in the transformed material tensors. To this end, we employ a full-vectorial finite-difference BPM that can handle anisotropic media [10] for modeling the straight waveguide structure in the (ξ, η, ζ)-space as obtained from the TO approach. Since the transformed waveguide does not have any bend, we can safely rely on a paraxial implementation of BPM, which is computationally faster than its wide-angle counterpart [12]. Note also that the BPM is generally a forward-propagating method and does not account for reflections. For the moderately bent PWB waveguides with rather a small index contrast ($n_{\text{core}} = 1.53$ and $n_{\text{clad}} = 1.36$ at $\lambda = 1.55\,\mu\text{m}$), this approximation works very well.

The TO approach relies on a coordinate transformation that can be described by a differentiable function and that maps the original 3D freeform PWB in (x, y, z)-space into its straight counterpart in (ξ, η, ζ)-space. To find an appropriate coordinate transformation, we use the technique of curvilinear meshing, similar to the approach described in [14], see Fig. 1(c), which can accurately capture complex geometries with curved boundaries. More specifically, we discretize our original structure in (x, y, z)-space into a multitude of curvilinear hexahedra which are then mapped to equivalent rectangular hexahedra in the transformed (ξ, η, ζ)-space by a specific quadratic function individually defined for each of the elements. The elements of the Jacobian matrix are then derived from this function and serve for calculating the transformed permittivity and permeability tensors ε' and μ', see [9] for details on the underlying calculations.

3 Results and Discussions

To test our TO-BPM algorithm, we use it for analyzing moderately bent PWB with trajectories featuring minimum radii of curvature in the range $R_{\min} = 2\,\mu\text{m}$ to $R_{\min} = 215\,\mu\text{m}$, Fig. 2(a). These PWBs are derived from a set of merely random 3D non-planar trajectories. We have labeled them with their minimum radii of curvature. The rectangular cross-section of all the PWBs is $2\,\mu\text{m} \times 1.8\,\mu\text{m}$ in the x and y directions, respectively. Core and cladding refractive indices are $n_{\text{core}} = 1.53$ and $n_{\text{clad}} = 1.36$, respectively. To define our curvilinear and rectilinear elements, we choose mesh sizes of $\Delta\xi = \Delta\eta = 90\,\text{nm}$ in the directions perpendicular to the trajectory and a step size of $\Delta\zeta = 2\,\mu\text{m}$ in the direction along the trajectory. In Fig. 2(b), we plot the simulated power transmission and power loss at $\lambda = 1.55\,\mu\text{m}$ obtained from the TO-BPM for each PWB trajectory characterized by $R_{\min}$ and compare the results to those obtained from a commercially available FDTD solver (Lumerical FDTD, Ansys Inc., Canonsburg, PA, USA) [15]. For practical low-loss PWB trajectories interconnecting PICs,

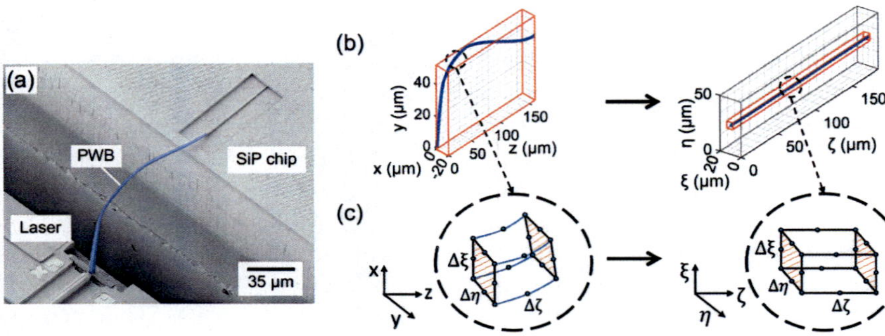

Fig. 1. Illustration of the TO-BPM method for modeling non-planar freeform photonic wire bonds (PWB). (a) PWB transition between a horizontal-cavity surface-emitting laser (HCSEL) and a silicon-on-insulator (SOI) waveguide [Figure reprinted from [16] © 2020 OSA]. (b) The exemplary freeform waveguide in the original (x, y, z)-space with a simulation volume of $20\times50\times160\,\mu\text{m}^3$, and transformed waveguide in the virtual (ξ, η, ζ)-space with a simulation volume of $5\times5\times180\,\mu\text{m}^3$. (c) The transformation from an original curvilinear element to a virtual rectangular element. The horizontal dimensions $\Delta\xi=\Delta\eta=90\,\text{nm}$ of the volume element are the same in both spaces, and longitudinal step size amounts to $\Delta\zeta=2\,\mu\text{m}$. (Color figure online)

the minimum radii of curvature R_{min} should be larger than $R_{\text{min}}=30\,\mu\text{m}$, and for this case, the TO-BPM demonstrates a mean percentage error (MPE) of 3.2%. This error represents the absolute deviation of the power transmission for the TO-BPM calculation relative to the FDTD reference values, averaged over all data points with $R_{\text{min}}\geq30\,\mu\text{m}$. At a bending radius of $30\,\mu\text{m}$, the bend loss of the PWB is 0.5 dB. The variations between neighboring transmission and loss values can be explained by the fact that the trajectories of PWB are random and may be vastly dissimilar from one another, containing, e.g., more than one bend with a bending radius close to the minimum radius of curvature depicted on the horizontal axis. For better comparison, the TO-BPM and the FDTD simulations were run on the same system (AMD EPYC 7713P 64-Core Processor, 512 GB DDR4-3200 Memory). As shown in Fig. 2(c), our implementation is up to 612 times faster than conventional FDTD simulations for PWBs with bending radii up to $30\,\mu\text{m}$. With increasing bending radius, the resulting reduction in simulation volume for the FDTD calculation becomes less prominent, and the disparity in performance between the two methods diminishes. Still, the TO-BPM exhibits a 96-fold performance improvement for bending radii beyond $100\,\mu\text{m}$, and the computation times stay consistently below 30 s, with a high potential to speed up the simulation further. Similarly, the memory requirements of the TO-BPM (typically smaller than 2 GB) are significantly lower than that of the FDTD solver (typically larger than 100 GB). Figure 2(d) depicts the simulated electric field intensity ($|\boldsymbol{E}|^2$) for the PWB trajectory with a minimum bending radius of $R_{\text{min}}=15\,\mu\text{m}$, which is marked with a green arrow in Fig. 2(a). We demonstrate the results in the virtual (ξ,η,ζ)-space for TE polarized light, having a

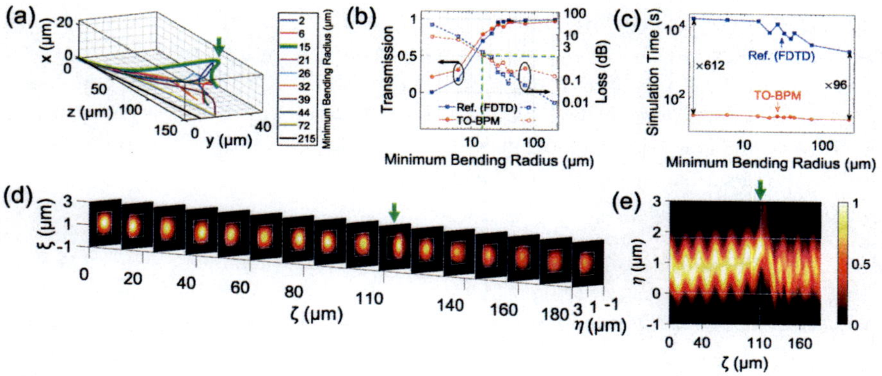

Fig. 2. TO-BPM simulation results for exemplary non-planar PWBs. (a) Selection of random exemplary PWBs with non-plane trajectories characterized by their respective minimum radii of curvature. (b) Simulated power transmission and power loss in dB of the PWB set, obtained by the TO-BPM algorithm and by reference FDTD simulations. For minimum radii of curvature $R_{\min}$ larger than 30 µm, the mean percentage error (MPE) amounts to 3.2%. (c) Comparison of total simulation times for the TO-BPM algorithm and the reference FDTD simulations, including the transformation from the physical PWB to the straightened waveguide. The TO-BPM can offer a more than 600-fold reduction in calculation time compared to the FDTD. (d) Simulated electric field intensity ($|\boldsymbol{E}|^2$) for the freeform waveguide with minimum bending radius $R_{\min} = 15\,\mu\mathrm{m}$ shown in (a) and marked with (a) green arrow, also in (d) and (e). (e) Electric field intensity in the (ξ, η, ζ)-space on the $\xi = 0$ plane. (Color figure online)

dominant electric field oriented along x in (x, y, z)-space and along ξ in (ξ, η, ζ)-space. The bending in the η-direction (y-direction) is severe, see Fig. 2(a), with the smallest bend radius of $R_{\min} = 15\,\mu\mathrm{m}$ at $\zeta = 110\,\mu\mathrm{m}$, which causes the mode to wiggle in this direction. This type of data provides insight into the bending losses along the trajectory and facilitates appropriate design and optimization strategies. Figure 2(e) illustrates the field distribution in (ξ, η, ζ)-space on the plane at $\xi = 0$. From this figure, it can be seen that the wave deviates from the transformed waveguide axis as it propagates along the curved waveguide trajectory. The smallest bend radius of $R_{\min} = 15\,\mu\mathrm{m}$ is found at $\zeta = 110\,\mu\mathrm{m}$ (green arrow). This bend causes a noticeable leakage of light from the core into the cladding and subsequently into the perfectly matched layer surrounding the simulation region.

4 Conclusions

We presented a computationally efficient TO-BPM algorithm based on a combination of TO and a full-vectorial BPM to model 3D-printed freeform PWBs. We verify the capability of our method by simulating freeform PWBs with non-plane trajectories for a wide range of bending radii. The method shows an up to 600-fold reduction in computational effort compared to the reference FDTD simulations in the original 3D space while obtaining a mean percentage error of 3.2% compared to the FDTD method. To the best of our knowledge, our work represents the first implementation of a fully vectorial BPM that can handle 3D freeform waveguides that have non-plane trajectories. We believe that the TO-BPM technique has the potential to facilitate the design and optimization of optical devices based on 3D freeform PWBs.

Acknowledgements. This work was funded by the Deutsche Forschungsgemeinschaft via the DFG Collaborative Research Center (CRC) WavePhenomena (SFB 1173) and via the Excellence Cluster '3D Matter Made to Order', by the European Research Council via the ERC Consolidator Grant TeraSHAPE (# 773248), and by the Karlsruhe School of Optics & Photonics (KSOP).

References

1. Dong, P., et al.: Silicon photonic devices and integrated circuits. Nanophotonics **3**(4–5), 215–228 (2014)
2. Shen, Y., et al.: Deep learning with coherent nanophotonic circuits. Nature Photon. **11**(7), 441–446 (2017)
3. Drayss, D., et al.: Non-sliced optical arbitrary waveform measurement (OAWM) using soliton microcombs. Optica **10**, 888–896 (2023)
4. Trocha, P., et al.: Ultrafast optical ranging using microresonator soliton frequency combs. Science **359**(6378), 887–891 (2018)
5. Kohler, D., et al.: Biophotonic sensors with integrated Si3N4-organic hybrid (SiNOH) lasers for point-of-care diagnostics. Light Sci. Appl. **10**(1), 64 (2021)
6. Xu, Y., et al.: Hybrid external-cavity lasers (ECL) using photonic wire bonds as coupling elements. Sci. Rep. **11**, 16426 (2021)
7. Nesic, A., et al.: Ultra-broadband polarisation beam splitters and rotators based on 3D-printed waveguides. Light Adv. Manuf. **4**, 23 (2023)
8. Xu, Y., et al.: 3D-printed facet-attached microlenses for advanced photonic system assembly. Light Adv. Manuf. **4**, 3 (2023)
9. Nesic, A., et al.: Transformation-optics modeling of 3D-printed freeform waveguides. Opt. Express **30**(21), 38856–38879 (2022)
10. Xu, C.L., et al.: A full-vectorial beam propagation method for anisotropic waveguides. J. Lightw. Technol. **12**(11), 1926–1931 (1994)
11. Lu, Y.Y., Ho, P.L.: Beam propagation modeling of arbitrarily bent waveguides. IEEE Photon. Technol. Lett. **14**(12), 1698–1700 (2002)
12. Kawamura, S., et al.: Design of tapered polarization splitter based on EC-CHFs by full-vectorial FE-BPM using coordinate transformation. J. Opt. Soc. Am. B **37**, 1075–1082 (2020)
13. Pedrola, G.L.: Beam Propagation Method for Design of Optical Waveguide Devices. John Wiley & Sons, Chichester (2015)
14. Viquerat, J.: Simulation of electromagnetic waves propagation in nano-optics with a high-order discontinuous Galerkin time-domain method, Ph.D. diss. Université Nice Sophia Antipolis (2015)
15. Ansys-Lumerical. https://www.lumerical.com. Accessed 14 June 2023
16. Billah, M.R., et al.: Hybrid integration of silicon photonics circuits and InP lasers by photonic wire bonding. Optica **5**, 876–883 (2018)

Wavelength Tunable, Polymer-Based Arrayed Waveguide Gratings for Hybrid Integration

Martin Kresse[1(✉)], Moritz Kleinert[1], David de Felipe[1], Tianwen Qian[1], Philipp Winklhofer[1], Madeleine Weigel[1], Klara Mihov[1], Jakob Reck[1], Crispin Zawadzki[1], Norbert Keil[1], and Martin Schell[1,2]

[1] Fraunhofer Institute for Telecommunications, Heinrich Hertz Institute, Einsteinufer 37, 10587 Berlin, Germany
martin.kresse@hhi.fraunhofer.de
[2] Technische Universität Berlin, Straße des 17. Juni 135, 10623 Berlin, Germany

Abstract. We present a hybrid integrated photonic circuit with tunable arrayed waveguide gratings (AWGs) as (DE-)MUX stages for optical modulators for use in parallel convolution processing. The simulated and fabricated AWGs have 16 outputs with 500 GHz channel spacing centered around 193.1 THz. A peak insertion loss of 4.5 dB is achieved with a 3 dB channel bandwidth of 380 GHz. Heater electrodes over the array waveguides enable tuning of more than 4 nm. The AWGs are hybridly integrated with indium phosphide-based modulators and amplifiers. This paves the way for integrated light modulation sources for the use in vector-matrix multiplication.

Keywords: hybrid photonic integration · arrayed waveguide grating · thermo-optic tuning

1 Introduction

Arrayed waveguide gratings (AWGs) are a popular means of multiplexing and demultiplexing optical signals in dense wavelength division multiplexing (DWDM) systems [1–3]. Their development revolutionized optical communication networks by enabling the simultaneous transmission of high-capacity data over multiple wavelengths.

More recently, AWGs have been used in a growing number of applications such as spectroscopy and astronomy [4, 5]. Another topic that is receiving increasing attention is integrated photonic tensor cores, where AWGs can serve as optical (DE-)MUX stages enabling parallel convolutional processing. This development has the potential to enable ultra-fast, energy-efficient computing and could transform artificial intelligence (AI) and machine learning technologies by leveraging the speed and parallelism inherent in photonic systems [6]. Such systems take advantage of lower latency, higher bandwidth, and lower power consumption compared to electronic counterparts by exploiting the intrinsic properties of light. The development of various material platforms for photonic integration has leveraged strengths in certain aspects of photonics, leading to the creation of passive platforms that specialize in tasks such as low-loss waveguiding, optical

J. Witzens et al. (Eds.): ECIO 2024, SPPHY 402, pp. 111–117, 2024.
https://doi.org/10.1007/978-3-031-63378-2_19

filtering and coupling. Active photonic platforms excel at tasks such as optical amplification and modulation. For optimum performance, a combination of both passive and active photonic components is desirable and has led to the creation of hybrid integration platforms.

In this work, polymer-based (DE-)MUX stages with wavelength tunable AWGs are designed and characterized. The hybrid integration concept with indium phosphide (InP) based components and assembly of the photonic integrated circuit is demonstrated.

2 Simulation and Design

2.1 Arrayed Waveguide Grating

In an arrayed waveguide grating (AWG), the light entering the input waveguide is split into several output waveguides depending on the wavelength. A free propagation region (FPR) distributes the light from the input waveguide to the array waveguides, where it undergoes a phase shift due to the different waveguide lengths. This leads to wavelength dependent interference. Constructive interference amplifies the signals of certain wavelengths at the output ports, while destructive interference suppresses others. This behavior is periodic, leading to equidistant peaks in the frequency spectrum spaced by the free spectral range (FSR) of the AWG. These peaks correspond to the different diffraction orders of the AWG.

The AWG presented has 16 channels with a center wavelength in the C-band. The channels are aligned around 193.1 THz and have a channel spacing of 500 GHz. The diffraction order of 15 was chosen, as this results in a FSR, large enough to cover the optical bandwidth of 7.5 THz. For tolerance reasons a flat-top channel response is desired. The wider channels relax the loss penalty that comes from shifts in the input frequencies with regard to the targeted channel frequencies. This flat-top performance is achieved by including a MMI (multi-mode interference) section in the input waveguide of the AWG. This section leads to a broadened field distribution at the beginning of the FPR compared to a single mode input waveguide and by convulsion with the output mode a broadening of the channel shape is achieved. The increase of the 3 dB-width from 210 to 380 GHz, or 1.7 to 3.0 nm, is paid for by a 1.5 dB reduction of the peak transmission. The geometric parameters and spectral response of the AWG were simulated and numerically optimized using the photonic CAD tool EPIPPROP. Figure 1 shows the simulated spectra of the 16 channels for TE polarization. A peak transmission value of −3.5 dB and cross talk level between channels of −33 dB are achieved.

The optimized AWG has 90 array waveguides and a Rowland radius of 1250 μm which results in a footprint of 9.5 mm × 2.8 mm.

2.2 Tunability

Tunable AWGs provide the ability to dynamically adjust the wavelength channels. For the device reported in this work, the vision is to combine multiple of these modules with 20 GHz separation. Therefore the polymer material is attractive, because of its high thermo-optic coefficient and low thermal conductivity compared to other inorganic

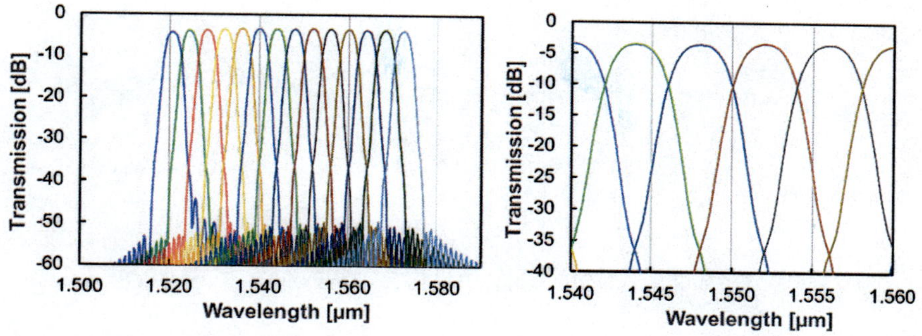

Fig. 1. Simulated AWG spectra for 16 channels with flat-top channel shape.

optical materials such as silicon nitride or silicon [7]. This reduces heat loss and improves tuning efficiency.

Tuning of the AWG is achieved by heating the array waveguides, which changes their optical length, and thus the center wavelength. The channel spacing remains largely unaffected at 500 GHz. By placing meandering electrodes over the array waveguide region, the effective index of the waveguides is changed homogeneously. The design of the AWG waveguides and heating structures is shown in Fig. 2.

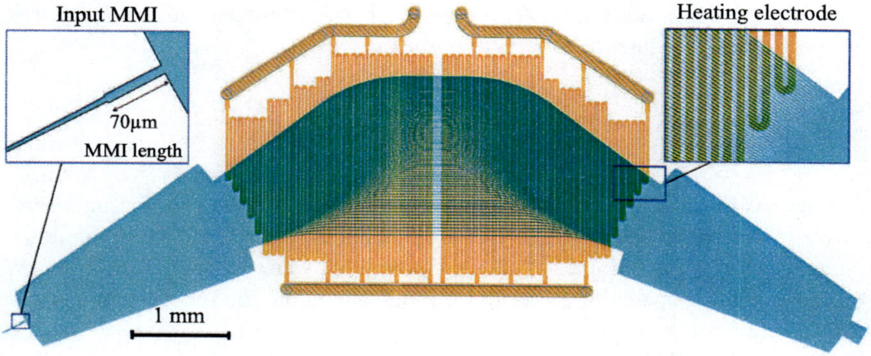

Fig. 2. CAD layout of AWG waveguide structures (blue) and meandering heating electrodes for tunability (orange). The input MMI serves to increase the channel width (Color figure online).

3 Characterization and Assembly

3.1 Design of MUX/DEMUX-PICs

The complete structure of the future modulation unit is shown in Fig. 4. Key components are the (DE-)MUX stages based on the designed tunable AWGs.

The layout of the DEMUX stages includes the input waveguide and alignment loops for fibre array coupling on the west side of the chip. On the east side of the chip are

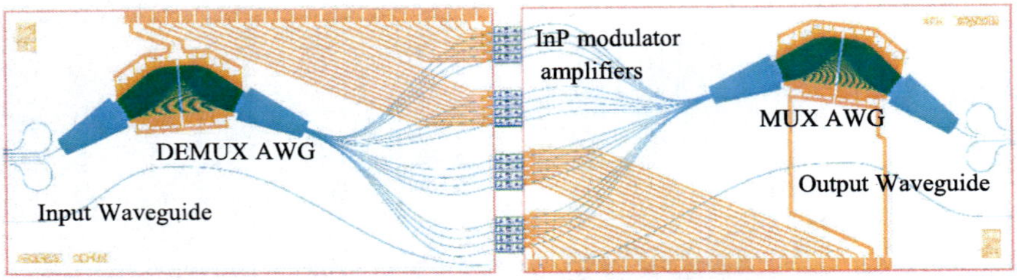

Fig. 3. CAD of the PHOENICS (DE-)MUX stages with flat-top AWGs.

the waveguides for coupling to the InP modulator amplifiers. The outputs are angled to reduce reflections. They are arranged in groups of four, with each group to be coupled to one 4-channel SOA-EAM-SOA array. In addition to the waveguides, an electrode layer is placed on the chip. Pads and supply lines to interface the driving electronics are placed on the chip connecting the north side of the chip to the InP chips. Two additional pads are used to drive the heating electrode on top of the AWG (Fig. 3).

A total of 20 (DE-)MUX stages can be produced on a 4-inch wafer. The fabricated chips were optically characterized by measuring the spectral transmittance through all 16 channels. The measurement was performed with a tunable laser source with polarization control. The insertion loss of the device under test was recorded using a fiber-coupled photodetector for each of the 16 output waveguides. The measurements were carried out on a temperature-controlled chuck. A heating current was applied to the tuning electrodes of the AWGs to investigate the tunability of the devices.

3.2 Characterization and Assembly

Figure 4(a) shows the measured optical transmission through the flat top AWG. The performance is in good agreement with the simulation results. The minimum insertion loss in the fabricated device is 4.5 dB. The additional loss compared to the simulation can be explained through additional loss contributions that were not included in the AWG modelling, e.g. the propagation losses of the input and output routing waveguides which together account for 0.9 dB of loss.

The wavelength shift of all 16 channels versus the applied heating power is shown in Fig. 5(a). The tuning efficiency was −4.5 nm/W. Figure 5(b) shows the peak insertion loss of the channels as a function of tuning. Up to 800 mW heating power, corresponding to 4 nm tuning, no effect is observed. Losses increase at higher tuning conditions. We believe that this is due to an unwanted temperature gradient at high tuning temperatures. Increasing the heated area of the electrodes could mitigate this effect, but at the expense of lower tuning efficiency.

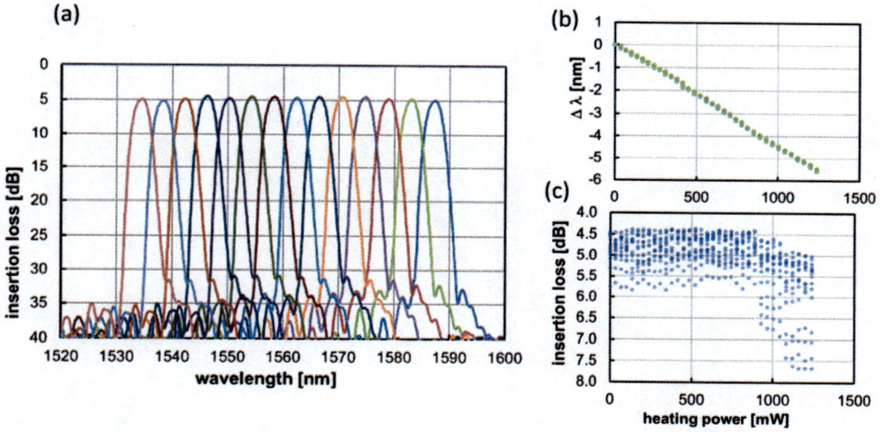

Fig. 4. (a) Transmission spectra of 16 channel AWG for TE-polarization; (b) wavelength shift $\Delta\lambda$ (note that at each heating power 16 values are plotted) (c) peak insertion loss

Based on the characterisation data, pairs of polymer DEMUX and MUX chips with matching performance were selected for hybrid integration into the input modulation units. The InP-based modulator array was installed between the two polymer chips using butt coupling. The modulator array is first bonded to a high thermal conductivity sub-mount to enable heat dissipation during operation. Alignment to the DEMUX chip is accomplished using 6-axis piezo-controlled motion stages with sub-micron accuracy. Coupling feedback for active alignment is provided by monitoring the photocurrent of the SOA of the upper and lower InP-array-waveguides while the input of the AWG is coupled to a broadband light source. After alignment, the two dies are permanently bonded using index-matched and UV-curable adhesive. In the subsequent assembly step, the MUX-chip is coupled in a similar way. By illuminating the polymer waveguides through the MUX-AWG with the broadband light source and simultaneously reading out the induced photocurrent in the DEMUX-modulator-subassembly, the optimum position can be found and the coupling is completed. The photonic circuit is placed on a common carrier for mechanical stability. Fiber arrays with polarization-maintaining fibers are attached as optical input and output. Their alignment is achieved using looped waveguides for coupling feed-back. The final assembly is shown in Fig. 5. It should be noted that a reduced number of 2 instead of 4 InP-arrays were populated in the device to form a unit with lower channel count.

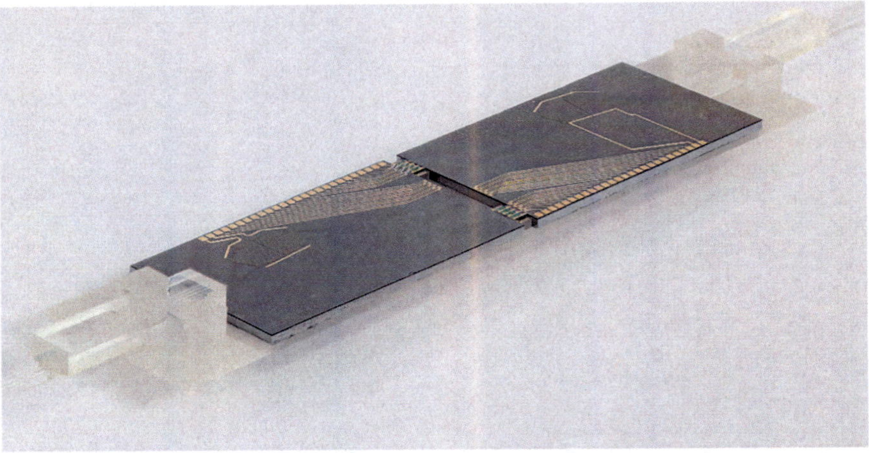

Fig. 5. Assembled 4 channel input modulation unit

4 Summary and Conclusion

Our study demonstrates a hybrid photonic integrated circuit with tunable polymer-based arrayed waveguide gratings (AWGs) as (DE-)MUX stages, designed to be combined with arrays of indium phosphide amplifiers and modulators. The AWGs provide 16 channels with 500 GHz spacing. They are simulated, designed and fabricated and achieve a peak insertion loss of 4.5 dB and significant thermo-optic tunability. The integration of the modulation unit demonstrates a pioneering approach to improving photonic convolution processors for vector matrix multiplication and promises advances in energy-efficient and high-speed computing. The successful fabrication and characterization of the AWG confirms the device's potential for future applications in AI and machine learning technologies, leveraging the advantages of hybrid photonic integration for improved performance in optical signal processing.

Acknowledgment. This work was supported by the European Union's Horizon 2020 research and innovation program (grant no. 101017237, PHOENICS project).

References

1. Smit, M.K.: Progress in AWG design and technology. In : Proceedings of 2005 IEEE/LEOS workshop on fibres and optical passive components, 2005. 2005 IEEE/LEOS workshop on fibres and optical passive components, pp. 26–31. Mondello, Italy IEEE (2005)
2. Keil, N., Yao, H.H., Zawadzki, C.: Athermal polarisation-independent arrayed- waveguide grating (AWG) multiplexer using an all-polymer approach. Appl. Phys. B **73**(5–6), 619–622 (2001). https://doi.org/10.1007/s003400100706
3. Okamoto, K., Sugita, A.: Flat spectral response arrayed-waveguide grating multiplexer with parabolic waveguide horns. Electron. Lett. **32**(18), 1661 (1996). https://doi.org/10.1049/el:19961108

4. Imran, A.B., Christopher, R.D.: Miniaturized spectrometers using integrated optics. In: Tromberg, B.J., Yodh, A.G., Sevick-Muraca, E.M., Alfano, R.R. (eds.): Optical tomography and spectroscopy of tissue XII. SPIE BiOS. San Francisco, California, United States, Saturday 28 January 2017: SPIE (SPIE Proceedings), p. 1005922 (2017)
5. Stoll, A., Zhang, Z., Haynes, R., Roth, M.: High-resolution arrayed-waveguide-gratings in astronomy: design and fabrication challenges. Photonics **4**(4), 30 (2017). https://doi.org/10.3390/photonics4020030
6. Feldmann, J., Youngblood, N., Karpov, M., Gehring, H., Li, X., Stappers, M., et al.: Parallel convolutional processing using an integrated photonic tensor core. Nature **589**(7840), 52–58 (2021). https://doi.org/10.1038/s41586-020-03070-1
7. Zhang, Z., Keil, N.: Thermo-optic devices on polymer platform. Opt. Commun. **362**, 101–114 (2016). https://doi.org/10.1016/j.optcom.2015.08.026

Nonlinear Photonics

Observation of Fundamental Charge Noise in Electro-Optic Photonic Integrated Circuits

Junyin Zhang[1,2], Zihan Li[1,2], Johann Riemensberger[1,2,3(✉)], Grigory Lihachev[1,2], Guanhao Huang[1,2], and Tobias J. Kippenberg[1,2(✉)]

[1] Institute of Physics, Swiss Federal Institute of Technology, Lausanne (EPFL), CH-1015 Lausanne, Switzerland
junyin.zhang@epfl.ch

[2] Center of Quantum Science and Engineering, EPFL, CH-1015 Lausanne, Switzerland
tobias.kippenberg@epfl.ch

[3] Department of Electronic Systems, Norwegian University of Science and Technology, 7491 Trondheim, Norway

Abstract. We observed that Lithium Niobate and Lithium Tantalate photonic integrated microresonators exhibit an unexpected Flicker type (i.e. $1/f^{1.2}$) scaling in the intrinsic refractive noise spectrum, significantly deviating from the well-established thermorefractive noise theory. We show that this noise is consistent with thermodynamical charge noise, which leads to electrical field fluctuations that are transduced via the strong Pockels effects of electro-optic materials. We demonstrated a Lithium Niobate self-injection locked integrated photonics laser with 56 Hz intrinsic linewidth in a 200 GHz FSR microresonator, achieving the charge-noise limit. Our results establish electrical Johnson-Nyquist noise as the fundamental limitation for Pockels integrated photonics, crucial for determining performance limits for classical and quantum devices. Equally, this observation offers an novel approach to probe mesoscopic electric field fluctuations with exceptional precision.

Keywords: integrated photonics · Lithium Niobate · integrated laser · thermal-charge-carrier-refractive noise · thermorefractive noise

1 Introduction

Refractive noise sets the fundamental performance limits of many photonic integrated devices, i.e., the signal-to-noise ratio of microresonator-based sensors, the linewidth of integrated lasers, and the pulse repetition rate noise of dissipative Kerr solitons. Recently emerged photonic platforms based on ferroelectric thin films [1] such as lithium niobate-on-insulator have revolutionized integrated photonics with efficient and ultrafast modulation capabilities. The intrinsic refractive noise and its impact for photonic devices, in particular those based on high-Q

J. Witzens et al. (Eds.): ECIO 2024, SPPHY 402, pp. 121–125, 2024.
https://doi.org/10.1007/978-3-031-63378-2_20

optical microresonators has not been investigated so far. Here, we discover that unexpectedly charge carrier density fluctuations give rise to a novel noise process [2] in recently emerged electro-optic photonic integrated circuits (Fig. 1.(a-b)). We show that photonic integrated microresonators based on thin-film Lithium Niobate and Lithium Tantalate exhibit an unexpected Flicker-type (i.e. $1/f^{1.2}$) scaling in their frequency noise spectra, significantly deviating from the well-established thermorefractive noise theory [3,4] (Fig. 1.(c)). This noise stems from thermodynamical charge fluctuation [5,6], similar to electrical Johnson-Nyquist noise, which leads to electrical field fluctuations that are transduced to optical frequency noise via the strong Pockels effects of electro-optic materials. Our results establish electrical Johnson-Nyquist noise as the fundamental limitation for the intrinsic noise of Pockels integrated photonics, crucial for determining performance limits for both classical and quantum devices, ranging from ultra-fast tunable and low-noise lasers, Pockels soliton microcombs, to quantum transduction, squeezed light or entangled photon-pair generation [7–9]. Moreover, our work offers novel optical methods to probe mesoscopic charge fluctuations with exceptional precision.

2 Fundamental Charge Refractive Noise via Pockels Effect

The refractive noise is measured by utilizing high Q/V (quality factor - volume ratio) microresonators via the balanced-homodyne method (Fig. 1.(d)). In this approach, a stable, low-noise probe laser, filtered by a narrow-linewidth Fabry-Pérot cavity, is locked into the microresonator's resonance, converting the fundamental refractive index fluctuations into phase noise of the transmitted laser light. The measured frequency noise for 1 THz, 200 GHz (FSR, free-spectral range) Z-cut $LiNbO_3$, 80 GHz X-cut $LiNbO_3$ and 80 GHz X-cut $LiTaO_3$ is shown in Fig. 1.(e). In each, we observed a Flicker-type scaling of $S_\nu(f) \propto f^{-1.2}$. The prediction of the thermorefractive noise model underestimates the low frequency noise below 10 kHz offset, while overestimating noise at higher frequencies. We built a theoretical model that explains the measured Flicker-type scaling based on charge diffusion, i.e., the thermal-charge-refractive noise theory originally developed in the context of gravitational wave detectors [5,6]:

$$S_{\nu,\mathrm{TCCR}}(\omega, T) = \frac{n^4 r^2 \nu^2 k_B T}{V_{\mathrm{eff}}} \frac{\sigma(\omega, T)}{\epsilon_0^2 \epsilon_r^2 \omega^2}, \tag{2}$$

wherein $\omega \gg 1/\tau_d$, and with the dielectric relaxation time $\tau_d = \epsilon_r \epsilon_0 / \sigma(0, T)$. For insulating ferroelectric crystals, $\tau_d > 1\mathrm{s}$ due to the low conductivity. Here, n is the refractive index, r is the Pockels coefficient for the specific polarization, k_B is the Boltzmann constant, T is the absolute temperature, V_{eff} is the effective optical mode volume, ν is the optical frequency, ϵ_0 is the vacuum permittivity, ϵ_r is the relative permittivity and $\sigma(\omega, T)$ is the temperature and frequency-dependent conductivity which is proportional to the charge-carrier

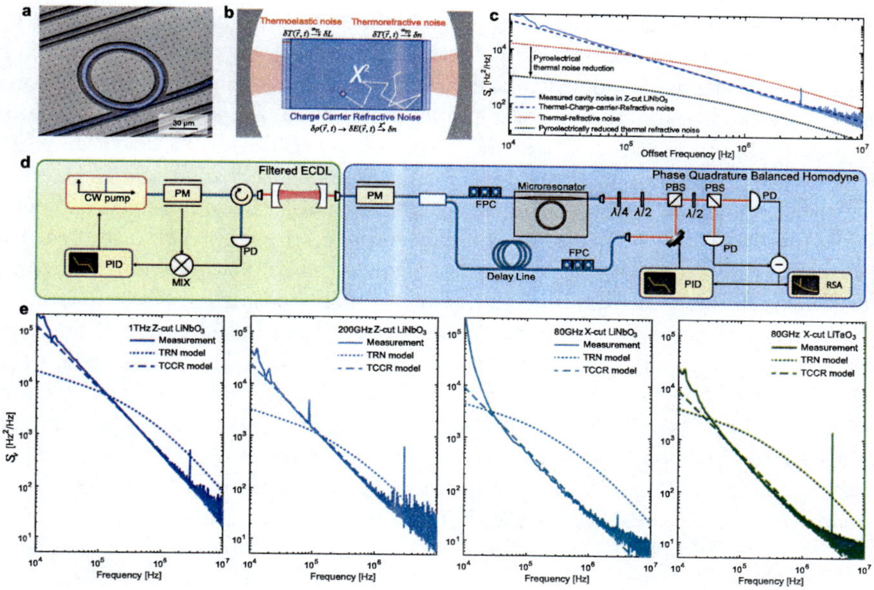

Fig. 1. Refractive noise from charge fluctuations via the Pockels effect. (a) False-colored SEM image of integrated microresonators on thin-film Lithium Niobate. **(b)** Schematic of various noise processes in optical resonators. **(c)** Schematic of intrinsic cavity noise in electro-optic microresonators. **(d)** Balanced homodyne method for refractive noise measurement. **(e)** Single-sided power spectral density (PSD) comparing intrinsic frequency noise across Z-cut, X-cut $LiNbO_3$, and X-cut $LiTaO_3$ microresonators at different free-spectral ranges.

diffusion coefficient. Hence, the conventionally accepted refractive noise model (thermorefractive noise) that has been proven to correctly describe the frequency noise in inversion symmetric semiconductors and amorphous dielectrics [3] hence is insufficient to capture the rich physics of refractive noise in ferroelectric photonics. The complex nature of the interplay of charge and temperature fluctuations in lithium niobate and tantalate does not only lead to excess noise at low offset frequency but also to a cancellation of noise at high frequencies due to the pyroelectricity of the crystal contributing an effective negative thermo-optic effect.

3 Charge Noise Limited Integrated Self-injection-locked Lasers

One particular application sensitive to fundamental refractive noise are integrated lasers, where the fundamental refractive noise limits the minimum achievable linewidth [10]. We experimentally verify that the charge fluctuation-induced refractive noise gives the accurate estimation of the minimum achievable frequency noise limit for self-injection locked lasers built on electro-optic materials. A self-injection locked laser is constructed using a high-quality microresonator with a Sagnac mirror for enhanced frequency-selective back-reflection that is butt-coupled to a distributed-feedback (DFB) laser diode (Fig. 2.(a)). The frequency noise of the integrated laser is measured with the heterodyne spectroscopy (Fig. 2.(b)) with the cavity-filtered laser. The measured frequency noise is aligned with the fundamental charge-induced refractive noise measured previously (Fig. 2.(c-d)), confirming that charge noise is the accurate estimation of the minimum attainable linewidth of integrated lasers. A comparison of the frequency noise between Z-cut $LiNbO_3$ and Si_3N_4-based lasers (Fig. 2.(d)) shows a reduction of up to five times in high-offset frequencies. This suggests that $LiNbO_3$-based integrated frequency-agile lasers [7] offer not just rapid tunability but can also feature a noise floor below the thermo-refractive noise limit at high-offset frequencies due to pyroelectric modified thermo-optical response. In our z-cut LN laser, we achieve a white frequency noise limit of 20 Hz^2/Hz which is record-low for LNOI-based hybrid integrated lasers. This advantage comes at the trade-off of excess noise at low offset frequencies due to the interplay of thermally driven charge fluctuations and the electro-optical Pockels effect.

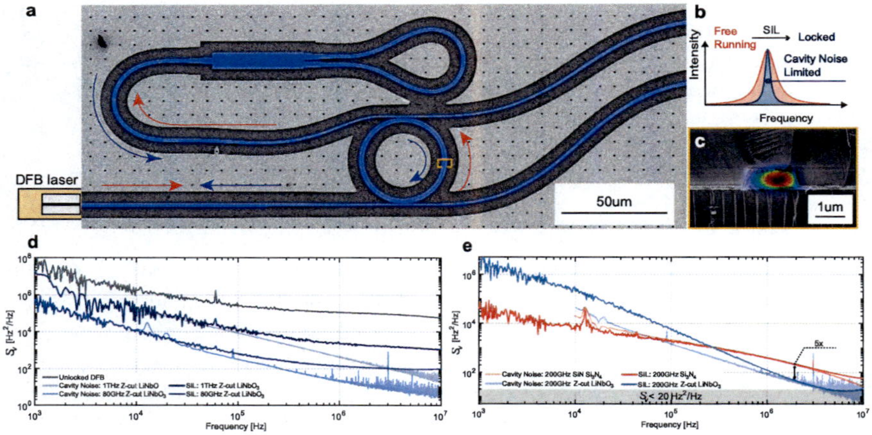

Fig. 2. Frequency noise of self-injection-locked lasers limited by charge noise. **(a)** False-colored SEM images of microresonators featuring Sagnac mirrors for self-injection locked (SIL) lasing. **(b)** Schematic illustrating the noise reduction mechanism in self-injection locked. **(c)** Optical mode profile within tightly bent waveguides. **(d-e)** Single-sided power spectral density comparing frequency noise of SIL lasers to intrinsic charge noise across different free-spectral ranges in Z-cut $LiNbO_3$ and in non-electro-optic material Si_3N_4.

4 Discussion and Conclusion

In summary, we observed that the charge fluctuation induced refractive noise via the Pockels effect as the fundamental noise limit for electro-optic photonics devices, showing significantly different behavior compared with the thermorefractive noise induced by temperature fluctuations, commonly observed in non-electro-optic materials such as Si_3N_4. The significance of this type of noise for photonic applications is exemplified in the self-injection locked laser. An intrinsic linewidth of 56 Hz was achieved in microresonators with a 200 GHz FSR, whose frequency noise spectra exhibit a distinct charge-noise-limited behavior, characterized by noise suppression at high-offset frequencies and pronounced excess noise at low-offset frequencies. Broadly viewed, our measurements in high-Q $LiNbO_3$ and $LiTaO_3$ photonic integrated circuit-based microresonators carry substantial implications for the ultimate and noise limitations for a variety of applications from quantum transduction, quantum electro-optic experiments to coherent communication, LiDAR, and sensing. We expect this charge-carrier noise might also be a dominating noise in future doped-silicon based [11] and other ferroelectric-based electro-optic integrated photonics [12]. Equally, the methodology of homodyne spectroscopy of the phase noise of ferroelectric microresonators offers an direct optical method to probe mesoscopic electric field fluctuations with exceptional precision.

Acknowledgments. The samples were fabricated in the EPFL Center of MicroNanoTechnology (CMi) and the Institute of Physics (IPHYS) cleanroom.

Funding Information. This work has received funding from the European Research Council (ERC) under the EU H2020 research and innovation programme, grant agreement No. 835329 (ExCOM-cCEO). This material is based upon work supported by the Air Force Office of Scientific Research under award number FA9550-19-1-0250.

References

1. Boes, A., et al.: Lithium niobate photonics: unlocking the electromagnetic spectrum. Science **379**, eabj4396 (2023)
2. Zhang, J., Li, Z., Riemensberger, J., Lihachev, G., Huang, G., Kippenberg, T.J.: arXiv preprint arXiv:2308.15404 (2023)
3. Braginsky, V., Gorodetsky, M., Vyatchanin, S.: Phys. Lett. A **271**, 303–307 (2000)
4. Levin, Y.: Phys. Lett. A **372**, 1941–1944 (2008)
5. Bruns, F., et al.: Phys. Rev. D **102**, 022006 (2020)
6. Siegel, H., Levin, Y.: Phys. Rev. D **107**, 022002 (2023)
7. Snigirev, V., et al.: Nature **615**, 411–417 (2023)
8. Stokowski, H.S., et al.: Nat. Commun. **14**, 3355 (2023)
9. Sahu, R., et al.: Science **380**, 718–721 (2023)
10. Jin, W., et al.: Nat. Photonics **15**, 346–353 (2021)
11. Xu, Q., Schmidt, B., Pradhan, S., Lipson, M.: Nature **435**, 325–327 (2005)
12. Abel, S., et al.: Nature Mater. **18**, 42–47 (2019)

An Integrated Gallium Phosphide Optical Parametric Amplifier

Nikolai Kuznetsov[1,2](✉), Alberto Nardi[1,3], Alisa Davydova[1,2], Mikhail Churaev[1,2], Johann Riemensberger[1,2,4], Paul Seidler[3], and Tobias J. Kippenberg[1,2]

[1] Institute of Physics, Swiss Federal Institute of Technology Lausanne (EPFL), 1015 Lausanne, Switzerland
[2] Center of Quantum Science and Engineering (EPFL), 1015 Lausanne, Switzerland
[3] IBM Research Europe, Säumerstrasse 4, Rüschlikon, 8803 Zurich, Switzerland
[4] Department of Electronic Systems, Norwegian University of Science and Technology, Trondheim, Norway

{nikolai.kuznetsov,tobias.kippenberg}@epfl.ch,
johann.riemensberger@ntnu.no
pfs@zurich.ibm.com

Abstract. Optical parametric amplification is a promising alternative to traditional erbium-doped amplifiers, offering unidirectional, low-noise, and broadband amplification. Here we present an integrated continuous-traveling-wave parametric amplifier that surpasses existing technologies, achieving up to 35 dB of gain and up to 25 dB of net gain over a 140-nm-wide wavelength window centered at 1550 nm, thus covering S, C, and L telecommunication bands simultaneously. Our device utilizes a 5.55-cm-long spiral waveguide made of gallium phosphide, renowned for its high Kerr nonlinearity and large linear refractive index. These properties are key factors for efficient parametric amplification.

Keywords: integrated photonics · parametric amplification · gallium phosphide

1 Continuous-Wave Parametric Amplification

The advancement of optical fibers and broadband amplification, particularly through erbium-doped fiber amplifiers (EDFAs) [1,2], has played a crucial role in modern science and technology. This progress, especially in the field of optical communications for long-distance information transmission, led to a significant breakthrough with intercontinental optical fiber networks. In contrast to doped amplifiers, optical parametric amplifiers (OPAs) utilize the intrinsic nonlinearities of optical materials for unidirectional gain generation (Fig. 1(a)). OPAs possess unique properties [3,4], allowing them to amplify signals across a broad range of wavelengths limited only by the transparency window of optical materials. Notably, optical traveling-wave parametric amplifiers (TWPAs) achieved

J. Witzens et al. (Eds.): ECIO 2024, SPPHY 402, pp. 126–130, 2024.
https://doi.org/10.1007/978-3-031-63378-2_21

exceptional performance breakthroughs with the emergence of highly nonlinear fibers (HNLF) in 1986 [3,5]. These optical TWPAs, operating near the quantum noise limit, offer noiseless phase-sensitive amplification, potentially extending the span length of long-haul fiber optic links [6]. However, continuous operation with substantial gain has so far been demonstrated only in bulky systems, such as those utilizing hundreds of meters of HNLF [7–9] or mechanically-diced lithium-niobate waveguides [10,11]. Despite recent advancements, the development of a continuous-wave optical TWPAs on photonic integrated circuits (PICs) remained a challenging frontier until very recently when meter-long Si_3N_4 waveguides demonstrated CW amplification due to reduced losses [12–14]. However, parametric amplifiers in Si_3N_4 waveguide spirals still require multiple meters in length and achieve only a 2 dB narrowband gain from input to output fiber despite high pump powers exceeding 6 W [13], limiting their suitability for optical communication systems.

2 Gallium Phosphide Spiral Waveguide

Here, we present the first practical time-continuous traveling-wave parametric amplifier. We use a gallium phosphide (GaP) on-insulator spiral waveguide with a compact length of 5.55 cm and a footprint of $500 \times 500\ \mu\text{m}^2$ (Fig. 1(b, c)). GaP, recognized as a promising material for nonlinear integrated photonics [15], combines a high optical refractive index ($n > 3.1$) with a strong Kerr nonlinearity ($n_2 = 1.1 \times 10^{-17}\ \text{m}^2\ \text{W}^{-1}$) and a sufficiently large indirect bandgap of $E_\text{g} = 2.24$ eV [16] to mitigate two-photon absorption at telecom wavelengths. GaP's properties have facilitated low threshold power frequency comb generation [15] and dissipative Kerr soliton generation [17] in GaP ring and photonic crystal optical microresonators, respectively. Continued and significant advances in GaP waveguide nanofabrication techniques [15,17] have resulted in optical propagation losses below 1 dB cm^{-1} and high bonding yield with low defect density. These advancements are pivotal for successfully fabricating centimeter-long waveguide spirals for optical TWPAs. In strongly confining integrated waveguides it is possible to engineer GVD parameter β_2 and the zero dispersion wavelengths of over a wide range by varying the cross-sectional geometry. Therefore, we perform dispersion simulations considering a straight bus waveguide for a range of film thicknesses and waveguide widths. We use a commercially available FEM solver (COMSOL Multiphysics®) to evaluate the dispersion of rectangular GaP waveguides fully cladded with silica. For a prepared thin film layer thickness of 299 nm, we designed a waveguide with a width of 790 nm to operate in the anomalous dispersion regime; we also use waveguides with a smaller width, varied in steps of 5 nm. We measure the propagation loss of the waveguides using optical frequency domain reflectometry and find uniform losses in a wavelength range from 1500 nm to 1600 nm. Input and output coupling of the light is enabled by inverse tapers with a design tip width of 180 nm. The inverse tapers are revealed via stealth dicing to achieve clean vertical chip facets to achieve efficient fiber-to-chip coupling to reduce the optical insertion loss.

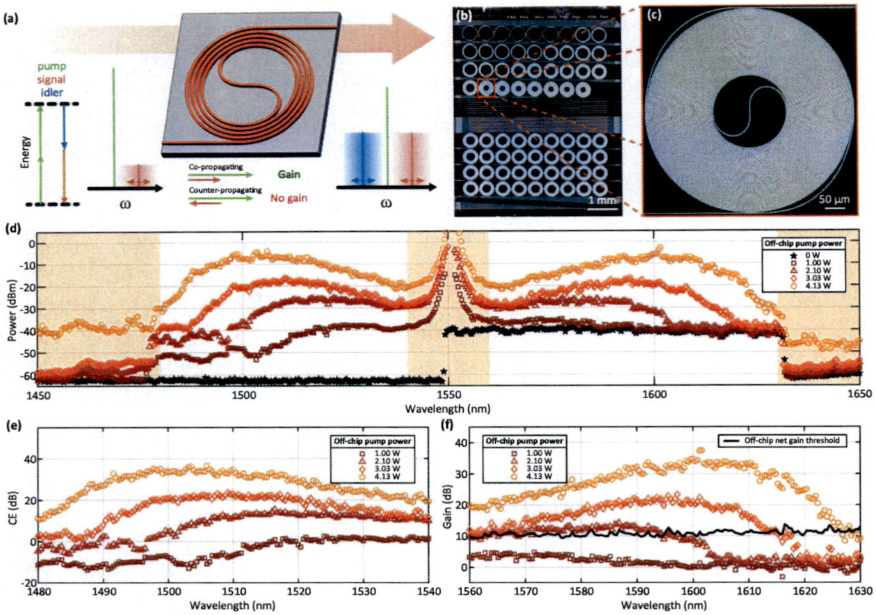

Fig. 1. Gallium phosphide optical parametric amplifier. (a) Principle of parametric amplification using the degenerate four-wave mixing process. **(b, c)** Microscope images of the whole GaP chip with multiple devices and a separate 5.55-cm-long spiral waveguide used in our experiments. **(d)** Measurements of parametric gain using "Max Hold" mode of the optical spectrum analyzer (OSA), i.e., recording the maximum power at every scan of the OSA while simultaneously scanning the signal wavelength. **(e, f)** Calculated idler conversion efficiency and signal gain spectra, respectively, for different pump power levels. The off-chip input signal power is 1.5 μm.

3 Optical Parametric Gain Measurements

The continuous-wave amplification spectrum is measured and evaluated following the procedures outlined in [13]. A dispersion-engineered spiral waveguide with a designed width of 780 nm is employed. We operate the OPA with a single pump to amplify signals via the degenerate four-wave-mixing (DFWM) relying on the optical Kerr effect. The optical energy is redistributed by annihilating two pump photons and amplifying a signal, while simultaneously generating a phase-conjugated idler at an equidistant but opposite frequency offset to the pump laser. Figure 1(d) shows the power-calibrated amplification spectrum. In the spectral region around the pump wavelength, we observed a spurious signal due to the leakage of the amplified spontaneous emission from the EDFA installed in the pump optical path. The frequency conversion and amplification spectra (Fig. 1(e, f)) are determined by comparing the measurements of the signal laser sweep spectra recorded at various pump power levels (red, orange) with the measurement obtained when the pump is switched off (black). Increasing the input power from 1 W to 4.13 W, we observe increasing gain, conversion efficiency, and amplification bandwidth following established theory [3,18]. The generated idler spectrum spans from 1550 nm to approximately 1480 nm at the highest pump power. In Fig. 2(c), spurious signals in the shaded regions outside the signal and idler laser scan bandwidth originate from higher-order (non-degenerate) FWM. Hence, we only calculate the internal idler conversion efficiency (Fig. 2(d)) from 1480 nm to 1550 nm and the internal signal gain from 1560 nm to 1630 nm (Fig. 2(e)). We separately measure the passive system transmission loss bypassing the lensed fibers and the chip, plotted as a black line in Fig. 2(e). We calculate the off-chip net gain by subtracting the 10-12 dB total transmission loss, revealing a net-gain of up to 25 dB, exceeding 10 dB over a bandwidth of 70 nm. Assuming a symmetric gain spectrum on the idler side, we estimate the 10 dB gain region to exceed 140 nm, or 17 THz. This bandwidth is larger by almost 10 times compared to the prior work on Si_3N_4 [13], and nearly 3 times larger than the bandwidth of a single C-band EDFA.

4 Conclusion

GaP photonic integrated circuits enable both high net gain and broadband operation, attaining a fiber-to-fiber net gain above 20 dB and a combined signal and idler gain bandwidth of 140 nm, exceeding the bandwidth of EDFAs. Further reducing the optical propagation losses will greatly reduce the required pump power and may even allow direct pumping of the OPA with semiconductor lasers. The photonic integrated GaP optical parametric amplifier solves the most pertinent challenges that have prevented the widespread adoption of fiber OPA systems. Firstly, the dispersion of the waveguides is defined lithographically. A short length prevents a bandwidth reduction [19] and reduces sensitivity to fabrication imperfections compared to the optical fibers that are hundreds of meters long. Secondly, the low Brillouin gain of integrated waveguides eliminates the technical complexity of adding additional pump phase-modulators [6,8,9]. In principle, the pump frequency can be freely chosen in a wide frequency range since it is not restricted as in doped amplifiers. Our results signal the emergence of practical optical Kerr TWPA that could be employed in future optical communication systems or other applications.

Acknowledgments. The sample was fabricated at the Binnig and Rohrer Nanotechnology Center (BRNC) at IBM Research Europe, Zurich.

Funding Information. This work was supported by the European Union's Horizon 2020 research and innovation programme under the Marie Skłodowska-Curie grant agreement No 812818 (MICROCOMB), by the Swiss National Science Foundation (SNF) under grant number 192293, and by the Air Force Office of Scientific Research under award number FA9550-19-1-0250.

References

1. Mears, R., Reekie, L., Jauncey, I., Payne, D.: Optical Fiber Communication Conference, p. WI2. Optical Society of America (1987)
2. Desurvire, E., Simpson, J.R., Becker, P.C.: Opt. Lett. **12**, 888 (1987)
3. Hansryd, J., Andrekson, P.A., Westlund, M., Li, J., Hedekvist, P.-O.: IEEE J. Sel. Top. Quantum Electron. **8**, 506 (2002)
4. Marhic, M.E., et al.: Laser Photonics Rev. **9**, 50 (2015)
5. Sudo, S., Hosaka, T., Itoh, H., Okamoto, K.: Electron. Lett. **16**, 833 (1986)
6. Tong, Z., et al.: Nat. Photonics **5**, 430 (2011)
7. Marhic, M., Kagi, N., Chiang, T.-K., Kazovsky, L.: Opt. Lett. **21**, 573 (1996)
8. Hansryd, J., Andrekson, P.A.: IEEE Photonics Technol. Lett. **13**, 194 (2001)
9. Torounidis, T., Andrekson, P.A., Olsson, B.-E.: IEEE Photonics Technol. Lett. **18**, 1194 (2006)
10. Umeki, T., Tadanaga, O., Takada, A., Asobe, M.: Opt. Express **19**, 6326 (2011)
11. Kishimoto, T., Inafune, K., Ogawa, Y., Sasaki, H., Murai, H.: Opt. Lett. **41**, 1905 (2016)
12. Ye, Z., et al.: Sci. Adv. **7**, eabi8150 (2021)
13. Riemensberger, J., et al.: Nature **612**, 56 (2022)
14. Zhao, P., et al.: 2023 Asia Communications and Photonics Conference/2023 International Photonics and Optoelectronics Meetings (ACP/POEM), pp. 1–5. IEEE (2023)
15. Wilson, D.J., et al.: Nat. Photonics **14**, 57 (2020)
16. Lide, D.R.: CRC handbook of chemistry and physics, vol. 85. CRC press (2004)
17. Nardi, A., et al.: arXiv preprint arXiv:2304.12968 (2023)
18. Stolen, R., Bjorkholm, J.: IEEE J. Quantum Electron. **18**, 1062 (1982)
19. Ayan, A., Liu, J., Kippenberg, T.J., Brès, C.-S.: Opt. Express **31**, 40916 (2023)

Lithium Tantalate Photonic Integrated Circuits (Student Paper)

Chengli Wang[1,2](✉), Zihan Li[1], Johann Riemensberger[1], Grigory Lihachev[1], Mikhail Churaev[1], Xinru Ji[1], Xin Ou[2], and Tobias J. Kippenberg[1](✉)

[1] Institute of Physics, Swiss Federal Institute of Technology Lausanne (EPFL), CH-1015 Lausanne, Switzerland
{chengli.wang,tobias.kippenberg}@epfl.ch

[2] State Key Laboratory of Functional Materials for Informatics, Shanghai Institute of Microsystem and Information Technology, Chinese Academy of Sciences, Shanghai 200050, China

Abstract. We demonstrate a wafer-scale lithium tantalate photonic platform with loss as low as 5.5 dB m^{-1}. Mach-Zehnder modulators with a 40 GHz electro-optical bandwidth and a 1.96 V·cm voltage-length product, along with microresonators supporting dissipative Kerr soliton generation.

Keywords: thin-film lithium tantalate · microresonator · photonic circuit · electro-optical modulator

1 Introduction

Lithium tantalate ($LiTaO_3$, LT), like lithium niobate ($LiNbO_3$, LN), is a traditional ferroelectric crystal with superior electro-optic and piezo-electric properties. For photonic applications, $LiNbO_3$ is widely considered to be a more excellent material [1]. In contrast, although $LiTaO_3$ is comparable to LN not only in terms of some physical coefficients, such as nonlinear-optic, electro-optic, and piezo-electric coefficients, but also in maturity and commercial availability of bulk materials, its application in the domain of photonic devices has been rarely reported. One of the reasons is that the Curie temperature of $LiTaO_3$ (610–700 °C) is much lower than the temperatures needed for the fabrication of optical waveguides by ion diffusion (typically above 1000 °C), which hinders the use of $LiTaO_3$ for the initial progress of the ion diffused waveguide [2]. However, akin to the development of $LiNbO_3$-on-insulators (LNOI) and SOI platforms, thin-film-on-insulator structure combined with advanced nanofabrication techniques can make high-performance waveguides possible without using ion diffusion process, which would bring excellent prospects for exploiting the advantages of $LiTaO_3$-based integrated circuits.

J. Witzens et al. (Eds.): ECIO 2024, SPPHY 402, pp. 131–135, 2024.
https://doi.org/10.1007/978-3-031-63378-2_22

Many studies based on $LiTaO_3$ bulk materials clearly show that $LiTaO_3$ exhibits a higher optical duration [3], a weaker birefringence effect and wider transparency window (0.24–5.5 μm) [4] compared with LN. However, there are few studies on photonic integrated $LiTaO_3$ thin films. Here we show the first $LiTaO_3$ photonic integrated circuits (PICs) fabricated on a 4-inch $LiTaO_3$-on-insulator (LTOI) wafer with optical loss as low as 5.5 dB m^{-1} and high-yields at the wafer scale.

2 Results

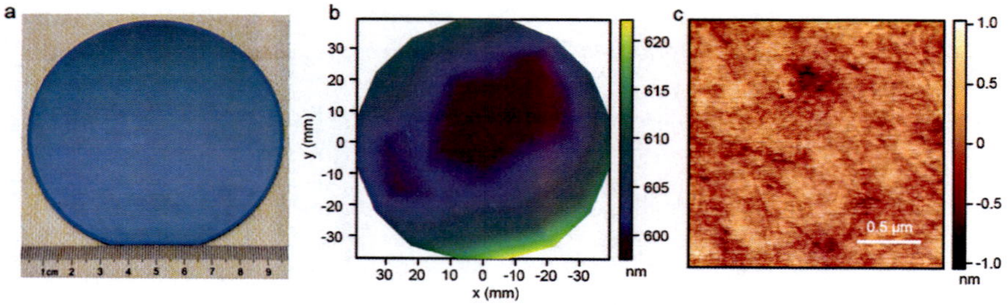

Fig. 1. (a) Photograph of 4-in. thin-film lithium tantalate wafer. (b) Measurement of LT thickness uniformity. (c) Atomic force microscopy (AFM) image of the LT layer surface.

In analogy to the fabrication of SOI and LNOI, an "ion-slicing" process is used to fabricated LTOI wafer [5]. The $LiTaO_3$ layer is transferred from a commercial x-cut $LiTaO_3$ bulk wafer via H+ ion implantation and wafer bonding. After $LiTaO_3$ layer transferring, the LTOI wafer is performed in a chemical mechanical polished (CMP) process. A top-view photograph of the fabricated LTOI wafer is shown in Fig. 1(a). The thickness of $LiTaO_3$ thin film after CMP is mapped as shown in Fig. 2(b), where the thickness uniformity is characterized as ±2.05%. The surface roughness is less than 0.25 nm. LTOI wafers with good uniformity and smooth surface roughness are the prerequisites for high-yield and low-loss $LiTaO_3$ PICs.

Figure 2(a) depicts the schematic process flow of the fabrication of our $LiTaO_3$ PICs. Key to our process flow is the utilization of diamond like carbon (DLC) as a hard mask for $LiTaO_3$ etch, which was reported in detail in our previous work on the LNOI PICs [6]. Briefly, a 300 nm thick DLC film is deposited on LTOI wafer by plasma-enhanced chemical vapor deposition (PECVD). The $LiTaO_3$ PICs pattern is defined in the DLC layer via DUV stepper photolithography and DLC dry etching. Then, the $LiTaO_3$ PICs forms via argon ion beam etching (IBE), followed by the removal of the redeposition and hard mask.

Figure 2(b) depicts several scanning electron micrographs of the racetrack microresonator with a bending radius of 100 μm and a straight arm length of 500 μm. The SEM images of the typical waveguide reveals a smooth sidewall. The rib optical waveguide manufactured by our DCL-base process has a sidewall angle of about 70°. To characterize the resonance linewidths of the microresonator, we use frequency-comb-assisted,

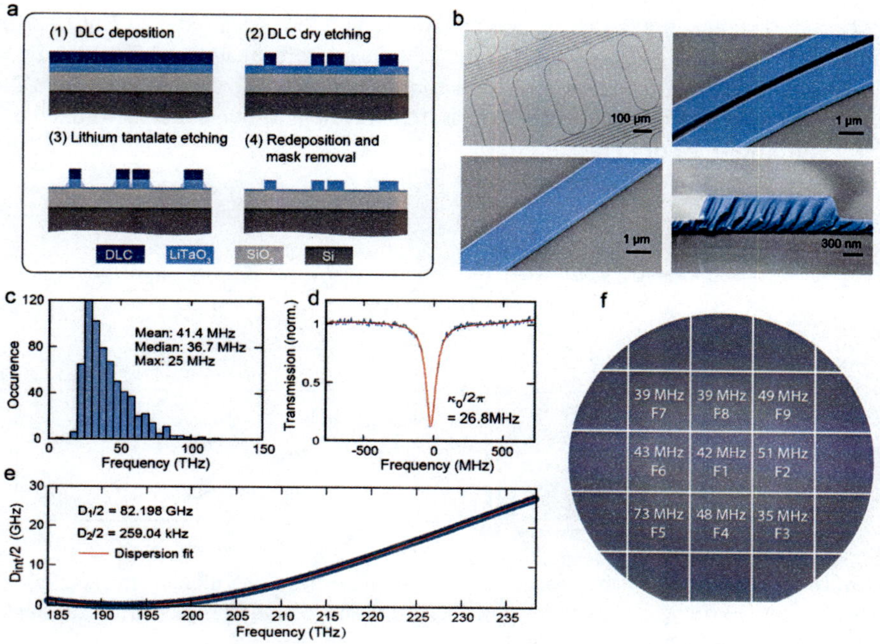

Fig. 2. (a) The wafer-scale $LiTaO_3$ PICs process flow, including DLC deposition, DUV stepper lithography (not shown), preform DLC and $LiTaO_3$ dry etching, redepositon cleaning, and hard mask removal. (b) False-colored scanning electron microscope (SEM) images of $LiTaO_3$ PICs, including racetrack resonantors, close-up of the pulley coupling section, crosssection of the partially etched ridge waveguide. False-colored scanning electron microscope image of a LNOD waveguide cross-section. (c) Histogram indicating intrinsic decay rate distribution of 692 TE_{00} resonances of a LTOI racetrack with a free spectral range (FSR) of 82 GHz. The median of the distribution is 36.7 MHz, which corresponds to a quality-factor of 5.3×10^6 and a linear propagation lass of 7.6 dB m^{-1}. (d) A recorded resonance with linewidth fitting at 1552.2 nm. The intrinsic loss rate $\kappa_0/2\pi$ is extracted as 26.8 MHz, corresponding to a quality factor of 7.2×10^6 and a linear propagation lass of 5.5 dB m^{-1}. (e) Measured integrated dispersion of a 2 μm width racatrack resonator, the fitted $D_2/2\pi = 259.04$ kHz shows anomalous dispersion. (f). DUV stepper exposure layout, and the most probable intrinsic loss rate $\kappa_0/2\pi$ of the 9 chips at different fields.

cascaded diode laser spectroscopy covering the entire telecommunication E- to L-band (1350 to 1630 nm). Here, we mainly study the fundamental transverse electric (TE_{00}) mode. Figure 2(c) shows the histogram of the intrinsic decay rate distribution of 692 TE_{00} resonances. The median of the distribution is 36.7 MHz, which corresponds to an intrinsic quality-factor of 5.3×10^6 and a linear propagation lass of 7.6 dB m^{-1}. A recorded resonance with linewidth fitting at 1552.2 nm is presented in Fig. 2(e). The intrinsic loss rate $\kappa_0/2\pi$ is extracted as 26.8 MHz, corresponding to a quality factor of 7.2×10^6 and a linear propagation loss of 5.5 dB m^{-1}. We achieve the LTOI PICs with a wafer-scale yield. Figure 2(f) shows a DUV stepper reticle layout containing tens of fields, each fields comprising 16 5×5 mm chips are exposure uniformly over the full 4-inch LTOI wafer. The characterization of wafer-scale loss over 9 fields is depicted in

Fig. 2(d), the average loss is 46 MHz. Figure 2(e) depicts the measured integrated dispersion $D_{int}/2\pi = \omega(\mu) - \omega_0 - \mu D_1/2\pi$ of the $LiTaO_3$ microresonator with FSR $D_1/2\pi =$ 82.2 GHz featuring anomalous microresonator dispersion of $D_2/2\pi = 259.04$ kHz. No observable mode mixing occurs at the measured wavelength range thanks to the weak inherent birefringence of $LiTaO_3$.

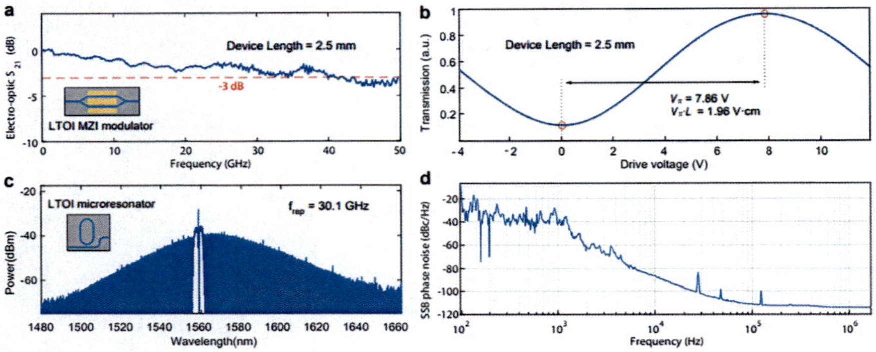

Fig. 3. Demonstration of electro-optical and Kerr nonlinearities using lithium tantalate. (a) EO bandwidths and (b) voltage tuning curve of MZMs. (c) Measured optical spectrum and (d) single-side band phase noise power spectral density.

To demonstrate the utility of the LTOI platform for electro-optics, we fabricated Mach-Zehnder modulator (MZM) at either end and a push-pull optical waveguide phase shifter pair with a length of 2.5 mm. The waveguide width is 1.2 μm and the gap between the LT waveguide sidewalls and the gold electrode is 2 um on each side. The measured 3 dB EO modulation bandwidth is beyond 40 GHz (cf. Fig. 3 (a)). The half-wave voltages V_π is 7.86 V under a 10 kHz triangular voltage sweep (cf. Fig. 3 (b)), corresponding a modulation efficiency of V_π L = 1.96 V·cm. The V_π is expected to be less than 1 V by simply increasing the device length to 2 cm. Pumping the high-Q resonance of a racetrack at a high power of 90 mW we achieve dissipative Kerr soliton (DKS) formation. Single soliton generation at pulse repetition rates 30.1 GHz with a bandwidth of 4.0 THz centered at 1556 nm and a pulse duration of 71 ps is shown in Fig. 3 (c). We detect the DKS pulse train on a fast photodetector and generate a 30.1 GHz microwave beat note with phase noise of −86 dBc/Hz at an offset frequency of 10 kHz, higher than earlier measurements using Si_3N_4 optical microresonators and in Z-cut LN. It is notable that here DKS generation was achieved in an x-cut ferroelectric crystal sample for the first time [1] facilitated by the low Raman response of LT.

3 Discussion

We show that the $LiTaO_3$ can be fabricated at thin-film- on-insulator structure, and the PICs fabricated based on LTOI achieved low optical loss with good uniformity on wafer-scale yield. Compared with $LiNbO_3$, $LiTaO_3$ has the advantages of wider window, weaker birefringence and higher power tolerance, which can compete with $LiNbO_3$ to enable a wide range of applications, including optical communications, classical and quantum computing networks.

Acknowledgements. This work was supported by the European Research Council grant no. 835329 (ExCOM-cCEO) and the EU Horizon Europe research and innovation program no. 101113260 (HDLN). J.R. acknowledges funding from the SNSF through an Ambizione Fellowship no. 201923. X.O. acknowledges the National Key R&D Program (No. 2022YFA-1404601), and National Natural Science Foundation of China (62293520). The samples were fabricated in the EPFL Center of MicroNanoTechnology (CMi). The LTOI wafers were fabricated in Shanghai Novel Si Integration Technology (NSIT) Co., Ltd. and the SIMIT-CAS.

References

1. Boes, A., et al.: Science **379**, 6627 (2023)
2. Tormo, V., et al.: Opt. Express **27**, 8696 (2019)
3. Tangonan, T., et al.: Appl. Phys. Lett. **30**, 238–239 (1977)
4. Meyn, J., Fejer, M.: Opt. Lett. **22**, 1214 (1997)
5. Yan, Y., et al.: ACS Appl. Electron. Mater. **1**, 1660–1666 (2019)
6. Li, Z., et al.: Preprint at http://arxiv.org/abs/2208.05556 (2022)

Voltage-Controllable Second-Order Susceptibility in Arsenic Sulfide Film

Laurids Wardenberg[1](✉), Benito Bunk[1], Georg von Freymann[2,3], and Jörg Schilling[1]

[1] Department of Physics, Martin-Luther-Universität Halle-Wittenberg, 06120 Halle, Germany
{laurids.wardenberg,joerg.schilling}@physik.uni-halle.de
[2] Department of Physics and Research Center OPTIMAS, RPTU Kaiserslautern-Landau, 67663 Kaiserslautern, Germany
[3] Fraunhofer Institute of Industrial Mathematics ITWM, 67663 Kaiserslautern, Germany

Abstract. Electric field induced second harmonic (EFISH) generation on arsenic sulfide films deposited via thermal evaporation is investigated, demonstrating a strong voltage-controllable second-order optical susceptibility ($\chi^{(2)}$). Surprisingly, even at 0 V, a non-EFISH bulk $\chi^{(2)}$ is also revealed. Supplementary polarization- and angle-dependent SHG measurements on as-deposited films show Maker fringe patterns from which the components of the $\chi^{(2)}$ tensor are determined. Calibrating the observed second harmonic generation (SHG) intensity from the arsenic sulfide against the SHG from a classic beta barium borate (BBO) crystal, a main component of $\chi^{(2)}_{zzz} = 0.22$ pm/V at 0 V is ascertained. Adding a dc-electric field during the EFISH-measurements the $\chi^{(2)}$ rises and reaches a maximum value of $\chi^{(2)}_{zzz} = 13{,}5$ pm/V at the electric breakdown field strength of 102 MV/m. This EFISH-enhanced $\chi^{(2)}_{zzz}$-value surpasses the $\chi^{(2)}$ main components of traditional nonlinear crystals, such as KDP, LBO or BBO significantly. This and a high intrinsic third order susceptibility of $\chi^{(3)} \approx 5\times10^{-20}\,\frac{\mathrm{m}^2}{\mathrm{V}^2}$ makes arsenic sulfide an interesting material for nonlinear hybrid photonic applications.

Keywords: second harmonic generation · electric field induced second harmonic generation · arsenic sulfide · second-order optical nonlinearity · third-order optical nonlinearity

1 Introduction

Due to its specific optical properties, such as a high refractive index [1] and low losses in the IR spectral region [2], arsenic sulfide is an interesting material platform for integrated hybrid photonic applications. In addition, the photosensitivity of freshly deposited arsenic sulfide films allows the direct lithographic structuring of the material, so that 2D and 3D micro- and nanostructures (e.g. waveguides) can be easily created [3, 4]. However, its amorphous structure and the resulting absence of bulk $\chi^{(2)}$ primarily limit its use. To broaden the applicability of arsenic sulfide for nonlinear optical devices, we investigated therefore the possibility to create an "artificial" $\chi^{(2)}$ in arsenic sulfide by applying a strong electric dc-field. We utilize the electric field induced second harmonic

J. Witzens et al. (Eds.): ECIO 2024, SPPHY 402, pp. 136–144, 2024.
https://doi.org/10.1007/978-3-031-63378-2_23

generation effect to demonstrate the resulting instantly changing, voltage-controllable $\chi^{(2)}$. This artificial $\chi^{(2)}$ depends on the applied electric field strength E_{DC} and the third-order susceptibility $\chi^{(3)}$ of the material:

$$\chi^{(2)}_{EFISH} = 3\chi^{(3)} E_{DC} \tag{1}$$

As a result the SHG intensity is proportional to the square of the applied electric field strength:

$$I_{SHG} \sim \left|\chi^{(2)}_{EFISH}\right|^2 \sim E_{DC}{}^2 \tag{2}$$

With a $\chi^{(3)} \approx 5 \times 10^{-20} \frac{m^2}{V^2}$ [5] arsenic sulfide exhibits a significantly higher third order susceptibility in the near IR than many other glasses (e.g. silicon dioxide $\chi^{(3)} \approx 2 \times 10^{-22}\ m^2/V$ [6]), promising a great potential for strong EFISH-signals and efficient second and third order nonlinear processes in general.

2 Sample Preparation

The sample for the EFISH measurement is designed in a sandwich structure (see Fig. 1). For that a fused silica substrate was coated with transparent 4 nm chromium bottom contact followed by the thermally evaporated arsenic sulfide layer and transparent 4 nm chromium top contact. The temperature controlled evaporation of arsenic sulfide was carried out at 320 °C, so that an approximately 5,5 μm thick layers could be deposited within 25 min. Note, that this sample design allows us to apply a homogeneous static electric field E_{DC} localized in the investigated film without effecting the substrate material while measuring the resulting SHG intensity simultaneously.

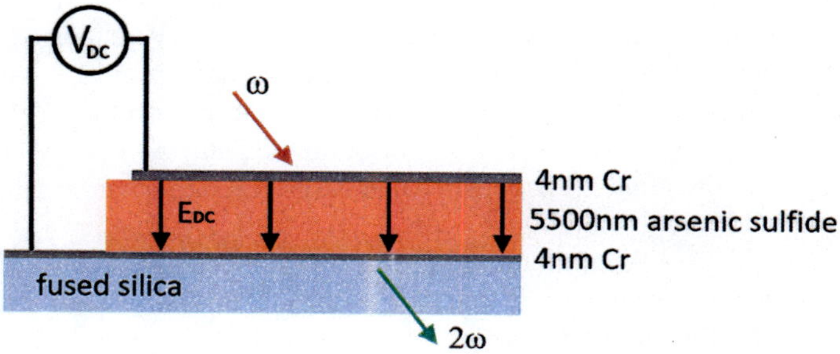

Fig. 1. Schematic representation of a sample for EFISH measurements.

3 Free-Space Measurements

3.1 EFISH Measurements

All EFISH and SHG measurements were conducted in transmission mode (see Fig. 2), driven by a Yb:KGW laser at a wavelength of 1030 nm (1 kHz repetition rate, 180 fs pulse duration). The average power is modulated between 50 mW and 200 mW, and the beam diameter at the sample position is approximately 3 mm. A zero-order half-wave plate at the output of the laser enables to adjust the polarization orientation of the laser. The sample is positioned on a motorized rotation mount, allowing changes in the angle of incidence during the measurement by tilting the sample. A polarizer in front and another polarizer behind the sample are used to control the polarization of the input beam and the corresponding SHG signal, respectively. In this way all fours polarization configurations s-s, s-p, p-s and p-p could be measured, where the first letter describes the polarization of the input beam and the second letter represents the polarization of the SHG beam (e.g. s-p indicates that the input beam is s-polarized while the SHG beam is p-polarized). To separate both beams, long and short pass filters (LPF, SPF) are placed in front and behind the sample, and a photomultiplier tube (PMT) is used to detect the SHG signal.

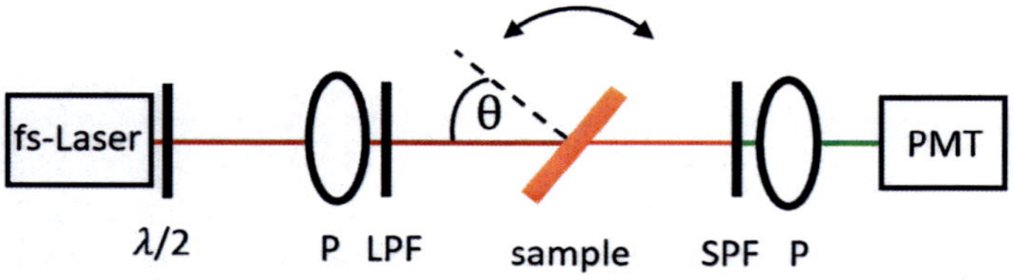

Fig. 2. Schematic representation of the free-space transmission SHG measurement setup. The combination of a half-wave plate polarizer (P) enables the adjustment of intensity, while the long-pass filter (LPF) and short-pass filter (SPF) function to separate pump and SHG beam before directing them through an analyzer (P) into the detector (PMT).

The EFISH measurements are conducted at a fixed angle of incidence of $\theta = 22.5°$. Depending on the polarization configuration, we observed distinct behaviors. While no SHG was observed for p-s and s-s configuration, the measured SHG intensity displayed a quadratic dependence on the applied electric field strength for s-p and p-p configuration (see Fig. 3), as expected according to Eq. 2. The fact that the resulting SHG signal is always p-polarized proves the *in-plane isotropy* of the film and the vertical (z) direction as the distinguished direction - as anticipated due to the perpendicular orientation of the applied electric field.

According to Eq. 1, the minimum of the EFISH intensity is predicted at 0 V/m. However, unexpectedly the experimental results show a deviation with a shift of the minimum to negative voltages. To explain this, the consideration of a quiescent bulk $\chi^{(2)}$ is necessary - similar as it has been observed in other amorphous materials, such as silicon nitride [7]. The resulting total $\chi^{(2)}$ is then the sum of the bulk-$\chi^{(2)}$ and the EFISH-$\chi^{(2)}$:

$$\chi^{(2)}_{\text{total}} = \chi^{(2)}_{\text{bulk}} + \chi^{(2)}_{\text{EFISH}} \tag{3}$$

The sign of the EFISH-$\chi^{(2)}$ depends on the polarity (sign) of the applied electric field E_{DC} (see Eq. 1). Therefore, the bulk- and the EFISH-$\chi^{(2)}$ can either add up or counteract each other. If the signs differ, the minimum of the EFISH intensity is found at the electric field strength where $\chi^{(2)}_{\text{bulk}}$ and $\chi^{(2)}_{\text{EFISH}}$ balance each other out.

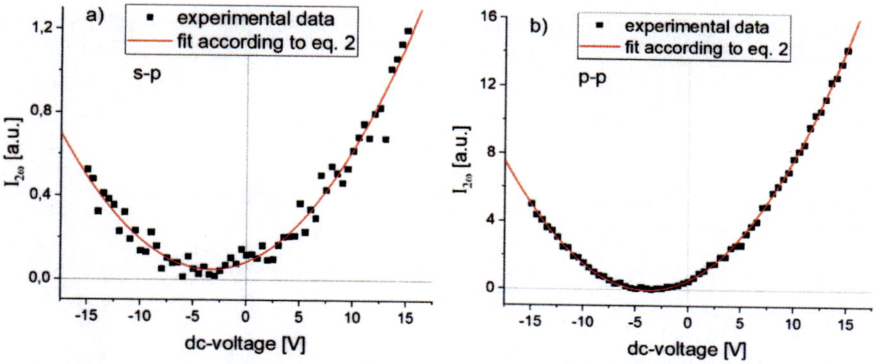

Fig. 3. a) Experimental results from an EFISH measurement with s-p polarization configuration. b) Experimental results from an EFISH measurement with p-p polarization configuration.

3.2 Bulk-SHG Measurements and $\chi^{(2)}$-Determination

To validate the existence of a bulk-$\chi^{(2)}$ in the arsenic sulfide film, further non-EFISH SHG measurements are required. For that purpose arsenic sulfide films were deposited on fused silica substrates without chromium electrodes. To avoid interference effects by multiple reflections within the film, we created films with thicknesses between 180 nm and 600 nm.

To confirm that the observed SHG signal indeed originates from the bulk of the film and is not a pure surface effect, we investigated the SHG from different thicknesses. The result is shown in Fig. 4. Given the good parabolic dependence of the SHG-intensity with the layer thickness, we conclude that the SHG has a bulk and not a surface origin.

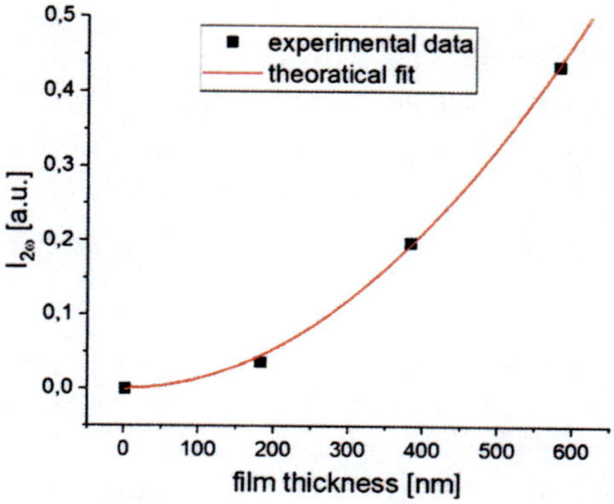

Fig. 4. Experimental data of the SHG intensity depending on the thickness of the arsenic sulfide film.

Further investigation on the bulk second-order nonlinear properties of arsenic sulfide are carried out on a film with 582 nm thickness. For that the polarization configuration was varied and the angle of incidence was modulated in a range from −60° to +60° while measuring the SHG-intensity (see Fig. 5a, b).

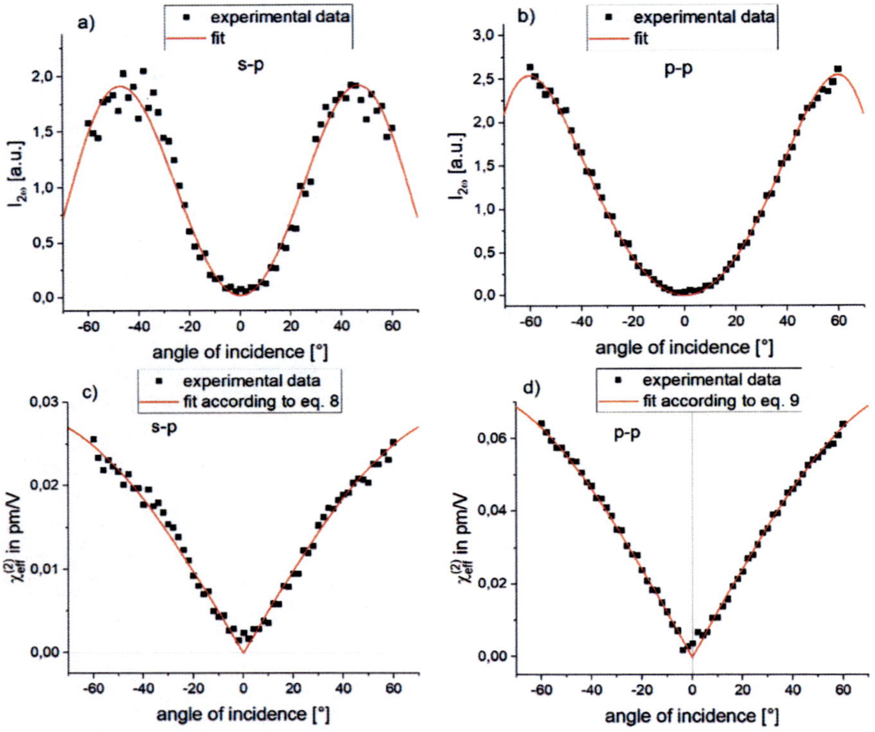

Fig. 5. a) Angle-dependent SHG intensity with s-p polarization configuration. b) Angle-dependent SHG intensity with p-p polarization configuration. c) Determined angle-dependent $\chi^{(2)}_{eff}$-values according to Eq. 4 based on the experimental data represented in a). d) Determined angle-dependent $\chi^{(2)}_{eff}$-values based on the experimental data represented in b)

Similar to the EFISH measurements, a SHG signal was detected for s-p and p-p configuration, but non for s-s and p-s configuration. This leads again to the conclusion that the film exhibits an in-plane isotropy and vertical polar axis, which coincides with the symmetry of the EFISH-geometry. For in-plane isotropy or symmetry group $C_{\infty v}$ the tensor components of the corresponding $\chi^{(2)}$-tensor are reduced to $\chi^{(2)}_{xxz} = \chi^{(2)}_{xzx} = \chi^{(2)}_{yyz} = \chi^{(2)}_{yzy}$, $\chi^{(2)}_{zxx} = \chi^{(2)}_{zyy}$ and $\chi^{(2)}_{zzz}$ while all other components vanish [8]. Note that here, z is parallel to the surface normal and x and y are any two orthogonal in-plane directions. To evaluate the $\chi^{(2)}$-tensor components quantitatively we apply the air-film-substrate-air model suggested by Herman and Hayden describing the transmitted SHG intensity $I_{2\omega}$ from a thin film by [9]:

$$I_{2\omega}(\theta) = \frac{128\pi^3}{c} \frac{\left(t_{\mathrm{af}}^{\omega}\right)^4 \left(t_{\mathrm{fs}}^{2\omega}\right)^2 \left(t_{\mathrm{sa}}^{2\omega}\right)^2}{\left(n_{2\omega}\cos\theta_{2\omega}\right)^2} \left(\frac{2\pi L}{\lambda}\right)^2 I_{\omega}^2 \left(\chi_{\mathrm{eff}}^{(2)}\right)^2 \times \frac{\sin^2(\psi) + \sinh^2(\delta_\omega - \delta_{2\omega})}{\psi^2 + (\delta_\omega - \delta_{2\omega})^2} \exp(-2(\delta_\omega + \delta_{2\omega})) \quad (4)$$

In this model t_{af}^{ω}, $t_{\mathrm{fs}}^{2\omega}$ and $t_{\mathrm{sa}}^{2\omega}$ are standard Fresnel transmission coefficients at the air-film (af), film-substrate (fs), and substrate-air (sa) interface, respectively, which depend on the polarization and frequency of the fundamental and the SHG beam. Other parameters in Eq. 4 are the thickness of the film L, $n_{2\omega}$ as the refractive index of the film at frequency 2ω, I_ω is the intensity of the fundamental beam and $\chi_{\mathrm{eff}}^{(2)}$ is the effective susceptibility tensor component. ψ describes the angle dependent phase mismatch due to dispersion of the material of fundamental and second harmonic waves:

$$\psi = \frac{2\pi L}{\lambda}(n_\omega \cos\theta_\omega - n_{2\omega}\cos\theta_{2\omega}) \quad (5)$$

leading to possible Maker fringe patterns in the measured SHG intensity $I_{2\omega}$. The δ's are correlated with absorption and contain the imaginary parts of the corresponding refractive indices κ_ω and $\kappa_{2\omega}$:

$$\delta_\omega = \frac{2\pi L}{\lambda}\frac{\kappa_\omega}{\cos\theta_\omega} \quad (6)$$

$$\delta_{2\omega} = \frac{2\pi L}{\lambda}\frac{\kappa_{2\omega}}{\cos\theta_{2\omega}} \quad (7)$$

Fitting the experimental values in Fig. 5 with the model (Eq. 4, red curves in Fig. 5) shows a very good agreement. Comparing the SHG-intensities of the sample with the SHG-intensities obtained from a 10 μm thick phase-matched BBO-crystal with a known $\chi^{(2)}$ as a reference, we determine the absolute values of $\chi_{\mathrm{eff}}^{(2)}$ of our sample (see Fig. 5c, d). These angle dependent $\chi_{\mathrm{eff}}^{(2)}$-values are related to the nonvanishing $\chi^{(2)}$-tensor components as follows [10]:

$$\text{s-p}: \chi_{\mathrm{eff}}^{(2)} = \chi_{zxx}^{(2)} \sin\theta_{2\omega} \quad (8)$$

$$\text{p-p}: \chi_{\mathrm{eff}}^{(2)} = \chi_{xxz}^{(2)} \sin 2\theta_\omega \cos\theta_{2\omega} + \sin\theta_{2\omega}(\chi_{zxx}^{(2)} \cos^2\theta_\omega + \chi_{zzz}^{(2)} \sin\theta_\omega) \quad (9)$$

Fitting the angle-dependent $\chi_{\mathrm{eff}}^{(2)}$-values according to Eq. 8 and 9 (see Fig. 5c, d) leads to the $\chi_{xxz}^{(2)}$-, $\chi_{zxx}^{(2)}$- and $\chi_{zzz}^{(2)}$-values of the investigated arsenic sulfide film at zero applied electric field (Table 1).

3.3 Evaluation of EFISH Results

Now we can utilize these zero-field $\chi^{(2)}$-tensor components to obtain absolute $\chi_{\mathrm{eff}}^{(2)}$-values at higher applied electric fields from the SHG intensities of the EFISH measurements in Fig. 3. The fact that the zero field $\chi_{\mathrm{eff}}^{(2)}$-value obtained from the 582 nm thick

Table 1. Zero-field $\chi^{(2)}$-tensor components of as-deposited arsenic sulfide

$\chi^{(2)}_{zxx}$ [pm/V]	$\chi^{(2)}_{xxz}$ [pm/V]	$\chi^{(2)}_{zzz}$ [pm/V]
0.07	0.05	0.22

film corresponds to the EFISH-$\chi^{(2)}_{\mathrm{eff}}$ at 0 V/m of the 5,5 μm thick sample (when both are measured under the same angle of incidence) enables an easy conversion of the EFISH intensities into applied field $\chi^{(2)}_{\mathrm{eff}}$-values. The results of this transformation are shown in Fig. 6.

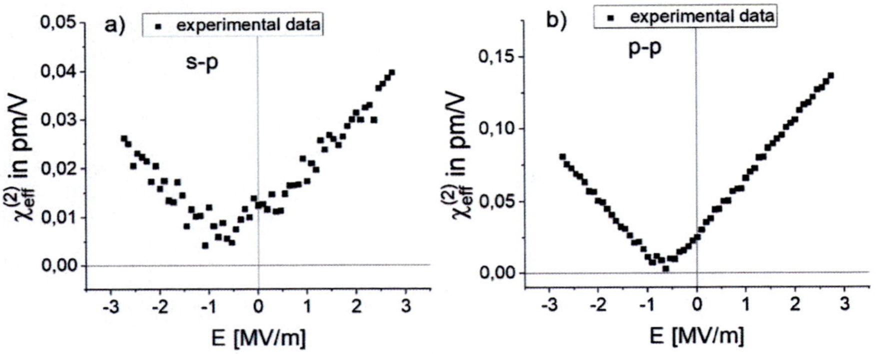

Fig. 6. a) Determined $\chi^{(2)}_{eff}$-values based on the experimental data in Fig. 3a depending on applied electric field. b) Determined $\chi^{(2)}_{eff}$-values based on the experimental data in Fig. 3b depending on applied electric field.

The final step of our investigations on EFISH in arsenic sulfide is to test the limit of the artificially achievable $\chi^{(2)}$. For this purpose, the electric field strength in the film was increased from 0 V/m to the breakdown field strength of 102 MV/m (see Fig. 7). Assuming that all $\chi^{(2)}$-tensor components are affected linearly by the applied electric field, we can first connect the SHG-intensity at 0 V/m to the zero field $\chi^{(2)}$-tensor components $\chi^{(2)}_{zzz} = 0{,}22$ pm/V - the main component of $\chi^{(2)}$. Subsequently we can perform again an easy transformation from EFISH intensities into electric field dependent $\chi^{(2)}_{zzz}$-values. This results in a maximum $\chi^{(2)}_{zzz}$-value of 13,5 pm/V. According to Eq. 1, from this the $\chi^{(3)}$ of the arsenic sulfide film can be derived as $\frac{\Delta\chi^{(2)}_{zzz}}{3\Delta E} = 4{,}4 \cdot 10^{-20} m^2/V^2$. This aligns very well with $\chi^{(3)}$-values reported in the literature, where $\chi^{(3)} \approx 5\times10^{-20}$ m^2/V^2 [5] and reassures the overall treatment.

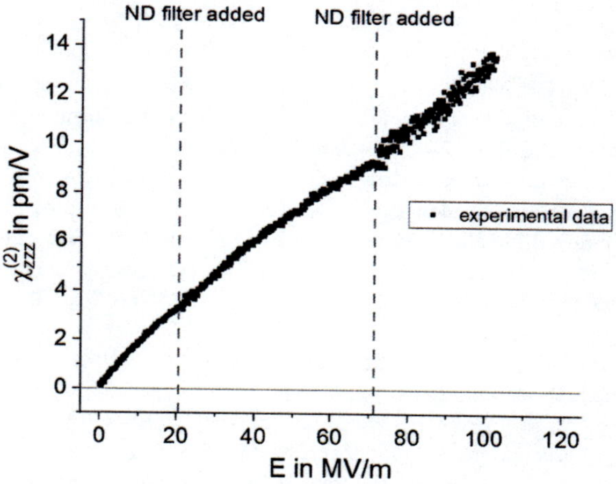

Fig. 7. Electric field-dependent $\chi^{(2)}_{zzz}$-values with a maximum value of 13,5 pm/V. The different noise levels appear due to the application of additional neutral density filters at higher intensities to protect the detector.

4 Conclusion

Free-space SHG measurements reveal an unexpected bulk-$\chi^{(2)}$ in arsenic sulfide films deposited using the thermal evaporation technique. We demonstrated that this effect exhibits in-plane isotropy and determined the values of the $\chi^{(2)}$-tensor components according to the matching air-film-substrate-air model. A possible explanation might be that the bulk $\chi^{(2)}$ is caused by fixed near-surface charges, inducing a permanent static electric field across the material thickness, which is comparable to the effect caused by near-surface charges in poled glasses [11–13]. However, as the origin of this nonlinear bulk effect is not yet clarified, further investigations are required. In addition, we performed EFISH measurements on arsenic sulfide films and determined the instantly changing, voltage-controllable second-order susceptibility. We demonstrated that this artificial $\chi^{(2)}$ can be widely tuned depending on the applied static electric field, giving great control over the second-order nonlinearity of the material. The limit of this effect is given by the breakdown field strength of 102 MV/m leading to a maximum value of the $\chi^{(2)}$ main component of $\chi^{(2)}_{zzz} = 13{,}5$ pm/V. This surpasses the $\chi^{(2)}$ main component of traditional nonlinear crystals such as BBO ($\chi^{(2)}_{yyy} = 2{,}2$ pm/V [14]) significantly. Therefore, in combination with its easy structuring in 2D and 3D, arsenic sulfide may enable the realization of novel or enhanced nonlinear hybrid photonic applications on-chip.

References

1. McCloy, J.S., Riley, B.J., Sundaram, S.K., Qiao, H.A., Crum, J.V., Johnson, B.R.: Structure-optical property correlations of arsenic sulfide glasses in visible, infrared, and sub-millimeter regions. J. Non-Cryst. Solids (25–27), 1288–1293 (2010)
2. Devyatykh, G.G., et al.: Low-loss infrared arsenic-chalcogenide glass optical fibers (2000)
3. Wong, S.H., et al.: Highly selective wet etch for high-resolution three-dimensional nanostructures in arsenic sulfide all-inorganic photoresist. Chem. Mater. **17**, 4213–4221 (2007)
4. Schwarz, C.M., et al.: Fabrication and characterization of microstructures created in thermally deposited arsenic trisulfide by multiphoton lithography. J. Micro/Nanolith. MEMS MOEMS (2), 23508 (2017)
5. Tanaka, K., Saitoh, A.: Pulsed light effects in amorphous As_2S_3: review. J. Mater. Sci. Mater. Electron. (28), 22029–22052 (2022)
6. Guillet de Chatellus, H., Freysz, E.: Measurement of the third-order susceptibility of glasses by EFISH of femtosecond pulses. Opt. Express (11), 586–591 (2001)
7. Lin, H.-H., et al.: On the observation of dispersion in tunable second-order nonlinearities of silicon-rich nitride thin films. APL Photonics (3), 36101 (2019)
8. Ning, T., Pietarinen, H., Hyvärinen, O., Simonen, J., Genty, G., Kauranen, M.: Strong second-harmonic generation in silicon nitride films. Appl. Phys. Lett. (16), 161902 (2012)
9. Herman, W.N., Hayden, L.M.: Maker fringes revisited: second-harmonic generation from birefringent or absorbing materials. J. Opt. Soc. Am. **B**(3), 416 (1995)
10. Sutherland, R.L.: Handbook of Nonlinear Optics, 2 nd edn., Revision and Expanded. Marcel Dekker, New York (2003)
11. Lepicard, A., Adamietz, F., Rodriguez, V., Richardson, K., Dussauze, M.: Demonstration of dimensional control and stabilization of second harmonic electro-optical response in chalcogenide glasses. Opt. Mater. Express (6), 1613 (2018)
12. Dussauze, M., Zheng, X., Rodriguez, V., Fargin, E., Cardinal, T., Smektala, F.: Photosensitivity and second harmonic generation in chalcogenide arsenic sulfide poled glasses. Opt. Mater. Express OME (1), 45–54 (2012)
13. Guo, H., et al.: Large second-order nonlinearity in thermally poled Ge–Sb–Cd–S chalcogenide glass. Opt. Mater. (6), 865–869 (2009)
14. Shoji, I., et al.: Absolute measurement of second-order nonlinear-optical coefficients of β-BaB2O4 for visible to ultraviolet second-harmonic wavelengths. J. Opt. Soc. Am. B JOSAB (4), 620–624 (1999)

Second Harmonic Generation and $\chi^{(2)}$ Cascading in Periodically Poled MgO:LiNbO$_3$ Photonic Wires

Halvor R. Fergestad[1](✉), Daiheng Fu[1], Muhammed Alqedra[1], Kore Hasse[2], Detlef Kip[2], Val Zwiller[1], and Katia Gallo[1]

[1] KTH Royal Institute of Technology, Roslagstullsbacken 21, 10691 Stockholm, Sweden
halvorf@kth.se

[2] Faculty of Electrical Engineering, Helmut Schmidt University, 22043 Hamburg, Germany

Abstract. We study second harmonic generation in LiNbO$_3$ nanowaveguides, operated with sub-pJ pulses at 1424 nm in the picosecond regime generating 0.5 μW at 712 nm. Spectra recorded at both wavelengths provide evidence for enhanced $\chi^{(2)}$ cascading effects.

Keywords: lithium niobate on insulator · integrated nonlinear optics · second harmonic generation · quadratic cascading

The high intrinsic nonlinearity of high-confinement LiNbO$_3$ on insulator (LNOI) nanophotonic waveguides and their unique dispersion engineering capabilities afford $\chi^{(2)}$ frequency conversion and parametric generation devices with unprecedented performance in terms of footprint, bandwidth and power, enhanced by chip-level integration and multiplexing [1, 2]. Moreover, the combination of engineered $\chi^{(2)}$ nonlinearities achieved by periodic poling (PP) and dispersion control in the PPLNOI nanophotonic platform is ideal to leverage quadratic cascading towards applications beyond mere frequency conversion, encompassing all-optical signal processing, pulse compression and solitary wave generation [3]. Here we report on second harmonic generation (SHG) in PP:MgOLNOI waveguides with sub-μm cross-sections and poling periods in the 3.5 μm range, tailored for high-efficiency frequency conversion between the 1400 and 700 nm wavelength ranges (FF and SH, respectively) in the picosecond pulse regime. Walk-off-free devices with a length of 4.8 mm efficiently generate a SH centered at 712 nm with sub-pJ pumping levels and exhibit a rich spectral dynamics at both FF and SH underpinned by the interplay of $\chi^{(2)}$ cascading and waveguide dispersion.

Figure 1 illustrates the guided-wave PPLNOI devices designed for SHG and fabricated in commercial 600 nm-thick 5% MgO-doped LNOI chips (NANOLN). Simulations (Fig. 1b) performed with a commercial mode solver (Lumerical) for representative waveguide geometries yield effective areas $A_{eff} < 0.7\ \mu m^2$ and poling periods $\Lambda \backsim 3.5\ \mu m$ for SHG between TE_{00} guided modes at FF ($\lambda_{FF} \backsim 1400$ nm) and SH ($\lambda_{SH} \backsim 700$ nm).

J. Witzens et al. (Eds.): ECIO 2024, SPPHY 402, pp. 145–148, 2024.
https://doi.org/10.1007/978-3-031-63378-2_24

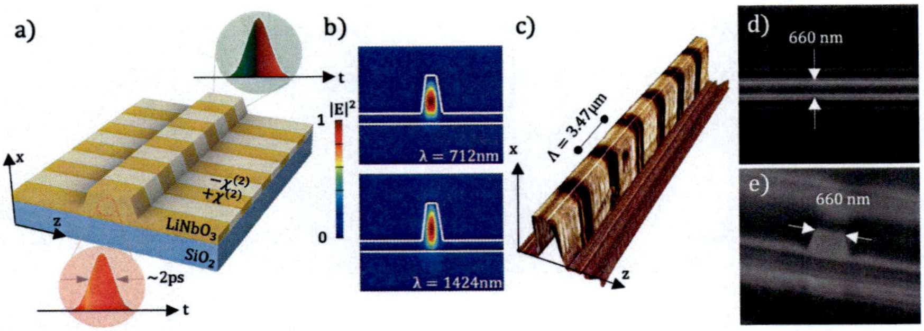

Fig. 1. a) A SHG waveguide device made in a 600 nm-thick film of x-cut MgO:PPLNOI film pumped by ~ 2 ps pulses at $\lambda_{FF} \backsim 1400$ nm. b) TE_{00} modes intensity computed at $\lambda_{FF} = 1424$ nm and $\lambda_{SH} \backsim 712$ nm for a rib waveguide with $w = 660$ nm, $h = 460$ nm, $\theta = 65°$, c) Overlayed atomic and piezo response force microscopy images revealing the rib waveguide and PPLNOI grating structures, respectively in a fabricated waveguide with the properties of Fig. 1b and poling period $\Lambda = 3.47$ μm. SEM images of the same waveguide: d) top view and e) side-view of the end-facet.

Periodic poling of the x-cut MgOLN thin film [1] was performed with 8 mm-long and 30 μm-wide patterns of periods ranging from 3.45 to 3.55 μm in steps of 20 nm, prior to the fabrication of optical waveguide circuitry by means of e-beam lithography and reactive ion etching, as described elsewhere [4]. The chip was finally cladded with an 800 nm layer of cured resist (Medusa 82) with high storage stability and refractive index akin to silica, followed by dice-polish and focused ion beam trimming of the waveguide end facets for optical edge coupling [5]. The finished PPLNOI chip consisted of sets of 20 straight nanowire rib waveguides (4 for each PPLNOI band) of length $L = 4.8$ mm, side-wall angle $\theta \sim 65°$ and height $h = 460$ nm, with widths w ranging from 650 to 850 nm, of which Fig. 1 c–e is an example.

The SHG response was studied with the setup sketched in Fig. 2a, employing a fiber-coupled FF source with output tunable in the spectral range around 1400 nm, delivering $\backsim$2 ps pulses at a repetition rate of 320 MHz (APE picoEmerald). The FF pump polarization was adjusted by a fiberized polarization controller and then filtered by a polarizing film in free space to align it to the z-axis of the LNOI waveguide for maximum nonlinear efficiency (using the $d_{33} = 25$ pm/V coefficient). The fiber-outcoupling system also included a collimator (Thorlabs TC12FC-1550), and a long pass filter (Thorlabs FELH1000) to remove any spurious signal in the SH range prior to FF coupling into the PPLNOI waveguide with a 50x objective (Mitutoyo plan-apo). The FF and SH signals at the waveguide output were collected with a lensed fiber (OZ Optics smf28) connected to an optical spectrum analyzer or a power meter, to conduct spectral and power characterizations at varying FF excitation powers (P_{FF}) and central wavelengths (λ_{FF}).

Figure 2b–c illustrate experimental results from the PPLNOI waveguide of Fig. 1, corresponding to a SHG quasi-phase matching wavelength of 712 nm. Figure 2b shows the spectrum recorded at the waveguide output, highlighting both FF and SH components. The inset illustrates the SHG power response in terms of generated SH (P_{SH}) versus FF

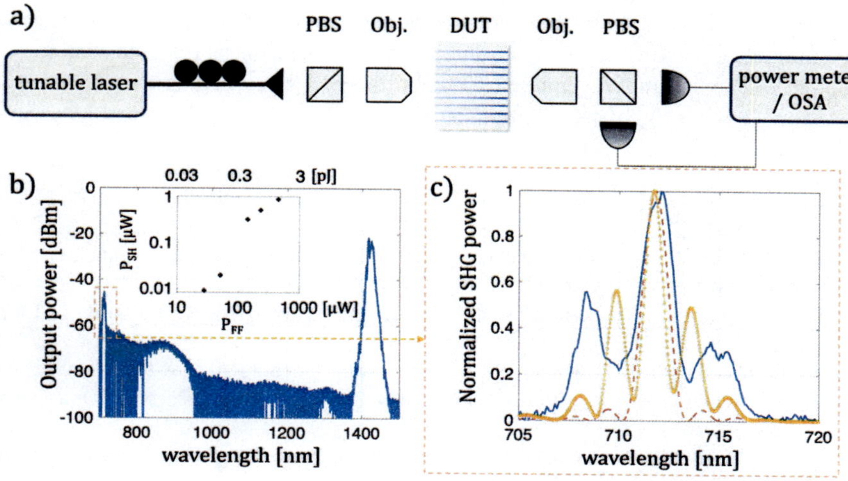

Fig. 2. a) Illustration of the setup for SHG excitation with a tunable picosecond laser around 1400 nm. b) Output spectrum recorded measured for the PPLNOI waveguide of Fig. 1, with an FF peaking around 1420 nm and SH centered at 712 nm. Inset: Measured SHG average powers (P_{SH}) as a function of the FF input power (P_{FF}) and pulse energy in log-scale. c) SH spectra measured for $P_{FF} = 230$ µW (blue line), and simulations in undepleted-pump (dashed red line) and high-conversion $\chi^{(2)}$ cascading (yellow line) regimes.

excitation in the waveguides. The maximum SH power of $P_{SH} = 0.9$ µW is obtained at an FF pulse energy of 1.4 pJ ($P_{FF} = 450$ µW). Spectral measurements at varying FF excitations revealed the emergence of spectral features well beyond the estimated ~3 nm-extent of the low-power SHG bandwidth, as illustrated by the comparison of the experiments (blue solid line) and simulations in the undepleted-pump approximation (red dashed line) in Fig. 2c, for $P_{FF} = 230$ µW. A pump-dependent spectral broadening spanning the range of 1410–1440 nm is also apparent in the measured FF throughput, consistently with the rich nonlinear wave dynamics expected in the strong cascading regime, underpinned by the interplay of dispersion, $\chi^{(2)}$:$\chi^{(2)}$ self-phase modulation and cascaded sum- and difference frequency intermixing in the proximity of the SHG phase-matching point [6]. Such insights are further confirmed by simulations based on advanced nonlinear models developed to rigorously account for SHG and broadband cascading effects in strongly nonlinear and broadband regimes [7]. This is illustrated by the qualitatively good agreement of the SH spectral coverage observed in the experiments with the one obtained through preliminary simulations using the pyChi software [8] with the specific FF input spectral distributions and waveguide dispersion properties of our experiments (Fig. 2c, thicker yellow line).

In conclusion, we demonstrated efficient SHG in lumped PPLNOI waveguides pumped with sub-pJ picosecond pulses in the 1400 nm band, generating 0.5 µW of guided-wave powers around 700 nm. The experiments provide evidence for a rich spectral dynamics indicative of $\chi^{(2)}$ cascading effects in a strongly nonlinear and dispersive regime, confirmed by preliminary simulations of supercontinua from mixed and cascaded nonlinearities in the considered waveguide geometry. The results pave the way for the demonstration of a broad range of all-optical devices for e.g. pulse compression, supercontinuum generation and even solitary wave generation across the visible and infrared ranges, leveraging the unique functionalities of $\chi^{(2)}$ cascading for ultrafast and low-power chip-scale signal processing.

References

1. Jankowski, M., et al.: Ultrabroadband nonlinear optics in nanophotonic periodically poled lithium niobate waveguides. Optica **7**(1), 40–46 (2020)
2. Zhu, D., et al.: Integrated photonics on thin-film lithium niobate. Adv. Opt. Photonics **13**(2), 242–352 (2021)
3. Stegeman, G.I., Hagan, D.J., Torner, L.: $\chi(2)$ cascading phenomena and their applications to all-optical signal processing, mode-locking, pulse compression and solitons. Opt. Quant. Electron. **28**(12), 1691–1740 (1996)
4. Prencipe, A., Baghban, M.A., Gallo, K.: Tunable ultranarrowband grating filters in thin-film lithium niobate. ACS Photonics **8**(10), 2923–2930 (2021)
5. Suntsov, S., Hasse, K., Kip, D.: Efficient watt-level 0.775 μm SHG and 3.3 μm DFG in annealed and reverse proton exchanged diced PPLN ridge waveguides. EPJ Web Conf. **255**, 06001 (2021)
6. Lee, K.J., et al.: Analysis of acceptable spectral windows of quadratic cascaded nonlinear processes in a periodically poled lithium niobate waveguide. Opt. Express **19**(9), 8327–8335 (2011)
7. Conforti, M., Baronio, F., De Angelis, C.: Nonlinear envelope equation for broadband optical pulses in quadratic media. Phys. Rev. A **81**(5), 053841 (2010)
8. Voumard, T., et al.: Simulating supercontinua from mixed and cascaded nonlinearities. APL Photonics **8**(3) (2023)

Fiber Communication Systems

Sustainable Pbps Co-packaged Optics Using Passively Assembled, Flip-Chip Evanescent Couplers

Drew Weninger[1](✉), Samuel Serna[2], Luigi Ranno[1], Lionel Kimerling[1], and Anuradha Agarwal[1]

[1] Massachusetts Institute of Technology, Cambridge, MA 02139, USA
drewski@mit.edu

[2] Bridgewater State University, Bridgewater, MA 02324, USA

Abstract. A passively assembled chip-to-interposer evanescent coupler between silicon nitride and silicon was experimentally demonstrated with a 0.33 ± 1.03 dB coupling loss at 1550 nm, a 160 nm 1-dB wavelength tolerance (1480–1640 nm), and a 1-dB lateral alignment tolerance of ± 1.56 μm. The thermal stability was evaluated from 23–60° C with average coupling loss and alignment tolerance varying by less than ± 0.35 dB and ± 30 nm, respectively. Finally, repeatability was evaluated across four packaged systems, demonstrating a coupling loss range of 1.5 dB. Together, these results show this coupler can help achieve future Pbps co-packaged optics input/output.

Keywords: silicon photonics · co-packaged optics · passive assembly · evanescent coupling · sustainability

1 Introduction

Historically, data center top of rack (ToR) switch packages have doubled total bandwidth capacity every two years to meet demand, with commercial packages reaching 51.2 Tbps using 106.25 Gbps per lane [8], and standardization of 200 Gbps per lane already underway [14]. However, the power consumed by board level copper traces and the limited pitch of pluggable transceivers has forced a transition to co-packaged optics (CPO) [11]. Scaling using CPO has challenges also, such as the active alignment of single mode fiber (SMF) arrays to silicon photonic integrated circuits (PICs). Additionally, 10^3-10^4 SMFs will be required to scale to the Pbps bandwidth capacities required by 2035 according to current trends [10]. However, standard telecom SMF arrays have minimum pitches of 127 μm, meaning a density of only 8 fibers/mm is possible. A sustainable solution

This work was funded by NSF ITE Convergence Accelerator Track I: Building a Sustainable, Innovative Ecosystem for Microchip Manufacturing, Award Number ITE-2236093 and carried out in part through the use of MIT.nano's cleanroom and characterization facilities.

J. Witzens et al. (Eds.): ECIO 2024, SPPHY 402, pp. 151–159, 2024.
https://doi.org/10.1007/978-3-031-63378-2_25

is to use fine pitch waveguide-to-waveguide couplers to connect die to interposers and interposers to boards, enabling optical fan-out. Such an architecture would permit massive numbers of parallel optical connections and passive flip-chip assembly using automated pick-and-place technology. To this end, the interposer-to-die evanescent coupler design outlined in [15] was fabricated, packaged, and tested to demonstrate feasibility.

2 Device Fabrication, Die Packaging, and Testing

The proposed coupler included two overlapping, double tapers, and was distinct from other evanescent couplers in two ways: 1) it was an inter-chip coupler as opposed to an intra-chip coupler and 2) it used a fused silica interposer with 220 nm silicon nitride (Si_3N_4) waveguides. The Si_3N_4-on-glass interposer maintained CMOS compatibility, lowered cost, had the prospect of low propagation losses, and increased minimum feature sizes to > 200 nm due to a lower index contrast [15]. By optimizing a double taper coupler geometry (as seen in the callout of Fig. 1(a)) using 3D FDTD simulations, footprint was minimized ($< 520\ \mu m$ total length) for wide alignment tolerances ($> \pm\ 2.5\ \mu m$) compared to linear inverse tapers [15]. Prototypes were fabricated using 150 mm fused silica wafers with a 220 nm low pressure chemical vapor deposited Si_3N_4 film and 150 mm silicon-on-insulator (SOI) wafers (250 nm device layer, 5 μm bottom oxide (BOX)). Waveguides were patterned using electron beam lithography followed by inductively coupled plasma reactive ion etching (ICP-RIE), and were cladded with 500 nm SiO_2 using plasma enhanced chemical vapor deposition (PECVD). This prototype was for testing the evanescent coupler only, so SOI was used as the "interposer" to allow for pick-and-place alignment accuracy to be easily verified by measuring through the back of the glass die. Moreover, the fabrication of high quality SOI edge facets was done through a combination of ICP-RIE and deep reactive ion etching (DRIE). After DRIE, wafers were diced and cleaned thoroughly using a combination of an O_2 plasma clean and isopropyl alcohol. The bonding was then accomplished using a Finetech Lambda2 high speed pick-and-place die bonder through passive alignment with machine vision only. The bonding process consisted of alignment, dispensing a NOA61 epoxy droplet ($n = 1.54$) using a 27 gauge needle, application of 40–100 N of force, and a 2 min in situ UV exposure. A pair of test die bonded using this recipe demonstrated epoxy bond line thicknesses of 0.48–0.63 μm as measured using a SEM, indicating a taper-to-taper distance of greater than 1 μm when including the SiO_2 cladding. A Vernier waveguide layout comprised of 14 Si channels with a pitch of 127 μm and 14 Si_3N_4 channels with a pitch of 127.4 μm was used to measure lateral alignment tolerance to within 400 nm precision. Additionally, Si spirals of length 1–5 cm were included to determine the facet and propagation loss for the Si waveguides. To determine the Si_3N_4 propagation loss, a separate fused silica die containing paperclips of lengths 1–3 cm was fabricated using an identical process flow to that listed above, except that the 70 μm deep edge facets were defined using magnetically enhanced reactive ion etching instead of

ICP-RIE and DRIE. With this information, the following equation was used to extract the loss of a single pair of evanescent tapers:

$$P_{\text{chip-to-chip}} = \frac{(P_{\text{output}} - P_{\text{facets}} - P_{\text{Si}} - P_{\text{SiN}})}{2} \tag{1}$$

The testing setup consisted of a 1480–1640 nm wavelength laser injecting 0 dBm of light into a SMF. The SMF connected to a tapered, lensed polarization maintaining fiber (8 μm core, 125 μm cladding, 3 μm working distance) which went through polarization controlling paddles before being connected to a piezo-electric arm for sub-micron fiber-to-chip alignment. As shown in Fig. 1(a), after edge coupling the light traveled from the lower die's Si waveguides to the upper die's Si_3N_4 waveguides using the vertical coupler and then back down using a second vertical coupler where it was collected using an edge coupled, lensed SMF and fed into a photodetector. The aforementioned testing was repeated across four sets of packaged die to determine process and device repeatability. Finally, because the stage that the packaged die were positioned on could be temperature controlled from 23–60 °C, measurements were repeated for the best performing package at temperatures of 23 °C, 40 °C, and 60 °C to determine thermal stability.

3 Results

A plot showing the raw measured output power for a completely Si reference channel and a well aligned chip-to-chip channel can be found in Fig. 1(b). The Si propagation and facet losses (P_{Si} and P_{facets}) were determined by measuring optical output power for the spirals, averaging within a ± 2.5 nm wavelength window, and fitting power versus spiral length for each wavelength window using linear regression, with the results being shown in Fig. 1(f). The R^2 values for all fits of the Si spirals from 1480–1640 nm wavelengths were above 0.94. The same process was used for the Si_3N_4 paperclips on the separate fused silica die to determine P_{SiN}, with the results being shown in Fig. 1(e). The R^2 values for all fits of the Si_3N_4 paperclips from 1480–1640 nm wavelengths were above 0.98. The coupling loss ($P_{\text{chip-to-chip}}$) for the best-aligned channel is shown in Fig. 2(b), with a value of 0.33 ± 1.03 dB at 1550 nm wavelength. The average coupling loss was calculated to be 0.56 ± 0.90 dB (± 2σ) across the S, C, and L bands (1480–1640 nm wavelengths). The 1-dB excess loss wavelength tolerance, also termed the 1-dB bandwidth, was determined to be 160 nm based on Fig. 2(b), extending from 1480 nm to 1640 nm wavelength. Likewise, the 1-dB and 3-dB excess loss lateral alignment tolerances were determined for each wavelength window using the Vernier layout, with results shown in Fig. 2(c). The 1-dB alignment tolerance at 1550 nm was ± 1.56 μm and the 3-dB alignment tolerance was ± 2.41 at 1550 nm. Similarly, the maximum 1-dB tolerance was ± 1.60 μm at 1545 nm while the average 1-dB tolerance from 1480–1640 nm was ± 1.38 μm, with 95% of the data falling within ± 0.24 μm of this value for a vertical taper-to-taper spacing of greater than 1 μm.

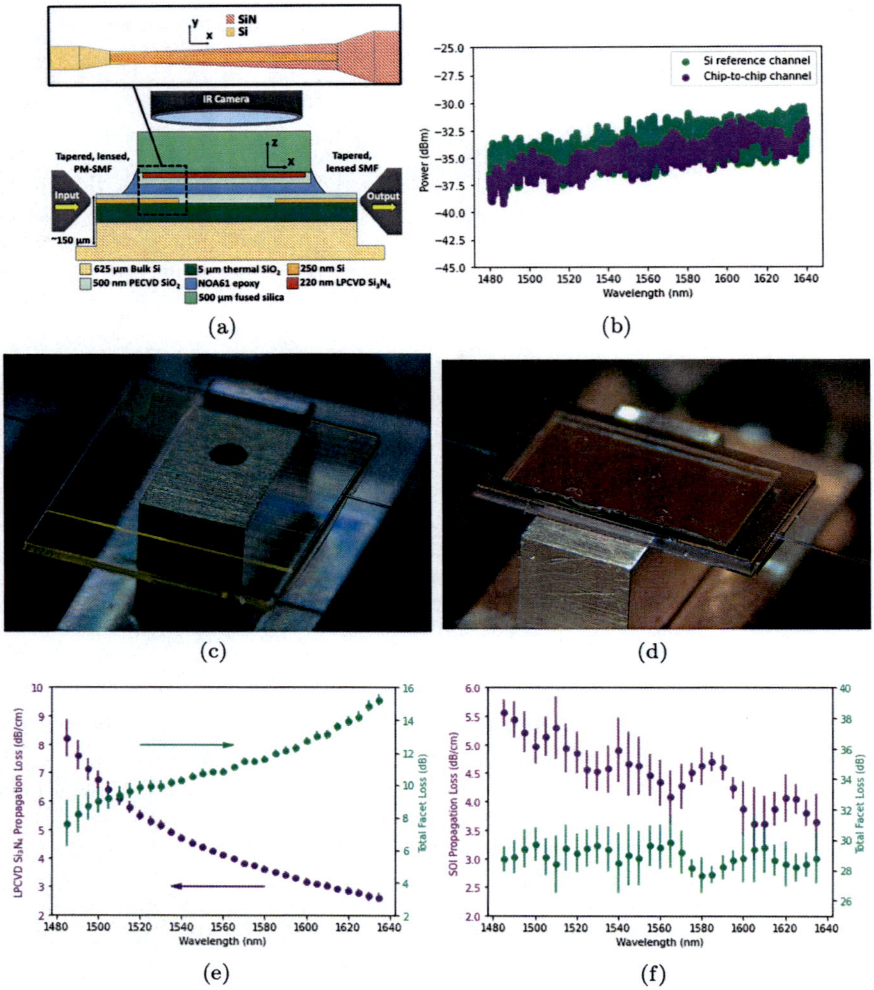

Fig. 1. In (a) a cross sectional view of the experiment with a callout showing the double taper design. In (b) raw data of a wavelength sweep for a reference Si channel and a chip-to-chip channel. In (c) and (d), the fabricated glass die with Si_3N_4 waveguides and the fully packaged system are shown on the testing stage, respectively. In (e) and (f), the propagation and facet loss from the SiN paperclips and Si spirals, respectively.

Furthermore, the coupling loss varied by less than ± 0.35 dB from 1480–1640 nm for temperatures from 23 °C to 60 °C as shown by the results in Fig. 3(a). Likewise, data in Fig. 3(c) shows that the 1-dB lateral alignment tolerance varied by less than ± 30 nm from 23 °C to 60 °C, demonstrating that the coupler will operate nearly identically, even if misaligned, for a wide temperature window.

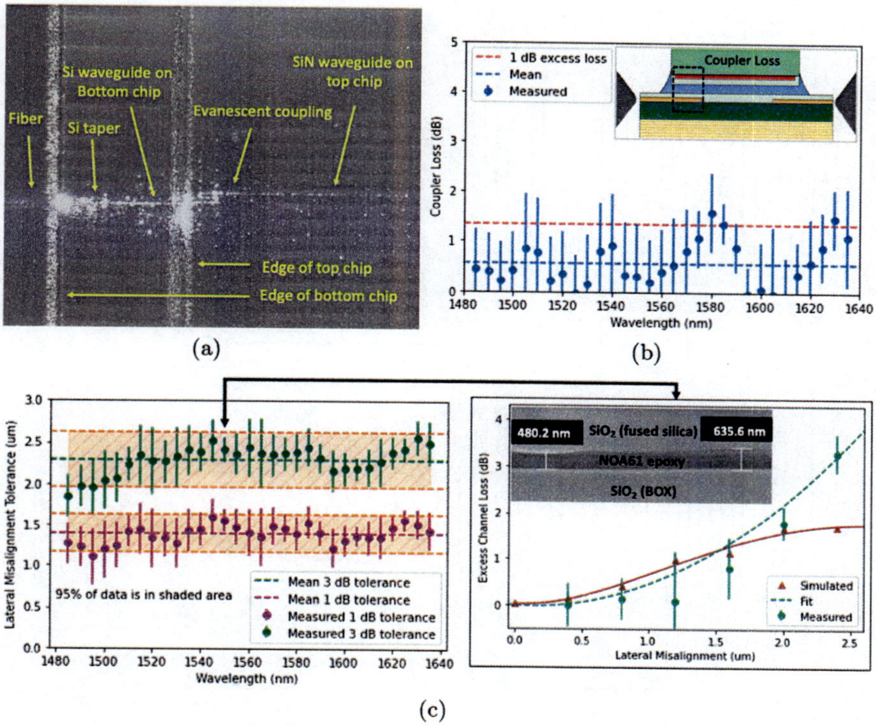

Fig. 2. The image in (a) shows coupling near 1550 nm from the SMF to the SOI interposer and into the separate fused silica die using a mounted infrared camera. In (b) $P_{\text{chip-to-chip}}$ is shown for the best aligned channel. The inset shows schematically the plotted quantity. In (c), the 1-dB and 3-dB lateral alignment tolerances from 1480–1640 nm. The callout shows measured and 3D-EME results at 1550 nm. The inset is a cross sectional SEM image of a bonded package showing an epoxy thickness of 0.48–0.64 μm.

Additionally, the average coupling loss from 1480–1640 nm had a range of less than 1.5 dB across the four sets of packaged die as shown by Fig. 3(b), from a minimum of 0.34 ± 1.32 dB (Package 3) to a maximum 1.83 ± 1.33 dB (Package 1). The coupling loss when combining the data from all four packages was 1.11 dB with a standard deviation of ± 0.99 dB. Similarly, Packages 2–4 showed a tightly bound 1-dB lateral alignment tolerance with the average across all three packages being ± 1.05 μm with a standard deviation of ± 0.35 as shown in Fig. 3(d). The lower alignment tolerance of ± 0.24 μm for Package 1 can be explained by this package having a 0.7–0.9 μm larger epoxy thickness than Packages 2–4 based on how 3D-EME simulations correlated with the measured data.

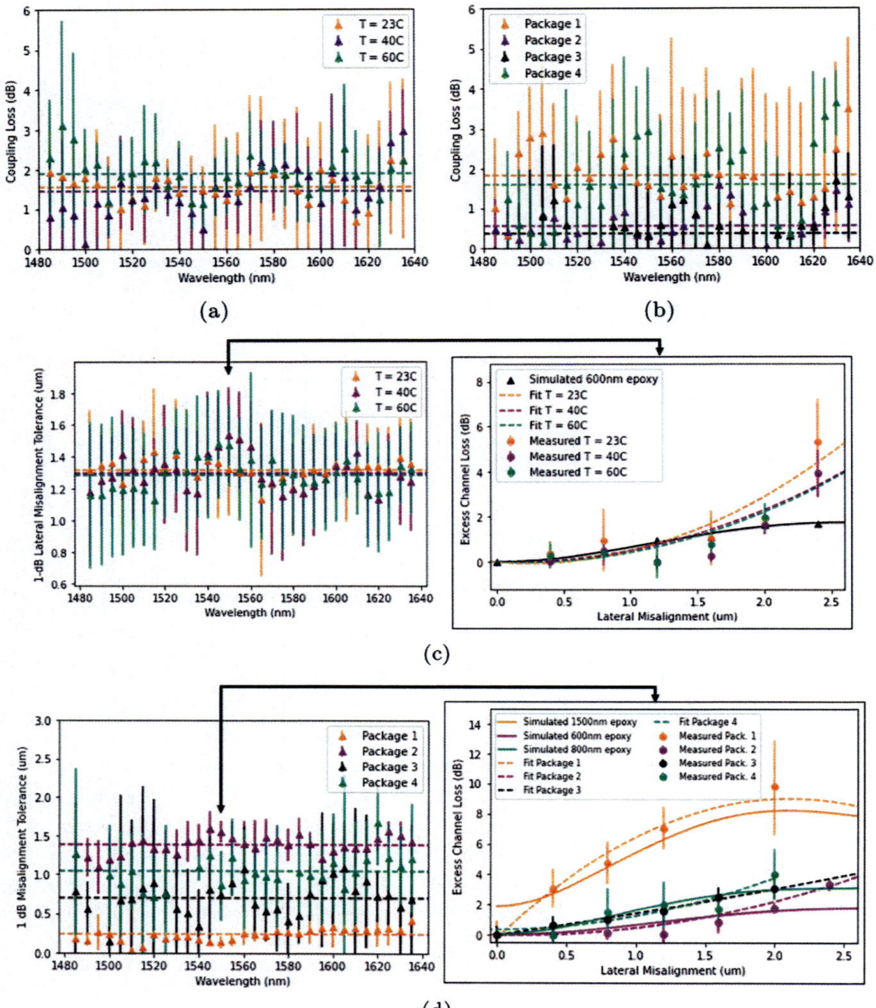

Fig. 3. In (a) $P_{\text{chip-to-chip}}$ is shown for temperatures between 23 °C and 60 °C. In (b) $P_{\text{chip-to-chip}}$ is shown for four separately packaged systems. In (c) the 1-dB lateral alignment tolerance is shown for temperatures between 23 °C and 60 °C, while in (d) the 1-dB lateral alignment tolerance is shown for Packages 1–4.

Table 1. Summary of silicon photonic waveguide-to-waveguide evanescent couplers in chronological order of publication date. The rows shaded blue indicate only simulation results were reported. Only evanescent couplers designed for inter-chip coupling were included. Note that this table provides a comparison using total insertion loss values.

Package	IL_{1550} (dB)	1-dB TOL (μm)		1-dB BW (nm)	LxW(μm)	Ref.
		L	V			
SOI to polymer	0.4 (TE)	± 3	/	± 100 (1550)	200x5	[12]
SOI to polymer	0.4 (TE/TM)	2	1.5	100 (1470-1570)	1750x5.915	[3,4]
Si_3N_4 to Si_3N_4	3.1	< ± 1	< 0.6	/ (1550)	500x3	[16]
Si_3N_4 to Si_3N_4	0.76	± 2	< 1	/ (1550)	500x3	[16]
SOI to polymer	1.25/0.5 (TE/TM) < 0.5 (TE/TM)	± 2 /	/	± 60 (1310) ± 70 (1550)	1500x6.5	[7,9, 13]
SOI to polymer	0.2	> ± 5 ± 1.5° (yaw)	0.5	± 100 (1550)	200x15.3	[2]
SOI to IOX	< 1 (TE/TM)	± 4	3	± 30 (1550)	1500x12	[6]
Si_3N_4 to IOX	0.7	> ± 4	< 2	75 (1515-1590)	2000x11	[5]
Si_3N_4 to Si_3N_4	1.31*	< ± 2	< 0.8	400 (1200-1600)	1000x3	[17]
SOI to Si_3N_4	**0.33 (TE)**	**± 1.56**	**> 1.1**	**160 (1480-1640)**	**520x1**	**this work**

IL = total insertion loss
TOL = misalignment tolerance
BW = bandwidth
L = lateral
V = vertical
/ = not reported
* = total insertion loss including 0.54 dB coupling loss and 0.77 dB scattering loss

4 Discussion and Conclusion

The device presented in this work was the first demonstration of an inter-chip evanescent coupler directly connecting SOI waveguides to other Si based waveguides (in this case Si_3N_4). In comparing the coupling loss of this device to the other inter-chip evanescent couplers shown in Table 1, the 1550 nm coupling loss of 0.33 ± 1.03 dB is the lowest reported coupling loss for an entirely Si based coupler. The maximum 1-dB lateral alignment tolerance of ± 1.60 μm at 1545 nm was also on par with other experimentally demonstrated inter-chip evanescent couplers and was shown to be large enough to enable passive assembly using automated pick-and-place tools [1]. Additionally, the longitudinal footprint of the coupler was roughly a third of the typical length of other models, and the lateral footprint of this coupler occupied only 33% of the space of the next thinnest design (and roughly 10%-20% of the space of typical designs), showing the potential for scaling to Pbps input/output counts in integrated photonic systems. Finally, the vertical taper-to-taper distance measured in this study was greater than 1 μm which will aid significantly in making the design compatible with typical BEOL layer thicknesses.

In summary, an evanescent interposer-to-die coupler was fabricated, packaged, and tested using standard CMOS foundry materials and processes and showed low loss, broadband, and alignment tolerant operation. This coupler can enable Pbps CPO systems by reducing the packaging inefficiencies associated with die-to-SMF connections.

References

1. Mrsi-s-hvm high speed, flexible, 0.5-micron flip-chip die bonder for high volume manufacturing. https://mrsisystems.com/mrsi-s-hvm/
2. Bandyopadhyay, S., Englund, D.: Alignment-free photonic interconnects (2021)
3. Barwicz, T., et al.: A novel approach to photonic packaging leveraging existing high-throughput microelectronic facilities. IEEE J. Sel. Top. Quantum Electron. **22**(6), 455–466 (2016). https://doi.org/10.1109/JSTQE.2016.2593637
4. Barwicz, T., Taira, Y.: Low-cost interfacing of fibers to nanophotonic waveguides: design for fabrication and assembly tolerances. IEEE Photonics J. **6**(4), 1–18 (2014). https://doi.org/10.1109/JPHOT.2014.2331251
5. Brusberg, L., et al.: Glass platform for co-packaged optics. IEEE J. Selected Topics Quantum Electr. **29**(3: Photon. Elec. Co-Inte. and Adv. Trans. Print.), 1–10 (2023) https://doi.org/10.1109/JSTQE.2023.3247245
6. Brusberg, L., et al.: Glass substrate with integrated waveguides for surface mount photonic packaging. J. Lightwave Technol. **39**(4), 912–919 (2021) https://doi.org/10.1109/JLT.2020.3033295
7. Dangel, R., et al.: Polymer waveguides enabling scalable low-loss adiabatic optical coupling for silicon photonics. IEEE J. Sel. Top. Quantum Electron. **24**(4), 1–11 (2018). https://doi.org/10.1109/JSTQE.2018.2812603
8. Fathololoumi, S., et al.: 1.6 tbps silicon photonics integrated circuit and 800 gbps photonic engine for switch co-packaging demonstration. J. Lightwave Technol **39**(4), 1155–1161 (2021). https://doi.org/10.1109/JLT.2020.3039218
9. Lamprecht, T., et al.: Electronic-photonic board as an integration platform for tb/s multi-chip optical communication. IET Optoelectr. **15**(2), 92–101 (2021). https://doi.org/10.1049/ote2.12017, https://ietresearch.onlinelibrary.wiley.com/doi/abs/10.1049/ote2.12017
10. Minkenberg, C., Krishnaswamy, R., Zilkie, A., Nelson, D.: Co-packaged datacenter optics: Opportunities and challenges. IET Optoelectr. **15**(2), 77–91 (2021). https://doi.org/10.1049/ote2.12020, https://ietresearch.onlinelibrary.wiley.com/doi/abs/10.1049/ote2.12020
11. Piehler, D.: Optical interconnects in enterprise and hyperscale datacenters. In: Schröder, H., Chen, R.T. (eds.) Optical Interconnects XX, vol. 11286, p. 1128602. International Society for Optics and Photonics, SPIE (2020). https://doi.org/10.1117/12.2550150
12. Shu, J., Qiu, C., Zhang, X., Xu, Q.: Efficient coupler between chip-level and board-level optical waveguides. Opt. Lett. **36**(18), 3614–3616 (2011). https://doi.org/10.1364/OL.36.003614, https://opg.optica.org/ol/abstract.cfm?URI=ol-36-18-3614
13. Soganci, I.M., Porta, A.L., Offrein, B.J.: Flip-chip optical couplers with scalable i/o count for silicon photonics. Opt. Express **21**(13), 16075–16085 (2013). https://doi.org/10.1364/OE.21.016075, https://opg.optica.org/oe/abstract.cfm?URI=oe-21-13-16075
14. Wang, X., He, X., Ren, H.: Advanced fec for 200 gb/s transceiver in 800 gbe and 1.6 tbe standard. IEEE Commun. Standards Mag. **7**(3), 56–62 (2023) https://doi.org/10.1109/MCOMSTD.0001.2200031

15. Weninger, D., et al.: High density vertical optical interconnects for passive assembly. Opt. Express **31**(2), 2816–2832 (2023). https://doi.org/10.1364/OE.475726, https://opg.optica.org/oe/abstract.cfm?URI=oe-31-2-2816
16. Zhang, Y., Shang, K., Zhang, Y., Ben Yoo, S.J.: Low-loss wafer-scale silicon photonic interposer utilizing inverse-taper coupler. In: 2018 IEEE Photonics Conference (IPC), pp. 1–2 (2018) https://doi.org/10.1109/IPCon.2018.8527256
17. Zhang, Y., Shang, K., Zhang, Y., Li, S., Lin, Y.C., Yoo, S.J.B.: Low-loss and broadband wafer-scale optical interposers for large-scale heterogeneous integration. Opt. Express **32**(1), 40–51 (2024). https://doi.org/10.1364/OE.504032, https://opg.optica.org/oe/abstract.cfm?URI=oe-32-1-40

Visible Photonics and Sensing

Visible-Light Optical Phased Arrays with a Convex Grating Emitter on Implantable Neural Probes for Spatially Targeted Deep Brain Optogenetics

Ankita Sharma[1,2(✉)], Fu-Der Chen[1,2], Alperen Govdeli[1,2], Xianshu Luo[3], Hongyao Chua[3], Guo-Qiang Lo[3], Wesley D. Sacher[1], and Joyce K. S. Poon[1,2(✉)]

[1] Max Planck Institute of Microstructure Physics, Weinberg 2, 06120 Halle, Germany
ank.sharma@mail.utoronto.ca, joyce.poon@mpi-halle.mpg.de

[2] Department of Electrical and Computer Engineering, University of Toronto, 10 King's College Road, Toronto, ON M5S 3G4, Canada

[3] Advanced Micro Foundry Pte Ltd., 11 Science Park Road, Singapore Science Park II, 117685 Singapore, Singapore

Abstract. We demonstrate single-lobe beam-steering at $\lambda = 460$–478 nm using an end-fire optical phased array coupled to a free-propagation region slab and convex grating for out-of-plane emission. Our design achieves sidelobe-free steering over 57% of the total maximum steering range 100 μm away from the OPA. The devices were implemented in silicon nitride waveguides and integrated on implantable silicon neural probes for targeted optogenetics.

Keywords: integrated optical phased arrays · visible-light · neural probes

1 Introduction

Combining optogenetic techniques with patterned light is important for selectively stimulating targeted areas of brain tissue to study neural circuits [1]. To effectively implement patterned photostimulation for optogenetics, it is crucial that light can reach tissues deep within the brain, necessitating technologies to overcome the attenuation length of light in brain tissue of about 1 mm [1, 2]. Implantable silicon neural probes with integrated SiN waveguides and grating emitters excel in delivering precisely tailored beam profiles deep within tissue compared to other minimally invasive implants like tapered multimode fibers and silicon probes with integrated μLEDs, which typically emit diverging beams [3, 4]. In our previous works we have developed neural probes with grating emitters that can produce low-divergence beams, focused beams, light sheets or steerable beams [5–10]. We created steerable beams using passive optical phased arrays (OPAs), which permitted the directional control of emitted beams within tissue by wavelength tuning [8, 9].

In the visible spectrum, due to fabrication constraints and inter-waveguide crosstalk innovative methods are needed to achieve single-lobe OPA emission [10]. In [8], we

J. Witzens et al. (Eds.): ECIO 2024, SPPHY 402, pp. 163–167, 2024.
https://doi.org/10.1007/978-3-031-63378-2_26

developed a neural-probe-compatible OPA using a 1D end-fire phased array that emitted into a free-propagation region (FPR) slab waveguide to eliminate side-lobes. A single beam was radiated out of the plane of the probe by a concave grating emitter at the end of the FPR. Our method allowed us to steer over 16° in the far-field and the photonic circuit occupied a large footprint of 0.018 mm^2 on the probe shank. In [9] using a straight instead of concave-curved grating, we were able to reduce the OPA footprint by 2.5x and extend sidelobe-free beam steering over a > 3x wider range in the far-field (~50°). However, at propagation distances close to the implanted OPAs ($\leq$100 μm) the improvement in the single-lobe angular steering range was minimal ($\leq$1.5x). Considering that the microelectrodes on silicon probes detect within ranges <140 μm, it is critical to enhance beam-steering at distances close to the implanted OPAs [11]. Here we report an OPA design featuring a novel slab grating emitter with a convex shape, enabling single-lobe light emission over an arc-length of 180.2 μm at a propagation distance of 100 μm; effectively doubling the single-lobe angular steering range at this distance compared to our previous works. Assuming the maximum steering range to be an arc-length corresponding to 180° (100π μm), our newest design achieves single-lobe beam steering over 57% of this total arc-length.

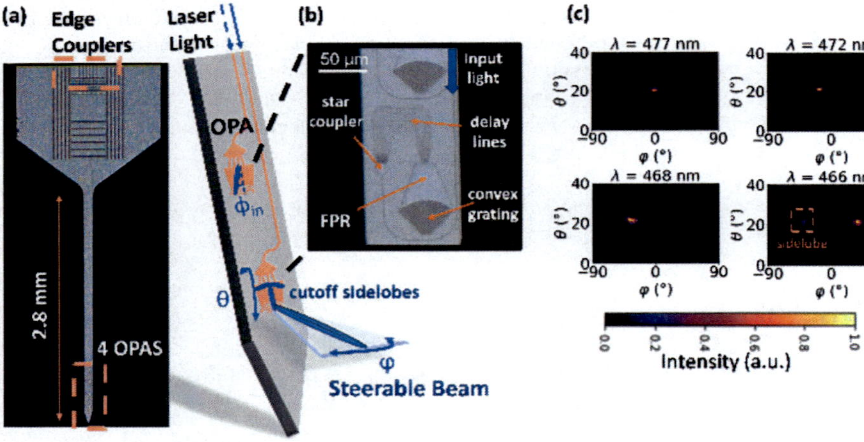

Fig. 1. (a) Optical micrograph of the neural probe and illustration of the OPA operation. A steerable beam forms in the-plane of the probe in the free-propagation region (FPR) and is then emitted out-of-plane by a grating. (b) Annotated optical micrograph of the OPA design. Beam-steering is enhanced because of the convex grating emitter. (c) 3D FDTD far-field simulations of the OPA design at various wavelengths. A single beam is emitted over most of the steering range.

2 Neural Probe Design and Fabrication

Figure 1(a) shows an implantable neural probe with 4 OPAs integrated on 103 μm-wide shank. Each device occupied an area of 0.011 mm^2. The 4 OPAs spanned ~0.6 mm along the shank's 2.8 mm length. The neural probes were fabricated at Advanced Micro

Foundry on 200 mm silicon (Si) wafers using the processes described in [5]. The photonic circuits were defined in a 150 nm thick silicon nitride (SiN) layer formed from plasma-enhanced chemical vapor deposition (PECVD). As outlined in Fig. 1(b), the OPAs employed a star coupler to distribute input light from a waveguide into 16 separate delay lines with a 6 μm differential path length. The delay lines converged to form an end-fire phased array with a pitch of 400 nm between waveguide emitters. The emission from the end-fire array was directed into an FPR slab, designed to eliminate sidelobes over most of the steering range by causing the undesired grating lobes to intersect with the slab sidewalls. Within the 40 μm-long FPR, a single steerable beam formed, which was radiated out of plane by the convex grating at the end of the slab (period = 480 nm, duty cycle = 50%).

In this design, beam-steering was enhanced by two distinct mechanisms: (1) the lateral translation of the beam in the FPR and (2) the dependence of the out-of-plane emission angle (φ) on the incident angle of the in-plane beam on the convex grating (ϕ_{in}). We verified the second mechanism using three dimensional (3D) finite difference time domain (FDTD) far-field simulations shown in Fig. 1(c). In simulation, the design achieved a wide single-lobe steering range of 72° in a medium with refractive index of n = 1.46, matching that of SiO_2; nevertheless with the tradeoff of an increased beam size (mean FWHM of 7.1°) and larger variation in beam profile. Nearing the ends of the steering range two lobes were present, for example at $\lambda = 466$ nm.

3 Characterization

To fully characterize the OPA design, the fabricated neural probes were attached with epoxy to a custom multi-core fiber. Each core in the fiber was addressed using a MEMS mirror-based laser scanning system, and connected to a single OPA as in [6]. The light source inputted to the scanning system was a supercontinuum laser with a tunable band-pass filter. We inserted a fiber-attached neural probe into fluorescein solution and positioned the probe to ensure that the beams were emitted parallel to the surface of the liquid.

Figure 2(a) shows top-down fluorescence images of the beam profiles at different wavelengths. The cross-sectional cuts in Fig. 2(b) are taken along an arc of 100 μm radius. These line cuts show that a single beam with a mean full width at half maximum (FWHM) of 35 μm ($\pm$9.7 μm) could be steered over a lateral distance of 180.2 μm at a distance of 100 μm from the OPA. This length corresponds to 57% of the total arc length. Over the 180.2 μm, the main-lobe-to-background ratio was $\geq$4 dB. In Fig. 2(c) and (d) we plot the single-lobe steering arc length and FWHM as a function of propagation distances between 50–140 μm from the OPA. The OPA emitted neuron-sized beams (FWHM of ~23–25 μm) 50–60 μm away from the inserted probe. Since microelectrodes on the neural probe can capture electrophysiology activity up to about 140 μm from the probe surface [11], in future designs, to improve the number of resolvable neurons, the OPA emission should be less divergent. In this way, the optical beam diameter is less than or equal to the average size of a neuron across all propagation distances $<$140 μm.

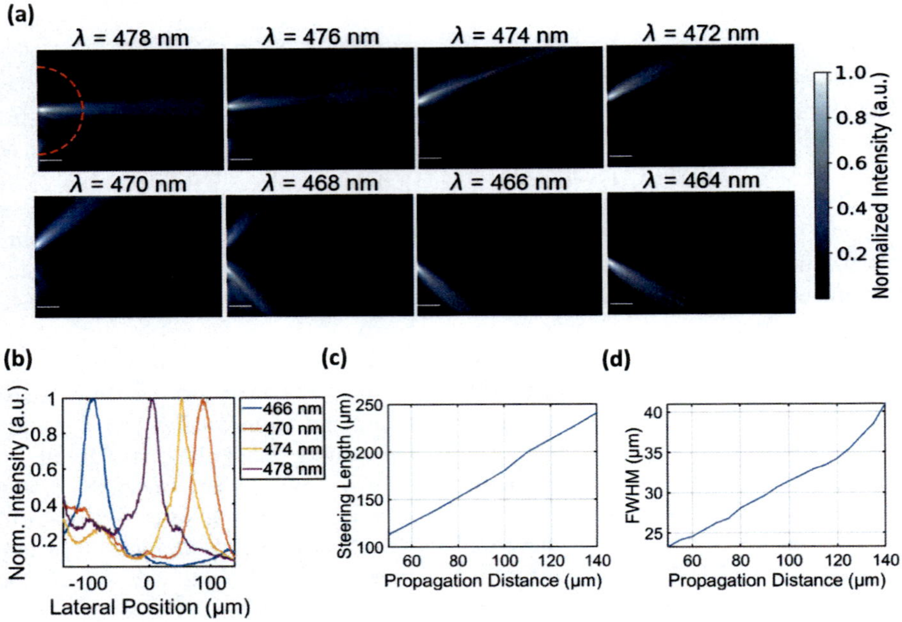

Fig. 2. (a) Top-down beam profiles measured in fluorescein solution at several wavelengths. The scale bars are 50 μm. The red dashed line corresponds to an arc of 100 μm centered on the OPA. (b) Cross-sectional cuts along the red dashed line in (a). (c) Single-lobe steering arc length as a function of propagation distance away from the OPA. (d) FWHM for the beam emitted at $\lambda = 478$ nm as a function of propagation distance away from the OPA.

4 Conclusion

In conclusion, we have demonstrated silicon neural probes with OPAs that through an innovative slab grating emitter design, can steer a single beam over 180.2 μm, 100 μm away from the OPA. This development in visible-light OPAs introduces new capabilities to silicon neural probes that could enable precisely targeted optogenetic stimulation.

References

1. Ronzitti, E., et al.: Recent advances in patterned photostimulation for optogenetics. J. Opt. **19**(11), 113001 (2017)
2. Emiliani, V., et al.: Optogenetics for light control of biological systems. Nat. Rev. Methods Primers **2**(55) (2022)
3. Pisanello, M., et al.: Tailoring light delivery for optogenetics by modal demultiplexing in tapered optical fibers. Sci. Rep. **8**(1), 4467 (2018)
4. Voroslakos, M., et al.: Hectorstar μLED optoelectrodes for large-scale, high precision in vivo opto-electrophysiology. Adv. Sci. **9**(18), 2105414 (2022)
5. Sacher, W.D., et al.: Visible-light silicon nitride waveguide devices and implantable neurophotonic probes on thinned 200 mm silicon wafers. Opt. Exp. **27**(26), 37400–37418 (2019)

6. Chen, F.D., et al.: Implantable neural probe system for patterned photostimulation and electrophysiology recording. In: CLEO 2022, JTh6A.7. Optica Publishing Group (2022)
7. Xue, T., et al.: Out-of-plane focusing grating on implantable neural probes for spatially targeted optogenetic stimulation. In: CLEO/Europe 2023, cl_1_3. Optica Publishing Group (2023)
8. Chen, F.D., et al.: Sidelobe-free beam steering using optical phased arrays for neural probes. In: CLEO 2021, SW3B.2. Optica Publishing Group (2021)
9. Sharma, A., et al.: Wide-angle single-lobe beam-steering using optical phased arrays on implantable neural probes. In: CLEO 2023, SF2E.5. Optica Publishing Group (2023)
10. Sacher, W.D., et al.: Optical phased array neural probes for beam-steering in brain tissue. Opt. Lett. **47**(5), 1073–1076 (2022)
11. Henze, D.A., et al.: Intracellular features predicted by extracellular recordings in the hippocampus in vivo. J. Neurophysiol. **84**(1), 390–400 (2000)

Focusing Optical Phased Array for Optically Enabled Probing of the Retina with Subcellular Resolution

Pedram Hosseini[1](✉), Prachi Agrawal[1], Alireza Tabatabaei Mashayekh[1], Sandra Johnen[2], Jeremy Witzens[1], and Florian Merget[1]

[1] Institute of Integrated Photonics, RWTH Aachen University, 52074 Aachen, Germany
phosseini@iph.rwth-aachen.de

[2] Department of Ophthalmology, University Hospital RWTH Aachen, 52074 Aachen, Germany

Abstract. We present a silicon-nitride-based optical phased array with built-in focusing and steering capability, that operates at 522 nm and is aimed at complementing a micro-electrode array for joint electrical and optical probing of retinal tissue. It achieves subcellular resolution with a beam diameter of 1.4 μm at a focal point located above the chip. Targeted cellular excitation can be achieved by steering the beam through a combination of wavelength tuning and simplified thermo-optical phase shifters with a single electrical input for each of transverse beam steering and selection of the focal plane.

Keywords: Optical phased arrays · silicon nitride photonic integrated circuits · bio-photonics

1 Introduction

Ex-vivo recording of electrical retinal activity performed with microelectrode arrays (MEAs) and time-varying blanket optical illumination [1] can be augmented by spatially selective optical excitation. The ability to target single photoreceptor cells appears a desirable characteristic for light delivery systems, together with the ability to span large fields of view (FOV) covering a substantial portion of the retina. The joint or sequenced excitation of selected photoreceptor cells would also be a powerful tool to gain new biological insights.

State-of-the-art approaches like micro-LEDs [2, 3] and optical holographic setups [4] have limitations, such as the dissipation of heat generated by micro-LEDs into the tissue or limited resolution and FOV. Moreover, focusing light onto the tissue on a micrometer scale with objective lenses results in bulky free space setups. Using a photonic integrated circuit (PIC) as a substrate for the MEA enables the co-integration of programmable optical emitters with a small footprint [5]. Beam shaping and steering can be obtained at the chip scale with optical phased arrays (OPAs) [6]. Even though OPAs have been intensely studied at longer wavelengths for free-space optical communication or light detection and ranging (LiDAR) [7, 8], there has been more limited research

J. Witzens et al. (Eds.): ECIO 2024, SPPHY 402, pp. 168–175, 2024.
https://doi.org/10.1007/978-3-031-63378-2_27

in the visible spectrum, which is more applicable to biophotonics applications such as optogenetics [9, 10] or flow cytometry [11]. The fabrication of PICs on silicon substrates with established industrial CMOS-manufacturing processes reduces the fabrication cost and process variability. The large transparency window and low losses of silicon nitride (SiN) technology makes it ideal for short wavelength applications [12, 13].

Surface emitting OPAs are promising integrated devices for replacing free-space beam shapers, that enable beam scanning [8] as well as advanced beam shaping, such as focusing [14, 15], multi-beam emission or patterned illumination, by emitting light from an array of emitters with programmable phases. These can be statically set, tuned with thermo-optic or micro-mechanical phase shifters, or set by optical delay lines and tuning of the light source wavelength [16]. Two-dimensional (2D) OPAs require a complex actuation system and a large number of control signals for phase tuning. Reducing the complexity of such systems is an active field of research [17]. For system simplification, using the wavelength dependency of the grating coupler (GC) emission angle can be used to steer the beam in one direction. Using a one-dimensional OPA with a combination of wavelength tuning and thermo-optical phase shifters leads to 2D beam steering with considerably less complexity [18].

Here, we present a one-dimensional (1D) phased array with built-in focusing capability fabricated in the SiN PIC platform (Sect. 2). Light can be steered by 3° in the longitudinal direction by means of input wavelength tuning in a range from 510 nm to 530 nm. In the transverse direction, along which the GCs are arranged, focusing is achieved above the chip, with good consistency with the simulation results. In Sect. 3, we describe a thermal tuner concept with supporting simulations, by which beam steering in the transverse direction and variation of the focal spot position can be each achieved with a single tuning signal.

2 Focusing One-Dimensional Optical Phased Array

2.1 Design

Figure 1(a) describes the architecture of the SiN-PIC-based focusing OPA. It is implemented in the double-stripe SiN waveguide platform from LioniX International BV with layer thicknesses optimized for operation in the visible range, deposited on an 8 μm buried oxide (BOX) layer [13, 19]. It consists of a core with a 50 nm silicon dioxide (SiO_2) layer sandwiched between two SiN layers that are 27 nm (bottom) and 80 nm (top) thick. Finally, the core is encapsulated within a 4 μm thick top cladding SiO_2 layer over which metal heaters are fabricated for phase tuning. The PIC comprises an edge coupler for light in-coupling from a laser source with a 520 nm central wavelength. The light is evenly distributed by a 1×16 passive light distribution network that employs 4 levels of identical 1×2 multi-mode interferometers (MMIs). These have a length of 15.3 μm, a width of 3 μm, and an access waveguide width of 0.8 μm (Fig. 1 (c)). At the output of the MMI, the waveguide centers are offset by $\pm$ 0.75 μ m relative to the center of the MMI. For optical focusing, delay lines of varying length are implemented in the resulting waveguide array (Fig. 1(a)), that introduce a phase delay depending on the waveguide index in a parabolic manner (Fig. 1(b)). The GCs, with a periodicity of 600 nm and a fill factor of 0.5, are uniformly spaced and arranged with a 2.4 μm pitch

(Fig. 1 (d)). Each grating emitter has a 2 μm width and 20 trenches that are etched in the top SiN layer only. A 10-μm long adiabatic taper prevents higher-order mode excitation while transitioning from a single-mode waveguide (0.4 μm, fully etched, with an effective index $n_{eff} = 1.68$) to the multi-mode waveguide width at the grating section. The GCs are simulated to scatter 51% of the light towards the top of the chip along an angle oriented 44° from the surface normal.

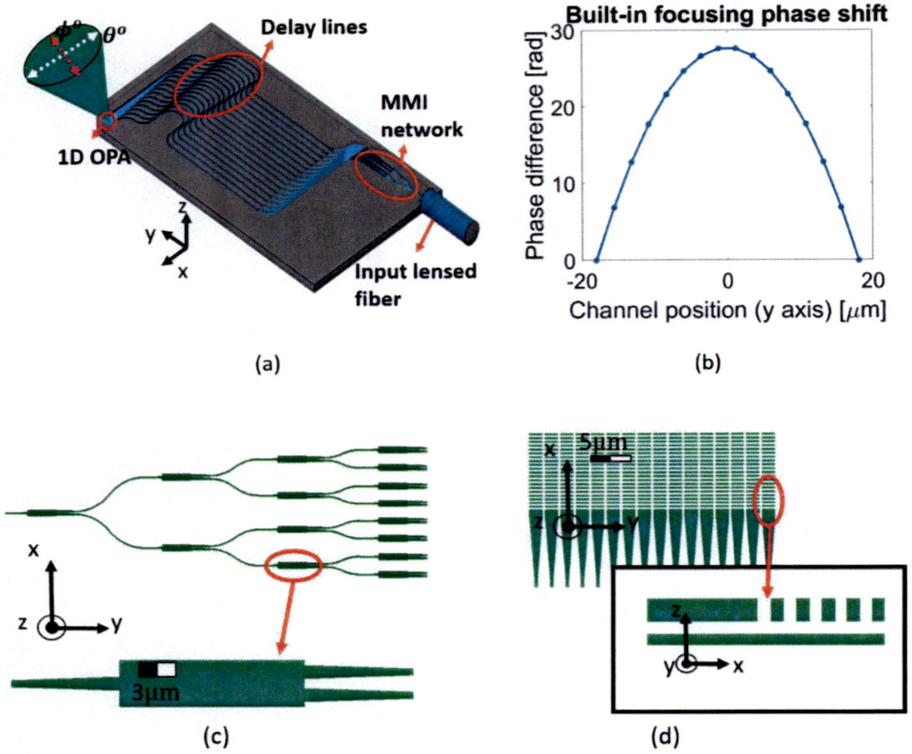

Fig. 1. Schematic of the designed 1D OPA with built-in focusing. It includes an edge coupler for in-coupling of the light, a 1 × 16 passive light distribution network based on 1 × 2 MMIs, delay lines for generating the phase delays required for focusing the light above the chip, and optical GC emitters. θ and ϕ describe the beam propagation in the longitudinal and transverse directions, respectively. (b) Phase delay induced in each of the OPA waveguide channels. (c) Layout of the passive light splitter network based on 1 × 2 MMIs. (d) Top and side views of the 1D OPA layout (including the layer stack of the double stripe SiN platform).

2.2 Simulation

Figure 2 depicts the simulated far-field emission of the OPA. The near-field of a single GC is first simulated and recorded by utilizing the 3D finite-difference time-domain

(FDTD) method. The field emitted at the surface of the PIC by the 1×16 GC array is then generated from it, by tiling the GC emission after applying the expected emission phase. A far-field transform is then applied in Fourier space to predict the emitted field profile at different heights above the chip. Three beams are generated with a spatial separation of 12.2° consistent with the GC pitch and a spacing of 15.3 μm from each other at the focal plane. The center beam and each of the two side-beams carry 38% and 26% of the total, top-emitted power, respectively. As expected from the built-in delay lines, the main beam focuses at a height of 50 μm above the surface of the chip (measured along z), that allows to accommodate the thickness of a biological sample, and propagates relative to the z-direction with the GC propagation angle of 44°, as depicted in Fig. 2(a). Figure 2(b) shows the illumination pattern along the (y, u_z) plane oriented along the direction of propagation of the beam (u_z) and Fig. 2(c) shows the illumination pattern in the (u_x, y) plane perpendicular to its direction of propagation, at the focal point of the beam. The full width at half maximum of the beam (FWHM), recorded along the y-direction along which it is focused, is 1.4 μm and is small enough to target a single cell. Along u_x the beam forms an extended sheet with a FWHM equal to 6.62 μm, and the FWHM of the beam along its direction of propagation (u_z) is 23 μm.

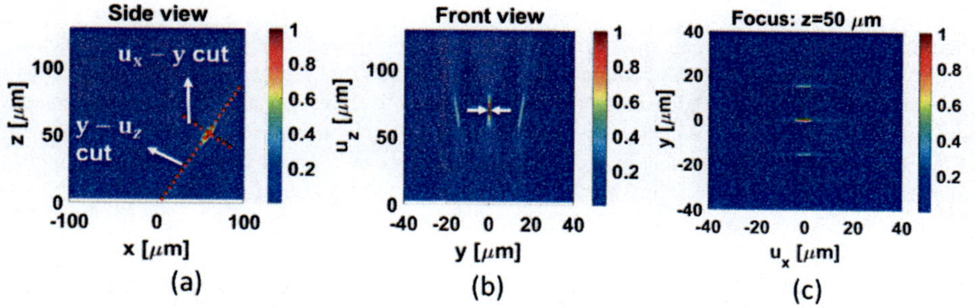

Fig. 2. Simulated field intensities resulting from 3D FDTD followed by a far-field projection at wavelength 520 nm. (a) Side view of the illumination pattern revealing the GC emission angle 44°; illumination patterns along (b) the (y, u_z) plane along the direction of propagation of the beam and (c) the (u_x, y) plane perpendicular to its direction of propagation.

2.3 Measurement and Optical Characterization

To characterize the OPA, we orient a CMOS camera equipped with an objective lens such that its optical axis coincides with the direction of propagation of the central beam. The position of the camera is then varied to record images of the out-coupled light at different distances above the surface of the PIC. Figure 3(a) depicts the field intensity recorded at a height at which the three beams reach their focal points. It reveals a FWHM in the direction of focusing (y) of less than 2 μm, limited by the resolution of the camera. The FWHM recorded along u_x is 7 μm, in good agreement with simulations. There is also a good consistency between the measured positions of the side lobes, at ±15.5 μm from the central beam, and the simulations.

Several such images are then recorded at different camera heights and stitched together to generate the intensity profile in the (y, u_z) plane shown in Fig. 3(b). In the (x, z) plane, the beam propagates with an angle of 38° relative to the surface normal. The 6° difference in the propagation angle is likely caused by fabrication tolerances or uncertainties in the measurement setup resulting from the alignment of the CMOS camera.

Wavelength dependent beam steering in the longitudinal direction was also measured in the wavelength range 510 nm–530 nm and was found to be 3°, in good agreement with simulations.

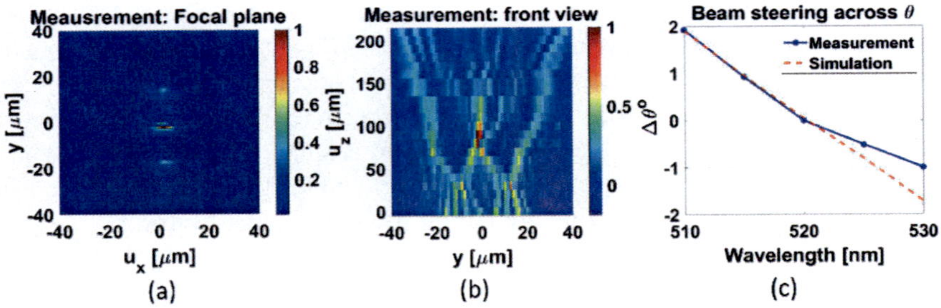

Fig. 3. Beams imaged with the optical axis of the camera aligned in the direction of propagation of the central beam. In (a), the camera was positioned such that the beam profiles in the focal plane of the PIC are projected onto the imager by its objective. Several such recordings are taken at different camera heights, recentered and stitched together to obtain the front view profile shown in (b). (c) Steering of the beams in the longitudinal direction (θ) as a function of wavelength, relative to the angle of propagation at 520 nm. (a) and (b) were recorded at 522 nm.

3 Thermo-Optical Phase Shifter for Beam Steering

Incorporating thermo-optical phase shifters into the PIC enables the application of a tunable linear phase shift across the OPA and thus transverse beam steering (ϕ) in addition to the longitudinal beam steering (θ) achieved through wavelength tuning. However, equipping each channel with a separate heater requires a large number of control signals and significantly enlarges the device footprint once the required pads are included.

Consequently, adopting a single heating element capable of inducing a linear phase shift across the waveguide array with a reduced number of control inputs significantly increases the practicability of the system [20]. Figure 4(a) shows a meandered heater configuration that allows applying a linear phase shift across the waveguide array with a single electrical current and a single pair of pads. To steer the light in both directions, that require linear phase shifts with increments of opposite signs, two such meandered heaters are implemented, with the top one decreasing the phase shift and the bottom one increasing the phase shift as one moves down the waveguide array.

These meandered phase tuners are simulated in a 3D model considering the 4.3 Ω/sq sheet resistance of the heater layer, resulting in a 270 Ω resistance for their 20 μm width and 1.25 mm length. Cascading several such heater elements along the length of the waveguide array, larger cumulative deflections can be achieved. Considering two cascaded elements and applying a 55-mA current in each, the beam can be deflected by 3° taking the thermo-optic coefficients of SiN and SiO_2 into account, see Fig. 4(b).

A similar concept can be used to apply a tunable parabolic phase shift across the waveguide array, in addition to the one that is already passively set, in order to tune the height at which the beam reaches its focal spot (Fig. 5(a)). This meandered heater has a length of 1.12 mm and a resistance of 241 Ω. Cascading again two such elements and applying a 50 mA current to each shifts the focal spot from the passively built-in 50 μm height (measured along the z-direction) to 38 μm. Considering the direction of propagation of the beam, this corresponds to a displacement of 16.7 μm along u_z, which substantially exceeds the half width at half maximum of the beam in that direction. Adding meandered heaters with the opposite convexity would allow tuning the focal spot position in both directions.

Experimental results for these meandered heaters will be reported in a later publication.

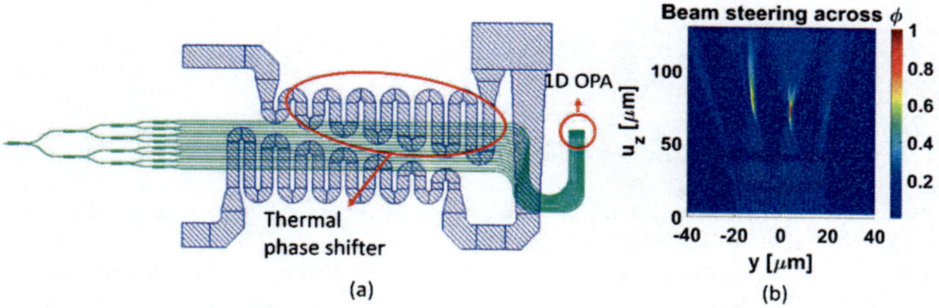

Fig. 4. (a) Meandered phase tuner design for linear phase shifting. (b) Simulation result after applying 55 mA current over two cascaded 1.55 mm long meandered heaters.

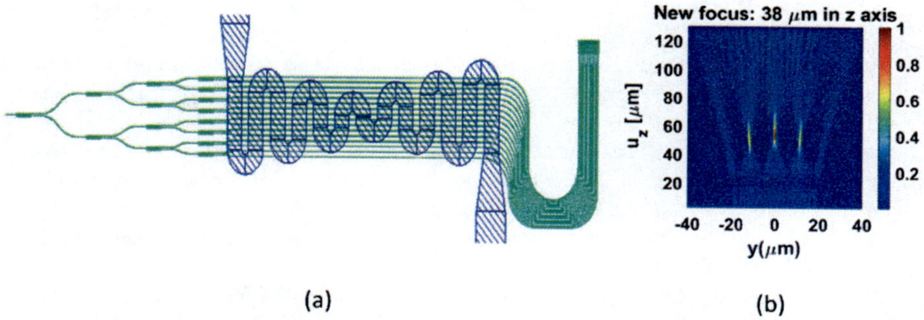

Fig. 5. (a) Meandered phase tuner design for shifting of the focal plane. (b) Simulation result after applying 50 mA current over two cascaded 1.12 mm long meandered heaters.

4 Conclusion

In this work, we experimentally demonstrate a focusing 1D optical phased array enabling the optical excitation of single cells located above the surface of the chip, with a spot FWHM of ~1.4 μm in the transverse direction along which the grating emitters are arranged. The PIC is implemented in SiN technology and aimed at simultaneous optical excitation of single photoreceptor cells and electrical recording in ex-vivo setups for the analysis of retinal activity. Three degrees of beam steering are achieved in the longitudinal direction by tuning the input wavelength between 510 nm and 530 nm, enabling targeted cell addressability. A concept for compact and easy to handle thermal actuation of beam steering in the transverse direction and shifting of the focal plane position was simulated. Meandered heater configurations allow the implementation of phase profiles of predetermined shape and tunable overall amplitude across the OPA, allowing steering or variable focusing with a single control signal. This significantly decreases the OPA's complexity, while enabling steering of the interrogation point in all three directions.

Acknowledgements. Funded by the Deutsche Forschungsgemeinschaft (DFG) under Research Training Network 2610 "InnoRetVision," project B3.

References

1. Kireev, D., et al.: N^3-MEA probes: scooping neuronal networks. Front. Nanosci. **13**, Article no. 320 (2019)
2. Ji, Y., et al.: An in vivo and in vitro photopolar array for precision optogenetic stimulation. AIP Adv. **13**(8), Article no. 085202 (2023)
3. Pisano, F., et al.: Large scale matching of function to the genetic identity of retinal ganglion cells. Sci. Rep. **7**(1), Article no. 15395 (2017)
4. Goetz, G.A., et al.: Holographic display system for restoration of sight to the blind. J. Neural Eng. **10**(5), Article no. 056021 (2013)
5. Welkenhuysen, M., et al.: An integrated multi-electrode optrode array for in vitro optogenetics. Sci. Rep. **6**, Article no. 056021 20353 (2016)
6. Heck, M.J.R.: Highly integrated optical phased arrays: photonic integrated circuits for optical beam shaping and beam steering. Nanophotonics **6**(1), 93–107 (2017)
7. Bhandari, B., et al.: Dispersive silicon–nitride optical phased array incorporating arrayed waveguide delay lines for passive line beam scanning. Sci. Rep. **12**(1), Article no. 18759 (2022)
8. Poulton, C.V., et al.: Optical phased array with small spot size, high steering range and grouped cascaded phase shifters. In: Integrated Photonics Research, Silicon and Nanophotonics, IW1B.2, Vancouver, Canada. Optica Publishing Group (2016)
9. Xue, T., et al.: Implantable photonic neural probes with out-of-plane focusing grating emitters. arXiv:2401.04525 (2024)
10. Mohanty, A., et al.: Reconfigurable nanophotonic silicon probes for sub-millisecond deep-brain optical stimulation. Nat. Biomed. Eng. **4**, 223–231 (2020)
11. Tabatabaei Mashayekh, A., et al.: Multi-color flow cytometer with PIC-based structured illumination. Biomed. Opt. Express **13**, 5098–5115 (2022)
12. Romero-García, S., et al.: Silicon nitride CMOS-compatible platform for integrated photonics applications at visible wavelengths. Opt. Express **21**(12), 14036–14046 (2013)

13. Roeloffzen, C.G.H., et al.: Low-loss Si_3N_4 triplex optical waveguides: technology and applications overview. J. Lightwave Technol. **24**(4), Article no. 4400321 (2018)
14. Notaros, J., et al.: Near-field-focusing integrated optical phased arrays. J. Lightwave Technol. **36**(24), 5912–5920 (2018)
15. Sacher, W.D., et al.: Optical phased array neural probes for beam-steering in brain tissue. Opt. Lett. **47**(5), 1073–1076 (2022)
16. Sharma, A., et al.: Wide-angle single-lobe beam-steering using optical phased arrays on implantable neural probes. In: 2023 Conference on Lasers and Electro-Optics (CLEO), SF2E.5. IEEE (2023)
17. Agrawal, P., et al.: Optical phased array with 2D beam steering based on nonlinear beam mixing. Opt. Lett. **47**(16), 3996–3999 (2022)
18. Wang, Q., et al.: Silicon nitride assisted 1×64 optical phased array based on a SOI platform. Opt. Express **29**, 10509–10517 (2021)
19. Tabatabaei Mashayekh, A., et al.: Silicon nitride PIC-based multi-color laser engines for life science applications. Opt. Express **29**(6), 8635–8653 (2021)
20. Rabinovich, W.S., et al.: Free space optical communication link using a silicon photonic optical phased array. In: Proceedings of the SPIE 9354, Article no. 93540B (2015)

Beyond the Free Spectral Range: On-Chip Spectrometer with Multi-color Cascaded Colloidal Quantum-Dot Photodiodes

Chao Pang[1,2,3](✉), Raúl López March[1], Ezat Kheradmand[2,3], Yu-Hao Deng[2,3], Luis Moreno Hagelsieb[4], Lukas Elsinger[1,2,3], David Cheyns[4], Pieter Geiregat[2,3], Zeger Hens[2,3], and Dries Van Thourhout[1,2]

[1] Photonics Research Group, Ghent University - imec, 9052 Gent, Belgium
Chao.Pang@Ugent.be
[2] NB Photonics, Ghent University, 9052 Gent, Belgium
[3] Physics and Chemistry of Nanostructures Group, Ghent University, 9000 Gent, Belgium
[4] IMEC, 3001 Leuven, Belgium

Abstract. Colloidal quantum dots (QDs), as a low-cost, flexibly tunable and fabrication scalable semiconductor material, have demonstrated exceptional capabilities in many complex applications including high-quality displays, high-resolution infrared imaging, advanced spectrometry and integration into photonic circuits. This study explores combining the tunable spectral response features of QD-based photodiodes (QDPDs) with dispersive photonic integrated circuits (PICs), demonstrating a spectrometer with operational spectral range extended beyond its original free spectral range (FSR). Experimentally, two types of PbS QDPDs, with different absorption features, were integrated in cascade on the output channels of a planar concave grating (PCG) with a 90 nm FSR. The differential responses of these QDPDs to two adjacent diffraction orders of the PCG enabled the creation of a spectrometer with a spectral range of approximately 180 nm, effectively decoupling two FSRs of the PCG. The proposed cascaded QDPDs, with diverse spectral photodetection capabilities, present great potential when integrated into complex optical systems.

Keywords: Photonic integrated circuits · colloidal quantum dots · spectrometer

1 Introduction

While complex devices such as arrayed waveguide grating (AWG) or planar concave gratings (PCG) are commonly used in silicon photonics as dispersive components [1–4] for on-chip spectrometers, a fundamental trade-off exists between their spectral range and spectral resolution [5]. Achieving better spectral resolution within a similar footprint requires larger optical dispersion, typically implying a larger diffraction order and, consequently, a smaller FSR and a narrower spectral range. This traditional trade-off persists because a single photodetector is used to detect light in each channel, making it challenging to distinguish optical signals from adjacent diffraction orders. If we place

J. Witzens et al. (Eds.): ECIO 2024, SPPHY 402, pp. 176–181, 2024.
https://doi.org/10.1007/978-3-031-63378-2_28

multiple, different types of photodetectors on each channel in cascade, each with distinct responses to different diffraction orders, we are able to separate signals from more than one FSR, thereby eliminating the aforementioned trade-off. The flexibility in absorption tuning offered by QDs positions them as promising candidates for implementing this approach.

In this work, we propose and demonstrate the integration of the tunable features of QDs with dispersive PICs, pushing the spectral range beyond a single FSR. In our demonstration, we integrated two distinct types of QDs with excitonic absorption peaks centered around the wavelengths of two adjacent diffraction orders of the PCG. These QDs were integrated as waveguide-coupled photodiodes in cascade on the output channels of the PCG, each exhibiting a unique spectral response. By carefully designing the response of the QDPDs, we demonstrated a spectral range of approximately 180 nm using an eight-channel PCG with a FSR of 90 nm, breaking the FSR limit of the spectrometer.

2 Principle

In an optical dispersive component, such as a PCG, light with different wavelengths is deflected to distinct angles and collected by separate channels, as shown in Fig. 1(a). However, each channel contains light from different diffraction orders, separated by the FSR. In conventional dispersive spectrometers, a single photodetector is placed on each channel, converting optical power into an electrical signal for readout, as shown in Fig. 1(b). The presence of a single photodetector on each channel implies that optical signals from different diffraction orders are indistinguishable, limiting the working spectral range to one FSR.

Contrastingly, by incorporating multiple photodetectors with distinct wavelength responses for each channel, as shown in Fig. 1(c), different responsivities at various diffraction orders are achieved. This innovative approach provides additional information, enabling the separation of signals from different diffraction orders. In detail, considering an input signal with a power spectral density of $P_{in}(\lambda)$, the power density measured by the PCG is discretized into $P_{in}(\lambda_{i,k})$, where $\lambda_{i,k}$ is the peak transmission wavelength of channel i within diffraction order k. For two diffraction orders and two cascaded QDPDs at channel i, the power density at each order is linked to the photocurrents of these QDPDs by a 2×2 matrix $\boldsymbol{M_i}$:

$$\begin{bmatrix} I_{i,1} \\ I_{i,2} \end{bmatrix} = \begin{bmatrix} m_{i,11} & m_{i,12} \\ m_{i,21} & m_{i,22} \end{bmatrix} \begin{bmatrix} P_{in}(\lambda_{i,1}) \\ P_{in}(\lambda_{i,2}) \end{bmatrix} \tag{1}$$

Here, $I_{i,j}$ is the photocurrent measured by QDPD of type j at channel i. For any unknown input spectrum, we can retrieve it by measuring $I_{i,j}$ and solving this linear problem. The matrix elements $m_{i,jk}$ are determined by the channel responsivity $R_{i,j}(\lambda)$ of QDPD j around the peak transmission wavelength $\lambda_{i,k}$ of channel i, diffraction order k:

$$m_{i,jk} = \int_{\lambda_{i,k}-\frac{\Delta}{2}}^{\lambda_{i,k}+\frac{\Delta}{2}} R_{i,j}(\lambda) d\lambda \tag{2a}$$

$$R_{i,j}(\lambda) = R_{cas,j}(\lambda) T_{PCG_i}(\lambda) \tag{2b}$$

Here, $R_{cas,j}(\lambda)$ is the responsivity of the cascaded QDPD of type j (Fig. 1(c)), $T_{PCG_i}(\lambda)$ is the transmission spectrum of channel i of the PCG, Δ is the width of the channel spectral responsivity peak, as shown in Fig. 1(c). In practice, $R_{i,j}(\lambda)$ can be measured by sweeping a monochromatic light source and recording the photocurrent of each QDPD.

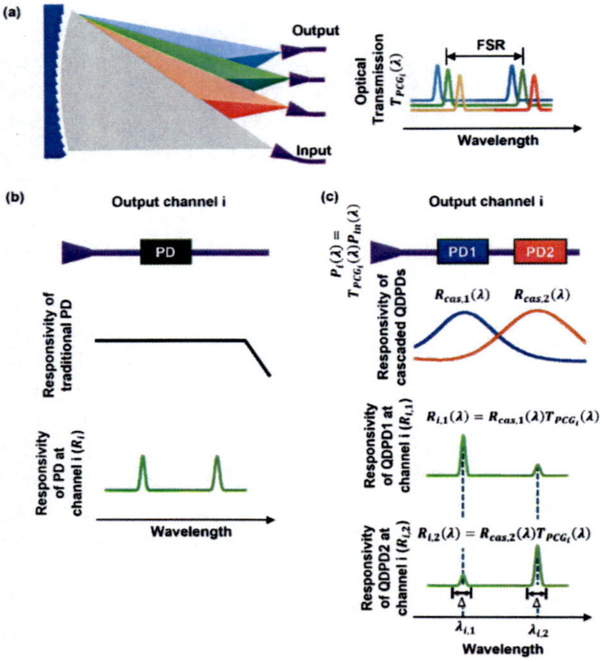

Fig. 1. Principle of breaking the FSR limit. (a) Planar concave grating and optical transmission at different channels. (b) Traditional photodetection solution for a spectrometer. Signals from different diffraction orders are mixed, so working range is limited to one FSR. (c) Cascaded QDPDs for photodetection. Different QDPDs have different response to different diffraction orders, extra information is achieved for separate signals beyond single FSR.

3 Experimental Results

The fabrication took place in the cleanroom facilities of Ghent University. Initially, the PCG, excluding the reflectors, was fabricated on a 300 nm SiN layer deposited atop 3 μm buried oxide substrates, utilizing E-beam lithography (EBL) and reactive ion etching (RIE) dry etching. Subsequently, another round of EBL and RIE was employed to define etched facets, serving as retroreflectors. A metal coating of 5 nm Ti/100 nm Au/5 nm Ti was then evaporated onto the facets, enhancing reflection and reducing insertion loss. The remaining metal on the chip was removed through lift-off techniques. The chip was then planarized with a 370 nm flowable oxide (hydrogen silsesquioxane,

HSQ) as a top cladding, providing a smooth interface for subsequent integration. This cladding also reduced optical power density in the QDPD integrated on top, mitigating power saturation and improving the linear response range.

Next, the QDPD stacks were integrated, starting from the planarized waveguide, as depicted in Fig. 2(a). Initially, a 20 nm thick indium tin oxide (ITO) layer was sputtered as the bottom electrode, followed by HCl-based wet etching to confine the ITO to the target QDPD regions as shown in Fig. 2(b). Subsequently, a 50 nm ZnO layer was deposited using sol-gel chemistry, patterned with dilute HCl solution, as shown in Fig. 2(c). We then place 20 nm Ti/100 nm Au n-contact pads on the side using the lift-off technique, as shown in Fig. 2(d). On top of the ZnO layer, we placed QD stacks for QDPD 1 and QDPD 2 using the lift-off method sequentially, as shown in Fig. 2(e) and (f). For QDPD 1, PbS QDs with a band-gap transition of 1250 nm were spin-coated as the absorption layer. This layer was treated with tetra-n-butylammonium iodide (TBAI) in a methanol solution to remove as-synthesized long surface ligands, turning the film n-type and enhancing the carrier mobility. On top, PbS QDs with a 940 nm band-gap transition were spin-coated and treated with ethanedithiol (EDT) in methanol solution for p-type doping. Layers of 1250 nm PbS-TBAI QDs and 940 nm PbS-EDT QDs were deposited, reaching a thickness of 60 nm each, and then lifted off simultaneously to achieve the desired pattern. QDPD 2 was processed similarly to QDPD 1, but with a 1355 nm transition bandgap for the absorption layer. Finally, 100 nm Au was placed on both types of QD stacks as p-contact pads. In the fabricated QDPDs, the ZnO/PbS-TBAI/PbS-EDT structure forms an n-i-p hetero-junction, aiding in carrier extraction and

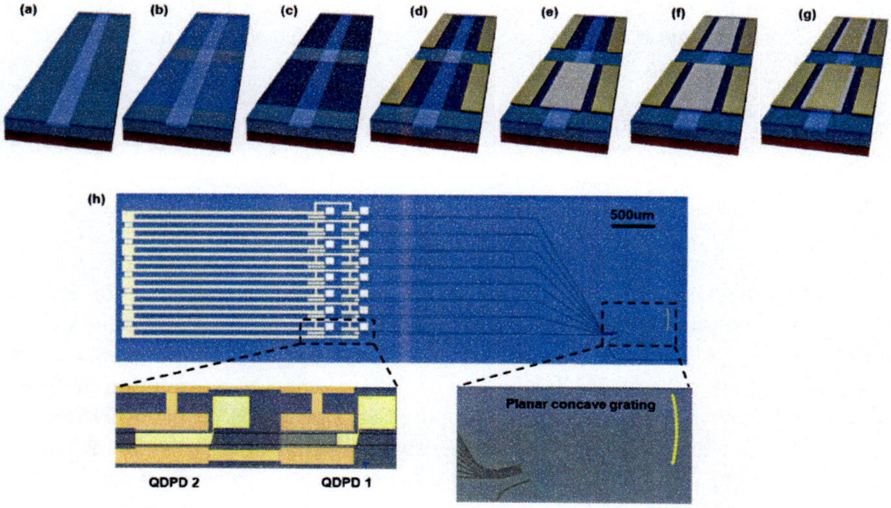

Fig. 2. Integration steps to fabricate cascaded QDPDs. (a) Optical passive structures fabrication and planarization. (b) ITO sputtering and patterning with wet etching. (c) ZnO deposition with solgel method and patterning with wet etching. (d) N-contact evaporation and pattering with liftoff. (e) Type 1 QDPD deposition and patterning with liftoff. (f) Type 2 QDPD deposition and patterning with liftoff. (g) P-contact evaporation and patterning with liftoff. (h) Top view of the fabricated spectrometer.

reducing dark current. The completed spectrometer, shown in Fig. 2 (h), features eight channels, each equipped with two cascaded QDPDs. All n-contacts were interconnected as a common ground to simplify measurements.

The wavelength dependent channel responsivity $R_{i,j}(\lambda)$ for both types of QDPDs was measured with a tunable laser, as shown in Fig. 3(a) and (b). As expected, QDPD 1 exhibits a stronger response to order 1, while QDPD 2 shows a more pronounced response to order 2. The capability of the spectrometer was demonstrated using amplified spontaneous emission (ASE) from an O-band semiconductor amplifier. Photocurrents were measured on both QDPDs at all eight channels for two different ASE spectra. With the 16 measured photocurrents and the transmission matrix M_i, input spectra were reconstructed, fitting well with those obtained from commercial benchtop spectrometers (Fig. 3(c)). Notably, both test spectra covered two FSRs of the PCG, making conventional spectrometers incapable of accurately reading the input spectrum.

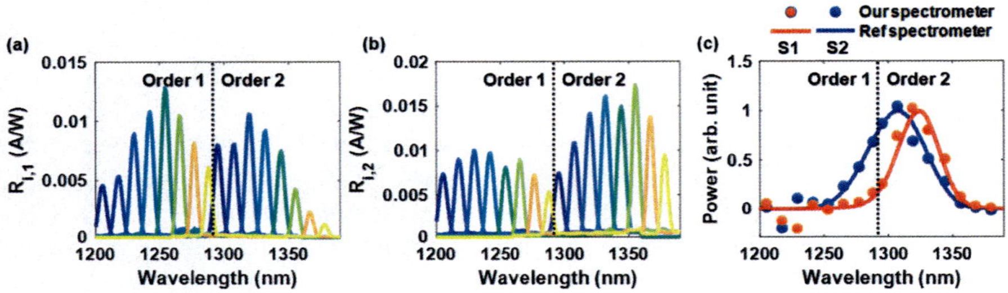

Fig. 3. Spectrum reconstruction. (a) and (b) Channel spectral responsivity of QDPD 1 and QDPD 2. (c) Reconstructed spectrum of spectrum S1 and S2 and their reference spectrum measured from a commercial spectrometer.

4 Conclusion

In this work, we have demonstrated an integrated spectrometer scheme using a PCG and cascaded QDPDs on a SiN platform. Leveraging the different wavelength response within the two consecutive QDPDs, we successfully decoupled two diffraction orders of the PCG. Our implementation achieved a broad operational range of approximately 180 nm on an eight-channel PCG with a 90 nm FSR, achieving a resolution of 12 nm. The tunable wavelength response of the proposed cascaded QDPDs opens avenues for their integration with various optical configurations, promising enhanced resolution, broader bandwidth, and increased robustness for on-chip computational spectroscopy applications.

References

1. Brouckaert, J., Bogaerts, W., Selvaraja, S., Dumon, P., Baets, R., Van Thourhout, D.: Planar concave grating demultiplexer with high reflective Bragg reflector facets. IEEE Photonics Technol. Lett. **20**, 309–311 (2008)
2. Brouckaert, J., Bogaerts, W., Dumon, P., Van Thourhout, D., Baets, R.: Planar concave grating demultiplexer fabricated on a nanophotonic silicon-on-insulator platform. J. Lightwave Technol. **25**, 1269–1275 (2007)
3. Pathak, S., Van Thourhout, D., Bogaerts, W.: Design trade-offs for silicon-on-insulator-based AWGs for (de)multiplexer applications. Opt. Lett. **38**, 2961 (2013)
4. Bogaerts, W., et al.: Silicon-on-insulator spectral filters fabricated with CMOS technology. IEEE J. Sel. Top. Quantum Electron. **16**, 33–44 (2010)
5. Kyotoku, B.B.C., Chen, L., Lipson, M.: Sub-Nm resolution cavity enhanced microspectrometer (2010)

Characterization of Grating Out-Couplers in the Ultraviolet-C Wavelength Range for On-Chip Spectroscopy

Chenming Su[1,2](✉) and Nicolas Le Thomas[1,2]

[1] Photonics Research Group, INTEC Department, Ghent University-imec, Technologiepark-Zwijnaarde, 9052 Ghent, Belgium
Chenming.Su@Ugent.be

[2] Center for Nano- and Biophotonics, Ghent University, Ghent, Belgium

Abstract. On-chip optical spectroscopy has been applied from visible to mid-infrared wavelength, but not yet in the ultraviolet-C (UVC, λ = 200 – 280 nm) spectral region. Since most of the biomolecules exhibit strong absorption in the UVC, on-chip spectral analysis of their autofluorescence excited by an UVC pump is envisioned to be promising for healthcare applications. Here, we report the optical far-field characterization of the first grating out-coupler operating in the UVC region (λ = 266 nm). The simulated effective index of the fundamental mode propagating in the grating is in line with the retrieved experimental one.

Keywords: Integrated Photonic Circuits · Ultraviolet-C · Biosensing · Grating Coupler

1 Introduction

Ultraviolet light, especially the ultraviolet-C region (UVC, λ = 200 – 280 nm), is particularly interesting for biosensing [1] and pharmaceutical applications [2] due to its unique interaction with biomolecules. When excited in the UVC spectral region, most biomolecules exhibit strong absorption and some of them even produce auto-fluorescence at longer wavelengths. This phenomenon is notable in amino acids such as tryptophan, tyrosine, and phenylalanine, which absorb strongly at approximately 260 nm [3]. These three amino acids are also characterized by autofluorescence peaks in the UVA (λ = 300 – 400 nm) wavelength range, for instance the tryptophan peak is located at around 360 nm [4]. To analyze the autofluorescence spectrum of biomolecules, the utilization of integrated photonic circuits (PICs) presents a compelling approach due to their enhanced compactness, sensitivity, resolution, and reduced power consumption. Our research leverages our newly low-loss suspended SiO_x waveguide in the UVC region in view of analyzing the spectral signature of biomolecules located near the waveguide. As equipment such as tunable lasers, high numerical aperture (NA) microscope objectives, and highly sensitive detectors are difficult to find in the UVC wavelength range, our methodology prioritizes achieving the best detection limit by employing a low numerical aperture (NA = 0.12) UV-objective and a low-cost UV-sensitive camera (340UV-USB, Thorlabs).

J. Witzens et al. (Eds.): ECIO 2024, SPPHY 402, pp. 182–186, 2024.
https://doi.org/10.1007/978-3-031-63378-2_29

Figure 1 shows the overall design of our integrated sensing approach. The waveguide spiral is intended to efficiently collect the autofluorescence of the biomolecules by offering a long interaction length with a single mode propagating in the waveguide. The single mode nature is important to enable a simple mapping between the angular spectrum of the fluorescence signal that is diffracted out-plane by the grating out-coupler and its frequency spectrum. As a first step, we investigate here the diffraction properties of the grating coupler by imaging the diffracted signal in the Fourier space.

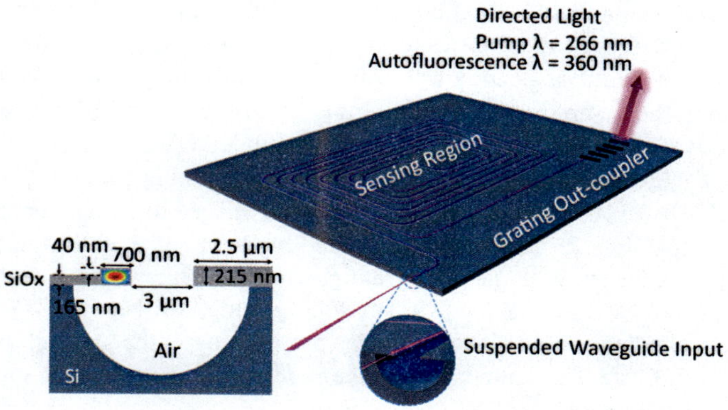

Fig. 1. Schematics of the suspended SiO_x waveguide with sensing region and grating out-coupler. The inset depicts the size of the suspended waveguide.

2 Result

We focus on engineering grating coupler for the pump light ($\lambda = 266$ nm). The suspended waveguides are patterned by a two-step electron beam lithography (EBL) and etched by an inductively coupled plasma-reactive ion etching (ICP-RIE) tool. Utilizing four-inch wafers with 300 nm thermally oxidized SiO_x on silicon substrate, the first EBL patterning and anisotropic ICP-RIE etching generate a 3-μm-wide trench, while the second EBL step is used to produce a shallow etch that defines the waveguide profile. The silicon below the waveguide is isotopically etched away through the trench generated by the first step, leaving a partially suspended SiO_x waveguide. The waveguide width is chosen to be 700 nm, which is the largest width ensuring single mode propagation at the wavelength of 266 nm.

The optimal pitch (Λ) of the grating coupler is determined from the phase matching condition between the mode guided in the grating with a wave vector $\boldsymbol{k}_g$ and the field diffracted in the far-field with a wave vector $\boldsymbol{k}_0$ at a diffraction angle θ with respect to the vertical direction:

$$k_g = k_0 \sin\theta + m \cdot G \tag{1}$$

where m is the diffraction order m and G is the modulus of reciprocal vector of the grating. For the first diffraction order $m = 1$, and using $G = 2\pi/\Lambda$, $k_0 = 2\pi/\lambda$ and

$k_g = (2\pi/\lambda)n_{eff}$ with n_{eff} is the effective index of mode propagating in the grating, the expression to retrieve the pitch from Eq. (1) is given by $\Lambda = \frac{\lambda}{n_{eff} - \sin\theta}$.

Utilizing a UV objective with NA = 0.12 and n_{eff} values obtained from simulation (MODE solver in Lumerical software), we processed gratings with pitches of $\Lambda = 200$ nm. The etch depth and the waveguide height are both 40 nm. The duty cycle of the grating is 0.3, i.e. the non-etched part corresponds to 30% of the unit cell of the grating.

Figure 2 (a) and (d) show the scanning electron microscope (SEM) images of the two fabricated grating couplers. The under-etching boundaries are distinguishable and annotated as SiO_x/Air and SiO_x /Si. The schematics in Fig. 2 (b) and (e) help to illustrate the two different designs, where the taper in Fig. 2 (b) has currently to be asymmetric to fit in the design of the suspended waveguide and fabrication constrains. For optical characterization, the pump laser ($\lambda = 266$ nm) is TE polarized and edge-coupled to the input of the waveguide via an aspherical lens. The coupled UVC light propagates along the spiral waveguide and is diffracted in free space by the grating coupler. The diffracted light is collected by the UV objective (NA = 0.12) and imaged by the camera in a Fourier plane. As shown in Fig. 2 (c), a single diffracted mode is collected in the aperture of the collecting lens for the 700 nm wide grating, while four lines corresponding to four different modes are present for the 1.5 μm wide grating, see Fig. 2 (f). Considering that the effective indices of the different modes propagating in the gratings are given by $n_{eff} = \sin\theta + \lambda/\Lambda$, the experimental effective index of the Bloch mode guided in the grating coupler is given by:

$$n_{eff} = \frac{k_y}{k_{max}} \cdot NA_{max} + \frac{\lambda}{\Lambda} = \frac{N_y}{N_{max}} \cdot NA_{max} + \frac{\lambda}{\Lambda} \tag{2}$$

Where k_y is the propagation constant for a given peak in the Fourier space in y direction, k_{max} is the largest propagation constant for the given NA = 0.12, N_y is the pixel number from the k_y peak to the center of the exit pupil (red circle) and N_{max} is the pixel number corresponding to the radius of the exit pupil.

As a result, the measured and simulated effective index of the single mode propagating in the 700 nm wide grating are in good agreement with values equal to 1.409 and 1.406, respectively. As regard the 1.5 μm wide grating, the angular spectrum reveals several peaks labelled P_i (i = 1 to 4). The value of the effective index associated with P_1 is 1.409 that is close to the simulated one (1.408) for the fundamental TE mode. This favors P_1 as being the peak associated with the fundamental TE mode. The simulated effective index values of the first and second higher order modes are 1.406 and 1.403, respectively. These values are far from those corresponding to P_2 and P_3, namely 1.383 and 1.358 respectively. It seems that the asymmetry of the taper that links the 700 nm wide waveguide to the 1.5 μm wide grating does not induce the excitation of higher order mode. The value 1.436 of the effective index of P_4 is larger than that of a mode propagating in the waveguide with similar dimensions as the grating (1.424). The origin of the P_2, P_3 and P_4 peaks are currently unclear, but propagating plasmon modes at the silicon sidewall might be an explanation for their presence.

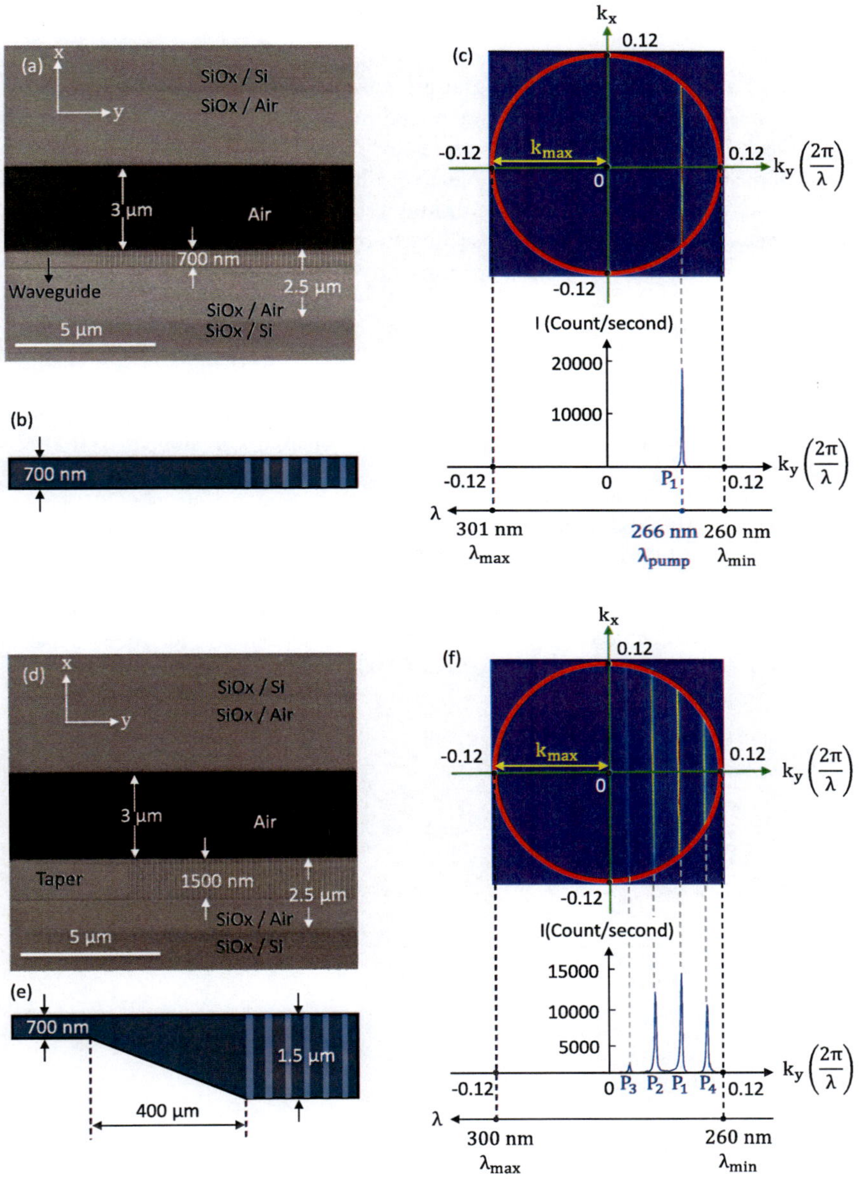

Fig. 2. (a) SEM images of the fabricated 700 nm-wide grating out-coupler, which is also depicted in the schematics (not to scale) of the design in (b). (d) and (e) are the SEM image and schematics (not to scale) for the fabricated 1500 nm-wide grating out-coupler with a taper. (c) and (f): images in the Fourier space for the two grating couplers. The position of peaks in the Fourier plane correspond to different wave numbers k_y in the y direction. The operating wavelength ranges are indicated for the two grating coupler designs taking into account the wavelength dispersion of the effective index.

3 Conclusion

We designed and fabricated the first grating out-coupler working in the UVC region. Using Fourier space imaging, we characterized the grating couplers with two different designs. The 700 nm wide grating coupler exhibits a single diffracted mode in the aperture of the collecting microscope objective, which validates our design on suspended structures. As regard the 1.5 μm wide grating coupler, unexpected lines are observed. In future, new grating couplers with parameters optimized for a wavelength of 360 nm will allows us to implement on-chip spectroscopy in the UVC region.

References

1. Asher, S.A.: UV resonance Raman spectroscopy for analytical, physical, and biophysical chemistry, Part 2. https://pubs.acs.org/doi/pdf/https://pubs.acs.org/doi/pdf/10.1021/ac00052a001
2. Görög, S.: Ultraviolet-visible spectrophotometry in pharmaceutical analysis (2018)
3. Mahalingam, V., Hazra, C., Samanta, T.: A resonance energy transfer approach for the selective detection of aromatic amino acids. J. Mater. Chem. C **2** (2014). https://doi.org/10.1039/C4TC01954G
4. Joshi, D., Kumar, D., Maini, A., Sharma, R.: Detection of biological warfare agents using ultra violet-laser induced fluorescence LIDAR. Spectrochim. Acta Part A Mol. Biomol. Spectrosc. **112** (2013). https://doi.org/10.1016/j.saa.2013.04.082

Microwave Photonics

Simultaneous Notch Filtering and True Time Delay RF Photonic Front-End

Shangqing Shi[1,2(✉)], Kaixuan Ye[1], M. T. van den Berg[1], Okky Daulay[1], Gaojian Liu[1], and David Marpaung[1]

[1] Nonlinear Nanophotonics Group, University of Twente, Enschede, NB 7522, The Netherlands
s.shi-1@utwente.nl

[2] Advanced Photonics Center, Southeast University, Nanjing 210096, China

Abstract. We experimentally demonstrate an integrated, multi-functional RF photonic front-end capable of simultaneous notch filtering and true time delay. Both the notch filter and the true time delay functions can be controlled independently. For the notch filter part, the central frequency can be tuned from 5 to 10 GHz, with the rejection exceeding 40 dB. In the meantime, the true time delay can be tuned from 100 ps to 600 ps.

Keywords: Integrated microwave photonics · RF front-end · Notch filter · True time delay

1 Introduction

With the rapid advancement of wireless communication systems, the performance requirements for RF front-ends are continuously escalating. To address this, integrated optical and microwave photonic (MWP) technologies have been combined to develop integrated photonic RF front-ends that offer advantages such as high-speed transmission, wide bandwidth, reconfigurability, and strong anti-interference capabilities [1–3]. In the intricate electromagnetic environment, the RF front-end necessitates simultaneously processing signals with multiple functions, including filtering [4, 5], phase shifting [6], delay [7, 8], and frequency measurement [9]. In our previous work [10], we demonstrate a programmable MWP circuit that can be configured as a high-performance notch filter or a phase shifter. Building upon this, we also realized an RF front-end capable of simultaneous notch filtering and phase shifting [11].

Here, we report a multi-functional integrated RF photonic front-end that can simultaneously perform tunable filtering and true time delay. The tunable notch filter demonstrates a rejection exceeding 40 dB while tuning from 5 GHz to 10 GHz. Additionally, a maximum delay tuning range of from 100 ps to 600 ps, with a corresponding bandwidth of 6 GHz to 1 GHz.

J. Witzens et al. (Eds.): ECIO 2024, SPPHY 402, pp. 189–192, 2024.
https://doi.org/10.1007/978-3-031-63378-2_30

2 Chip Structure and Working Principle

The chip structure and working principle of the experiment are shown in Fig. 1. The vector network analyzer (VNA) emits continuous frequency-scanning signal from ω_0 to ω_2, which is modulated by the intensity modulator (IM) to the optical carrier with an angular frequency of ω_c. Subsequently, the modulated double-sideband signal undergoes amplification by an erbium-doped fiber amplifier (EDFA) and is subsequently injected into the chip. The chip consists of an 80-GHz-FSR spectral de-interleaver, and four cascaded all-pass 50-GHz-FSR ring resonators [10]. The on-chip de-interleaver [12] is utilized to selectively eliminate the lower sideband of the modulated signal, thereby generating a single-sideband (SSB) modulated signal that preserves the optical carrier and upper sideband for subsequent processing.

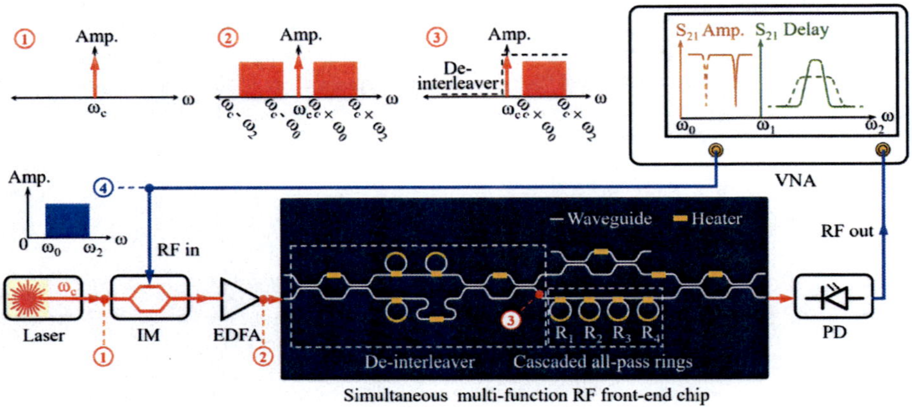

Fig. 1. Chip structure and working principle of the experiment. The VNA emits a continuous frequency-scanning signal (point ①), which is modulated onto the optical carrier by the IM. Subsequently, the amplified double-sideband (DSB) signal (point ②) is injected into the chip. The on-chip de-interleaver filters out the lower sideband of the DSB signal (point ③) for further filtering and delay processing. The photodetector (PD) converts the optical signal back to the RF domain, which is subsequently received by the VNA.

The first all-pass micro-ring resonator R_1 is precisely tuned to operate at its critical coupling state for effective notch suppression. We achieve efficient filtering of potential interference signals within the frequency range of ω_0 to ω_1 by tuning the resonant position of R_1. We further use R_2 and R_3 at their over coupling states to achieve tunable true time delays within the frequency range of ω_1 to ω_2. Additionally, R_4 serves as a separate carrier tuning unit [13] to compensate for the phase errors of the optical carrier signal.

3 Experimental Results

The functionalities of the multi-function integrated RF photonic front-end are verified through measurements conducted with both fixed group delay and varied notch positions, as well as measurements performed with fixed notch position and varied group delays.

The amplitude and group delay responses of the multi-function integrated RF photonic front-end at a fixed true time delay and different notch frequencies are shown in Fig. 2(a) and Fig. 2(b). The notch filter is precisely tuned from 5 to 10 GHz with an average attenuation of about 46 dB, while the true time delay is fixed at 500 ± 3 ps with an average bandwidth of about 1.3 GHz. Figure 2(c) and Fig. 2(d) depict the amplitude and group delay responses of the chip at a fixed notch position and varied group delays. The maximum group delay of 600 ps, 500 ps, 400 ps, 300 ps, 200 ps, and 100 ps are achieved for respective delay bandwidth settings of approximately 1 GHz, 1.3 GHz, 1.8 GHz, 2.2 GHz, 3.3 GHz and 6 GHz. At the same time, the notches are placed at exactly 7 GHz with a precision of ±0.01 GHz.

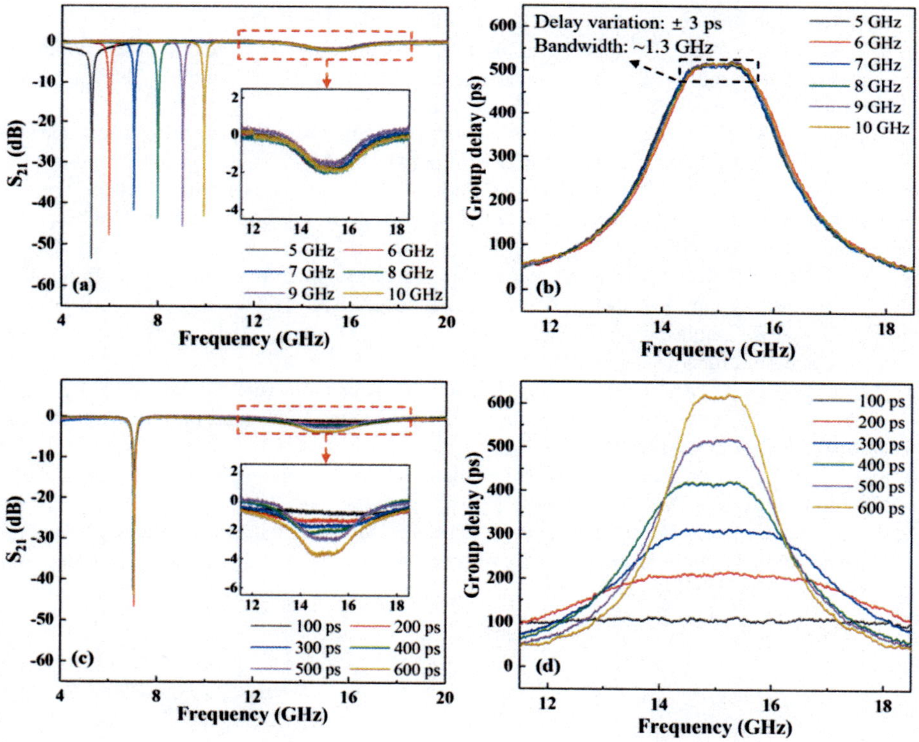

Fig. 2. (a) Amplitude and (b) group delay response at fixed 500 ps true time delay with notch frequency tuned from 5 GHz to 10 GHz; (c) Amplitude and (d) group delay response at fixed notch frequency at 7 GHz with varying true time delays from 100 ps to 600 ps.

4 Conclusion

We present the first integrated RF photonic front-end that enables simultaneous tunable filtering and true time delay, which shows the application potential of integrated RF photonic front-ends in practical RF environments.

References

1. Marpaung, D., Yao, J., Capmany, J.: Integrated microwave photonics. Nat. Photonics **13**(2), 80–90 (2019)
2. Akyildiz, I.F., Kak, A., Nie, S.: 6G and beyond: the future of wireless communications systems. IEEE Access **8**, 133995–134030 (2020)
3. Feng, H., Ge, T., Guo, X., et al.: Integrated lithium niobate microwave photonic processing engine. Nature **627**, 80–87 (2024)
4. Liu, Y., Choudhary, A., Marpaung, D., Eggleton, B.J.: Integrated microwave photonic filters. Adv. Opt. Photon. **12**(2), 485–555 (2020)
5. Daulay, O., Liu, G., Ye, K., et al.: Ultrahigh dynamic range and low noise figure programmable integrated microwave photonic filter. Nat. Commun. **13**, 7798 (2022)
6. Lin, D., Xu, X., Zheng, P., Hu, G., Yun, B., Cui, Y.: A high-performance microwave photonic phase shifter based on cascaded silicon nitride microrings. IEEE Photonic Technol. Lett. **32**(19), 1265–1268 (2020)
7. Lin,D., Xu, X., Zheng, P., et al.: A tunable optical delay line based on cascaded silicon nitride microrings for Ka-band beamforming. IEEE Photonics J. **11**(5), 1–10 (2019)
8. Liu, Y., Lu, L., Ni, Z., Chen, J., Zhou, L., Poon, A.W.: Continuously tunable silicon optical true-time delay lines with a large delay tuning range and a low delay fluctuation. Opt. Express **32**(5), 7848–7864 (2024)
9. Tao, Y., Yang, F., Tao, Z., et al.: Fully on-chip microwave photonic instantaneous frequency measurement system. Laser Photonics Rev. **16**(11), 2200158 (2022)
10. Liu, G., Ye, K., Daulay, O., Tan, Q., Yu, H., Marpaung, D.: Linearized integrated microwave photonic circuit for filtering and phase shifting. APL Photonics **8**(5), 051303 (2023)
11. Ye, K., Daulay, O., Liu, G., Marpaung, D.: Simultaneous multi-function integrated RF photonic front-end. In: Conference on Lasers & Electro-Optics 2023, Jose, pp. SW3O.3. Optica Publishing Group San (2023)
12. Guo, X., Liu, Y., Yin, T., et al.: Versatile silicon microwave photonic spectral shaper. APL Photonics **6**(3), 036106 (2021)
13. Burla, M., Marpaung, D., Zhuang, L., et al.: On-chip CMOS compatible reconfigurable optical delay line with separate carrier tuning for microwave photonic signal processing. Opt. Express **19**(22), 21475–21484 (2011)

Large-Scale Photonic Chip Based Pulse Interleaver for Low-Noise Microwave Generation

Zheru Qiu[1,2], Neetesh Singh[3], Yang Liu[1,2], Xinru Ji[1,2], Rui Ning Wang[1,2,4], Franz X. Kärtner[3], and Tobias Kippenberg[1,2(✉)]

[1] Swiss Federal Institute of Technology Lausanne (EPFL), 1015 Lausanne, Switzerland

[2] Center for Quantum Science and Engineering, EPFL, 1015 Lausanne, Switzerland
tobias.kippenberg@epfl.ch

[3] Center for Free-Electron Laser Science, Deutsches Elektronen-Synchrotron, 22607 Hamburg, Germany
franz.kaertner@desy.de

[4] Currently with Luxtelligence SA, 1015 Lausanne, Switzerland

Abstract. We demonstrate a 64-times multiplication of a mode-locked laser's 217 MHz pulse repetition rate to 14 GHz using an integrated Si_3N_4 interleaver. Six cascaded stages of Mach-Zehnder interferometers with up to 33 cm long optical delay lines are fully integrated into a very compact footprint of 8.5 mm × 1.7 mm. By overcoming photodetector saturation during the high power peak, the generated microwave power was improved by 36 dB, which reduces the phase noise floor by more than 10-fold to −160 dBc/Hz on the 14 GHz carrier.

1 Introduction

High-purity microwave signals have a wide range of applications in communications, radar, metrology, and analog-to-digital conversion. A powerful method of generating high-purity microwave is optical frequency division (OFD) that has resulted in exceptionally low-phase-noise microwaves reaching below -170 dBc/Hz at offsets > 10 kHz [1,2]. In this technique, solid-state or fiber-based mode-locked lasers (MLLs) are referenced to ultra-stable lasers and generate optical pulses with a regular temporal interval.

The stable repetition rate is then converted to a high-purity electronic signal by photodetection. A challenge in practical applications of OFD is the gap between the pulse repetition rate and the microwave frequencies desired for utilization. Most important applications such as communication systems, radar, and instrumentation require high-purity microwaves in the range of a few to tens of GHz. At the same time, the repetition rate of the low-noise MLLs typically falls in the range of a few hundred MHz, determined by the cavity length

Z. Qiu and N. Singh—These authors contributed equally to this work.

J. Witzens et al. (Eds.): ECIO 2024, SPPHY 402, pp. 193–197, 2024.
https://doi.org/10.1007/978-3-031-63378-2_31

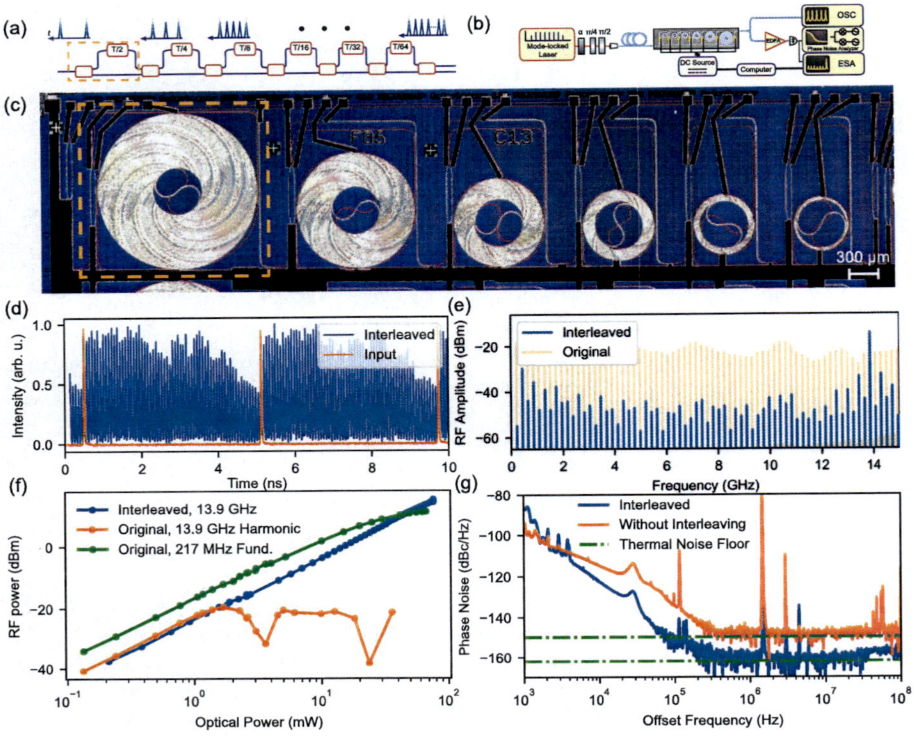

Fig. 1. (a) Principle of the cascaded unbalanced MZI pulse interleaver. (b) Setup for the operation of the interleaver and microwave generation. OSC: sampling oscilloscope, ESA: radio frequency (RF) spectrum analyser, EDFA: erbium doped fiber amplifier. (c) Optical micrograph of the fabricated 6-stage interleaver chip. (d) Pulse train at the input and after interleaving. (e) Spectrum of the generated microwave. (f) Generated microwave power as functions of the average optical power at the PD. (g) Phase noise of the generated microwave, in comparison to the thermal noise floor.

and usually limited by the gain coefficient for fundamental mode-locked fiber lasers. The short photocurrent pulses naturally contain high-order harmonics of the repetition rate, which can be used to bridge the gap in frequency. However, the signal-to-noise ratio (SNR) of the generated microwave will be limited by the photocurrent saturation in the photodiodes (PDs), even with state-of-the-art modified uni-traveling-carrier (MUTC) PDs [3]. The PDs saturate with the incident of the high-peak-power fs-pulses from MLLs, slowing down the response and thus limiting the achievable power of high-order harmonics.

A common strategy to alleviate this problem is multiplying the optical pulse repetition rate using mode-filtering cavities [4, 5] or interleavers [6]. The pulse train with multiplied repetition rates have a lower peak power at the PD, increasing the microwave power by reducing saturation and improve the signal to noise ratio limited phase noise floor. However, the fiber-based and free-space interleavers come with a substantial footprint, are prone to mechanical noise and temperature fluctuations, and necessitate careful polishing or adjustment for achieving the desired delay lengths [7].

In this work, we demonstrate a miniaturized 6-stage pulse interleaver on a low-loss silicon nitride photonic integrated circuit (Fig. 1(d)) fabricated at wafer scale, suitable for photonic generation of microwaves in portable devices. By incorporating up to 33 cm long low-loss delay lines, this chip can multiply the repetition rate f_{rep}=216.7 MHz of a fiber-based 1550 nm MLL by 2^6 times. The device footprint of 8.5 mm×1.7 mm device is significantly smaller than fiber-based or free-space counterparts and is $<$ 1/25th of the size of the reported 4-stage silica waveguide mode-filter [8]. This allows for low-cost mass production of compact ultra-low-phase-noise photonic microwave sources well suited for fiber-based or future on-chip integrated MLLs [9,10]. We note that by inserting more stages with shorter delay lengths of a few mm or hundreds of μm, impossible with fiber delay lines but can be simply implemented on chip, the repetition rate of the generated pulse train can be easily extended to $>$ 100 GHz [11], enabling ultra-low phase noise millimeter wave generation with a simple setup.

Here, we demonstrate a miniaturized 6-stage pulse interleaver on a low-loss silicon nitride photonic integrated circuit (Fig. 1(a), (b)). This chip can multiply the f_{rep}=216.7 MHz repetition rate by 2^6-times. The 8.5 mm×1.7 mm device footprint is much smaller than any fiber or free space interleavers and is $<$ 1/25 of the 4-stage silica waveguide device in [8]. Delay lengths in the device are precisely defined by lithography, which eliminates the need for the adjustment of delay lengths. This allows for low-cost mass-production of compact ultra-low-phase-noise photonic microwave generators with fiber-based or future integrated mode-locked lasers.

2 Device Design, Fabrication and Pulse Interleaving Demonstration

We adopt a cascaded unbalanced Mach-Zehnder interferometer (MZI) design (Fig. 1(a)) for the interleaver. In the nth stage of the interleaver, each input pulse is equally (ideally) divided into two halves by a coupler; one half is then delayed for $1/(2^n f_{\mathrm{rep}})$ and combined with the other half by another coupler. To compensate for the fabrication imprecision of the coupling ratio and the loss in the delay lines (up to 32 cm), we utilize balanced MZI couplers with Ti/Pt micro-heaters as tunable couplers. The Si_3N_4 waveguides have a 1.5 μm ×700 μm cross-section to reduce coupling to high-order modes and are fabricated with the photonic damascene process [12].

Errors in the time delay can corrupt the regular timing between the pulses and negate the phase noise reduction by shot noise correlations of the short pulses [13]. On the interleaver chip, the time delay is controlled by the waveguide length and group index, which are precisely defined by deep ultraviolet lithography. We characterized the time delay of each interleaver stage on fabricated chips by performing a Fourier transform of the transmission spectrum of the interleaver (see Methods). For the longest delay time of 2308 ps (33 cm), the error is < 2 ps (0.12%), comparable to or better than the hand-polished fiber delay lines and can minimize microwave power loss and increase in phase noise [14].

We demonstrate the pulse interleaving by sending the pulses from a fiber-based MLL (OneFive Origami) to a short fiber delay line and then to the interleaver chip. The pulses enter from the 6th stage to avoid excessive nonlinear broadening in the long delay lines. The resulting pulse is characterized by a sampling oscilloscope (Fig. 1(d)) and sent to a MUTC PD (Freedom Photonics FP1015A, 5.3 V bias) to generate the microwave. We use the Nelder-Mead method to maximize the suppression of the undesired harmonics on the RF spectrum by optimizing the heating power of the 7 tunable couplers. These undesired side modes can be suppressed by 14 dB (Fig. 1(e)) relative to the generated microwave. Figure 1(f) shows the RF power as a function of the average optical power at the PD. While the non-interleaved pulses saturate the PD at < 1mW average optical power and limit the 13.87 GHz power to < -19 dBm, the amplified interleaved pulses can generate more than 16 dBm microwave power with 76 mW average optical power. The microwave phase noise is characterized by a phase noise analyzer (R&S FSWP). With the interleaver, we reach the best phase noise performance of -160 dBc/Hz at 1 MHz offset (without EDFA, ~ 1.2 mA photocurrent) for the generated -15 dBm 13.87 GHz microwave signal. This result is limited by the shot and thermal noise floor and ~ 12 dBc/Hz lower than the non-interleaved case.

3 Summary

We present a large-scale integrated pulse interleaver on a Si_3N_4 photonic integrated circuit platform. With low-loss waveguide delay lines, this 8.5 mm×1.7 mm device efficiently multiplies the pulse repetition rate from 217 MHz to 14 GHz on a single chip. Applying this photonic multiplier to a free running femtosecond laser allows for an increase in the generated microwave power by 37 dB and a reduction of the phase noise floor by 12 dB. This interleaver holds potential for future miniaturized ultra-low-phase-noise photonic microwave generators with fiber-based or integrated mode-locked lasers.

References

1. Fortier, T.M., et al.: Nat. Photonics **5**(7), 425 (2011)
2. Xie, X., et al.: Nat. Photonics **11**(1), 44 (2017)
3. Fortier, T.M., et al.: Opt. Lett. **38**(10), 1712 (2013)
4. Diddams, S.A., Kirchner, M., Fortier, T., Braje, D., Weiner, A., Hollberg, L.: Opt. Express **17**(5), 3331 (2009)
5. Jiang, H., Taylor, J., Quinlan, F., Fortier, T., Diddams, S.A.: IEEE Photonics J. **3**(6), 1004 (2011)
6. Haboucha, A., et al.: Opt. Lett. **36**(18), 3654 (2011). https://doi.org/10.1364/OL.36.003654. https://opg.optica.org/ol/abstract.cfm?URI=ol-36-18-3654

7. Matlis, N.H., et al.: Opt. Express **31**(26), 44424 (2023).https://doi.org/10.1364/OE.503480. https://opg.optica.org/oe/abstract.cfm?URI=oe-31-26-44424
8. Sander, M.Y., Frolov, S., Shmulovich, J., Ippen, E.P., Kärtner, F.X.: Opt. Express **20**(4), 4102 (2012)
9. Singh, N., Ippen, E., Kärtner, F.X.: Opt. Express **28**(15), 22562 (2020)
10. Singh, N., et al.: Nature Photonics, pp. 1–7 (2024). https://doi.org/10.1038/s41566-024-01388-0
11. Lo, M.C., Guzmán, R., Gordón, C., Carpintero, G.: Opt. Lett. **42**(8), 1532 (2017). https://doi.org/10.1364/OL.42.001532. https://opg.optica.org/ol/abstract.cfm?URI=ol-42-8-1532
12. Liu, J., Huang, G., Wang, R.N., He, J., Raja, A.S., Liu, T., Engelsen, N.J., Kippenberg, T.J.: Nat. Commun. **12**(1), 2236 (2021)
13. Quinlan, F., et al.: Nat. Photonics **7**(4), 290 (2013)
14. Quinlan, F., et al.: Opt. Lett. **39**(6), 1581 (2014)

Check for updates

Agile Spectral Multiplication of Narrow-Band Comb using Integrated InP Multi-wavelength Laser

Pablo Marin-Palomo, Shahab Abdollahi, Mathieu Ladouce, and Martin Virte(✉)

Brussels Photonics Team (B-PHOT), Vrije Universiteit Brussel, Pleinlaan 2, 1050 Brussel, Belgium
martin.virte@vub.be

Abstract. Chip-scale frequency comb sources, due to their reduced footprint, low cost, and energy efficiency, have revolutionized, among others, the fields of communications, metrology, and sensing. The generated frequency combs, however, suffer from a trade-off between bandwidth and controllability of its envelope and tone spacing. Here, we present an approach to overcome this trade-off by enabling agile spectral multiplication of a narrowband comb. By leveraging the nonlinear dynamics of a single InP distributed Bragg reflector (DBR) laser under modulated optical injection, we can achieve spectral multiplication of an injected narrowband comb at frequency offsets of up to 1.3 THz. We also demonstrate nanosecond-scale feedback-controlled tuning speed of the laser emission by a monolithically integrated feedback cavity. Lastly, we propose a technique to phase lock the multiplied combs through cascaded phase locking and show that our approach is scalable and potentially covering a range of several THz with a suitable laser design, compatible with generic foundry platforms.

Keywords: InP DBR laser · Multi-wavelength laser · Frequency comb · spectral multiplication

1 Introduction

Integrated photonics technology has opened the path towards compact, energy efficient and low-cost frequency comb sources. On-chip frequency comb sources have become key building blocks for many applications ranging from spectroscopy [1, 2] and lidar [3] to optical communications [4]. Over the last two decades, a wide variety of integrated frequency comb sources have been demonstrated on different photonic platforms [5, 6]. The main approaches for on-chip comb generation involve either mode-locking in InP quantum well (QW) lasers [7], or optical nonlinearities in Kerr resonators [8] or electro-optic (EO) modulators [9]. While there has been progress towards increasing the bandwidth and power of the associated combs, there is, however, an inherent trade-off of current chip-scale comb sources between bandwidth and tunability. Large bandwidths typically come at the price of no or limited tunability and the system is optimized for one specific use case. On one side, mode-locked lasers (MLL) and Kerr resonators enable

J. Witzens et al. (Eds.): ECIO 2024, SPPHY 402, pp. 198–203, 2024.
https://doi.org/10.1007/978-3-031-63378-2_32

the generation of broad combs, however, the corresponding bandwidth and FSR are intrinsically fixed and FSR values of few GHz or lower represent a challenge with these approaches. On the other side, EO combs provides higher flexibility in terms of mode spacing and spectral shape but reaching mm-wave frequencies becomes challenging and requires the use of more complex systems involving, e.g., cascaded EO modulators or highly nonlinear waveguides or microresonators [10, 11].

In this work, we demonstrate frequency-agile spectral multiplication of a narrow-band EO-comb by an InP multiwavelength laser (MWL). Our scheme combines the flexibility of the EO-comb with the large bandwidth of a MWL to obtain a versatile comb with a flexible spectral envelope over THz-range frequencies [12]. The approach is based on injecting a comb into a semiconductor laser, which has been demonstrated to lead to comb broadening [13]. Here we demonstrate spectral comb multiplication at frequency offsets of 26 GHz, 54 GHz, and up to 1.3 THz. Our approach incorporates on-chip optical feedback control that allows for high suppression ratio of over 30 dB and tuning of the offset frequency at the nanosecond time scale. Our system is scalable and, with proper laser design, can reach several THz enabling, e.g., low-cost, compact, and power-efficient sources for agile THz comb generation.

2 Device Structure and Characterization

The multi-wavelength laser (MWL) is a dual-cavity laser [14] with two detuned distributed Bragg reflectors (DBR) placed in a row, a shared semiconductor optical amplifier (SOA), and a broadband reflector (BR), see Fig. 1(a).

The DBRs are detuned by approximately 10 nm (1.3 THz) with corresponding Bragg wavelengths at approximately 1537 nm and 1547 nm, where multiple longitudinal modes can emit. The laser threshold current is at 21 mA, and we operate at 30 mA. In this configuration, the free-running laser emits at $\lambda 4 = 1547.58$ nm, see Fig. 1(b). Three additional residual modes can be observed at $\lambda_1 = 1536.70$ nm, $\lambda_2 = 1536.92$ nm, and $\lambda_3 = 1537.13$ nm. When increasing the feedback strength by setting $i_{FB} = 21$ mA, switching between emission at λ_1, λ_2 and λ_4 can be achieved by adjusting the voltage V_{FB} applied to the electro-optic phase modulator (EOPM), see Fig. 1(c) and (d).

We characterize the required time by the MWL to change the wavelength at which it emits by sending a voltage step to the EOPM and measuring the time-domain evolution of the output optical signal from the MWL. The corresponding setup is depicted in Fig. 2(a), where we employ an arbitrary waveform generator (AWG) to create the step voltage and use a DC power supply and a bias tee to shift the voltage level to the value where switching occurs. The output optical signal from the MWL is monitor by a high-speed oscilloscope and displayed in Fig. 2(b). The time at which the laser starts reacting, $\tau_r = 1.6$ ns, and the time at which the switching is done, $\tau_s = 4.0$ ns, i.e., when the optical power emitted at λ_1 is reduced by more than 10 dB.

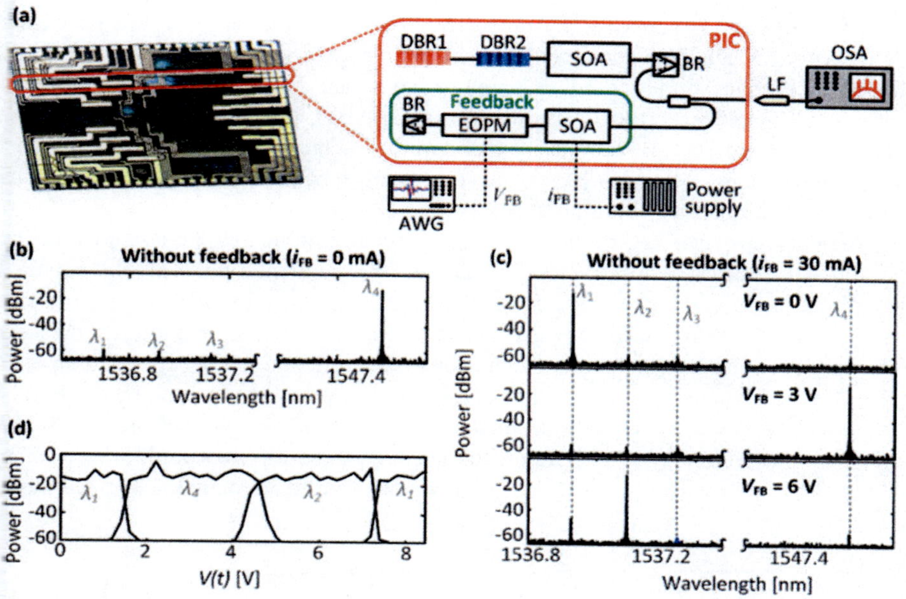

Fig. 1. Multi-wavelength laser photonic integrated circuit (PIC). (a) Image of the PIC containing eight distinct MWL structures and scheme describing the components of our MWL. (b) Optical spectrum of MWL output without feedback (FB) and (c) with different FB phases. (d) MWL output power as a function of the FB phase. OSA: Optical spectrum analyzer. LF: Lensed Fiber.

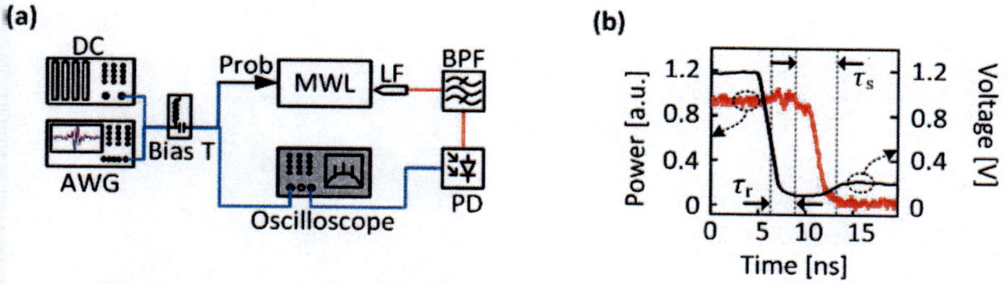

Fig. 2. High-speed feedback-controlled mode switching in our MWL. (a) Experimental setup for the switching speed measurement. AWG: Arbitrary waveform generator. PD: Photodiode. LF: Lensed fiber. BPF: Tunable bandpass filter. (b) In black: Voltage signal sent to the EOPM. In red: Signal recorded by the oscilloscope, associated with the power emitted at λ_1. The vertical dotted lines indicates the reaction time, τ_r, and the switching time, τ_s, of the laser.

3 THz-Range Spectral Multiplication and Phase Locking of Narrow-Band EO-Comb

The spectral multiplication of a narrow-band EO comb is achieved by optical injection on our MWL. The setup for comb generation and optical injection is depicted in Fig. 3(a). The optically injected signal consists of a frequency comb featuring 5 tones with an

FSR of 1 GHz and is generated by modulating the light from an external-cavity laser (ECL) with a Mach-Zehnder modulator (MZM). A polarization controller (PC) and erbium-doped fiber amplifier (EDFA) are used to adjust the polarization and power of the injected signal respectively. The output from the MWL is sent to an optical spectrum analyzer (OSA) to measure the spectrum and towards a band-pass filter (BPF) followed by a photodetector (PD) and electrical spectrum analyzer (ESA) to record the individual beating of each spectrally multiplied sub-comb.

We inject the center of the comb around $\lambda 3$ with a detuning of 2.3 GHz and injection strength of 9.5 dBm. The spectral multiplication of the modulated signal at $\lambda 4$ is obtained through the modulation of the carrier population. Because the injected mode experiences amplitude modulation induced by the frequency comb, the carrier population of the laser will experience similar fluctuations. Then, due to the carrier coupling between all modes of the MWL, frequency combs emerge at the other wavelengths emitted by the MWL, in this case, $\lambda 4$. We thus obtain multiplied combs separated by up to 1.3 THz. With iFB = 21 mA and tuning VFB, we can switch between three different multiplied combs, see Fig. 3(b) and (c). The re-generated comb which is emitted around the injected mode

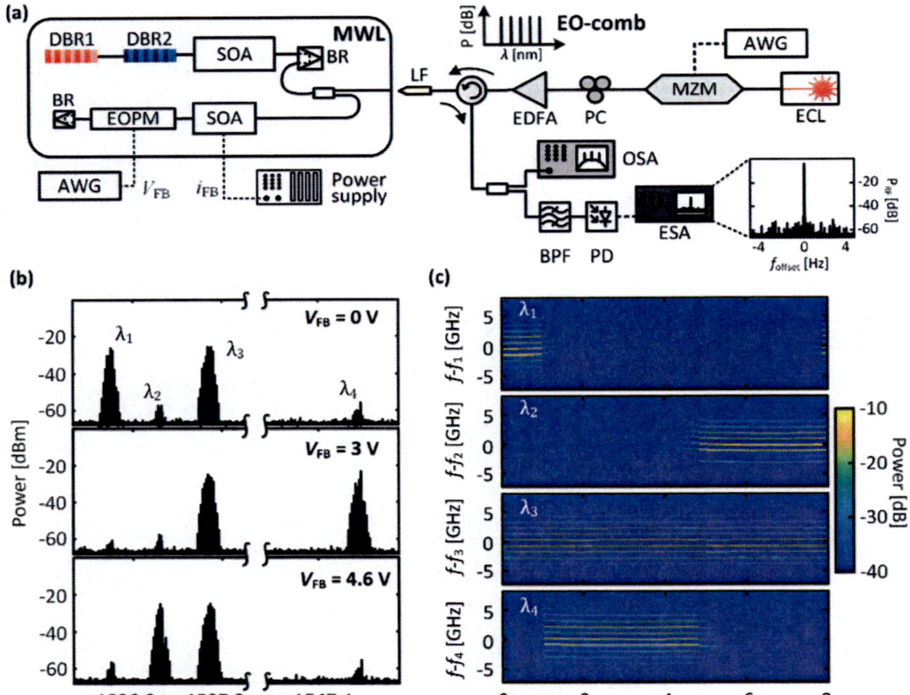

Fig. 3. MWL-based THz-range spectral multiplication of narrow-EO frequency comb. **(a)** Experimental setup for spectral multiplication of injected comb. Inset: beating of a single sub-comb. **(b)** Optical spectral of the output of the MWL for different FB phase values. **(c)** Spectral evolution of the four sub-combs as a function of the feedback phase, adjusted via V_{FB}.

always remains present. In this MWL configuration, we achieve sequential emission of three multiplied combs in the mm-wave and THz range with an FSR of 1 GHz and central frequencies around 26 GHz, 54 GHz, and 1.3 THz, We achieve an extinction ratio, for each multiplied comb, of above 30 dB.

The sub-combs are not inherently phase correlated. By adding an extra tone to the EO-comb near an active mode of the MWL, we actively phase-lock the subcombs at λ_3 and λ_2. By carefully adjusting the position of the additional tone, the beating of the sub-combs at λ_3 and λ_2 locks the mode λ_1 which is not directly affected by the injection. We believe that such cascading phase-locking can be extended to modes at larger frequency offsets, with a suitable MWL configuration.

4 Conclusion

We demonstrate spectral multiplication of a narrow-band EO-comb over more than 10 nm (1.3 THz) with controllable frequency offset using an integrated InP multiwavelength laser. The MWL features nanosecond timescale switching and the spectral multiplication is solely controlled by a single parameter, namely the feedback phase. Further, we discuss a technique to trigger a cascaded mechanism leading to phase-locking of the sub-combs. Our approach is scalable in frequency and has the potential to bridge the gap between broadband and dense but narrow combs.

References

1. Myoung-Gyun, S., et al.: Microresonator soliton dual-comb spectroscopy. Science **354**, 600–603 (2016)
2. Trocha, P., et al.: Ultrafast optical ranging using microresonator soliton frequency combs. Science **359**, 887–891 (2018)
3. Riemensberger, J., et al.: Massively parallel coherent laser ranging using a soliton microcomb. Nature **581**(7807), 164–170 (2020)
4. Marin-Palomo, P., et al.: Microresonator-based solitons for massively parallel coherent optical communications. Nature **546**, 274–279 (2017)
5. Chang, L., et al.: Integrated optical frequency comb technologies. Nat. Photonics **16**(2), 95–108 (2022)
6. Gaeta, A.L., et al.: Photonic-chip-based frequency combs. Nat. Photon **13**, 158–169 (2019)
7. Liu, S., et al.: High-channel-count 20 GHz passively mode-locked quantum dot laser directly grown on Si with 41 Tbit/s transmission capacity. Optica **6**, 128–134 (2019)
8. Del'Haye, P., et al.: Optical frequency comb generation from a monolithic microresonator. Nature **450**, 1214–1217 (2007)
9. Ren, T., et al.: An integrated low-voltage broadband lithium niobate phase modulator. IEEE Photonics Technol. Lett. **31**, 889–892 (2019)
10. Wang, C., et al.: Integrated lithium niobate electro-optic modulators operating at CMOS-compatible voltages. Nature **562**, 101–104 (2018)
11. Buscaino, B., et al.: Design of efficient resonator-enhanced electro-optic frequency comb generators. J. Lightwave Technol. **38**, 1400–1413 (2020)
12. Abdollahi, S., et al.: Agile THz-range spectral multiplication of frequency combs using a multi-wavelength laser. Nat. Commun. **15**, 1305 (2024)

13. Doumbia, Y., et al.: Two polarization comb dynamics in VCSELs subject to optical injection. Photonics **9**(2), 115 (2022)
14. Abdollahi, A., et al.: Mode-coupling effects in an optically-injected dual-wavelength laser. Opt. Express **30**, 32447–32458 (2022)

Microwave-Optical Transduction Using High-Overtone Bulk Acoustic Resonators on Silicon Nitride Photonics

Terence Blésin[1,2], Wil Kao[1,2], Anat Siddharth[1,2], Rui N. Wang[1,2], Alaina Attanasio[3], Hao Tian[3], Sunil A. Bhave[3], and Tobias J. Kippenberg[1,2(✉)]

[1] Institute of Physics, Swiss Federal Institute of Technology Lausanne (EPFL), 1015 Lausanne, Switzerland
tobias.kippenberg@epfl.ch
[2] Center of Quantum Science and Engineering (EPFL), 1015 Lausanne, Switzerland
[3] OxideMEMS lab, Purdue University, West Lafayette, IN, USA
bhave@purdue.edu

Abstract. We present compact microwave-optical transducers between the microwave S-band and optical C-band based on monolithic integration of piezoelectric actuators atop silicon nitride photonic circuits. The silica cladding of the optical waveguide core is suspended below the actuators, forming a bulk acoustic resonator that mediates interaction between microwave and optical fields through piezoelectric and stress-optical effects. An off-chip photon conversion efficiency of −48 dB is measured over a bandwidth of 25 MHz for both up- and down-conversion. Microwave pulses are synthesized on-chip through down-conversion and multiplexed transduction is demonstrated, indicating its potential as a photonic interconnect between cryogenic circuits.

Keywords: Microwave-optical conversion · Silicon nitride photonics · Bulk acoustic waves · Cryogenic interconnect · MEMS

1 Introduction

Transducers for microwave-optical frequency conversion are expected to become essential building blocks of future quantum networks, bridging stationary microwave qubits to flying optical photons [1]. Beyond enabling quantum coherent conversion, transducers also fulfil a pivotal role in reducing the heat load and cabling constraints in cryostats by replacing conventional microwave wiring with optical fibers for tasks such as readout and control of superconducting qubits.

2 Design and Fabrication

In this context, we have made a compact microwave-optical transducer based on monolithic integration of piezoelectric actuators atop silicon nitride photonic circuits [2]. Optical microring resonators with 22 µm radius are fabricated using the

J. Witzens et al. (Eds.): ECIO 2024, SPPHY 402, pp. 204–206, 2024.
https://doi.org/10.1007/978-3-031-63378-2_33

photonic Damascene process for Si_3N_4 waveguides cladded by SiO_2. Piezoelectric actuators made of AlN and Mo electrodes are subsequently integrated on top of the photonic circuitry using sputtering and ion beam etching. AlN is removed from the feedline to reduce actuator capacitance and suppress parasitic modes. In addition, shorting between the top and bottom electrodes is avoided by etching back the bottom electrode with XeF_2 before the top electrode is deposited. The Si substrate is remove from below the actuators using the Bosch process, thereby confining acoustic waves within the SiO_2 cladding [4]. The microwave ports excite high-overtone bulk acoustic resonances (HBAR) through piezoelectricity over a bandwidth extending from roughly 2 GHz to 4 GHz with a free spectral range of 320 MHz. Photoelastic and moving boundaries effects induce optomechanical interaction between the acoustic wave and light circulating in the microresonators. Strong evanescent field coupling between adjacent optical cavities is engineered to form a photonic molecule. The resulting hybrid mode splitting is controlled by applying a DC bias voltage to the piezoelectric actuator, thereby realizing the triply resonance configuration.

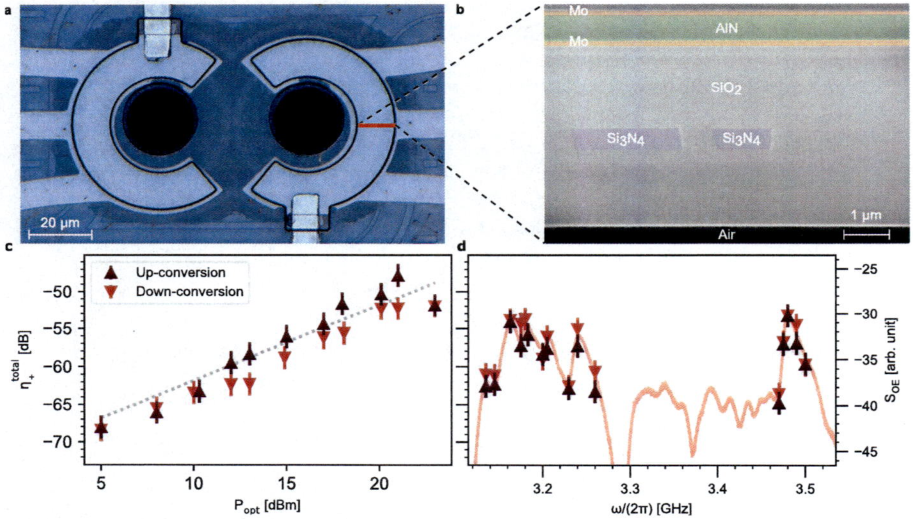

Fig. 1. Bidirectional, multimode transduction with a continuous-wave pump. Conversion between microwave and optical frequencies using the HBAR-silicon nitride photonic transducer. **a** Optical micrograph of the transducer. **b** False-color cross section of the device imaged by focused ion beam scanning electron microscopy showing the top and bottom electrodes, piezoelectric layer, suspended cladding, and optical waveguides. **c** Off-chip photon number transduction efficiencies as a function of off-chip input pump power in the triply resonant configuration. **d** Bidirectional transduction with a detuned pump leveraging multiple HBAR modes, as a function of input microwave frequency and input optical detuning for an input optical pump power of 20 dBm. Up- and down-conversion efficiencies are denoted by triangles and inverted triangles, respectively, with the acoustic-optic response spectrum superimposed.

3 Results

A calibrated measurement of the total conversion efficiency is performed by connecting the transducer microwave ports to a vector network analyzer using high frequency probes and the optical ports to an optical heterodyne detection setup using optical fibers. As shown in Fig. 1(c), bidirectional conversion is demonstrated with photon number transduction efficiencies higher than −50 dB over a bandwidth of 25 MHz for both Stokes and anti-Stokes processes. Microwave pulses with powers up to −95 dBm are synthesized directly on chip from a purely optical input [3]. Furthermore, the transducer is operated in the pulsed regime with a pulse duration of 1 µs and a pulse repetition rate of 100 kHz, replicating the conditions that would be used for addressing superconducting qubits optically in a cryostat. Finally, the multimode nature of the transducer can be exploited for multiplexed transduction, as demonstrated by converting signals separated by hundreds of MHz using distinct overtones without compromising the conversion efficiency, as depicted in Fig. 1(d).

Acknowledgements. This work was supported by the Air Force Office of Scientific Research under award no. FA8655-20-1-7009, Swiss National Science Foundation under grant agreement no. 204927, and European Research Council (ERC) under the EU H2020 research and innovation programme, grant agreement no. 835329 (ExCOM-cCEO). This work was further supported by United States Air Force Research Laboratory Award no. FA8750-21-2-0500, as well as United States National Science Foundation's International Collaboration Supplements in Quantum Information Science and Engineering Research for RAISE-TAQS Award no. 18-39164. A.S. acknowledges support from the European Space Technology Centre with ESA Contract No. 4000135357/21/NL/GLC/my. Samples were fabricated in the Center of MicroNanoTechnology (CMi) at EPFL and Birck Nanotechnology Center at Purdue University.

References

1. Awschalom, D., et al.: Development of Quantum Interconnects (QuICs) for Next-Generation Information Technologies. PRX Quantum **2**(1), 017,002 (2021)
2. Blésin, T., Tian, H., Bhave, S.A., Kippenberg, T.J.: Quantum coherent microwave-optical transduction using high-overtone bulk acoustic resonances. Phys. Rev. A **104**(5), 052,601 (2021)
3. Blésin, T., et al.: Bidirectional microwave-optical transduction based on integration of high-overtone bulk acoustic resonators and photonic circuits (2023). http://arxiv.org/abs/2308.02706. ArXiv:2308.02706 [physics, physics:quant-ph]
4. Tian, H., et al.: Magnetic-free silicon nitride integrated optical isolator. Nature Photonics, pp. 1–9 (2021)

Phonon-Photon Interaction in Integrated Photonics

Brillouin-Active Subwavelength Silicon Membrane Waveguides

Paula Nuño Ruano[1](✉), Jianhao Zhang[2], David González-Andrade[1], Hiba El Batoul Ferhat[1], Daniele Melati[1], Eric Cassan[1], Pavel Cheben[2], Laurent Vivien[1], Norberto Daniel Lanzillotti-Kimura[1], and Carlos Alonso-Ramos[1]

[1] Centre de Nanosciences et de Nanotechnologies, CNRS, Université Paris-Saclay, 91120 Palaiseau, France
paula.nuno-ruano@c2n.upsaclay.fr
[2] National Research Council Canada, 1200 Montreal Road, Bldg. M50, Ottawa, ON K1A 0R6, Canada

Abstract. On-chip Brillouin scattering plays a key role in numerous applications in the domain of signal processing and microwave photonics due to the coherent bidirectional coupling between near-infrared optical signals and GHz mechanical modes. Here, we propose and demonstrate a novel subwavelength waveguide design to achieve narrow linewidth and strong optomechanical interaction. We experimentally demonstrate optomechanical coupling with 5–8 MHz linewidth and a coupling strength of $G_B = 1360\ W^{-1}m^{-1}$. A Stokes gain of 1.5 dB and anti-Stoke attenuation of −2 dB is observed for a 6 mm-long waveguide with 35.5 mW of input power. We show tuning of the mechanical frequency between 5 and 8 GHz by geometrical optimization without loss of the optomechanical coupling strength.

Keywords: Brillouin Scattering · Subwavelength · Silicon Photonic

1 Introduction

Brillouin scattering refers to the three-wave coherent interaction between the optical field and mechanical vibrations of the bulk material. Contrary to other optoacoustic interactions, such as those we find in acousto-optics modulators, in Brillouin scattering, the optical fields excite and affect the mechanical field. Currently, Brillouin interactions are being exploited in optical fibers and integrated devices to implement a wide range of devices, including ultra-narrow linewidth lasers [1], radiofrequency (RF) filters [2], and quantum applications [3].

Traditionally, Brillouin scattering was considered as an intrinsic nonlinearity and, thus, determined only by the material properties. However, this picture fails for nanoscale devices due to the importance of boundaries in these structures. There, four main mechanisms govern the Brillouin interaction with comparable strength: photoelastic effect, electrostriction forces, moving boundary effect, and radiation pressure forces. Generally,

J. Witzens et al. (Eds.): ECIO 2024, SPPHY 402, pp. 209–213, 2024.
https://doi.org/10.1007/978-3-031-63378-2_34

these four mechanisms are partnered up by pairs (electrostriction forces and photoelasticity, radiation pressure forces and moving boundary effect) with one being the action of the mechanical motion upon the optical modes, and the other, the optical action upon the mechanical modes [4].

Because of their strong optomechanical interactions and compatibility with standard CMOS technology, silicon-based devices, since their first experimental observation of Brillouin scattering in 2013 [5], have garnered significant attention among the various platforms used to harness this effect. However, achieving tight phonon confinement and strong photon-phonon coupling in nanoscale silicon waveguides, simultaneously, is a significant challenge to achieving efficient Brillouin scattering in this platform. The primary obstacle to using the Brillouin nonlinearities in silicon-on-insulator waveguides is the strong phonon leakage towards the silica cladding [6]. Isolating the active silicon optomechanical region by complete or partial removal of the silica cladding is the most extended strategy to observe strong optomechanical interactions. Exploiting this strategy, large Brillouin gain has been demonstrated in silicon membrane rib waveguides [7], fully suspended silicon nanowires [8], and pedestal waveguides [9]. These structures, however, exhibit poor fabrication tolerance, their mechanical stability might be jeopardized, or they depend on very low optical loss to compensate for a moderate Brillouin gain.

Here, we propose a suspended waveguide design that produces a large Brillouin gain while making fabrication easier by taking advantage of a single-etch subwavelength structure (Fig. 1). To expand the design space for CMOS-compatible integrated devices, subwavelength silicon waveguides have been thoroughly investigated [10]. Remarkably, since GHz phonons and near-infrared photons have similar wavelengths (around 1 μm) in Si waveguides, the same periodic structure can operate in the subwavelength regime for both [11]. Furthermore, intramodal forward Brillouin scattering, where both optical waves propagate in the same waveguide mode and the same direction, relies on longitudinally propagating photons and transversally propagating phonons [10]. Thus, independent control of photonic and phononic modes is possible through subwavelength engineering of the longitudinal and transversal geometries.

2 Design and Simulation

The proposed membrane waveguide (Fig. 1) consists of a phononic crystal barrier and a subwavelength lattice cladding surrounding the central core. The latter is made up of a square lattice of air-cross unit cells, preventing mechanical leakage towards the silicon slab and localizing the mechanical modes inside the core and inner cladding. The fundamental transverse-electric (TE) optical mode propagates in the subwavelength regime due to the periodic arrangement of silicon arms in the inner part. Its objective is to tune the frequency of the mechanical mode that couples to the light by adjusting the cavity length formed by the phononic crystal and to separate the phononic crystal from the optical mode to achieve low-loss optical propagation.

Based on COMSOL simulations, this geometry couples two co-propagating fundamental TE optical modes to a breathing mechanical mode (Fig. 2) with a frequency within the bandgap of the phononic crystal between 5 and 8 GHz. Assuming a mechanical Q-factor of 1000, a Brillouin gain coefficient G_B of up to 2000 $W^{-1}m^{-1}$ can be achieved.

Fig. 1. Proposed optomechanical waveguide. a) Geometry of the structure, with core width of $W_c = 600$ nm, a period of $\Lambda_{SWG} = 200$ nm for the subwavelength arms, and thickness of $t = 220$ nm. The lateral subwavelength lattice has a length of $\ell_{SWG} = 2$ μm and a duty cycle of 40% of silicon. The phononic crystal period is twice the SWG period, and the arms of the air cross have dimensions of 350 nm × 140 nm.

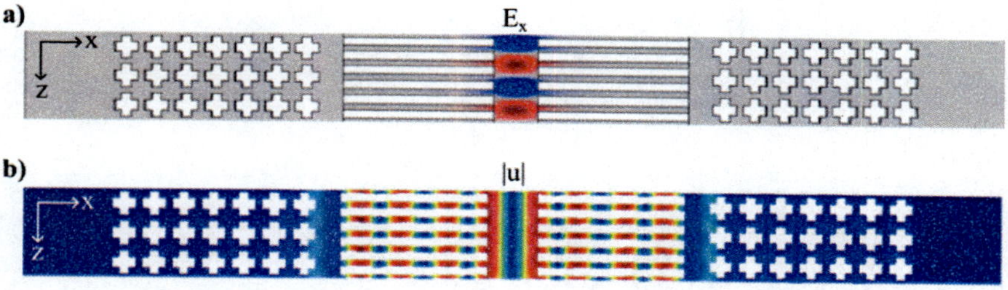

Fig. 2. a) Normalized x-component of the TE-polarized optical mode with a wavelength of 1554.9 nm in vacuum. Here, blue means a negative value, and red means a positive value. **b)** Normalized displacement for the breathing mode with a frequency of 7.184 GHz. Here, red means maximum displacement, and blue, a zero-displacement region.

3 Experimental Results

In intramodal forward Brillouin scattering, optical dispersion can be generally ignored due to the five orders of magnitude frequency difference between near-infrared photons (~200 THz) and GHz phonons. Besides, the phase-matching condition with co-propagating photons leads to an interaction mediated by mechanical modes with very small wavevectors. As a result, anti-Stokes and Stokes waves interact through the same phonon mode, and the spontaneous, thermally initiated optomechanical effect is negligible. A stimulated process is, therefore, required to obtain measurable Brillouin gain.

We implement the three-tone gain scheme illustrated in Fig. 3a. We sweep the driving frequency. When the modulating frequency equals the Brillouin shift, the optical beating note between the pump and the blue- and red-detuned sidebands coherently drives the acoustic field, leading to observable gain and loss for the Stokes and anti-Stokes sidebands (Fig. 3b). A Lorentzian shape centered at the phononic frequency and with linewidth given by the phonon lifetime in the system characterizes Brillouin interactions. After fitting the experimental measurements with the amplitude coupled-mode equations that modeled the process [12], we obtain a mechanical frequency of $\Omega/2\pi = 7.0453$ GHz and a linewidth of $\Gamma/2\pi = 6.3$ MHz, or often used in the context of Brillouin scattering, a mechanical quality factor of $Q = 1100$. This linewidth corresponds to a phonon lifetime of around 25 ns. A Brillouin gain coefficient of $G_B = 1360\ W^{-1}\ m^{-1}$ is achieved in this structure, leading to a maximum attenuation of 2 dB and gain of 1.5 dB for anti-Stokes and Stokes sidebands, respectively, with an input power of 35.5 mW in a 6 mm-length waveguide. Net Brillouin gain may be obtained with further reduction of the optical propagation loss, e.g. through post-processing to reduce sidewall roughness.

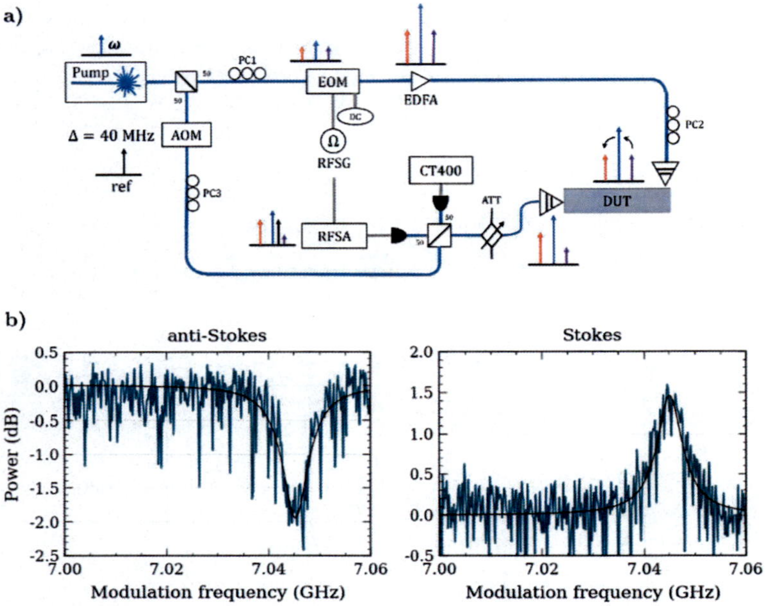

Fig. 3. Experimental characterization of the optomechanical waveguide. **a)** Three-tone gain setup. PC: polarization controller, EOM: electro-optical modulator, RFSG: radio-frequency signal generator, EDFA: erbium-doped fibre amplifier, DUT: device under test, RFSA: radio-frequency signal analyzer, AOM: acousto-optical modulator, ATT: variable optical attenuator. **b)** Normalized RF spectrum, i.e., power spectral density (PSD) as a function of the modulating frequency for an input power of 35.5 mW. In solid black line, theoretical fitting to amplitude coupled-mode equations.

4 Conclusions

We propose and demonstrate a novel geometry that effectively confines the optical and mechanical fields within the waveguide core by combining the benefits of phononic crystals and subwavelength cladding. This geometry results in a strong, narrow-linewidth optomechanical interaction ($G_B \approx 1360\ W^{-1}m^{-1}$, $\Gamma/2\pi \sim 6$ MHz) between the breathing mechanical modes ($\Omega/2\pi \sim 7$ GHz) and the fundamental TE optical mode. These findings provide intriguing new avenues for the application of narrow-linewidth notch filters, which are essential for microwave-photonics applications.

Acknowledgments. The fabrication was performed with the C2N micro nanotechnologies platforms and partly supported by the RENATECH network and the General Council of Essonne. This work has been funded by ANR BRIGHT (ANR-18-CE24-0023-01), the European Union (ERC-SPRING, under the Grant Agreement 101087901), and European Union's Horizon Europe (Marie Sklodowska-Curie grant agreement N° 101062518). Views and opinions expressed are, however, those of the author(s) only and do not necessarily reflect those of the European Union or the European Research Council. Neither the European Union nor the granting authority can be held responsible for them.

References

1. Gundavarapu, S., et al.: Sub-hertz fundamental linewidth photonic integrated Brillouin laser. Nature Photonics **13**(1), 60 (2019)
2. Kittlaus, E.A., Kharel, P., Otterstrom, N.T., Wang, Z., Rakich, P.T.: *RF-Photonic Filters via* "On-Chip Photonic-Phononic Emit–Receive Operations." J. Lightwave Technol. **36**(13), 2803 (2018)
3. Merklein, M., Stiller, B., Vu, K., Ma, P., Madden, S.J., Eggleton, B.J.: On-chip broadband nonreciprocal light storage. Nanophotonics **10**(1), 75–82 (2021)
4. Wolff, C., Steel, M.J., Eggleton, B.J., Poulton, C.G.: Stimulated Brillouin scattering in integrated photonic waveguides: Forces, scattering mechanisms, and coupled-mode analysis. Phys. Rev. A **92**(1), 013836 (2015)
5. Shin, H., et al.: Tailorable stimulated Brillouin scattering in nanoscale silicon waveguides. Nat. Commun. **4**(1), 1944 (2013)
6. Wiederhecker, G.S., Dainese, P., Mayer Alegre, T.P.: Brillouin optomechanics in nanophotonic structures. APL Photonics **4**(7), 71101 (2019)
7. Kittlaus, E.A., Shin, H., Rakich, P.T.: Large brillouin amplification in silicon. Nat. Photonics **10**(7), 463–467 (2016)
8. Laer, R.V., Bazin, A., Kuyken, B., Baets, R., Thourhout, D.V.: Net on-chip brillouin gain based on suspended silicon nanowires. New J. Phys. **17**(11), 115005 (2015)
9. Van Laer, R., Kuyken, B., Van Thourhout, D., Baets, R.: Interaction between light and highly confined hypersound in a silicon photonic nanowire. Nat. Photonics **9**(3), 199–203 (2015)
10. Cheben, P., Halir, R., Schmid, J.H., Atwater, H.A., Smith, D.R.: Subwavelength integrated photonics. Nature **560**, 565–572 (2018)
11. Safavi-Naeini, A.H., Thourhout, D.V., Baets, R., Laer, R.V.: Controlling phonons and photons at the wavelength scale: integrated photonics meets integrated phononics. Optica **6**(2), 213–232 (2019)
12. Kharel, P., Behunin, R.O., Renninger, W.H., Rakich, P.T.: Noise and dynamics in forward Brillouin interactions. Phys. Rev. A **93**(6), 063806 (2016)

Quantum Technologies

High-Visibility Interference for Time Bin Encoded Entanglement on Silicon Integrated Platform

Yue Qin, Hongnan Xu, Gaolei Hu, and Hon Ki Tsang(✉)

The Chinese University of Hong Kong, Shatin, NT, Hong Kong, SAR of China
hktsang@ee.cuhk.edu.hk

Abstract. Time bin entangled photon pairs are robust against transmission fluctuations and suitable for long-distance quantum entanglement distribution. Here we report the first monolithically integrated silicon time-bin encoding entanglement system, which integrates the photon sources, wavelength demultiplexers, and Franson interferometers on a single chip. We measured quantum interference visibility of 98.98% $\pm$ 0.55%, which is among the highest yet reported for monolithic integrated time-bin entanglement systems.

Keywords: Quantum Entanglement · Time Bin Encoding · Silicon Photonics · Integrated Photonics

1 Introduction

In recent years integrated quantum photonics has advanced considerably [1], and quantum entanglement networks with enhanced security have been demonstrated [2, 3]. Among the different quantum encoding methods, such as polarization-, mode-, path-, and frequency-bins, time-bin encoding is well suited for long-distance entanglement distribution because of its robustness to fluctuations in transmission media [4, 5]. Time-bin encoding entanglement was first proposed by Franson [6] for the test of Bell's inequality [7]. Time-bin entanglement generation using bulk optics [17] and fiber optics [16] are now well established. With the advance of silicon photonics, there has been increasing interest on integrated platforms for time-bin entanglements that can make the systems more compact, better in performance and more robust to environment disturbances. Unbalanced Mach-Zender interferometers (UMZIs) with path time delays of the order of nanoseconds are key components in time-bin entanglement generation. Planar light circuits (PLC) provide a robust platform for UMZIs and have been widely used in time-bin entanglement generation [12]. Recently, a large channel capacity was demonstrated with wavelength demultiplexers [10]. However, silica PLCs have low nonlinearity, which makes the silica PLC ill-suited for nonlinear multi-photon generation by spontaneous four wave mixing, and the broad spontaneous Raman emission spectrum in silica can degrade the quality of silica based single photon sources. Silicon photonics has attracted much attention for integrated quantum photonics because it has higher

J. Witzens et al. (Eds.): ECIO 2024, SPPHY 402, pp. 217–221, 2024.
https://doi.org/10.1007/978-3-031-63378-2_36

nonlinearity than silica and its Raman emission is limited to a very narrow spectral of only ~105 GHz. Time-bin entanglement using silicon waveguides and discrete Franson interferometers was reported recently [20]. Previously reported monolithic integration for time-bin entanglement include the use of the Triplex (silicon nitride) platform [9, 13], which yielded quantum interference visibility of around 88%. Compared to the Triplex platform, silicon waveguides have higher index contrast which enable tighter waveguide bends and smaller footprint, and silicon photonics can potentially be lower cost when scaled to large volume production, while the higher third-order nonlinearity of silicon improves the efficiency of generating photon pairs from spontaneous four wave mixing. The main disadvantage of silicon waveguides are their typically higher linear losses than silica platforms, but we mitigate this in this paper by using multimoded waveguides in the long delay lines needed for the Franson interferometer.

In this paper, we report the first silicon integrated monolithic time-bin entanglement system and achieve a record high quantum interference visibility for a monolithic system. The entangled photon sources, wavelength demultiplexers, and Franson interferometers are all monolithically integrated on a single chip. The quantum interference visibility of the system was $98.98\% \pm 0.55\%$. This result represents the highest visibility yet reported for monolithic time-bin entanglement systems. Our work also shows the potential for practical implementation of a time-bin entanglement system that is fully integrated with the pump lasers [19] and single-photon detectors [18].

2 System Design

The photonic circuit is schematically shown in Fig. 1(a). A spiral single-mode waveguide is used as the nonlinear photon source (labelled as part I) which produces entangled photon pairs by spontaneous four wave mixing. The single-mode waveguide has a cross section area of 450×220 nm^2 and a length of 1 cm. Two asymmetric MZIs with different free-spectral ranges are combined as a wavelength demultiplexer, which rejects the pump in part II and separates signal and idler photons in part III.

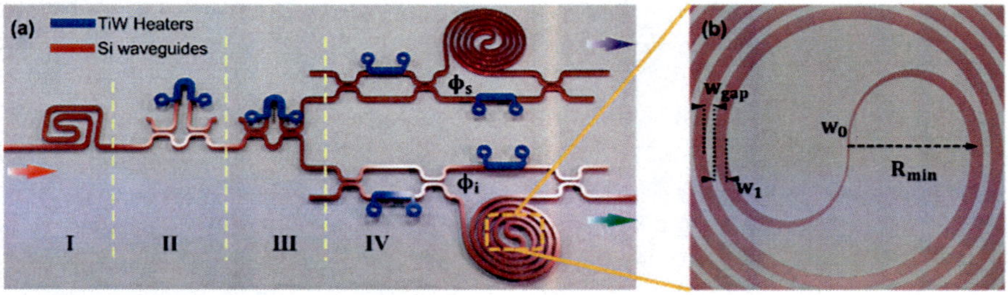

Fig. 1. (a) The schematic of our proposed systems, which involves photon sources (I), wavelength filters (II & III) and Franson interferometers (IV) for time-bin entanglement generation. (b) The design schematic of the low loss multimode waveguides used as delay lines.

Part IV are Franson-type interferometers built with two UMZIs where the path delay difference is longer than the photon coherence time of ~100ps (determined by the spectral bandwidth of the off-chip bandpass filters) so that classical interference is not present. The path delay difference between the long and short paths is designed to be 380 ps and was implemented with a 3-cm long spiraled multimoded waveguide (MMW). The MMW reduced the optical overlap with sidewalls and hence minimize the linear scattering propagation loss [14]. We set the width of the MMW as $w_1 = 2.7\ \mu\text{m}$ and the gap width as $w_{\text{gap}} = 1.7\ \mu\text{m}$ to avoid inter-waveguide coupling. The minimum radius is $R_{\min} = 27.5\ \mu\text{m}$ and the central width of the S-bend is $w_0 = 400$ nm. The loss of the spiral waveguide was measured to be about −0.9 dB, which is much lower than the 4.5 dB loss of a comparable delay single-mode waveguide. To compensate for the loss imbalance in between the long path and short path in the UMZIs, two variable beam splitters are used to replace the input 50-50 directional couplers.

After the two UMZIs, the signal and idler photon states can be described as $\frac{|L\rangle_s + e^{i\phi_s}|S\rangle_s}{\sqrt{2}}$ and $\frac{|L\rangle_i + e^{i\phi_i}|S\rangle_i}{\sqrt{2}}$, in which $|L\rangle$ and $|S\rangle$ respectively denote the long and short paths available to the photons and the subscripts indicate signal and idler photons. After post selecting in coincidence basis, the bi-photon states become a Bell state as follows:

$$|\Psi\rangle = \frac{|L_s L_i\rangle + e^{i(\phi_s+\phi_i)}|S_s S_i\rangle}{\sqrt{2}}$$

Since the cases when the two photons take either both long paths $|L_s L_i\rangle$, or both short paths $|S_s S_i\rangle$, are indistinguishable, the quantum interference result can be used to assess the violation of Bell's inequality and verify quantum entanglement [6].

3 Experiment Result

The whole experimental setup is shown in Fig. 2(a). The device under test (DUT) was wire-bonded and attached to a copper heatsink connected to a thermal electric cooler (TEC) for temperature control, as shown in Fig. 2(b). Light was coupled into the chip via an array of ultra-high numerical aperture fibers (UHNA7). Three commercial dense wavelength division demultiplexing (DWDM) single channel band-pass filters with 200 GHz bandwidth were connected to the system. The center wavelength of DWDM filters were 1558.98 nm, 1552.52 nm and 1546.12 nm, respectively used for signal, pump, and idler filtering. The electrodes were routed through a customized PCB board and connected to a home-made multi-channel voltage source (MCVS) with current feedback. The applied phase to the photon states were normalized by the measured half-wave power P_π ($\approx$ 20.8 mW) of the heaters.

The coincidence results are shown in Fig. 2(C) and (D). The three peaks represent bi-photon state $|L_s S_i\rangle$, $\frac{|L_s L_i\rangle + e^{i(\phi_s+\phi_i)}|S_s S_i\rangle}{\sqrt{2}}$, and $|S_s L_i\rangle$, respectively. Here, the time-bin histogram resolution is 10 ps and the coincidence integration time is 1 s. The actual path delay can be characterized from the time-bin distance between each pair of adjacent peaks. The measured results of time delay (~390 ps) are in good accord with simulations. The single photon rates of each channel are shown in Fig. 2(f). The curves stable over time, without noticeable single-photon interference pattern.

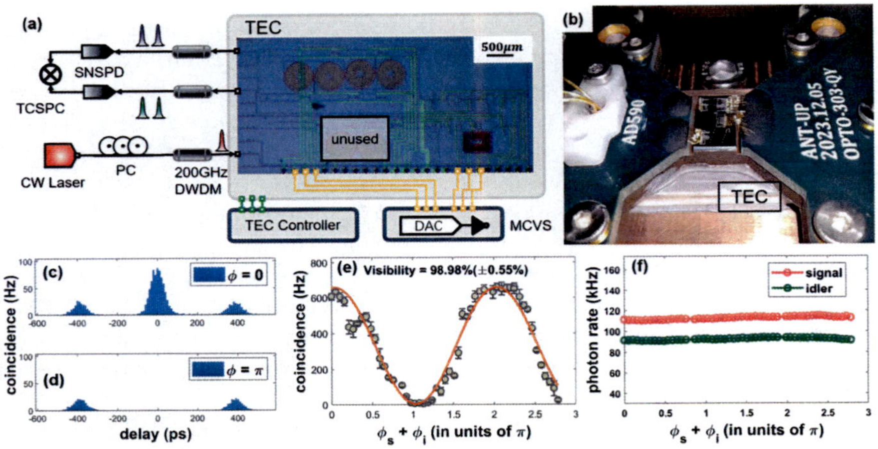

Fig. 2. (a) Experimental setup for system test (left) and microscope image of the chip (right). (b) Wire-bounded DUT on copper heatsink and TEC. (c)–(d) Measured time-bin histogram of photon correlations with varying channel delay between detectors. Here, the central peak denotes the entanglement state $|\Psi\rangle = \frac{|L_sL_i\rangle + e^{i(\phi_s+\phi_i)}|S_sS_i\rangle}{\sqrt{2}}$. (e) Total coincidence counts measured in the central peak with different phase applied to phase shifters. (f) Measured single-photon count rates of two channels with varying phase shift applied.

Figure 2(e) shows the result of our bi-photon interference. We record the coincidence counts of the central peak with a time window of 110 ps, corresponding to the FWHM of the central peak. The measured visibility is 98.98% ± 0.55% using a integration time of 1s. We think linear optical losses in our measurement system (including 4.5 dB in edge couplers, 1.5 dB in DWDM filters and 1 dB in fiber connections) are the main impairments that degrade the visibility of this integrated system. The optical losses degrade the photon brightness and thus require longer integration time, resulting in the accumulation of environment fluctuations. The error bars represent the calculated one standard deviation of five independent measurements. Compared to the other integrated time-bin entanglement generation systems listed in Table1, the demonstrated system features the highest visibility and highest photon coincidence rate.

Table 1. Comparison of visibilities in integrated time-bin systems.

Year	Platforms	Best visibility	Coincidence rate
2016 [13]	Triplex Si_3N_4 + Silicon source	88.4%	~2.3 Hz
2018 [9]	Triplex Si_3N_4	88%	~2 Hz
2022 [8]	Silicon + Silicon source	93.13%	NA
2023 [10]	Silica + Silicon source	96.329 ± 0.760%	~3.75 Hz
Our work	**Silicon**	**98.98% ± 0.55%**	**~600Hz**

4 Conclusion

We have proposed and demonstrated the first monolithically integrated silicon time-bin entanglement system with built-in photon source, wavelength demultiplexers and Franson interferometers. The low loss delay line and high nonlinearity enabled a record high visibility up to 98.98% ± 0.55%.

Acknowledgement. This work was supported by Hong Kong RGC Research Matching Grant RMG-01. YQ also thank RGC for funding support from the Hong Kong PhD Fellowship scheme.

References

1. Pelucchi, E., et al.: The potential and global outlook of integrated photonics for quantum technologies. Nat. Rev. Phys. **4**, 194–208 (2021)
2. Koduru, S., et al.: A trusted node–free eight-user metropolitan quantum communication network. Science Advances (2020)
3. Wengerowsky, S., et al.: An entanglement-based wavelength-multiplexed quantum communication network. Nature **564**, 225–228 (2018)
4. Kim, J.H., et al.: Quantum communication with time-bin entanglement over a wavelength-multiplexed fiber network. APL Photonics 7 (2022)
5. Wen, W., et al.: Realizing an entanglement-based multiuser quantum network with integrated photonics. Phys. Rev. Appl. **18**, 024059 (2022)
6. Franson, J.D.: Bell inequality for position and time. Phys. Rev. Lett. **62**, 2205–2208 (1989)
7. Bell, J.S.: On the einstein podolsky rosen paradox. Physics **1**, 195–200 (1964)
8. Chemnitz, M., et al.: Telecom-compatible, on-chip generation and processing of complex photon states in time and frequency. In: Proc. SPIE 12004, Integrated Optics: Devices, Materials, and Technologies XXVI, 1200409, 5 March 2022. https://doi.org/10.1117/12.2607224
9. Zhang, X., et al.: Integrated silicon nitride time-bin entanglement circuits. Opt. Lett. **43**, 3469–3472 (2018)
10. Ren, S.-Y., et al.: Photonic-chip-based dense entanglement distribution. PhotoniX **4**, 12 (2023)
11. Zheng, Y., et al.: Multichip multidimensional quantum networks with entanglement retrievability. Science **381**, 221–226 (2023)
12. Takesue, H., Inoue, K.: Generation of 1.5 μm band time-bin entanglement using spontaneous fiber four-wave mixing and planar light-wave circuit interferometers. Phys. Rev. A **72** (2005)
13. Xiong, C., et al.: Compact and reconfigurable silicon nitride time-bin entanglement circuit. Optica **2**, 724–727 (2015)
14. Hong, S. et al.: Ultralow-loss compact silicon photonic waveguide spirals and delay lines. Photonics Res. **10** (2021)
15. Fujiwara, M., Wakabayashi, R., Sasaki, M., Takeoka, M.: Wavelength division multiplexed and double-port pumped time-bin entangled photon pair generation using Si ring resonator. Opt Express **25**, 3445–3453 (2017)
16. Reimer, C., et al.: Generation of multiphoton entangled quantum states by means of integrated frequency combs. Science **351**, 1176–1180 (2016)
17. Jayakumar, H., et al.: Time-bin entangled photons from a quantum dot. Nat. Commun. **5**, 4251 (2014)
18. Akhlaghi, M.K., Schelew, E., Young, J.F.: Waveguide integrated superconducting single-photon detectors implemented as near-perfect absorbers of coherent radiation. Nat. Commun. **6**, 8233 (2015)
19. Liang, D., Bowers, J.E.: Recent progress in lasers on silicon. Nature Photon **4**, 511–517 (2010)

Industrial Ion Trap Chips with Integrated Optics

Alexander Zesar[1,2](✉), Jakob Wahl[1,3], Sofia Cano Castro[1,4], Klemens Schüppert[1], Yves Colombe[1], Silke Auchter[1], Max Glantschnig[1,5], Clemens Rössler[1], Bernhard Lamprecht[6], Philipp Hurdax[6], Marco Schmauser[3], Marco Valentini[3], Philipp Schindler[3], Thomas Monz[3], Oscar Jimenez Gordillo[4], Andrea Melloni[4], and Joachim Krenn[2]

[1] Infineon Technologies Austria AG, Siemensstraße 2, 9500 Villach, Austria
alexander.zesar@infineon.com
[2] Institut für Physik, Universität Graz, Universitätsplatz 5, 8010 Graz, Austria
[3] Institut für Experimentalphysik, Universität Innsbruck, Technikerstraße 25, 6020 Innsbruck, Austria
[4] Photonics Devices Group, Polytecnico di Milano, Via Ponzio 34, 20133 Milan, Italy
[5] Physikalisch-Technische Bundesanstalt, Bundesallee 100, 38116 Braunschweig, Germany
[6] Joanneum Research Materials, Franz-Pichler-Strasse 30, 8160 Weiz, Austria

Abstract. Trapped ions are among the most researched and advanced quantum computing hardware platforms. The use of free-space optics does not scale to large, high-fidelity ion trap processors due to beam pointing errors and spatial restrictions. Therefore, future ion trap chip architectures require integrated waveguiding and focusing for stable beam placement in ion ensembles. This paper gives an overview of photonics and optics integration schemes developed in the industrial environment of Infineon. Challenges that come with femtosecond laser written waveguides as well as thin film slab waveguides in conjunction with focusing grating couplers, including fiber-to-chip coupling and integration density are discussed.

Keywords: Ion Trap Chip · Optics Integration · Industrial Microfabrication

1 Introduction

Quantum computing (QC) will lead to a paradigm shift in computation enabling the solution of yet intractable problems. Useful quantum computers will require a high number of quantum bits (qubits). Trapped ions provide a scalable QC platform in which all ions are physically identical and well isolated quantum systems acting as qubits. Ion traps chips are able to host thousands of ions in two-dimensional arrangements via electromagnetic fields that originate from electrodes on the chip surface. Individual ion manipulation and readout requires laser beams with wavelengths ranging from the ultraviolet to near infrared [1]. Numerous laser beams must be placed in a precise and stable manner inside the two-dimensional ion ensemble to achieve low error rates and

J. Witzens et al. (Eds.): ECIO 2024, SPPHY 402, pp. 222–229, 2024.
https://doi.org/10.1007/978-3-031-63378-2_37

a high qubit connectivity (see Fig. 1). Optical integration in ion traps chips provides laser beams pointed at specific processing zones inside the ion trap chip fulfilling the above-mentioned requirements for precision and stability.

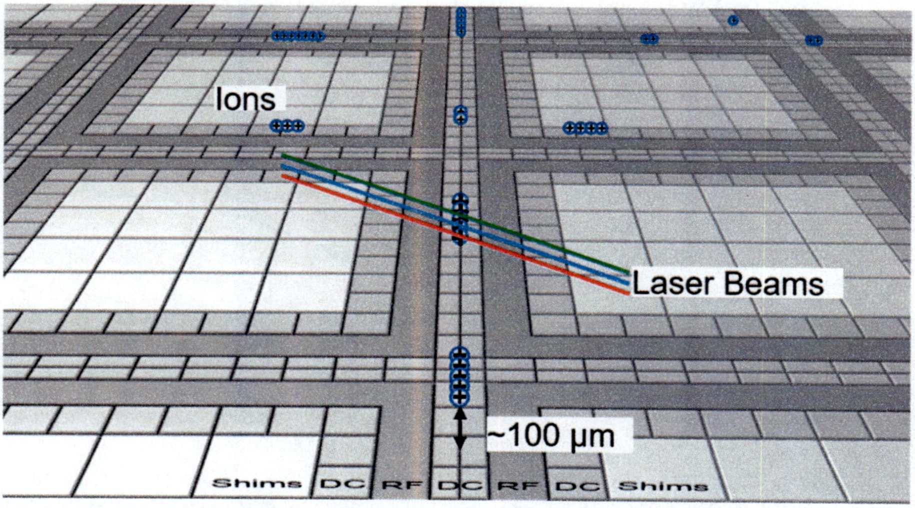

Fig. 1. Schematic of a quantum charge-coupled device (QCCD) architecture [2]. Electrode types are marked as *RF* for radio frequency (radial confinement), *DC* for slowly varying potentials (axial confinement and ion shuttling) and *Shims* for stray charge compensation. Laser beams are symbolically placed over single ions inside an ion chain in the middle of the graph.

This paper motivates industrial microfabrication before presenting an overview over two different approaches to optics integration into ion trap chips developed at Infineon Austria in cooperation with our academic partners. The first variant uses femtosecond laser written waveguides in the bulk of a borosilicate glass piece bonded onto the surface of an ion trap chip. The second approach relies on structured thin film photonics based on silicon nitride monolithically integrated in the ion trap chip below the surface electrodes. Finally, an optics testing setup for chip characterization as well as fiber-to-chip assembly is discussed.

2 Industrial Microfabrication of Ion Trap Chips

Ion trap chips have been produced in university cleanrooms [3, 4] as well as in larger microchip factories [5]. Main advantages of a smaller production facility are the flexibility in terms of process changes and tool modifications as well as faster overall processing times for single wafers or chips.

On the industrial scale ion trap dedicated fabrication facilities are not economical, which necessitates a shared use with other products. This leads to restrictions in process and tool modification, materials used and prolongs the overall processing time. However, the availability of a large park of well-maintained state-of-the art production tools including automated inspection (see Fig. 2) and metrology machines enable the fabrication of complex ion trap chips including multi metal layers, through substrate vias, multi-wafer stacks and optics integration. Furthermore, an experiment management system allows automated trial runs with many wafers in parallel for fast process validation.

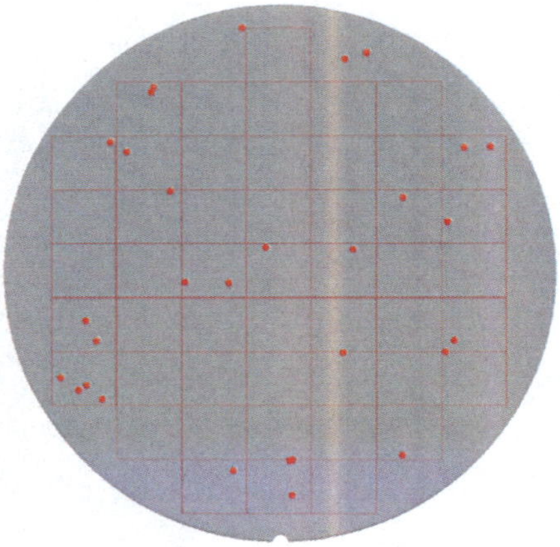

Fig. 2. Defect map created via an automated defect inspection tool in the cleanroom of Infineon Villach. Here a 200 mm silicon wafer with a silicon nitride layer on top was inspected. A total amount of 30 defects including scratches and dust particles with a size >100 nm has been found. The grid indicates the placement of individual devices.

3 Optics Integration

3.1 Femtosecond Laser Written Waveguides in Borosilicate Glass

A highly focused femtosecond laser beam is scanned over a borosilicate glass substrate. The focus volume is fully inside the glass bulk and the trajectory of the movement relative to the glass describes the waveguide. A modification in the glass caused by the laser pulses leads to a rise in refractive index relative to the unmodified material [6].

We inscribe such waveguides in glass pieces (nose like structures) which are bonded on top of an ion trap chip (see Fig. 3A). On the chip edge a single mode fiber (Thorlabs 630HP) for 730 nm light is attached to the waveguide with index matched glue (Norland NOA61), rendering a polishing of the mechanically diced sidewall unnecessary. Depending on the waveguide writing parameters, most importantly the pulse energy, the mode field diameter for 730 nm light is in the range of 3.5 to 6.0 μm.

Including coupling losses at the in- and output of the inscribed waveguide, the overall transmission at 730 nm light over a length of 1 cm is 0.85 ± 0.05 (corresponds to a loss of 0.5 to 1 dB). As described in Fig. 3, a flat end facet at the tip of the nose is achieved via cleaving, which leads to an unperturbed Gaussian beam near the at the ion location with an NA of 0.03–0.05 depending on the mode field diameter. The distance between ion and nose tip is about 1 mm to decrease the influence of stray charges on the nose leading to ion micromotion. A transparent conductive coating on the glass piece would allow the nose tip to be placed closer to the operation zone decreasing the beam diameter at the ion position.

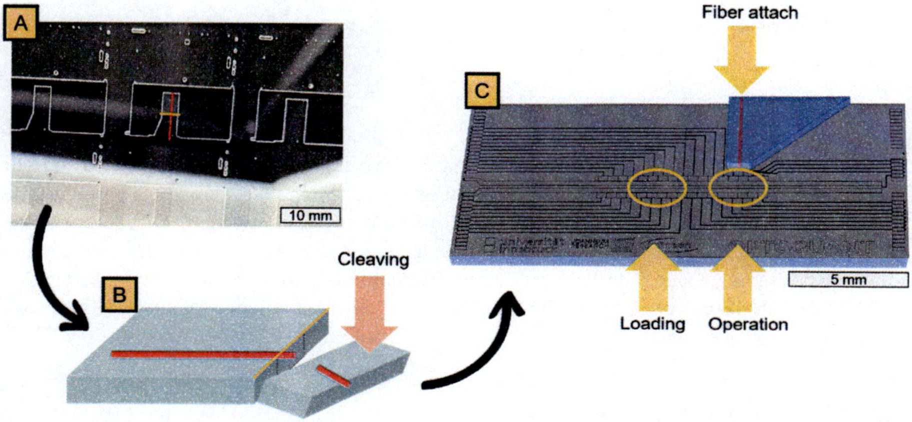

Fig. 3. Overview of the processing steps of the optics part in the "Nosetrap" ion trap. **A**: One or several waveguide(s) are written in nose-like structures in a glass wafer (shown in red). The orange line indicates the position of an ablation trench. **B**: The tips of the nose-like structures is removed via a flat cleave along the ablation line. **C**: The thus fabricated glass wafer is bonded on top of a second wafer containing the surface electrodes to complete the ion trap.

The integration of multiple waveguides in one chip coupled to a fiber array requires bend waveguides due to the mismatch in fiber pitch and ion to ion distance. Bends in waveguides lead to radiative losses as well as increased scattering Fig. 4. A shows the loss of a curved waveguide relative to a straight waveguide of the same length. Below a bending radius of 10 mm the relative transmission drops significantly below 0.8. Furthermore, our waveguides exhibit considerable amounts of stray light guided in the bulk of the material due to scattering in straight and bent sections (see Fig. 4B). Therefore, we conclude that the herein proposed optics integration scheme is suitable for small and intermediate scale ion trap architectures, where a high integration density is not an issue – for example in optical clocks [7].

3.2 Thin Film Waveguides and Grating Couplers

Ion traps with integrated thin film photonics have already been demonstrated for the silicon nitride (SiN) [3] and aluminium oxide platforms [5]. SiN photonics allows for waveguides with high integration densities (minimal bending radii on the order of 50 μm) as well as grating couplers with tight focusing on the order of a few μm spot diameter [8]. Our mission is to establish an industrial photonics platform for ion trap chips.

Image 5A and B show sketch cross sections through a waveguide as well as a grating coupler. Typical SiN core layer thicknesses are between 100 and 200 nm, whereas the

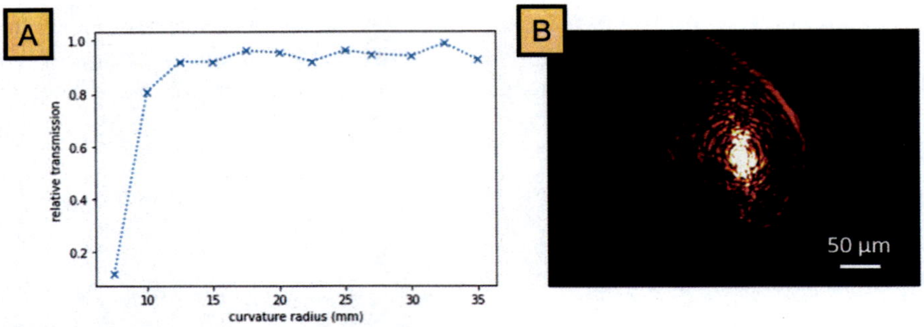

Fig. 4. A: Transmission of a curved femtosecond laser written waveguide relative to a straight waveguide of the same length as a function of curvature radius at a wavelength of 730 nm. The minimal admissible bending radius is on the order of 10 mm or larger. **B**: Microscope image of the cleaved end facet of a bend waveguide. Around the intensity peak due to the guided mode in the middle one can see a considerable amount of stray light stemming from scattering at the waveguide core-cladding interface. Straight waveguides exhibit considerable amounts of stray light as well (not shown).

width of waveguides is in the range of 200–1000 nm. Grating couplers for wavelengths between 400–800 nm require duty cycles well below 200 nm. Both gratings and waveguides are structured in one step putting high demands on the lithography and etching processes. With our current lithography and etching process we could fabricate structures with 200 nm feature sizes in 150 nm thick SiN on silicon oxide (see Fig. 6B and C).

Low waveguide losses are required for large and complex ion trap chips to keep stray light due to scattering and heat dissipation as small as possible. The roughness of the core top and bottom as well as the sidewalls are main contributors to scattering loss. Chemical mechanical polishing helps to remove top and bottom roughness, optimization in lithography and etching processes will reduce sidewall roughness (Fig. 5).

A first silicon waveguide test chip was produced in the cleanrooms at Infineon Villach, containing waveguide structures with varying lengths and widths as well as grating couplers and ring resonators to benchmark the limits of lithography and etching (see Fig. 6A). The lower cladding was formed by thermal oxidation of a 100-silicon wafer with a thickness of >1500 nm followed by a silicon nitride deposition via plasma enhanced chemical vapour deposition (PECVD) with thicknesses ranging from 100 to 200 nm. A resist was exposed with a 248 nm KrF excimer stepper. After reactive plasma etching an upper cladding was deposited via PECVD. Automated defect inspection, critical dimension measurement and dielectric thickness measurement complement the fabrication for process monitoring.

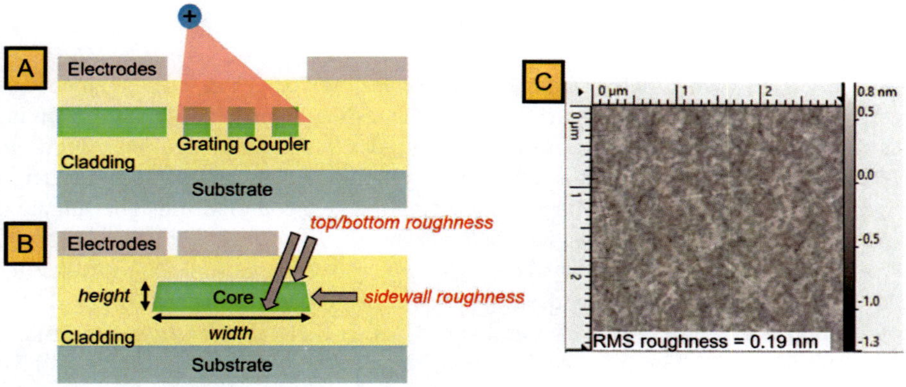

Fig. 5. **A**: Sketch cross section of a waveguide leading to a grating coupler. The light coming from the waveguide is diffracted and focused onto the ion through a gap in the metal electrode layer. **B**: Sketch cross section through a waveguide showing a core with top/bottom as well as sidewall roughness. Typical heights for SiN cores are 100–200 nm whereas the widths are within 200–1000 nm depending on the wavelength. **C**: Atomic force microscopy image of a PECVD silicon nitride surface after chemical mechanical polishing resulting in an RMS roughness of 0.19 nm.

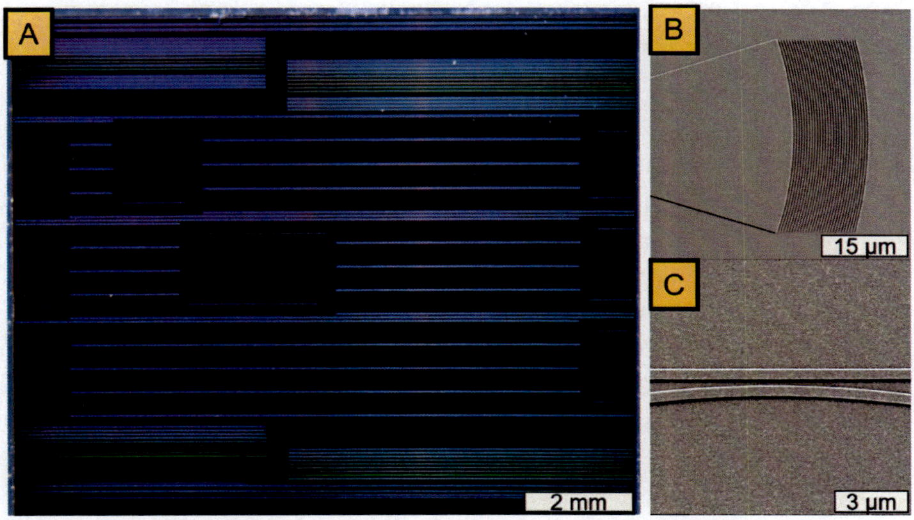

Fig. 6. A first test chip was produced at Infineon Austria containing waveguides of different widths and lengths for power transmission analysis. **A**: Optical micrograph of a chip section containing waveguides of variable lengths. **B**: Electron micrograph of a grating coupler with a period of 400 nm. **C**: Electron micrograph of the gap in a ring resonator.

4 Testing of Industrial Ion Traps with Integrated Optics

The fabrication of photonic structures requires testing for process development and screening as well as for design validation. Our testing setup features an incoupling assembly, which couples light into SiN waveguides at the chip edge via a high NA objective (Nikon LU Plan 100x NA 0.9). The system will be upgraded to enable active alignment of single fibers and fiber arrays using the hexapod platform. The sample is mounted on a holder fixed onto a Z-translation stage.

An analysis assembly is mounted on a XY-translation stage clamped onto an elevated table. A microscopic image produced by an objective (Nikon LU Plan 50x NA 0.8) in conjunction with a tube lens is captured via a CMOS. Polarization analysis is possible via a rotating polarization filter. A photodiode mounted on top of a beam splitter allows power transmission measurement as well as fast automatic incoupling beam alignment or fiber alignment. The entire analysis assembly can be rotated and moved on the optical table to facilitate grating analysis (Fig. 7).

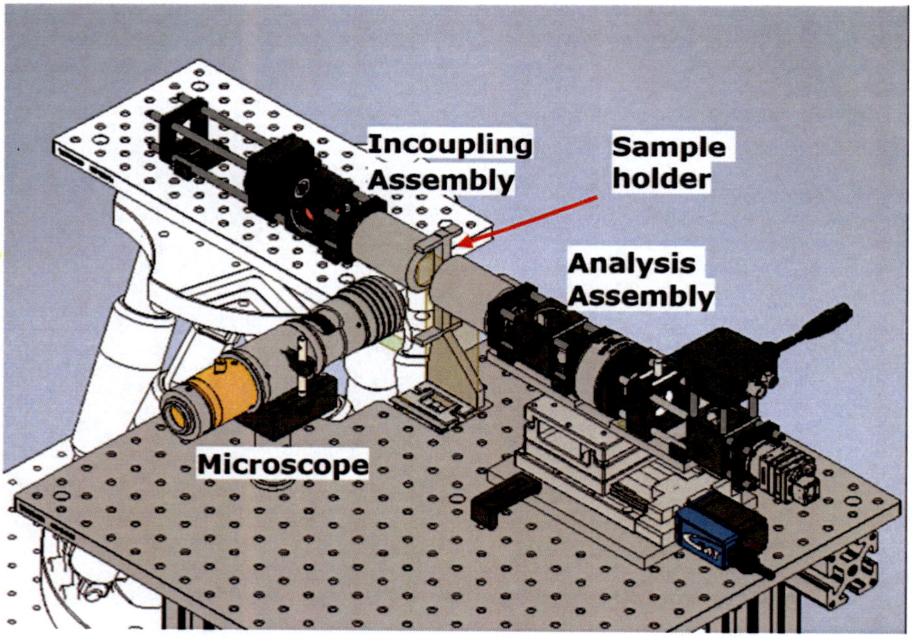

Fig. 7. CAD image of a photonics chip testing setup. The incoupling assembly is mounted on a hexapod with 6 degrees of freedom and allows for coupling of light into the chip via a tightly focused beam. The analysis assembly consists of a microscope with a magnification of 50x and an NA of 0.8. A beam splitter allows us to capture the image via CMOS sensor as well as photodiode for fast response as well as power measurement. An additional microscope looks at the chip from the side and helps in coarse alignment.

5 Outlook

Optics integration of femtosecond laser written waveguides in ion traps has been demonstrated. Further possible improvements mainly concern stray light management via precise fiber alignment and waveguide writing process optimization. Nosetrap chips will be tested in a cryogenic setup at the university of Innsbruck to demonstrate quantum operations in trapped ions.

Thin film photonics requires additional development, especially in loss reduction. Core sidewall roughness analysis will be done to estimate their loss contribution which will be compared to waveguide transmission measurements in the characterization setup. Furthermore, mode sizes for standard fibers and SiN waveguides differ significantly. Spot-size converters adiabatically transform the spot size [9] while introducing additional fabrication complexity. Lastly, effective gratings require duty cycles below 200 nm [10] which will be reached by using 193 nm ArF exposure tools.

References

1. Wineland, D.J., et al.: Quantum information processing with trapped ions. Philosophical Trans. Roy. Soc. London. Series A: Math. Phys. Eng. Sci. **361** (2003). edited by P. L. Knight, E. A. Hinds, and M. B. Plenio, 1349
2. Kielpinski, D., Monroe, C., Wineland, D.J.: Architecture for a large-scale ion-trap quantum computer. Nature **417**, 709 (2002)
3. Niffenegger, R.J., et al.: Integrated multi-wavelength control of an ion qubit. Nature 586.7830, 538-542 (2020)
4. Mordini, Carmelo, et al.: Multi-zone trapped-ion qubit control in an integrated photonics QCCD device.arXiv preprint arXiv:2401.18056 (2024)
5. McGuinness, H.J., et al.: Integrated photonics for trapped ion quantum information experiments at Sandia National Laboratories. Quantum Nanophotonic Materials, Devices, and Systems 2022, vol. 12206. SPIE (2022)
6. Tan, D., et al.: Fabricating low loss waveguides over a large depth in glass by temperature gradient assisted femtosecond laser writing. Optics Lett. **45**(14), 3941–3944 (2020)
7. Poli, N., Oates, C.W., Gill, P., Tino, G.M.: Optical atomic clocks. La Rivista del Nuovo Cimento **36**(12), 555–624 (2013). https://doi.org/10.1393/ncr/i2013-10095-x
8. Song, J.H., et al.: Adiabatically bent waveguides on silicon nitride photonics for compact and dense footprints. IEEE Photonics Technol. Lett. 28(20), 2164–2167 (2016)
9. Cegielski, P.J., et al.: Silicon nitride waveguides and spot size converters with< 1.76 dB loss over broad wavelength range from 1010 nm to 1110 nm for OCT applications. In: Proc. Passive Photonic Integr. Struct. New Platforms, Eur. Conf. Integr. Opt.(ECIO) (2020)
10. Beck, G.J., Home, J.P., Mehta, K.K.: Grating design methodology for tailored free-space beam-forming. arXiv preprint arXiv:2306.09220 (2023)

Photonic Platform for Industrially Microfabricated Ion Traps

Sofia Cano Castro[1,2](✉), Alexander Zesar[1,3], Max Glantschnig[1,4], Oscar A. Jimenez Gordillo[2], Silke Auchter[1], Yves Colombe[1], Clemens Rössler[1], and Andrea Melloni[2]

[1] Infineon Technologies Austria AG, Siemensstraße 2, 9500 Villach, Austria
sofia.canocastro@infineon.com
[2] Politecnico di Milano, P.za Leonardo da Vinci, 32, 20131 Milan, Italy
[3] University of Graz, Universitätsplatz 3, 8010 Graz, Austria
[4] Physikalisch-Technische Bundesanstalt, Bundesallee 100, Braunschweig 38116, Deutschland

Abstract. On chip light delivery has proven to be key for scaling up the number of qubits in a trapped ion quantum processor and enable large scale quantum computing. This can be achieved with structured thin films dielectrics, but its fabrication comes with challenges including a broad wavelength range and thermal budget constraints. At Infineon, we are studying photonic structures suitable for the visible spectrum, their fabrication processes and material properties. We present results for low surface roughness silicon nitride films and loss measurements on waveguides as a first step to integrate photonic layers in our microfabricated ion traps.

Keywords: quantum computing · ion traps · silicon nitride · industrial microfabrication

1 Introduction

The high prospect for quantum computing has pushed developments for different physical implementations, including trapped-ion based systems. To enable applications of practical interest there is an important challenge related to the scalability of ion traps [1] and Infineon is tackling this issue leveraging precision and repeatability of state-of-the-art industrial microfabrication.

A key element for the control of ions in Paul traps is laser beam delivery. Operation, readout, and cooling is achieved with coherent light in resonance with dipole and quadrupole transitions of the chosen ion species that need to reach different trapping sites in a scalable and robust manner. Passive integrated photonic elements including waveguides, splitters and grating out-couplers that route the light and focus it at the ion position have already been demonstrated [2]. This integration approach meets the need to increase the number of qubits on a single chip and leads to performance gains, paving the way for error-corrected universal quantum computation. It suppresses beam pointing

J. Witzens et al. (Eds.): ECIO 2024, SPPHY 402, pp. 230–234, 2024.
https://doi.org/10.1007/978-3-031-63378-2_38

errors usually arising from the usage of free space optics, provides precise focusing and additional beam shaping capabilities that benefit certain qubit quantum interactions [3].

This work presents a comprehensive development of a photonic platform, addressing the need of optical integration for ion traps to enable large scale quantum computing. The specific requirements for ion qubit applications include:

- Light guiding for a broad wavelength window spanning from the near infrared (NIR) to the ultraviolet (UV) on a single chip. Gate operation wavelengths, for instance 729 nm for calcium, pose the biggest constraint. They require tight focusing spots to deliver high power, that set the limit for effective index and geometry of the guiding structures for that wavelength [4].
- A photonic integration scheme compatible with trap fabrication. Ion confinement is driven by DC and RF electrodes placed on top of the chip that need to scale up as well. Multi-layer devices where photonic routing is embedded in a stack with metal lines for electrode routing need to be fabricated. This is a challenge since many mature photonic fabrication technologies rely on high temperature processes that do not comply with the temperature budget of commonly used metals such as aluminum.
- Scaling of the number of optical inputs that require packaging compatible with cryogenic temperatures and ultra-high vacuum. Both mode matching on chip and fiber attach needs to be addressed along with solutions to increase the number of inputs.

2 Material and Fabrication Development

We studied the material properties and fabrication schemes for photonic structures in silicon nitride (Si_3N_4), because of its transparency window, index contrast and loss mechanisms well understood within both academic and industrial environments [5].

2.1 Thin Films

Critical aspects for the optical behavior of the fabricated devices are refractive index, thin film absorption, roughness, and uniformity. Because of the need for alternative low temperature processes, we are comparing low pressure chemical vapor deposition (LPCVD) and plasma enhanced chemical vapor deposition (PECVD) techniques. To improve the film quality, a planarization process (chemical mechanical polishing, CMP) was tested achieving a root mean square roughness (Rq) below 0.2 nm for LPCVD and three different variants of PECVD films, suitable for low loss applications. The highest improvement observed was for one of the PECVD films, reducing RMS roughness from $Rq = 2.13$ nm to $Rq = 0.19$ nm as shown in Fig. 1. From the Payne Lacey Model [6] we can predict a reduction on surface scattering of almost two orders of magnitude from this CMP step leading to a contribution to losses of less than 0.01 dB/cm.

Ellipsometry is used to monitor the refractive index and the thickness uniformity of the films on every wafer. Uniformity can compromise efficiency of grating out-couplers and a deviation of the effective index value leads to a drift from the focusing spot. A change in the effective index of 5% can lead already to a few micrometers difference from the expected focus position at the ion distance (~100 um), from the dependence of the emission angle as: $\theta = \sin^{-1}(n_{eff} - \lambda_0/\Lambda)$ [4].

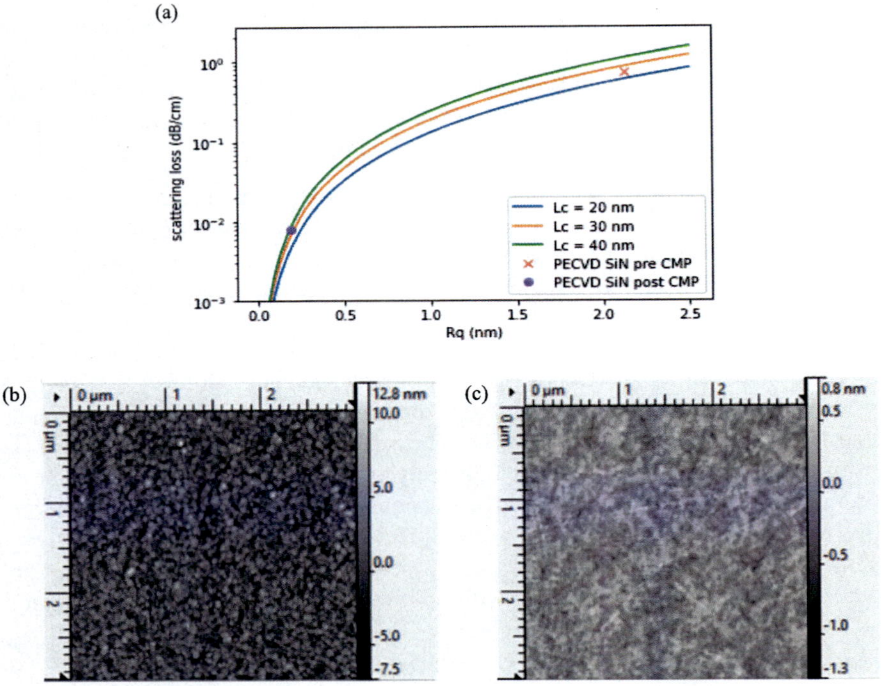

Fig. 1. a) Surface scattering as a function of Rq from Payne Lacey Model [6]. Corresponding value for measured Rq on pre CMP sample with coherence length Lc = 25 nm and post CMP sample with Lc = 36 nm. b) Pre CMP atomic force microscope (AFM) image with Rq = 2.13 nm. c) Post CMP AFM image with Rq = 0.19 nm.

Absorption on the other hand cannot be solely determined from ellipsometry since the sensitivity of these tools is at least an order of magnitude below relevant quantities [7]. Instead, measurements on structured waveguides have been conducted within the expected transparency window of the material, observing transmission for up to 5 cm long waveguides at wavelengths of $\lambda = 633$ nm and $\lambda = 450$ nm.

2.2 Structuring and Losses

Understanding additional loss mechanisms and identifying the main sources requires a more careful study. We designed a test chip including suitable structures for the transition wavelengths of the most prominent ion species and different mode confinements in both single and multi-mode regimes. Results from fabrication and characterization are relevant to the further development of our photonic platform.

We report on the structuring of LPCVD films including waveguides and gratings. For our first fabrication run we used photolithography with 248 nm exposure wavelength and the first chip obtained is shown below.

We measured propagation losses for single mode waveguides with aspect ratios $\frac{w_{core}}{h_{core}}$ of 30, 8 and 3. Analysis was done with cutback measurements and vertical scatter imaging, finding for all the structures losses that remain suitable for the application and in line with literature [7]. Lowest propagation losses were observed for $\frac{w_{core}}{h_{core}} = 30$, below 0.5 dB/cm, as one would expect from the minimized sidewall interaction area [8].

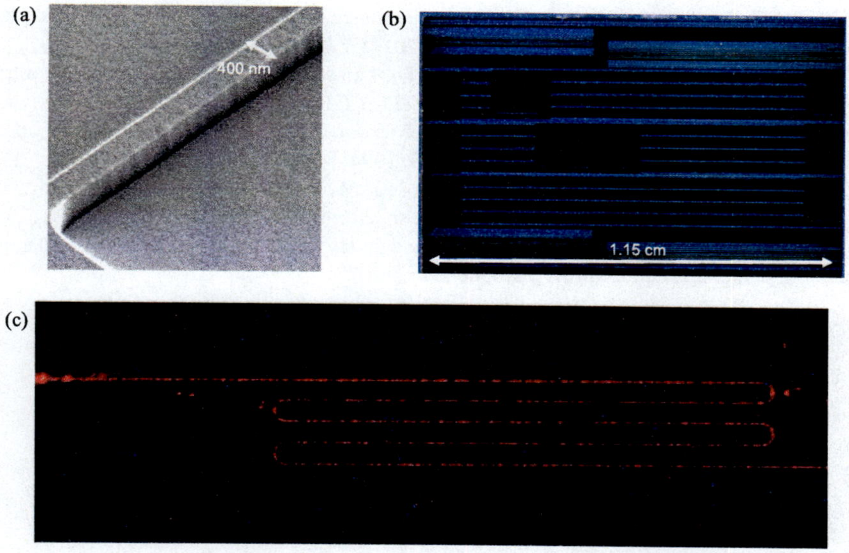

Fig. 2. a) Scanning electron microscope image of structured waveguide before cladding deposition. b) Cutout of first fabricated characterization test chip with waveguides of different lengths and widths. c) Vertical scattering visible in a long exposure photograph for a 4 cm long waveguide with aspect ratio $\frac{w_{core}}{h_{core}} = 30$.

3 Discussion and Outlook

Along with SiN, additional materials with a broader transparency window into the UV are also being investigated. Next steps include structuring of PECVD polished films for loss comparison and optimization of the optical lithography with 193 nm exposure wavelength to enable small pitch gratings beyond 100 nm line/space.

We are looking towards multi-layer integration and active components will eventually become important for further scaling. We are confident that together with innovative microfabrication processes for the traps, integrated photonics will enable the next generation of quantum processing units.

References

1. Bruzewicz, C.D., et al.: Trapped-ion quantum computing: progress and challenges. Appl. Phys. Rev. **6**(2), 021314 (2019)

2. Mehta, K., Bruzewicz, C., McConnell, R., et al.: Integrated optical addressing of an ion qubit. Nature Nanotech **11**, 1066–1070 (2016)
3. Peshkov, A.A., Jordan, E., Kromrey, M., Mehta, K.K., Mehlstäubler, T.E., Surzhykov, A.: Excitation of Forbidden electronic transitions in atoms by Hermite-Gaussian modes. Ann. Phys. **535**, 2300204 (2023)
4. Mehta, K.K.: Integrated optical quantum manipulation and measurement of trapped ions, Ph.D. thesis, Massachusetts Institute of Technology (2017)
5. Blumenthal, D.J., Heideman, R., Geuzebroek, D., Leinse, A., Roeloffzen, C.: Silicon nitride in silicon photonics. Proc. IEEE **106**(12), 2209–2231 (2018)
6. Payne, F.P., Lacey, J.P.R.: A theoretical analysis of scattering loss from planar optical waveguides. Opt. Quant. Electron. **26**, 977–986 (1994)
7. Corato-Zanarella, M., Ji, X., Mohanty, A., Lipson, M.: Absorption and scattering limits of silicon nitride integrated photonics in the visible spectrum. Opt. Express **32**, 5718–5728 (2024)
8. Nitesh, C., et al.: Ultra-low loss visible light waveguides for integrated atomic, molecular, and quantum photonics. Opt. Express **30**, 6960–6969 (2022)

Programmable Photonics

Programmable Integrated Photonics with Phase-Change Materials

Niloy Acharjee and Carlos A. Ríos Ocampo(✉)

University of Maryland, College Park, MD 20742, USA
riosc@umd.edu

Abstract. Phase-change materials (PCMs) have emerged as a promising platform for nonvolatile light modulation. The reversible switching between their stable amorphous and crystalline states leads to an impressive refractive index contrast (*Δn, Δk ~ 1–3*). Integrated photonics has benefited from the progress of PCMs such as Sb_2Se_3 and $Ge_2Sb_2Te_5$ for ultra-compact phase and amplitude modulators using all-optical and electro-thermal approaches. These devices allow small-form-factor silicon photonics with zero-static power yet reconfigurability, which are crucial properties in applications such as in-memory computing, optical synapses, switches, trimming, etc. Here, we review the various switching mechanisms and integration methods of PCMs in integrated photonic platforms.

Keywords: phase change materials · photonic integrated circuits · photonic materials

1 PCM Cells Embedded on Cladding

Due to the contrast between the optical properties of stable crystalline and amorphous phases, PCMs are considered ideal candidates for nonvolatile modulation in integrated photonics. One of the most common integration methods of PCMs has been to deposit them directly onto the waveguide and switch either via an evanescent coupling from within the waveguide or by a waveguide-integrated microheater (e.g. doped silicon, graphene, etc.). Chalcogenide-based PCMs like tellurides (e.g. $Ge_2Sb_2Te_5$), selenides (e.g. Sb_2Se_3), and sulfides (e.g. Sb_2S_3) have all been tested in such a platform, as they can switch between two physical states in response to appropriate heat stimuli: annealing (crystallization) or melt-quenching processes (amorphization) [1]. They can be switched on a sub-microsecond timescale with high reproducibility, enabling ultrafast operation over 10^6 switching cycles [1]. In addition, at normal operating temperatures both states of the PCMs are highly stable for years, which is a key requirement for a truly non-volatile response. PCMs like $Ge_2Sb_2Te_5$ (GST) show a significant change in refractive index in the near-infrared wavelength regime, the spectral region of choice for telecommunication applications. Because of its data retention capabilities [2] and high state discrimination [3] down to nanoscale cell sizes, which enables dense packaging and low-power memory switching, researchers have demonstrated fast and repeatable all-optical, multi-level, multi-bit, non-volatile memory operations, with wavelength division multiplexed

J. Witzens et al. (Eds.): ECIO 2024, SPPHY 402, pp. 237–241, 2024.
https://doi.org/10.1007/978-3-031-63378-2_39

(WDM) access on a chip at telecommunication wavelengths compatible with on-chip optical interconnects. These all-optical memories with GST cell placed directly on top of the nanophotonic waveguide are not restricted in size by the diffraction limit of the input light and can hence be miniaturized to nanoscale dimensions. Figure 1(A) shows how a PCM on top of a waveguide can be programmed to set and reset states and subsequently read out as a change in transmission of the waveguide [4]. The use of microheaters has gained more space in PCM integrated photonics since it enables a scalable platform to integrate many PCM cells and accessing them individually. Moreover, the all-optical evanescent method shown in Fig. 1(A) only works with absorptive PCMs, such as GST. For low- or zero-loss PCMs in the telecommunication bands, such as Sb_2Se_3, microheaters are the only option. Hence, the field has seen a growing number of proposals for microheater materials and geometries aiming to build phase shifters based on low-loss PCMs and amplitude modulation, mainly for computing applications, based on lossy PCMs (e.g. GST) [1].

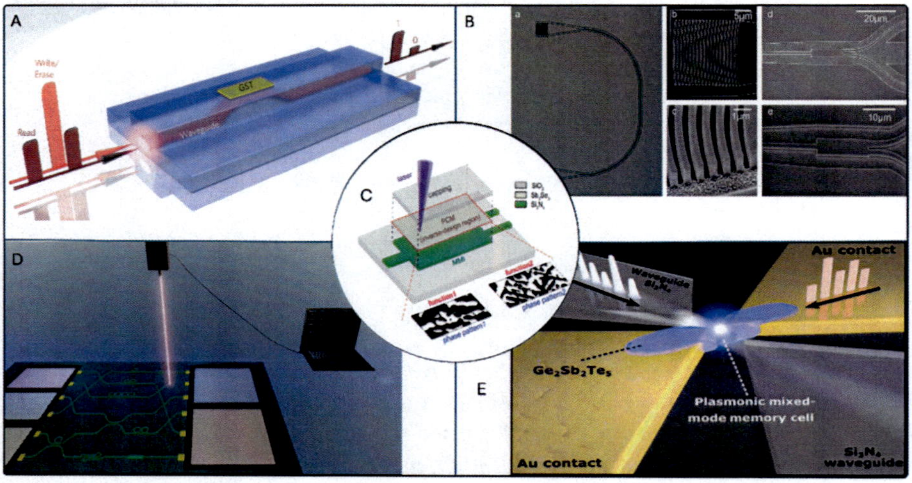

Fig. 1. (A) Illustration of a nanophotonic waveguide with GST programmable cell placed on top, (B) Optical and SEM images of PIC components on GSST platform, (C) Illustration of transferring inverse-designed phase pattern to Sb_2Se_3 thin-film on top of a multi-functional photonic MMI using DLW technique, (D) Illustration of freeform writing and rewriting PICs on Sb_2Se_3 thin film using simple, fast-turnaround, and low-cost approach, (E) Illustration of a device concept where light is delivered to the nanoscale device via a photonic waveguide, while the Au contacts serve as both device electrodes and plasmonic nanogap to focus incoming light to change the phase of GST cell.

2 Monolithic PCM Integrated Photonics

While PCMs are incorporated as a thin film in traditional photonic platforms like Silicon-on-Insulator (SOI) and Silicon Nitride for different applications, they have not been considered a photonic platform on their own. Recently, amorphous $Ge_2Sb_2Se_4Te$ (am-GSST)—a PCM featuring a large refractive index at 1550 nm ($n = 3$) and a 1.5 change in index upon crystallization—has been demonstrated as a monolithic platform for photonic integrated circuits (PICs). Building such a platform with PCMs allows monolithic active photonic devices since changing the phase of the material enables tuning the refractive index and the electrical conductivity. Thus, post-fabrication trimming and writing of photonic structures to control light and even the integration of microheaters for thermo-optical modulation can all be achieved on the same material platform via localized crystallization. A writing mechanism based on electron-beam-induced crystallization was also demonstrated in the study. Figure 1(B) shows the optical image of a straight waveguide with input-output grating couplers and SEM images of the top and tilted view of a grating coupler and a 1 × 2 MMI fabricated on am-GSST platform [5].

3 Inverse-Design in Cladding-Embedded PCM Films

In some other studies, to realize programmable photonics, PCMs, including GST, GSST, and Sb_2Se_3, have been incorporated in Mach–Zehnder interferometers (MZIs) [6] and directional couplers [7] for phase modulation and in ring resonators for resonance tuning. A variety of sophisticated design optimization techniques like genetic algorithms [8], particle swarm optimization, semi-analytical approaches, and inverse design [9] have been implemented in PCM-integrated photonic devices to enable programmability by exploring the design spaces, and yielding devices with an exceptionally compact footprint for superior performance metrics. Figure 1(C) shows such a programmable, multi-functional photonic device achieved by integrating the low-loss phase-change material, Sb_2Se_3, on a multi-mode interferometer (MMI) where the device's functionality can be switched between wavelength-division multiplexing (WDM) and mode-division multiplexing (MDM) by transferring the inverse-designed phase pattern to the Sb_2Se_3 thin-film using direct laser writing (DLW) technique while preserving the integrity of other components [10].

4 Direct-Laser-Written Waveguides in PCM Films

Researchers are also exploring flexible, reliable, and cost-effective techniques for directly writing and rewriting photonic circuits on low loss phase-change thin films. Traditionally, PICs are fabricated in thin-film materials, including silicon, silicon nitride, indium phosphide, and lithium niobates, using top-down nanofabrication processes including electron-beam or photolithography, dry or wet etching, sputtering, and deposition on tools installed in cleanroom facilities. Thus, photonics research faces the barriers of the limited accessibility to fabrication facilities, the high costs associated with the fabrication process, and the extended design-to-device turnaround time. Also, there is a growing concern regarding the environmental impact and chemical waste generated by the top-down fabrication process of PICs. The combination of these barriers impedes widespread innovation and throttles the broader impact of photonics research and education. Moreover, there is a growing interest in programmable PICs to realize highly flexible photonic networks for emerging applications, including optical computing, optical interconnect for neural network accelerators, and quantum computing [11]. This directly writing and rewriting technique of PICs on PCMs simplifies the complex nanofabrication processes typically required for fabricating integrated photonic devices down to one-step laser writing and enables reusing the same chip/die. This can be done with a much lower-cost laser writing system, which incorporates a laser diode, a focusing objective, a high-precision motion stage, and a computer control system, and can achieve the same results as a fabricated PIC in a cleanroom facility. This simple, fast-turnaround, and low-cost approach to creating and reprogramming PICs could shift the paradigm in photonics research, prototyping, and education. Figure 1(D) shows an illustration of how one group of researchers used DLW on PCMs to demonstrate PICs consisting of a full package of elementary photonic components, including waveguides, grating couplers, ring resonators, MZIs, programmable optical switch fabrics, reconfigurable photonic crossbar array, and tunable optical filters [12]. Soon, the phase-change thin films can also be integrated on $LiNbO_3$-on-insulator substrates, enabling the development of programmable electro-optical or acoustic-optical circuits. These advantages underscore the potential of the DLW technique in enabling the rapid prototyping and testing of innovative photonic circuits using only a low-cost tool that is affordable to a wide range of research and education communities.

5 Combining Plasmonics with PCM-Based Integrated Photonics

In the field of computing, exploiting optical signaling requires new on-chip devices that work seamlessly in both electrical and optical domains, without the need for repeated electrical-to-optical conversion. Phase-change devices can, in principle, provide such dual electrical-optical operation, but assimilating both functionalities into a single device were proved elusive owing to conflicting requirements of size-limited electrical switching and diffraction-limited optical response. Combining plasmonics with PCMs is a particularly promising approach for satisfying such stringent requirements, since the dimensions of such devices can be reduced to tens of nanometers and smaller—significantly below the diffraction limit of traditional optical devices. Researchers have demonstrated nonvolatile nanoscale electro-optic devices by exploiting both the nanoscale dimensions and strong field confinement of a plasmonic nanogap, where they enabled both electrical and optical nonvolatile switching of GST within the gap, allowing for full mixed-mode operation of a PCM memory cell. This kind of device enables both electrical and optical programming and readout using the synergetic combination of PCMs and nanoplasmonics. The combination of plasmonics, photonics, electronics and PCMs delivers an integrated phase-change memory cell that can be electrically or optically switched between binary or multilevel states. This unprecedented demonstration of an integrated, reversible, and nonvolatile phase-change memory cell fully bridges the gap between electro-optic mixed-mode operations. The plasmonic design simultaneously reduces the footprint of the device, enhances light-matter interaction [13], and reduces the separation between electrical contacts, creating a compact and highly sensitive device. Figure 1(E) shows the micrograph of such a device with tapered Si_3N_4 waveguide and Au contacts with GST cell at the junction to exploit the plasmonic effect [14].

The past few years have witnessed rapid expansion of research explorations in PCM-based photonics with diverse applications spanning optical switching, photonic memory, optical computing, plasmonics, active metamaterial/metasurface, reflective display, thermal camouflage, tunable filters, phase/amplitude modulators and color pixels. Demonstrating PICs on PCM as a monolithic platform is opening a new dimension to integrated photonics making it fast, affordable, and low-energy reconfigurable. We envision the synergy between PCMs and photonics to drive programmable and reconfigurable technology to a next level in the near future.

References

1. Rios Ocampo, C.A., Youngblood, N.: Configuring phase-change materials for photonics. Elsevier, Phase Change Materials-Based Photonic Computing (2024)
2. Wuttig, M., Yamada, N.: Phase-change materials for rewriteable data storage. Nature Mater. **6**, 824–832 (2007)
3. Rios, C., Hosseini, P., Wright, C.D., Bhaskaran, H., Pernice, W.H.P.: On-chip photonic memory elements employing phase-change materials. Adv. Mater. **26**, 1372–1377 (2013)
4. Ríos, C., Stegmaier, M., Hosseini, P., et al.: Integrated all-photonic non-volatile multi-level memory. Nature Photon **9**, 725–732 (2015)
5. Acharjee, N., Sun, H., Ghosh, S., Ríos, C., $Ge_2Sb_2Se_4Te$ Phase Change Material Monolithic. Platform for Integrated Photonics, CLEO 2024
6. Yang, X., et al.: Non-Volatile Optical Switch Element Enabled by Low-Loss Phase Change Material., Adv Funct Mater 2304601 (2023)
7. Xu, P., Zheng, J., Doylend, J.K., Majumdar, A.: Low-loss and broadband nonvolatile phase change directional coupler switches. ACS Photonics **6**, 553–557 (2019)
8. Ren, Y., et al.: Genetic-algorithm-based deep neural networks for highly efficient photonic device design. Photonics Res **9**, B247–B252 (2021)
9. Molesky, S., et al.: Inverse design in nanophotonics. Nat. Photonics **12**, 659–670 (2018)
10. Wu, C., Jiao, Z., Ríos, C.: Reconfigurable inverse designed phase-change photonics, arXiv2403.05649
11. Shen, Y., et al.: Deep learning with coherent nanophotonic circuits. Nat Photonics. **11**, 441–446 (2017)
12. Wu, C., et al.: Freeform direct-write and rewritable photonic integrated circuits in phase-change thin films. Sci. Adv. **10** (2024)
13. Rodriguez-Hernandez, G., Hosseini, P., Ríos, C., Wright, C.D., Bhaskaran, H.: Mixed-mode electro-optical operation of Ge2Sb2Te5 nanoscale crossbar devices. Adv. Electron. Mater. **3**, 1700079 (2017)
14. Farmakidis, N., et al.: Plasmonic nanogap enhanced phase-change devices with dual electrical-optical functionality. Sci. Adv. **5** (2019)

Sb_2Se_3 Based Non-volatile Memory for Photonic Matrix-Vector Multiplications

Mukta Janpandit[1](✉), Rakshitha Kallega[1], K. Ramesh[2], Chetan Singh Thakur[3], and Shankar Kumar Selvaraja[1]

[1] Centre for Nano Science and Engineering, Indian Institute of Science Bangalore, Bengaluru, India
muktaj@iisc.ac.in

[2] Department of Physics, Indian Institute of Science Bangalore, Bengaluru, India

[3] Department of Electronic Systems Engineering, Indian Institute of Science Bangalore, Bengaluru, India

Abstract. This work integrates nonvolatile antimony selenide (Sb_2Se_3) phase shifters onto a Mach-Zehnder interferometer (MZI) for reconfigurable 2×2 photonic matrix-vector multiplications (MVMs). The Sb_2Se_3 phase shifters enable programming via a one-time 500 mW power input, offering significant potential for reducing power consumption when matrix configurations remain fixed. Consequently, it offers promise for the development of energy-efficient photonic accelerators for machine learning applications. Through comprehensive device characterization and genetic algorithm fitting, we achieved a normalized mean square error of 0.0494 between measured data and the ideal MZI model, validating the reliable implementation of programmable matrix multiplications on this platform.

Keywords: Photonic Computing · Phase-Change material · Photonic Matrix-Vector multiplication · Photonic Memory

1 Introduction

In the era of exponential computational growth, exemplified by Moore's Law, the quest for efficient computing hardware has become increasingly pressing. However, this pursuit is countered by the persistent bottleneck posed by the von Neumann architecture, where the separation of memory and processing units leads to significant energy inefficiencies and limits overall system performance. As a result, there is a critical need for innovative approaches to computing hardware that address these limitations and strive for energy efficiency to meet the demands of modern computing applications.

In this context, photonic computing emerges as a promising alternative, leveraging the inherent properties of light, such as high propagation speed, strong robustness, and multiple degrees of freedom to encode information, thereby overcoming the limitations of electronic systems. Photonic Matrix-Vector Multiplication (MVM) presents a key operation in many computational tasks, particularly prevalent in machine learning and

J. Witzens et al. (Eds.): ECIO 2024, SPPHY 402, pp. 242–248, 2024.
https://doi.org/10.1007/978-3-031-63378-2_40

neural network inference. By exploiting the unique properties of light, such as high-speed propagation and parallel processing capabilities, photonic MVM offers the potential for significant performance gains while reducing energy consumption.

Multi-port Mach-Zehender interferometers can be used to perform matrix multiplications on vectors encoded in the amplitude of coherent light beams/pulses. To see this, consider the matrix Multiplication unit highlighted in Fig. 1a. It consists of 2 phase shifters and 2 splitters. If the coupling coefficient of the splitters is exactly 0.5, then the transfer function of this device is given by,

$$T_{ij}(\theta, \phi) = \frac{1}{2}\begin{bmatrix} e^{i\phi}\left(e^{i\theta} - 1\right) & i\left(e^{i\theta} + 1\right) \\ ie^{i\phi}\left(e^{i\theta} + 1\right) & 1 - e^{i\theta} \end{bmatrix}$$

where θ and ϕ are the phase shifts introduced at p1 and p2. This expression represents any 2x2 unitary matrix. Thus, MZIs can be used to implement MVMs on input electric field amplitudes. MZI based matrix multiplication units have been proposed for low-power Deep Learning accelerators [1]. Section 3.3 discusses the general case when the coupling coefficient is not 0.5.

Currently, photonic integrated circuits use thermo-optic effect to tune the phase shifters, which raises concerns regarding power consumption and compute-speed limitations. Though MEMS-based approach offers lower power consumption, the post-processing and packaging such circuits is still challenging for practical application. An energy-efficient and scalable alternative is phase change material (PCM)-based phase shifters, which can be tuned either electrically or optically. However, the commonly used GST-based PCMs employ an absorption mechanism for tuning light. This increases optical power loss and limits the scaling to larger circuits. In this work, we demonstrate novel Sb_2Se_3 based circuits for matrix multiplications. The optical absorption by Sb_2Se_3 increases only by 0.1 dB upon crystallization, compared to a 0.64 dB increase in loss due to the crystallization of GST [2].

2 Device Fabrication

We fabricated a balanced Mach Zehnder modulator on a silicon-on-insulator platform with two electrically tunable Sb_2Se_3 phase shifters. The fabrication process starts with taking a SOI wafer with a device layer thickness of 220 nm. The waveguide and fiber grating couplers are patterned using electron beam lithography followed by a 70 nm silicon etch. A thin layer of SiO_2 (40 nm) is deposited on top of the silicon device using a plasma-enhanced chemical vapor deposition process to isolate silicon and Sb_2Se_3. Next, we pattern the metal layer consisting of resistive microheaters by optical lithography. Metal layers of 10 nm thick Ti followed by 100 nm thick Pt are deposited using metal sputtering. This is followed by a lift-off process. On top, we deposited 30 nm thick and 20 μm wide Sb_2Se_3 using a thermal evaporator. Figure 1 shows the microscope image of the fabricated test device. The Sb_2Se_3 deposited using this process is amorphous [3].

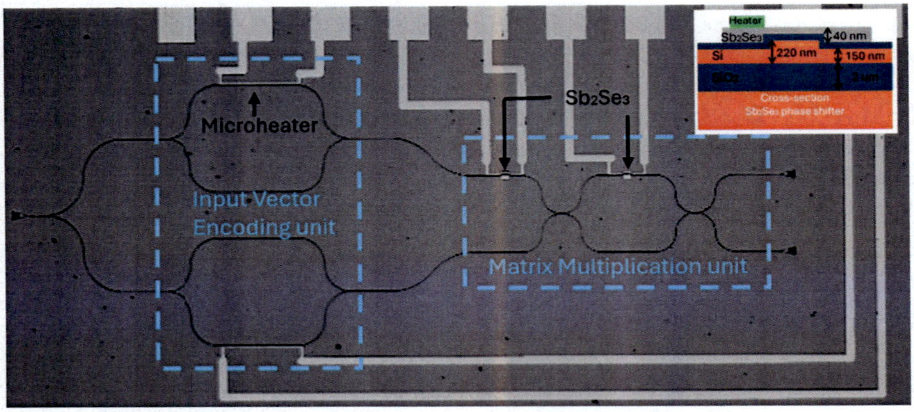

Fig. 1. Microscope image of the fabricated device highlighting the Input encoding and MVM units. The inset shows the cross-section schematic of the Sb_2Se_3 deposited region.

3 Results

We report the results in three sections. First, we characterize the input stage thermo-optic phase shifter to find the phase shift per milliwatt of electrical power. Then, we show the characterization results of the non-volatile phase shift observed in Sb_2Se_3. Finally, we demonstrate matrix-vector multiplications using our device in Sect. 3.3.

3.1 Input Encoding

An on-chip intensity modulator translates electrical data to intensity modulated optical signal. We have used a single input single output Mach-Zehender interferometer with a thermo-optic phase shifter on one of the arms to act as a intensity modulator. The thermo-optic phase shifter was characterized using an unbalanced MZI device, as shown in Fig. 2(a). The phase shift is approximately 0.04 rad/ mW. The characterization data of the thermo-optic phase shifter is as shown in Fig. 2(b).

3.2 Switching Characterization

To assess the non-volatile switching characteristics of Sb_2Se_3 phase shifters, we conducted experiments on a Mach Zehnder Interferometer (MZI) in a single-input-multi-output configuration (see Fig. 3a). Electrical power was supplied to a metal microheater positioned beneath the Sb_2Se_3 layer, situated 4 μm away from the waveguide. The microheater is used to crystallize the Sb_2Se_3. Importantly, this phase change process is non-volatile in nature, meaning that once the material has undergone crystallization, it remains in this state without the need for additional electrical power to maintain it.

We conducted characterization across a range of input electrical powers. This involved applying various levels of electrical power to the input for a brief duration, followed by discontinuation of the electrical supply. Subsequently, we measured the optical response, focusing on changes in optical output power at 1560 nm wavelength

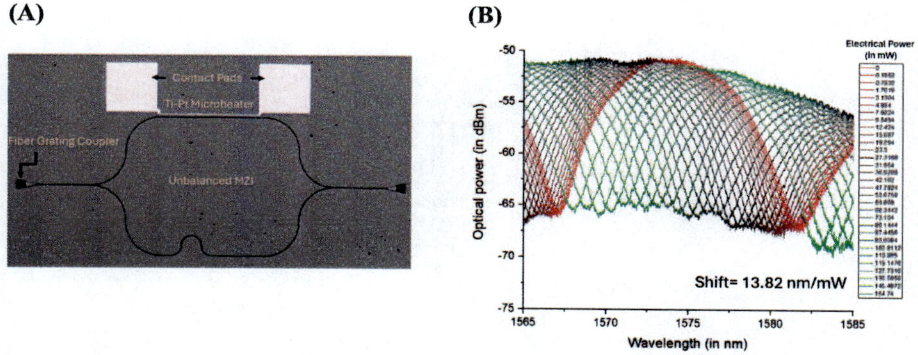

Fig. 2. (a) Test device for characterizing thermo-optic phase shifter. (b) Device response for different electrical powers applied to the Ti-Pt microheater.

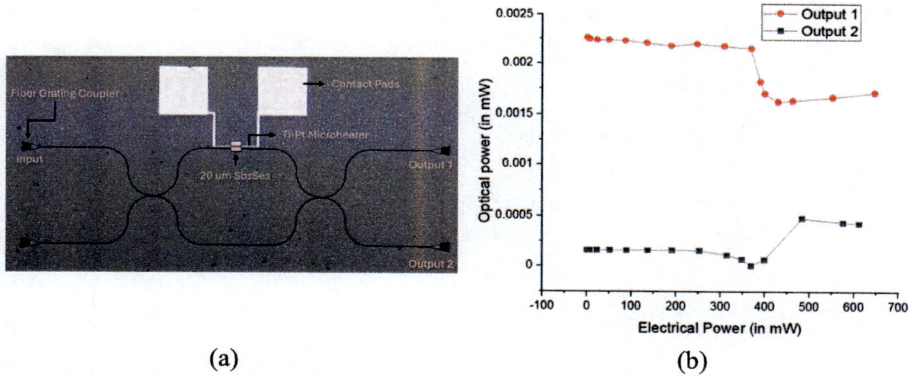

Fig. 3. a) Microscope image of a test device used for characterizing Sb_2Se_3 switching and b) Non-volatile change in Optical Power as a function of electrical power.

in both arms. The relationship between the supplied electrical power and the resulting changes in optical output power in arm 1 and arm 2 is illustrated in Fig. 3b. We observe a change in output power in the input range of 300–500 mW. This observation suggests that our material undergoes a phase transition within this specific range.

3.3 Measurements and Data Analysis

Figure 4 shows the measurement setup used for the 2X2 matrix multiplication characterization of the circuit depicted in Fig. 1.

The device testing steps are as follows: First, we tune the non-volatile Sb_2Se_3 phase shifters by supplying DC power corresponding to the desired phase shift. This sets up our MZI to a particular matrix configuration. Next, we use volatile thermo-optic phase shifters to encode input vectors. In the final step, the optical power output of both outputs is recorded. We record multiple input-output data for the same matrix configuration. This data is then fitted to a matrix model to predict the obtained matrix.

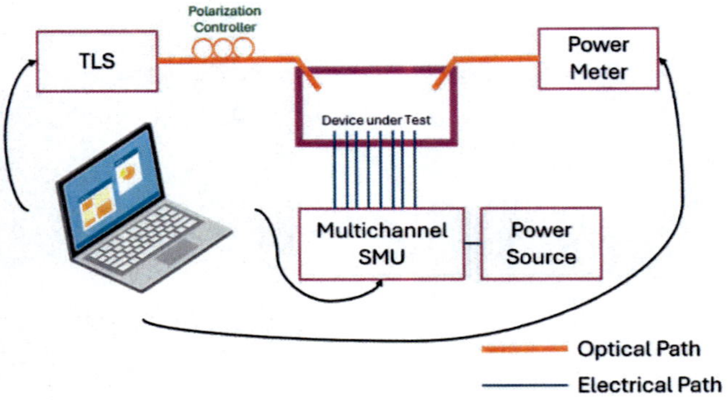

Fig. 4. Experimental Setup

To obtain a matrix model for the MZI, we split the MZI into its basic components- a phase shifter, then a directional coupler, followed by another phase shifter and a directional coupler. Consider the transfer function of a directional coupler with coupling coefficient c.

$$DC = constant \times \begin{bmatrix} \sqrt{1-c}e^{i\delta} & i\sqrt{c} \\ i\sqrt{c} & \sqrt{1-c}e^{-i\delta} \end{bmatrix}$$

c can take values from 0 to 1, 0 denotes no power transfer whereas 1 denotes complete power transfer. Therefor, the transfer matrix of an MZI is given by,

$$MZI = \begin{bmatrix} \sqrt{1-c}e^{i\delta} & i\sqrt{c} \\ i\sqrt{c} & \sqrt{1-c}e^{-i\delta} \end{bmatrix} \times \begin{bmatrix} e^{i\theta} & 0 \\ 0 & 1 \end{bmatrix} \times \begin{bmatrix} \sqrt{1-c}e^{i\delta} & i\sqrt{c} \\ i\sqrt{c} & \sqrt{1-c}e^{-i\delta} \end{bmatrix} \times \begin{bmatrix} e^{i\varphi} & 0 \\ 0 & 1 \end{bmatrix}$$

$$MZI = \begin{bmatrix} (1-c)\left(e^{i(\theta+\varphi+2\delta)}\right) - ce^{i\varphi} & i\sqrt{c(1-c)}(e^{i(\theta+\delta)} + e^{-i\delta}) \\ i\sqrt{c(1-c)}(e^{i(\theta+\varphi+\delta)} + e^{i(\varphi-\delta)}) & -ce^{i\theta} + (1-c)e^{-i2\delta} \end{bmatrix}$$

We utilize the genetic algorithm (GA) solver available in the optimization toolbox of MATLAB to fit our input-output data to the matrix described earlier. The algorithm employed closely resembles the one outlined in reference [4]. We defined a nonlinear loss function that accounted for the abovementioned matrix product with the input vector. The input vector was expressed as follows,

$$Input\ Vector = \begin{bmatrix} e^{i(b\times in1+c)} + e^{id} \\ e^{i(b\times in2+c)} + e^{id} \end{bmatrix}$$

where, in1 and in2 are electrical powers that encode input information. The output electric fields obtained after matrix-vector product were converted to optical intensities using the following equation.

$$I = \frac{1}{2}\epsilon_0 E^2 c$$

We can successfully fit the data with a normalized mean square error of 0.0494. Figures 5a and 5b show sample fitted outputs, and Fig. 5c and 5d show the variation of

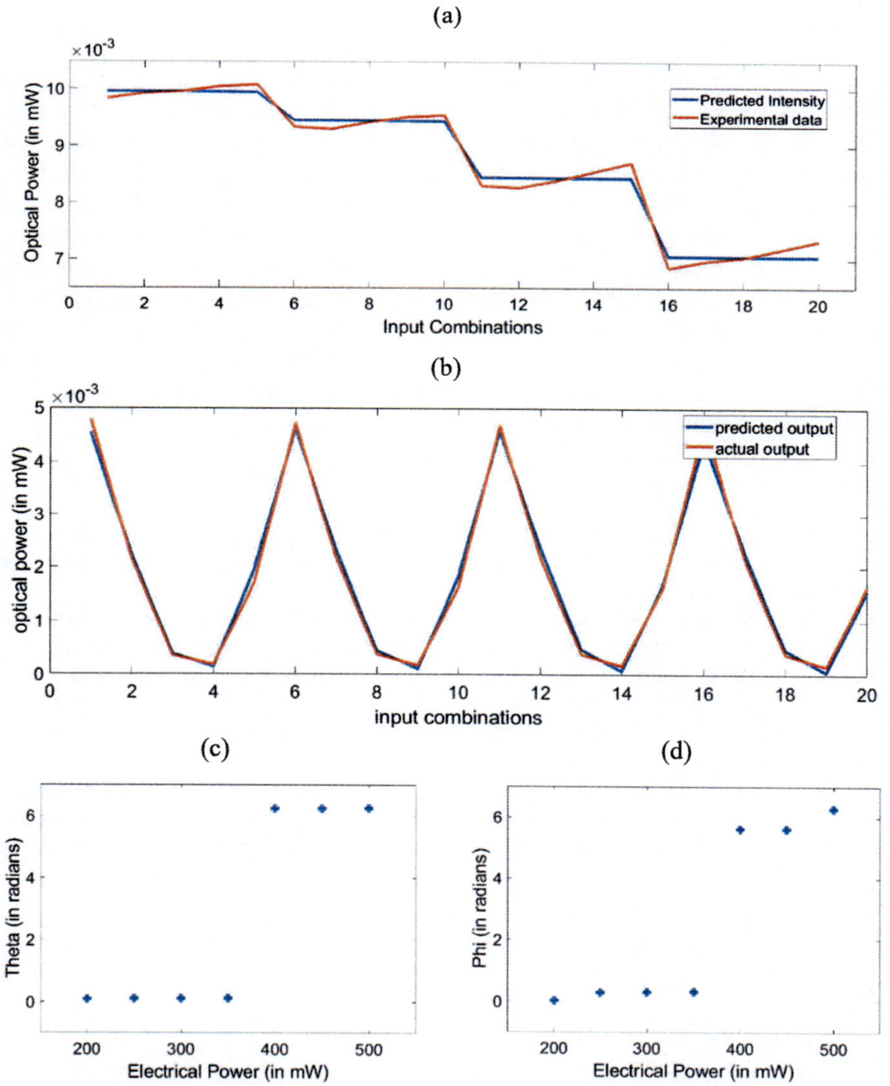

Fig. 5. a) Sample fitted outputs at output 1 for a particular (θ, φ) combination. X-axis represents various inout (in1, in2) combinations, b) similar samples for output 2, c) Theta (θ) obtained from the fitted model as a function of input power, and d) Phi (φ) obtained from the fitted model as a function of input power.

θ and , respectively, with supplied electrical power. We observe a sharp phase change at around 350–400 mW of applied electrical power due to phase change. However, the phase values are obtained from model fitting (so θ is same as $(2\pi + \theta)$), and thus, more data points are necessary to predict the absolute phase shift.

4 Conclusion

This work demonstrates the integration of nonvolatile antimony selenide (Sb_2Se_3) phase shifters onto a 2x2 Mach-Zehnder interferometer (MZI) for performing reconfigurable photonic matrix-vector multiplications (MVMs). The Sb_2Se_3 phase shifters enable programming the MZI matrix configuration using a one-time 300–500 mW electrical pulse, after which MVMs operate passively without additional power input - a critical advantage over conventional approaches. Comprehensive device characterization was performed, including quantifying the nonvolatile Sb_2Se_3 phase change dynamics. Input-output measurements were fitted to an ideal MZI model using genetic algorithm optimization, achieving a normalized mean square error of 0.0494. Thus, the integrated Sb_2Se_3-on-silicon photonics platform constitutes a pioneering solution for realizing nonvolatile, energy-efficient programmable nanophotonic processors targeted at accelerating computational workloads like deep learning inference.

References

1. Shen, Y., Harris, N., Skirlo, S., et al.: Deep learning with coherent nanophotonic circuits. Nature Photon **11**, 441–446 (2017)
2. Kallega, R., Shekhawat, R., Bhat K.U., Karuppannan, R., Selvaraja, S.K.: Controlled crystallization of thermal evaporated GST-on-SOI for photonic neuromorphic application. In: Yang, B., (eds.) OSA Advanced Photonics Congress 2021, OSA Technical Digest, paper IF3A.1 (2021)
3. Delaney, M., et al.: Non-volatile programmable silicon photonics using an ultralow-loss Sb2Se3 phase change material. Sci. Adv. **7**, 1–7 (2021)
4. Jain, P., Gupta, G., Kwon, H.-K., Vukasin, G.D., Chandorkar, S.A.: Machine learning based nonlinear resonator system identification. In: 2024 IEEE Electromechanical l Conference on Micro Electro Mechanical Systems (MEMS), Austin, TX, USA, pp. 541–544 (2024)

Reconfigurable Silicon Polarization Rotator in the O-Band Using Sb_2Se_3 on the Silicon-on-Insulator Platform

Jorge Parra[1], Juan Navarro-Arenas[1,2], Amadeu Griol[1], and Pablo Sanchis[1(✉)]

[1] Nanophotonics Technology Center, Universitat Politècnica de València, Camino de Vera s/n, 46022 Valencia, Spain
pabsanki@ntc.upv.es

[2] Institute of Materials Science (ICMUV), Universitat de València, Carrer del Catedràtic José Beltrán Martinez 2, 46980 Valencia, Spain

Abstract. We report a reconfigurable polarization rotator (PR) for the O-band on the silicon-on-insulator (SOI) platform. The proposed PR consists of an asymmetric Sb_2Se_3/Si waveguide. The polarization of the optical mode can be maintained or switch to the orthogonal state by changing the Sb_2Se_3 between the amorphous and crystalline states, respectively. Our fabricated device exhibits a polarization conversion efficiency (PCE) as low as −0.25 dB and a polarization extinction ratio (PER) of 12.5 dB at 1310 nm. Moreover, our device is just 21 μm long and exhibits broad spectral response.

Keywords: reconfigurable · non-volatile · polarization rotator · silicon photonics

1 Introduction

Silicon-on-insulator (SOI) technology has emerged as the leading platform for the development of photonic integrated circuits (PICs) [1]. Yet, one of its primary benefits—the significant refractive index contrast—also presents a notable challenge by inducing strong birefringence. This characteristic renders compact silicon photonic devices highly susceptible to polarization states, thereby narrowing their application scope and diminishing their efficacy [2]. Consequently, managing polarization is crucial in SOI-based PICs. Additionally, implementing on-chip polarization control can enhance data transmission rates through the utilization of polarization-division multiplexing (PDM), a method that allows data rates exceeding 200 Gb/s in silicon PICs [3].

Various strategies have been reported for incorporating polarization-diversity schemes into silicon PICs [2]. Among these, polarization rotators (PRs) are particularly effective for converting to orthogonal polarization states within straight silicon nanowires by altering the waveguide's cross-sectional symmetry and maintaining compact device footprints [4, 5]. However, these devices are passive, limiting the ability to reconfigure polarization states dynamically.

J. Witzens et al. (Eds.): ECIO 2024, SPPHY 402, pp. 249–252, 2024.
https://doi.org/10.1007/978-3-031-63378-2_41

In this work, we report experimental results on a reconfigurable (with zero static power consumption) PR in the SOI platform by using a phase-change material (PCM) such as Sb_2Se_3. Our design is based on our previous proposal based on numerical simulations and developed for the telecom band (1550 nm) [6]. Here, our device is fabricated and experimentally demonstrated for working in the datacom O-band (1310 nm). The working principle relies on the dramatic change and non-volatile change in the refractive index featured by Sb_2Se_3 [7]. By changing the state of the PCM, our PR can maintain the input polarization or switch to the orthogonal with a non-volatile response, i.e., zero holding power.

2 Results

Figure 1 (a) illustrates the working principle of the proposed PR. Our reconfigurable PR is based on a hybrid Sb_2Se_3/Si waveguide with an asymmetric cross-section in order to support hybrid optical modes. The cross-section asymmetry is achieved by fully etching a certain width of the silicon waveguide. On the other hand, reconfigurability is enabled by depositing a thin layer of Sb_2Se_3 on top of the etched waveguide. In the amorphous state, the PR does not affect the polarization of the input mode, whereas in the crystalline state, the polarization is rotated to the orthogonal optical mode.

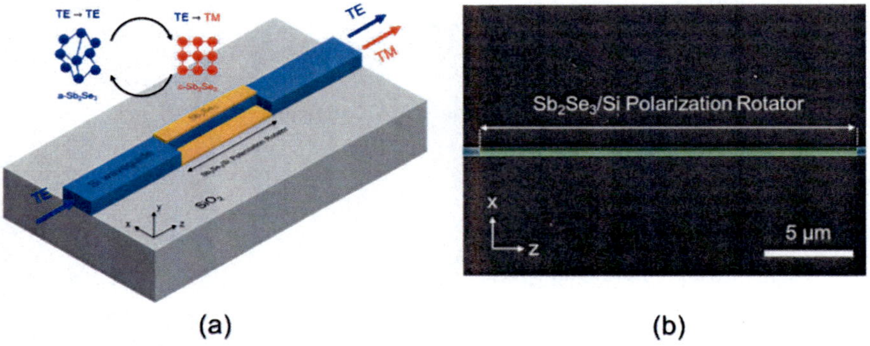

Fig. 1. 3D illustration of the proposed reconfigurable PR based on a hybrid Sb_2Se_3/Si waveguide. The polarization is reconfigurable by changing the Sb_2Se_3 between the amorphous (a-Sb_2Se_3) and crystalline (c-Sb_2Se_3) states. (b) Top-view false-color SEM image of the fabricated 21-μm-long Sb_2Se_3/Si PR.

The considered dimensions of the silicon waveguide are 500 × 220 nm. In the PR region, the etched width of the Si waveguide is 250 nm. The Sb_2Se_3 layer deposited on top of the etched waveguide is 500 nm wide and 40 nm thick. On the other hand, according to our numerical simulations, the length of the PR should be 21 μm to achieve almost full polarization conversion based on the effective refractive indices of the hybrid modes and the resulting beat length. Figure 1(b) shows a scanning electron microscope (SEM) image of the fabricated Sb_2Se_3/Si PR. The patterning of the photonic structures was made using e-beam lithography on an SOI chip, and the Sb_2Se_3 layer was deposited by e-beam evaporation.

Figure 2 shows an optical micrograph of the polarization diversity circuit used for characterizing the Sb_2Se_3/Si PR at 1310 nm. We used a TE grating coupler to inject light from the fiber to the Sb_2Se_3/Si PR, and then the output goes through a polarization splitter based on a directional coupler to split between TE and TM polarization. Finally, both polarizations are collected by using respective grating couplers. Before injecting the optical power to the polarization diversity circuit, we polarized the light to TE using an external polarization controller and a reference straight waveguide with TE grating couplers. The Sb_2Se_3/Si PR was first characterized in the as-deposited amorphous state. To crystallize the Sb_2Se_3, the sample was heated at 250 °C for 10 min. in an Ar atmosphere.

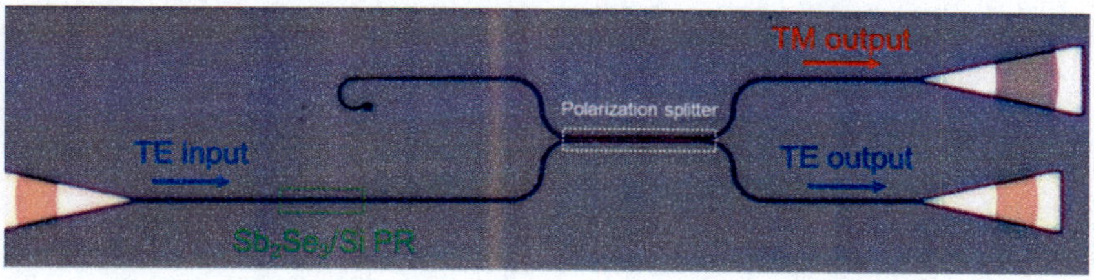

Fig. 2. Optical micrograph of the polarization diversity circuit used for characterizing the Sb_2Se_3/Si PR at 1310 nm.

Figure 3 shows the experimental results of our reconfigurable Sb_2Se_3/Si PR in the O-band. The polarization conversion efficiency (PCE) is calculated as $P^{TE}/(P^{TM} + P^{TE})$ and $P^{TM}/(P^{TM} + P^{TE})$ for the amorphous and crystalline state, respectively, where P^{TE} and P^{TM} are the normalized optical power at the TE and TM output, respectively. On the other hand, the polarization extinction ratio (PER) is calculated as $|P^{TM} - P^{TE}|$. At 1310 nm, the PCE can be as low as −0.25 dB together with a PER of 12.5 dB.

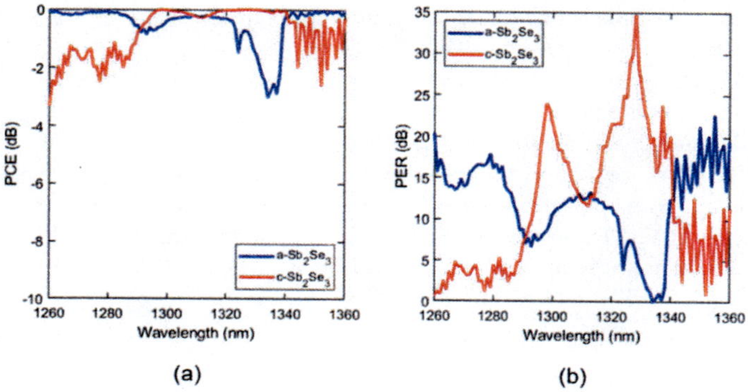

Fig. 3. Experimental performance of the Sb_2Se_3/Si PR. (a) Polarization conversation efficiency (PCE) spectrum. (b) Polarization extinction ratio (PER) spectrum.

3 Discussion

In summary, we have reported a reconfigurable PR in SOI and for the O-band that can maintain the polarization of the optical mode or switch to the orthogonal based on a hybrid Sb_2Se_3/Si waveguide. Our non-volatile device has a footprint of just 21 μm, it does not consume static power consumption and features a broad spectral response showcasing a PCE as low as −0.25 dB and a PER of 12.5 dB at 1310 nm. On the other hand, on-chip thermo-electrical switching of our device could be implemented using transparent heaters based on graphene [8] or indium tin oxide [9].

Acknowledgements. Agencia Estatal de Investigación (TED2021-132211B-I00, PID2022-137787OB-I00); Generalitat Valenciana (CIPROM/2022/14); Advanced Materials programme supported by MCIN with funding from European Union NextGenerationEU (PRTR-C17.I1) and by Generalitat Valenciana; Ministerio de Universidades (MS21–037); Universitat Politècncia de València (Grant PAID-06-23, PAID-10-23). The authors would like to thank Miroslavna Kovylina for the deposition of the Sb_2Se_3.

References

1. Lipson, M.: The revolution of silicon photonics. Nat. Mater. **21**, 974–975 (2022)
2. Dai, D., Liu, L., Gao, S., Xu, D., He, S.: Polarization management for silicon photonic integrated circuits. Laser Photonics Rev. **7**, 303–328 (2013)
3. Liao, P., et al.: A 260 Gb/s/λ PDM link with silicon photonic dual-polarization transmitter and polarization demultiplexer. In: 2021 European Conference on Optical Communication (ECOC), pp. 1–4. (IEEE, Bordeaux, France, 2021)
4. Aamer, M., et al.: CMOS compatible silicon-on-insulator polarization rotator based on symmetry breaking of the waveguide cross section. IEEE Photon. Technol. Lett. **24**, 2031–2034 (2012)
5. Xie, A., Zhou, L., Chen, J., Li, X.: Efficient silicon polarization rotator based on mode-hybridization in a double-stair waveguide. Opt. Express **23**, 3960 (2015)
6. Parra, J., Navarro-Arenas, J., Sanchis, P.: Sb2Se3/Si tunable rotator with ultra-compact footprint. In: 2023 IEEE Silicon Photonics Conference (SiPhotonics), pp. 1–2. (IEEE, Washington, DC, USA, 2023)
7. Delaney, M., Zeimpekis, I., Lawson, D., Hewak, D.W., Muskens, O.L.: A new family of ultralow loss reversible phase-change materials for photonic integrated circuits: Sb2S3 and Sb2Se3. Adv Funct Materials **30**, 2002447 (2020)
8. Fang, Z., et al.: Ultra-low-energy programmable non-volatile silicon photonics based on phase-change materials with graphene heaters. Nat. Nanotechnol. **17**, 842–848 (2022)
9. Taghinejad, H., et al.: ITO-based microheaters for reversible multi-stage switching of phase-change materials: towards miniaturized beyond-binary reconfigurable integrated photonics. Opt. Express **29**, 20449 (2021)

A Photonic Convolution Processor Based on Matched Pair of Arrayed Waveguide Gratings

Caiyue Zhao[1], Dan Yi[1], Zunyue Zhang[2], Hongnan Xu[1], and Hon Ki Tsang[1](✉)

[1] Department of Electronic Engineering, The Chinese University of Hong Kong, Shatin, NT, Hong Kong SAR
hktsang@ee.cuhk.edu.hk

[2] School of Precision Instrument and Opto-Electronics Engineering, Tianjin University, Tianjin, China

Abstract. We demonstrate a photonic convolution processor that uses the spectral shifting in output waveguides for adjacent input waveguides of an arrayed waveguide grating to implement the convolution sliding operation and generate convolution output maps without repetitive matrix calculations. We implement a balanced pair of arrayed waveguide gratings to realize for the first time both negative and positive weighting of photonic convolution kernels. We demonstrate the convolution processor experimentally, integrating a pair of arrayed waveguide gratings with a broadband Mach-Zehnder interferometer on the silicon photonics platform. Experimental results demonstrate the effectiveness of the proposed convolution processor in edge detection.

Keywords: convolution processor · arrayed waveguide grating · parallel computation

1 Introduction

Convolution is a basic operation applied in machine learning for image processing [1]. Previous research on implementing integrated optical convolution based on metamaterial and diffraction [2, 3] suffers the disadvantage of difficulty in reconfigurability. The use of tunable delay lines to interleave the signals temporally [4] requires precise fabrication and thermal control. The spectral response of arrayed waveguide grating (AWG) has also been previously explored for parallel computation using discrete packaged components [5]. In this paper, we present an integrated convolution processor (CP) based on the shifted output spectral response of AWG for different input ports. We demonstrate the first integrated AWG photonic processor capable of achieving both negative and positive weights. The bipolar weights are achieved by integrating a broadband Mach-Zehnder interferometer (MZI) with a pair of AWGs. The proposed photonic processor is used in edge detection to process and extract features from the original image, conducting convolutions with parallel kernels.

J. Witzens et al. (Eds.): ECIO 2024, SPPHY 402, pp. 253–257, 2024.
https://doi.org/10.1007/978-3-031-63378-2_42

2 Device Design

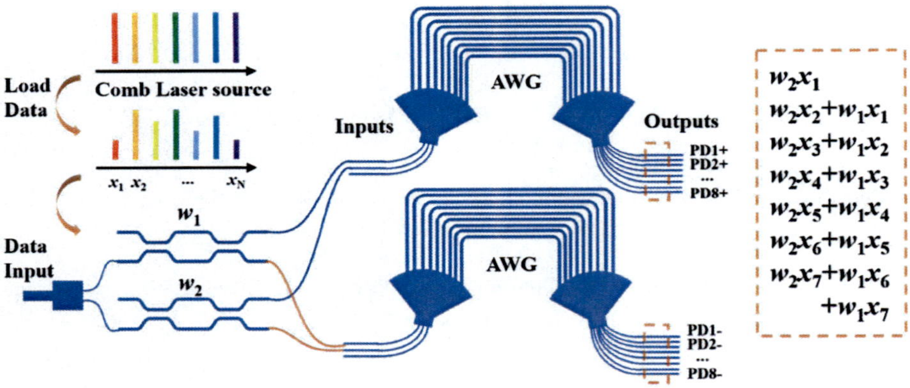

Fig. 1. The principle of the convolution operation with balanced AWGs.

Figure 1 exhibits the basic principle of the one-dimensional convolution using balanced 3 × 8 AWGs. The input image is encoded on different wavelength channels. The convolution kernel window slides by accessing data from adjacent input ports and utilizing the characteristic routing operation of the AWG. The routing of wavelengths, carrying distinct kernel weights, through AWG combines different wavelength from different inputs at the output ports to obtain convolutional results.

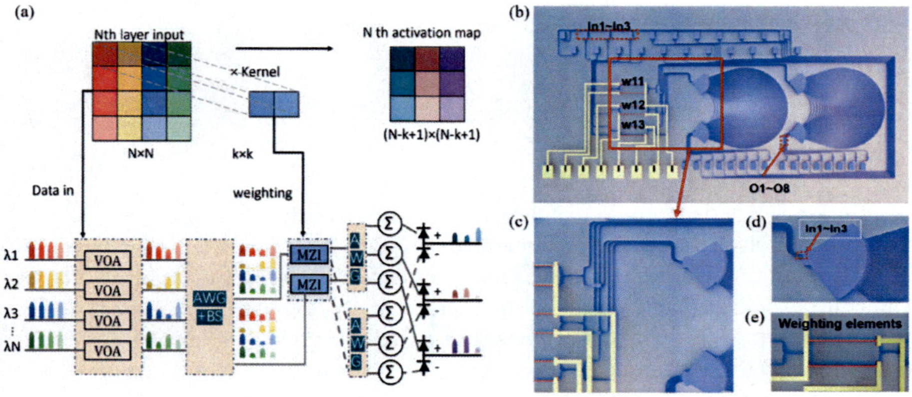

Fig. 2. (a) Data encoding and flow. VOA, variable optical attenuator; AWG, arrayed waveguide grating; BS, beam splitter. (b) Schematic of the CP using 3 × 8 AWGs. (c) Zoomed-in view of routing signals via the MZIs to the different AWGs. (d) The multi-input star coupler of the AWG. (e) The broadband MZI for balanced detection.

Figure 2(a) shows how the data is encoded temporally in different wavelengths by using the array of variable optical attenuators, which adjust the optical power in the time domain to represent the various intensities of a sequence of individual pixels within the grayscale image. The combined outputs are routed via the broadband MZIs to a pair of identical AWGs. The two MZIs also weight the data before processing by the final pair of AWGs, which sum the optical power of the different wavelengths before they are converted back to the electrical domain by the balanced photodetector array. The schematic of the proof-of-concept CP using 3 common wavelength inputs by 8 wavelength channel outputs AWGs is shown in Fig. 2(b). Figure 2(c) exhibits a zoomed-in view of routing signals from MZIs to different AWGs. Signals from each MZI upper branch are routed to the input ports of the right AWG in Fig. 2(b), while those from the lower branch are routed to the left AWG. Figure 2(d) depicts the star coupler within the AWG, while Fig. 2(e) illustrates the broadband MZI, which employs a bent directional coupler for broadband splitting [6] and enables parallel processing by applying a uniform weight for different wavelengths. Thermal phase shifters are integrated to adjust the output of the two waveguides. The proposed optical CP is fabricated on a silicon-on-insulator (SOI) platform with a 220-nm-thick fully-etched Si and a 2-μm-thick buried oxide layer. We experimentally demonstrate the CP with two MZIs carrying the weights of a 1×2 kernel matrix.

3 Experimental Results and Discussion

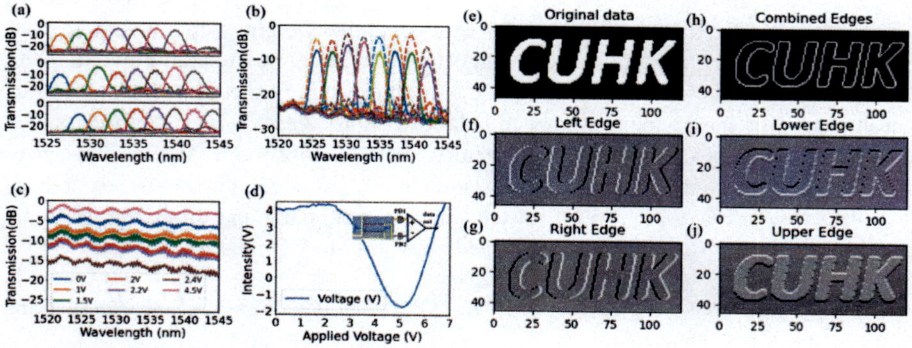

Fig. 3. Transmission spectra of (a) AWG for various input ports, (b) two AWGs outputs, and (c) the broadband MZI when applied with different voltages. (d) The MZI can be programmed to positive and negative weights when working in the balanced detection. Experimental results of edge detection: (e) original image, (f) left edge, (g) right edge, (h) combined edges, (i) lower edge, (j) upper edge.

The performance of devices is measured. The AWG is designed with 2.5 nm channel spacing across 1520–1545 nm. The experimental transmission spectra of an AWG are shown in Fig. 3(a). The transmission spectra of the output channels shift along the spectral channels when the light is launched from different ports. The insertion loss (IL) of AWG is measured to be below 4 dB, and the crosstalk is less than -15 dB. The transmission spectra of the two AWGs are depicted in Fig. 3(b). The spectral responses of the two AWGs match well, with identical center wavelengths. Figure 3(c) exhibits the transmission spectra of the broadband MZI, with a variation within 0.5 dB over 20-nm wavelength. The weights programmed by MZIs are adjustable to be both positive and negative as shown in Fig. 3(d), ranging from -2 to + 4 with a power consumption of approximately 22 mW. In the experiment, a 2 $\times$ 2 convolutional kernel is implemented by conducting performance tests twice on a 1 $\times$ 2 kernel. Specific kernel matrices can be realized by programming MZIs. We convolve 8-bit grayscale images of the letters "*CUHK*" to perform edge detection. Taking the lower edge as an example, the kernel is set as $\begin{bmatrix} -1 & 1 \\ -1 & 1 \end{bmatrix}$, thereby computing the discrete first-order derivative of the image. By programming two MZIs with voltages of [4.5, 0] V, the convolutional kernel [-1,1] is implemented. The resulting images of the four types of edge detection after convolution are shown in Fig. 3(f), (g), (i) and (j). The edges of the four types are superimposed to obtain the combined edge in Fig. 3(h). Figure 3(h) highlights the sharp edges of the pattern compared to the original image in Fig. 3(e). The edge detection images obtained through optical computation correspond well to the original image.

4 Conclusion

In summary, we demonstrated a novel photonic CP utilizing programmable MZIs and AWGs. The CP was employed for image processing. The system shows high programming resolution and broadband operation. The experimental results show that balanced AWGs can be utilized to configure positive and negative weights without further post-processing. The spectral response shift of AWG increases the computation parallelism. For high-speed convolution processing, the photonic circuit implementation can be further improved by integrating high-speed balanced photodetector arrays and high-speed microring modulators instead of the thermally tuned Mach-Zehnder interferometers to encode data at ~ 100Gb/s. The results presented in this paper demonstrated the potential of using a pair of matching AWGs for convolution processing and the scheme may be suited for future low-latency and high-speed convolution computation.

Acknowledgment. This work was funded by RGC GRF grant 14204623. The authors thank Hang Liu for helpful discussions.

References

1. Feldmann, J., et al.: Parallel convolutional processing using an integrated photonic tensor core. Nature **589**, 52–58 (2021)
2. Liao, K., Gan, T., Hu, X., Gong, Q.: AI-assisted on-chip nanophotonic convolver based on silicon metasurface. Nanophotonics **9**(10), 3315–3322 (2020)
3. Huang, Y., Fu, T., Huang, H., Yang, S., Chen, H.: Sophisticated deep learning with on-chip optical diffractive tensor processing. Photonics Res. **11**(6), 1125–1138 (2023)
4. Xu, X., et al.: 11 TOPS photonic convolutional accelerator for optical neural networks. Nature **589**(7840), 44–51 (2021)

5. Zhang, S., et al.: Redundancy-free integrated optical convolver for optical neural networks based on arrayed waveguide grating. Nanophotonics **13**(1), 19–28 (2024)
6. Chen, S., Dai, D.: Ultra-broadband low-loss 2×2 MZI (Mach- Zehnder interferometer)-based thermo-optic switch with bent directional couplers on silicon. In: Optical Fiber Communication Conference, pp. 1–3. Optica Publishing Group (2016)

Two Modes Unscrambling Using a Single Micro-Ring Resonator

Dan Yi and Hon Ki Tsang(✉)

Department of Electronic Engineering, The Chinese University of Hong Kong, Shatin, NT, Hong Kong SAR
hktsang@ee.cuhk.edu.hk

Abstract. We propose and demonstrate the use of one single micro-ring resonator (MRR) for unitary matrix operations. We proposed that a single MRR with two thermal tunable phase shifters can be used for 2×2 unitary conversion. For a proof-of-concept demonstration, a network composed of a 2d grating coupler and a MRR was designed and fabricated for polarization mode unscrambling. Specifically, we show that two arbitrarily mixed modes can be unscrambled to crosstalk levels below −19 dB. This paper shows how a single MRR may be used for matrix multiplications to process high-speed polarization-multiplexed communications signals with more compact footprint and modest power consumption.

Keywords: Micro-ring resonator · Waveguide devices · Polarization unscrambling

1 Introduction

Coherent networks based on a lattice of Mach-Zehnder Interferometers (MZI) can perform universal matrix transformations [1–3] and have found applications in deep learning [4], boson sampling and other applications in integrated quantum photonics [5], and optical communication systems [6]. Having no recirculating propagation, the MZI-based feedforward configurations do not inherently have optical feedback and are less efficient for implementing functions that need recursion [7]. Micro-ring resonators (MRR) are attractive because of their compact size, high quality factor, power-efficient phase tuning, and intrinsic infinite impulse response [8]. Previously, we proposed that the MRR-based coherent network can be used to realize both finite impulse response and infinite impulse response [7]. Therefore, the coherent network can be applied in two distinct ways: incoherent and coherent applications. Apart from tunable filters, optical switches, optical delay lines, and photonic differentiators, we also proposed the MRR-based network as alternative to MZI for matrix transformation. Based on that, we demonstrated coherent MRR can be used in a polarization processor for polarization unscrambling and polarization state monitoring [9]. Here, we propose and demonstrate that instead of using one phase shifter (PS) in MRR and one outside the ring, we can use two PSs within one MRR so that the extra PS outside the MRR can be omitted to implement unitary conversion as indicated in Fig. 1(a1).

J. Witzens et al. (Eds.): ECIO 2024, SPPHY 402, pp. 258–262, 2024.
https://doi.org/10.1007/978-3-031-63378-2_43

2 Simulation Results

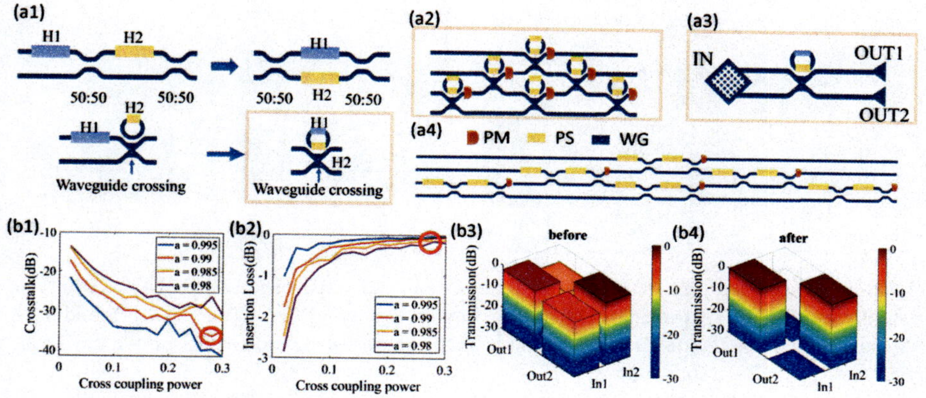

Fig. 1. (a1) Schematic of the MZI and MRRs for unitary conversion. (a2) Scalability of the MRRs for unitary conversion with matrix dimension of four. PM and WG denote power monitor and waveguide. (a3) Schematic of the device using MRR for polarization unscrambling. (a4) Schematic of MZIs for unitary conversion. (b1) Simulation results indicate the relationship between the unscrambling mode crosstalk level and the cross-coupling power/the roundtrip loss of MRR. (b2) Simulation results of insertion loss dependence on cross-coupling power and roundtrip loss. Simulated scattering matrix of the two polarization channels before (b3) and (b4) after unscrambling by the MRR when the (cross-coupling power, the roundtrip loss) were designed to be (0.25, 0.99).

The basic element of the proposed coherent network is the MRR shown in Fig. 1(a1) with two PSs on the ring waveguide. It comprises a dual-bus MRR and a waveguide crossing. The metal heaters were embedded in the top oxide cladding. The sum of the two PSs can be used to tune the resonance of the MRR (variable beam splitter) while the difference of the two PS can be used to tune the relative phase difference between two input waveguides. This MRR therefore can be regarded as a 2 by 2 lattice element providing a 2 by 2 unitary conversion. Here, we design the power coupling ratio of the MRR to be 0.25. Using the MRR-based lattice element, we implemented a photonic circuit to manipulate the two orthogonal polarizations from a single mode fiber which is coupled into the silicon chip via the polarization diversity two-dimensional grating coupler (2DGC) input shown in Fig. 1(a3). Simulation indicated that if we assume the round-trip loss and the cross-coupling power of the MRR to be 0.99 and 0.25, the mode crosstalk level can be recovered to be lower than −30 dB.

3 Device Fabrication and Experimental Setup

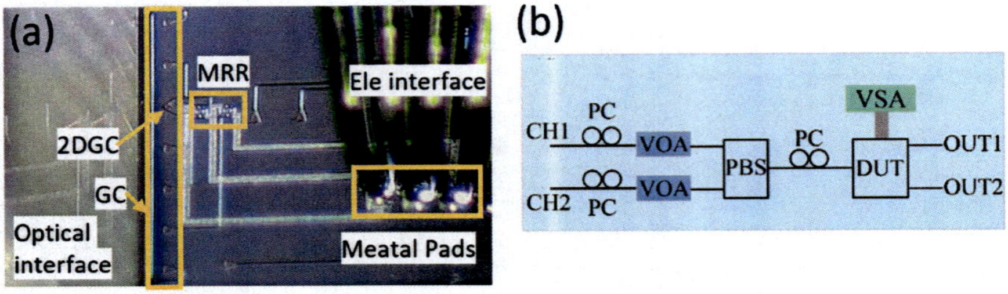

Fig. 2. (a) Micrograph image of the fabricated MRR and Experimental setup. GC represents grating coupler. (b) Experimental setup of the proposed system. PC, VOA, PBS, DUT, and VSA represent polarization controller, variable optical attenuator, polarization beam splitter, the device under test, and voltage source array.

The experimental setup is shown in Fig. 2(b). Two continuous-wave laser was used as an optical source for two polarization channels (CH1 and CH2). On each beam path, there is a polarization controller (PC) was used to adjust the polarization before combined with the fiber-optic packaged polarization beam splitter (PBS). In each path, there is an additional variable optical attenuator (VOA) which was used to measure the scattering matrix. In this way, two independent data channels were coupled to the orthogonal modes in the single mode fiber. After the PBS, another PC enabled the setting of different test input polarizations orientations before coupling into the device under test (DUT). Schematic of the DUT is shown in Fig. 1(a3).

4 Experimental Results

Performance of the MRR with PS were characterized. Figure 3(a1) and (a2) show the transmission spectra of the fabricated MRR, the resonance of the MRR was estimated to be around 14k. The tunability of the MRR is shown in Fig. 3(a3) and (a4). The power consumption to tune the MRR over one free-spectral range was 33mW. After testing the independent devices, we performed the experiments to demonstrate the polarization mode unscrambling. To achieve mode unscrambling, we used adapted gradient descendent method to optimize the applied voltages on the PSs of MRR. During the optimization process, we set the FOM = min (|S12-S11|, |S21-S22|). The value of FOM evolved with optimization epochs were shown in Fig. 3(b1). Scattering matrix Sij evolved with epochs were shown in Fig. 3(b2). FOM value was optimized from 2 dB to 19 dB. The scattering matrix after optimization is plotted in Fig. 3(b4) with crosstalk level < -19 dB. Figure 3(c1) and (c2) show the transmission spectra before and after unscrambling. Figure 3(c3) and (c4) show the mode unscrambling can be established automatically after disturbance (red arrows).

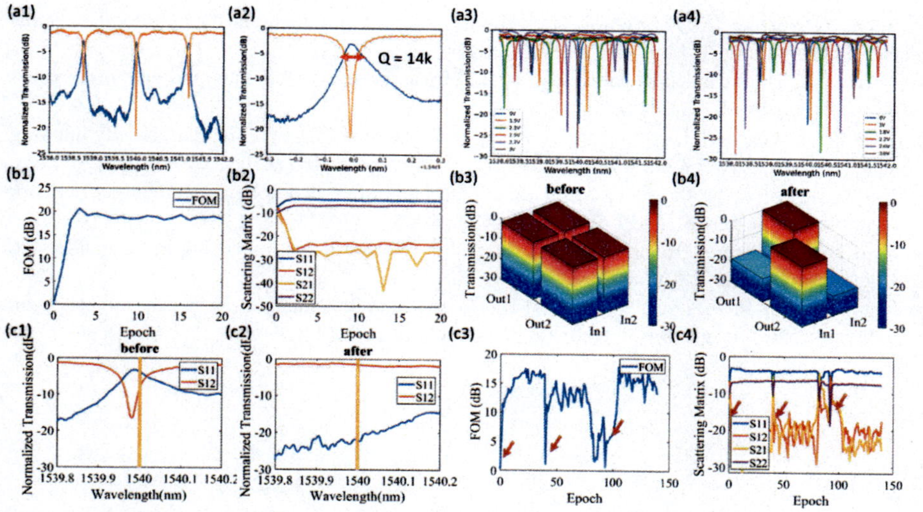

Fig. 3. Experimental results. (a) Device characterization. (b) & (c) Experimental results for two polarization unscrambling. (a1) Transmission spectra of the through port (orange) and drop port (blue) of the fabricated MRR. (a2) Zoomed-in figure of the transmission spectra of the MRR. The MRR has an estimated Quality factor of 14k. Thermal tunability of the two heaters (a3) HT1 and (a4) HT2 when applied with voltage from 0V to 3V. The power efficiency is 33mW per FSR or 2π. (b1) The training process of mode unscrambling. FOM value evolved with epoch number. (b2) Scattering matrix evolved with epoch number. Scattering matrix before (b3) and (b4) after unscrambling. Transmission spectra (c1) before and (c2) after unscrambling. (c3) The training process of mode unscrambling. (c4) Scattering matrix evolved with epoch number.

5 Conclusion

This paper reports the experimental demonstration of polarization unscrambling using one MRR with two thermal PSs. Specifically, we show that the two mixed polarization mode can be unscrambled to crosstalk level lower than -19 dB. This paper shows how a MRR may be used for matrix multiplications to process high-speed polarization-multiplexed communications signals with compact footprint and modest power consumption.

Acknowledgement. This work was funded by RGC GRF grant 14204623.

References

1. Miller, D.A.: Self-configuring universal linear optical component. Photon. Res. **1**(1), 1–15 (2013)
2. Miller, D.A.: Self-aligning universal beam coupler. Opt. Express **21**(5), 6360–6370 (2013)
3. Clements, W.R., Humphreys, P.C., Metcalf, B.J., Kolthammer, W.S., Walmsley, I.A.: Optimal design for universal multiport interferometers. Optica **3**(12), 1460–1465 (2016)

4. Shen, Y., Harris, N.C., Skirlo, S., et al.: Deep learning with coherent nanophotonic circuits. Nat. Photonics **11**(7), 441–446 (2017)
5. Harris, N.C., Carolan, J., Bunandar, D., et al.: Linear programmable nanophotonic processors. Optica **5**(12), 1623–1631 (2018)
6. Annoni, A., Guglielmi, E., Carminati, M., et al.: Unscrambling light—automatically undoing strong mixing between modes. Light: Sci. Appl. **6**(12), e17110 (2017)
7. Yi, D., Wang, Y., Tsang, H.K.: Multi-functional photonic processors using coherent network of micro-ring resonators. APL Photonics **6**(10), 100801 (2021)
8. Madsen, C.: General IIR optical filter design for WDM applications using all-pass filters. J. Lightwave Technol. **18**(6), 860–868 (2000)
9. Yi, D., Luan, J., Wang, Y., Tsang, H.K.: Reconfigurable polarization processor based on coherent four-port micro-ring resonator. Nanophotonics **12**(22), 4127–4136 (2023)

Photo-Thermal Plasmonic Sensors for Transparent Detection and Control of Integrated Photonic Devices

Alessandro di Tria[1(✉)], Andrés Martínez[1], Francesco Zanetto[1], Manuel Kohli[3], Juerg Leuthold[3], Alexandre Bouhelier[4], Francesco Morichetti[1], Andrea Melloni[1], Giorgio Ferrari[2], and Marco Sampietro[1]

[1] Department of Electronics, Information and Bioengineering, Politecnico Di Milano, 20133 Milano, Italy
alessandro.ditria@polimi.it
[2] Department of Physics, Politecnico Di Milano, 20133 Milano, Italy
[3] Institute of Electromagnetic Fields, ETH Zürich, 8092 Zürich, Switzerland
[4] Laboratoire Interdisciplinaire Carnot de Bourgogne, CNRS UMR 6303, Université de Bourgogne, 21000 Dijon, France

Abstract. This work presents novel photo-thermal plasmonic sensors as inline quasi-transparent candidates for optical detection and control of photonic integrated circuits. The good sensitivity and resolution proved them successful in the stabilization of photonic devices against external perturbations. Thanks to their sub-wavelength dimensions and simple fabrication, these sensors are attractive in any platform, offering a solution for technologies that do not feature active optoelectronics.

Keywords: plasmonics · optical detectors · transparent sensors · control loop

1 Introduction

Photonic Integrated Circuits (PICs) have emerged as a viable solution for a wide range of applications, including free-space telecommunication [1], data center interconnects [2] and quantum computing [3], among many others. As such photonic systems expand, the increasing number of integrated devices is pushing the need for stabilizing the operating point of each component in presence of thermal drifts and process tolerances. In order to implement such control, it is necessary to measure the light power in multiple points of the PIC, using photodetectors such as photodiodes, photoresistors, CLIPPs and others. However, some of these solutions require specific technological steps [4], which may be unavailable in certain technologies, or may be too large and prevent the operation on scaled devices [5].

J. Witzens et al. (Eds.): ECIO 2024, SPPHY 402, pp. 263–269, 2024.
https://doi.org/10.1007/978-3-031-63378-2_44

A solution that targets a generic photonic platform is offered by photo-thermal plasmonic sensors. These devices rely on the resistance variation of a metal strip placed in contact with the waveguide, caused by the scattering of Surface Plasmon Polaritons (SPP) modes.

In this work, two designs are presented: a pure plasmonic sensor in a TiO_2 photonic platform and a hybrid plasmonic sensor in Silicon technology. Experimental characterization of both devices is shown, along with the validation of their performances in the closed-loop stabilization of a Micro Ring Resonator, in presence of perturbations of the input light wavelength and temperature fluctuations.

2 Photo-Thermal Plasmonic Sensor

2.1 Working Principle

The photo-thermal plasmonic sensor consists of a strip of metal placed in contact with a dielectric waveguide. When the two materials are in direct contact, we speak of pure plasmonic structure. When the contact is made through an intermediate dielectric layer, we call it hybrid plasmonic (or conductor-gap-dielectric, CGD) structure.

We designed and fabricated two different sensors, as shown in Fig. 1a and 1b:

- a pure plasmonic structure targeting a Titanium Dioxide (TiO_2) platform, where the TiO_2 waveguide is in direct contact with the gold strip;
- a hybrid plasmonic structure targeting a Silicon (Si) platform, where the intermediate layer between Silicon and gold is made of alumina (Al_2O_3).

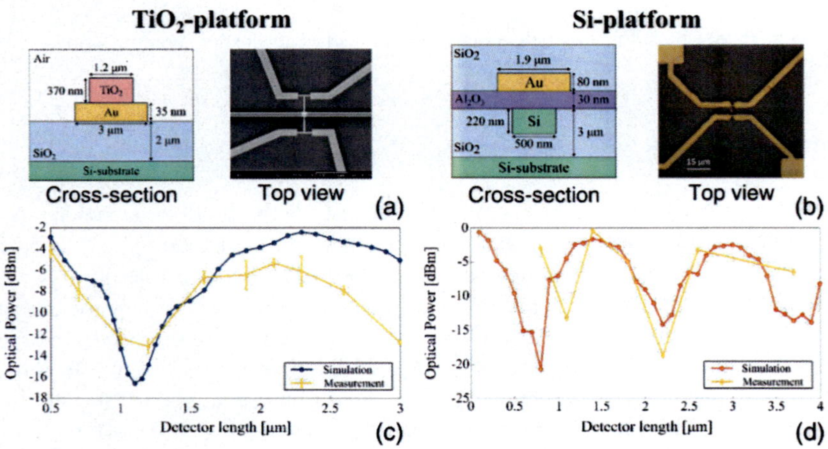

Fig. 1. Technological stack and top-view photograph of the sensor in TiO_2 platform (a) and in Silicon platform (b). Simulated and measured transmittance of the TiO_2 sensor (c) and Silicon sensor (d), as a function of the device length.

When the dielectric TM mode propagating in the waveguide reaches the sensor, it excites two Surface Plasmon Polariton (SPP) modes, propagating either at the metal-dielectric interface (pure plasmonic), or confined in the intermediate layer (hybrid plasmonic). The two modes travel with different phase velocities along the sensor, causing an interference at the metal strip end which depends on the device length [6]. With a correct sizing of the device, the interference is constructive and light couples back into the dielectric waveguide with minimal losses. Figure 1c and 1d show the measured transmittance of the two sensors as a function of the device length, highlighting the periodic interference of the SPP modes and showing insertion losses as low as 6 dB for the TiO_2 sensor and lower than 1 dB for the Silicon one.

When the SPP modes propagate along the sensor, they cause a local heating of the metal due to scattering phenomena. The generated heat changes the metal resistance, according to the formula:

$$R_{sens}(T) = R_0[1 + \alpha(T - T_0)] \tag{1}$$

where R_0 is the sensor resistance at room temperature T_0 and α is the metal temperature coefficient. The light-dependent resistance variations are measured with a four-probe lock-in detection scheme, schematized in Fig. 2a. A sinusoidal current I_{sens} is forced through two pads, producing a voltage variation across the sensor which is read through the other two pads by means of an Instrumentation Amplifier (INA). The output of the amplifier is demodulated and filtered by a Low-Pass Filter (LPF) in order to extract the modulated resistance value. The adoption of the four-probe lock-in detection ensures a high-resolution measurement and the rejection of strays.

In presence of light, the measured voltage changes proportionally to the optical power, according to the formula:

$$\Delta V_{sens} = I_{sens}\Delta R = I_{sens}R_0\alpha\Delta T \tag{2}$$

where ΔT is the light-induced temperature change.

2.2 Electro-Optical Characterization

Experimental characterizations have been carried out for both the TiO_2 sensor and the Si sensor. Figure 2b shows the measured sensitivities for both devices, demonstrating excellent linearity in response to the input optical power over the measured range.

The TiO_2 sensor exhibits a sensitivity of about 5 μV/mW, while the Silicon one shows a higher sensitivity of 30 μV/mW. The difference in sensitivity is attributed to the different structures of the two sensors: the Silicon one, being a hybrid plasmonic structure, exhibits greater confinement and interaction with the sensor compared to the pure plasmonic TiO_2 platform, resulting in increased heating of the metal and consequently a higher sensitivity. The better sensitivity of the Silicon sensor ensures also a higher dynamic range and resolution, reaching a minimum measured optical power of about −28 dBm, 8 dB lower than the TiO_2 sensor.

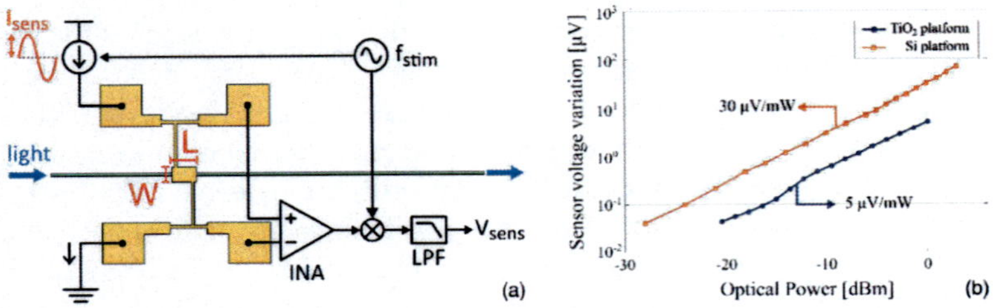

Fig. 2. (a) Schematic of the sensor readout. (b) Measured sensitivities for the TiO_2 sensor (blue) and Si sensor (red).

3 Closed-Loop Control of Micro Ring Resonators

3.1 Control Strategy

The good performances of the Silicon sensor in terms of sensitivity and resolution have been exploited in a control experiment to stabilize the working point of a Micro Ring Resonator (MRR). The schematic of the control loop is depicted in Fig. 3.

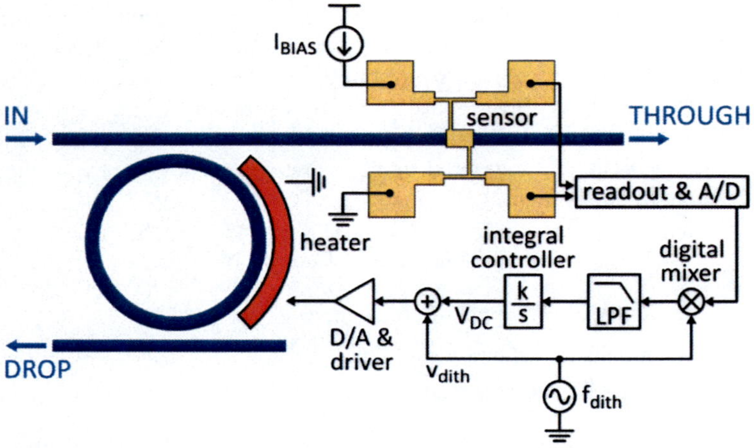

Fig. 3. Schematic of the control loop of a Micro Ring Resonator, employing a heater actuator and a photo-thermal plasmonic sensor.

The control strategy is based on the dithering technique [7]. A sinusoidal perturbation v_{dith} is added to the DC working point of the heater actuator, causing a sinusoidal oscillation of the optical power at the through port proportional to the first derivative of the MRR transmittance. This signal is readout by the photo-thermal plasmonic sensor, digitized and fed to the Digital Signal Processing (DSP) chain. The value of the oscillation is extracted with a digital mixer and a Low-Pass Filter (LPF) and fed to an integral controller. The latter updates the DC voltage V_{DC} applied to the heater until the oscillation of the optical power – and therefore the first derivative of the MRR transmittance – is zeroed, locking the MRR at resonance even in presence of external perturbations.

3.2 Control Experiment

The real-time control loop was tested by trying to keep the MRR at resonance in presence of input light wavelength steps.

Figure 4 shows the measurement of the through port output power when the control is active and the input beam is perturbed with incremental wavelength steps of 100 pm. The control loop is able to compensate for the perturbations updating the heater voltage in order to keep the MRR at resonance in a response time of about 10 ms

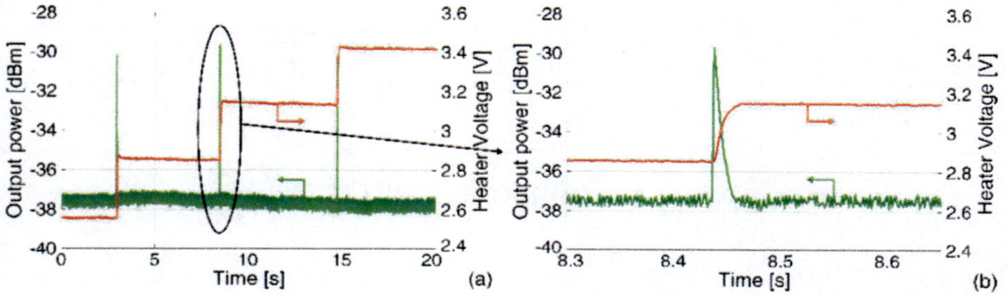

Fig. 4. (a) Measured optical power (green) and heater voltage (orange) at the through port of the MRR with the control activated and with incremental wavelength steps of 100 pm. (b) Zoom of the same measurement around one perturbation. A control response time of about 10 ms can be appreciated.

The control loop was then tested against global temperature variations. The temperature of the PIC was changed using a Peltier cell glued under the chip. Figure 5 shows the measurement of the sensor signal:

- when there is no input light. The curve shows a linear dependence of the sensor signal on the global chip temperature, as reported by Eq. 1 and 2;
- when there is light and the control is OFF. The ring is at resonance at a temperature of 25°C. The measured signal drifts from the linear reference, due to red resonance shifts when the temperature is above 25°C and to blue shifts when it is under 25°C;
- when there is light and the control is ON. The curve matches quite well the reference, meaning that the control loop is able to keep the MRR at resonance.

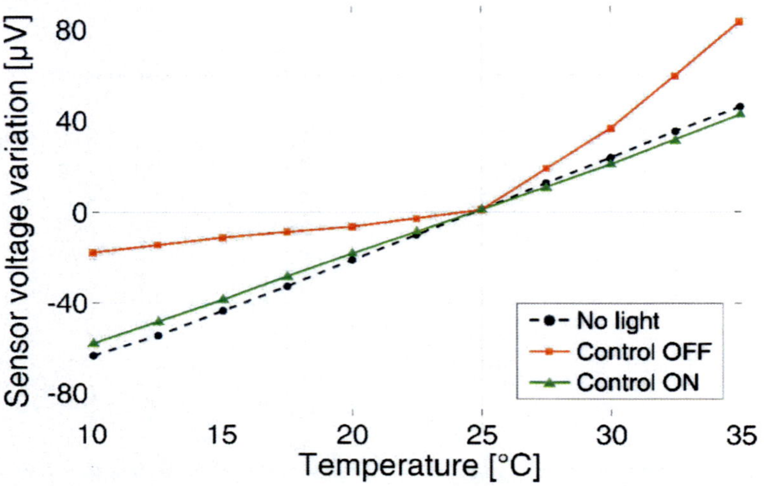

Fig. 5. Measurement of the sensor readout voltage as a function of the global chip temperature when there is no light (dashed blue line), when there is light and the control is off (orange line) and when the control is activated (green line).

4 Conclusions

The proposed photo-thermal plasmonic sensors have been designed and fabricated in two different photonic technologies: a pure plasmonic sensor in TiO_2 platform and a hybrid plasmonic sensor in Silicon technology. Both sensors have been characterized, demonstrating low insertion losses (less than 1 dB), high sensitivity (up to 30 μV/mW) and good resolution (down to -28 dBm). The hybrid plasmonic sensor exhibits better performances thanks to the higher confinement of optical power in the intermediate layer.

The Silicon sensor has been used in a control experiment to lock a Micro Ring Resonator at resonance in presence of external perturbations. The control proved successful in stabilizing the MRR against input wavelength steps and global temperature variations, effectively mitigating the resonance shifts with a response time of 10ms.

Therefore, photo-thermal plasmonic sensors emerge as promising candidates for inline optical detection within Photonic Integrated Circuits, enabling high-resolution sensing and control for any photonic platform.

References

1. Billault, V., et al.: Free space optical communication receiver based on a spatial demultiplexer and a photonic integrated coherent combining circuit. Opt. Express **29**, 33134–33143 (2021)
2. Wohlfeil, B., et al.: Photonic integrated circuits for data center interconnects. In: 2018 Photonics in Switching and Computing, pp.1–3 (2018)
3. Madsen, L.S., et al.: Quantum computational advantage with a programmable photonic processor. Nature **606**, 75–81 (2022)

4. Roelkens, G., et al.: Integration of InP/InGaAsP photodetectors onto silicon-on-insulator waveguide circuits. Opt. Express **13**(25), 10102–10108 (2005)
5. Xu, Y., Jinbiao, X.: Compact silicon hybrid plasmonic microring resonator-based polarization demultiplexer. Jpn. J. Appl. Phys. **54**(8), 082201 (2015)
6. Martinez, A., et al.: Inline photothermal surface plasmon detector integrated in titanium dioxide waveguides. IEEE J. Quantum Electron. **59**(3), 1–8 (2023)
7. Zanetto, F., et al.: Dithering-based real-time control of cascaded silicon photonic devices by means of non-invasive detectors. IET Optoelectron. **15**(2), 111–120 (2021)

Hybrid SiN– Al(Ga)N Optical Phase Shifter at 1550 nm

Cyrille Barrera[1,2,3](✉), Yohan Désières[2], Amélie Dussaigne[2], Eva Kempf[1,4], Ludovic Dupré[2], Pierre Ferret[2], Carlos Alonso-Ramos[3], Frédéric Barbier[2], Jean-Gabriel Mattei[1], Laurent Clément[1], Laetitia Adelmini[2], Laurent Vivien[3], and Frédéric Boeuf[1]

[1] STMicroelectronics, 38920 Crolles, France
cyrille.barrera@steerlight.com
[2] CEA LETI, 38000 Grenoble, France
yohan.desieres@cea.fr
[3] C2N, Paris-Saclay, 91120 Paris, France
[4] INL, 69100 Lyon, France

Abstract. A phase shifter based on GaN/$Al_xGa_{1-x}N$ superlattice is experimentally evaluated, achieving a phase-shift efficiency of 7.2V·cm at 1550 nm. The phase-shifter relies on a SiN-rich rib waveguide with a 13% overlap with an underlying vertical GaN/$Al_xGa_{1-x}N$ *p-i-n* junction. Additionally, we engage in a discussion about the role of charge depletion and Pockels effect on the measured phase-shift.

Keywords: modulators · gallium nitride · nonlinear optics · silicon nitride

1 Introduction

Integrated photonic circuits predominantly rely on silicon platforms, benefiting from the versatility offered by Si/SiN guiding, high-speed & low-consumption Si modulators, and Ge-based photodetectors. However, CMOS platforms are constrained to the infrared domain ($\lambda > 1100$ nm) and necessitate III-V laser source co-integration or advanced packaging for autonomous systems. To address applications across shorter wavelength ranges from NIR to visible, the SiN platform has demonstrated significant capabilities. Nevertheless, it lacks low-power phase modulation options, while the lithium niobate platforms present a promising one but encounters challenges in co-integrating with silicon. III-N bulk materials have also a large transparency window and can be more compatible with CMOS foundries but suffer from lower Pockels coefficient of ~ 1pm/V, and both moderate thermo-optic and dispersion plasma coefficients [1]. However, recent studies tend to show that enhanced electro-optic effect may exists in III-N superlattice [2, 3], most probably due to DC-Kerr effect [4] coupled with the strong polarization-induced internal electric fields in superlattice [5]. In this study, we have designed, fabricated and measured an integrated phase modulator based on silicon-rich SiN rib waveguide on top of an GaN/$Al_xGa_{1-x}N$ superlattice grown on sapphire/AlN substrate.

J. Witzens et al. (Eds.): ECIO 2024, SPPHY 402, pp. 270–275, 2024.
https://doi.org/10.1007/978-3-031-63378-2_45

2 Epitaxial Growth

A $GaN/Al_xGa_{1-x}N$ superlattice similar to the one used in [3] was grown using Metal Organic Chemical Vapor Deposition (MOCVD) on an AlN-on-sapphire (0001) substrate. Prior to the growth of the superlattice, 3 layers of $Al_xGa_{1-x}N$, with decreasing Al fraction x from 70% to 30% along the growth direction, were deposited in order to accommodate the strain between the substrate and the $Al_xGa_{1-x}N$ layers of the superlattice, as depicted in Fig. 1 (b)-(c). The superlattice comprises 5 periods of 4 nm GaN / 8 nm $Al_xGa_{1-x}N$ quantum wells, as illustrated in Fig. 1 (e). To evaluate the superlattice growth influence on the whole stack, a first structure configuration has been grown with only the AlGaN layers, i.e. without the superlattice, and with the last $Al_{0.3}Ga_{0.7}N$ layer doped at n~$5.10^{18}cm^{-3}$ (sample **A**). The superlattice and a top p-GaN layer doped at p~$10^{17}cm^{-3}$ have been added to the A stack on another sample (B) to form a vertically polarized *p-i-n* junction. The establishment of this junction was aimed at dual objectives: firstly, to prevent the occurrence of free carriers within the superlattice, and concurrently, to modulate the electric field within the III-N layers instead of relying on distant electrodes. X-ray diffraction confirmed pseudomorphic III-N layers on sample A, while showed strain relaxation in sample B due to the strain added by the p-GaN and the superlattice. The surface of sample A presented a R_q of 3.4 nm whereas sample B showed a R_q of 6.8nm. The presence of micronic islands (which can be seen in Fig. 1 (f)) with density ~ $3.10^{-4}/\mu m^2$ is also observed on sample B.

3 Design and Fabrication of SiN Waveguide and Hybrid Modulator

The PECVD deposition and patterning of an Si-rich SiN layer [6] with an index of ~ 2.51 allowed to make a 320 nm-thick, 1.2 μm-wide waveguide on top of the surface. The waveguide was designed to allow a TM0 mode at 1550 nm with significant lateral confinement while keeping an overlap of ~ 13% with the $GaN/Al_xGa_{1-x}N$ superlattice for sample B. The TM rather than TE polarization was chosen to be able to interact with the electro-optic coefficient of the III-N layers whether it is the linear r_{33} and/or quadratic s_{33} component. These SiN photonic devices were realized by e-beam lithography and reactive ion etching (RIE). A transmission electron microscopy (TEM) image of the epitaxial stack coupled with the electromagnetic field distribution simulated for the TM0 mode is shown in Fig. 1 (c). A set of waveguides of different lengths and 910 μm-long micro-ring resonators were fabricated in both sample A and B to measure propagation losses. On sample B, electrical contacts have been added on p-doped GaN and n-doped $Al_{0.3}Ga_{0.7}N$ layer as presented in Fig. 1 (a)-(b), to evaluate the phase-shift and loss variations with applied voltage. Electrical contacts were achieved through a sequential process involving standard photolithography, RIE, wet etching, gold deposition, and lift-off techniques. The final cross-sectional and top-view representations of the devices are illustrated in Fig. 1 (c)-(d)-(f), showcasing both transmission electron microscopy (TEM) and scanning electron microscopy (SEM) images.

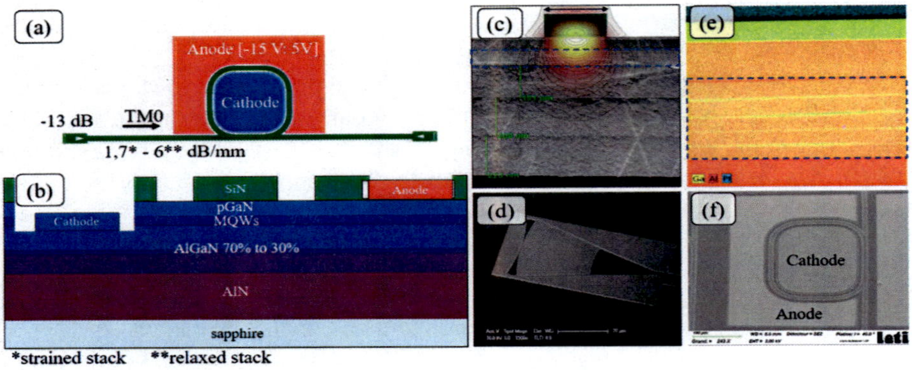

Fig. 1. (a) Top-view design of the ring modulator surrounded by dedicated contact and TM0 guided mode. (b) Cross-section design of the ring modulator with GaN/$Al_xGa_{1-x}N$ multiple quantum wells (MQWs) stack below the SiN waveguide. (c) TEM image of the GaN/$Al_xGa_{1-x}N$ superlattice with optical mode simulated profile overlapping both SiN and III-N layers. (d) SEM image of grating coupler and waveguide. (e) TEM - EDX image of the GaN/$Al_xGa_{1-x}N$ superlattice (f) SEM image of the (Al)GaN SiN modulator(sample **B**).

4 Experimental Results

Transmission measurements were conducted first on sample A between 1500nm and 1600nm, in TM polarization as shown in Fig. 2 (a). Grating insertion losses of -13 dB have been measured whereas ring quality factors at critical coupling were estimated at ~ 25000 with extinction ratios up to ~ 14 dB were measured around 1550nm. From these measurements, propagations losses of ~ 17 dB/cm have been calculated.

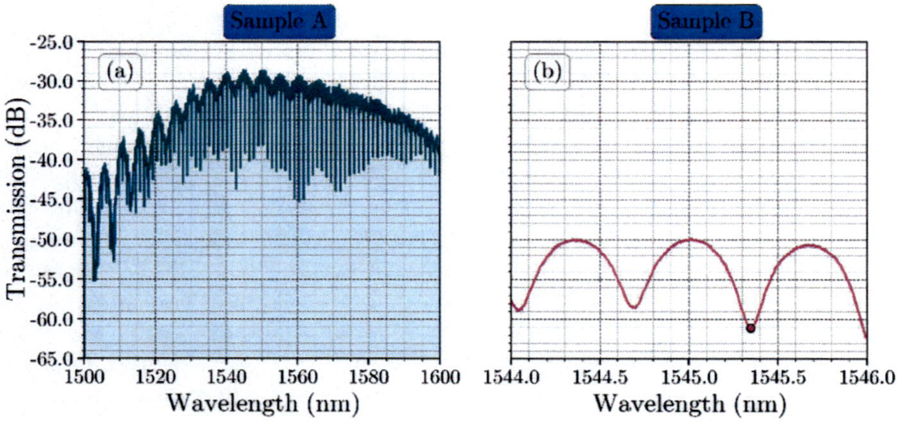

Fig. 2. Sample A: Ring resonator spectra, Q factor at critical coupling is calculated at ~ 25000. Sample B, Ring modulator spectra, Q factor is calculated at ~ 7000

Similar measurements were made on sample B on Fig. 2 (b). The presence of micronic islands at the III-N surface led to a decrease of quality factors down to ~ 7,000. Nevertheless, interference patterns with ER ~ 10dB were still observed on micro-rings with smaller gaps. Spectra were measured with voltage in both forward and reverse voltage and showed a 73 pm resonance shift toward higher wavelength from -8V to -15V, as shown in Fig. 3 (a) whereas no significant shift could be detected in forward bias up to 15V. The evolution of the phase-shift with voltage below -8V was found to be rather linear and, when normalized to the FSR of 1.3 nm, allowed us to evaluate a $V_{\pi}L$ of ~ 7.2V·cm (Fig. 3 (c)).

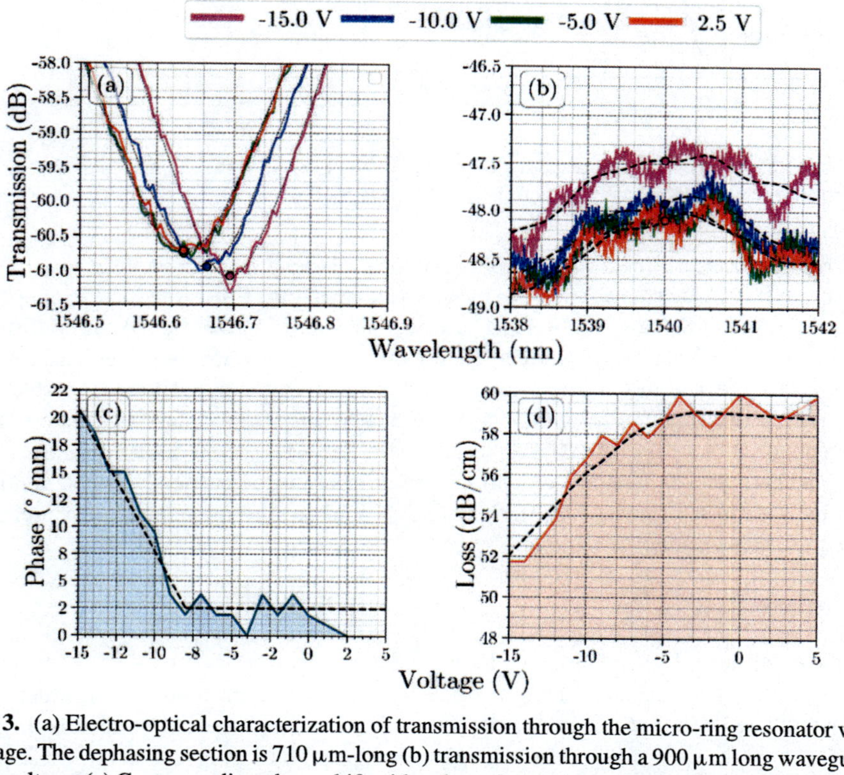

Fig. 3. (a) Electro-optical characterization of transmission through the micro-ring resonator with voltage. The dephasing section is 710 µm-long (b) transmission through a 900 µm long waveguide with voltage. (c) Corresponding phase-shift with voltage from Fig. 2. (a)-(d) propagation loss with voltage (evaluated from Fig. 2. (b) and propagation loss measurements using the cut back method).

I-V measurements in Fig. 4 (b) showed the presence of 5V-8V Schottky barriers in reverse polarization and with some current leakage (around -30 µA at -15V). Transmission line method (TLM) confirmed the presence of Schottky contacts on p-GaN and n-AlGaN layers in Fig. 4 (a). Despite the current leakage observed in reverse bias, the displacement of the peak to longer wavelengths is not consistent with a current injection but with a carrier depletion. We therefore expect the current leakage observed for reverse voltage remains localized below the p-GaN contact, where the serial resistance to reach the n-type layers is expected to be the lowest.

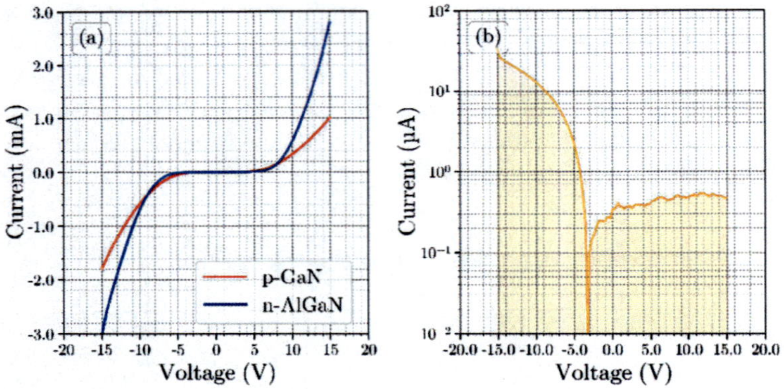

Fig. 4. (a) TLM measurement of both anode on p-GaN layer and cathode on n-AlGaN showing Schottky contact layer. (b) I-V characteristics from ring modulator.

Transmissions of linear phase-shifters of different lengths (410 μm, 910 μm) were measured to evaluate propagation losses with voltage in Fig. 3 (b). A reduction of propagation losses of -7dB/cm was observed between -5V and -15V in Fig. 3 (d) and confirmed that a carrier depletion in III-N layers below the waveguide contributes, at least, to the phase-shift observed in our micro-rings. At this stage, Pockels and Kerr DC effect [2, 3] could not be separated from the plasma effect. In our samples, both effects produce different variation of refractive index ($\Delta n > 0$ for DC Kerr/carrier depletion and $\Delta n<0$ for Pockels effect) and electro-optical modeling of both free carriers' distribution and electric fields in our GaN/$Al_xGa_{1-x}N$ samples would require meticulous measurement and calibration of material parameters.

5 Conclusion

We have fabricated an integrated phase modulator with a $V_\pi L$ of ~7.2 V.cm on a GaN/$Al_xGa_{1-x}N$ superlattice using a rib SiN waveguide. The modulation effect was attributed, at least in part, to carrier depletion. Furthermore, one can notice that in this first device, only 13% of the mode intensity overlaps with III-N layers. While these preliminary results mark the first demonstration, to the best of our knowledge, of integrated modulators using III-N epitaxial layers, and despite the high linearity of the phase-shift with voltage, we have not yet been able to unambiguously confirm a contribution of either the Pockels or DC Kerr effect on the measured dephasing. Second-harmonic generation (SHG) experiments on samples of higher epitaxial quality, such as those grown on AlN bulk substrates, might help elucidate this effect in integrated devices. We believe that having a low-power modulator based on III-N could also be advantageous for III-N photonic platforms, given their demonstrated capabilities in various photonic devices [7], laser sources in the VIS-NIR domain [8], nonlinear properties [9], and successful industrial co-integration with silicon.

Acknowledgements. This work is part of the IPCEI Microelectronics and Connectivity and was supported by the French Public Authorities within the frame of France 2030.

References

1. Soltani, M., Soref, R.: Free-carrier electrorefraction and electroabsorption in Wurtzite GaN. Opt. Express **23**, 24984–24990 (2015)
2. Avakyants, P., et al.: Linear electro-optic effect in electroreflectance spectra of AlGaN/InGaN/GaN light emitting diodes structure. Jpn. J. Appl. Phys. **52**(8), 08JK11 (2013)
3. Jiang, W., et al.: Electro-optic coefficient enhancement of AlGaN via multiple field modulations. ACS Appl. Mater. Interfaces **7**(32), 17:707–17:712 (2015)
4. Jonathan, P., et al.: High-speed silicon photonic electro-optic Kerr modulation. Photon. Res. **12**, 51–60 (2024)
5. Bretagnon, T., et al.: Radiative lifetime of a single electron-hole pair in GaN/AlN quantum dots. Phys. Rev. B **73**, 113304 (2006)
6. Kempf, E., et al.: Low loss SiN films for integrated photonics deposited by PVD at low temperature. Opt. Mater. Express **13**, 1353–1360 (2023)
7. Suraj, S.R., Raghavan, S., Selvaraja, S.: Polarization independent grating in a GaN-on-sapphire photonic integrated circuit. Opt. Expr. **31**(14), 23350 (2023). https://doi.org/10.1364/OE.487389
8. Kamei, T., et al.: Research toward a heterogeneously integrated InGaN laser on silicon. Phys. Status Solidi A **217**, 1900770 (2020)
9. Zheng, Y., et al.: Integrated gallium nitride nonlinear photonics. Laser Photonics Rev. **16**, 2100071 (2021)

Optical MEMS

Monolithically Integrated Visible-Light MEMS Switch

Alperen Govdeli[1,2](✉), Hong Chen[1], Saeed S. Azadeh[1], John N. Straguzzi[1], Hongyao Chua[3], Guo-Qiang Lo[3], Joyce K. S. Poon[1,2], and Wesley D. Sacher[1](✉)

[1] Max Planck Institute of Microstructure Physics, Weinberg 2, 06120 Halle, Germany
{agovdeli,wesley.sacher}@mpi-halle.mpg.de

[2] Department of Electrical and Computer Engineering, University of Toronto, 10 King's College Road, Toronto, ON M5S 3G4, Canada

[3] Advanced Micro Foundry Pte. Ltd., 11 Science Park Road, Science Park II, Singapore 117685, Singapore

Abstract. We demonstrate a monolithically integrated, electrostatic MEMS-actuated optical switch for visible light with >7 dB extinction ratio and 2.5 dB loss at a wavelength of 540 nm. We measured rise and fall times of 5 μs and 28 μs, respectively, and a low static power dissipation of <1 nW.

Keywords: Integrated photonics · optical switches · MEMS actuation

1 Introduction

The integration of silicon nitride (SiN) waveguides into silicon (Si) photonic integrated circuits (PICs) and the emergence of SiN-on-Si platforms [1] have led to recent reports of increasingly advanced on-chip photonic devices for visible (VIS) and near-infrared (NIR) light (wavelengths, λ, of 400–1000 nm), including phase shifters [2–6], photodetectors [1, 7], and co-packaged lasers [8] to support applications spanning quantum information [9], neurotechnology [10] and display [11]. A critical missing building block in existing VIS integrated photonic platforms is a low-power and fast optical switch.

The simplest and most-widely reported integrated photonic phase shifters / switches in the VIS are based on the thermo-optic (TO) effect. SiN TO phase shifters typically exhibit large power dissipations (20–30 mW [2]), and recent efforts have demonstrated power consumptions of ~1 mW via resonant enhancement with SiN microdisks [3] or via a suspended and folded-waveguide structure as in our previous work in Ref. [4]. However, these device improvements come at the expense of wavelength sensitivity (in the case of resonant enhancement) or long, ~1 ms, rise/fall times (for suspended TO phase shifters). Recent efforts to overcome these limitations of TO VIS optical switches and phase shifters have explored thin-film lithium niobate [12] and liquid crystal clad [5] waveguides, with ultralow power dissipations reported and, in the case of lithium niobate, large, >25 GHz, modulation bandwidths. However, integrating lithium niobate into a VIS Si photonic platform is inherently challenging and remains an open problem,

J. Witzens et al. (Eds.): ECIO 2024, SPPHY 402, pp. 279–283, 2024.
https://doi.org/10.1007/978-3-031-63378-2_46

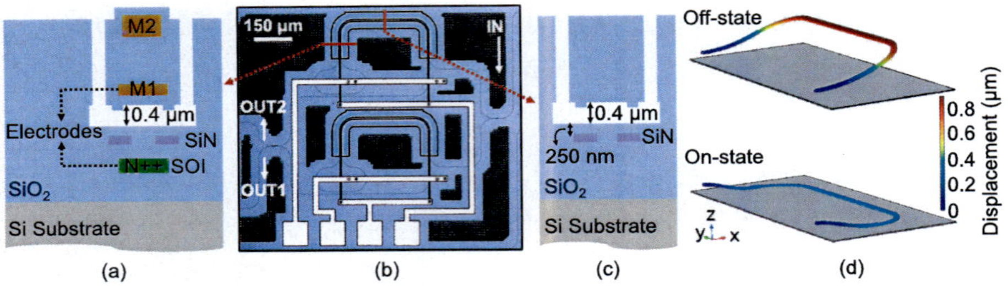

Fig. 1. Visible-light electrostatic MEMS-actuated optical switch. (a) Cross-section schematic of the electrostatic actuator section of the phase shifter (without deflection from internal stress). (b) Annotated optical micrograph of the switch; the brightness and contrast have been adjusted for enhanced visibility of the waveguides. (c) Cross-section schematic of the suspended beam section of the phase shifter. (d) Simulated displacement of the cantilever in the off-state ($V_{app} = 0$ V) and on-state ($V_{app} = 60$ V).

and liquid crystal modulators require post-processing steps and are not significantly faster than TO phase shifters.

An alternative strategy to design high-performance integrated photonic VIS switches and phase shifters is to incorporate microelectromechanical systems (MEMS) technology. Along these lines, aluminum nitride piezo-actuated phase shifters with SiN waveguides have been reported for NIR, achieving MHz-scale bandwidths with a voltage for a π rad. phase shift, V_π, of ~15 V. However, due to the reliance on the strain optic effect, a relatively large waveguide length (>3 mm) was required, adding excess propagation losses and corresponding to a phase shifter footprint of 350 μm $\times$ 325 μm [6]. An initial investigation of an electrostatically actuated MEMS phase shifter relying on a deformable Si-SiO_2 membrane was presented in Ref. [13], but the device was integrated into an otherwise passive platform (without photodetectors or other active devices); the modulation speed and optical spectrum were not reported.

In this work, we report an electrostatic MEMS-actuated visible-light switch monolithically integrated in our second-generation (Gen-2) VIS Si photonic platform, which includes waveguide-coupled photodetectors (PDs), TO phase shifters, and advanced passive devices (with 2 waveguide layers). In contrast to our Gen-1 platform, which was fabricated on bulk Si wafers, Gen-2 is fabricated on 200-mm silicon-on-insulator (SOI) wafers featuring high-efficiency and low-noise PDs, which outperform their Gen-1 counterparts [14], in addition to similar passive device and TO phase shifter capabilities as Gen-1. Here, we experimentally demonstrate an optical switch at $\lambda = 540$ nm with an extinction ratio (ER) > 7 dB, insertion loss of 2.5 dB, and low static power dissipation (<1 nW). The switching speeds were roughly an order of magnitude faster than our previously reported suspended TO switches.

2 Design, Fabrication, and Characterization

Figure 1 shows the device design and operating principle. The MEMS phase shifter has a U-shaped structure consisting of 2 suspended electrostatic actuators, Fig. 1(a), with a suspended SiO_2 beam in between, Fig. 1(c). With an applied voltage across the top (M1) and bottom (N++ doped SOI) electrodes of the actuators, the suspended beam is lowered onto the 120 nm-thick SiN waveguides (looped twice through the phase shifter), Fig. 1(d), changing the effective index of the waveguides resulting in an optical phase shift. Through design of the metal (M2 and M1) widths and resulting internal stress of the MEMS structure, a small upward deflection was achieved with no applied voltage, enabling an electrically controllable gap between the suspended beam and the waveguides. The electrostatic actuator sections were each 100 μm-long, and the suspended beam section was 250 μm-long. The suspended beam included mechanical stoppers to avoid stiction. The phase shifter was integrated into a 1 × 2 Mach Zehnder interferometer (MZI), forming an optical switch. The MZI included 1 × 2 and 2 × 2 multimode interference couplers and a path length mismatch of 8μm between its arms to facilitate characterization.

The MEMS-actuated optical switch was fabricated in our Gen-2 visible-light integrated photonics platform at Advanced Micro Foundry (AMF) on 200-mm SOI wafers. The fabrication follows our previous reports in Refs. 1 and 14 with the additional definition of a patterned amorphous Si sacrificial layer and subsequent undercut etching to the define the suspended MEMS structure.

To characterize the device, transverse magnetic (TM) polarized laser light was coupled into and out of the chip through cleaved single-mode optical fibers using on-chip edge couplers. The optical transmission was characterized using a supercontinuum laser (NKT Photonics SuperK Fianium) with a wavelength-tunable optical filter (NKT Photonics LLTF) and a fiber-coupled photodetector. To characterize the switching performance, the applied voltage was delivered by a piezoelectric actuator controller module via tungsten probes. The static power dissipation was measured using a source-meter (Keithley 2470).

3 Experimental Results

The characterization of the optical switch is summarized in Fig. 2. With increasing applied voltage, the phase shifter and MZI switched from their "off" to "on" states, Figs. 2(a) and (b). The optical transmission of the device with increasing and decreasing applied voltage at $\lambda = 540$ nm is shown in Fig. 2(c). The device reached its on-state with an applied voltage of 60 V. The switch turned off [Fig. 2(a)] when the bias voltage was reduced to <10 V. An extinction ratio (ER) of >7 dB at $\lambda = 540$ nm between on/off-states was observed in both the transmission vs. voltage, Fig. 2(c), and optical transmission spectra, Fig. 2(d). After several minutes with an applied voltage of 60 V, we observed a gradual increase in the required turn-on voltage for the switch. To circumvent this effect, we switched the polarity of the applied voltage between subsequent measurement trials. This effect is currently under investigation.

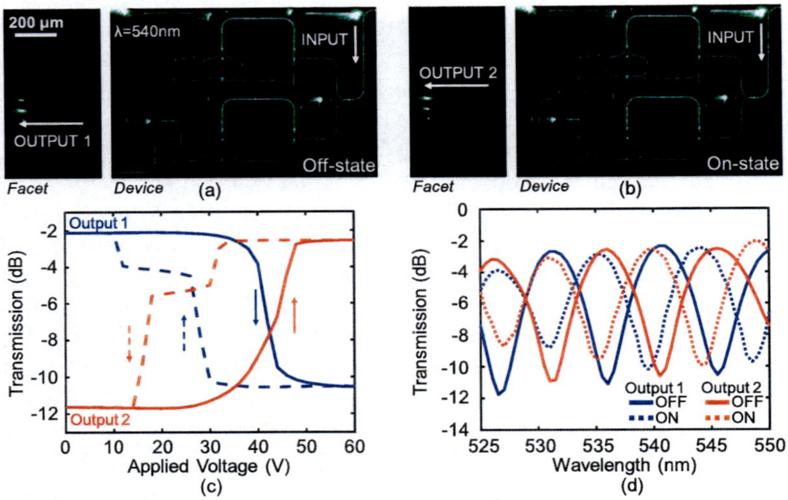

Fig. 2. MEMS-actuated optical switch characterization. Optical micrograph of the chip facet and device at $\lambda = 540$ nm in (a) the off-state ($V_{app} = 0$ V) and (b) the on-state ($V_{app} = 60$ V). (c) Measured optical transmission of the switch with increasing (solid trace) and decreasing (dashed trace) applied voltages for Output-1 (blue) and Output-2 (red) for TM-polarized input light at $\lambda =$ 540 nm; switching between the outputs is observed. (d) Optical transmission spectra of the switch for TM-polarized input light spanning $\lambda = 525$–550 nm measured at Output-1 (blue) and Output-2 (red). Transmissions in (c) and (d) were normalized to the transmission of a straight waveguide reference structure with nominally identical edge couplers to the switch.

The rise and fall times were measured to be 5 μs and 28 μs for a 60-V square-wave voltage waveform, respectively. In addition, low static power consumption (<1 nW) was observed – an inherent property of the electrostatic actuation mechanism.

4 Conclusion

In summary, we have demonstrated a visible-light electrostatic MEMS-actuated optical switch in an active foundry-fabricated visible-light integrated photonics platform. The switch operated at green wavelengths for TM-polarized light, and exhibited ER > 7 dB with an insertion loss of 2.5 dB at $\lambda = 540$ nm and a switching voltage of 60 V. The switching speed of the device reached tens of kHz, which is about an order of magnitude faster than our previously reported SiN TO phase shifters [4], while the static power dissipation was <1 nW. Characterization of the dynamic power dissipation and mechanical resonance frequency is ongoing.

References

1. Sacher, W.D., et al.: An active visible-light integrated photonics platform on 200-mm Si. In: 2023 Optical Fiber Communication Conference, (pp. Tu3C-5), Optica Publishing Group (2023)

2. Mohanty, A., et al.: Reconfigurable nanophotonic silicon probes for sub-millisecond deep-brain optical stimulation. Nat. Biomed. Eng. **4**(2), 223–231 (2020)
3. Liang, G., et al.: Robust, efficient, micrometre-scale phase modulators at visible wavelengths. Nat. Phot. **15**(12), 908–913 (2021)
4. Yong, Z., et al.: Power-efficient silicon nitride thermo-optic phase shifters for visible light. Opt. Express **30**(5), 7225–7237 (2022)
5. Notaros, M., et al.: Integrated visible-light liquid-crystal-based phase modulators. Opt. Express **30**, 13790 (2022)
6. Dong, M., et al.: Piezo-optomechanical cantilever modulators for VLSI visible photonics. APL Photonics **7**, 051304 (2022)
7. Yanikgonul, S., et al.: Integrated avalanche photodetectors for visible light. Nat. Comm. **12**(1), 1834 (2021)
8. Corato-Zanarella, M., et al.: Widely tunable and narrow-linewidth chip-scale lasers from near-ultraviolet to near-infrared wavelengths. Nat. Photon. **17**(2), 157–164 (2023)
9. Mehta, K.K., et al.: Integrated optical multi-ion quantum logic. Nature **586**(7830), 533–537 (2020)
10. Sacher, W.D., et al.: Optical phased array neural probes for beam-steering in brain tissue. Opt. Lett. **47**(5), 1073–1076 (2022)
11. Raval, M., et al.: Integrated visible light phased array system for autostereoscopic image projection. Opt. Lett. **43**(15), 3678–3681 (2018)
12. Li, C., et al.: High modulation efficiency and large bandwidth thin-film lithium niobate modulator for visible light. Opt. Express **30**(20), 36394–36402 (2022)
13. Figeys, B., et al.: Compact optomechanical phase shifter for visible light integrated photonics. In CLEO: Science and Innovations, (pp. JW2A-32), Optica Publishing Group (2018)
14. Govdeli, A., et al.: Broadband, efficient, and low dark current SiN-on-SOI waveguide-coupled photodetectors for visible light. In: Optical Fiber Communications (OFC), (pp. W2A.37), Optica Publishing Group (2024)

Photonic Integrated Piezo-MEMS (Pip-MEMS) Device for 1D Beam Scanning

P. Venkatachalam[1,2](✉), Daniel Yumnam[1], Sushma Gali[1], Mruthyunjaya K. Swamy[2], and Shankar Kumar Selvaraja[1](✉)

[1] Centre of Nano Science and Engineering, Indian Institute of Science, Bengaluru, India
{venkatachala,shankarks}@iisc.ac.in

[2] Department of Mechanical Engineering, National Institute of Technology, Surathkal, India

Abstract. We demonstrate a 1D beam scanning device with a maximum scan angle of 6.6° at 18.91 kHz for a 0.5 mm long PZT-on-SiN waveguide integrated Piezo-MEMS cantilever.

Keywords: Cantilever · Waveguide · Beam Scanning

1 Introduction

Optical beam scanning is indispensable for a wide range of applications, including displays, microscopy, LiDAR, and biomedical imaging [1]. One could achieve beam scanning by systematically tuning the angle or position of electromagnetic radiation, such as light or radio waves. Beam scanning enables precise control of signal transmission, tracking moving objects, profile shape of objects, laser-based manufacturing processes, and environmental monitoring. While bulk beam scanning is achieved through bulk mechanical schemes, its significance lies in the minaturisation of the system on a chip. Some of the key performance improvements include improving detection and imaging capabilities, achieving intricate designs in manufacturing, enabling autonomous systems, and facilitating precise data capture and reproduction. However, the challenge of shrinking the size of integrated optical beam scanning systems while simultaneously reducing their power consumption persists as a formidable obstacle. Current solutions like Galvo scanners, MEMS mirror scanners, and optoacoustic deflectors face limitations in miniaturisation and integration potential due to discrete integration of source, optics, and scanning elements [2, 3]. Moreover, optical phase arrayed systems necessitate complex and high-power consumption methodologies for effective phase control [2, 4].

This work proposes a novel solution by employing a monolithic photonic integrated piezo-MEMS (Pip-MEMS) platform for beam scanning [5]. Leveraging the silicon nitride photonic circuit platform and lead zirconium titanate (PZT)-based piezo-MEMS, our PipMEMS device achieves efficient beam scanning capabilities while addressing the challenges of size reduction and power consumption. This innovative approach holds promise for advancing the field of integrated optical beam scanning, facilitating enhanced performance and versatility across various applications.

J. Witzens et al. (Eds.): ECIO 2024, SPPHY 402, pp. 284–288, 2024.
https://doi.org/10.1007/978-3-031-63378-2_47

2 Design and Fabrication

The silicon nitride waveguide dimensions have been optimised for single-mode operation at a wavelength of 780 nm. We use 220 nm thick silicon nitride (SiN) as the waveguide layer. A ridge waveguide with a width of 550 nm and an etch depth of 90 nm is obtained as an optimal waveguide parameter. These dimensions ensure efficient single-mode operation and light confinement within the waveguide. Figure 1(a) shows the cross-section schematic and the cantilever beam steering device with cantilever stack details.

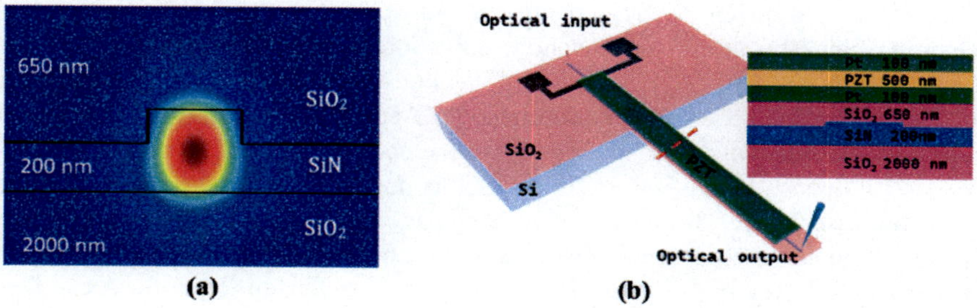

Fig. 1. (a). Cross-section of SiN waveguide with material thickness and (b) Cross-sectional view and Schematic Representation of the Proposed Device

The waveguide is embedded into the microcantilever structures for beam steering. Figure 1(b) shows the schematic of a piezo-MEMS cantilever with an integrated SiN waveguide. We use sputter-deposited PZT as the piezoelectric material. The proposed structure consists of a rectangular cantilever together with piezoelectric actuation. COMSOL Multiphysics finite element methods (FEM) is used to simulate the electro-mechanical behaviour of the MEMS cantilever structures.

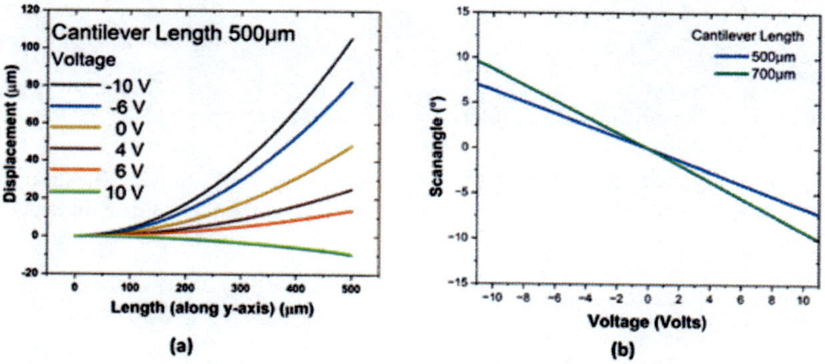

Fig. 2. (a) Simulated cantilever defection along the length of a 0.5 mm cantilever for different applied voltages (b) Simulated scan angle for 0.5 mm and 0.7 mm cantilever.

The MEMS module is used to analyse the static and dynamic deflection of the cantilevers under applied voltages. When applying an AC drive voltage across the platinum

(Pt) electrodes (Fig. 1(b)), the induced electric field generates mechanical stress within the sandwiched PZT, which leads to out-of-plane cantilever deflection. Figure 2(a) shows the cantilever displacement for an applied voltage and scan angle for two different cantilever lengths, and Fig. 2(b) shows the estimated scan angle of the cantilever device for lengths 0.5 and 0.7 mm for a maximum applied voltage of 10 V.

The proposed beam scanner is fabricated on a 500 μm thick silicon wafer as the substrate. Initially, a 2 μm thick layer of silicon dioxide is deposited using plasma-enhanced chemical vapor deposition (PECVD). This silicon dioxide layer serves as cladding and provides isolation from the substrate silicon. Following this, a 220 nm of PECVD silicon nitride (SiN) is deposited. Subsequently, the SiN layer undergoes patterning using e-beam lithography and reactive ion etching to create waveguides and grating couplers. A 650 nm thick PECVD silicon dioxide layer is deposited as top cladding, which also serves as isolation from PZT and prevents lead diffusion into the guiding layer. 100 nm platinum is sputtered as the bottom electrode and buffer layer for piezoelectric deposition. A 500 nm thick PZT is deposited using a sputter deposition process onto the Pt layer, followed by wet etch pattering of the PZT. The as-disposited PZT is amorphous; the patterned PZT is annealed to achieve the desired ferroelectric phase. This is followed by 100 nm of Pt deposition to form the top electrode. A cantilever pattern is defined using optical lithography, and the complete stack, including the buried 2 μm silicon dioxide under the waveguides, is etched using reactive ion etching. The sacrificial handle layer of Si is removed using an isotropic deep reactive ion etching to suspend the waveguide along with the cantilever. Figure 3(a) and (b) shows the optical and SEM images of the fabricated device. Figure 3(c) shows the output grating section of the device and the sandwiched PZT layer.

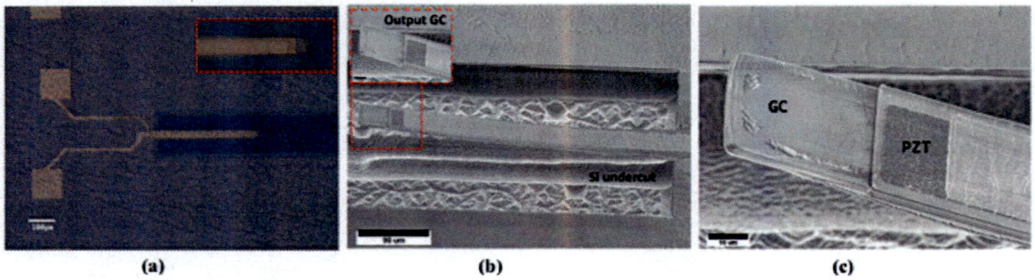

Fig. 3. (a) Optical Micrograph of the 0.5 mm Cantilever Device. (b) Scanning Electron Microscope (SEM) Image of the 0.5 mm Cantilever Device. (c) SEM image of cantilever tip with Output Grating Coupler.

3 Measurement Results

The characterisation of the fabricated device involves coupling light from a 780 nm laser into the waveguide through a single-mode fibre and measuring the out-coupling from the cantilever grating. Grating couplers are employed to couple light in and out of the waveguide efficiently. PipMEMS beam scanning device was tested for both static and dynamic characteristics. Figure 4(a) shows the non-contact optical surface profiler map of 0.5 mm cantilever. The upward deflection of the tested device indicates a final stress of 150 MPa, which shows good agreement with their simulated results; this can be attributed to repeated thermal processing and residual stress during stages of fabrication.

Laser Doppler vibrometer (LDV) is used to assess the resonant frequency of the fabricated cantilever. Electrical signals were sourced from a lock-in amplifier, covering a frequency spectrum ranging from 1 kHz to 400 kHz. Figure 4(b) illustrates a comparison between the measured and simulated resonance of the cantilever. We measured a resonance frequency of 18.91 kHz for a 0.5 mm long cantilever. We consistently measured approximately 2X difference between the estimation and measurement. This deviation can be attributed to variations in material properties and fabrication tolerances inherent in the device fabrication process.

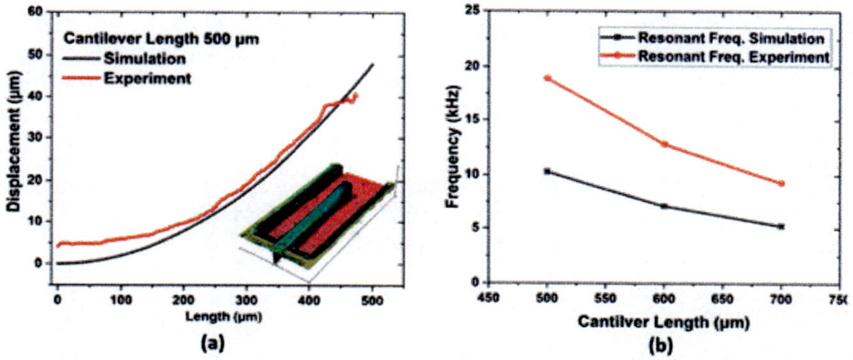

Fig. 4. (a) Cantilever tip position long the length of 0.5 mm cantilever extracted form the 3D topographic map from a optical profilometer in comparison with simulated results. (b) Resonant frequency of the measured devices in comparison with simulation results.

Despite the observed deviation, the characterisation provides valuable insights into the performance and behaviour of the fabricated device, paving the way for further refinement and optimisation to achieve desired operational parameters. To accurately measure the scan angle, a specialised test setup employing an optical camera is utilised, as depicted in Fig. 5(a). The setup includes a primary camera (Camera-1) positioned to capture images perpendicular to the device's plane, providing a field of view of 10.47°. Additionally, Camera-2 serves as a reference, placed outside the field of view of Camera-1, as illustrated in Fig. 5(b). An electrical signal sourced from a lock-in amplifier, spanning a voltage range of 1 V to 10 V with a frequency range of 1 kHz to 18 kHz, is applied to drive the device. The drive signal from a lock-in amplifier is applied across the electrodes, which induces an electric field and generates mechanical stress within the sandwiched lead zirconate titanate (PZT) material, resulting in cantilever deflection along the Z-axis and subsequent beam deflection. Figure 5(c) shows the captured image from camera-1. The camera captures the light from the output grating coupler and records the scanning action, as shown in Fig. 5(b). The recorded response, obtained for a full-scale measurement of the scanning action, is processed using a Python image processing program. This program tracks the change in position of the optical output beam across captured image frames and the estimation of the scan angle.

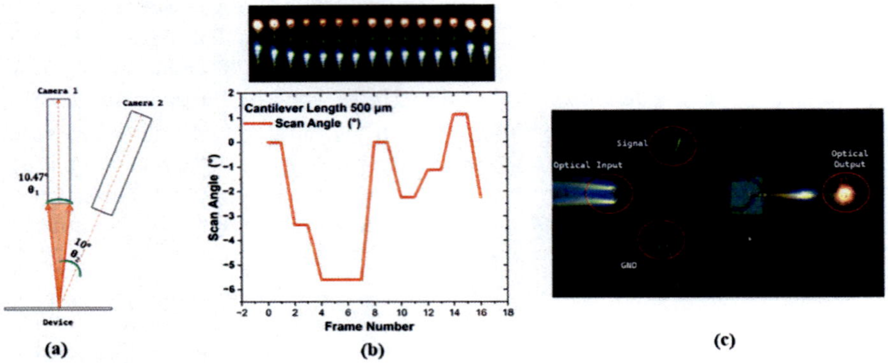

Fig. 5. (a) Optical measurement setup. (b) Scan angle esistmate for selected induviual image frames. (c) PipMEMS device under operation.

For a 500 μm cantilever device, we estimate a scan angle of 6.6 degrees, achieved with an applied signal of 10 V peak-to-peak (Vpp). The scanning range could be improved through PZT thickness and cantilever stress optimisation.

4 Conclusion

We have proposed and demonstrated one-dimensional beam scanning at 780 nm using a piezo-integrated photonic MEMS (PipMEMS) device. We demonstrated a sputter-deposited PZT-on-SiN waveguide cantilever as a scanning device. A maximum scan angle of 6.6^0 is achieved for 0.5 mm long cantilever. This demonstration shows the possibility of monolithic integration of piezoelectric MEMS with silicon nitrate photonics for its potential applications in various light modulation and scanning applications, including LiDAR and imaging. With its compact design and low power consumption, the PipMEMS device allows for the further development of integrated beam scanning systems.

References

1. Kim, I., et al.: Nanophotonics for light detection and ranging technology. Nat. Nanotechnol. **16**, 508–524 (2021)
2. Azadeh, S.S., et al.: Microcantilever-integrated photonic circuits for broadband laser beam scanning. Nat. Commun. **14**, 1–8 (2023)
3. McManamon, P.F., Ataei, A.: Progress and opportunities in optical beam steering (2019)
4. Li, N., et al.: A progress review on solid-state lidar and nanophotonics-based lidar sensors. Laser Photonics Rev. **16**, 2100511 (2022)
5. Mere, V., et al.: Photonics integrated piezomems (PipMEMS): a scalable hybrid platform for next-generation mems. IEEE Sens. Lett. **4**, 1–4 (2020)

Nanomechanical Phase Shifting on a Gallium Arsenide Platform

Celeste Qvotrup[1(✉)], Rodrigo Thomas[1], Zhe Liu[1], Marcus Albrechtsen[1], Arne Ludwig[2], Andreas D. Wieck[2], and Leonardo Midolo[1]

[1] Niels Bohr Institute, University of Copenhagen, Blegdamsvej 17, 2100 Copenhagen, Denmark
celeste.qvotrup@nbi.ku.dk

[2] Lehrstuhl für Angewandte Festkörperphysik, Ruhr-Universität Bochum, Universitätsstrasse 150, 44780 Bochum, Germany

Abstract. Integrated photonics on membranes containing quantum dots constitute a promising platform for quantum information processing. On-chip photonic information processing requires reconfigurable active elements such as phase shifters and beamsplitters. We report the design, fabrication and characterization of a slot waveguide phase shifter manufactured on a suspended gallium arsenide (GaAs) membrane with embedded indium arsenide (InAs) quantum dots (QDs). Using a comb drive we electromechanically actuate the gap of a slot waveguide, demonstrating a phase shift of π in a device with a slot waveguide length of 10 µm.

Keywords: gallium arsenide · quantum photonic integrated circuits · phase shifting · nano-opto-electro-mechanical systems

1 Introduction

Indistinguishable single photons generated in photonic integrated circuits (PICs) are a promising choice of qubits for quantum information processing [1]. Due to its direct bandgap and high refractive index, gallium arsenide (GaAs) membranes with epitaxially grown integrated indium arsenide (InAs) quantum dots (QDs) provide an excellent platform for generating and controlling single photons. The manipulation and routing of the on-chip single photons requires development of low-loss optical elements such as beam splitters [2] and phase shifters that reliably can be tuned.

Quantum dots require a cryogenic environment ($< 10\,\text{K}$) to freeze out vibrations in the QDs, i.e. phonons, which otherwise degrade the quality of the generated single photons. As Franz-Keldysh electroabsorption in GaAs renders electrooptic modulation unviable at typical InAs emission wavelengths (930nm) [3], this leads to mechanical actuation being the preferred strategy for controlling and routing photons in GaAs PICs [4].

J. Witzens et al. (Eds.): ECIO 2024, SPPHY 402, pp. 289–292, 2024.
https://doi.org/10.1007/978-3-031-63378-2_48

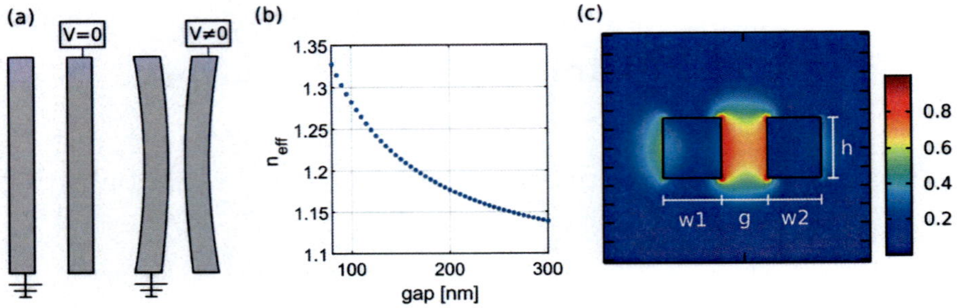

Fig. 1. a: Schematic of displacement on two parallel beams, fixed on each end, with a potential difference. By connecting one beam to one cantilever of a slot waveguide, actuation of the effective mode can be carried out. b: Effective index of slot mode versus gap of a GaAs slot waveguide suspended in air. c: Electric field norm of a quasi-TE supported in a GaAs slot waveguide with a gap $g = 120\,\mathrm{nm}$, $w1 = 150\,\mathrm{nm}$, $w2 = 140\,\mathrm{nm}$, $h = 160\,\mathrm{nm}$.

In this paper, we present a nano-opto-electro-mechanical system (NOEMS) constisting of a phase shifter integrated on GaAs. By exploiting the strong gap dependence of the effective refractive index, n_{eff}, of a mode in a slot waveguide, where the mode is guided in the low-index gap between two high-index cantilevers [5], the accumulated phase of light in one arm of an integrated Mach-Zender interferometer (MZI) can be actively controlled.

Former works has demonstrated slot waveguide couplers utilized as efficient phase shifters on materials such as silicon [6] and silicon nitride [7]. However, we wish to integrate this directly with our cryogenic GaAs QD platform to route the monolithically generated single photons.

2 Fabrication and Experiment

The devices are fabricated on 160-nm-thick GaAs membranes grown by molecular beam epitaxy (MBE) on a 1150 nm $Al_{0.75}Ga_{0.25}As$ sacrificial layer. First, Cr-Au wires and electrodes are defined via 125 keV electron-beam lithography and electron-beam metal evaporation followed by liftoff. Next, Cr-Ni-Au strips, which provide an intrinsic stress that stiffens the fragile device, are deposited using a similar procedure. Shallow etch gratings (SEGs) [8] are defined, after which waveguides, isolation trenches and optical elements are defined utilizing ICP. Finally, the chip is undercut in a 5% HF solution, which is highly selective to AlGaAs over GaAs. Finally, the chip is dried with critical point drying to prevent stiction.

A scanning electron micrograph of the finished chip is shown on Fig. 2(a). After fabrication, the chip is wirebonded and placed in a cryostat with optical fiber access, as shown in Fig. 2(b). The finalized device consists of an MZI containing two passive beam splitters, the upper arm of the interferometer contains the phase shifter. Actuation is carried out by a comb drive, which allows larger

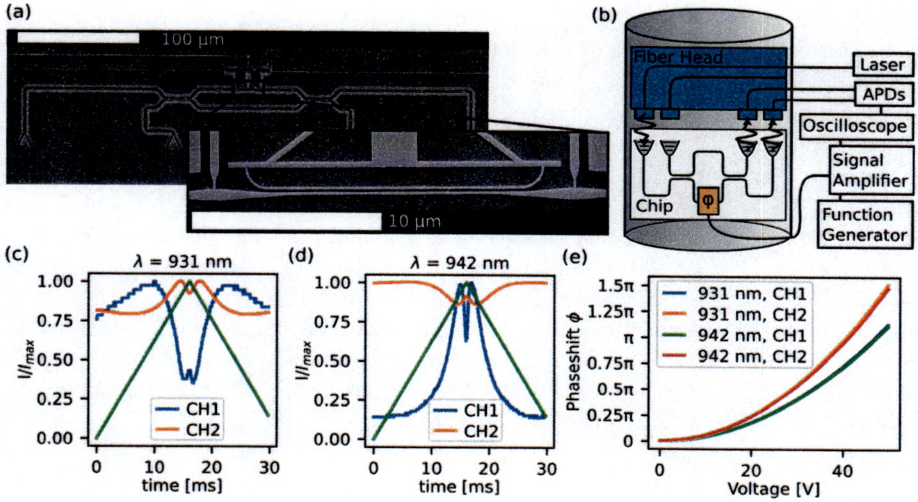

Fig. 2. Experimental setup and results. a: Scanning electron micrograph of the fabricated device, which is symmetric around a vertical middle axis. Insert: Close up of slot waveguide. b: Diagram of the experimental setup. c (d): Normalized transmission data for a wavelength of $\lambda = 931(942)$ nm. e: Calculated phase shift for the two different outcoupling gratings at respectively 931 and 942 nm.

displacements without risking collapse due to the pull-in effect [9], using a novel design that pulls the cantilevers of the slot waveguide closer, rather than apart, when actuated. To characterize the phase shifter, we measure the response of the interferometer to when a phase shift is applied to the upper arm. We excite the structure with a continuous wave laser and collect from both output ports, labeled CH1 and CH2, simultaneously while sweeping both wavelengths and applied potential difference. We use a signal amplifier to supply potential differences > 10 V over the comb drive. The result at two different wavelengths can be seen in Fig. 2(c) and (d). Electrostatic actuation originates in the capactive energy stored, which scales quadratically with the voltage. Therefore, the phase shift can be extracted from a fit of the intensity against V^2, as plotted in Fig. 2 (e). Smallest actuation needed to obtain a π phase shift is for $\lambda = 931$ nm in channel 2, where $V_{\pi,931}^{CH2} = 39.7$ V is needed, equivalent to a half-wave-voltage-length product $V_{\pi,931}^{CH2} L_s = 0.00397\,\text{V} \cdot \text{cm}$. Largest actuation needed to obtain a π phase shift is for $\lambda = 942$ nm in channel 1, where $V_{\pi,942}^{CH1} = 47.7$ V is needed, equivalent to a half-wave-voltage-length product $V_{\pi,942}^{CH1} L_s = 0.00477\,\text{V} \cdot \text{cm}$.

3 Discussion and Conclusion

We proposed a phase shifter integrated on GaAs utilizing a suspended slot waveguide with a controllable gap. After fabricating the device, initial characterization shows that the device is capable of inducing a π phase shift. The data analysis is carried out using a simplified model assuming that n_{eff} varies linearly with the gap between the cantilevers of the slot waveguide. As can be seen from Fig. 1(b), this only holds true over very small displacements. We attribute the inconsistent phase shift between the two channels in part to this simplification.

Although the device exhibits a π phase shift, full switching between the arms of the MZI over π phase shift is not observed. MZIs utilizing balanced beam splitters (R=T=0.5), with one lossless arm and one lossy arm can exhibit a visibility less than one, even in the case of two lossless beam splitters. Reducing the loss by designing a better slab-to-slot waveguide coupler, or alternatively balancing the loss by inserting a passive slot waveguide on the other arm of the phase shifter can yield a higher visibility. As an alternative, the beam splitters could be replaced with a 1×2 and 2×1 multimode interferometer (MMI), respectively. By fine-tuning the properties of the comb drive, a better displacement per applied potential difference can be obtained.

References

1. Uppu, R., Midolo, L., Zhou, X., Carolan, J., Lodahl, P.: Quantum-dot-based deterministic photon-emitter interfaces for scalable photonic quantum technology. Nature Nanotechnol. **16**(12), 1308–1317 (2021)
2. Papon, C., et al.: Nanomechanical single-photon routing. Optica **6**(4), 524–530 (2019)
3. Wang, Y., et al.: Electroabsorption in gated GaAs nanophotonic waveguides. Appl. Phys. Lett. **118**(13), 131106 (2021)
4. Midolo, L., Schliesser, A., Fiore, A.: Nano-opto-electro-mechanical systems. Nature Nanotechnol. **13**(1), 11–18 (2018)
5. Almeida, V.R., Qianfan, X., Barrios, C.A., Lipson, M.: Guiding and confining light in void nanostructure. Opt. Lett. **29**(11), 1209–1211 (2004)
6. Van Acoleyen, K., Roels, J., Mechet, P., Claes, T., Van Thourhout, D., Baets, R.: Ultracompact phase modulator based on a cascade of NEMS-operated slot waveguides fabricated in silicon-on-insulator. IEEE Photonics J. **4**(3), 779–788 (2012)
7. Grottke, T., Hartmann, W., Schuck, C., Pernice, W.H.P.: Optoelectromechanical phase shifter with low insertion loss and a 13pi tuning range. Opt. Express **29**(4), 5525–5537 (2021)
8. Zhou, X., Kulkova, I., Lund-Hansen, T., Hansen, S.L., Lodahl, P., Midolo, L.: High-efficiency shallow-etched grating on GaAs membranes for quantum photonic applications. Appl. Phys. Lett. **113**(25), 251103 (2018)
9. Legtenberg, R., Groeneveld, A.W., Elwenspoek, M.: Comb-drive actuators for large displacements. J. Micromech. Microeng. **6**(3), 320 (1996)

Photonic Integrated Circuits for Free-Space Communications and Sensing

Cutting-Edge Integrated Photonics in Space

Caterina Ciminelli(✉)

Optoelectronics Laboratory, Politecnico di Bari, Bari, Italy
caterina.ciminelli@poliba.it

Abstract. The space industry is currently experiencing unprecedented transformation and growth on a global scale. Significant technological advances have propelled the space sector towards global innovation. Integrated photonics technology serves as a key driver of innovation in the New Space domain. This paper explores the integration of photonics into space missions, highlighting its transformative potential in telecommunications, Earth observation and space exploration. It also analyses the financial trajectory of the space industry, with the significant investments made by the European Union and NASA to promote the development of space-grade photonic components, and the role of space companies in advancing photonic technologies.

Keywords: Photonic Integrated Circuits · Space Optical Communication · Earth Observation · Space Exploration

1 Introduction

Photonics technology has become a catalyst for driving new capabilities and opportunities across various applications within the New Space sector. Photonics enables the miniaturization of space instruments, enhances navigation precision, facilitates high-speed data transfer in free space, and empowers advanced imaging and sensing technologies for Earth observation, leading to size, weight, and power consumption reductions alongside enhanced performance. PICs emerge as transformative forces in space applications, offering chip-scale integration of multiple optical elements critical for mission success. PICs are expected to extend beyond space exploration, influencing the defence, security, and telecommunications sectors [1].

Research and Markets Global Space Economy Market Report 2022–2026 offers a compelling perspective on the financial trajectory of the space industry. Analysts anticipate that as launch costs continue to decrease, the space industry is poised to achieve an annual revenue of $1 trillion by 2040 [2].

Space companies play a crucial role in advancing integrated photonics technologies, leveraging their expertise to push the boundaries of innovation in space exploration and telecommunications. Incorporating PICs into its satellites can enhance the performance and efficiency of communication systems, enabling high-speed data transfer and connectivity across vast distances, as in the case of SpaceX constellation [3]. Airbus introduced the Space Photonics Unified Roadmap [4], a strategic initiative aimed at identifying

J. Witzens et al. (Eds.): ECIO 2024, SPPHY 402, pp. 295–300, 2024.
https://doi.org/10.1007/978-3-031-63378-2_49

shared photonic requirements across future space products. Another example is Lockheed Martin, which has pioneered the integration of photonics into space-based sensing and imaging systems, enhancing the precision and capabilities of its Earth observation satellites and facilitating critical tasks such as environmental monitoring and disaster response. Additionally, companies like Blue Origin and OneWeb are exploring the integration of photonics into their space infrastructure to enable next-generation satellite communication networks and advanced sensing capabilities.

Besides the companies' thrust towards photonics, institutional funding for the development of space technologies in Europe reached approximately 935 million euros in 2021 [5].

Furthermore, the EU has allocated funding for numerous research projects within the Horizon 2020 (H2020) program, exemplified by initiatives like INPHOMIR [6] and SIPhoDiAS [7], aimed at fostering the development of new space-grade technological platforms.

NASA has been actively advancing integrated photonics for space applications, with significant funding allocated to support these efforts.

Optoelectronic components have become vital in telecommunications satellites. Optopyrotechnics, utilizing laser ignition, offer safety and cost advantages for launchers. LIDAR systems, such as ADM Aeolus and EarthCare, play key roles in scientific and environmental applications. Microwave photonics enhances RF payloads' performance through optical manipulation. Key components include lasers, optical amplifiers, and photonic integrated circuits.

In the following sections, the important role of integrated photonics in space optical communication technologies, Earth observation, and space exploration is highlighted.

2 Space Optical Communication and Earth Monitoring Photonic Technologies

In the domain of space-based optical communication technologies, the pursuit of higher frequencies stands as a paramount imperative to meet the escalating data rate requirements of next-generation systems. In this framework, while free-space optical communications, harnessing laser- and fiber- technologies, are well-recognized for their efficacy in addressing these challenges head-on, space agencies are diligently crafting photonic payloads tailored for the demands of next-generation telecommunication satellites and earth monitoring missions.

Regarding space laser communications technologies, pioneering studies have been carried out by the ESA since 1977, notably with the Artemis satellites [8], demonstrating with the deployment of the first commercially operational optical data relay satellite, European Data Relay System (EDRS)-A [9], high data rates of up to 1.8 Gbit/s with remarkable reliability, as well as optical inter-satellite and satellite-to-ground links with data rates from hundreds of Gbps to Terabits per second. Also, this sector has been impacted by the migration towards integrated photonic components since the first decade of the new millennium, spurred by the ESA's Scy-Light program (HydRON initiative) and the European Commission's VERTIGO project [10, 11].

Satellite communication transceivers are increasingly demanding flexible telecom payloads that can be reconfigured post-launch, with operating bandwidth shifted from traditional low-frequency bands (C/Ku) to higher-frequency Ka or Q/V bands. Microwave Photonic (MWP) technology has emerged as a promising alternative to standard RF components, as demonstrated by ESA within the Artes program, offering broader bandwidth, enhanced routing flexibility, and reduced mass and size. To accomplish the market requirements, a novel TLC payload architecture has been proposed, including optical signal generation, distribution, and switching functionality. Although a migration towards a fully photonic payload is envisaged and expected in the upcoming missions, several studies have been conducted, including industrial research, with Technology Readiness Level (TRL) ≤ 4 [12, 13], and experimental developments, with TRL ≥ 5 [4, 14, 15], mainly focused on the development of the key functionalities of the proposed photonic repeater, as the optical distribution of high-frequency microwave local oscillator signals, photonic frequency conversion of RF signals, and routing via photonic switching. Miniaturized X- and Ka-band oscillators with very low phase noise (-91 dBc/Hz @1MHz and < -100 dBc/Hz @10 kHz from the carrier, respectively) have been demonstrated [12, 13].

Upon recognizing the benefits –23% for both power consumption and unit saving and 3% for mass saving - space companies and agencies have pushed this transition through experimental developments with high TRL. As an example, an ESA-funded project, SODaH, has contributed to advancing the TRL of photonic components, including photonic modulators, routing, and digitalization units, to TRL 5 [14]. Similarly, the ORIONAS program has developed miniaturized optical transmitters and receiver amplifiers compatible with next-generation small high-speed satellite constellations, achieving remarkable miniaturization compared to conventional free space optics receivers [15]. The SIPhoDIAS project has further enhanced photonic component reliability at TRL 7, with developments including a high-performance photodiode (0.8 W/A with a bandwidth of 38 GHz), electro-optic modulators (45 GHz with losses less than 7 dB), and an on-board optics transceiver module with a small form factor [4].

The integration of photonics into microwave spaceborne operations introduces new functionalities and capabilities also for Synthetic Aperture Radar (SAR) systems for Earth observation. A SAR payload typically comprises essential components such as a linearly chirped microwave waveform (LCMW) generator, frequency converters (FC), an analog-to-digital converter (ADC), and a beamforming network (BFN) [16]. Moreover, recent preliminary studies on photonic Fast-Fourier-Transform processors, have been initiated, guaranteeing 300 MHz sampling spacing of the upconverted signal from the RF band, 8-bit, and real-time on-board processing [17]. Integrated delay lines with fast reconfigurability (of the order of ns), low power consumption (of the order of mW), and a maximum delay time of hundreds of ps, that ensure a steering angle of tens of degrees, have been demonstrated [18, 19]. The RETINA program started in 2013, is the pioneer in the development of a high TRL (5–6) reconfigurable multibeam photonic BFN with centralized processing [20]. Recently, in the framework of the European Commission's project SCORE-SAR the first-ever photonics-assisted receiver module for scan-on-receive synthetic aperture radar has been demonstrated [21].

3 Space Exploration Photonic Technologies

In preparation for a long-term human stay in extraterrestrial space, it is necessary to characterize and then explore possible planets in the solar system that could host human settlements.

The aims of robotic exploration activities are mainly twofold, the identification and characterization of possible extraterrestrial habitats, and the development of technologies for robotic exploration of potential extraterrestrial habitats.

Research and development activities in support of future human missions have objectives such as the definition of requirements and architectures of new missions, with particular attention to the human factor, and the identification and study of infrastructures to create suitable habitats to support long-term human permanence.

Photonics technology has long been crucial in the exploration and study of outer space. Several space agencies are intensively engaged in planetary exploration activities, with missions to the Moon and Mars.

A key enabling technology in planetary exploration applications, such as the approach and landing of celestial bodies, rendezvous, docking, and identification of space debris, as well as in CubeSat constellations, is represented by LIDARs [22].

Current LIDAR systems used in space are based on discrete optical components, which makes them very bulky. This negative feature is typically compounded by high power consumption. Considerable miniaturization efforts have been made over the years.

Integrated photonics now promises, even in these systems, high levels of functionality, and a potential co-integration with electronics and micro-optics, with the ultimate target of higher miniaturization, compactness, reliability, and lower size, weight, and power (SWaP), all critical factors for these applications.

A key element of LIDAR is the phased array, which when realized with photonic integrated circuits, represents a robust and non-mechanical approach to beam steering.

In addition to lunar and planetary exploration, space photonics also holds promise for the study of distant celestial bodies.

4 Conclusions

This paper provides a comprehensive overview of state-of-the-art developments in space photonics across various sectors, including space-based communication, Earth monitoring, and exploration missions. PICs have emerged as pivotal components, by offering chip-scale integration capabilities and unlocking new possibilities and functionalities in space applications, exhibiting superior performance compared to competing technologies. PICs offer inherent advantages such as reduced sensitivity to external disturbances, compactness, and lower power consumption, making them ideal for space payloads, especially in the context of the expected proliferation of micro- and nanosatellites in the coming years. Analysis of present and future European and American photonic space missions underscores the significance of space-grade programs in shaping future initiatives. Despite advancements, challenges persist in finding a single integrated photonic platform meeting all modern system requirements. To address this, targeted actions driven by space companies, agencies, and R&D centers have been established to promote PIC technology and reduce production costs, paving the way for widespread and well-established use of photonics in space.

References

1. Brunetti, G., et al.: Integrated photonics for NewSpace. In: International Conference on Applications in Electronics Pervading Industry, Environment and Society, pp. 294–299. Springer, Cham (2022). https://doi.org/10.1007/978-3-031-30333-3_39
2. Global Space Economy Market: Analysis By Client Type, By Application, By Value Chain, By Region Size and Trends with Impact of COVID-19 and Forecast up to 2028, https://www.researchandmarkets.com/reports/5574887/global-space-economy-market-analysis-by-client. Accessed 18 Mar 2024
3. Carrizo, C., Knapek, M., Horwath, J., Gonzalez, D.D., Cornwell, P.: Optical inter-satellite link terminals for next generation satellite constellations. In: Free-Space Laser Communications XXXII vol.11272, SPIE, pp. 8–18 (2020).
4. Fernandez, V. Mariojouls, S., Kokkinos, D.: Space Photonics Unified Roadmap Airbus Space Systems Initiative SPUR Team, EPIC Meeting on Photonics for Space: Opening New Horizons (2023)
5. European Commission Press Release: European Investment Fund announces €300 Million of Space Sector Finance with New Investments into Orbital Ventures and Primo Space Brussels (2021)
6. INPHOMIR project: https://ec.europa.eu/info/funding-tenders/opportunities/portal/screen/how-to-participate/org-details/922504533/project/101135749/program/43108390/details. Accessed 18 Mar 2024
7. Sourikopoulos, I., et al.: The H2020-SPACE-SIPHODIAS project: space-grade optoelectronic interfaces for photonic digital and analogue very-high-throughput satellite payloads. In: International Conference on Space Optics, ICSO, vol. 11852. SPIE (2021)
8. Alonso, A., Reyes, M., Sodnik, Z.: Performance of satellite-to-ground communications link between ARTEMIS and the optical ground station. In: Optics in Atmospheric Propagation and Adaptive Systems VII, vol. 5572, pp. 372–383. SPIE (2004)
9. Berretta, G., De Agostini, A., Dickinson, A.: The European data relay system: present concept and future evolution. Proc. IEEE **78**, 1152–1164 (1990)
10. Perdigues, J., et al.: HYDRON: The ESA initiative towards optical networking in space. In: 2021 European Conference on Optical Communication (ECOC), pp. 1–4. IEEE (2021)
11. Le Kernec, A., et al.: The H2020 VERTIGO project towards Tbit/s optical feeder links. In: International Conference on Space Optics, ICSO, vol. 1185. SPIE (2021)
12. Tang, J., et al.: Integrated optoelectronic oscillator. Opt. Express **26**, 12257–12265 (2018)
13. Brunetti, G., Marocco, G., Armenise, M.N., Ciminelli, C.: High performance chirped microwave generator for space applications. In: International Conference on Space Optics, ICSO, vol. 11852. SPIE (2021)
14. Polo, V., Piqueras, M.A., Mati, J.: Photonic payloads for next generation 5G satellite networks. IEEE Future Netw. Tech Focus **3**, 1–8 (2019)
15. Stampoulidis, L., et al.: The H2020-SPACE-ORIONAS project: Lasercom-on-chip" for high-speed satellite constellation interconnectivity. In: International Conference on Space Optics, ICSO, vol. 11852. SPIE (2020)
16. Moreira, A.: Synthetic Aperture Radar (SAR): principles and applications. In: Proceedings of 4th Advanced Training Course in Land Remote Sensing, ESA, Athens, Greece (2013)

17. Armenise, M.N., di Toma, A., Brunetti, G., Saha, N., Ciminelli, C.: Flexible photonic integrated circuits: a new paradigm to process data on-board satellites. In: 2023 23rd International Conference on Transparent Optical Networks (ICTON). IEEE (2023)
18. Brunetti, G., Conteduca, D., Dell'Olio, F., Ciminelli, C., Armenise, M.N.: Design of an ultra-compact graphene-based integrated microphotonic tunable delay line. Opt. Express **26**, 4593–4604 (2018)
19. Wang, X., Zhao, Y., Ding, Y., Xiao, S., Dong, J.: Tunable optical delay line based on integrated grating-assisted contradirectional couplers. Photonics Res. **6**, 880–886 (2018)
20. Chmielak, B., et al.: High-efficiency grating coupler for an ultralow-loss Si3N4-based platform. Opt. Lett. **47**, 2498–2501 (2022)
21. Wang, H., Zhang, H., Dai, S., Sun, Z.: Azimuth multichannel GMTI based on Ka-band DBF-SCORE SAR system. IEEE Geosci. Remote Sens. Lett. **15**, 419–423 (2018)
22. Avraam, C., et al.: Advanced concept of a photonic integrated circuit microlidar for navigation, landing and debris detection. In: International Conference on Space Optics, ICSO, vol. 12777. SPIE (2022)

Automatic Real-Time Configuration of a 15-Channel Free-Space Optical Receiver with a Fully Integrated CMOS Controller

Emanuele Sacchi[1](✉), F. Zanetto[1], A. Martinez[1], S. SeyedinNavadeh[1], F. Morichetti[1], A. Melloni[1], M. Sampietro[1], and G. Ferrari[2]

[1] Department of Electronics, Information and Bioengineering, Politecnico di Milano, 20133 Milano, Italy
emanuele.sacchi@polimi.it
[2] Department of Physics, Politecnico di Milano, 20133 Milano, Italy

Abstract. We present a fully integrated CMOS electronic control system for automatic configuration of integrated free-space optical receivers fabricated on a Silicon Photonics platform. The electronic chip consists of 8 identical and independent channels, dissipating roughly 10 mW each. By connecting via direct wire-bonding two chips to the photonic integrated circuit, our closed-loop architecture succeeded in automatically configuring, starting from a random working point, a mesh of 15 interferometers in less than 10 ms. The chips also compensate in real-time atmospheric turbulence with a bandwidth up to 300 Hz, providing a scalable solution for enabling reliable free-space optical communications.

Keywords: silicon photonics · free-space optics · CMOS integrated circuit

1 Introduction

Photonic integrated circuits (PICs) are currently experiencing an increasing demand in the field of free-space optics (FSO), where large wireless bandwidth, low latency and interference-free communication are needed to provide efficient solutions for Internet-of-Things technologies in 5G networks and beyond [1]. In FSO links, the main obstacle for the propagation and reconstruction of the signal at receiver side is due to the presence of atmospheric turbulence along the beam path. Such turbulence, which appears as a random phase and amplitude modulation with a bandwidth of hundreds of Hz, results in scintillation and fading of the received beam spot, thus impairing the overall coupling efficiency [2].

This issue can be tackled directly in the optical domain by employing two design solutions, both implemented in this work. First, a multi-aperture receiver was used to sample multiple portions of the distorted wavefront by means of an integrated optical phase array (OPA), thus increasing the amount of power available for signal reconstruction. Second, a programmable photonic processor, consisting in a binary-tree mesh of Mach-Zehnder Interferometers (MZI), reconstructs the received beam by coherently summing the sampled inputs according to their time-varying phase and amplitude [3].

J. Witzens et al. (Eds.): ECIO 2024, SPPHY 402, pp. 301–308, 2024.
https://doi.org/10.1007/978-3-031-63378-2_50

The control of each MZI needs to be carried out by a dedicated electronic system operating in real time. This approach was previously validated by means of a custom-designed electronic platform [4], which is however hardly scalable to PIC of increasing complexity due to its large area occupation (>cm^2) and power dissipation (≈W). In this work, we present a scalable approach to the problem of the dedicated control electronics, realized here as an 8-channel CMOS application specific integrated circuit (ASIC). By using two ASICs in parallel, our solution proved successful in automatically stabilizing, without any prior calibration, the working point of up to 15 MZIs, as well as dynamically reconfiguring their relative phase shift to compensate the effect of atmospheric turbulence.

2 Architecture of the System

2.1 Free-Space Receiver

A detailed scheme of the targeted PIC and its micrograph are shown in Fig. 1a and 1b. The FSO receiver was built on AMF Silicon Photonics (SiP) platform and comprises 15 MZIs forming a 16×1, 4-layers binary-tree mesh. Each device is equipped on one output branch with an integrated Ge photodiode (PD) with sensitivity $S_{21} \approx 0.85$ A/W, feeding the controller with the information about the working point of the interferometer. Two integrated thermal actuators (*heaters*), modelled with a resistance $R_H \approx 400\ \Omega$ and a thermal transient response $\tau_H \approx 1\ \mu s$, are used to unbalance the optical lengths of the two branches and control the output power splitting ratio of the MZI. By driving the heaters in the 0–6 V range, it is possible to induce a phase shift between the arms up to $\varphi_H \approx 3\pi$. The control electronics normally operates to minimize the optical power impinging on each integrated sensor, thus steering all the light towards the other branch of the MZI, eventually allowing for a coherent sum of all the sampled input beams at the output of the PIC. The SiP die was mounted on a custom-designed PCB holder (Fig. 2a) and connected to two CMOS electronic controllers via direct wire-bonding.

2.2 Electronic Controller Architecture

Each channel of the ASIC, whose architecture is detailed in Fig. 2b, closes a negative feedback control loop around the interferometric device. The loop relies on the dithering technique [5] to generate the error signal that drives a calibration-free integral controller. The pair of DACs that determines the working point of the actuators superimposes on top of their bias voltages two orthogonal square-wave modulations (*dithering* signals), with $\approx 1-100$ mV programmable amplitude, at a frequency $f_{dith} \approx$ 8.3 kHz. The photogenerated current, which is modulated accordingly, is amplified and filtered by an analog front-end, digitized by a 10-bit ADC and finally demodulated at digital level. The extracted dithering components, which are easily discriminated due to their orthogonality, are proportional to the partial derivatives of the interferometer's transfer function around the bias point set by the heaters. The integration is performed in the digital domain by accumulating the dithering signals over a period $T_{dith} = 120\ \mu s$, after which both heaters' working points are updated. Thus, the circuit works either

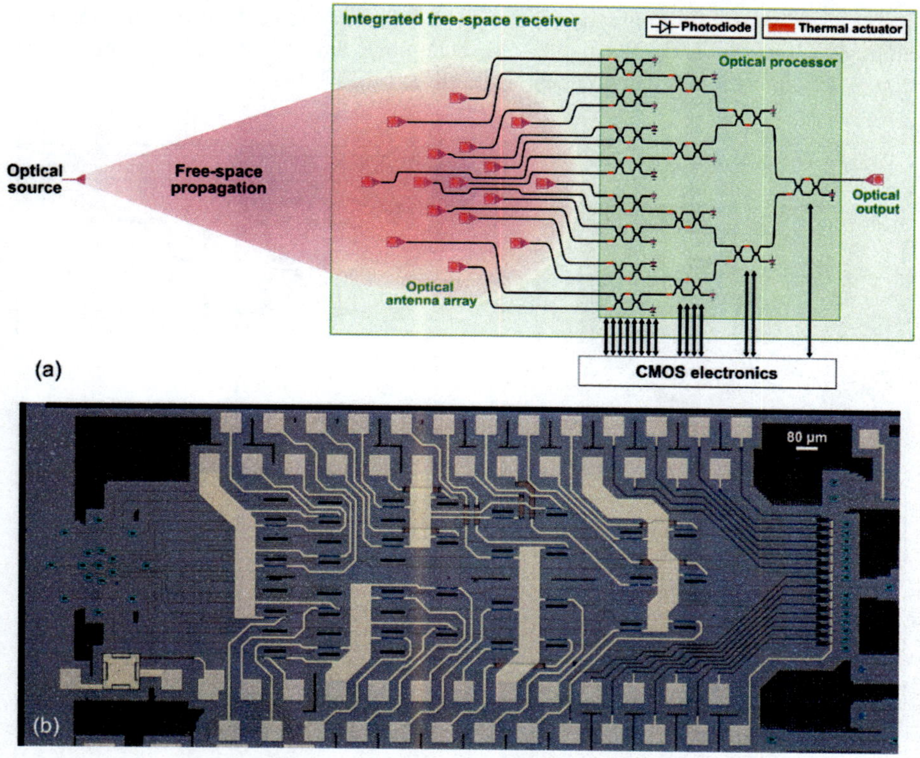

Fig. 1. a) Schematic view and b) microscope photograph of the integrated free-space receiver, made of a mesh of 15 MZIs.

to minimize or maximize the amount of power impinging on the PD, according to the user-defined sign of the overall loop gain.

The analog front-end features a variable transimpedance gain, which is adjusted automatically to ensure that the ADC always works over its full-scale range, regardless of the total input optical power. With this solution, the ASIC handles input signals between 0 and −50 dBm. The digital part is equipped with a dedicated logic that performs a square-root operation on the digital words fed to the DACs. This provides a compact solution to mitigate the quadratic non-linearity that intrinsically arises when controlling the voltage V_H of the actuators, that induces phase shift φ_H proportional to $V_H{}^2$. The communication between the ASIC and an external personal computer is provided by a serial interface that allows the user to define several parameters, including the bandwidth of the control loop and the amplitude of each dithering modulation, thus optimizing the operation of the FSO receiver for different values of total input optical powers. Furthermore, the user can turn the loop off to either fix the actuators working point to the state reached by the controller or to manually set a new voltage for the heaters. This latter feature was used to characterize the transfer function of each interferometer in the PIC. Finally, an additional 2-wire interface monitors both the photogenerated current flowing in the analog read-out and the output voltage of the DACs.

The chip, whose micrograph is reported in Fig. 2c, was fabricated with a 350 nm feature size. Overall, each 8-channel ASIC occupies 12 mm^2, while dissipating only $\approx$ 10 mW per channel, less than the heaters required by each MZI.

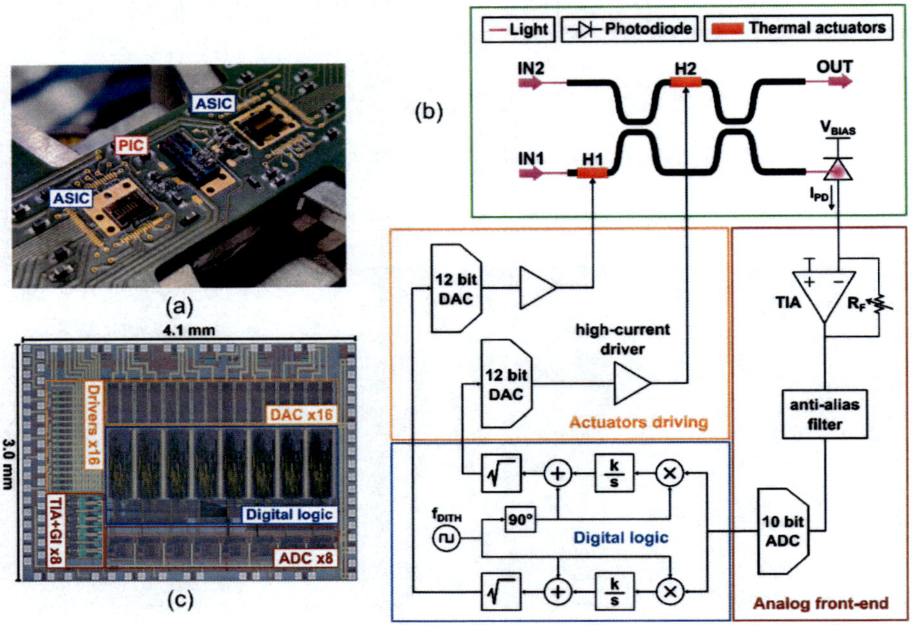

Fig. 2. a) Photograph of the PIC connected to 2 ASICs. b) Microscope photograph of the ASIC. c) Schematic view of the architecture of a single ASIC channel, connected to a MZI.

3 Experimental Results

A detailed scheme of the experimental setup employed to validate the electro-optical system is shown in Fig. 3a. The optical signal is generated by a laser with $\lambda = 1550$ nm and sent into free-space by a single-mode fiber. The collimated beam is expanded up to a diameter of $\approx$ 5 cm by a pair of lenses after the fiber source, reflected by a mirror after $\approx$ 1 m of propagation to form a folded path and finally collimated onto the SiP chip OPA by a telescopic system, allowing to simulate an overall optical link of 100s of meters [4]. A thermal gun placed along the path was used to locally inject free-space turbulence, comparable in bandwidth ($\approx$300 Hz) to the real-world ones reported in literature [6]. A photograph of the complete laboratory setup is shown in Fig. 3b.

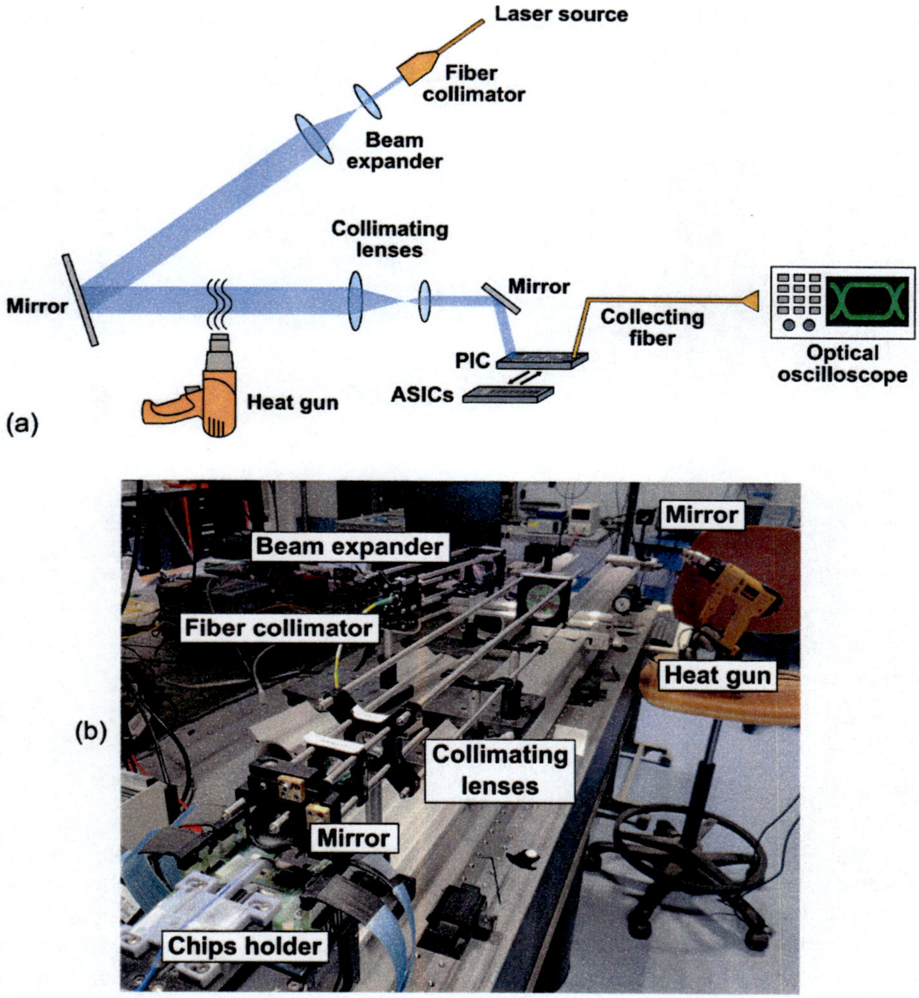

Fig. 3. a) Schematic view of the experimental setup for FSO transmission and b) its photograph.

3.1 Automatic Reconfiguration of the Receiver

An initial experiment was carried out to estimate the bandwidth of the controller, and the results are reported in Fig. 4. First, with the control loop turned off, each MZI of the PIC was set to a random working point, far from the optimal configuration, so that only a fraction (<1%) of the input optical power was correctly transferred from the OPA to the fiber coupled to the output of the receiver. Around t = 10.85 s, the control loop was turned on, letting the ASIC drive all the heaters simultaneously in order to minimize the power impinging on each integrated PD, consequently maximizing the available power at the output of the mesh (Fig. 4a). As a result, the free-space receiver is automatically driven to the optimal configuration in less than 10 ms, increasing the power delivered at the output by ≈ 23 dB, as highlighted in Fig. 4b. Figure 4c provides the trace over time of all the photogenerated currents read by the two ASICs, showing that the input interferometers reach a stable equilibrium point even within few ms.

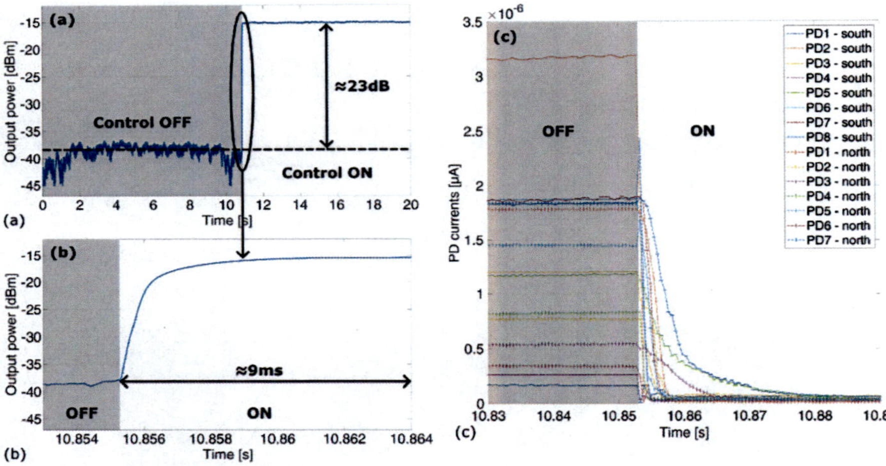

Fig. 4. a), b) Time evolution of the optical power at the output of the receiver and c) of the photocurrents detected by the integrated PDs, before and after the mesh configuration.

3.2 Atmospheric Turbulence Phase-Compensation

A second experiment was conducted to assess the robustness of the control loop against free-space turbulences, which were artificially introduced along the beam path by activating the thermal gun. Figure 5 reports the results of the measurements taken in three different conditions. A reference measurement (yellow curves) was taken by activating the control loop while leaving the heat gun off: in this ideal case, the average output power is −20.1 dBm, with a standard deviation of 0.0554 dB, as derived by the histogram plot in Fig. 5b. At this point, the control loop was suspended, fixing the FSO receiver in the configuration previously reached with no turbulence, and the gun turned on: the average received optical power dropped down to −21 dBm, with a particularly detrimental effect on the standard deviation, which becomes as large as 0.3429 dB (blue curves). Finally, the ASICs were activated to compensate the incoming turbulence: measurements show that the average received optical power recovers to −20.4 dBm, with a standard deviation shrunk down to 0.1216 dB (red curves).

Figure 5a highlights the different levels of stability and average intensity of the optical signal over time in the three different conditions, whereas Fig. 5c provides a characterization in the frequency domain of the injected turbulence and the effectiveness of the compensation performed by our ASICs. In particular, it is possible to observe that a good rejection (up to 10 dB) of disturbances is provided up to few hundreds Hz, demonstrating that the system is suitable to be employed in real outdoor FSO links.

After this frequency, the three spectra are almost perfectly superimposed, meaning that no turbulence is present beyond such bandwidth, as expected [6]. Disturbs due to power supply interference are present in all spectra, at multiples of f = 100 Hz, and can be removed by better shielding the setup.

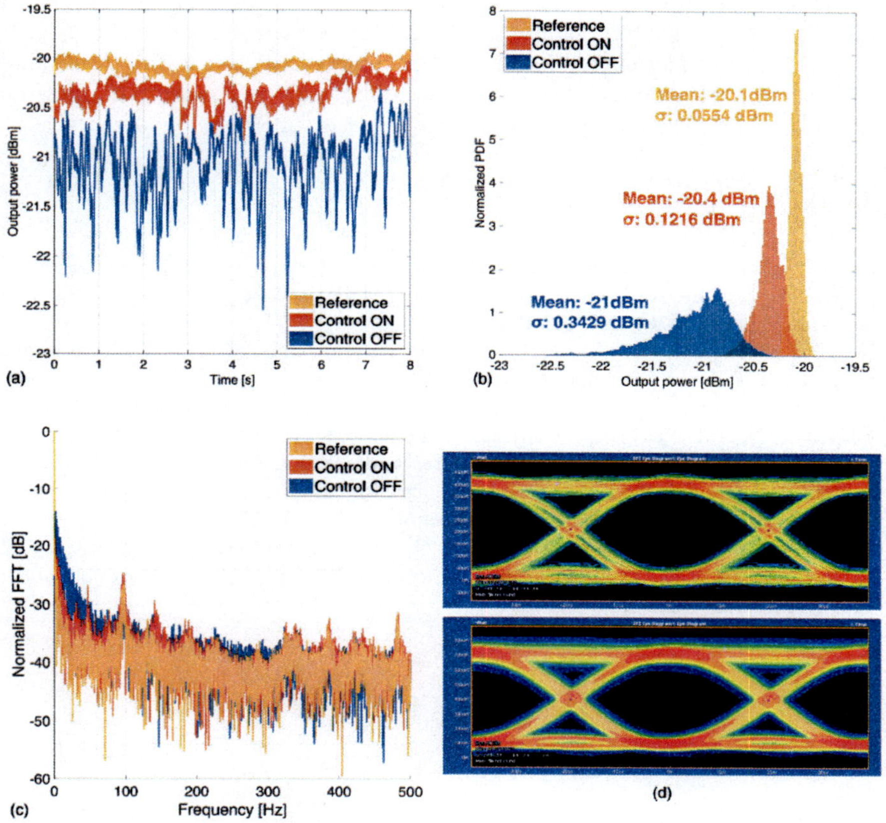

Fig. 5. a) Time evolution of the optical power at the output of the mesh and b) corresponding probability density histogram and c) frequency spectra, before the activation of the turbulence and after, with the ASIC on and off. d) Eye diagrams at the transmitter side of the FSO link and at the output of the PIC, demonstrating the correct operation of the electro-optical receiver.

Figure 5d finally shows the eye diagram obtained when transmitting a 25 Gbps on-off keying modulated signal through the FSO link enabled by the integrated optical receiver. The eye diagram captured at the output of the receiver does not show significant degradation when compared to the one measured at transmitter side (right after the light modulator), thus certifying the correct operation of the PIC as well as the proper tuning and control of the mesh performed by the CMOS electronics.

4 Conclusions

The fully integrated CMOS controller presented in this work provides a viable solution to implement scalable dedicated control electronics needed by PICs of growing complexity. In particular, experimental results show that our solution matches, when applied to a FSO receiver, the performance of other discrete components architectures [4], while consuming only a fraction of their power and area. Future ASICs will be designed targeting even more compact solutions, ideally matching the area occupied by one CMOS electronic channel to the area of the controlled SiP device. This would allow the fabrication of very compact, flip-chip connected photonic-electronic co-designs, thus introducing a chiplet-based approach also in the industry of PICs.

References

1. Dhasarathan, V., Singh, M., Malhotra, J.: Development of high-speed FSO transmission link for the implementation of 5G and Internet of Things. Wireless Netw. **26**(4), 2403–2412 (2020). https://doi.org/10.1007/s11276-019-02166-5
2. Zhu, X., Kahn, J.M.: Free-space optical communication through atmospheric turbulence channels. IEEE Trans. Commun. **50**(8), 1293–1300 (2002)
3. Miller, D.A.B.: Self-aligning universal beam coupler. Opt. Express **21**, 6360–6370 (2013)
4. Morichetti, F., et al.: Mitigation of atmospheric turbulence in an optical free space link with an integrated photonic processor. In: OFC Conference, pp. W3I-7, San Diego, CA, USA (2023)
5. Zanetto, F., et al.: Dithering-based real-time control of cascaded silicon photonic devices by means of non-invasive detectors. IET Optoelectron. **15**(2), 111–120 (2021)
6. Cox, M.A., Mphuthi, N., Nape, I., Mashaba, N., Cheng, L., Forbes, A.: Structured light in turbulence. IEEE J. Selected Top. Quantum Electron. **27**(2), 1–21 (2021). https://doi.org/10.1109/JSTQE.2020.3023790

Integrated Mode-Selective Repeater for Free-Space Optical Communications

SeyedMohammad SeyedinNavadeh(✉), Alessandro di Tria, Francesco Zanetto, Giorgio Ferrari, Marco Sampietro, Andrea Melloni, and Francesco Morichetti

Dipartimento di Elettronica, Informazione e Bioingegneria (DEIB), Politecnico di Milano, Via Ponzio 34/5, 20133 Milan, Italy
seyedmohammad.seyedinnavadeh@polimi.it

Abstract. A silicon photonic integrated circuit is designed and fabricated to couple and regenerate arbitrary beams for free-space optical communications. The mode-selectivity is achieved through self-configuration of the Mach-Zehnder interferometer diagonal. A transmission experiment using 25 Gbps modulated data demonstrates the advantages of using this free-space repeater in the case of arbitrary overlapping beams.

Keywords: Free-Space Optical Communications · Photonic Integrated Circuits · Mach-Zehnder Interferometer Mesh

1 Introduction

Transmission of spatially multiplexed channels, or spatial modes, has been considered for years as a strategy to increase the capacity of Free-Space Optical (FSO) links [1,2]. However, in free-space, mode orthogonality can be impaired by the presence of obstacles as well as by the inherent time-varying nature of environment (e.g., turbulence). As a result, it can be difficult, or even impossible, to identify a-priori the mode sets to be used at the transmitter side that is more robust against mode mixing effects. In this work, we experimentally demonstrate that a pair of self-configuring Mach-Zehnder (MZI) meshes, used for coupling and regenerating FS beams, can be advantageous in the case of arbitrary overlapping beams. Two incident beams are selectively coupled into one of the apertures of the PIC. On the other side, the re-transmitting aperture emits an arbitrary beam that is the most suited for the specific channel. The eye diagram of the modulated signal at 25 Gbps can be retrieved once the input MZI mesh is configured to the desired transmitter (Tx), even in the presence of another interfering Tx.

2 Reconfigurable Free-Space Beam Coupler and Regenerator Chip

The proposed device for performing arbitrary FS beam coupling and generation is shown schematically in Fig. 1(a). The photonic circuit consists of two apertures

J. Witzens et al. (Eds.): ECIO 2024, SPPHY 402, pp. 309–314, 2024.
https://doi.org/10.1007/978-3-031-63378-2_51

on either side realized using arrays of grating couplers as optical antennas; the one on the left, for example, is used to sample the incoming beam into the input single-mode waveguides of Mesh_{in} (coupling), and the one on the right couples back the beam out of the output mesh Mesh_{out} into the free space (regeneration). The photonic integrated circuit in between (i.e., Mesh_{in} and Mesh_{out}) for on-chip light processing is based on mesh waveguides of Mach-Zehnder Interferometers (MZIs) [Fig. 1(a)]. Actuating on 2 phase shifters integrated on the MZI, the two single-mode inputs can coherently sum up in one of the outputs, e.g., upper waveguide in this case. To do this, a photodetector (PD), making use of a 10% tap coupler (not shown in the schematic), is integrated at the output that is nulled so that a simple power minimization algorithm can be used to configure a single MZI. For summing up all the other sampled inputs, a cascaded diagonal row of MZIs can be configured in the same manner to couple all the power in the input waveguides into the top output waveguide of Mesh_{in}. The behavior of this part of the device is what is known as a self-aligning beam coupler [3]. The other part of the device (Mesh_{out}), instead, is a mirrored replica of the first part so that for the generator aperture at the right, the amplitude and the phase of the elements are provided by another row of MZIs [Fig. 1(a)].

A microphotograph of the designed and fabricated chip is shown in Fig. 1(b). The chip is fabricated in a standard 220-nm-thick waveguide Silicon photonics platform at Advanced Micro Foundry (MPW run). The components of the PIC are designed to work at a center wavelength of $\lambda = 1550$ nm. The total size of the chip is 6.4 mm $\times$ 1.9 mm. Each single balanced MZI [Fig. 1(b) zoomed] is realized using two 50/50 couplers, and the two thermo-optic phase shifters (TiN heaters) are realized on external and internal arms on the MZI. Deep trenches have also been utilized to reduce the thermal cross-talk effect. The grating couplers used as optical antennas for FS beam coupling have about $5° \times 9°$ divergence angle and a tilt angle of 12°. A 3×3 array of these elements, used as an antenna array (aperture of the chip), is also shown in Fig. 1(b). The distance between elements thus exceeds half-lambda, and multiple diffraction orders are expected to be present in the far field beam shape, as shown in Fig. 1(c) (see Sect. 4 for more details). In between the two meshes, S-like double-sided coupler taps are introduced for power monitoring [not shown in the schematic of Fig. 1(a)].

3 Experimental Setup and Control Electronics

The above-mentioned PIC is used to perform FS beam coupling and regeneration experiments. The detailed schematic of the optical setup is depicted in Fig. 2(a). Two fiber collimators are aligned to couple the FS beam into the aperture of the PIC on the left (and thus into the waveguides of Mesh_{in}), using a converging lens ($f = 50$ mm). The polarization of each branch is adjusted, making use of two polarization controllers, as the grating couplers work only with TE polarization. The modulated signal (25 Gbps OOK-NRZ, at a center wavelength of $\lambda_0 = 1550$ nm) is amplified using an Erbium-doped fiber amplifier (EDFA_1) and then split into two branches to inject the light to the fiber collimators, namely, **Tx_A** and

Tx_B as in Fig. 2(a). The FS beams, injected from the collimators, are combined in FS to couple to the aperture of the chip. After on-chip manipulation (coupling and regeneration), the beam from the aperture on the right is converged and Fourier-transformed, and another fiber coupler (**Rx**) is aligned to couple one of the diffraction orders of the far-field beam. A phase mask is also utilized in between the lens and fiber coupler to manipulate further the FS beam at the Rx side. The received signal after the coupler is then amplified once again ($EDFA_2$), and a high-bandwidth oscilloscope is used to measure the eye diagrams.

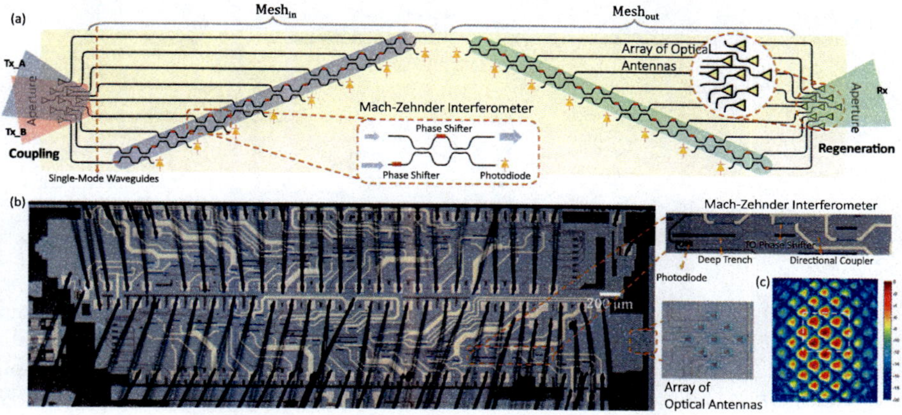

Fig. 1. Reconfigurable free-space beam coupler and regenerator chip, (a) schematic of the PIC consisting of two MZI diagonal rows for coupling and regeneration, and grating couplers as optical antenna arrays, (b) micro-photograph of the chip along with the zoomed detail of an MZI and antenna array, (c) a far-field beam shape radiated by the antenna array consisting of multiple diffraction orders.

Figure 2(b) shows a photograph of the optical setup in the laboratory. Some of the FS components, described in Fig. 2(a) are labeled in the picture. As can be seen in the zoomed panel on the right, the chip is mounted on a printed circuit board (PCB) consisting of trans-impedance amplifiers (TIA), which will be described in the following. A mirror is aligned on the aperture on the right (however, not shown in the schematic) to deflect the beam in the desired direction. An IR camera is also used to capture the far-field beam shapes of the two apertures, making use of a beam splitter in the path.

A custom electronic system has been developed to control the PIC. The schematic is depicted in Fig. 2(c). Making use of the interface PCB, the signals from the PIC are pre-amplified and routed to the main electronic board [Fig. 2(d)]. The platform has a modular structure made of a main control motherboard and six expansion modules. Each module is designed to perform a specific functionality, namely the acquisition of signals from integrated PDs and the generation of voltages to drive thermal actuators. The whole system, with

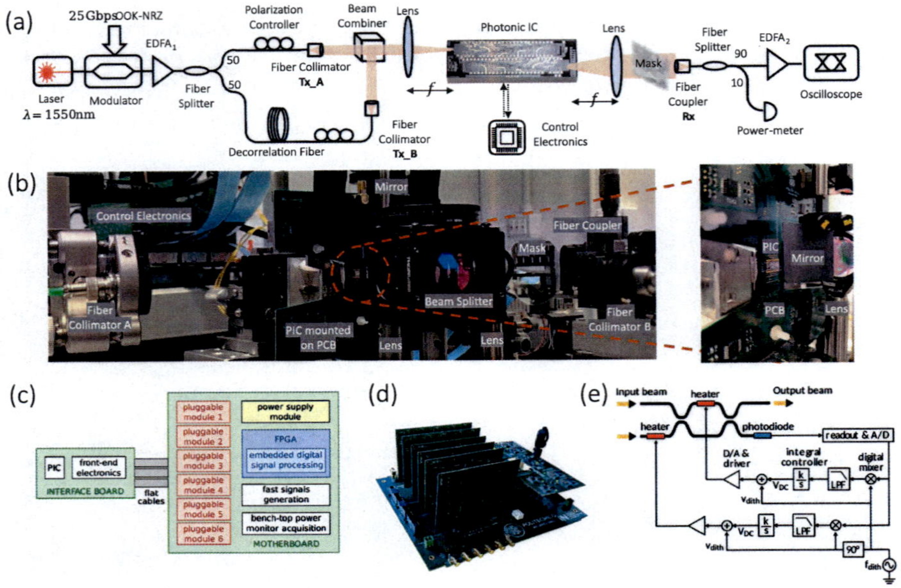

Fig. 2. Experimental setup and control electronics, (a) schematic of the optical setup for performing beam coupling and regeneration experiments, (b) photograph of the laboratory setup, (c)-(e) schematic of the custom control electronics system, photograph of the modular board, and control algorithm using the dithering technique, respectively

three acquisition modules and three actuation modules plugged into the motherboard, is able to read up to 48 PD channels and drive up to 72 heaters, enough to control both the MZI meshes.

An FPGA housed on the motherboard performs the lock-in detection and implements a feedback law in the digital domain [Fig. 2(e)], using the dithering technique [4]. For each MZI, the system drives the heater with a sinusoidal oscillation v_{dith} around the DC working point V_{DC}. This perturbation induces an oscillation in the MZI output optical power proportional to the first derivative of its transmittance. This oscillation is read out by the integrated PD and extracted with a digital lock-in comprised of a mixer and a Low-Pass Filter (LPF). This quantity is fed to an integral controller, which updates the DC voltage until the amplitude of the oscillation, and therefore, the first derivative of the MZI transmittance is nulled. This strategy automatically locks each device in a stationary point, allowing for parallel real-time control of the entire PIC. A custom graphical user interface allows one to set up the electronic system and visualize on a personal computer the state of the photonic processors.

4 Results

The setup explained in Sect. 3 is used to perform the beam coupling experiments. Initially, the light is injected only into **Tx_A**, and Mesh_{in} is configured to sum up all the power of this incident Gaussian beam at the left aperture of the PIC. The same procedure is done but instead injecting the light backward into **Rx**, so that Mesh_{out} is configured to adaptively capture all the power of this arbitrary beam, passing through the phase mask. Keeping the status of all the MZIs in Mesh_{in} and Mesh_{out}, the whole PIC in this way is configured to couple an incident beam from **Tx_A** and regenerate an arbitrary beam to best couple to the receiver fiber **Rx**. The beam shapes that are obtained after configuration, for each aperture on the left and right are shown in Fig. 3(a) and (b). The total end-to-end loss is about 47 dB, considering the geometrical loss of the antenna array with respect to the beam size, grating couplers as antenna elements, fiber splitters and equipment,

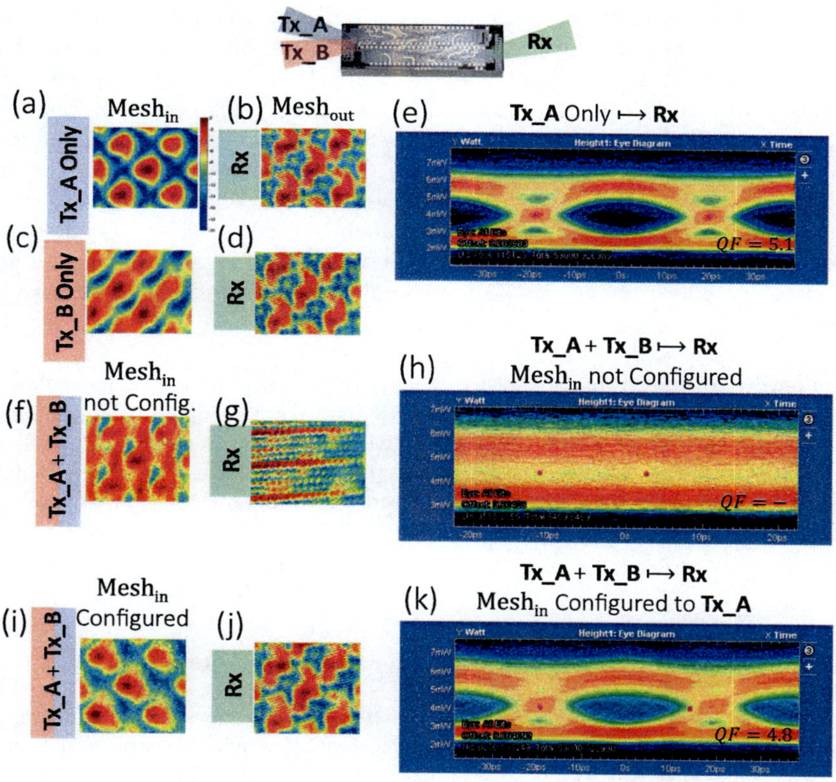

Fig. 3. Experimental results of the beam coupling and regeneration, (a)-(e) beam shapes along with the eye diagram of coupling only one Tx collimator, (f)-(h) same for the case of Mesh_{in} not configured and injecting light in both Tx collimators, (i)-(j) same for the case Mesh_{in} configured

FS components (beam splitter, lenses, etc.) and alignment tolerances. Note that Mesh_{in} is configured in such a way to reproduce a Gaussian-like beam, at least in each diffraction order as in Fig. 3(a), in order to maximize the power coupling to **Tx_A**. All the plots of the beam shapes, acquired using the IR camera, are normalized to the maximum intensity.

The same procedure is done in the case of **Tx_B**, however, the fiber collimator is intentionally aligned in such a way as only to have partially overlapped beams with **Tx_A**. Once Mesh_{in} is configured for **Tx_A**, injecting the light only into **Tx_B** shows about 10 dB of crosstalk at the **Rx**, maintaining Mesh_{out} fixed. This can be considered as the in-parallel component of the beam shone by **Tx_B** with respect to **Tx_A**, in the basis set defined by the antenna array aperture. The beam shapes for the case of **Tx_B**-only are shown in Fig. 3(c) and (d). Note that to show the beam shape of **Tx_B**, Mesh_{in} is configured to this Tx.

Figure 3(e) shows the eye diagram acquired for the case of the **Tx_A** only. Instead, if now both **Tx_A** and **Tx_B** are kept on at the same time, and if Mesh_{in} is kept in an arbitrary working point, i.e., randomizing phase shifters of the MZIs as shown in the beam shape of Fig. 3(f), a coherent combination of **Tx_A** and **Tx_B** is fed into Mesh_{out}. Although the beam shape of Mesh_{out} [Fig. 3(g)]is covered by the beam emitted from the monitor grating couplers after PDs (horizontal lines), the beam shape is similar to the previous cases, as Mesh_{out} is kept fixed. Even though the power in **Tx_B** is deliberately reduced from the initial situation by around 14 dB, the strong degradation in the eye diagram is evident as in 3(h). However, if now Mesh_{in} is configured again for the case of **Tx_A** while injecting at the same time into **Tx_B** as shown in the beam shapes of 3(i) and (j), the eye diagram of the modulated signal can be retained as seen in 3(k), which demonstrates the significant contribution of the mesh as a mode-selective repeater.

Acknowledgments. Authors acknowledge support from the Italian National Recovery and Resilience Plan (NRRP) of NextGenerationEU, partnership on "Telecommunications of the Future" (PE00000001 - program "RESTART", Structural Project "Rigoletto" and Focused Project "HePIC"). Part of this work was carried out at Polifab, the micro- and nanofabrication facility of Politecnico di Milano (https://www.polifab.polimi.it/).

References

1. David, A., Miller, B.: Reconfigurable add-drop multiplexer for spatial modes. Opt. Express **21**, 20220–20229 (2013)
2. SeyedinNavadeh, S.M., Milanizadeh, M., Zanetto, F., et al.: Determining the optimal communication channels of arbitrary optical systems using integrated photonic processors. Nat. Photon. **18**, 149–155 (2024). https://doi.org/10.1038/s41566-023-01330-w
3. David, A., Miller, B.: Self-aligning universal beam coupler. Opt. Express **21**, 6360–6370 (2013)
4. Zanetto, F., et al.: Dithering-based real-time control of cascaded silicon photonic devices by means of non-invasive detectors. IET Optoelectron. **15**, 111–120 (2021). https://doi.org/10.1049/ote2.12019

Silicon Nitride Metalenses at Near-Infrared Wavelengths Manufactured Using Deep-Ultraviolet Scanner Lithography

David De Vocht[1](✉), Luise Armbruster[1], Alonso Millan-Mejia[2], Angel Savov[2], Yuqing Jiao[1], and Erwin Bente[1]

[1] Eindhoven Hendrik Casimir Institute, Eindhoven University of Technology, 5612 AZ Eindhoven, The Netherlands
d.j.y.d.vocht@tue.nl
[2] SMART Photonics, 5656 AE Eindhoven, The Netherlands

Abstract. We present the Si_3N_4 metalenses fabricated on an InP wafer using deep-ultraviolet scanner lithography which is suitable for high-volume manufacturing. Despite limiting the pillar height to 750 nm, preventing full phase coverage, we demonstrate diffraction-limited focusing in the near-infrared regime. Furthermore, we showcase an achromatic metalens design that successfully reduces chromatic aberration by 40%. These achievements propose a promising alternative for mass-producing metalenses relevant to various applications such as optical sensing.

Keywords: Metalens · metasurface · DUV lithography · near-infrared · silicon nitride

1 Introduction

Recent studies have demonstrated the mass producibility of metalenses based on Si or SiO_2 nanopillars by manufacturing them using deep-ultraviolet (DUV) lithography [1–3]. These metalenses are microscopically thin lenses constructed from an array of cylindrical nanopillars (Fig. 1). SiO_2 emerges as a particularly intriguing material due to its wide transmission bandgap, spanning from the visible to the near-infrared (NIR) regime. However, its low refractive index requires 1.5 μm tall pillars for achieving full phase coverage at visible wavelengths and even taller at NIR wavelengths [3]. SiO_2-based metalenses also use SiO_2 as the substrate material, intensifying the coupling between neighbouring pillars. In contrast, Si_3N_4 shares a similar transmission bandgap but possesses a higher refractive index. This allows for more manageable pillar heights, specifically 0.7 at a wavelength of 633 nm and 1.6 μm at 1550 nm [4, 5]. However, the mass production of Si_3N_4-based metalenses has yet to be demonstrated.

J. Witzens et al. (Eds.): ECIO 2024, SPPHY 402, pp. 315–322, 2024.
https://doi.org/10.1007/978-3-031-63378-2_52

Only two years ago, the integration of metasurfaces on top of InP-based substrates was demonstrated (using e-beam lithography) to enhance avalanche photodiodes [6]. This makes Si_3N_4 an even more interesting material as it is compatible with InP foundry processes. Consequently, these metalenses can be combined with Generic InP or InP-membrane integration platforms to advance the miniaturization of optical devices such as sensors [7].

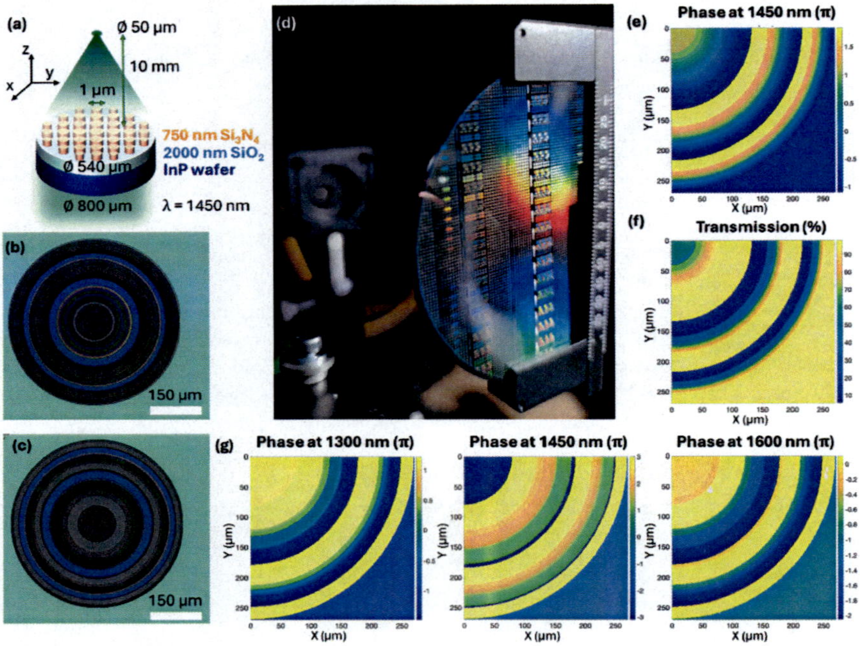

Fig. 1. (a) Schematic of our metalens. Optical Images showcasing a single (b) chromatic and (c) achromatic metalens. (d) InP wafer containing 5000 metalenses. (e) Phase and (f) transmission profile for the designed chromatic metalens at 1450 nm wavelength. (g) Phase profiles for the achromatic metalens at the wavelengths 1300 nm, 1450 nm and 1600 nm.

We present the manufacturing of 5000 Si_3N_4-based metalenses on an InP wafer using DUV scanner lithography (Fig. 1(a)). Using a high ion density Induction Coupled Plasma Reactive Ion Etching (ICP-RIE) tool, we have achieved precise etching of 750 nm tall Si_3N_4 pillars directly from the DUV resist mask. Although the pillars need to be 2-3x taller to achieve full phase coverage, we successfully demonstrate diffraction-limited focusing in the near-infrared regime. Furthermore, we introduce an achromatic metalens design tailored specifically for wavelengths 1300 nm, 1450 nm and 1600 nm, showcasing a reduction of chromatic aberration by 40%. This study marks a pioneering investigation into achromatic focusing for DUV scanner lithography metalenses. Our goal is to introduce these ultra-flat metalenses as replacements for conventional lenses in optical sensors.

2 Design

We specialize in miniaturizing NIR sensors where light focusing is required over centimeter distances. Specifically, we aim to develop a metalens optimized for achieving a focal spot size of around 45 μm at a distance of 10 mm. Our metalens, with a diameter of 540 μm, consist of a 2D array of 800 nm tall cylindrical Si_3N_4 nanopillars on a SiO_2 base layer. As illustrated in Fig. 1(a), the array follows a cartesian grid with a 1 μm pitch. We optimize the diameter D of the pillars across the metalens to enable the focusing behaviour. For simulation purposes, we assume that locally the pillars have similar shapes and dimensions, enabling us to apply infinite boundary conditions. We utilize the 3D finite-difference time-domain method (FDTD) of Lumerical to simulate the transmission $T(\lambda, D)$ and phase $\varphi(\lambda, D)$ profiles at different wavelengths λ. Due to the limited pillar height, the difference in phase $\varphi_{pillars}$ between the thinnest and thickest pillar is 48%, 45% and 36% of the desired 2π phase coverage at 1300 nm, 1450 nm and 1600 nm, respectively.

To focus a Gaussian beam at a focal length f, we require the following phase profile

$$\varphi(r, \lambda) = -\frac{2\pi}{\lambda}\left(\sqrt{r^2 + f^2}\right) + C(\lambda) \tag{1}$$

where r is the radial position, λ is the wavelength and C(λ) is a spectral degree of freedom. For each r, we identify the most optimal pillar m by minimizing the error between the desired and resulting electrical field near the lens plane as proposed in [8]

$$\min \sum_{\lambda} \left| \exp(i\varphi(r, \lambda)) - \sqrt{T_m(\lambda)} \exp(i\varphi_m(\lambda)) \right| \tag{2}$$

In other words, we want to approach as closely as possible the desired phase profile for the target wavelengths. Since we cannot obtain full phase coverage, certain locations on the metalens will have larger errors. By incorporating a Particle Swarm Optimization algorithm, we iteratively adjust the constant phase shift $C(\lambda)$ to maximize metalens efficiency. We calculate the far-field intensity profile of the metalens utilizing the Rayleigh-Sommerfeld diffraction formula.

This approach is instrumental in designing both chromatic and achromatic metalenses. During the optimization of the chromatic metalens, we solely consider the target wavelength of 1450 nm. However, for the achromatic metalens, we extend our considerations to wavelengths of 1300 nm and 1600 nm. Figure 1(e-g) shows the phase and transmission profile of the designed chromatic and achromatic metalens.

3 Fabrication

We have developed a fabrication process to manufacture Si_3N_4-based metalenses using DUV scanner lithography whilst eliminating the need for a metal lithography mask. The process begins with a 3-inch Dual Side Polished InP wafer onto which 2 μm of SiO_2 is deposited using Plasma-Enhanced Chemical Vapor Deposition (PECVD), followed by 750 nm of Si_3N_4. Next, we perform a spin coating process using a tri-layer configuration, comprising the TOK P6111 photoresist sandwiched between a bottom and a top

Table 1. Comparison of the measured and simulated beam diameters and focal lengths for chromatic and achromatic metalenses at wavelengths 1300 nm, 1450 nm, and 1600 nm. Δf indicates the variation in focal length across the wavelength range.

		Beam Diameter (μm)		
		1300 nm	1450 nm	1600 nm
Chromatic metalens	Measurement	44.8	45.6	47.5
Achromatic metalens	Measurement	44.9	47.0	47.1
Airy function	Theoretical	37.7	42.2	46.7
Fiber collimator	Measurement	830	888	954

		Focal Length (mm)			
		1300nm	1450nm	1600nm	Δf (mm)
Chromatic metalens	Simulation	10.9	9.8	8.9	2
	Measurement	10.1	9.2	8.6	1.5
Achromatic metalens	Simulation	11.2	9.6	9.6	1.6
	Measurement	10.4	9.6	9.5	0.9

anti-reflection coating, applied over the Si_3N_4 layer. We then expose this photoresist utilizing a 193 nm DUV scanner lithography tool (ASML Pass 5500/1100B scanner). This projection lithography employs a 4:1 reduction ratio with a DUV lens column of 0.75 NA under conventional illumination conditions. After developing the resist, we use a high ion density Induction Coupled Plasma Reactive Ion Etching (ICP-RIE) tool to etch the nanopillars into the Si_3N_4 layer. Finally, we strip away the resist. The Si_3N_4 pillars are slightly shorter than designed as the etching process was not optimized yet to achieve 800 nm. This method seamlessly translates to 4-inch InP wafers, employing identical equipment. Furthermore, our approach is fully compatible with other processing steps of the InP-based generic and InP-membrane integration technology [9, 10].

4 Performance

We characterize the chromatic and achromatic metalenses at the three target wavelengths of 1300 nm, 1450 nm and 1600 nm. In the experimental set-up, a collimated laser beam is incident on the metalens and the 3D far-field of the lens is measured by acquiring a series of 2D images at different distances from the lens. To generate the light at 1300 nm, 1450 nm and 1600 nm, we use three fibre-coupled lasers: Santec TSL-550, Keysight 81600B and ID Photonics CoBrite DX4. We then use a fibre collimator (PAF2-A4C collimator) to obtain a Gaussian beam with a beam diameter of 830 μm, 890 μm and 950 μm at the respective wavelengths. After transmitting this beam through our metalens, we capture the light by a 20x microscope objective (Mitutoyo Plan APO NIR using a tube lens with a focal length of 200 mm) and image it onto an InGaAs camera (Xeva 320 - Xenics). We mounted the objective and the camera on a motorized stage (CONEX-MFACC - Newport) to measure the far-field at different distances.

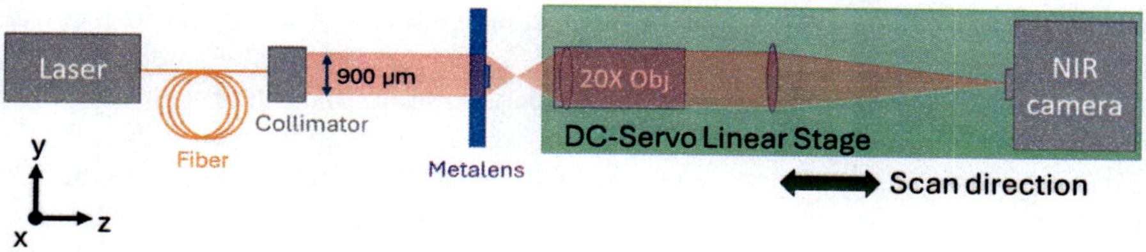

Fig. 2. Experimental set-up used to characterize the metalenses.

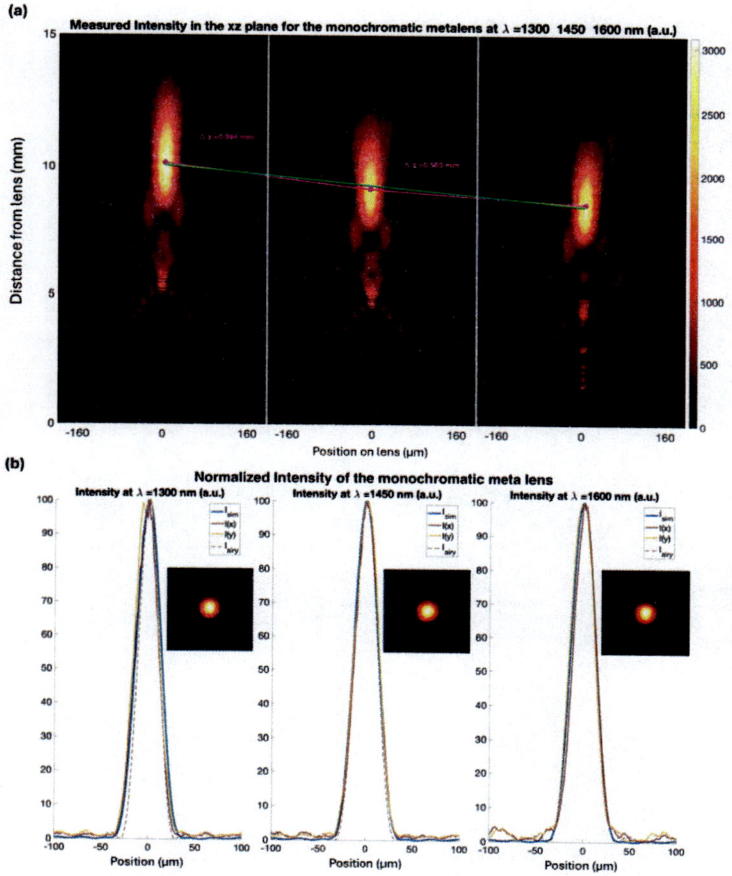

Fig. 3. (a) Measured far-field intensity distribution of the chromatic metalens in the axial plane for the wavelengths 1300 nm, 1450 nm and 1600 nm. (b) Comparison of the measured, simulated and theoretical expected intensity profile at the focal plane. The inset shows the focal spot.

Figure 3 shows the measured far-field intensity profiles of the chromatic metalens. To determine the focal point, we apply a 2D Gaussian fit across the region of high-intensity values in the xz plane and take the maximum of this Gaussian profile. At the designated wavelength of 1450 nm, we observe that the metalens focuses the beam to a 48 μm wide Gaussian spot at a distance of 9.2 mm from the lens surface. These values are within 10% of the target values. Figure 3(b) shows a comparison of our measured intensity profile with our simulation results and the theoretically expected result from an Airy disk. The main lobe closely aligns with the simulations, though it is slightly broader than the Airy disk. This divergence is because the Airy disk model assumes uniform illumination of the opening. The measured sidelobes are a few percent higher than our simulations predict. To investigate the chromatic aberration, we also measure the intensity profile at 1300 nm and 1600 nm, revealing a similar focusing behaviour. Table 1 compares the measured focal lengths and beam diameters with simulation results. As anticipated for chromatic metalenses, the focal length decreases with increasing wavelength, contrary to classical lenses. The measured focal lengths are generally smaller than the simulation results,

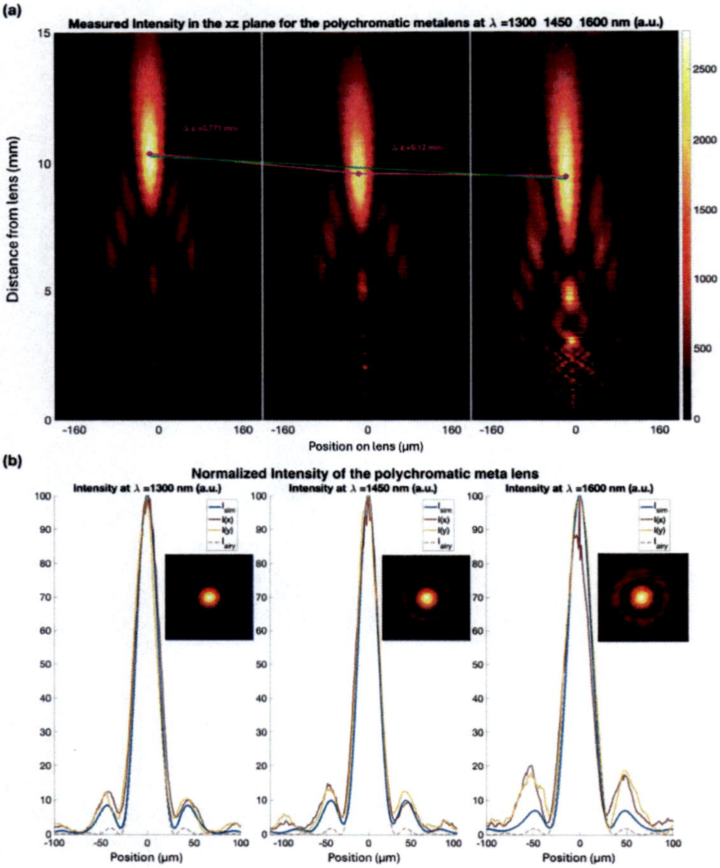

Fig. 4. (a) Measured far-field intensity distribution of the achromatic metalens in the axial plane for the wavelengths 1300 nm, 1450 nm and 1600 nm. (b) Comparison of the measured, simulated and theoretical expected intensity profile at the focal plane. The inset shows the focal spot.

particularly at 1300 nm. The overall chromatic aberration across the measured NIR region is 1.5 mm. We observe that the measured beam diameters vary more uniformly across the different wavelengths than the Airy disk model predicts. This observed uniformity is attributed to the dispersive behaviour of our fibre collimator (see Table 1). Specifically, when considering the varying input diameters across the wavelengths, the ratios between the measured and theoretical beam diameters exhibit only a minimal 3% difference.

Figure 4 shows the measured far-field intensity profiles of our achromatic metalens for the wavelength 1300 nm, 1450 nm and 1600 nm. Compared to the chromatic metalens, we notice a larger depth of field and prominent sidelobes. Table 1 gives an overview of the measured and simulated focal lengths and beam diameters, revealing nearly identical focal lengths for 1450 nm and 1600 nm. Surprisingly, the measured focal length is lower at 1300 nm compared to simulation results, resulting in a 40% improvement in overall chromatic aberration, surpassing the anticipated 20% reduction from simulation. The beam diameters align closely with the chromatic metalens results. As shown in Fig. 4(b), the measured sidelobes' peak intensity is 1.5x higher than the simulated results.

5 Conclusion

We successfully demonstrated the fabrication of Si_3N_4 metalenses using DUV scanner lithography. Despite the limited pillar height, resulting in only 35–50% phase coverage, the chromatic metalens exhibited measured focal lengths and beam diameters within 10% of the design targets. The achieved spots were near-diffraction-limited. The achromatic metalens displayed similar behaviour with an 40% improvement in chromatic aberration. As the fabrication process is compatible with InP foundries, this study facilitates the integration of metalenses on InP or InP-membrane integration platforms. To increase the pillar height and thereby phase coverage, an extra lithography step can be employed transferring the photoresist mask to a Chromium layer because Cr has an even better etching selectivity w.r.t. Si_3N_4 than photoresist. It is important to note that the current pillar height and minimum feature sizes are sufficient for operation in the visible regime, where Si_3N_4 metalenses are superior to glass- or Si-based metalenses. This work offers a pathway to more compact sensing solutions.

Acknowledgements. This publication is part of NWO HTSM 2019 project 17971 which is (partly) financed by NWO. We thank Irwan Setija for the insightful discussions.

References

1. She, A., Zhang, S., Shian, S., Clarke, D.R., Capasso, F., Paulson, J.A.: Large area metalenses: design, characterization, and mass manufacturing. Opt. Express **26**, 1573–1585 (2018)
2. Zhang, L., et al.: High-Efficiency, 80 mm Aperture Metalens Telescope. Nano Lett. **23**(1), 51–57 (2023)
3. Park, J.S., et al.: All-glass, large metalens at visible wavelength using deep-ultraviolet projection lithography. Nano Lett. **19**(12), 8673–8682 (2019)
4. Zhan, A., Colburn, S., Trivedi, R., Fryett, T.K., Dodson, C.M., Majumdar, A.: Low-contrast dielectric metasurface optics. ACS Photonics **3**(2), 209–214 (2016)

5. Zhan, A., Majumdar, A., Bayati, E., Zhelyeznyakov, M.V., Colburn, S.: Role of refractive index in metalens performance. Appl. Opt. **58**(6), 1460–1466 (2019)
6. Zhang, H., et al.: Metalens array for InGaAs/InP avalanche photodiodes at optical-communication wavelengths. Opt. Commun. **514**, 128159 (2022)
7. De Vocht, D., Millan-Mejia, A., Savov, A., Jiao, Y., Bente, E.: Dual focus wavefront tilting silicon nitride metalens manufactured using deep-ultraviolet scanner lithography. In: Conference on Lasers and Electro-Optics, Technical Digest Series, Optica Publishing Group, FM3L.4 (2024)
8. Shi, Z., et al.: Single-layer metasurface with controllable multiwave-length functions. Nano Lett. **18**(4), 2420–2427 (2018)
9. Smit, M., et al.: An introduction to InP-based generic integration technology. Semicond Sci. Technol. **29**(8), 083001 (2014). https://doi.org/10.1088/0268-1242/29/8/083001
10. Jiao, Y., et al.: Indium phosphide membrane nanophotonic integrated circuits on silicon. Phys. Status Solidi (A) **217**(3), 1900606 (2020). https://doi.org/10.1002/PSSA.201900606

Chirped Bragg Gratings for On-Chip Pulse Compression in the Mid-Infrared Wavelength Range

Annabelle Bricout[1](✉), Victor Turpaud[1], Thi-Hao-Nhi Nguyen[1], Hamza Dely[1], Natnicha Koompai[1], Stefano Calcaterra[2], Jacopo Frigerio[2], Samson Edmond[1], Jean-René Coudevylle[1], Etienne Herth[1], Carlos Alonso-Ramos[1], Laurent Vivien[1], Giovanni Isella[2], and Delphine Marris-Morini[1]

[1] Centre de Nanosciences et de Nanotechnologies, Université Paris-Saclay, CNRS, 91120 Palaiseau, France
annabelle.bricout@universite-paris-saclay.fr
[2] L-Ness, Dipartimento di Fisica, Politecnico di Milano, Polo di Como, Via Anzani 42, 22100 Como, Italy

Abstract. Integrated ultrashort pulsed lasers are widely used in the telecom wavelength range. However, the mid-infrared spectral range still lacks compact integrated ultrashort pulsed lasers. In this work, we present integrated chirped Bragg gratings for pulse compression working in the 7.8–8.2 μm wavelength range. It will be shown that group delay dispersion of a few tens of ps^2 can be obtained in such spectral range. Considering an input pulse of 11.8 ps, a compression factor of 9 can thus be obtained, with an increase of the peak power of a factor 9.

Keywords: pulse compression · mid-infrared · silicon · germanium

1 Introduction

Photonics integration in the mid-infrared range (6–15 μm wavelength) has garnered considerable interest for its potential applications in spectroscopy and sensing because of its correspondence with the so-called "fingerprint region". Integrated broadband sources are therefore substantial for sensing applications as they enable the possibility to detect numerous molecules in a single compact device. Indeed, on-chip two octave generation was obtained from 3 to 13 μm wavelength using Ge-rich graded index SiGe waveguides [1]. However, this demonstration required a bulky and expensive table-top high pulsed power laser source. There is currently a widespread interest around achieving a compact integrated photonics platform in the mid-infrared, requiring the supercontinuum generation to be triggered by compact laser sources. Herein, short mid-infrared pulses in the picosecond range have been demonstrated exploiting frequency modulated QCL with external pulse compression [2]. This work presents prospects to realize on chip pulse compression in the mid-infrared range, in order to be able to pump supercontinuum source by compact QCL.

J. Witzens et al. (Eds.): ECIO 2024, SPPHY 402, pp. 323–327, 2024.
https://doi.org/10.1007/978-3-031-63378-2_53

2 Results

2.1 Design

The main idea of this work is to compensate the group delay of an initially chirped pulse coming from a quantum cascade laser (QCL) frequency comb [2]. Thus a chirp filter must be designed and fabricated to compensate the chirp of the initial pulse. The chirp filter consists of a chirped Bragg grating with lateral corrugation where each wavelength is reflected at different distances (Fig. 1b) resulting in a wide stopband (from 7.5 to 8 μm) and a group delay of the reflected light that varies with the wavelength. To be used in practice, it will be necessary to be able to collect the reflected light. This can be done using a 2 by 2 multimode interferometer (MMI), as shown in Fig. 1c and presented in [3]. By injecting a pulse into the top left arm, the light will be split into the two output ports on the right where it will be reflected by the Bragg gratings. Therefore, the reflected light will be collected from the drop port. MMI operating in the mid-IR based on Ge-graded SiGe waveguide have already successfully been demonstrated [4]. This work thus focuses on the Bragg grating design, fabrication and characterization.

The Bragg grating is designed and fabricated on a 6 μm thick graded SiGe epitaxial layer waveguide, consisting in a graded SiGe epitaxial layer which allows refractive index gradient and thus mode confinement on top of the waveguide. The etching depth is 4 μm. The width of the waveguide is 6 μm (respectively 4 μm) in the wide (respectively narrow) part of the Bragg grating. The device is constructed as follows: it is composed of 15 sections, of N periods each. In each section i, the period of the Bragg grating is Λ_i, allowing to reflect a part of the optical spectrum. As a function of the section design, it is then possible to obtain either positive or negative group delay dispersion. In the following the Bragg grating starts with the smallest period, so that the shortest wavelength is reflected at the beginning of the grating, while the longest wavelength is reflected at the end (Fig. 1b). Negative group delay dispersion will thus be obtained. By varying N, we can tune the reflectivity and group delay dispersion by increasing the length of the Bragg grating. The higher N is, the longer is the device and the higher are the group delay dispersion and the reflectivity. Structures with N = 50, 100 and 200 have been simulated and fabricated.

2.2 Simulations

The Bragg gratings were simulated using an eigenmode expansion solver which allowed us to evaluate the transmission and the reflection of the Bragg gratings as well as the group delay. By using 15 sections, with periods varying from 1.068 to 1.128 μm, we can see that a wide stopband is obtained from 7.7 to 8.2 μm wavelength (Fig. 2a). For N from 50 to 200 periods the reflectivity varies from 90% to 100% (Fig. 2a) and the group delay dispersion from -11ps^2 to -50ps^2 (Fig. 2b). Again, the group delay dispersion (GDD) is negative here but it could be positive just by flipping the Bragg grating. Furthermore, we simulated the effect of such chirp filter on a 11.8ps wide chirped pulse with a chirp parameter properly chosen for each GDD value (Fig. 2c). A compression factor of 9 is obtained, with an increase of the peak power by the same factor 9.

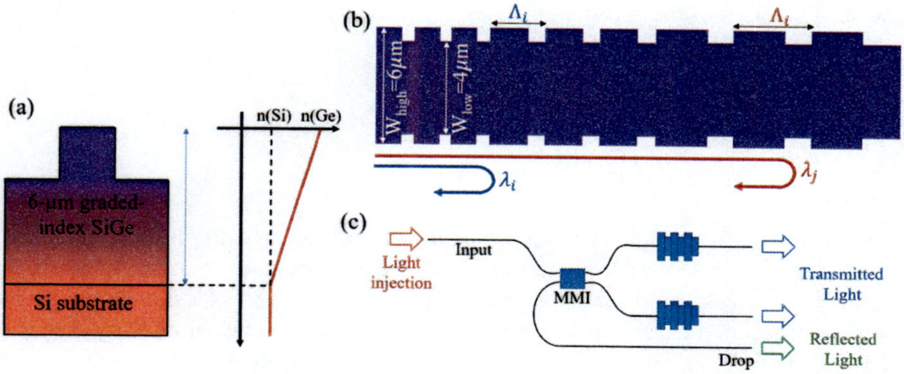

Fig. 1. **(a)** Schematic view of the waveguide cross section, **(b)** Schematic top view of the Bragg grating, **(c)** Device to measure the reflected pulse

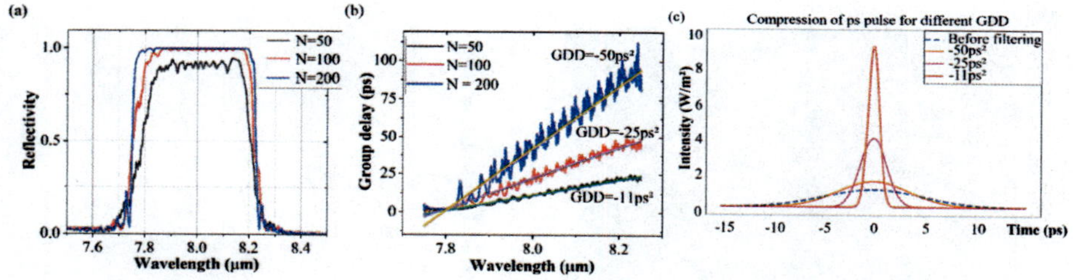

Fig. 2. **(a)** Simulated Bragg gratings reflectivity for different N, **(b)** Simulated Bragg gratings group delay for different N, **(c)** Simulated compression for the different group delay dispersions obtained

2.3 Fabrication and Characterization

Device fabrication starts with the epitaxial growth of the graded-index SiGe layer by using low-energy plasma-enhanced chemical vapor deposition (LEPECVD). The device fabrication is then carried out using e-beam lithography to pattern the device, followed by inductively coupled plasma reactive ion etching (ICP-RIE). Finally, waveguide input and output facets are diced to obtain a smooth facet to couple the light from the free-space laser beam. An SEM picture of the fabricated device can be seen in Fig. 3a. For optical characterization, we used a pulsed QCL emitting from 7.4 to 8.5 μm and measured the transmission of the device using a MCT detector. The transmission of the chirped Bragg gratings with N = 100 and 200 can be seen Fig. 3b. A low pass filter was used to smooth the transmission. It can be seen that the stopband is slightly shifted compared to the simulations. A rejection between −5 and −8 dB for N = 100 and −7 and −10 dB for N = 200 (Fig. 3b). We also measured the group delay of a chirped Bragg grating with N = 50. For this purpose, two Bragg gratings were put next to each other to form a Fabry-Perot cavity where the longer wavelengths travel a longer effective cavity length than the shorter wavelengths resulting in a free spectral range (FSR) that varies linearly with the wavelength. From the FSR measurements, it is possible to extract the group delay as a function of the wavelength using the following equations:

$$L_c = \frac{\lambda^2}{2n_g \Delta\lambda} \tag{1}$$

$$\frac{L_c n_g}{c} = \tau_g = \frac{\lambda^2}{2c\Delta\lambda} \tag{2}$$

where L_c is the effective length of the cavity, n_g is the group index, τ_g is the group delay and $\Delta\lambda$ is the FSR measured from the transmission spectrum. The group delay dispersion can thus be extracted from the slope of the curve. Figure 3c shows a simulation of this device considering the actual resolution of the experimental setup (limited by the laser linewidth), and Fig. 3d shows the real measurement of this device. We obtain a good agreement between the simulation and the measurement. We also obtain the same group delay dispersion calculated in the simulations in Sect. 2.2 of -1ps^2.

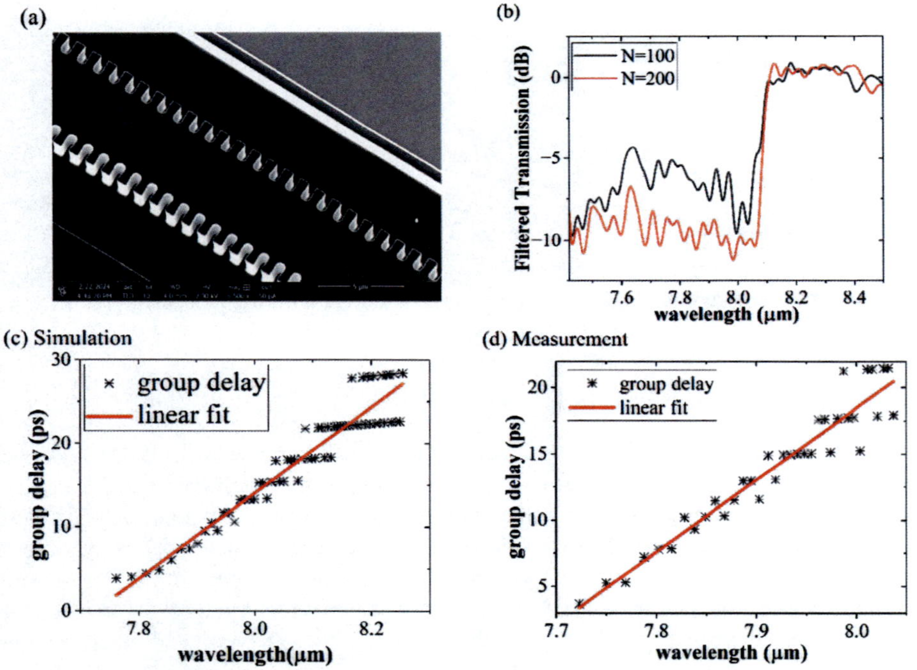

Fig. 3. **(a)** SEM image of a Bragg grating, **(b)** Reflectivity of Bragg gratings for N = 100 and 200, **(c)** Simulation of the Fabry-Perot cavity to measure the group delay for N = 50, **(d)** Measurement of the group delay of a Bragg grating with N = 50

3 Conclusion

We have demonstrated the integration of chirped Bragg gratings for the compression of mid-infrared pulses. Rejections of −10 dB in a stopband from 7.5 to 8 μm and group delay dispersions of few tens of ps^2 were measured enabling possible compression factors of up to 9 for 10ps wide pulses and a peak power increase with the same factor 9. These results hold promises for the integration of a compact ultrashort pulsed laser source in the mid-infrared wavelength range.

Acknowledgements. The fabrication of the device was partially performed at the Platform de MicroNanoTechnologie/C2N, which is partially funded by the "Conseil Général de l'Essonne". This work was partly supported by the French RENATECH network. Co-funded by the European Union (ERC-Electrophot, 101097569). Views and opinions expressed are however those of the author(s) only and do not necessarily reflect those of the European Union or the European Research Council. Neither the European Union nor the granting authority can be held responsible for them.

References

1. Montesinos-Ballester, M., et al.: On-chip mid-infrared supercontinuum generation from 3 to 13μm wavelength. ACS Photonics **7**, 3423–3429 (2020)
2. Täschler, P., et al.: Femtosecond pulses from a mid-infrared quantum cascade laser. Nat. Photonics **15**, 919–924 (2021)
3. Kashyap, R., Maxwell, G.D., Ainslie, B.J.: Laser-trimmed four-port bandpass filter fabricated in single-mode photosensitive Ge-doped planar waveguide. IEEE Photonics Technol. Lett. **5**(2), 191–194 (1993)
4. Vakarin, V., et al.: Ultra-wideband Ge-rich silicon germanium integrated Mach-Zehnder interferometer for mid-infrared spectroscopy. Opt. Lett. **42**, 3482–3485 (2017)

Monday Poster Session

CMOS Electronic Circuits in Standard Silicon Photonics

Monica Crico[1], Samuele De Gaetano[1], Andrés Martinez[1], Francesco Morichetti[1], Andrea Melloni[1], Giorgio Ferrari[2], Marco Sampietro[1], and Francesco Zanetto[1](✉)

[1] Department of Electronics, Information and Bioengineering, Politecnico di Milano, p.za Leonardo da Vinci 32, Milano, Italy
francesco.zanetto@polimi.it

[2] Department of Physics, Politecnico di Milano, p.za Leonardo da Vinci 32, Milano, Italy

Abstract. We present the integration of CMOS electronic circuits into a standard Silicon Photonics platform. The designed electronics, fabricated without modifying the fabrication process employed by commercial foundries, targets the detection and processing of electrical signals directly on the photonic chip. First, we show the operation of multiplexers and sample&hold circuits for sequentially controlling large-scale photonic systems with few electrical signals. Then, we present the integration of a transimpedance amplifier for monitoring light signals down to −40 dBm with a frequency up to 2 MHz. These examples represent a first step towards complex electronic processing and elaboration in standard Silicon Photonics.

Keywords: Silicon Photonics · CMOS electronics · monolithic integration · electronic multiplexers · transimpedance amplifier

1 Introduction

Silicon Photonics (SiP) has enabled the integration of multiple photonic devices in a compact silicon chip, opening the way to the design of large-scale optical systems for complex manipulation of light beams. In addition to its excellent optical properties, silicon is a popular photonic material thanks to its compatibility with the manufacturing process of the microelectronic industry. This latter aspect has stimulated the investigation of monolithic electronic-photonic chips, combining the unrivaled bandwidth of optical devices and the versatility of CMOS electronics [1]. The option of integrating photonic devices into mature microelectronic stacks, properly modified to maximize speed and energy performance, has been successfully pursued to design high-speed transceivers operating at hundreds of Gbps [2]. However, due to its high cost, this approach still struggles to gain popularity in those applications where photonic functionalities prevail and large-scale optical circuits are needed, as in optical computers, quantum photonic processors and neuromorphic systems [3]. To overcome this limitation, the opposite approach has been proposed, with the successful integration of complementary metal-oxide-semiconductor (CMOS) devices and digital circuits into standard SiP technologies

J. Witzens et al. (Eds.): ECIO 2024, SPPHY 402, pp. 331–336, 2024.
https://doi.org/10.1007/978-3-031-63378-2_54

[4]. Here, we further extend this concept and show that analog electronic circuits can also be integrated in purely photonic platforms, opening the way to on-chip amplification and processing of electrical signals.

2 Electronic Integration in Standard Silicon Photonics

2.1 CMOS Transistors

CMOS transistors have been integrated into the photonic chip with a zero-change approach, achieved by using only the fabrication steps already available in the standard technological stack of Advanced Micro Foundry (AMF, Singapore). Since in SiP technologies the metal layers are relatively far from the silicon, the transistors were designed with lateral silicon gates, by exploiting the minimum waveguide-to-waveguide distance of around 200 nm (Fig. 1a). Both nMOS and pMOS devices were realized, employing the same geometry while inverting the doping species of the diffusion regions. A reasonably low threshold voltage around 1.5 V has been achieved thanks to the fact that the native silicon layer of a photonic chip is naturally depleted of free carriers [4]. The two devices have been electrically characterized demonstrating successful operation (Fig. 1b) and used to integrate complex circuits on the photonic chip.

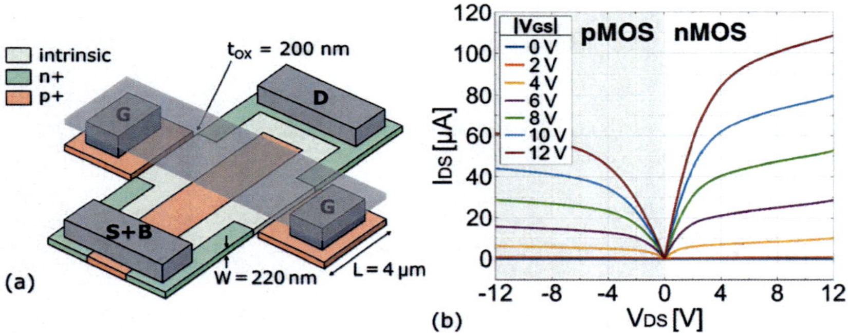

Fig. 1. a) Schematic structure of the designed nMOS transistor in standard Silicon Photonics. b) Experimental characterization of the nMOS and pMOS transistors, demonstrating their correct functionality.

2.2 On-Chip Multiplexers for Sequential Control of Photonic Circuits

Figure 2a shows an application enabled by embedding CMOS electronics into large-scale photonic circuits (PIC). As an example, a programmable mesh of Mach-Zehnder interferometers (MZI) is reported. To arbitrarily set the circuit functionality, sensors and actuators are needed for each MZI, operated by external electronic circuits. However, as the photonic complexity increases, the quantity of interconnections towards the control electronics explodes, limiting the scalability of this approach. This issue is overcome by serializing the control action with on-chip multiplexers (MUX). One MUX can be used to sequentially interrogate the on-chip sensors and route their signals towards a single readout circuit [5]. Similarly, the actuators can be sequentially driven through an on-chip demultiplexer (DEMUX). Since most of the photonic actuators are volatile, an on-chip electronic memory is needed to keep their operation stable. To this end, we designed a Sample&Hold (S&H) circuit, connected to the on-chip actuators current drivers. The S&H exploits the switches of the DEMUX and a bank of memory capacitors of 10 pF. In the sample phase, the actuator command is directly applied to the driver and memorized by the electronic memory. During the hold phase, the on-chip capacitor maintains the operation of the unselected actuators stable, thus providing a non-volatile behaviour. Figure 2b reports the measured variation of optical power at the output of an interferometer when operating its heater with the S&H. The output of the MZI is kept stable when the S&H refresh time is below 100 ms, which represents the maximum hold time of the memory capacitors.

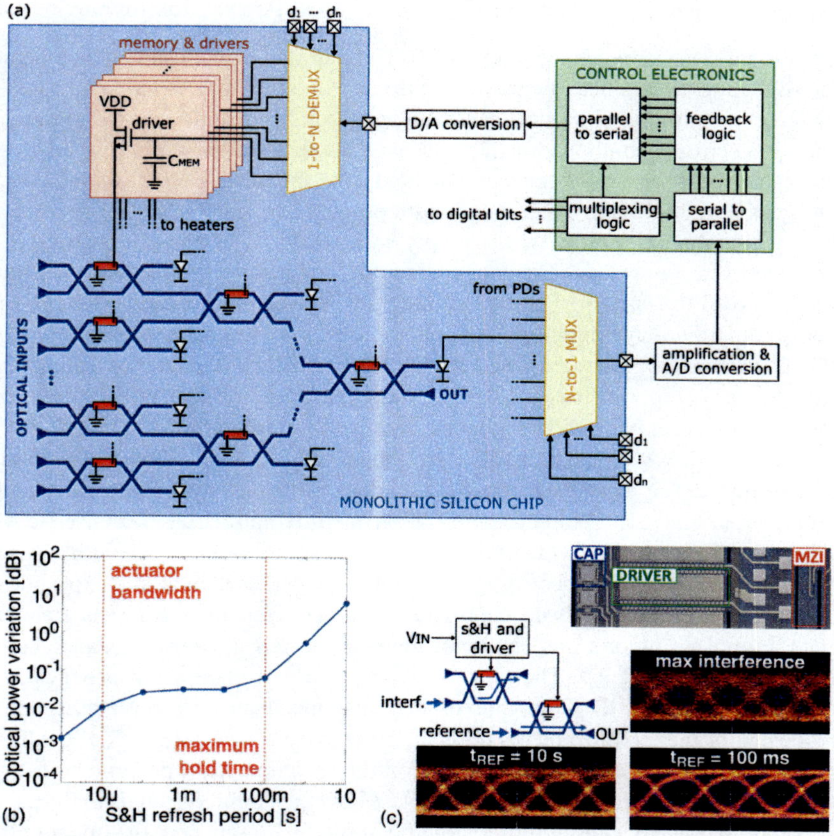

Fig. 2. a) Schematic view of a programmable PIC with on-chip MUXes and S&H to reduce the number of electrical connections. b) Optical power variation at the output of a MZI when controlling its heater with the on-chip S&H. c) Schematic view of a two-MZI structure designed for testing the S&H operation and eye diagrams at its output, for different S&H refresh times.

The impact of the S&H discharge, due to leakage currents, has been evaluated on the transmission of high-speed signals, with the circuit shown in Fig. 2c. Two signals at 10 Gbps have been injected into the input ports of the structure. The first MZI has been configured in cross state, while the second has been operated in bar state to route all the light to the output port. When the first MZI drifts from the cross condition, its input signal acts as an interference for the second interferometer, degrading the bit-error-rate (BER) at the output. The measured BER when the interference is off has a value around 10^{-11}. The same performance is obtained for refresh periods below 100 ms. Instead, at longer refresh times, the S&H discharge induces a shift of the first MZI bias point, thus producing a relevant interference at the output. This is confirmed by the eye diagrams measured for 10 s and 100 ms S&H refresh times, reported in Fig. 2c. This result translates into the possibility of multiplexing hundreds of actuators without relevant penalties on the chip performance.

2.3 Transimpedance Amplifier for On-Chip Optical Power Monitoring

An operational transconductance amplifier (OTA) has also been designed as a demonstration of analog electronics integrated in Silicon Photonics. OTAs and amplifiers are the fundamental building blocks of many analog circuits and could be used to integrate a variety of electronic components, like ADCs and DACs. As shown in Fig. 3a, a standard two-stage amplifier has been implemented. The OTA has been designed with an open-loop gain of 38 dB and a gain-bandwidth product of a few MHz. The circuit has a power consumption of 400 μW with a supply voltage of 5 V. In this demonstration, the amplifier has been connected in transimpedance configuration with a feedback resistance of 10 kΩ, to read the current signal from integrated germanium photodiodes (PD) and produce an output voltage directly on the photonic chip.

A test structure has been fabricated to characterize and validate the operation of the TIA, as shown in Fig. 3b. It is made of a straight waveguide that injects light to an integrated germanium photodiode connected to the on-chip amplifier. The PD, having a responsivity of 0.8 A/W and a dark current around 50 nA, is provided by the foundry process design kit. To precisely compare the readout performed by the integrated amplifier with conventional bench-top instrumentation, a 90–10 splitter has been used to route 10% of the injected light signal out of the chip. A laser source at 1550 nm, with a variable optical attenuator (VOA) connected to its output, has been used in all the experiments.

The fabricated TIA has been electrically characterized by testing its frequency response. Figure 3c reports its behaviour when connected in transimpedance configuration to the integrated PD. The plot shows that the gain-bandwidth product of the amplifier matches the −3 dB bandwidth of the transimpedance gain, thus confirming the correct design of the circuit. The resulting 2 MHz closed-loop bandwidth of the TIA is wide enough to perform effective monitoring of the on-chip optical power. Optical validation has then been performed in order to assess how the circuit responds to the current signals generated when shining light on the photodiode. To this end, the optical power injected into the chip has been progressively attenuated with the external VOA and the corresponding voltage variation at the output of the TIA has been monitored. The results are shown in Fig. 3d. The relation between optical power and output voltage variation is linear, in accordance with the theoretical behaviour estimated from the responsivity of the PD and the TIA feedback resistance. The lowest signal that can be detected by the circuit has been assessed by measuring the noise at the TIA output without light. A noise RMS value of 660 uV has been measured with a readout bandwidth of 1 kHz, corresponding to a minimum detectable light variation of −40 dBm. This performance level highlights the good sensitivity of the circuit, suitable for most photonic applications.

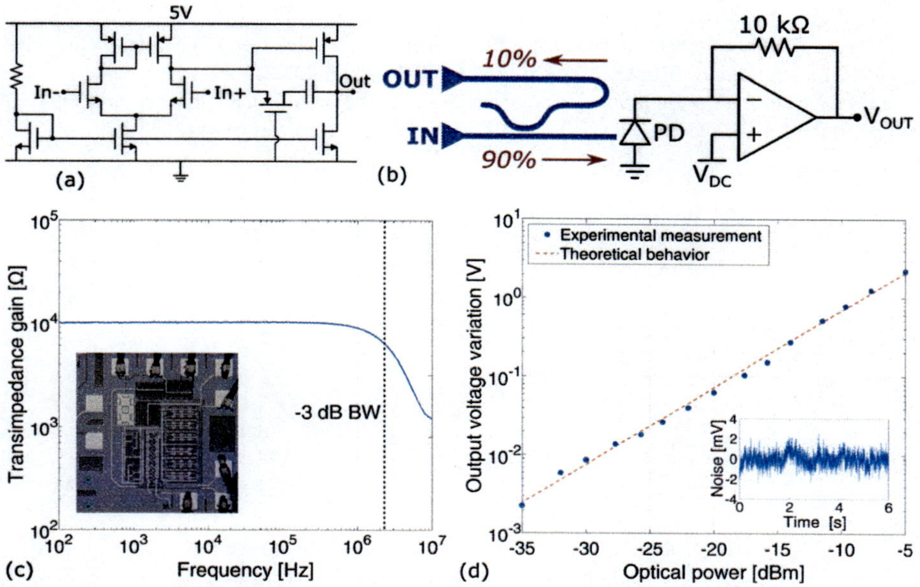

Fig. 3. Schematic view of a) the designed operational transconductance amplifier and b) the test structure fabricated to validate it. c) Measured frequency response of the TIA. The inset reports the microscope photograph of the chip. d) Voltage variation at the output of the TIA as a function of the optical power reaching the photodiode. The inset shows the noise at the TIA output with a 1 kHz readout bandwidth.

3 Conclusions

We demonstrated that it is possible to monolithically integrate CMOS electronic circuits on purely photonic platforms, manufactured by commercial foundries without modifications to the standard fabrication flow. First, we showed the possibility of controlling silicon photonic circuits by time-multiplexing on-chip the electrical signals needed to read the sensors and drive the actuators. Non-volatility has been provided to the actuators by integrating a monolithic S&H circuit, with a maximum hold time of 100 ms. Then, we integrated a transimpedance amplifier for on-chip optical power monitoring, suitable for reading current signals from integrated germanium photodiodes with a frequency up to 2 MHz and a power down to −40 dBm. This approach will be further developed by employing the realized circuits as a building block for more complex analog structures, aiming at bringing more and more electronic functionalities on the photonic chip.

References

1. Thomson, D., et al.: Roadmap on silicon photonics. J. Opt. **18**(7), 073003 (2016)
2. Rakowski, M., et al.: 45 nm CMOS - silicon photonics monolithic technology (45CLO) for next-generation, low power and high speed optical interconnects. In: Optical Fiber Communications Conference (OFC), Paper T3H.3 (2020)
3. Shen, Y., et al.: Deep learning with coherent nanophotonic circuits. Nature Photonics **11**(7), 441–446 (2017)
4. Zanetto, F., et al.: Unconventional monolithic electronics in a conventional silicon photonics platform. IEEE Trans. Electron Devices **70**(10), 4993–4998 (2023)
5. Zanetto, F., et al.: Time-multiplexed control of programmable silicon photonic circuits enabled by monolithic CMOS electronics. Laser Photonics Rev. **17**(11), 2300124 (2023)

Comparison of Thermo-Optic Phase Shifters in Imec's Silicon Photonics Platform

Yuxi Fang[1](✉), Hong Deng[1], Xiangfeng Chen[1], Halil Cuma[1], K. P. Nagarjun[1], Filippo Ferraro[2], Guy Lepage[2], Peter De Heyn[2], and Wim Bogaerts[1](✉)

[1] Department of Information Technology, Ghent University – IMEC, Photonics Research Group, Gent, Belgium
{yuxi.fang,wim.bogaerts}@ugent.be
[2] Imec vzw, Kapeldreef 75, Heverlee, Belgium

Abstract. Taking advantage of the high thermo-optic coefficient of silicon is an effective way to realize optical phase shifting in silicon photonic circuits. Based on the imec iSiPP200 silicon photonics platform, different types of integrated silicon photonic phase shifters are compared, including doped-silicon side heaters, diode-loaded heaters, doped waveguides, and metal top heaters. The effect of thermal insulation on phase shifter efficiency and bandwidth is discussed for different architectures with and without substrate undercut. By locally removing the silicon substrate underneath the heater, the efficiency of doped-silicon side heaters is improved by a factor of 2.7 for a π phase shift.

Keywords: Silicon Platforms · Thermo-optic Phase Shifter · Undercut · Efficiency · Bandwidth

1 Introduction

Silicon platform is becoming a key technology for photonic integrated circuits, due to its CMOS compatibility and small footprint. In many applications, such as optical switching, optical communication, and programmable photonic integrated circuits [1], thermo-optic phase shifters form a key building block. Taking advantage of the high thermo-optic coefficient of silicon [2], local heaters are an effective way to realize optical phase shifting in silicon platforms.

Using on the imec iSiPP200 silicon photonics platform [3], we implemented various phase shifters to illustrate the impact of thermal insulation on both efficiency and bandwidth, including doped-silicon side heaters, diode-loaded heaters, doped wave-guides, metal top heaters, and folded metal heaters. Due to the local removal of the silicon substrate underneath the heater, the efficiency of doped-silicon side heaters could be improved by a factor of 2.7 for a π phase shift. We compare doped waveguides with lateral and longitudinal current injection, the −3 dB bandwidths of the lateral-injected is 3.4 times larger than that of longitudinal-injected doped waveguides. Furthermore, the

Y. Fang and H. Deng—Contributed equally.

J. Witzens et al. (Eds.): ECIO 2024, SPPHY 402, pp. 337–342, 2024.
https://doi.org/10.1007/978-3-031-63378-2_55

asymmetric electrical response of diode-loaded heaters provides opportunities for utilizing pulse-width modulation (PWM) or multiplexed driving techniques, which offers valuable guidance in selecting appropriate phase shifters for large-scale programmable circuits. The most efficient heater we tested is a folded tungsten heater, with an efficiency of 1.97 mW for a π phase shift.

2 Design and Measurement Method

2.1 Different Designs of Phase Shifters

Figure 1 depicts three types of heaters in imec's platform. Doped-silicon side heaters with and without undercut are shown in Fig. 1(a_1) and (a_2), respectively. Based on typical shallow-etched rib waveguides, P-doped silicon is positioned as a resistor next to the waveguides. The dopants increase the conductivity of side heaters, allowing current to flow through them via electrical contacts. Using deep-etched rib waveguides, we propose two doped waveguides in Fig. 1(b_1) and (b_2), using waveguide itself as a resistor. The current flows through the waveguides directly in Fig. 1(b_1). Therefore, this type of heater is more efficient. As shown in Fig. 1(b_2), only one electrical contact is placed on each side, two side heaters and a central waveguide form a circuit. We utilize full-etched strip

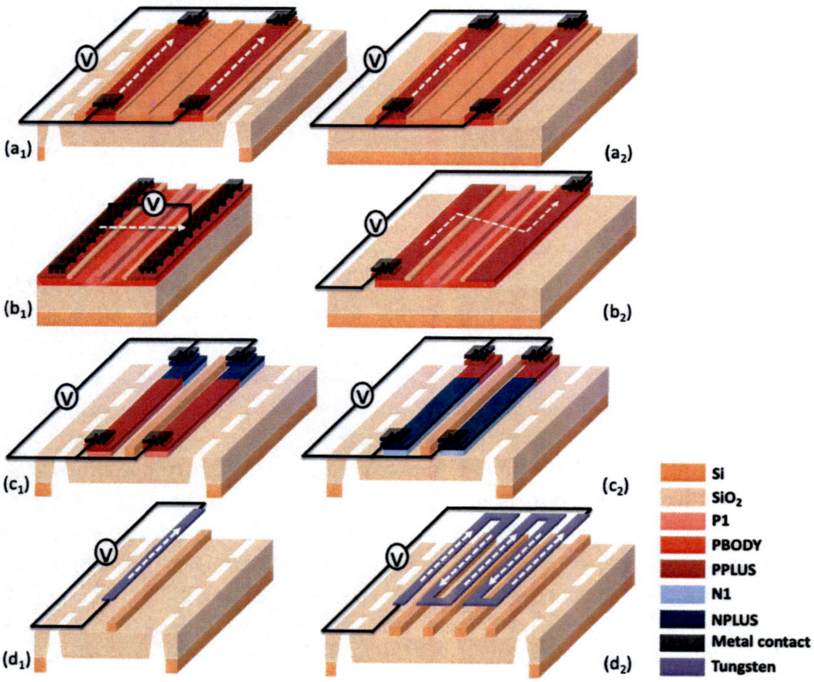

Fig. 1. Schematic diagram of doped-silicon side heaters (a_1) with and (a_2) without undercut, (b_1) lateral and (b_2) longitudinal-injected doped waveguides, (c_1) P and (c_2) N diode-loaded undercut heaters, (d_1) single and (d_2) 4 folded tungsten heaters.

waveguides to demonstrate the diode-loaded heaters. As shown in Fig. 1(c_1), we use P-type dopants to dope the main body of the heaters, and use N-type dopant to dope the area near one of the electrical contacts, creating a PN junction inside the heater. This is defined as P diode-loaded heater. Likewise, Fig. 1(c_2) shows a N diode-loaded heater. Another alternative is to use a metal resistor on top of the waveguide. Figure 1(d_1) shows the standard single-line heater. To improve the efficiency, we use folded tungsten lines above the waveguides, as shown in Fig. 1(d_2).

2.2 Measurement Methods

We use asymmetric Mach-Zehnder interferometers (MZIs) to demonstrate the phase shifters. Each type of heater is placed in such an MZI with standard 2 × 2 multimode interferometers (MMI). We could characterize electrical and optical response of the heaters using the method shown in Fig. 2 (a). Light is coupled into the MZI by grating couplers, and we drive the heaters by sweeping the direct current (DC) voltage, then monitor the current and the optical output power simultaneously. We could further measure the frequency response of the heaters, as depicted in Fig. 2(b). Using a function generator, we inject a radio frequency (RF) signal at a single frequency to the heaters. Light couples out of the fiber and enters the photodetector.

After adjusting the DC bias, we set the heaters to the quadrature point to get the maximum modulation amplitude on the oscilloscope. By sweeping different frequencies, a Bode plot could be constructed.

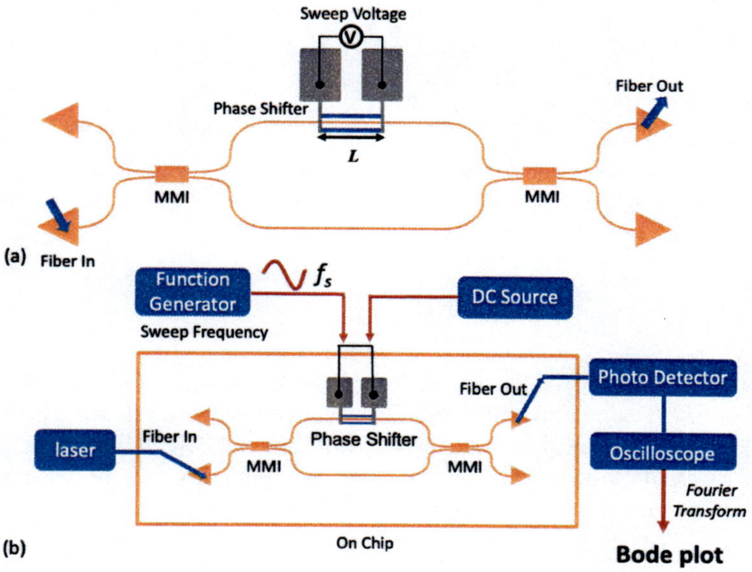

Fig. 2. Schematic diagram for (a) efficiency and (b) bandwidth measurement.

3 Experimental Results

3.1 Doped-Silicon Side Heaters

As shown in Fig. 3, the normalized optical transmission follows a sinusoidal curve, from which the thermal efficiency can be extracted. We first demonstrate the influence of the thermal insulation undercut. We take the P-doped silicon side heater as an example, with a length of 28 μ. For a π phase shift, the efficiency of side heaters with and without undercut is measured as a P_π (power to induce a π phase shift) of 13.20 mW and 35.16 mW, respectively (lower P_π is better). The doped-silicon side heater with undercut is approximately 2.7× more efficient than without undercut. As the length of the undercut heater increases from 28 to 53 μm, the P_π decreases from 13.2 mW to 8.92 mW, and the efficiency improves by a factor of 1.48×. As the heater length increases, the relative area of substrate removal increases, helping to trap the heat in the region around the waveguide. Therefore, the efficiency of the 53-μm heater is higher.

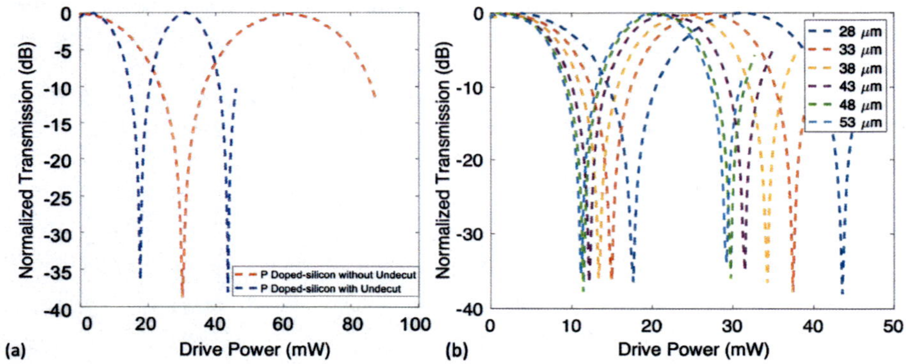

Fig. 3. Transmission as a function of electrical power for (a) 28-μm P-doped silicon side heaters with and without undercut, (b) P-doped silicon side undercut heaters with different length.

3.2 Diode-Loaded Heaters

From Fig. 4(a), one can see a clear diode I-V curve in 28-μm diode-loaded heaters. For P or N diode-loaded heater, the reverse current is negligible below the break-down voltage (−3 V). All diodes show a similar threshold around 0.8 V. As shown in Fig. 4(b), if we operate the heaters in the range of ±3 V, it will only drive one heater in a matrix addressing topology.

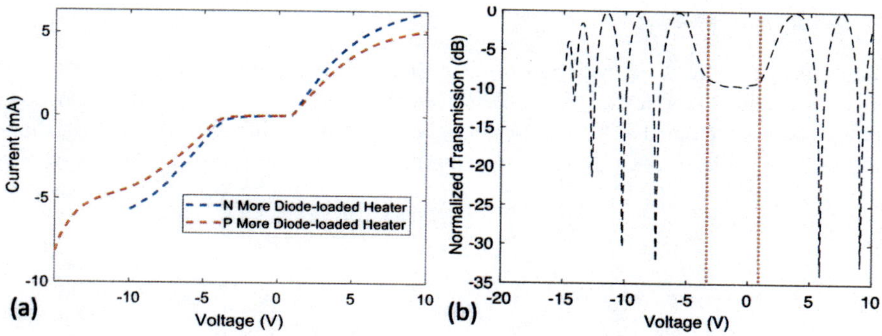

Fig. 4. (a) I-V curve of a 28-μm N and P diode-loaded heaters, (b) transmission as a function of voltage for a 28-μm P diode-loaded heater.

3.3 Doped Waveguides

Figure 1(b_1) shows the doped waveguides with lateral current injection (over the entire length) and without thermal insulation, with a P_π of 18.73 mW. The design in Fig. 1(b_2) is also a doped waveguide, but the current is injected only at the ends. Therefore, the driving power is distributed over three parts: two side heaters, and the waveguide itself. The result is a P_π of 32.43 mW, which is significantly larger than with the lateral injection, as a larger volume is heated. This difference is also reflected in the response time. The −3 dB bandwidths of the lateral-injected and longitudinal-injected doped waveguides are 139 and 41 kHz, respectively. We can use a simplified thermal resistor-capacitor (RC) model to explain it. Compared to longitudinal-injected doped waveguide, the lateral-injected doped waveguide has a smaller thermal resistance, which means the RC time constant is smaller, thus the bandwidth is larger. Also, as the heated volume is smaller, it has a lower thermal capacitance (Fig. 5).

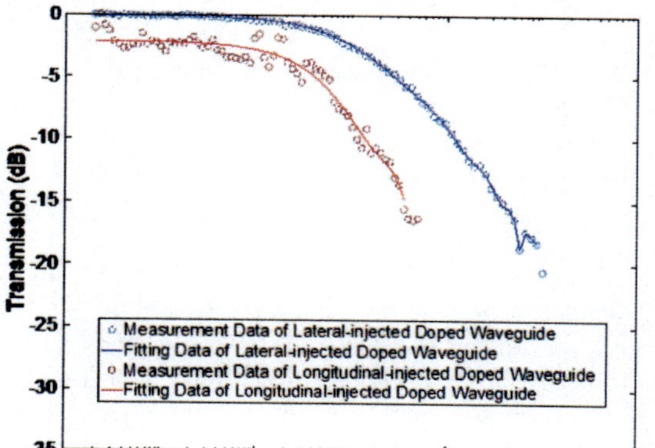

Fig. 5. Bandwidth of lateral-injected and longitudinal-injected P-type doped waveguides.

3.4 Metal Top Heaters

We propose two types of overhead tungsten heaters, including a single tungsten line and 4 folded lines above the 5× folded waveguides. For a π phase shift, the power consumption is 6.6 and 1.97 mW, respectively, showing the increased efficiency thanks to the folding, which passes the same waveguide multiple times through the same heated volume (Fig. 6).

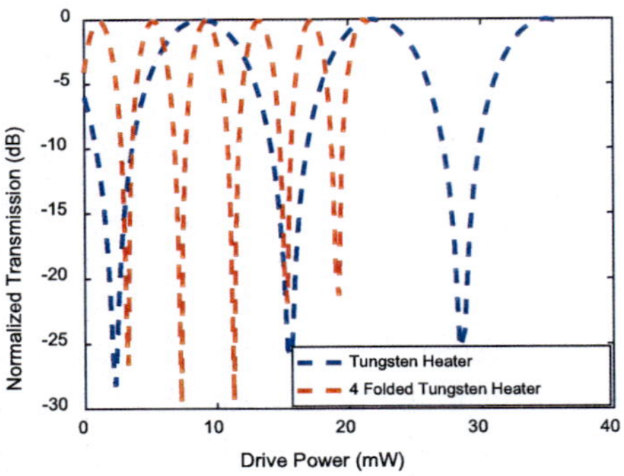

Fig. 6. Transmission as a function of electrical power for tungsten and 4 folded tungsten heaters.

4 Conclusion

In conclusion, substrate undercut makes the heater about 3× more efficient than before. Through using lateral current inject, the efficiency could be improved 1.73 times than longitudinal-injected doped waveguides, and the corresponding response time is faster. In addition, a 4 folded tungsten heater has the smallest P_{π}, which is below 2 mW.

Acknowledgement. Part of this work was funded by the European Union through the ERC grant PhotonicSWARM (grant 725555).

References

1. Bogaerts, W.: Programmable photonic circuits. Nature **586**(7828), 207–216 (2020)
2. Komma, J., Schwarz, C., Hofmann, G., Heinert, D., Nawrodt, R.: Thermo-optic coefficient of silicon at 1550 nm and cryogenic temperatures. Appl. Phys. Lett. **101**(4), 041905 (2012)
3. Ferraro, F.: Imec silicon photonics platforms: performance overview and roadmap. In: Next-Generation Optical Communication: Components. Sub-Systems, and Systems XII, pp. 22–28. SPIE, San Francisco, California, United States (2023)

Optomechanical Crystal Cavities for Mechanically-Enabled All-Optical Upconversion of 3GPP 5G NR Signals

Raúl Ortiz(✉), Vicente Fito, Maria Morant, Laura Mercadé, Roberto Llorente, and Alejandro Martínez

Nanophotonics Technology Center, Universitat Politècnica de València, Camino de Vera s/n, Building 8F, 46022 Valencia, Spain
{rortfer,amartinez}@ntc.upv.es

Abstract. Phonon lasing is attained by driving optomechanical crystal cavities (OMCCs) with blue-detuned lasers, producing narrow and stable microwave tones at integer multiples of the mechanical frequency. We evaluate all-optical frequency upconversion employing an OM frequency comb generated with a silicon OMCC targeting the generation of 3GPP 5G new-radio signals. The performance of the upconverted 5G signal from the low-band to higher frequency bands employing the first two harmonics and their phase noises are evaluated. The proposed OMCC technology is a compact and power-efficient solution for all-photonic processing of microwave signals, a crucial aspect in 5G NR and future beyond-5G systems.

Keywords: Silicon optomechanics · all-Optical 5G frequency conversion

1 Introduction

Cavity optomechanics investigates the interplay between confined light and mechanical waves in a semiconductor structure, which is an intriguing physical phenomenon [1]. When an optomechanical crystal cavity (OMCC) is driven by a laser with a frequency exceeding its optical resonance (blue-detuned), the optical mode interacts with mechanical motion, reducing damping and leading to self-sustained oscillations, ultimately reaching the so-called phonon lasing regime [2]. In this regime, the OMCC modulates the driving laser with a narrow tone at the mechanical resonance frequency and its harmonics, creating an optical frequency comb [3,4], which can serve as a multi-frequency microwave oscillator suitable for microwave photonics applications.

Phased antenna arrays are used in new 5G communications to increase the transmission capacity (users × bitrate per user) of cellular nodes [5]. This is necessary to meet the continuously increasing data traffic demand [6]. Operating phased antenna arrays in the millimeter frequency band is challenging, as a

J. Witzens et al. (Eds.): ECIO 2024, SPPHY 402, pp. 343–348, 2024.
https://doi.org/10.1007/978-3-031-63378-2_56

potentially large number of frequency conversion stages are required. These usually require bulky and costly electronics [7].

Optoelectronic oscillators (OEOs) are an effective solution for a flexible, scalable, compact, high bandwidth and power-efficient frequency conversion in the optical domain, with a phase noise independent of the oscillation frequency, allowing high frequencies to be reached without phase noise limitation [8–10]. In the lasing regime, the OMCC generates an OM frequency comb with several harmonics located at integer multiples of the mechanical resonance (f_m). Thus, the silicon OMCC can be used for all-optical frequency conversion at different target frequencies with the advantages of flexibility, scalability, and seamless integration with other silicon photonics and electronic subsystems presented in the phased array. In this paper, we evaluate experimentally the frequency upconversion performance of radio signals following the 3rd Generation Partnership Project (3GPP) 5G new-radio (NR) standard using the first and second harmonics generated by the silicon OMCC comb.

2 Silicon OMCC Design for Comb Generation

The silicon OMCC has been designed and fabricated using standard silicon fabrication toolchain and processes [11]. A scanning electron microscope (SEM) image of the device is included in the inset of Fig. 1(b) for reference. The OMCC has been designed to support several mechanical resonances in the GHz range [12], each exhibiting a distinct vacuum OM coupling rate with the optical mode. The mechanical mode at $f_m = 3.84$ GHz has the highest g_0 value $g_0 \approx 2\pi \cdot 200$ kHz [3] while the optical mode is centered at $\lambda_o = 1523$ nm, with quality factors $Q_m = 6.7 \times 10^3$ and $Q_o = 3.0 \times 10^3$, respectively. This leads to an optical decay rate of $\kappa = 65$ GHz, corresponding to the unresolved sideband regime ($\kappa \gg 2\pi f_m$). We characterize experimentally both the optical and mechanical responses of the cavity using the setup depicted in Fig. 1(a), coupling the light

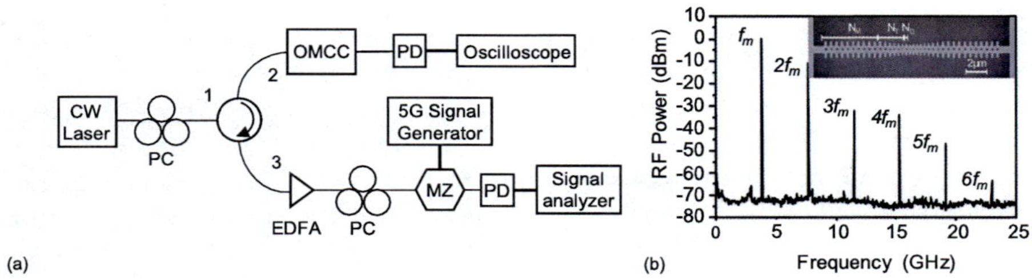

Fig. 1. (a) Experimental setup used for all-optical frequency conversion with the proposed OMCC. The transmission pathway measures the phonon lasing, while the reflection pathway is used for 5G NR transmission implementing frequency upconversion. (b) Measured RF spectrum displaying six harmonics of the OM frequency comb in the lasing regime (measured with $R_{BW} = 10$ MHz) and SEM image of the OMCC comprising $N_M = 10$ mirror cells, $N_T = 6$ transition cells, and $N_D = 1$ defect cell.

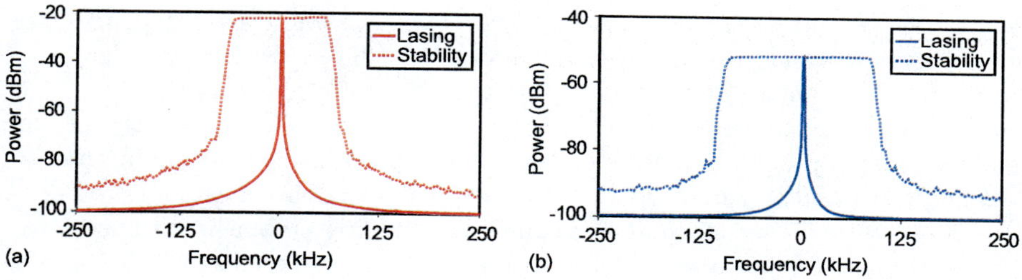

Fig. 2. Measured RF power of the OM frequency comb generated in the lasing regime (solid line) and mechanical frequency drift measured with max hold during 20 min (dashed line) for: (a) first and (b) second OMCC harmonics.

using a tapered fiber via a fiber loop [11]. The resulting optical resonance is monitored in transmission with an oscilloscope, while the mechanical response is measured in reflection using a circulator and modulated with a 5G NR signal. Polarization controllers (PC) are used to optimize the optical coupling to the OMCC and to the Mach-Zehnder (MZ) modulator. When pumping high power into the OMCC, it enters a dynamic state where mechanical self-sustained oscillations manifest, leading the cavity into the phonon lasing regime and generating an optomechanical frequency comb [4].

Figure 1(b) shows the measured RF spectrum of the OM comb, illustrating the first six harmonics spaced by the mechanical frequency. The harmonics provided by the OM frequency comb allow coverage of the microwave photonics RF spectrum to the K-band from 18 to 27 GHz, according to IEEE.

Figure 2 depicts the Voigt-like shape of the first and second harmonics in the lasing regime, including the maximum amplitude span and frequency drift measured over 20 min. It can be observed in Fig. 2 that the frequency offset of the second harmonic is approximately twice that of the first harmonic, yet both below 250 kHz meeting the specifications defined by 3GPP 5G NR TS 38.141-1 (Release 15) [13]. To comprehensively evaluate the system's stability, Fig. 3(a) depicts the phase noise of the first harmonic, along with their dependencies based

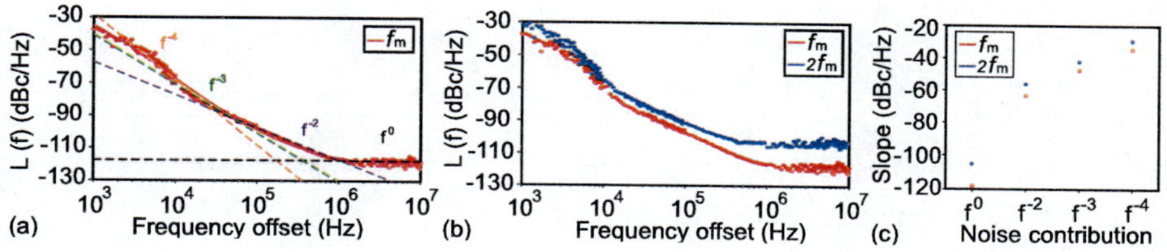

Fig. 3. Phase noise measurements of the GHz lasing regime including: (a) phase noise curve of the first harmonic and fitted slopes corresponding to the noise sources in the oscillator and (b) comparison of first and second harmonic phase noises. (c) y-Intercept of the fitted slopes for the first and second harmonics of the OMCC.

on Lesson's model [14]. The phase noise exhibits characteristic dependencies, including f^{-4} (random walk of frequency), f^{-3} (flicker frequency noise), f^{-2} (white frequency noise or random walk), and f^{0} (white phase noise).

Further analysis involves retrieving the mechanical linewidth of the harmonics from the white frequency noise dependence f^{-2}, resulting in an effective linewidth of $\Gamma_{eff}/2\pi = 74$ Hz and lower values of the phase noise of –97.1 dBc/Hz at 100 kHz for the first harmonic. This finding emphasizes the system's highly competitive performance as a stable free-running oscillator [15,16]. The jitter of the system for the two given harmonics –Fig. 3(b)– was also calculated, providing values as low as $J_{RMS} = 1.1$ ps at 10 kHz - 100 kHz for both comb lines, resulting in a thirtyfold reduction in jitter compared to other free-running oscillators. Figure 3(c) shows the intercepts at the y-axis obtained from fitting the slopes of the phase noise plots of both harmonics.

3 Experimental Evaluation of All-Optical OMCC Upconversion with 3GPP 5G NR Signals

The resulting RF spectrum for the frequency upconversion process is depicted in Fig. 4(a), including the original low-band 5G NR signal centered at $f_0 = 1$ GHz and the first OMCC harmonic located at $f_m = 3.84$ GHz. The double-sideband replicas corresponding to the 5G mid-band are centered at $f_m - f_0 = 2.84$ GHz and $f_m + f_0 = 4.84$ GHz. The second harmonic is found at $2f_m = 7.68$ GHz, with upconverted replicas of the original signal centered at $2f_m - f_0 = 6.68$ GHz and $2f_m + f_0 = 8.68$ GHz (corresponding to the 5G extended mid-band). In this study, the worst-case scenario is considered, measuring the error

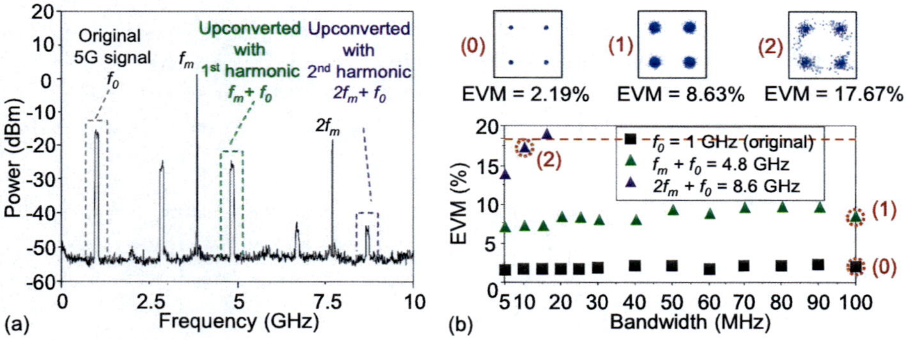

Fig. 4. (a) Received RF spectrum of the upconversion process of a 100-MHz BW 5G NR signal originally at $f_0 = 1$GHz (measured with $R_{BW} = 10$ MHz) and optical power received by the PD = -2 dBm. (b) Measured EVM of the PDSCH upconverted 5G signals at 4.86 GHz and 8.68 GHz using the first and second harmonics vs. BW, compared with the original 5G NR signal at $f_0 = 1$ GHz. Inset constellations and 3GPP 5G NR EVM recommendation included as a horizontal dashed line.

vector magnitude (EVM) of the higher sidebands located at $f_m + f_0 = 4.84$ GHz (upconverted using the first harmonic) and at $2f_m + f_0 = 8.68$ GHz (upconverted using the second harmonic).

The EVM compliance is evaluated by comparing with the 3GPP standard EVM recommendation for 5G signals using QPSK modulation in the physical downlink shared channel (PDSCH) at 18.5% [13]. Figure 4(b) presents the received EVM for each upconverted frequency and for different electrical BWs following the 3GPP 5G NR standard (ranging from 5 MHz to 100 MHz). Figure 4(b) includes insets with examples of received constellations of the 5G PDSCH using QPSK modulation comparing the original frequency at f_0 with the upconverted signal using the first and second harmonics.

The experimental results confirm that the upconverted signals to the 5G mid-band using the first OMCC harmonic meet the 3GPP EVM recommendation for the maximum 5G BW of 100 MHz. However, the upconversion to the extended mid-band using the second harmonic is limited to 5G NR signals with up to 10 MHz BW. The EVM degradation is more pronounced for the signals upconverted using the second harmonic due to its reduced signal-to-noise ratio (SNR), together with the higher frequency offset measured in the second harmonic reported in Fig. 2(b). For frequency-conversion of signals with higher BW, a higher SNR should be provided.

4 Conclusion

We have experimentally validated the concept of all-optical frequency conversion for 3GPP 5G NR signals using a OMCC designed and fabricated employing standard silicon toolchain and processes. The results obtained demonstrate that the OMCC can successfully generate an OM frequency comb at multiples of the first harmonic (3.8 GHz). The optical 5G NR frequency upconversion performance when considering the first and second harmonics produced by the OMCC in lasing regime mode has been evaluated in terms of amplitude span, frequency offset, and phase noise. The stability of the OMCC was measured over 20 min, obtaining a frequency drift smaller than 250 kHz for both harmonics, which meets the 3GPP specification for 5G NR signals. Low phase noise values of −97.1 dBc/Hz and −91.95 dBc/Hz at 100 kHz for the first and second harmonics of the OMCC, respectively, were measured.

The upconversion feasibility study has been performed experimentally employing full-standard 3GPP 5G NR signals with BWs up to 100 MHz using the first and second harmonics of the OMCC. The study reveals that while both harmonics achieve effective upconversion, using the second harmonic exhibits a 6% degradation in EVM compared to the first harmonic. Although our demonstration focuses on the frequency upconversion, the downconversion process would be identical but using a different frequency f_0 to obtain linear combinations of f_m and f_0 of a lower positive frequency than f_0. Ongoing work includes the implementation of a closed loop to optimize the oscillator stability and reduce the frequency offset in higher-order harmonics. These results allow us to conclude that silicon OMCCs can become an attractive solution for implementing on-chip microwave oscillators and nonlinear mixing functionalities suitable for processing 5G signals in the optical domain.

Acknowledgments. This work was supported in part by Generalitat Valenciana CIAICO/2021/201 TERAFLEX project, IDIFEDER/2021/061 grant; Spanish Ministry of Science and Innovation under grants ALLEGRO (PID2021-124618NB-C21), SLOMO (PDC2022-133267-C21) and MUSICIAN (CHISTERA IV Cofund 2021) and European Commission under grant MAGNIFIC (HORIZON-CL4-2022-RESILIENCE-01-10 101091968). Authors R. Ortiz and V. Fito contributed equally to this work, supported by GVA CIACIF/2021/006 and PAID-01-20 UPV grants, respectively. M. Morant work is supported by Spanish PID2019-106163RJ-I00 MULTICORE+ project.

References

1. Aspelmeyer, M., et al.: Cavity Optomechanics. Rev. Mod. Phys. **86**(4), 1391–1452 (2014)
2. Navarro-Urrios, D., et al.: Dynamical back-action at 5.5 GHz in a corrugated optomechanical beam. AIP Adv. **4**, 124601 (2014)
3. Mercadé, L., et al.: Microwave oscillator and frequency comb in a silicon optomechanical cavity with a full phononic bandgap. Nanophotonics **9**, 3535–3544 (2020)
4. Miri, M., et al.: Optomechanical frequency combs. New J. Phys. **20**, 043013 (2018)
5. Rao, L., et al.: 5G beamforming techniques for the coverage of intended directions in modern wireless communication: in-depth review. IJMWT **13**(10), 1039–1062 (2021)
6. Lin, T., et al.: Beamforming design for large-scale antenna arrays using deep learning. IEEE Wireless Commun. Lett. **9**(1), 103–107 (2020)
7. Gong, T., et al.: RF chain reduction for MIMO systems: a hardware prototype. IEEE Syst. J. **14**(4), 5296–5307 (2020)
8. Capmany, J., et al.: Microwave photonics combines two worlds. Nat. Photonics **1**(6), 319–330 (2007)
9. Zou, F., et al.: Optoelectronic oscillator for 5G wireless networks and beyond. J. Phys. D Appl. Phys. **54**(42), 423002 (2021)
10. Zhang, K., et al.: An all-optical ka-band microwave long-distance dissemination system based on an optoelectronic oscillator. IEEE Photonics J. **14**(4), 1–5 (2022)
11. Gomis-Bresco, J., et al.: A one-dimensional optomechanical crystal with a complete phononic band gap. Nat. Commun. **5**, 4452 (2014)
12. Mercadé, L., et al.: Engineering multiple GHz mechanical modes in optomechanical crystal cavities. Phys. Rev. Applied **19**(1), 014043 (2023)
13. 3GPP TS 38.141-1 version 15.15.0 Release 15, 5G NR Base Station (BS) conformance testing. Part 1: Conducted conformance testing (2023)
14. Rubiola, E.: Phase Noise and Frequency Stability in Oscillators. Cambridge University Press, The Cambridge RF and Microwave Engineering Series (2008)
15. Ghorbel, I., et al.: Optomechanical gigahertz oscillator made of a two photon absorption free piezoelectric III/V semiconductor. APL Photonics **4**, 116103 (2019)
16. Horváth, R., et al.: Sub-Hz closed-loop electro-optomechanical oscillator with gallium phosphide photonic crystal integrated on SoI circuitry. ACS Photonics **10**(8), 2540–2548 (2023)

Generating Photon Pairs in a Hybrid Si-BTO Platform

Daniel Marchant[1,2](✉), Imad Faruque[1], and Jorge Barreto[1]

[1] QET Labs, University of Bristol, Bristol, UK
d.marchant@bristol.ac.uk
[2] QECDT, University of Bristol, Bristol, UK

Abstract. Here we show photon pair generation in a hybrid Si-BaTiO_3 (Si-BTO) on insulator platform. Analysis of photon generation rates show that SFWM is comparable to SOI devices of similar characteristics with a γ_{eff} of (14.7 ± 1.3) and (2.0 ± 0.3) MHz/mW^2 for ring resonator and waveguide structures respectively.

Keywords: photon generation · SFWM · BTO · hybrid platform

1 Introduction

Photon pair sources play a crucial role in linear optical quantum computing (LOQC) [1, 2] and quantum communication [3, 4]. In LOQC, essential components include photon sources and switches. Integrated photon sources utilize waveguide or micro ring resonator structures, relying on the non-linear interaction of light with the material. Through spontaneous four-wave mixing (SFWM), two pump photons are annihilated to produce a signal and idler photon. Detecting the idler photon can then herald the presence of the signal photon. Switches, crucial for routing generated photons, typically consist of phase shifters in a Mach Zehnder configuration. A phase shifter for this purpose must be high-speed and low-loss [5]. A hybrid Si-BTO platform has been shown to fulfill both requirements [6, 7]. It is shown here that this hybrid platform is also suitable for photon generation.

2 Device Characterization

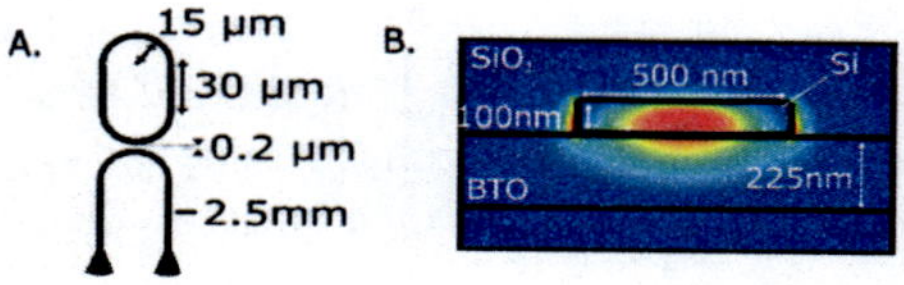

Fig. 1. **A.** Schematic of the device under test (DUT). **B.** Lumerical Mode simulation of device cross section.

J. Witzens et al. (Eds.): ECIO 2024, SPPHY 402, pp. 349–352, 2024.
https://doi.org/10.1007/978-3-031-63378-2_57

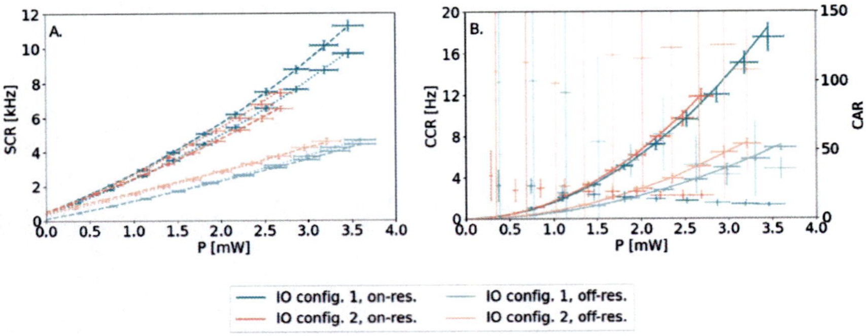

Fig. 2. SCR: single count rate; CCR: coincidence count rate; CAR: coincidence-to-accidental ratio. IO config. 1 and IO config. 2 denote the two possible grating coupler configurations that the device can be measured in. **A.** shows the raw data and fits for the signal (dashed line) and idler (dotted line) SCRs. **B.** shows the raw data and CCR fits along with the CAR values for each data point. The CCR values are represented by bold error bars while the CAR values are represented by light error bars.

We evaluated the photon generation characteristics of the devices using a 50 MHz pulsed laser centered at 1551 nm as the pump. Control over pump power was achieved through a variable optical attenuator and coupling to the Photonic Integrated Circuit (PIC) was facilitated by grating couplers. Employing DWDM filters before and after the chip ensured $\geq$ 80 dB rejection band suppression on each channel. The signal and idler photons, generated through Spontaneous Four-Wave Mixing (SFWM), were routed to their respective DWDM channels (each with a 0.5 nm bandwidth and 4.8 nm separation from the pump). Detection involved two superconducting nanowire single photon detectors and a time interval analyzer. Figure 1 shows a schematic of the device under test (DUT). The DUT was pumped off-resonance to eliminate the resonant contribution of the cavity and measure SFWM brightness from the waveguide. Figure 2 depicts single and coincidence generation rates, showcasing their dependence on input laser power for both on- and off-resonance scenarios. The DUT was measured in two configurations due to imbalanced grating coupler losses. Guided by the approach of Sinclair et al. [8], the data from both configurations was fitted to find the effective non-linearity from SFWM, γ_{eff}.

3 Device Performance

All the referenced results compared in Table 1 are derived from an SOI platform, utilizing a pulsed laser for excitation. Our determined value for γ_{eff} aligns favorably with comparable devices reported in literature where the same method of excitation was used.

Table 1. Comparison of SFWM brightness in integrated photon pair sources. WG: Waveguide; MRR: Micro-ring resonator.

Ref.	Structure	CAR	γ_{eff} [MHz/mW^2]
9	WG	14	1.1
This work	WG	108	2.0 ± 0.3
10	WG	25	2.5
11	MRR	>100	6.1
12	MRR	3000 ± 500	14.6
This work	MRR	32 ± 19	14.7 ± 1.3
13	MRR	-	19.9
14	MRR	10	30.6

4 Summary

Through measurements on hybrid waveguide and ring resonator structures, we've validated the potential to generate photon pairs on a Si-BTO on insulator platform. These devices exhibit SFWM brightness performance comparable to or surpassing similar SOI devices, while also offering the advantage of high-speed, low-power electro-optic modulation capabilities.

Acknowledgements. D.M. expresses gratitude to Ben Burridge for insightful discussions related to this study and acknowledges the support of the QECDT through the EPSRC training grant, EP/LO15730/1. J.B. extends thanks to Graham D. Marshall and Mark G. Thompson for valuable discussions. The authors collectively appreciate the support of Stefan Abel and Felix Eltes (Lumiphase AG) in facilitating this work.

References

1. Knill, E., Laflamme, R., Milburn, G.J.: A scheme for efficient quantum computation with linear optics. Nature **409**, 46–52 (2001)
2. Kok, P., Munro, W.J., Nemoto, K., Ralph, T.C., Dowling, J.P., Milburn, G.J.: Linear optical quantum computing with photonic qubits. Am. Phys. Soc. **79**, 135–174 (2007)
3. Gisin, N., Thew, R.: Quantum communication. Nat. Photon. **1** (2007)
4. Ma, X., Lo, H.: Quantum key distribution with triggering parametric down-conversion sources. New. J. Phys. **10**, 073018 (2008)
5. Rudolph, T.: Why I am optimistic about the silicon-photonic route to quantum computing. APL Photon. **2**, 030901 (2017)
6. Eltes, F., et al.: A BaTiO3-based electro-optic pockels modulator monolithically integrated on an advanced silicon photonics platform. J. Lightwave Technol. **37**, 1456–1462 (2019)
7. Eltes, F., et al.: Low-loss BaTiO3–Si waveguides for nonlinear integrated photonics. ACS Photon. **3**, 1698–1703 (2016)

8. Sinclair, G.F., Tyler, N.A., Sahin, D., Barreto, J., Thompson, M.G.: Temperature dependence of the Kerr nonlinearity and two-photon absorption in a silicon waveguide at 1.55 mm. Phys. Rev. Appl. **11**, 044084 (2019)
9. Faruque, I., et al.: Estimating the indistinguishability of heralded single photons using second-order correlation. Phys. Rev. Appl. **12**, 054029 (2019)
10. Sharping, J.E., et al.: Generation of correlated photons in nanoscale silicon waveguides. Opt. Express **14**, 12388–12393 (2006)
11. Faruque, I.I., Sinclair, G.F., Bonneau, D., Rarity, J.G., Thompson, M.G.: On-chip quantum interference with heralded photons from two independent micro-ring resonator sources in silicon photonics. Opt. Express **26**, 20379–20395 (2018)
12. Ma, C., Wang, X., Mookherjea, S.: Photon-pair and heralded single photon generation initiated by a fraction of a 10 GBps data stream. Opt. Express **26**, 22904–22915 (2018)
13. Zheng, Q., et al.: Bright 547-dimensional Hilbert-space entangled resource in 28-pair modes biphoton frequency comb from a reconfigurable silicon microring resonator. Chin. Phys. B **31**, 024206 (2021)
14. Silverstone, J.W., et al.: Qubit entanglement between ringresonatorphoton-pair sources on a silicon chip. Nat. Commun. **6** (2015)

Investigating the Spectral Response of a Taiji-CROW Device

Bülent Aslan(✉), Riccardo Franchi, Stefano Biasi, Salamat Ali, and Lorenzo Pavesi

Department of Physics, Nanoscience Laboratory, University of Trento, 38123 Povo, Italy
bulent.aslan@unitn.it

Abstract. We report a study on modeling, designing, and testing an integrated photonic structure based on a sequence of two taiji microresonators coupled with a middle link microresonator. The so called taiji-CROW device exhibits strong unidirectional reflection by design and a non-reciprocal intermodal coupling due to its non-Hermitian nature. Our study provides design constraints to preserve the ideal operation of the structure by quantifying a minimum ratio between the coupling coefficients and the backscattering coefficients. This ratio is valuable to avoid surface roughness problems in designing topological integrated photonic devices based on arrays of microresonators.

Keywords: Optical microresonator · Coupled resonators · non-Hermitian photonics

1 Introduction

Topological photonics, as an emerging field for manipulating light using optical structures, has attracted significant interest in recent years. Integrated optical systems like an array of coupled microresonators (MRs) provide a fertile environment for obtaining topological properties [1, 2]. One of the building blocks used in integrated two-dimensional topological lasers is the taiji-microresonator; an MR with an S-shaped waveguide inside [3, 4]. It allows the selection of a preferred chirality for the mode surface in topological laser systems and can exhibit robustness against backscattering [5]. Nevertheless, the implementation of topological features in integrated photonic devices is challenging because of the imperfections introduced by the fabrication processes. That considerably alters the spectral response of such devices.

In this work, we investigate the response of a photonic structure consisting of coupled resonator optical waveguides (CROW) that are intended to form a complex matrix for exploring the topological properties. The observed characteristics were analyzed in terms of the physical properties using the Temporal Coupled Mode Theory (TCMT) to describe the light propagation behavior in the taiji-CROW.

J. Witzens et al. (Eds.): ECIO 2024, SPPHY 402, pp. 353–359, 2024.
https://doi.org/10.1007/978-3-031-63378-2_58

2 Design and Theoretical Model

The studied CROW structure consists of two side taiji MRs coupled by a link MR in the middle as shown in Fig. 1 (left). The structure is sandwiched between two side bus waveguides where the incoming and outgoing light is coupled through the grating couplers. The embedded S-shaped waveguide in taiji MRs couples clockwise (CW) mode to counterclockwise (CCW) mode but not vice versa. This non-reciprocal intermodal coupling between the two counter-propagating modes makes taiji also a non-Hermitian structure [4, 6]. The taiji-CROW device can generate different transmission spectra at each output port without violating the Lorentz reciprocity theorem, regardless of which of the four input ports is excited. On the other hand, the reflection spectra exhibit strong unidirectionality depending on the specific input port. Notably, the matrix depicted in Fig. 1 (right) provides immediate insight into the exit ports based on the input port the light enters.

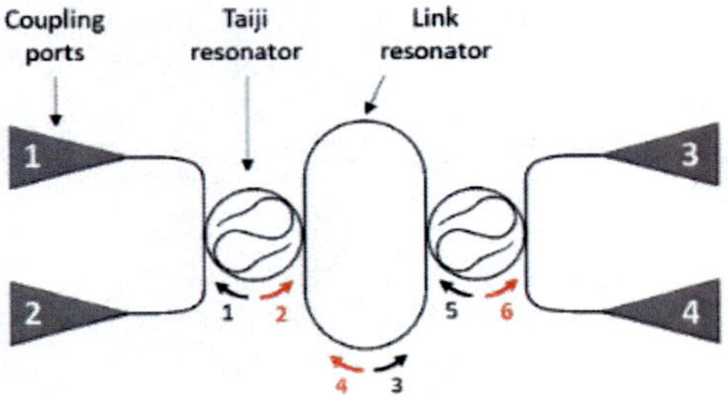

Coupling ports		output			
		1	2	3	4
input	1	X	√	√	X
	2	√	√	√	√
	3	√	√	√	√
	4	X	√	√	X

Fig. 1. (Left) The taiji-CROW device. Arrows with numbers indicate the propagating fields. Enumerated triangles represent grating couplers, i.e. input/output device ports. (Right) Device's input/output matrix.

To describe the light propagation behavior in the linear regime (i.e., low input intensities), we model the structure with TCMT. The propagating modes (α_j) in each MR are indicated in Fig. 1 by enumerated arrows. The dynamics of the field amplitudes can be written as [6]:

$$i\frac{d\boldsymbol{\alpha}}{dt} = \mathbf{H}\boldsymbol{\alpha} - \mathbf{E} \tag{1}$$

where $\mathbf{E}$ is a vector associated with the input electric fields. The Hamiltonian $\mathbf{H}$ of this system can be written as

$$\begin{bmatrix} \omega_1 - i\gamma_{t,1} & 0 & -i\beta_{31} & 0 & 0 & 0 \\ -i\beta_{12} & \omega_1 - i\gamma_{t,1} & 0 & -i\beta_{42} & 0 & 0 \\ -i\beta_{13} & 0 & \omega_3 - i\gamma_{t,3} & 0 & -i\beta_{53} & 0 \\ 0 & -i\beta_{24} & 0 & \omega_3 - i\gamma_{t,3} & 0 & -i\beta_{64} \\ 0 & 0 & -i\beta_{35} & 0 & \omega_5 - i\gamma_{t,5} & 0 \\ 0 & 0 & 0 & -i\beta_{46} & -i\beta_{56} & \omega_5 - i\gamma_{t,5} \end{bmatrix} \tag{2}$$

where ω_i are the resonance frequencies, $\gamma_{t,i}$ are the total loss rates (include intrinsic and extrinsic damping rates) and β_{ij} is the intermodal coupling coefficient. Here, the single digit subscripts 1, 3 and 5 are used to represent MRs (from left to right). Whereas two-digit subscripts refer to the coupling modes; for example, β_{12} is the energy exchange from α_1 to α_2 through the S-shape waveguide. Further details and a complete analysis of the model can be found elsewhere [7].

Figure 2 shows the computed output responses as a function of frequency detuning ($\Delta\omega = \omega - \omega_1$) for a single excitation when $\omega_1 = \omega_3 = \omega_5$. At the eigenfrequencies, spectra show three specific peaks: the separation depends on the coupling coefficients between MRs, while the coupling strength between bus waveguides and the taiji MRs and the intrinsic damping rates of the MRs determine the width of each peak. The model assumed that all inter-resonator mode couplings were Hermitian and identical for both directions of propagation at any interface. The non-Hermitian nature of intra-resonator mode couplings produces different transmission spectra at each output port for a single input and different reflection spectra at each port. Note that the same transmission spectra were obtained for each pair of reciprocal ports (e.g., 13: 1 → 3 and 31: 3 → 1).

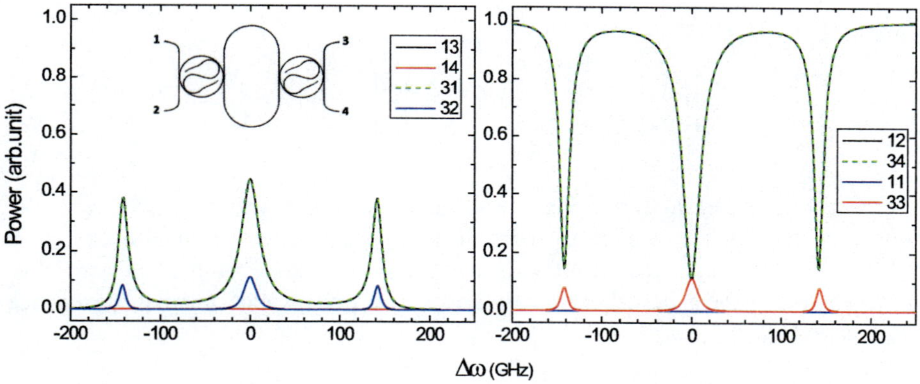

Fig. 2. Computed transmission and reflection responses of the taiji-CROW. Numbers in the legend correspond to the input → output ports reported in the inset. All MRs are assumed to have the same resonant frequency.

The spectral shape of the taiji-CROW is also affected by the individual resonance frequencies of the MRs: Fig. 3 illustrates how the spectra change when the isolated resonance frequency of the link MR differs from that of the site taiji MRs ($\omega_1 = \omega_5 \neq \omega_3$): 250 GHz and 500 GHz are chosen as this detuning frequency of the link MR. As the detuning increases, the spectral splitting of the taiji MRs decreases while the spectral response associated with the link MR moves away towards higher frequencies.

Imperfections on the surface of the microresonators scatter part of the light into the counter-propagating mode and reduce the performance of the applications [8, 9]. To account for the effect, we incorporate the following backscattering terms for each MR

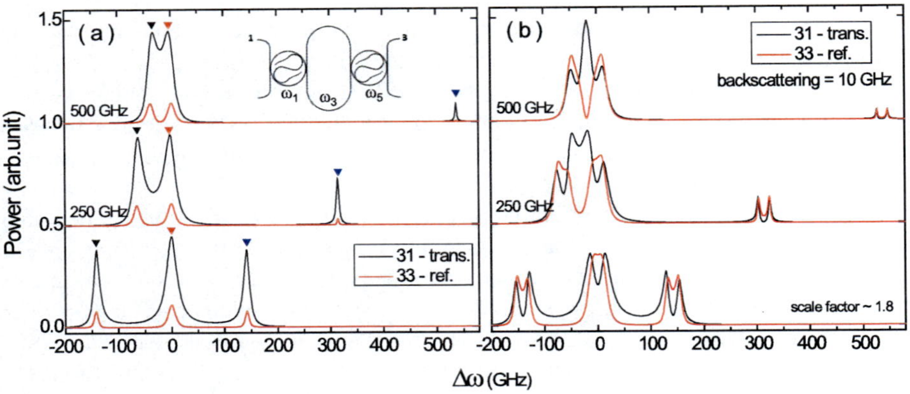

Fig. 3. Transmission (black lines) and reflection (red lines) spectra as a function of the frequency detuning with respect to the taiji MR resonant frequencies ($\Delta\omega = \omega_3 - \omega_1$) when the light is coupled into the device from port 3. Same spectra are plotted on the right when the backscattering is incorporated.

into the Hamiltonian:

$$\begin{bmatrix} 0 & -i\beta_{BS,21} \\ -i\beta_{BS,12} & 0 \end{bmatrix}\begin{bmatrix} \alpha_1 \\ \alpha_2 \end{bmatrix}, \begin{bmatrix} 0 & -i\beta_{BS,43} \\ -i\beta_{BS,34} & 0 \end{bmatrix}\begin{bmatrix} \alpha_3 \\ \alpha_4 \end{bmatrix}, \begin{bmatrix} 0 & -i\beta_{BS,65} \\ -i\beta_{BS,56} & 0 \end{bmatrix}\begin{bmatrix} \alpha_5 \\ \alpha_6 \end{bmatrix} \tag{3}$$

where $\beta_{BS,ij}$ are the backscattering coefficients that induce coupling from α_i to α_j. Figure 3b shows the transmission and reflection spectra when an equal backscattering coefficient is considered for all MRs. The interference between the CW and CCW modes excited by the backscattering causes a spectrally resolvable split in the resonance frequencies [6].

3 Experimental Results

Taiji-CROW devices were designed and fabricated with silicon waveguides. The perimeter of the link MR is selected as twice the perimeter of the taiji MRs. More details on the design parameters and the experimental setup can be found elsewhere [7].

Figure 4 shows the transmission and reflection spectra for single side excitations from port 1 and port 3 for three different devices having different resonance conditions. For each device (Fig. 4a, 4b, and 4c), the transmission spectra are identical while the reflection spectra differ. In an ideal taiji-CROW device, no reflection from port 1 is expected when the input signal is injected in port 1 because there is no intra-resonator mode coupling in action. However, the observed reflections indicate significant backscattering from the waveguide surface-wall roughness [10, 11]. This causes the splitting of the individual peaks as well, as backscattering creates coupling between counter-propagating modes that splits the eigenfrequencies. We fitted the model to the experimental spectra and extracted the physical parameters that control the properties of the structure. The good

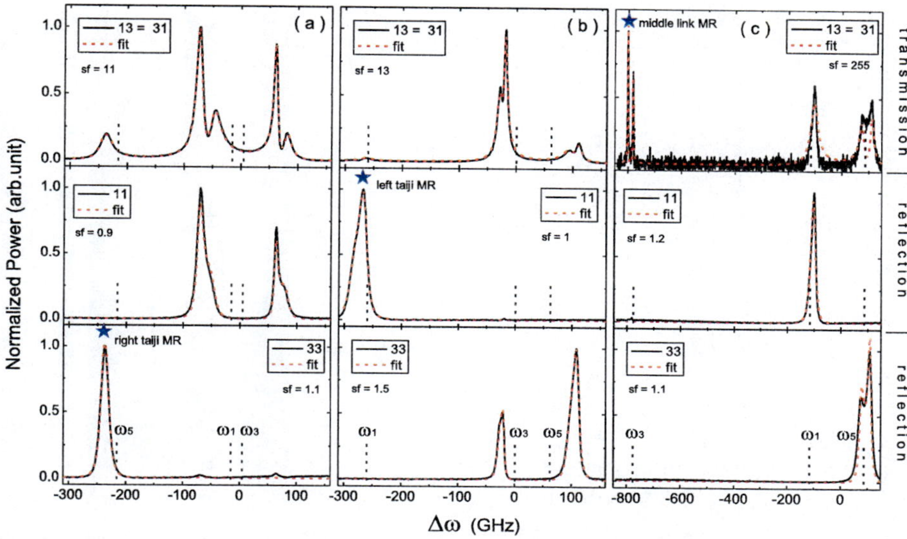

Fig. 4. Measured spectra (black line) for single side excitation and the corresponding fit (red dashed line) for three samples showing different resonance conditions: (a) ω_5, (b) ω_1 and (c) ω_3 is distinctly shifted from the others. Vertical dashed lines indicate the extracted resonant frequencies of each MR.

agreement with the experimental data and the consistency of the fitting validates our modelling.

Each device presented in Fig. 4 has one of the resonance frequencies slightly different from the others: the associated peak is separated from the others and corresponds to ω_5, ω_1 and ω_3 in the cases of Fig. 4a, Fig. 4b and Fig. 4c, respectively. The isolated MR resonance frequencies (vertical dashed lines) are different from the observed spectral peak positions since they correspond to the eigenvalues of **H** due to the mode couplings in the taiji-CROW device. Figure 4c shows the spectra for a device where the link MR is off-resonance with the taiji MRs: indeed, the weak exchange of energy reduces the signal-to-noise ratio of the transmission signals compared to other devices. The scale factor (sf) on each graph was written to indicate the amplitude relations between different panels of Fig. 4. Additional details on the estimation of the parameters as well as the fit parameter values can be found elsewhere [7].

Figure 5 shows the wavelength dispersion of the backscattering coefficients (Fig. 5a) and the $1 \rightarrow 3$ transmission spectra (Fig. 5b) for the device of Fig. 4a. Although it is present throughout the spectral range, backscattering is only strong enough to split resonance at certain mode orders. This is due to its random nature affecting the wavelength region where peak splitting is observed. In these cases, $\beta_{BS,1}$ is distinctly different from the other coefficients. Overall, the backscattering coefficients vary in the 2–20 GHz range for all the investigated devices as reported in [12].

To quantify the impact of the backscattering with respect to the experimentally controllable parameters, we calculated the transmitted fields for different inter-resonator coupling rates while keeping the backscattering coefficient fixed. In the analysis, we

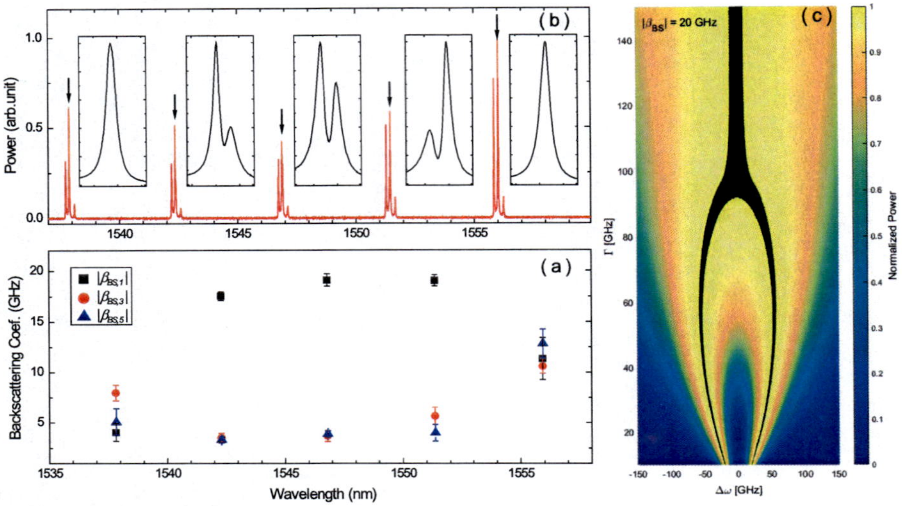

Fig. 5. Wavelength dispersion of the backscattering coefficients (a) and 1 → 3 transmission spectra (b) for the same device as Fig. 4a. Normalized intensity map the transmission as a function of the coupling rate for a given backscattering value (c).

considered (i) the link MR is off-resonance with respect to the taiji MRs, (ii) the same coupling rate for each point coupling (with $\Gamma = |\beta_{13}|^2/(2FSR)$ where FSR is the free spectral range in frequency), (iii) the same backscattering coefficient for each MR, and (iv) $\beta_{12} = 2\Gamma$. Figure 5c shows the intensity map of the transmission as a function of coupling rate for $\beta_{BS} = 20$ GHz. To make the peak positions of the split resonances visually traceable, each spectrum was normalized to 1 (black color in Fig. 5c) with respect to its maximum. The noticeable effect of backscattering on the spectral response (i.e., resonance splitting) becomes negligible at larger coupling rates. In the case of Fig. 5c where $\beta_{BS} = 20$ GHz, the threshold value was found to be $\Gamma \gtrsim 115$ GHz. This leads to a ratio $|\beta_{BS}|/\Gamma \lesssim 0.17$ ($\Gamma \gtrsim 6|\beta_{BS}|$) and indicates a regime where the backscattering is less powerful. In our structures, the ratio is 1.5, indicating the backscattering is predominant. Note that this threshold value depends on the backscattering coefficient, but the ratio remains constant. The corresponding high value of the coupling coefficient between MRs limits the quality factor of the device and restricts the parameters of topologically integrated photonic devices. Moreover, it leads to an increase in the width of the transmission resonances and to a decrease in the field enhancement of the microresonator.

4 Conclusion

In summary, we have demonstrated the modeling and measurement of the transmission and reflection properties of an integrated photonic structure consisting of two taiji MRs coupled with a middle link MR. The so-called taiji-CROW structure exhibits strong unidirectional reflection by design and non-reciprocal intermodal coupling due to its non-Hermitian nature. The physical parameters in the model were extracted by fitting to the experimental data, which allowed us to analyze the effect of these parameters on the spectral shape and the ultimate performance of the taiji-CROW structure. Backscattering terms are also incorporated into the model to account for the influence of waveguide surface-wall roughness which restrains the intended functionality. Our study provides a minimum value for coupling coefficient for a given backscattering coefficient, which is necessary to avoid surface roughness problems. The ratio was found to be $|\beta_{BS}|/\Gamma \lesssim 0.17$ at the onset of mode splitting which we use it as an indication of a strong backscattering.

Acknowledgements. We acknowledge funding from PAT through the Q@TN joint lab, from MIUR (PRIN PELM (20177 PSCKT)), from ERC Horizon 2020 (No 788793, BACKUP) and by European Union FSE-REACT-EU, PON Research and Innovation 2014-2020 DM1062/2021.

References

1. Parto, M.: Non-Hermitian and topological photonics: optics at an exceptional point. Nanophotonics **10**(1), 425–434 (2021)
2. Hafezi, M.: Imaging topological edge states in silicon photonics. Nat. Photon. **7**(12), 1001–1005 (2013)
3. Bandres, M.: Topological insulator laser: experiments. Science **359**(6381), eaar4005 (2018)
4. Calabrese, A.: Unidirectional reflection from an integrated “taiji” microresonator. Photon. Res. **8**(8), 1333–1341 (2020)
5. Ozawa, T.: Topological photonics. Rev. Mod. Phys. **91**(1), 015006 (2019)
6. Biasi, S.: Interferometric method to estimate the eigenvalues of a non-Hermitian two-level optical system. Photon. Res. **10**(4), 1134–1145 (2022)
7. Aslan, B.: On the spectral response of a taiji-CROW device. Opt. Express (submitted)
8. Heebner, J., Grover, R., Ibrahim, T.: Optical Microresonators: Theory, Fabrication, and Applications. Springer, London (2008). https://doi.org/10.1007/978-0-387-73068-4
9. Li, A.: Backscattering in silicon microring resonators: a quantitative analysis. Laser Photon. Rev. **10**(3), 420–431 (2016)
10. Biasi, S.: Hermitian and non-hermitian mode coupling in a microdisk resonator due to stochastic surface roughness scattering. IEEE Photon. J. **11**(2), 1–14 (2018)
11. McCutcheon, W.: Backscattering in nonlinear microring resonators via a Gaussian treatment of coupled cavity modes. APL Photon. **6**(6), 066103 (2021)
12. Franchi, R., Biasi, S.: On the response of the Taiji microresonator against small perturbation of the counter propagating mode. In: García-Blanco, S.M., Cheben, P. (eds.) SPIE OPTO 2022, vol. 12004, p. 13. SPIE, San Francisco (2022)

Multimodal Semi-analytical Model for Bound States in the Continuum and Unidirectional Guided Resonances in a Photonic Crystal

Thomas Delplace[1](✉), Tom van Loon[2], Minpeng Liang[2], Jaime Gómez Rivas[2], and Bjorn Maes[1](✉)

[1] Micro- and Nanophotonic Materials Group, Research Institute for Materials Science and Engineering, University of Mons, Place du Parc, 20, 7000 Mons, Belgium
{Thomas.Delplace,Bjorn.Maes}@umons.ac.be

[2] Department of Applied Physics and Science Education, Institute for Complex Molecular Systems, and Eindhoven Hendrik Casimir Institute, PO Box 513, Eindhoven 5600 MB, The Netherlands

Abstract. Recent attention has been directed towards optical bound states in the continuum (BICs) within photonic crystal slabs. Unidirectional guided resonances (UGRs) have also garnered interest, an associated phenomenon that involves intentionally broken symmetry and resulting in directional leakage. This study introduces a microscopic semi-analytical model to better understand these resonances. Building on a multimodal interference method used for BIC exploration, our approach extends to investigating UGRs, providing valuable insights into their distinct properties. Using this model, we aim to contribute to the design and comprehension of BICs and UGRs in photonic crystal slabs.

Keywords: BIC · photonic crystal · symmetry breaking

1 Introduction

In the last decade, the investigation of bound states in the continuum (BICs) has garnered significant attention because of their distinctive characteristics. Unlike typical confined modes, BICs exist alongside the radiation continuum, decoupling from it, and showcasing potential applications in areas such as spintronics, nanocavities, low-threshold lasing, sensing, and communication [1]. A closely related resonance, termed unidirectional guided resonance (UGR) [2], has recently been investigated. UGRs manifest in similar structures as BICs but exhibit broken symmetry, allowing radiation to leak exclusively in only one direction.

Various mechanisms explain the emergence of BICs, with the multimodal interference model being one such approach [3]. This model entails the interference of multiple fundamental modes, leading to destructive interference outside the structure and the creation of a BIC under specific conditions. This study demonstrates the applicability of the multimodal interference model to UGRs, offering a semi-analytical description of these

J. Witzens et al. (Eds.): ECIO 2024, SPPHY 402, pp. 360–364, 2024.
https://doi.org/10.1007/978-3-031-63378-2_59

resonances. This adaptability enables efficient exploration and interpretation of diverse geometries, providing valuable insights into UGR behavior and potential applications in photonic devices.

2 Method and Structure

The method employs multiple steps. First, we divide the crystal unit cell (Fig. 1a) into two halves, a top and bottom half (Fig. 1b), and create an infinite waveguide with the same dimension as the cell where it was cut (Fig. 1c). Subsequently, we calculate the dispersion of the infinite waveguide, providing insight into the coexisting modes in the vertical direction of the structure (Fig. 1d). Using a boundary mode analysis within the COMSOL simulation software, these modes are introduced into the two halves of the structure. This process enables the computation of two half-trip matrices, namely S_u for the upper part and S_d for the lower part, illustrating how the modes are reflected and mixed by the interfaces of the structure.

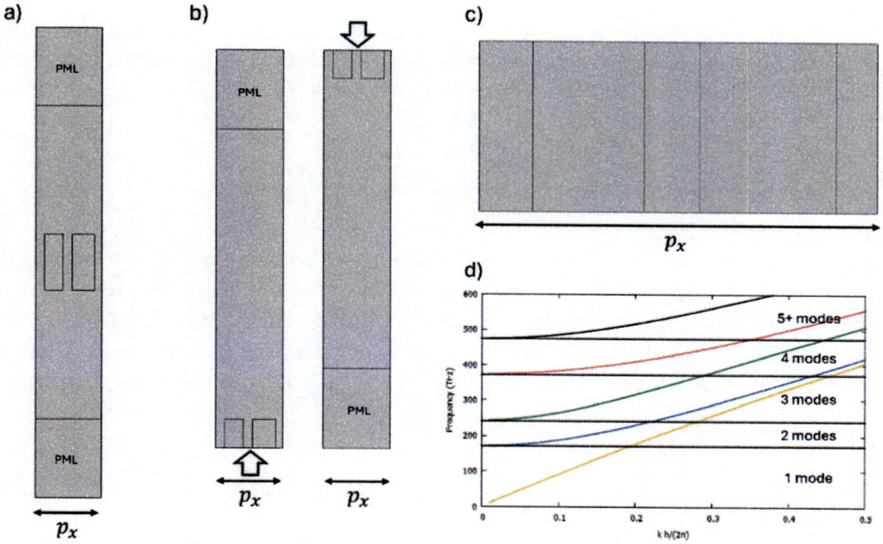

Fig. 1. a) The full unit cell of a photonic crystal, with period p_x along the horizontal direction (x). b) The two halves of the cell used to compute the half-trip matrices. c) A section of the waveguide that is infinite along the vertical direction (y). $p_x = 408$ nm, the width of the largest particle is 140 nm. The width of the smallest particle is varied to possibly break the symmetry between the two particles. The height is the same for the two particles and can be varied analytically via the propagation matrices (in this figure $L = 360$ nm). d) The TE (transverse electric) dispersion for the structure of c). Depending on the frequency, a different number of guided modes are available.

By combining the two half-trip matrices, we construct the round-trip matrix $S_d \times S_u$ which we then analyze by computing its eigenvalues and eigenvectors, similar as in Ref. [4]. For a system with 3 modes, one thus obtains:

$$S_d \times S_u \begin{bmatrix} C1 \\ C2 \\ C3 \end{bmatrix} = \lambda \begin{bmatrix} C1 \\ C2 \\ C3 \end{bmatrix}$$

This gives valuable information on the interaction of the modes in the structure. To have a resonance, the imaginary part of the eigenvalue must be zero and the real part must be positive (and close to 1 for a large quality factor). The associated eigenvector shows the contributions of each mode that construct this specific resonance. Note that the dimension of S_u, S_d and the number of eigenvalues is equal to the number of modes to model the structure, which is typically limited to the guided modes only. The number of guided modes is indicated in Fig. 1d.

3 Q-factor and Results

For true BICs, the eigenvalue alone suffices as we just need $\lambda_{real} = 1$ and $\lambda_{imag} = 0$. For UGRs, we have losses on one side of the cell, meaning that the real part of the round-trip eigenvalue is not sufficient to characterize the resonances. To solve this, we separate the Q-factors of the top and bottom parts of the cell:

$$Q_u = \frac{2\omega_0 L}{|v_g| T_u} \quad \text{and} \quad Q_d = \frac{2\omega_0 L}{|v_g| T_d}$$

where ω_0 is the angular frequency, L is the height of the particles, v_g the (averaged) group velocity, and T_u, T_d the transmission losses of upper and lower half, respectively. The transmissions are computed via the half-trip matrices and the specific eigenvectors. Figure 2 shows the results for the array of Fig. 1. We can find the precise position of BICs and UGRs in any parameter space (here, the angular frequency vs. the height

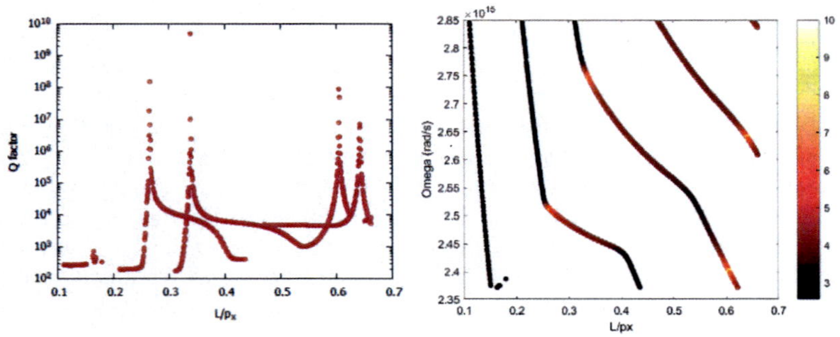

Fig. 2. (Left) The quality factor for the top and bottom sides of the structure (equal for the symmetric quasi-BIC). (Right) Placement of the resonances in the parameter space of frequency versus height (color indicates the Q-factor).

of the particles), and via the eigenvectors we obtain information e.g. about the near-field. Figure 3 shows results of the multimodal method for a structure with top-down asymmetry. We observe a good agreement between the multimodal analysis and an eigenmode solver, proving the viability of our method for high contrast UGRs.

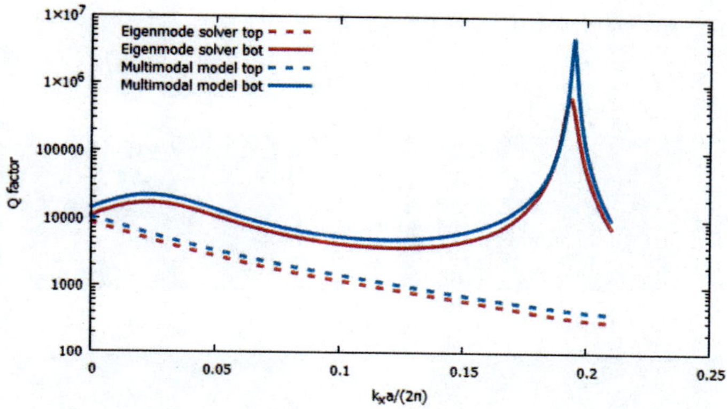

Fig. 3. Comparison of Q factors between the multimodal method and a rigorous eigenmode solver for a UGR structure.

4 Extension to 3D

The method is also useful in a 3D context. We are currently analyzing a structure used to demonstrate quasi-BICs for sensing via symmetry breaking effects [5]. We study a 2D array of cylindrical silicon particles (Fig. 4). The goal of this study is to understand how the symmetry breaking in size and position allows tuning the position of BICs. Connected with sensing experiments, we also examine the impact of different superstrate refractive indices on the resonances.

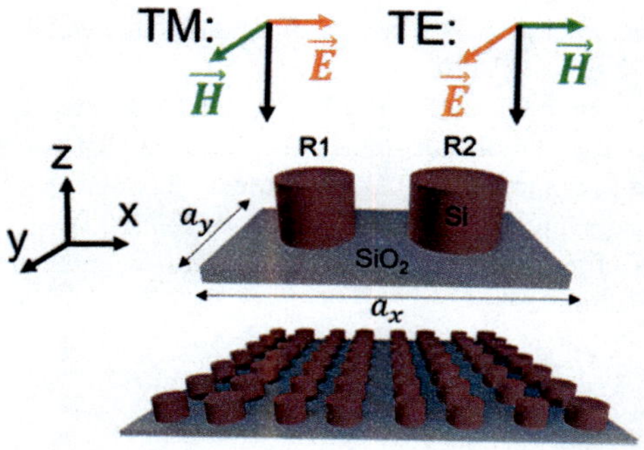

Fig. 4. Illustration of the array. $a_x = 408$ nm and $a_y = 204$ nm. The height of the particles is *90* nm. The radii are varied to tune the resonances.

We have initial results for the multimodal approach in this 3D structure. The electric field profiles obtained via the multimodal method (Fig. 5b) fit well with the results of a scattering simulation (Fig. 5a), showing that a decomposition in multiple fundamental modes can explain the resulting field in the array. The first quasi-BIC (on the left of Fig. 5) is mainly monomodal, while the second quasi-BIC (right of Fig. 5) is a composition of around 40% and 60% of 2 interacting modes. This shows that our model works for different types of BICs.

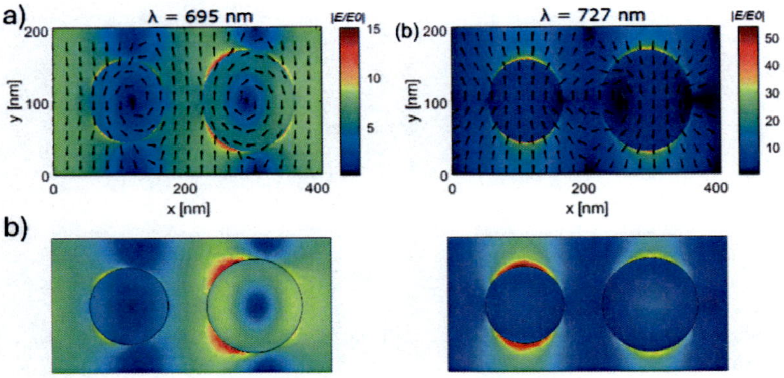

Fig. 5. Profile of the normalized norm of the electric field. a) Profile obtained by means of a scattering simulation on COMSOL. b) Profile obtained via the multimodal computation mixing waveguide simulations and analytical calculations.

References

1. Shereena, J., Saurabh, P., Swagato, S., Joby, J.: Bound states in the continuum in resonant nanostructures: an overview of engineered materials for tailored applications. Nanophotonics **10**(17), 4175–4207 (2021)
2. Yin, X., Jin, J., Soljačić, M., et al.: Observation of topologically enabled unidirectional guided resonances. Nature **580**, 467–471 (2020)
3. Ovcharenko, A.I., Blanchard, C., Hugonin, J.P., Sauvan, C.: Bound states in the continuum in symmetric and asymmetric photonic crystal slabs. Phys. Rev. B **101**, 155303 (2020)
4. Maes, B., Ibanescu, M., Joannopoulos, J.D., Bienstman, P., Baets, R.: Microcavities based on multimodal interference. Opt. Express **15**(10), 6268–6278 (2007)
5. Van Loon, T., et al.: Refractive index sensing using quasi-bound states in the continuum in silicon metasurfaces (submitted)

An Efficient Singlet-Triplet Spin Qubit to Fiber Interface Assisted by a Photonic Crystal Cavity

Kui Wu[1](✉), Sebastian Kindel[2], Thomas Descamps[2], Tobias Hangleiter[2], Jan Christoph Müller[2], Rebecca Rodrigo[1], Florian Merget[1], Hendrik Bluhm[2], and Jeremy Witzens[1]

[1] Institute of Integrated Photonics, RWTH Aachen University, 52074 Aachen, Germany
kwu@iph.rwth-aachen.de

[2] Quantum Technology Group, 2nd Institute of Physics, RWTH Aachen University, 52074 Aachen, Germany

Abstract. We introduce a novel optical interface between a singlet-triplet spin qubit and a photonic qubit which would offer new prospects for future quantum communication applications. The interface is based on a 220 nm thick GaAs/AlGaAs heterostructure membrane and features a gate-defined singlet-triplet qubit, a gate-defined optically active quantum dot, a photonic crystal cavity and a bottom gold reflector. All essential components can be lithographically defined and deterministically fabricated, which greatly increases the scalability of on-chip integration. According to our FDTD simulations, the interface provides an overall coupling efficiency of 28.7% into a free space Gaussian beam, assuming an SiO_2 interlayer filling the space between the reflector and the membrane. The performance can be further increased to 48.5% by undercutting this SiO_2 interlayer below the photonic crystal.

Keywords: quantum technology · gate-defined quantum dot · optical interface · singlet-triplet spin qubit · photonic crystal cavity · Gaussian beam

1 Introduction

Interconnecting distant stationary qubits with photonic qubits is essential for advancing the quantum Internet [1], distributing quantum entanglement across various quantum technology platforms for applications like secure quantum communication [2, 3] and distributed quantum computing [4]. Gate-defined quantum dots (GDQDs), which trap single electrons in tunable electrostatic potential minima within a two-dimensional electron gas (2DEG), emerge as promising candidates due to their compatibility with top-down fabrication techniques and the potential for integrating multiple locally interconnected qubits. Singlet-triplet qubits defined by GDQD based on gallium arsenide (GaAs) are particularly promising as they feature all necessary prerequisites for quantum information processing, including initialization, readout, all electrical control, and high-speed qubit-state manipulation [5–10]. The direct bandgap of GaAs further facilitates efficient photonic qubit to spin qubit conversion by direct optical absorption and emission.

J. Witzens et al. (Eds.): ECIO 2024, SPPHY 402, pp. 365–372, 2024.
https://doi.org/10.1007/978-3-031-63378-2_60

However, achieving coherent and efficient interfacing between photonic and spin qubits in GDQD systems poses challenges, notably due to the lack of hole confinement in conventional heterostructures. A solution involves integrating an optically active quantum dot (OAQD), such as a fully gate-defined electrostatic exciton trap [11], as an intermediary for coherently transferring photogenerated electrons to GDQD spin qubits [12]. This approach, complemented by advancements in optical interface technologies such as photonic crystal cavities (PCCs), enables enhanced light-matter interaction [13–17] and efficient photon emission [18–20].

In this paper and in line with our previous publications [21–23], we introduce a novel design of a singlet-triplet spin qubit to fiber interface incorporating a gate-defined OAQD at the center of a PCC. All the structures are fully lithographically defined on a 220 nm GaAs/AlGaAs heterostructure membrane. According to our theoretical modeling and numerical evaluation, this design ensures that more than 50% of the photons emitted by the OAQD are coupled into a narrow free space beam in the perpendicular direction, which in turn has an optical overlap greater than 50% with an ideal Gaussian beam, facilitating coupling to an optical fiber. This integration approach is not only compatible with electrode integration into the photonic cavity, but also promises to significantly advance scalability by means of top-down fabrication. It thus provides a practical path towards the efficient and scalable integration of GDQD and photonic cavities for quantum information applications.

2 Modeling of the Optical Interface

Figure 1(a) presents a top view of our optical interface, which consists of a PCC defined by a triangle lattice of holes (black circles), cavity openings with mini-stopband (light and dark blue circles) providing electrical connectivity to the surrounding 2DEG, an Au/Ti metal gate system defining the GDQD, the OAQD and a single electron transistor (SET) used to read out the state of the GDQD. The detailed gate structure is depicted in Fig. 1(b). The entire structure, including the PCC and the gate system, is lithographically defined on a 220 nm thick GaAs/AlGaAs membrane (Fig. 1(c)). The OAQD emission wavelength is 823 nm at cryogenic temperatures and can be tuned from 820 nm to 825 nm by adjusting the electrical voltage applied to the trap and the guard gate (quantum-confined Stark effect [24]). A 200 nm thick gold reflector (not shown) is attached to the bottom side of the membrane via an SiO_2 interlayer to recycle photons back to the desired positive z-direction.

The PCC is created by modifying a hexagonal H4 cavity [21], with a lattice constant $a_1 = 290$ nm and a hole radius $r_1 = 109$ nm (lattice of black circles). When such a lattice is etched in the 220 nm GaAs/AlGaAs heterostructure membrane (Fig. 1(c)), a transverse electric (TE) photonic bandgap is opened from 717 nm to 1021 nm.

The four cavity openings in the diagonal directions are created by removing a row of etched holes. Those cavity openings are required to ensure a continuous 2DEG in the GaAs quantum well layer located at the center of the stack, which is an essential prerequisite for the functionality of the SET and of the GDQD. The underlying reason is that the surface charges trapped on the side walls of the etched holes deplete the free carriers in their vicinity, which prohibits electron transport through the photonic crystal lattice except in the cavity openings.

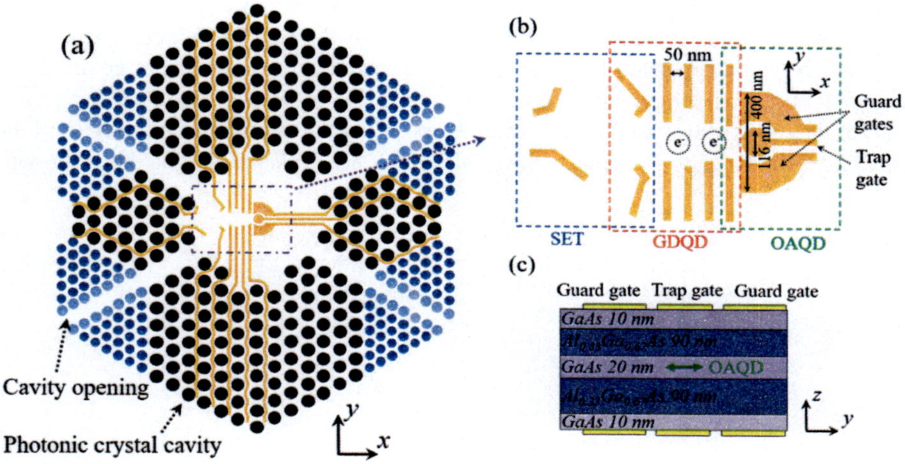

Fig. 1. (a) Top view of the optical interface including the PCC, the electrode system, and the cavity openings. (b) Detailed gate structure of the SET, the GDQD and the OAQD. The Au/Ti wires have a thickness of 9 nm and a width of 30 nm. Gates for the OAQD are fabricated on both sides of the membrane, while the SET and the GDQD only require top side electrodes. (c) Cross sectional view of the OAQD in the *yz*-plane through the center of the trap gate.

On the other hand, to maintain optical confinement inside the PCC, we modified the lattice constant and the hole radii of the PCC near the cavity openings. As shown in Fig. 1(a), the lattice marked in light and dark blue circles has a reduced lattice constant of $a_2 = 214$ nm (down from $a_1 = 290$ nm), and reduced radii of $r_2 = 90$ nm and $r_3 = 73$ nm, respectively (down from $r_1 = 109$ nm). These cavity openings support a TE mini-stopband [25] opening from 793 nm to 845 nm, i.e., no propagating mode is allowed to transmit though the cavity openings within this wavelength range.

3 Simulation Results

To evaluate the theoretical performance of the complete structure, we performed three-dimensional finite-difference time-domain (3D FDTD) simulations. All the structures shown in Fig. 1 are reproduced in the simulations (Lumerical FDTD). The OAQD is modeled by a dipole orientated in the *y*-direction located at the center of the PCC. Modeling the dipole this way is justified by considering that the in-plane polarization of the emitted photon is determined by the external magnetic field and by the heavy-hole nature of the participating hole [12]. The refractive indices of GaAs and $Al_{0.33}Ga_{0.67}As$ utilized in the simulations are approximated to be n_{4K}, GaAs = 3.59 and n_{4K}, AlGaAs = 3.38 by considering their temperature dependence and linearly extrapolating the room temperature values [26]. The absorption coefficients of GaAs and AlGaAs are assumed to be zero in the simulations. The thickness of the SiO_2 interlayer is optimized in such a way that a high extraction efficiency is achieved at the working wavelength (823 nm) of the OAQD.

Figure 2 shows the simulated cavity spectrum from 790 nm to 870 nm excited by a y-dipole located in the center of the PCC. Due to the relatively large size of the PCC, multiple resonant peaks are observed in this wavelength range. The target mode at 823 nm, whose wavelength coincides with the emission wavelength of the OAQD as a result of the optimization of the PCC parameters, is highlighted with a green arrow. For comparison to the target mode, another mode at 833 nm is also highlighted by a brown arrow.

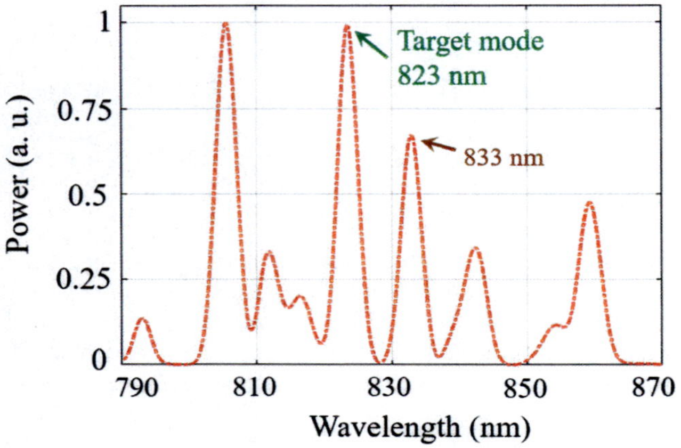

Fig. 2. Cavity power spectrum in the PCC. The target mode, which has a high extraction efficiency, is marked by a green arrow at 823 nm. For comparison, another cavity mode at 833 nm is also highlighted by a brown arrow.

To study the far-field emission properties, we record the electromagnetic field with a 5 μm by 5 μm 2D monitor located 20 nm above the top surface of the membrane and apply a far-field transform to the recorded near-field. The far-field patterns at a distance 1 m away from the membrane for both cavity modes are depicted in Fig. 3. It is clear that the target cavity mode at 823 nm features a pronounced single-beam vertical emission at $x = 0$ and $y = 0$. On the other hand, the far-field emission pattern of the cavity mode at 833 nm shows a complex profile not easily matched to that of an optical fiber.

The difference in the far-field emission pattern can be explained by applying a spatial 2D Fourier transformation to the recorded near-field electromagnetic waves and comparing the dominant k-space components with the reciprocal lattice points of the photonic crystal (black lattice in Fig. 1). For the target mode, we observe that the cavity mode features dominant *k*-space components at the reciprocal lattice points of the photonic crystal (Fig. 4(a)). As a result, these components experience a Bragg-scattering to the Γ-point ($k_x = k_y = 0$) and result in a pronounced vertical emission. On the other hand, the dominant k-space components of the cavity mode at 833 nm do not overlap with the reciprocal lattice points (Fig. 4(b)), which explains the complex far-field emission pattern seen in Fig. 3(b).

To rigorously evaluate the probability that a photon emitted by the OAQD couples into the fundamental mode of a single mode fiber, we first calculate the possibility of

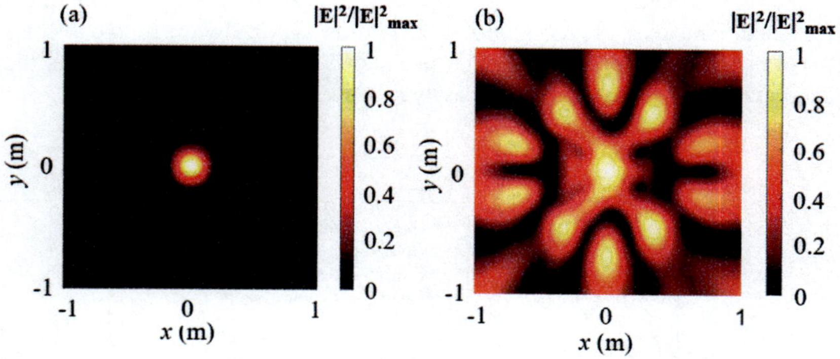

Fig. 3. Far-field patterns of the two highlighted cavity modes at (a) 823 nm and (b) 833 nm, at a distance 1 m away from the membrane. The intensities are independently normalized by the maximum value for both far-field patterns.

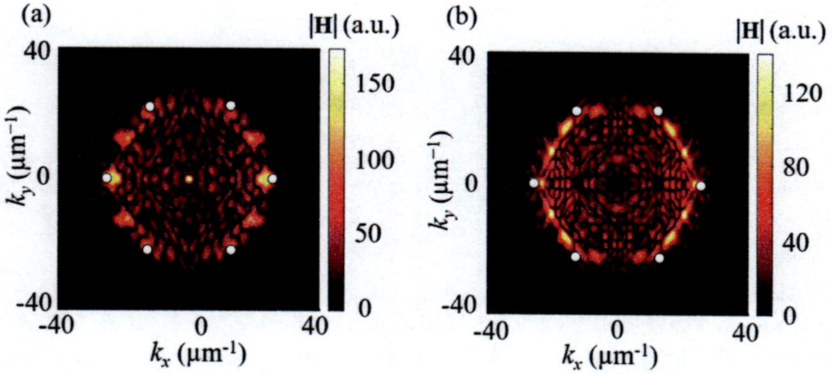

Fig. 4. Fourier space distribution of the cavity mode at (a) 823 nm and (b) 833 nm. The white dots represent the reciprocal lattice points of the photonic crystal lattice with a lattice pitch 290 nm.

the emitted photon escaping the semiconductor membrane and being recorded by the monitor (escape probability). This is done by integrating the normal component of the Poynting vector over the entire monitor surface and dividing the integrated power by the total power emitted by the dipole. Then, we calculate the vectorial optical overlap (OV) between the recorded near-field and an ideal 2D Gaussian beam according to [27]

$$\mathrm{OV}(\theta) = \frac{|\iint (\vec{E}_m \times \vec{H}^*_g(\theta) + \vec{E}^*_g(\theta) \times \vec{H}_m) \bullet d\vec{S}\,|^2}{4 \iint Re(\vec{E}_m \times \vec{H}^*_m) \bullet d\vec{S} \iint Re(\vec{E}_g(\theta) \times \vec{H}^*_g(\theta)) \bullet d\vec{S}} \tag{1}$$

where $\vec{E}_m$ ($\vec{H}_m$) is the electric (magnetic) near-field recorded by the monitor, and $\vec{E}_g(\theta)$ ($\vec{H}_g(\theta)$) represents the electric (magnetic) field of a linearly polarized Gaussian beam with an $1/e^2$ intensity half angle θ.

The multiplication of the escape probability with OV(θ) gives the overall efficiency, which is equal to the overall probability of the photon emitted by the dipole to couple to a Gaussian beam with a half angle θ. Figure 5 shows the overall efficiency and the OV as a function of θ for both highlighted cavity modes.

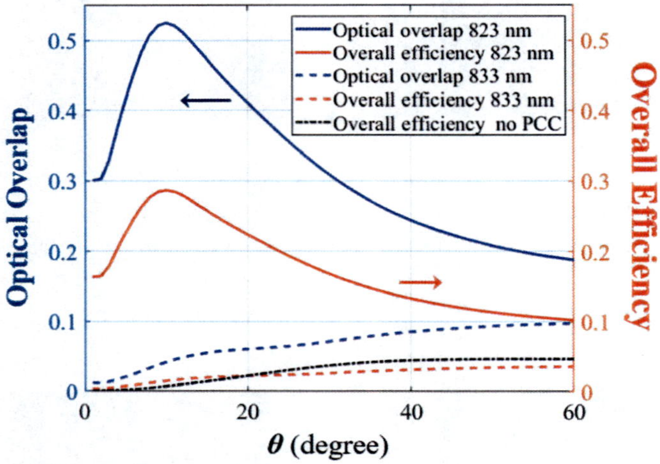

Fig. 5. OV and overall efficiency as a function of θ for both cavity modes. To show the effectiveness of the PCC, we also plot the overall efficiency when the PCC is removed (black dotted curve).

As we can see in Fig. 5, the maximum OV and overall efficiency for the target cavity mode are 0.526 and 28.7%, respectively, at a divergence angle of 10° with an escape probability of 54.6%. In comparison, the OV for the 833 nm cavity mode stays below 0.1, and the overall efficiency for the 833 nm cavity mode and for an electrode system without the PCC stays below 5% for all θ. In addition, we also investigated the performance when the SiO_2 interlayer is removed and the PCC fabricated in a suspended membrane. After re-optimizing the position of the bottom reflector, the maximum OV reaches 0.839 with a maximum overall efficiency of 48.5% at a divergence angle of 10°. While this configuration increases the fabrication complexity, it offers a path forward for a significant improvement of the device performance.

4 Conclusion

In this paper, we present an efficient optical interface, assisted by a photonic crystal cavity, between a singlet-triplet spin qubit and a photonic qubit. The entire structure can be lithographically defined and fabricated. By utilizing the Bragg scattering of the PCC, we obtain vertical Gaussian emission at the desired wavelength. According to our calculations, our design reaches an overall efficiency of 28.7% by emitting into a Gaussian beam with a divergence angle of 10°. The performance can be further increased by removing the SiO_2 interlayer. In this case, the overall efficiency is 48.5% at the same divergence angle. Fabrication of this device is currently in progress and we expect to be able to report first experimental results in the near future.

Acknowledgement. This project is funded by the Deutsche Forschungsgemeinschaft (DFG, German Research Foundation) under Germany's Excellence Strategy – Cluster of Excellence Matter and Light for Quantum Computing (ML4Q) EXC 2004/1 – 390534769.

References

1. Kimble, H.J.: The quantum internet. Nature **453**, 1023 (2008)
2. Gisin, N., Thew, R.: Quantum communication. Nature Photon. **1**, 165–171 (2007)
3. Gisin, N., Ribordy, G., Tittel, W., Zbinden, H.: Quantum cryptography. Rev. Mod. Phys. **74**, 145–195 (2002)
4. Jiang, L., Taylor, J.M., Sørensen, A.S., Lukin, M.D.: Distributed quantum computation based on small quantum registers. Phys. Rev. A **76**, 062323 (2007)
5. Cerfontaine, P., Botzem, T., Ritzmann, J., et al.: Closed-loop control of a GaAs-based singlet-triplet spin qubit with 99.5% gate fidelity and low leakage. Nat. Commun. **11**, 4144 (2020)
6. Cerfontaine, P., Botzem, T., DiVincenzo, D.P., Bluhm, H.: High-fidelity single-qubit gates for two-electron spin qubits in GaAs. Phys. Rev. Lett. **113**, 150501 (2014)
7. Petta, J.R., et al.: Coherent manipulation of coupled electron spins in semiconductor quantum dots. Science **309**, 2180 (2005)
8. Elzerman, J., et al.: Few-electron quantum dot circuit with integrated charge read out. Phys. Rev. B **67**, 161308(R) (2003)
9. Elzerman, J., Hanson, R., Willems van Beveren, L., et al.: Single-shot read-out of an individual electron spin in a quantum dot. Nature **430**, 431–435 (2004)
10. Bluhm, H., Foletti, S., Neder, I., et al.: Dephasing time of GaAs electron-spin qubits coupled to a nuclear bath exceeding 200 μs. Nature Phys. **7**, 109–113 (2011)
11. Descamps, T., et al.: Semiconductor membranes for electrostatic exciton trapping in optically addressable quantum transport devices. Phys. Rev. Applied **19**, 044095 (2023)
12. Joecker, B., Cerfontaine, P., Haupt, F., Schreiber, L.R., Kardynał, B.E., Bluhm, H.: Transfer of a quantum state from a photonic qubit to a gate-defined quantum dot. Phys. Rev. B **99**, 205415 (2019)
13. Lodahl, P., Floris van Driel, A., Nikolaev, I., et al.: Controlling the dynamics of spontaneous emission from quantum dots by photonic crystals. Nature **430**, 654–657 (2004)
14. Englund, D., et al.: Controlling the spontaneous emission rate of single quantum dots in a two-dimensional photonic crystal. Phys. Rev. Lett. **95**, 013904 (2005)
15. Luxmoore, I., Toro, R., Pozo-Zamudio, O., et al.: III–V quantum light source and cavity-QED on Silicon. Sci. Rep. **3**, 1239 (2013)
16. Kim, H., Sridharan, D., Shen, T.C., Solomon, G.S., Waks, E.: Strong coupling between two quantum dots and a photonic crystal cavity using magnetic field tuning. Opt. Express **19**, 2589–2598 (2011)
17. Kuruma, K., et al.: Strong coupling between a single quantum dot and an L4/3 photonic crystal nanocavity. Appl. Phys. Express **13**, 082009 (2020)
18. Daveau, R.S., et al.: Efficient fiber-coupled single-photon source based on quantum dots in a photonic-crystal waveguide. Optica **4**, 178–184 (2017)
19. Gröblacher, S., Hill, J.T., Safavi-Naeini, A.H., Chan, J., Painter, O.: Highly efficient coupling from an optical fiber to a nanoscale silicon optomechanical cavity. Appl. Phys. Lett. **103**(18), 181104 (2013)

20. Uppu, R., et al.: Scalable integrated single-photon source. Sci. Adv. **6**, eabc8268 (2020)
21. Wu, K., et al.: High-efficiency gate-defined quantum dot to single mode fiber interface assisted by a photonic crystal cavity. AIP Adv. **10**, 115016 (2020)
22. Wu, K., et al.: Highly efficient spin qubit to photon interface assisted by a photonic crystal cavity. Proc. SPIE Int. Soc. Opt. Eng. **11995**, 40–48 (2022)
23. Wu, K., et al.: Modeling of an efficient singlet-triplet spin qubit to photon interface assisted by a photonic crystal cavity. arXiv:2310.18690 (2023)
24. Schinner, G.J., et al.: Confinement and interaction of single indirect excitons in a voltage-controlled trap formed inside double InGaAs quantum wells. Phys. Rev. Lett. **110**, 127403 (2013)
25. Olivier, S., Benisty, H., Weisbuch, C., Smith, C.J.M., Krauss, T.F., Houdré, R.: Coupled-mode theory and propagation losses in photonic crystal waveguides. Opt. Express **11**, 1490–1496 (2003)
26. Talghader, J., Smith, J.S.: Thermal dependence of the refractive index of GaAs and AlAs measured using semiconductor multilayer optical cavities. Appl. Phys. Lett. **66**, 335–337 (1995)
27. Romero-García, S., Shen, B., Merget, F., Marzban, B., Witzens, J.: Alignment Tolerant Couplers for Silicon Photonics. IEEE J. Sel. Top. Quant. Electron. **21**(6), 765–778 (2015)

Photonic Integrated Filter in Silicon Nitride Technology for High-Performance Microwave Photonics Applications

Valentina Gemmato[1(✉)], Filippo Scotti[2], Federico Camponeschi[1], Luca Rinaldi[2], Manuel Reza[1,4], Marco Bartocci[3], Paolo Ghelfi[2], and Claudio Porzi[1]

[1] TeCIP Institute, Scuola Superiore Sant'Anna, via Moruzzi 1, 56124 Pisa, Italy
valentina.gemmato@santannapisa.it
[2] PNTLab, CNIT, via Moruzzi 1, 56124 Pisa, Italy
[3] Elettronica Spa, via Tiburtina km 13, 700, 00131 Rome, Italy
[4] IMEC, 3001 Leuven, Belgium

Abstract. We report the characterization of a waveguide optical filter designed for RF scanning operation. The filter is based on a 2nd-order microring resonator-assisted balanced Mach-Zehnder Interferometer (MZI) and realized in silicon nitride (SiN) technology. Through careful optimization of the design layout, filter transmission bandwidth down to 2 GHz with a flatness lower than 0.5 dB, and record values of free-spectral range and out-of-band rejection of 36 GHz and 57 dB, respectively, are observed.

Keywords: Microwave photonics · Integrated microwave photonics filter · Silicon Nitride

1 Introduction

By introducing photonic technology into conventional electronic systems, microwave photonics (MWP) techniques offer an array of advantages for the processing of radio-frequency (RF) signals [1, 2]. These include improved flexibility, high information transmission capabilities, low propagation losses, and heightened resilience against electromagnetic interference. By combining the different functions of a complex MWP processor within a photonic integrated circuit (PICs), significant improvements in the reliability of the operation and in the mechanical robustness can additionally be achieved, while minimizing, at the same time, the size, weight, and power consumption (SWaP) of the system. It has long since been pointed out, however, the lack of a unique photonic integration technology platform fulfilling altogether the necessary functionalities with the demanding target performance required for several modern RF systems applications. To this scope, hybrid PIC technologies [3] can provide a viable solution to the development of ultra-compact and high-performance photonics-based RF signal processing systems. Recently, we proposed a hybrid architecture aiming at combining the advantages of silicon nitride (SiN) and indium phosphide (InP) material platforms for realizing an

J. Witzens et al. (Eds.): ECIO 2024, SPPHY 402, pp. 373–377, 2024.
https://doi.org/10.1007/978-3-031-63378-2_61

advanced RF scanner suitable for modern radar applications [4]. While the InP circuit provides the active (e.g., light generation and amplification, fast electro-optic modulation and switching) functionalities, the low-loss property of SiN passive waveguides can be conveniently exploited to realize the strict filtering requirements [5, 6] that are demanded for the application. Additionally, the high power-handling capability of SiN waveguides permits to optimize the RF gain and noise figure of the microwave photonics processor. Here, we present the preliminary characterization results of the filter realized in SiN PIC technology and confirm its ability of simultaneously matching the target performance in respect to the ultra-high extinction ratio, narrow-bandwidth, wide free-spectral-range, and low-insertion-loss.

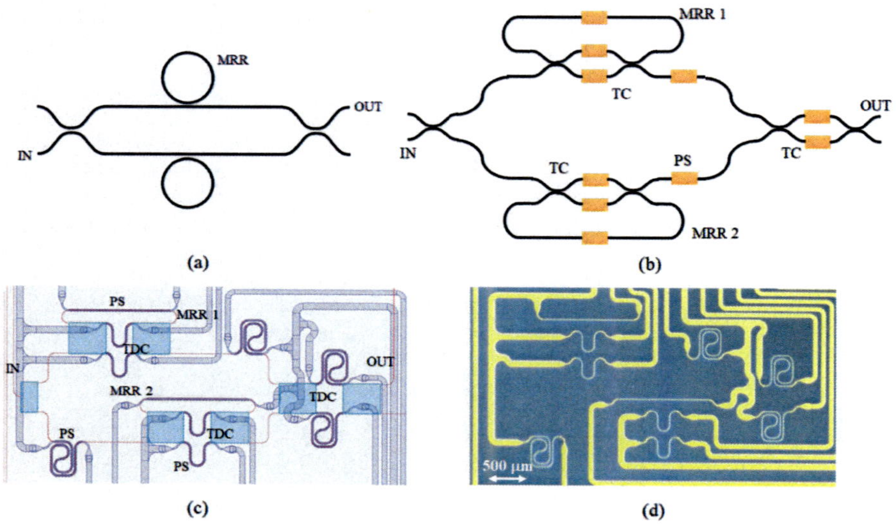

Fig. 1. (a) Schematic diagram, (b) functional layout, (c) mask layout and (d) micrograph of fabricated device.

2 Filter Architecture and Implementation

The schematic structure of the considered photonic integrated filter is shown in Fig. 1(a). A sum-difference filter is realized in a balanced Mach-Zehnder Interferometer (MZI) assisted by a microring resonator (MRR) all-pass filter (APF) onto each arm. A multiple bandpass transfer function can be obtained at the output coupler as the results of the interference of the periodic phase responses produced by the APFs [7]. Due to the pole-zero nature of the filter, a narrow bandwidth and sharp roll-off can be already achieved with this 2nd-order design. Practical filter realizations require controlling the relative detuning of the MRR resonances and of the phases in the MZI, which is typically accomplished through optical phase shifters (PS) embedded in the structure. Additionally, by implementing tunable coupling sections between the MZI arms and the MRRs, the shape of the phase response of the APFs can also be controlled, permitting to implement any second-order passband transfer function within the same structure. Despite this architecture has been extensively studied [7, 8], challenges remain in synthetizing filter responses simultaneously exhibiting GHz-range passband widths, large free-spectral ranges (FSRs) values of several tens of GHz, out-of-band rejection levels up to 40 dB or more, and low insertion loss at the peak transmission of few dB. The propagation losses in the MRRs indeed causes a smoothing of the phase response for a given microring coupling coefficient and the appearance of amplitude notches in the APF response. Both these effects contribute to degrading the filter transfer function, by reducing the roll-off and peak transmission, and ultimately limiting its selectivity [8]. For the application described in [5], an FSR of 36 GHz, a maximum passband width of 4 GHz, and a minimum attenuation of 40 dB at 17 GHz from the passband central frequency, are required.

The functional implementation of the realized device is then reported in Fig. 1(b), whereas Fig. 1(c) shows the corresponding produced mask layout for its fabrication through low-loss SiN-based TriPleX waveguides [9], using a standard multi-process wafer service. The amount of light coupled into/out each MRR is adjusted by means of tunable couplers (TCs) realized as balanced MZIs, where thermo-optic PSs are implemented through metal heaters placed above the waveguides. An additional TC is used for the output coupler of the parent MZI, which permits to counteract small fabrication inaccuracies and carefully balance the amplitude of the interfering fields at the device output for maximizing the extinction ratio of the filter. The need for keeping a sufficiently short MRR length for reaching the target FSR also collides with the ability of driving the heaters in the optical PSs with current levels below their damage threshold, and introduce an issue related to thermal crosstalk due to the proximity of the waveguides. To this scope, the layout has been optimized to keep a minimum distance between the waveguides of around 170 μm and by replicating the PS onto each of the MZI arms implementing the TCs, permitting to exploit the most favorable condition in terms of dissipated electrical power. A micrograph of the fabricated device is also shown in Fig. 1(d), and the overall circuit dimensions are ~5.2 $\times$ 2.5 mm^2.

3 Experimental Results

For the characterization of the filter, a packaged version of the PIC, illustrated in Fig. 2(a), has been employed. The measured spectral transmission of the filter under different setting configurations of the various PSs realizing various filter transfer functions all fulfilling the target specifications are then shown in Fig. 2(b). A FSR of ~36.2 GHz is observed, well in accordance with the target simulation. From the traces, a maximum out-of-band rejection as large as 57.5 dB is measured. To the best of our knowledge, this is the largest value reported for a SiN-based photonic integrated filter [6, 10]. The corresponding roll-off over the -1 dB - to- -40 dB transition band is of 4 dB/GHz. The passband top details of Fig. 3(a) reveal the different -1 dB bandwidths for the three filter settings of 2.2 GHz (green curve), 3.4 GHz (blue curve), and 4.4 GHz (red curve). The -3 dB bandwidths result to be 3 GHz, 4.2 GHz, and 5.5 GHz, respectively. As evident from Fig. 2(b), an extinction ratio larger than 40 dB is preserved for all the configurations, and the − 40dB stopband regions between adjacent passband windows are always greater than 10 GHz, with maximum value of 17 GHz (green trace). A low typical insertion loss of ~3.5 dB is also measured.

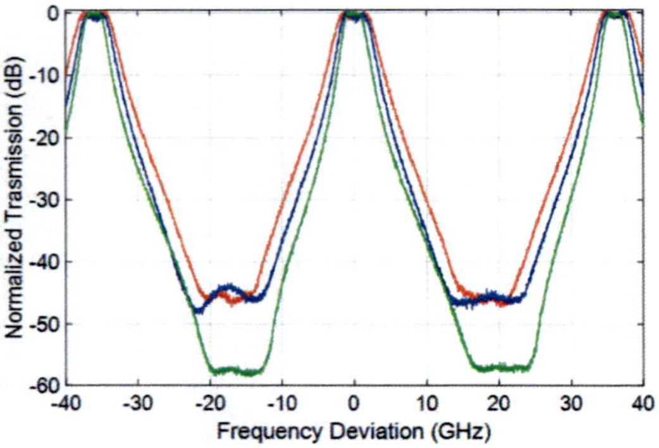

(a) (b)

Fig. 2. (a) Picture of the packaged PIC used for the experimental characterization; (b) Transmission spectra versus relative frequency detuning for different filter configurations.

The tuning of the photonic integrated filter central wavelength, which is also required for the application devised in [5], has then been verified, as reported in Fig. 3(b). To this scope, departing from an optimized filter shape (blue trace, *tuning 1*), only the resonances of the two MRRs have been adjusted through the relative PSs, leading to a filter replica which is shifted by more than 20 GHz toward shorter frequencies (red trace, *tuning 2*) and preserves the passband features, which confirms the reduced thermal crosstalk of the realized filter. The measured tuning efficiency is 10.2 mW/GHz.

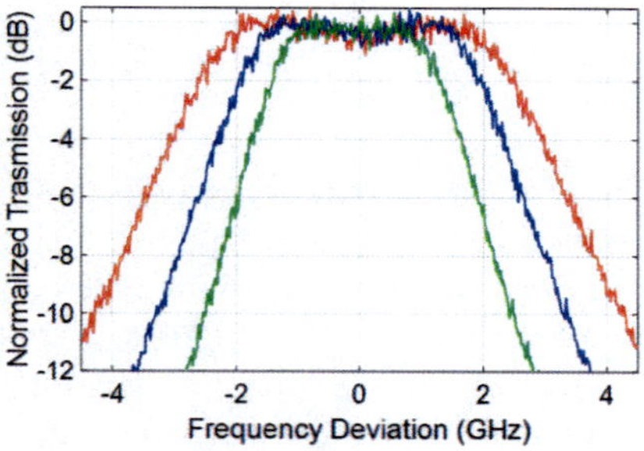

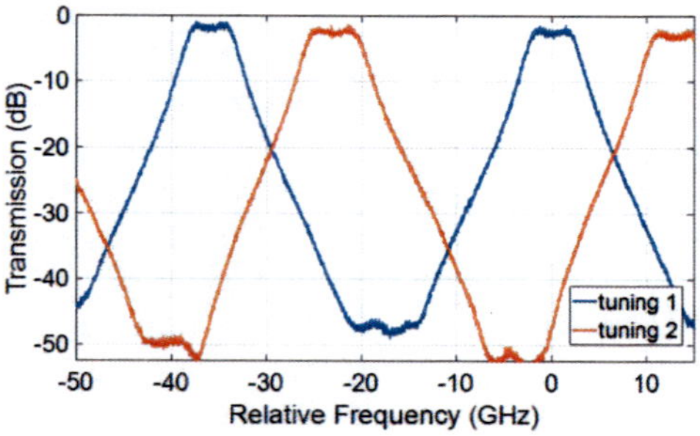

(a) (b)

Fig. 3. (a) Passbands details for the different filter settings of Fig. 2; (b) Tuning of the passband central frequency.

4 Conclusions

We reported the results of the characterization of a photonic integrated bandpass filter realized in silicon nitride technology designed for photonics-based advanced RF processing applications. The filter is implemented in a balanced Mach-Zehnder interferometer loaded with two micro-ring resonators and features a wide free-spectral-range of 36 GHz. Different filter configuration led to a tunable bandpass width down to 2 GHz, a −40 dB stopband region between adjacent resonances up to 17 GHz, and a maximum out-of-band rejection of nearly 60 dB. The results demonstrate the suitability of the fabricated circuit in silicon nitride technology for enabling the use of microwave photonics techniques in the most demanding RF applications.

Acknowledgements. This work is partially funded by the EU under the Italian National Recovery and Resilience Plan (PNRR) of NextGenerationEU partnership on "Telecommunications of the Future" (PE00000001 - program "RESTART") and by the Italian Defence Ministry within the PNRM project esaSCAN #37/2020. The authors acknowledge Lionix International for PIC manufacturing.

References

1. Serafino, G., et al.: Microwave photonics for remote sensing: from basic concepts to high-level functionalities. J. Lightwave Technol. **38**(19), 5339–5355 (2020)
2. Porzi, C., et al.: Frequency and bandwidth tunable integrated microwave photonic bandpass filter. In: 2022 International Workshop on Fiber Optics in Access Networks (FOAN), Valencia, Spain, pp. 1–5 (2022)
3. Mohammad, A.W., et al.: Design, fabrication, and characterization of a hybrid integrated photonic module for a synthetic aperture radar receiver. J. Lightwave Technol. **42**(2), 760–770 (2024)
4. Camponeschi, F., et al.: Integrated photonics for multi-channel RF scanning receivers. In: 2023 International Topical Meeting on Microwave Photonics (MWP), Nanjing, China, pp. 1–4 (2023)
5. Xiang, C., et al.: Silicon nitride passive and active photonic integrated circuits: trends and prospects. Photon. Res. **10**, A82–A96 (2022)
6. Taddei, C., Zhuang, L., Roeloffzen, C.G.H., Hoekman, M., Boller, K.-J.: High-selectivity on-chip optical bandpass filter with sub-100-MHz flat-top and under-2 shape factor. IEEE Photon. Technol. Lett. **31**(6), 455–458 (2019)
7. Madsen, C.K., Zhao, J.H.: Optical Filter Design and Analysis: A Signal Processing Approach, 1st edn. Wiley, New York (1999)
8. Rasras, M.S., et al.: Demonstration of a fourth-order pole-zero optical filter integrated using CMOS processes. J. Lightwave Technol. **25**(1), 87–92 (2007)
9. Roeloffzen, C., et al.: Low-loss Si3N4 TriPleX optical waveguides: technology and applications overview. IEEE J. Sel. Top. Quant. Electron. **24**(4), 1–21 (2018)
10. Alonso-Ramos, C., et al.: Bragg grating filter for suspended silicon waveguides. In: 2017 IEEE 14th International Conference on Group IV Photonics (GFP), Berlin, Germany, pp. 91–92 (2017)

Pulse Driving of Thermo-Optic Phase Shifters on an InP Membrane

Qiyuan Sheng(✉), Yi Wang, Kevin Williams, and Yuqing Jiao

Eindhoven Hendrik Casimir Institute (EHCI), Eindhoven University of Technology, 5600MB Eindhoven, The Netherlands
q.sheng@student.tue.nl, y.wang10@tue.nl

Abstract. Thermo-optic phase shifters (TOPS) have been widely applied in photonic integrated circuits (PIC), as they offer a straightforward way of phase tuning by locally heating the optical waveguide. Leveraging direct waveguide heating, an ultra-compact, efficient TOPS has been recently demonstrated on an InP membrane, that allows native integration with active components. In this paper, we apply a pulse-driving method on this InP-membrane-based TOPS to drastically boost the tuning speed by more than one order of magnitude, from 26.67 μs to 3.25 μs.

Keywords: Thermal-optic phase shifter · optical switch · InP photonics · membrane

1 Introduction

InP photonics holds a key position in the development of advanced PICs thanks to its comprehensive set of building blocks (BBs) that enable the monolithic construction of complex PICs structure [1]. Thermo-optic phase shifters (TOPS), crucial for phase tuning, significantly enhance photonics performance, with their integration proving instrumental in advancing high-speed optical switching and precision sensing tech-nologies [2]. Due to the low heat delivery efficiency and poor isolation caused by the bulky waveguide structure in traditional InP photonics, TOPS in InP materials has problems such as low power efficiency (~100 mW) and slow speed (miliseconds).

The Indium Phosphide Membrane on Silicon (IMOS) platform introduces an approach for integrating active and passive photonic functionalities into a single nanophotonic InP layer to provide high index contrast [3]. Therefore, tight light con-finement and comprehensive photonic features can be simultaneously attained. In 2023, an ultra-compact and efficient (2.26 mW/π) TOPS was demonstrated on an InP membrane platform with a leading figure of merit (FoM) in terms of the power speed product of 29 mW·μs [4]. Though being efficient in power consumption, its response time is still limited to tens of micro-seconds. Some previous studies have demonstrated a reduction in time response by advanced control methods without sacrificing power efficiency. Xu et al. [5] applied pre-emphasis in micro-ring electro-optic modulators to expand bandwidth and decrease response times, and similarly, Lino et al. [6] demonstrate same pulse driving method in MZI based TOPS.

J. Witzens et al. (Eds.): ECIO 2024, SPPHY 402, pp. 378–382, 2024.
https://doi.org/10.1007/978-3-031-63378-2_62

In this paper, to further push the performance of developed TOPS, we demonstrate a pulse-driving method and significantly reduce the response time by almost one order of magnitude. By overdriving the TOPS at the rising edge temporarily, this method quickly modulates the refractive index of the InP waveguide beyond the DC limit, reducing response times to 12.5%. Such advancements are crucial for applications requiring high-speed optical modulation, contributing to the development of faster, more efficient photonic systems. The fast-switching speed demonstrated in this paper combined with the low energy consumption may find applications in optical phased arrays, photonic analog computing, and switching.

2 Experimental Method and Result

Figure 1(a) shows the TOPS device's cross-section. On top of the i-InP waveguiding layer is an n-InP layer which acts as the conduction layer for Joule heating. This arrangement optimizes the overlap between thermal energy and the optical mode by minimizing the heat capacity of the device [4]. The semiconductor layers are encapsulated in polyimide (PI) and benzo-cyclobutene (BCB) for planarization and structural support. On the circuit-level, the TOPS cascaded and embedded in one arm of a balanced Mach-Zehnder interferometer (MZI), with the optical input and output connected to fiber grating couplers (FGC) for vertical coupling.

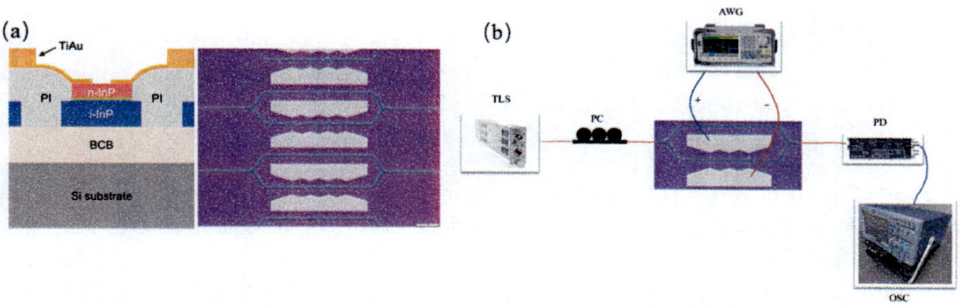

Fig. 1. Devices and experimental setup. (a) Cross-section of the TOPS and microscope imagine of 4 microheaters on MZI structure. (b) Experimental setup of the transient measurement.

The experimental setup is shown in Fig. 1(b). The TE-polarized laser from a tunable laser source (TLS, Agilent 81940A) with a wavelength of 1550 nm and 0 dBm power was coupled into the MZI with FGC and the output power was detected by a high-speed photodetector (PD, Femto OE-300-IN-01) after the MZI. Signals from an arbitrary waveform generator (AWG, Siglent SDG 1032X), with 100 ns rise time and bandwidth up to 60 MHz, were applied to the two metal contacts with DC probes. The transient electro-optical response of the TO-MZI is then measured using a real-time digital oscilloscope (Wavemaster 8600A). We utilize a 1 kHz square wave voltage, alternating between high (V_{on}) and low (V_{off}) levels, to dynamically modulate the output optical power of a thermo-optic Mach-Zehnder interferometer (TO-MZI). Here, V_{on} corresponds to the maximum constructive interference condition of the TO-MZI, while V_{off} corresponds to the maximum destructive interference condition. An additional voltage pulse is added to the rising edge or falling edge. Rise time (t_{rise}) is measured as the output's escalation from 10% to 90% of the 'on' state, and fall time (t_{fall}) as the reduction from 90% to 10%.

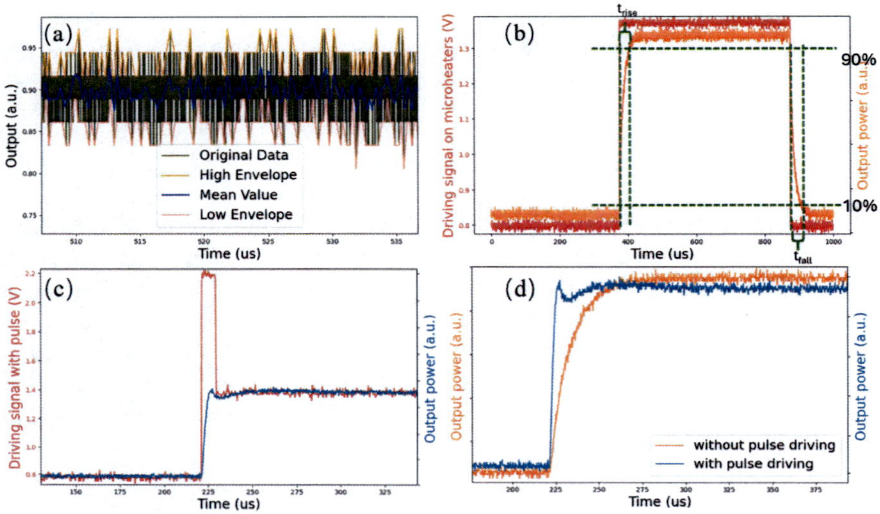

Fig. 2. (a) Data smoothing method. (b) Square wave signal without pulse-driving and TO-MZI switching time response. (c) TO-MZI switching time responses pulse-driving. (d) Comparison of switching time responses with or without pulse-driving on rising edge.

The obtained waveform contains rounding errors of oscilloscope, and high-frequency noise from both the oscilloscope and the photodetector (PD). This can lead to inaccuracies in determining the correct data points when calculating rise and fall time. To solve it, we smooth the data by calculating the average of the signal's envelope, as illustrated in Fig. 2(a). By averaging the constructed upper and lower envelopes from the signal's local peaks and troughs, we partially filter noise and extract the actual waveform, this method doesn't strictly filter out noise in all frequencies but is enough to determine the rise time. We first apply a 1 kHz square wave with voltage levels $V_{on} = 1.4$ V, $V_{off} = 0.815$ V to calculate the reference switching time without pulse-driving. The obtained waveform is presented in Fig. 2 (b), the 10%–90% rise and fall time are 26.67 μs and 25.99 μs, respectively.

The transient response with pulse-driving is measured on the rising edge of the square wave, in which a pulse voltage much larger than V_{on} was temporarily applied to the TOPS when switching from 'off' to 'on'. The pulsed waveform is shown in Fig. 2 (c). By trail and error, we apply a pulse voltage $V_p = 2.18$ V and pulse width $t_p = 5.9$ μs, and thus the rise time is reduced from 26.67 μs to 3.25 μs as shown in Fig. 2 (d), it is only 12.5% of the rise time without pulse-driving. Due to the output power of MZI and phase change introduced by microheaters following a cosine relation, we observe a rapid decrease in output power after it initially reaches its maximum, meaning the pulse we apply overshoots the π phase change. In the time response graphs shown in Fig. 2 (c) and (d), the voltage of the 'on' state is slightly lower than the maximum value. This is due to the experimental platform perturbation during the experiment, causing the fiber and FGC to be misaligned. Thus, the rise time here is calculated based on 90% of the maximum value rather than the steady 'on' state voltage. Since we are more concerned about the rising and falling edges of the TOPS, this misalignment here has no impact on the results.

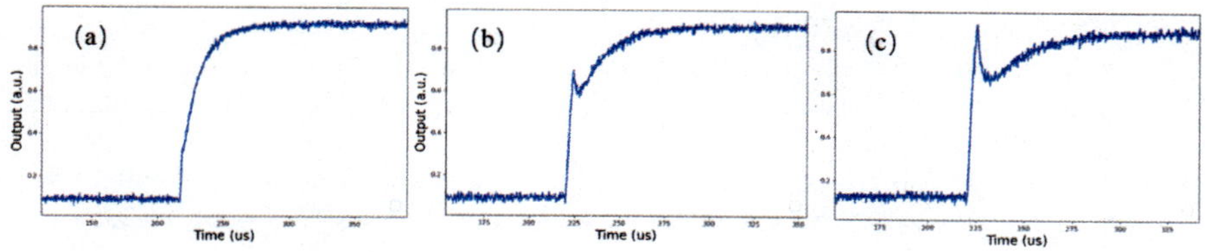

Fig. 3. (a) Case 1: $V_p = 2$ V, $t_p = 2.5$ μs. (b) Case 2: $V_p = 2.1$ V, $t_p = 4.5$ μs. (c) Case 3: $V_p = 2.28$ V, $t_p = 5.9$ μs.

To explore how the pulse affects the switching waveforms and response time, we varied the pulse widths and voltages. A few representative cases are shown in Fig. 3. In the first case, when a low pulse voltage and narrow pulse width is applied, the expected π phase shift cannot be achieved within the pulse period, resulting in a negligible improvement in rise time ($t_{rise} = 24.8$ μs), as shown in Fig. 3 (a). Case 2 exhibits an unexpected decrease right after the pulse ends, as seen in Fig. 3 (b). This is likely because of the thermal crosstalk between the MZI arms – the heat pulse transferring to the other arm reduces the phase difference and pulls back the optical signal. Case 3 indicates that an overly large pulse can overshoot the phase shift rather than speed up the time response. Given these insights, future efforts will focus on equipping both MZI arms with microheaters for finer control.

3 Conclusion

In this paper, we've explored the significant impact of pulse driving on the switching speed of an InP membrane-based TOPS. Our research demonstrated that, the response times can be shortened from 26.67 μs to 3.25 μs with pulse voltage $V_p = 2.18$ V and pulse width $t_p = 5.9$ μs. With $P_\pi = 2.33$ mW, an effective 7.6 mW·μs power-speed product can be achieved. This is beyond the state-of-the-art demonstrations, to the best of our knowledge. We further explored the influence of the pulse waveform on the response time.

References

1. Jiao, Y., et al.: Indium phosphide membrane nanophotonic integrated circuits on silicon. Phys. Status Solidi A **217**(3), 1900606 (2020). https://doi.org/10.1002/pssa.201900606
2. Liu, S., Feng, J., Tian, Y., et al.: Thermo-optic phase shifters based on silicon-on-insulator platform: state-of-the-art and a review. Front. Optoelectron. **15**(9) (2022). https://doi.org/10.1007/s12200-022-00012-9
3. van der Tol, J.J.G.M., Jiao, Y., Van Engelen, J.P., Pogoretskiy, V., Kashi, A.A., Williams, K.: InP membrane on silicon (IMOS) photonics. IEEE J. Quant. Electron. **56**(1), 1–7 (2020)
4. Wang, Y., Dolores-Calzadilla, V., Williams, K.A., Smit, M.K., Jiao, Y.: Ultra-compact and efficient microheaters on a submicron-thick InP membrane. J. Lightwave Technol. **41**(6), 1790–1800 (2023). https://doi.org/10.1109/JLT.2022.3225110
5. Xu, Q., Manipatruni, S., Schmidt, B., Shakya, J., Lipson, M.: 12.5 Gbit/s carrier-injection-based silicon micro-ring silicon modulators. Opt. Express **15**(2), 430–436 (2007). https://doi.org/10.1364/OE.15.000430
6. Lino, K., Kita, T.: Ultrafast operation of Si thermo-optic switch using differential control method. Jpn. J. Appl. Phys. **62**(1) (2022). https://doi.org/10.35848/1347-4065/aca268

Design of Silicon-Based Quantum Squeezer

Mouhamad Al-Mahmoud[1,2](✉) and Stéphane Clemmen[1,2,3,4]

[1] Photonics Research Group, INTEC-Department, Ghent University-IMEC, Ghent, Belgium
mouhamad.almahmoud@ugent.be
[2] Center for Nano- and Biophotonics (NB-Photonics), Ghent University, Ghent, Belgium
[3] OPERA-Photonique CP 194/5, Université Libre de Bruxelles, 1050 Brussels, Belgium
[4] Laboratoire d'Information Quantique, Université Libre de Bruxelles, 1050 Brussels, Belgium

Abstract. We introduce a silicon squeezer designed for high levels of squeezing and compatibility with state-of-the-art integrated balanced detection systems. Our approach combines linear elements, such as dispersion control and cavity design, with the strategic selection of pump parameters to enhance nonlinear performance. To optimize signal confinement and reduce phase mismatches in four-wave mixing (FWM), we meticulously adjust the waveguide dimensions, cladding materials, bend radii, and directional coupler settings. This comprehensive strategy is aimed at minimizing FWM discrepancies and ensuring effective signal confinement, demonstrating our commitment to addressing the complexities of high-performance squeezing in silicon platforms.

Keywords: quantum squeezer · nonlinear optics · FWM · silicon

1 Introduction

The field of quantum squeezing has significantly evolved in theory and now serves a wide range of applications, including gravitational wave detection [1], quantum simulation [2], and communication technologies [3]. Particularly for sensing and simulation, attributes like higher brightness, more significant squeezing levels, better bandwidth control, and tunability, along with the miniaturization and ability to produce identical squeezers on a single photonic chip, are highly valued [4]. Silicon photonics has emerged as a prime source for generating squeezed vacuum states, particularly through the utilization of microring cavities [5], proving effective in producing high quality photon pairs [6–8]. Silicon photonics is highlighted in the context of squeezing for its superior homodyne detection capabilities using integrated Germanium-based detectors [9–11]. However, inherent

J. Witzens et al. (Eds.): ECIO 2024, SPPHY 402, pp. 383–390, 2024.
https://doi.org/10.1007/978-3-031-63378-2_63

nonlinearities in silicon, such as two-photon absorption and subsequent free carrier absorption, pose challenges in generating states with multiple photons per mode [12].

Recent successes have been achieved with SiN microcavities and PPLN waveguides, the latter typically offering a narrow optical bandwidth [13,14] but capable of achieving high squeezing levels [15,16]. SiN has been a preferred material for generating squeezing for over two decades, including structures for guiding [17]. Recent research has explored squeezing in PPLN nanophotonic waveguides [18–20], although integrating small sizes with high spectral brightness remains a challenge. Applications requiring monomode squeezing, like gaussian boson sampling, often utilize pump non-degenerate four-wave mixing, especially in the absence of second-order optical nonlinearity [21]. In silicon photonics, adopting a dual pump configuration has been identified as an effective approach to significantly enhance the squeezing parameter beyond unity. However, the simplest spectrally degenerate pump configurations face limitations in spectral brightness due to two-photon absorption. While pumping at wavelengths longer than silicon's half-bandgap avoids this issue [22], many applications require telecom wavelength squeezing. A non-degenerate pump scheme, using one pump above and one below the half-bandgap, can generate monomode squeezing [23,24]. This configuration's efficiency depends on both pump beams' intensities, necessitating adjustments to avoid two-photon absorption and potentially leading to frequency comb generation at high long-wavelength pump powers [25,26], although this can be managed with dispersion engineering [27] and other strategies such as lowering Q-factor of neighboring resonances to prevent frequency comb generation while preserving squeezing [add reference].

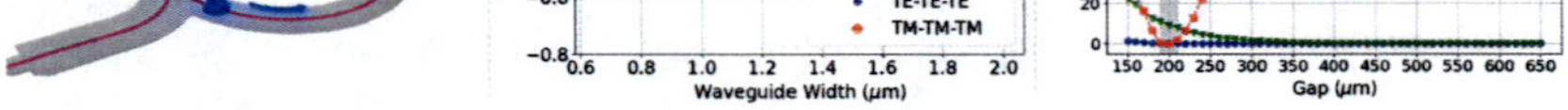

Fig. 1. Visualization of the quantum squeezer configuration, featuring: (1.a) the two-pump four-wave mixing scheme, (1.b) the micro-resonator outline, (1.c) the phase mismatch (Δk) against waveguide width under air and SiO_2 claddings, and (1.d–e) coupling factors for the two bus-to-waveguide directional couplers as function of the gap width (grey zones mark chosen parameters; see text)

Figure 1a illustrates the pump non-degenerate four-wave mixing process's wavelength setup. The targeted wavelengths are $\lambda_{pump} = 2350 \pm 30$ nm for the main pump, $\lambda_{sq} = 1550 \pm 30$ nm for the squeezed light, and $\lambda_{aux} = 1156 \pm$

40 nm, adhering to the relationship $2\lambda_{sq}^{-1} = \lambda_{aux}^{-1} + \lambda_{pump}^{-1}$. The challenge lies in designing a cavity that maintains high Q at these three wavelengths while ensuring significant overcoupling at the squeezed wavelength and proper phase matching within the cavity. A significant nonlinear phase shift, primarily from Kerr nonlinearity and potentially from photogenerated carriers due to cross two-photon absorption, complicates the design [24,28,29].

Despite its strengths, the four-wave mixing process in silicon is challenged by nonlinear phenomena like two-photon and free carrier absorption in extended waveguides. Micro-ring resonators (Fig. 1.b) offer an efficient alternative by enhancing light interaction in a confined space, avoiding the need for long waveguides [30]. Integrating a PN junction within the cavity addresses these nonlinear issues by enabling carrier evacuation, underscoring the importance of comprehensive design considerations for minimizing FWM phase mismatching and ensuring effective signal confinement across all wavelengths.

2 Device Optimization

Preliminary numerical studies on phase mismatch in silicon waveguides, incorporating specific dimensions and partial etching for enhanced carrier removal, have been performed. The phase mismatch Δk dependence to the waveguide width is depicted in Fig. 1.c for waveguides clad in both Air and SiO_2. These findings highlight the superiority of Air cladding compared to SiO_2, and the preference for TE modes over TM modes. Although perfect phase matching is attainable in TM modes with both types of cladding for waveguide widths exceeding 0.8 μm, narrower widths are preferred to facilitate effective coupling at the Aux wave.

Optimizing the coupling between the cavity and its external environment poses a significant challenge, as achieving critical coupling at the wavelengths of the auxiliary and pump waves is essential for maximizing intracavity power. Conversely, overcoupling at the squeezing wavelength is desirable to ensure a high escape probability for the squeezed wave. To address this, we've selected a design depicted in Fig. 1b, featuring distinct coupler for the squeezed and auxiliary waves and another for introducing the main pump. For the pump coupler (Fig. 1d), the design strategy involves creating a relatively wide gap between the bus waveguide and the cavity. This allows the pump's coupling intensity to be fine-tuned to optimal levels ranging from 1–10%, while ensuring that the coupling for the other two waves remains significantly lower, below 1/1000, especially for a waveguide width of 600 nm in TE polarization. The optimal parameters, highlighted by the grey-shaded area, are chosen to be 500 nm for gap size and 0 μm for the straight length, result in an coupling $\kappa_{\text{pump}} = 7.5\%$.

The coupling for the squeezed and auxiliary waves (Fig. 1.e), however, requires a more nuanced approach due to the need for weaker coupling at λ_{pump} while maintaining stronger coupling at λ_{sq} and λ_{aux}. This can be achieved by exploiting the directional coupler's periodic response to variations in its length. Through a detailed parameter optimization process, including adjustments to waveguide width, gap size, and coupling section length, we identified an effective configuration for this coupler, indicated by the grey-shaded area in Fig. 1e (200 nm gap and a 10 μm coupling section), resulting in $\kappa_{\text{aux}} = 2\%$ and $\kappa_{\text{sq}} = 10\%$. These simulations, conducted using Finite-Difference Time-Domain (FDTD) methods, also consider coupling within the directional coupler's curved sections.

A racetrack-shaped cavity, characterized by a L_{cav} =200 μm length and a 20 µm bend radius with air as the top cladding, has been analyzed. This model takes into account propagation losses of $\alpha_{\text{aux}} = 2\,\text{dB/cm}$ for the auxiliary wavelength, $\alpha_{\text{sq}} = 1.4\,\text{dB/cm}$ for the squeezed wavelength, and $\alpha_{\text{pump}} = 2.5\,\text{dB/cm}$ for the pump wavelength, alongside the coupling coefficients previously discussed. The expected cavity resonances in the linear operation mode are depicted in Fig. 2(left), and can be calculated as follows [31]

$$\frac{I_{\text{cavity}}}{I_{\text{input}}} = \frac{\kappa^2}{1 + (\sigma\tau)^2 - 2\sigma\tau\cos\phi}, \tag{1}$$

where κ^2 is the coupling, $\tau^2 = 1 - \kappa^2$ is the through transmission, and $\sigma^2 = \exp(-\alpha L)$ is the single-trip transmission, with α and L depict the linear loss and the cavity length, respectively. Here, Eq. 1 describes the case of an All-pass ring resonator, since each wave has a neglected coupling coefficient at one of the couplers.

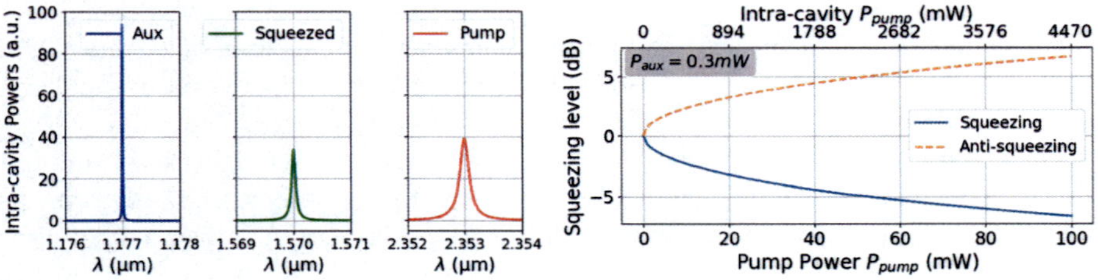

Fig. 2. (Left) Intra-cavity power enhancement factors for the three waves. (Right) Predicted levels of squeezing and anti-squeezing (dB) [see text for the assumed parameters].

3 Squeezing Level

Assuming, phase matching is indeed achieved, the monomode amplitude squeezing can easily be expressed in the photon pair regime when both the main pump and the auxiliary are such that of small enough so that $\gamma F_{aux}^2 P_{aux} L_{cav} \ll 1$ and $\gamma F_{aux}^2 P_{aux} L_{cav} \ll 1$, where γ, F, P and L are the nonlinear coefficient, the field enhancement, the pump power and the cavity length, respectively. The prediction of the squeezing levels lies in the calculation of photon generation through the FWM process, quantified by the spectral brightness $\mathcal{B}$ (photons/Hz/s) [32], as

$$\mathcal{B} = \gamma^2 L_{\text{cav}}^2 F_{aux}^2 F_{pump}^2 F_{sq}^4 P_{aux} P_{pump}. \tag{2}$$

Integrating the spectral brightness $\mathcal{B}$ across both the time and frequency domains yields the average photon number $\langle n \rangle$, which then allows for the determination of the squeezing factor r, using the relationship $\langle n \rangle = \sinh^2(r)$ [33]. This factor r significantly impacts the levels of observable squeezing ($R = e^r$) and its counterpart, anti-squeezing. These effects are respectively quantified by the squeezing and anti-squeezing variances in their respective quadratures, as

$$V_s = \langle \Delta X^2 \rangle = \frac{1}{2R^2} \quad \text{and} \quad V_{as} = \langle \Delta P^2 \rangle = \frac{R^2}{2}. \tag{3}$$

Hence, the measurement of squeezing levels is expressed in decibels (dB) as $10 \log_{10}(2V_{s,as})$. Our analysis, shown in Fig. 2 (right), reveals the dynamics between squeezing and anti-squeezing levels in relation to the main pump power, with the intra-cavity auxiliary pump power held constant at $P_{\text{aux}} = 0.3\,\text{mW}$. This efficient approach highlights the silicon platform's capability for quantum squeezing applications and provides a straightforward path toward their practical implementation, as demonstrated by the configurations in Fig. 1(d,e).

4 Effect of Nonlinear Losses

In this system, additional nonlinear loss effects emerge from nonlinear processes such as two-photon absorption (2PA), three-photon absorption (3PA), and cross-photon absorption (XPA). The 2PA and 3PA phenomena involve the absorption of 2 or 3 photons, respectively, of identical frequency by the nonlinear medium, which promotes the system from its ground state to a virtual state. XPA is akin to 2PA but involves photons of differing frequencies. A significant issue with these processes is the liberation of charge carriers—electrons and holes—due to the absorption of photons, leading to increased loss and changes in the refractive index through the Kerr effect. Within the context of the silicon-based quantum squeezer at hand, we confront all these parasitic nonlinear effects concurrently: 2PA predominantly induced by the auxiliary pump, 3PA by the main pump, and XPA by both pumps. To measure the nonlinear loss caused by these induced carriers, we analyze how the quantity of free carriers "N" changes over time, which can be described by the following equation

$$\frac{dN(t,z)}{dt} = \frac{\beta_{2PA}}{2\hbar\omega}\left(\frac{P}{A_{eff}}\right)^2 + \frac{\beta_{3PA}}{3\hbar\omega}\left(\frac{P}{A_{eff}}\right)^3 - \frac{N(t,z)}{\tau_{eff}}, \tag{4}$$

where, $\beta_{2PA,3PA}$ are the absorption coefficients, P and ω are respectively the pump power and frequency, A_{eff} is the effective area, and τ_{eff} is the effective carrier lifetime. We conduct separate analyses for each nonlinear process, taking into account two different effective carrier lifetimes: $\tau_{eff,min} = 200\,\text{ps}$, which is typical with a $p-n$ junction present, and $\tau_{eff,min} = 2000$ ps, which is characteristic in its absence. Beginning with 2PA and 3PA, these are driven by the first and second terms on the right-hand side of Eq. 4, respectively. As depicted in

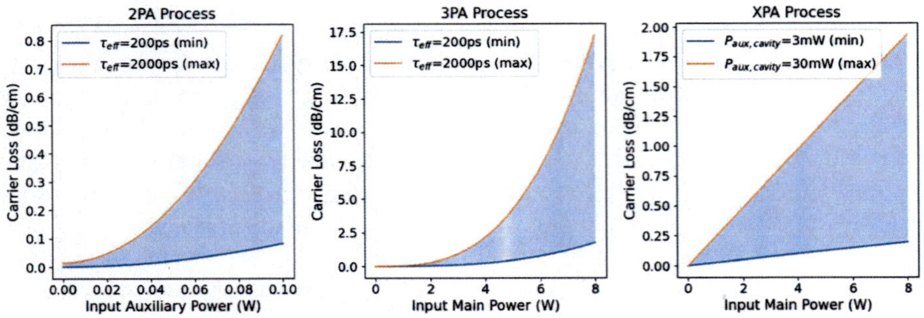

Fig. 3. Carrier losses, measured in dB/cm, associated with nonlinear processes—2PA, 3PA, and XPA—as they relate to the power of input pumps. The first two plots account for two different effective carrier lifetimes. The final plot displays carrier losses for two levels of intra-cavity auxiliary power, specifically 3 mW and 30 mW.

Fig. 3, the resulting carrier nonlinear losses (in dB/cm) exhibit an exponential growth. Notably, substantial carrier loss is associated with 3PA due to the main pump power being on the order of several watts. The defined limits for both τ_{eff} values illustrate that incorporating a $p-n$ junction can effectively curtail these losses. Turning to the final graph, we examine the effects at two intra-cavity auxiliary power levels, 3 mW and 30 mW. Here, carrier loss is computed using the 2PA component from Eq. 4, but at the squeezed frequency. This calculation is pertinent due to the involvement of both pumps where one photon from each is absorbed. With this configuration, the carrier loss is observed to increase linearly with the Main pump power. Again, the predicted effect of the presence of the $p-n$ junction is very clear.

A 3 dB level of squeezing is achievable at a pump power of $P_{pump} = 20\,\text{mW}$ and an auxiliary power of $P_{aux} = 0.3\,\text{mW}$, as shown in the right-hand plot of Fig. 2. The carrier losses (α) at these threshold powers for each nonlinear process, considering both the minimum and maximum effective lifetimes, are detailed as follows: for the 2PA process, $\alpha_{2PA,min} = 7 \times 10^{-7}\,\text{dB/cm}$ and $\alpha_{2PA,max} = 1.53 \times 10^{-2}\,\text{dB/cm}$; for the 3PA process, $\alpha_{3PA,min} = 7 \times 10^{-7}\,\text{dB/cm}$ and $\alpha_{3PA,max} = 1.56 \times 10^{-2}\,\text{dB/cm}$; and for the XPA process, $\alpha_{XPA,min} = 5 \times 10^{-4}\,\text{dB/cm}$ and $\alpha_{XPA,max} = 5 \times 10^{-3}\,\text{dB/cm}$. In the most extreme case—absent the $p-n$ junction which corresponds to a longer carrier lifetime—the summed losses amount to 0.035 dB/cm for the aforementioned threshold pump powers. However, with the $p-n$ junction in place, indicative of a shorter carrier lifetime, these losses are considerably reduced to $\alpha = 6.5 \times 10^{-4}\,\text{dB/cm}$. Consequently, it is inferred that a high degree of squeezing is attainable even in the presence of these nonlinear parasitic processes, particularly with a reduced effective carrier lifetime, which can be effectively managed by incorporating a $p-n$ junction.

5 Conclusion

Our research focused on developing a silicon-based quantum squeezer. We carefully selected cavity parameters, including waveguide dimensions, bend radius, and cladding materials. Additionally, we justified our choice of particular parameters for the directional couplers. We provided a thorough prediction of the expected squeezing levels in both linear and nonlinear operating conditions, emphasizing the role of a $p-n$ junction in mitigating carrier losses. Our conclusions demonstrate the feasibility of attaining high levels of squeezing despite the challenges posed by nonlinear losses.

References

1. Schnabel, R., Mavalvala, N., McClelland, D.E., Lam, P.K.: Quantum metrology for gravitational wave astronomy. Nat. Commun. **1**(1), 121 (2010)
2. Hamilton, C.S., Kruse, R., Sansoni, L., Barkhofen, S., Silberhorn, C., Jex, I.: Gaussian boson sampling. Phys. Rev. Lett. **119**(17), 170501 (2017)
3. Gottesman, D., Preskill, J.: Secure quantum key distribution using squeezed states. Quant. Inf. Continuous Variables, 317–356 (2003)
4. Arrazola, J.M., et al.: Quantum circuits with many photons on a programmable nanophotonic chip. Nature **591**(7848), 54–60 (2021)
5. Clemmen, S., Huy, K.P., Bogaerts, W., Baets, R.G., Emplit, Ph., Massar, S.: Continuous wave photon pair generation in silicon-on-insulator waveguides and ring resonators. Opt. Express **17**(19), 16558–16570 (2009)
6. Engin, E., et al.: Photon pair generation in a silicon micro-ring resonator with reverse bias enhancement. Opt. Express **21**(23), 27826–27834 (2013)
7. Yang, Z., et al.: A squeezed quantum microcomb on a chip. Nat. Commun. **12**(1), 4781 (2021)
8. Savanier, M., Kumar, R., Mookherjea, S.: Photon pair generation from compact silicon microring resonators using microwatt-level pump powers. Opt. Express **24**(4), 3313–3328 (2016)
9. Yamada, K., et al.: Silicon-germanium-silica monolithic photonic integration platform for high-performance optical data communication systems. ECS Trans. **64**(6), 749 (2014)
10. Hajomer, A.A.E., et al.: Continuous-variable quantum key distribution at 10 gbaud using an integrated photonic-electronic receiver. arXiv preprint arXiv:2305.19642 (2023)
11. Lischke, S., et al.: Ultra-fast germanium photodiode with 3-dB bandwidth of 265 GHz. Nat. Photon. **15**(12), 925–931 (2021)
12. Helt, L.G., Steel, M.J., Sipe, J.E.: Parasitic nonlinearities in photon pair generation via integrated spontaneous four-wave mixing: critical problem or distraction? Appl. Phys. Lett. **102**(20) (2013)
13. Reimer, C., et al.: Generation of multiphoton entangled quantum states by means of integrated frequency combs. Science **351**(6278), 1176–1180 (2016)
14. Ramelow, S., et al.: Silicon-nitride platform for narrowband entangled photon generation. arXiv preprint arXiv:1508.04358 (2015)
15. Dutt, A., Luke, K., Manipatruni, S., Gaeta, A.L., Nussenzveig, P., Lipson, M.: On-chip optical squeezing. Phys. Rev. Appl. **3**(4), 044005 (2015)

16. Vaidya, V.D., et al.: Broadband quadrature-squeezed vacuum and nonclassical photon number correlations from a nanophotonic device. Sci. Adv. **6**(39), eaba9186 (2020)
17. Kanter, G.S., Kumar, P., Roussev, R.V., Kurz, J., Parameswaran, K.R., Fejer, M.M.: Squeezing in a Linbo 3 integrated optical waveguide circuit. Opt. Express **10**(3), 177–182 (2002)
18. Mondain, F., et al.: Chip-based squeezing at a telecom wavelength. Photon. Res. **7**(7), A36–A39 (2019)
19. Stokowski, H.S., et al.: Integrated quantum optical phase sensor in thin film lithium niobate. Nat. Commun. **14**(1), 3355 (2023)
20. Chen, P.-K., Briggs, I., Hou, S., Fan, L.: Ultra-broadband quadrature squeezing with thin-film lithium niobate nanophotonics. Opt. Lett. **47**(6), 1506–1509 (2022)
21. Zhang, Y., et al.: Squeezed light from a nanophotonic molecule. Nat. Commun. **12**(1), 2233 (2021)
22. Rosenfeld, L.M., et al.: Mid-infrared quantum optics in silicon. Opt. Express **28**(25), 37092–37102 (2020)
23. Vernon, Z., et al.: Scalable squeezed-light source for continuous-variable quantum sampling. Phys. Rev. Appl. **12**(6), 064024 (2019)
24. Zhao, Y., Okawachi, Y., Jang, J.K., Ji, X., Lipson, M., Gaeta, A.L.: Near-degenerate quadrature-squeezed vacuum generation on a silicon-nitride chip. Phys. Rev. Lett. **124**(19), 193601 (2020)
25. Griffith, A.G., et al.: Silicon-chip mid-infrared frequency comb generation. Nat. Commun. **6**(1), 6299 (2015)
26. Kuyken, B., et al.: An octave-spanning mid-infrared frequency comb generated in a silicon nanophotonic wire waveguide. Nat. Commun. **6**(1), 6310 (2015)
27. Ramelow, S., et al.: Strong nonlinear coupling in a si3n4 ring resonator. Phys. Rev. Lett. **122**(15), 153906 (2019)
28. Poulvellarie, N., Ciret, C., Kuyken, B., Leo, F., Gorza, S.-P.: Highly nondegenerate two-photon absorption in silicon wire waveguides. Phys. Rev. Appl. **10**(2), 024033 (2018)
29. Chembo, Y.K.: Quantum dynamics of kerr optical frequency combs below and above threshold: Spontaneous four-wave mixing, entanglement, and squeezed states of light. Phys. Rev. A **93**(3), 033820 (2016)
30. Sharping, J.E., et al.: Generation of correlated photons in nanoscale silicon waveguides. Opt. Express **14**(25), 12388–12393 (2006)
31. Bogaerts, W., et al.: Silicon microring resonators. Laser Photon. Rev. **6**(1), 47–73 (2012)
32. Helt, L.G., Liscidini, M., Sipe, J.E.: How does it scale? Comparing quantum and classical nonlinear optical processes in integrated devices. JOSA B **29**(8), 2199–2212 (2012)
33. Foundations, S., Lvovsky, A.I.: Squeezed light. Photon. Technol. Appl. **1**, 121–163 (2015)

All Dielectric Integrable Optical Isolators

Sevag Abadian(✉), Getúlio Souza, Stanislav Winkler, Marian Bogdan Sirbu, Michail Symeonidis, and Tolga Tekin

Fraunhofer-Institut für Zuverlässigkeit und Mikrointegration (IZM), Gustav-Meyer-Allee 25, 13355 Berlin, Germany

sevag.abadian@izm.fraunhofer.de

Abstract. Integrable optical isolators, play a crucial role in fortifying the overall performance of Photonic Integrated Circuits (PICs). In this paper, we propose a device functioning in Transverse Magneto-Optic Kerr Effect (TMOKE) configuration, where a magneto-optic material is sandwiched in between parallel dielectric slab waveguides. The coupled optical modes hosted by such a geometry undergo an electromagnetic profile transformation that can be exploited to confine the input mode in one waveguide during forward propagation and in the other during backward propagation. Together with a ring resonator, such a system can provide a 20dB isolation ratio with negligible insertion losses.

Keywords: Optical Isolator · PIC · TMOKE · Waveguides · Coupled Modes

1 Introduction

The increasing reliance on online services, streaming platforms, and IoT devices is straining data centers, emphasizing the need for efficient hardware. Photonic Integrated Circuits (PICs) are crucial for data processing with minimal power consumption, relying on the seamless integration of photonic components for optimized data transmission. Integrable isolators contribute to overall circuit performance by protecting lasers from reflected light. The pursuit of isolators has been ongoing for decades, with researchers exploring various designs and mechanisms like magneto-optics (MO) [1], electro-optics [2], acousto-optics [3], optomechanical [4], photonic transition [5], photonic crystal [6] and piezoelectric interactions [7]. This paper demonstrates a novel dielectric-based MO isolator, with high isolation ratio, low insertion losses and back reflections that could be integrable in PICs.

2 Dielectric Isolator Design

The isolator constitutes parallel dielectric waveguides (*Wa* and *Wb*) with a magneto-optic (MO) material placed in between, all immersed in cladding (Fig. 1). Such a design provides a difference in the path that the optical mode undergoes in the forward and backward senses. Additionally, a ring resonator on one side of the coupled waveguides diverts solely the backward propagating light in *Wb* to a drop port.

J. Witzens et al. (Eds.): ECIO 2024, SPPHY 402, pp. 391–400, 2024.
https://doi.org/10.1007/978-3-031-63378-2_64

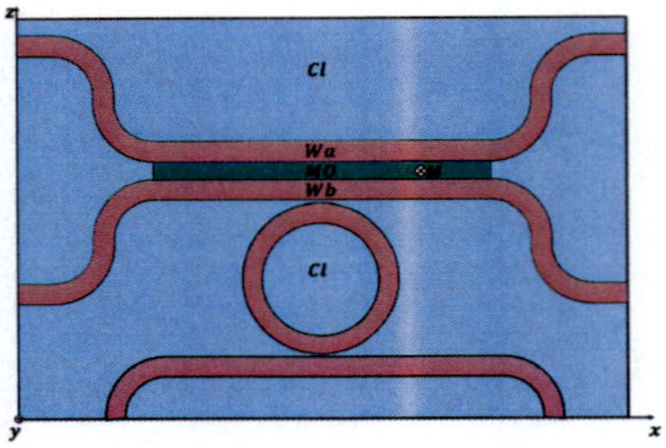

Fig. 1. Top view of the optical isolator constituting a MO material placed in between parallel waveguides, coupled to a ring resonator on one side.

We would like to show the theoretical derivations required to characterize the modal distribution of the coupled optical modes in the parallel waveguides. Moreover, we use Coupled Mode Theory to convey the interaction of these modes during simultaneous propagation. For simplicity, throughout the derivations we consider a 2D case.

2.1 Modal Analysis

Using Maxwell's equations, the generalized dispersion relations are derived. The MO material has width $2a$ and the dielectric *Wa* and *Wb* are of width d (Fig. 2). Such a structure hosts even and odd coupled modes.

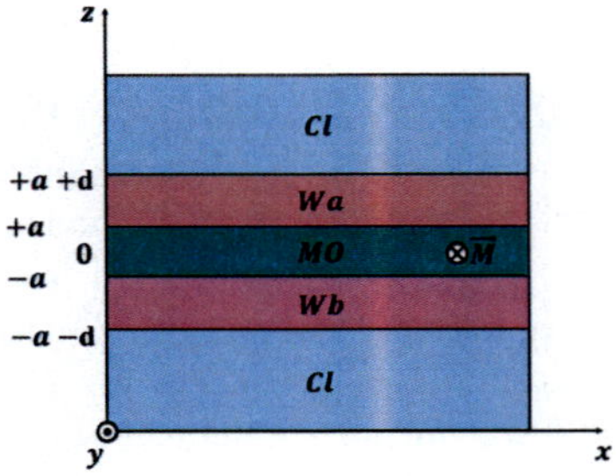

Fig. 2. Top view of the isolator core where the MO material is placed between parallel waveguides, all inserted within the cladding.

The dielectric tensor of a MO material can be represented in the form [8]:

$$\varepsilon_d = \begin{pmatrix} \varepsilon_{xx} & \varepsilon_{xy} & \varepsilon_{xz} \\ \varepsilon_{yx} & \varepsilon_{yy} & \varepsilon_{yz} \\ \varepsilon_{zx} & \varepsilon_{zy} & \varepsilon_{zz} \end{pmatrix} \tag{1}$$

The off-diagonal element is given by Eq. (2).

$$\varepsilon_{xy} = i \cdot g = i \cdot n_{MO} \cdot \lambda \cdot \theta_F / \pi \tag{2}$$

g is the MO gyrotropic parameter, ε_d is the dielectric constant, λ is the wavelength of operation and θ_F is the associated Faraday rotation coefficient.

In a cartesian coordinate system the electromagnetic fields can be represented:

$$\vec{E} = \vec{E_0} e^{i(\omega t - \beta x)} \tag{3}$$

$$\vec{H} = \vec{H_0} e^{i(\omega t - \beta x)} \tag{4}$$

$\vec{E_0}$ and $\vec{H_0}$ are the amplitude vectors. β is the propagation constant of the traveling waves and $\omega = 2\pi f$ is the angular frequency.

The magnetic field components for the structure in Fig. 2, can be expressed as follows:

$$H_y(\mathrm{z}) = \begin{cases} A_1 \exp[-\mathrm{K}_1(\mathrm{z}-\mathrm{a}-d)]\mathrm{z} \geq a+d \\ A_2 \cos[\mathrm{K}_2(\mathrm{z}-\mathrm{a}-d/2)]a \leq z \leq a+d \\ A_3 \exp[\mathrm{K}_3 z] + A_4 \exp[-\mathrm{K}_3 z] - \mathrm{a} \leq \mathrm{z} \leq a \\ A_5 \cos[\mathrm{K}_4(\mathrm{z}+\mathrm{a}+d/2)] - a - d \leq \mathrm{z} \leq -a \\ A_6 \exp[\mathrm{K}_5(\mathrm{z}+\mathrm{a}+d)]\mathrm{z} \leq -a-d \end{cases} \tag{5}$$

By replacing H_y of each layer into the wave equation, one can find the expression for the wavevectors in each layer.

$$\mathrm{K}_{1,3,5} = \sqrt{(\beta_{e,o}^2 \gamma_{xx}^{1,3,5} - k_0^2)/\gamma_{xx}^{1,3,5}},\ \mathrm{K}_{2,4} = \sqrt{(k_0^2 - \beta_{e,o}^2 \gamma_{xx}^{2,4})/\gamma_{xx}^{2,4}} \tag{6}$$

where:

$\gamma_{zx}^3 = ig/(n_3^4 - g^2)$, $\gamma_{xx}^3 = n_3^2/(n_3^4 - g^2)$

$$\gamma_{xx}^{1,5} = 1/n_{1,5}^2,\ \gamma_{xx}^{2,4} = 1/n_{2,4}^2 \tag{7}$$

Finally, applying the continuity of the electromagnetic fields E_x, H_y and E_z at the different boundaries ($-a-d$, $-a$, $+a$, $+a+d$), for this wave, the transcendental relation is written:

$$\mathrm{K}_2 \mathrm{d} = \tan^{-1}\left(\frac{\gamma_{xx}^1 \mathrm{K}_1}{\gamma_{xx}^2 \mathrm{K}_2}\right)$$
$$+\tan^{-1}\left(\frac{A_3 \exp[\mathrm{K}_3 a](\beta_{e,o}\gamma_{zx}^3 - i\gamma_{xx}^3 \mathrm{K}_3) + A_4 \exp[-\mathrm{K}_3 a](\beta_{e,o}\gamma_{zx}^3 + i\gamma_{xx}^3 \mathrm{K}_3)}{-i\gamma_{xx}^2 \mathrm{K}_2 (A_3 \exp[\mathrm{K}_3 a] + A_4 \exp[-\mathrm{K}_3 a])}\right) \tag{8}$$

where:

$$A_3 = \frac{(-2\beta\gamma_{zx}^3 \cosh(2\mathrm{K}_3 \mathrm{a}) A_4 \pm \sqrt{(2\beta_{e,o}\gamma_{zx}^3 \cosh(2\mathrm{K}_3 a) A_4)^2 - 4(\beta_{e,o}\gamma_{zx}^3 - i\gamma_{xx}^3 \mathrm{K}_3)(\beta_{e,o}\gamma_{zx}^3 + i\gamma_{xx}^3 \mathrm{K}_3) A_4^2}}{2(\beta_{e,o}\gamma_{zx}^3 - i\gamma_{xx}^3 \mathrm{K}_3)} \tag{9}$$

In Eq. (9), the two solutions obtained from the opposite signs correspond to β_e and β_o, respectively. Notably, the relation between A_3 and A_4 reveals distinct amplitudes of the $H_y(\mathrm{z})$ component for each of the coupled modes, A_2 and A_5 in the waveguides *Wa* and

Wb, respectively. This shows that the even and odd coupled modes adopt asymmetric and anti-asymmetric distribution around the z-axis. By solving both sides of the dispersion relation for the two cases of A_3, one can determine β_e and β_o.

Upon increasing the width of the MO layer, the system can be effectively partitioned into two sections (Fig. 3).

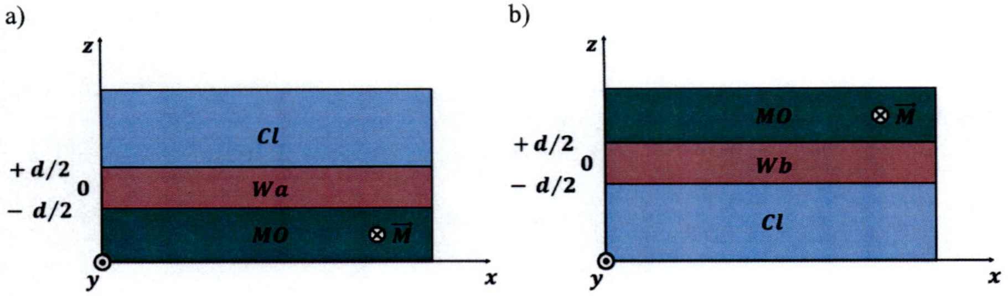

Fig. 3. Top view of the two partitions representing waveguide a) *Wa* and b) *Wb*.

For *Wa* in Fig. 3a, the magnetic field components in written:

$$H_y(\mathrm{z}) = \begin{cases} A_1\exp[-\mathrm{K}_1(\mathrm{z} - d/2)]\mathrm{z} \geq d/2 \\ A_2\cos[\mathrm{K}_2(\mathrm{z})] - d/2 \leq z \leq d/2 \\ A_3\exp[\mathrm{K}_3(\mathrm{z} + d/2)]\mathrm{z} \leq -d/2 \end{cases} \tag{10}$$

The dispersion relation can be expressed as:

$$\mathrm{K}_2\mathrm{d} = \tan^{-1}\left(\frac{\gamma_{xx}^3\mathrm{K}_3}{\gamma_{xx}^2\mathrm{K}_2}\right) + \tan^{-1}\left(\frac{\beta_a\gamma_{zx}^1 + i\gamma_{xx}^1\mathrm{K}_1}{i\gamma_{xx}^2\mathrm{K}_2}\right) \tag{11}$$

β_a, represents the propagation constant of the fundamental mode of *Wa*.

For *Wb* in Fig. 3b, the magnetic field components in *Wb* is written:

$$H_y(\mathrm{z}) = \begin{cases} A_1\exp[-\mathrm{K}_1(\mathrm{z} - d/2)]\mathrm{z} \geq d/2 \\ A_2\cos[\mathrm{K}_2(\mathrm{z})] - d/2 \leq z \leq d/2 \\ A_3\exp[\mathrm{K}_3(\mathrm{z} + d/2)]\mathrm{z} \leq -d/2 \end{cases} \tag{12}$$

The dispersion relation can be expressed as:

$$\mathrm{K}_2\mathrm{d} = \tan^{-1}\left(\frac{\gamma_{xx}^1\mathrm{K}_1}{\gamma_{xx}^2\mathrm{K}_2}\right) + \tan^{-1}\left(\frac{\beta_b\gamma_{zx}^3 - i\gamma_{xx}^3\mathrm{K}_3}{-i\gamma_{xx}^2\mathrm{K}_2}\right) \tag{13}$$

β_b, represents the propagation constant of the fundamental mode of *Wb*.

The obtained relations reveal distinct propagation constants in the two scenarios, with $\beta_a \neq \beta_b$. . To illustrate the principles discussed, we have chosen the following values for material and geometrical parameters (Table 1):

In all the gyrotropy cases, for narrow gap widths, the even and odd coupled modes exhibit strong coupling, resulting in distinct propagation constants. However, as the width increases, β_e and β_o gradually approach the propagation constants of the individual waveguides *Wa* and *Wb*, denoted as, β_a and β_b, respectively (Fig. 4).

Table 1. Material and geometrical parameters used in calculations for $\beta_{e,o}$ and $\beta_{a,b}$.

Variable	Value
n_{Cl}	1.45
$n_{Wa,b}$	3.48
n_{MO}	1.45
g	0. 1,0.05,0.01
$2a$	[0.2–1.7] μm
d	0.3 μm

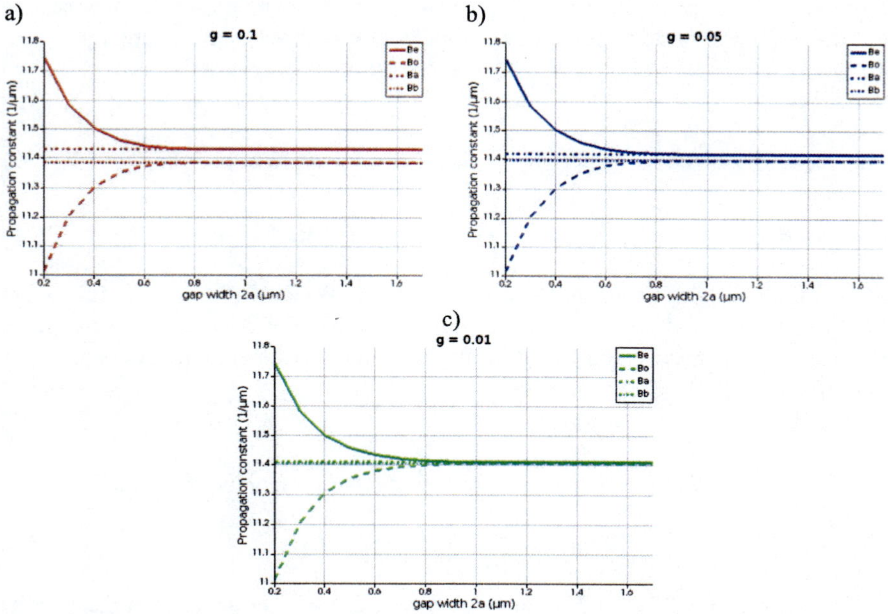

Fig. 4. $\beta_{e,o}$ and $\beta_{a,b}$ values for the material parameters represented in Table 1, with respect to variation of MO layer width $2a$.

2.2 Coupled Mode Theory (CMT)

The CMT proves valuable for analyzing the interaction between coupled modes. The relevant equations can be found in [9], expressed in matrix form (Eq. (14)).

$$\begin{pmatrix} P_1(x) \\ P_2(x) \end{pmatrix} = \begin{pmatrix} \cos(\psi x) - i\frac{\Delta}{\varphi}\sin(\psi x) & i\frac{K_{ab}}{\varphi}\sin(\psi x) \\ i\frac{K_{ba}}{\psi}\sin(\psi x) & \cos(\psi x) + i\frac{\Delta}{\psi}\sin(\psi x) \end{pmatrix} e^{\frac{i(\beta_a+\beta_b)}{2}x} \begin{pmatrix} P_1(0) \\ P_2(0) \end{pmatrix} \tag{14}$$

The two eigenvalues β_e and β_o of the even and odd modes:

$$\beta_{e,o} = \frac{\beta_a + \beta_b}{2} \pm \psi \tag{15}$$

$$\Delta = \frac{\beta_a - \beta_b}{2} \tag{16}$$

$$\psi = \sqrt{\Delta^2 + K_{ab}K_{ba}} \tag{17}$$

The coupling length to decouple the even and odd modes can be written as a function of the propagation constants:

$$\mathrm{L}_c = \frac{\pi}{\beta_e - \beta_o} \tag{18}$$

To understand the power exchange mechanism of the coupled modes, we assume that at the input $x = 0$, the optical power incident on both waveguides are equal, i.e., $P_1(0) = P_2(0) = 0.5$. The power at any point along the propagation axis can be extracted from Eq. (14).

$$P_1(x) = 0.5\left[\cos(\psi x) - i\frac{\Delta}{\psi}\sin(\psi x) + i\frac{K_{ab}}{\psi}\sin(\psi x)\right]e^{\frac{i(\beta_a+\beta_b)}{2}x} \tag{19}$$

$$P_2(x) = 0.5\left[\cos(\psi x) + i\frac{\Delta}{\psi}\sin(\psi x) + i\frac{K_{ba}}{\varphi}\sin(\psi x)\right]e^{\frac{i(\beta_a+\beta_b)}{2}x} \tag{20}$$

The maximum power transfer between waveguide *Wa* and *Wb* occurs at L_c, where $\psi x = \pi/2$. Assuming the power transfer between the parallel waveguides is complete, then the ratio of the normalized power difference can be written as equal to one.

$$Ratio = \frac{P_2(x)^2 - P_1(x)^2}{P_2(x)^2} = 1 \tag{21}$$

Replacing each with its equivalent term and after simplification we get:

$$Ratio = \frac{\beta_a - \beta_b}{\beta_e - \beta_o} = 0.707107 \tag{22}$$

When the *Ratio* is 0.707107, complete power transfer occurs between the parallel waveguides. Notably, the numerator, remains constant as the propagation constants of the partitioned *Wa* and *Wb* do not depend on the MO material width. Conversely, the denominator decreases with the increase of this width leading to an increase in the quantity *Ratio*. Finding the width corresponding to a *Ratio* = 0.707107 is essential to identify the optimal solution. After calculating the propagation constants $\beta_{a,b}$, $\beta_{e,o}$ through modal analysis, we determine L_c and the *Ratio* using CMT (Fig. 5).

The table below summarizes the MO gap layer width (2*a*) and the coupling length L_c required in each gyrotropy case to achieve complete power transfer.

As gyrotropy decreases, achieving the desired *Ratio* requires larger gap widths. While this may lead to an increase in coupling length, the *IL*s are confined to the waveguide material absorption, which remains below 1 dB/cm. This design presents a notable advantage: even with low gyrotropy MO material, it is feasible to create an integrable component that achieves non-reciprocal transmission with minimal *IL* s.

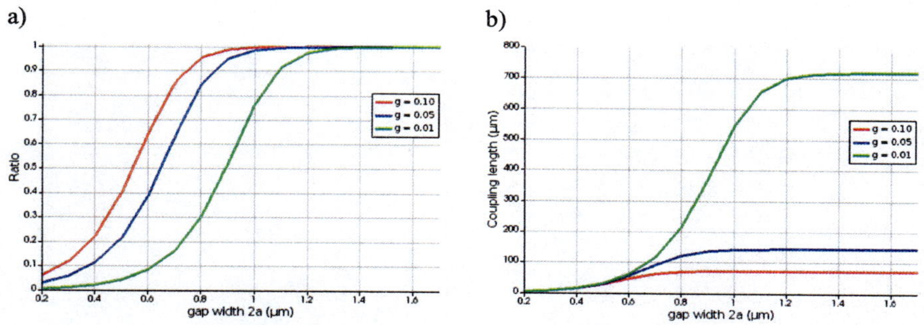

Fig. 5. a) Ratio b) Coupling length plots, calculated for different gyrotropy values with respect to variation of the MO layer width.

Table 2. MO layer width ($2a$) and coupling length (L_c) required in each gyrotropy case to achieve complete power transfer.

Gyrotropy	$2a$(µm)	L_c (µm)
0.1	0.62	50.7
0.05	0.73	101.4
0.01	0.97	506.9

3 All Dielectric Isolator

In an isolator, the transmission in the forward direction should be $T_f \approx 1$, whereas in the backward direction should be $T_b \approx 0$. Additionally, the component should not induce any back reflections *BR* to the ports.

$$IR = 10\log(T_f / T_b) \tag{23}$$

$$IL = 10\log_{10}(T_f) \tag{24}$$

In the forward case, following a distance of L_c, the coupled modes separate and the even mode appears in *Wa*. This mode remains unaffected by the ring resonator and upon traversing another L_c excites the coupled modes. Whereas, in the backward case, after covering a distance of L_c, the coupled modes separate and the even mode emerges in *Wb*. This mode is diverted by the ring resonator and coupled to the drop port ensuring that no power reaches the output ports (Fig. 6).

2D FDTD simulations were performed at $\lambda_0 = 1.55\mu m$, for the parameters presented in Table 2. We choose the case where $g = 0.1$ to maintain a low computational cost, and ensure better visibility of the pictures.

In the forward direction, (Fig. 7a), the light from the input ports reaches the output ports unperturbed. The normalized transmission is $T_f = 0.98$, with an estimated $IL = -0.08$ dB and $BR = -22$ dB. In the backward direction, (Fig. 7a), the light from the input

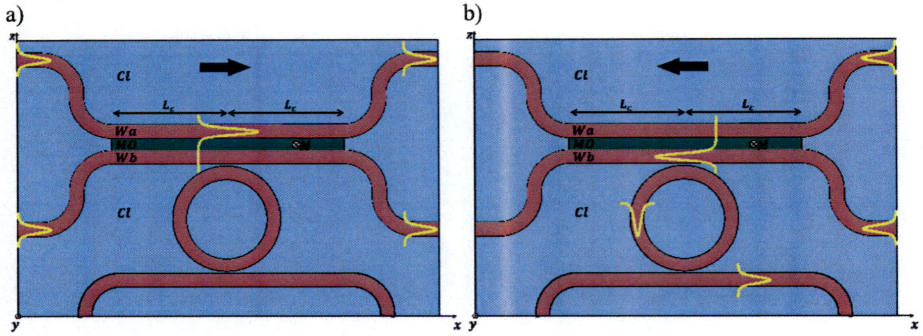

Fig. 6. Top view of the all dielectric optical isolator. Propagation of light in a) forward and b) backward directions are depicted.

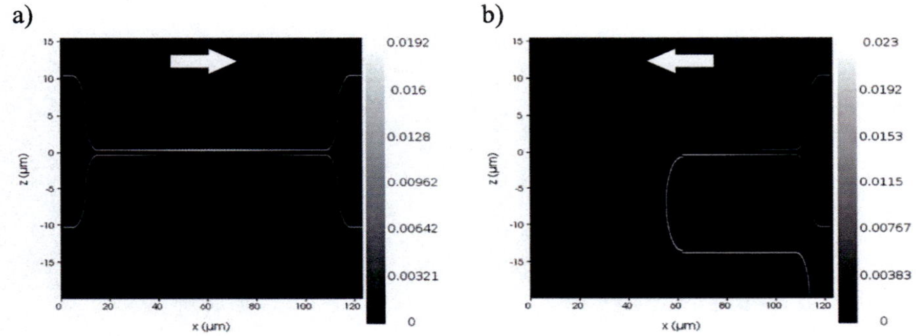

Fig. 7. Power distribution in the isolator as computed by the FDTD method (represented in greyscale). Propagation of light in a) forward and b) backward directions.

ports is redirected to the drop port and doesn't reach the output ports. The normalized transmission is $T_b = 0.006$, and $BR = -19$ dB.

Hence, the proposed isolator provides $IR = 22.13$ dB, $IL = -0.08$ dB and $BR = -22$ dB.

Simulations were conducted to ascertain the bandwidth by varying the wavelength from $1.54\mu m$ to $1.555\mu m$, without adjusting the MO layer width, the coupling length, and the ring resonator radius (Fig. 8).

For the optimized wavelength centered around $1.55\mu m$, high *IR* is obtained with negligible *IL*. The short bandwidth of the component is attributed to the utilization of the ring resonator. To address this issue, one can introduce a metal next to *Wb* to induce plasmonic absorption [10]. Alternatively, employing subwavelength gratings at the center of *Wb* with specific pitch and grating length, can diffract light outwards [11].

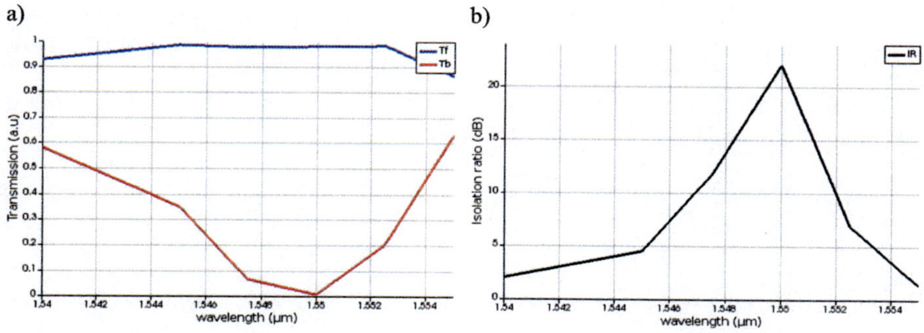

Fig. 8. a) Forward (blue) and backward (red) transmission as computed by the FDTD method. b) The calculated isolation ratio.

4 Conclusion

In conclusion, our study presents a thorough analysis of a dielectric isolator utilizing a system of parallel waveguides with MO material. The design achieves effective non-reciprocal transmission. Notably versatile, the isolator selectively routes light paths based on external factors and exhibits high isolation, low insertion losses, and minimal back reflections.

These insights contribute to advancing functionality of PICs and their integration into everyday applications.

References

1. Srinivasan, K., Stadler, B.J.: Review of integrated magneto-optical isolators with rare-earth iron garnets for polarization diverse and magnet-free isolation in silicon photonics [invited]. Opt. Mater. Express **12**(2), 697 (2022)
2. Yu, M., et al.: Integrated electro-optic isolator on thin-film lithium niobate. Nat. Photon. **17**(8), 666–671 (2023)
3. Corcoran, V.J.: A new acousto-optic isolator for the stabilization of optically pumped submillimetre lasers. Infrared Phys. **18**(5–6), 635–645 (1978)
4. Xia, C.-C., et al.: Ideal optical isolator with a two-cavity optomechanical system. Opt. Commun. **451**, 197–201 (2019)
5. Yu, Z., Fan, S.: Complete optical isolation created by indirect interband photonic transitions'. Front. Opt. 2008/Laser Science XXIV/Plasmonics and Metamaterials/Optical Fabrication and Testing (2008, Preprint)
6. Goswami, K., et al.: Design and analysis of all-optical isolator based on linear photonic crystal. Brazil. J. Phys. **52**(3) (2022)
7. Tian, H., et al.: Magnetic-free silicon nitride integrated optical isolator. Nat. Photon. **15**(11), 828–836 (2021). https://doi.org/10.1038/s41566-021-00882-z
8. Zvezdin, A.K., Zvezdin, A.K., Kotov, V.A.: Modern Magnetooptics and Magnetooptical Materials. CRC Press, Taylor & Francis Group, Boca Raton (2020)
9. Chuang, S.L.: Physics of Optoelectronic Devices. Wiley, New York (1995)

10. Boriskina, V.S., et al.: Losses in plasmonics: from mitigating energy dissipation to embracing loss-enabled functionalities. Adv. Opt. Photon. **9**(4), 775–827 (2017). Optical Society of America
11. Sychugov, V.A.: Optimization and control of grating coupling to or from a silicon-based optical waveguide. Opt. Eng. **35**(11), 3092 (1996)

Waveguide Undercut in High-Confinement Silicon Nitride Platform for Enhanced Phase-Shifters Performance

Anton Stroganov(✉), Henry Francis, Camiel Op de Beeck, Davide Sacchetto, Ozren Petrovic, Antoine Brimont, and Michael Geiselmann

LIGENTEC SA, EPFL Innovation Park, 1024 Ecublens, Switzerland
anton.stroganov@ligentec.com

Abstract. We demonstrate a reproducible, wafer-scale waveguide undercut process in an industrial silicon nitride platform for enhanced phase-shifters efficiency. In particular, this approach allows up to tenfold reduction in power consumption with respect to standard thermo-optical tuners.

Keywords: Photonic Integrated Circuits · Silicon Nitride · Undercut · Phase-shifters

1 Introduction

Silicon nitride (SiN) gets a lot of attention as a platform for photonic integrated circuits (PICs) due to its CMOS compatibility, maturity and excellent optical performance: it is widely considered a promising material for next generation PIC-based products. One of the key advantages of a SiN platform over silicon-on-insulator (SOI) platforms is its thermal stability. However, this can also be a limitation for thermal-optical phase shifters. The smaller thermo-optic coefficient of SiN with respect to Si can lead to lower heater efficiency for thermo-optical tuning of Mach-Zehnder interferometers (to realize switches or similar functionality), ring resonators and other structures. Besides, thermal dissipation into neighboring areas is a significant factor that limits the efficiency overall. One of the simplest solutions are on-chip trenches for thermal isolation around phase-shifter elements, which typically allow for 25%-50% improvement [1, 2]. However, trenches alone cannot isolate the main heat leakage channel through the high thermal conductivity Si substrate. A more advanced (but also more complicated) approach is to undercut the waveguide/heater - a process that not only forms trenches around the region of interest, but also removes the silicon directly under the structure, providing full isolation from the heat-sinking substrate. Such an undercut process was previously demonstrated by some research groups in Si [3] and thin SiN [4] platforms. In this work we implement the process in an industrial foundry setting for thick-film SiN waveguides. The AN800 platform used in this work is known to provide low-loss performance <0.1 dB/cm (with <0.02 dB/cm demonstrated for wider waveguides) in C-band, with on-demand access to dispersion engineering and nonlinear effects. It is widely used in

J. Witzens et al. (Eds.): ECIO 2024, SPPHY 402, pp. 401–404, 2024.
https://doi.org/10.1007/978-3-031-63378-2_65

both advanced research and product development in a variety of domains such as quantum, narrow-linewidth external cavity or self-injection-locked lasers for transceivers, fiber sensing and other applications, where power efficiency is an important metric.

2 Implementation and Performance

To perform the undercut we first used vertical trenches on both sides of the structure to open the oxide cladding and reach to the Si substrate. This was followed by an isotropic silicon etching step to release the structure from the bottom (see Fig. 1 right panel). Such process shows good within-wafer and wafer-to-wafer silicon etch rate uniformities, which is critical for practical applications. As the structures become suspended, it is important to explore design limitations on the length and width of trenches used for the process. We converged to the optimal approach of using several trenches alongside the waveguides to leave oxide bridges for mechanical stability (see Fig. 1 left panel). Careful design of those is required as the width of the suspended beam, as well as number and the dimensions of the supportive oxide bridges, directly affect the power efficiency gain for heater structures.

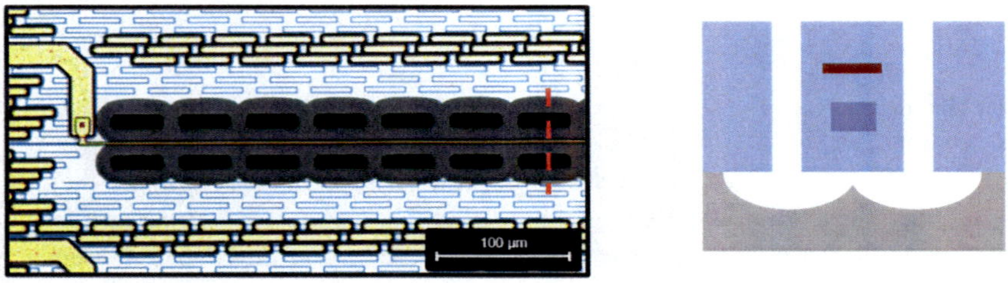

Fig. 1. An optical microscope image showing an undercut waveguide and a heater in an MZI arm (left) and a sketch of a corresponding cross-section (right).

The thermo-optical characterization to assess the improvement was performed using asymmetric Mach-Zehnder interferometer (MZI) structures with 750 µm-long heaters on one or both arms. The heaters were driven by varying values of direct current and the corresponding transmission traces from a tunable narrow-linewidth laser were recorded. The phase shift values related to each current level were extracted by assessing the shift of a MZI transmission dip position with respect to the reference measurement (at zero driving current). The obtained traces and extracted phase shifts for the tested structure can be seen in Fig. 2 below.

The measurements were performed for three types of structures: (i) without any thermal isolation, (ii) with trenches and (iii) with undercut, to obtain the improvement for various thermal isolation cases. The Table 1 below summarizes the obtained values and indicates the corresponding level of improvement with respect to the reference (no undercut, no trenches) device. The reference MZI devices were laid out using standard single-mode 1 µm-wide waveguides. Trenches for thermal isolation, as well as those

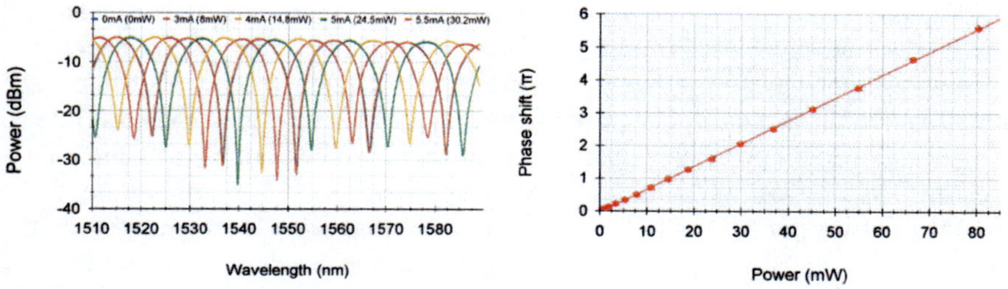

Fig. 2. Overlaid transmission spectra of an undercut device for various DC current levels (left) and corresponding extracted phase shift as a function of heater power (right).

used for the undercut process are placed at the same distance from the waveguides. All devices have been measured in the extended C-band, in TE polarization. It can be seen that the presence of trenches already gives ~30% improvement with respect to the reference structure, while the undercut boosts the power efficiency by almost tenfold. After improvement, the power required for a 2π phase shift $P_{2\pi}$ is only around 30 mW, which makes it comparable to the heater efficiencies in SOI platforms [5]. Besides, the lower current levels required for the same tuning give an opportunity to achieve larger phase shifts up to 6π without reaching critical current density values, which may otherwise lead to more complex packaging and limited device lifetimes.

Table 1. Power consumption for 2π phase shift for various heater configurations

Structure type	$P_{2\pi}$ [mW]	Improvement w.r.t reference
Reference	~296.5	–
With trenches	~229.5	~1.3x
With undercut	~30	~9.9x

3 Conclusion and Outlook

In conclusion, we have demonstrated an undercut process in a mature SiN photonic platform with high-confinement waveguides. The process has been performed on several manufacturing batches to confirm its stability and repeatability. All the steps have been performed at LIGENTEC standard manufacturing lines, using commercially available processes. A significant improvement of thermo-optical phase shifters power efficiency has been observed on dedicated MZI test structures in the telecom C-band wavelength window, without making compromises on the manufacturing process and optical performance. This approach may be followed irrespective of the silicon nitride layer thickness, is independent of the operational wavelength and may be equally applied to visible and NIR wavelength ranges. This makes it universally suitable for a variety of applications.

In this demonstration, the widths of the oxide bridges and suspended beams were still designed conservatively, biased towards mechanical robustness. Further optimization of the thermal resistance of the mechanical support without trading off on mechanical stability is still possible, providing an outlook at even higher heater efficiencies. Additional efficiency gains are still possible with optimized layouting, for example by having multiple passages of a waveguide within the same suspended beam [4]. Furthermore, in the context of the growing interest and demand for fast heterogeneously integrated (Lithium Niobate, BTO, etc.) phase shifters on Si/SiN platforms, the developed process flow can be directly reused for the next generation of such electro-optic phase shifters. For such devices, an undercut has been shown to reduce the losses experienced by microwave driving signals [6], allowing high-speed modulation, with bandwidths above 100 GHz.

References

1. Dong, P., et al.: Low power and compact reconfigurable multiplexing devices based on silicon microring resonators. Opt. Express **18**, 9852–9858 (2010)
2. Alemany, R., Muñoz, P., Pastor, D., Domínguez, C.: Thermo-optic phase tuners analysis and design for process modules on a silicon nitride platform. Photonics **8**, 496 (2021)
3. Liu, S., et al.: Thermo-optic phase shifters based on silicon-on-insulator platform: state-of-the-art and a review. Front. Optoelectron. **15**(1), 9 (2022)
4. Yong, Z., et al.: Power-efficient silicon nitride thermo-optic phase shifters for visible light. Opt. Express **30**, 7225–7237 (2022)
5. Jacques, M., Samani, A., El-Fiky, E., Patel, D., Xing, Z., Plant, D.: Optimization of thermo-optic phase-shifter design and mitigation of thermal crosstalk on the SOI platform. Opt. Express **27**, 10456–10471 (2019)
6. Wang, Z., et al.: Silicon-lithium niobate hybrid intensity and coherent modulators using a periodic capacitively loaded traveling-wave electrode. ACS Photon. **9**(8), 2668–2675 (2022)

Tuesday Poster Session

An Ultra-Small InP Microdisk Laser Diode for Programmable Non-linear Activation Functions in Neuromorphic Photonics

Christos Pappas[1(✉)], Andrea Demarchi[2], Ioannis Roumpos[1], Guilhem Madiot[3], Miltiadis Moralis-Pegios[1], George Giamougiannis[1], Apostolos Tsakyridis[1], Alexandre Bazin[2], Francesco Manegatti[2], Grégoire Beaudoin[2], Konstantinos Pantzas[2], Isabelle Sagnes[2], Fabrice Raineri[2,3], and Nikos Pleros[1]

[1] Centre for Interdisciplinary Research and Innovation, Informatics Department, Aristotle University of Thessaloniki, Thessaloniki, Greece
chripapp@csd.auth.gr

[2] Centre de Nanosciences et de Nanotechnologies, CNRS, Univ. Paris Saclay, Univ. Paris Cité, Palaiseau, France

[3] Institut de Physique de Nice, CNRS, Univ. Côte d'Azur, Nice, France

Abstract. The increased demands of Deep Learning (DL) stress electronic computing hardware, prompting researchers in new computing paradigms. Neuromorphic photonic, emerged as a candidate, offering high throughput and energy efficiency, by harnessing light's advantages. Here, we propose and experimentally demonstrate a Microdisk laser as a programmable all-optical activation function (AF) unit for photonic neural networks (PNNs). The device with a footprint of only 44.2 μm^2 produced three non-linear AFs at 2Gbaud i.e., Inverse-ELU, Sigmoid and Clipped-GeLU at input power envelopes as low as 11.5 μW, exhibiting energy efficiency of 1.89 pJ/Symbol.

Keywords: Neuromorphic Computing · Optical Integrated Circuit · Non-linear Activations

1 Introduction

The constantly growing energy and compute power requirements of next-generation Deep Learning (DL) workloads is putting tremendous pressure in the underlying computing hardware [1], calling for orders of magnitude improvements in energy and footprint efficiency. In this context, researchers have focused on developing non-Von Neumann computing hardware, with neuromorphic photonic arising as a potential candidate capable of offering the required efficiency breakthroughs, through leveraging light's proven credentials [1]. Driven by the scaling laws of computational complexity, dictating a N^2 scaling for the linear part, previous efforts have mainly focused on the linear part of NN [2–5], developing photonic implementations that can accelerate the required matrix multiplication calculations at unprecedented computational speeds, with the non-linear part usually implemented in digital hardware.

J. Witzens et al. (Eds.): ECIO 2024, SPPHY 402, pp. 407–412, 2024.
https://doi.org/10.1007/978-3-031-63378-2_66

This discrepancy has fueled research interest in non-linear activation function (NLAF) deployments during the last years, with previous demonstrations moving across three main directions: (i) Electronic deployments of the NLA in the digital domain following an Analog-to-Digital conversion, that comes with the price of increased energy consumption and latency [3] (ii) Hybrid optical-electrical deployments relying on a cascade of photonic and electronic components, typically inducing significant energy and footprint overhead to the photonic NN implementations [6, 8] (iii) All-optical approaches that leverage non-linear phenomena in photonic devices, hitting the sweet spot in terms of latency and operational overhead [9–13], yet requiring high optical input powers to trigger the required non linearities and typically the use of bulky Semiconductor Optical Amplifiers (SOA) [9, 10]. To this end, meeting the cascadability, energy and footprint requirements of next generation photonic NN necessitates an integrated all-optical activation module that can operate at the required performance envelope and yet offer a versatile set of NLAFs.

In this paper we propose and experimentally demonstrate a photonic integrated all-optical AF module based on a InP-on-SOI Microdisk laser diode (LD), capable of implementing a set of widely used, in NN, AFs at footprint and energy efficiencies of 44.2×10^3 Gsymbols/mm^2 and 1.89 pJ/Symbol, respectively. The Microdisk LD leverages an active region of 4-QW InGaAsP/InGaAsP heterostructures and achieves electrical injection operation with a threshold current as low as 1.35 mA by leveraging a novel PIN junction approach [14], while also having an ultra-compact radius of 3.75 um. We experimentally define the non-linear operational regime of the Microdisk LD at 10 kHz and proceed by showcasing its high-speed credentials at 2 Gbaud. Controlling the detuning of the injected optical beam as opposed to the laser free-running mode, enables the on-chip realization of three distinct NLAF i.e., inverse-ELU, Sigmoid and clipped GELU, at input power envelopes as low as 11.5 μW, highlighting the potential of our device as a next-generation all-optical integrated AF module.

2 Microdisk Laser Diode Fabrication and Characterization

Figure 1 (a) illustrates a 3D schematic of the proposed electrical injection Microdisk LD, comprising a 450 nm-thick PIN junction embedding an active region of 4-InGaAsP QW heterostructures. The PIN junction is patterned as a Microdisk resonator of 3.75 μm radius. The n-doped layer on top is contacted thanks to circular metal pad whereas the p-doped layer is contacted on the side of the microdisk using a circular metal crown. The InP-based device is evanescently coupled to the underlying SOI circuitry. The electrical injection scheme and optical confinement was optimized through simulations in Lumerical software suite, taking into account the difference in mobility between electrons and holes, as well as the absorption losses coming from the contact metal. The 450 nm-thick III-V epitaxy grown by MOCVD coated with MgF_2 and SiO_2 is adhesively bonded on the SOI circuitry using BCB polymer, as depicted in Fig. 1 (b). The fabrication of the final prototype is based on multiple steps of e-beam lithography (EBL), with a SEM of the fabricated chip prior etching illustrated in Fig. 1 (c). Finally, Fig. 1(d) illustrates the emitted power versus current (L-I) curve of the Microdisk LD, alongside the measured spectral linewidth, revealing an emission threshold of 1.35 mA, where the laser power consumption and emitted power are 1.56 mW and 8 μW, respectively.

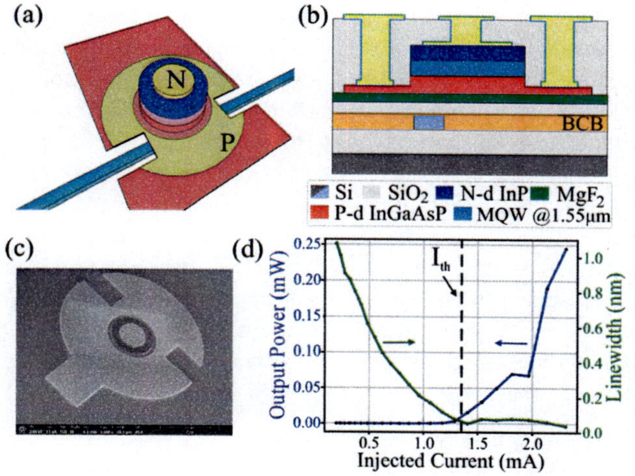

Fig. 1. (a) InP-on-SOI Microdisk Laser schematic (b) Vertical Scheme of the co-integrated laser diode: the Silicon circuitry is adhesively bonded on the III-V (c) SEM image of the fabricated Microdisk (d) Laser output power and linewidth versus DC current injection.

3 Microdisk Laser as an All-Optical Activation Function: Experimental Setup and Results

The experimental setup used for the characterization of the transfer function (TF) of the Microdisk LD is illustrated in Fig. 2 (a). A Tunable Laser Source (TLS) was used to generate an optical beam, henceforth referred to as λinj, that was injected into a GHz bandwidth electro-optical modulator (MOD). The MOD was responsible for imprinting the electrical waveform generated by the Arbitrary Waveform Generator (AWG) into the optical domain. The resulting modulated signal was then split via a 50:50 coupler into two constituents, with the first beam being injected to the Microdisk LD chip through a TE-optimized grating coupler (GC) and the second beam utilized for input signal monitoring purposes. Electrical access to the PIN diode of the fabricated laser was achieved through the use of a Ground-Singal-Ground electrical tip, while the injected current was controlled through a high-precision source measurement unit (SMU). The signal emitted by the Microdisk LD at a wavelength λMD = 1569.982 nm was then coupled out of the chip through an additional GC and was amplified in an Erbium Doped Fiber Amplifier (EDFA) prior getting filtered in a 1-nm Optical Band Pass Filter (OBPF). The filtered signal was split into a 70:30 coupler and monitored at both an Optical Spectrum Analyzer (OSA) and a digital sampling oscilloscope (OSC), after being converted into the electrical domain through a GHz-bandwidth photodiode (PD). Polarization controllers (PC) where employed at different stages of the setup to ensure proper polarization state of the signal.

Figure 2 (b) depicts a qualitative diagram of the Microdisk operational regions, showing two main operational regimes depending on the wavelength detuning Δλ of the injected optical beam compared to the free running mode of the Microdisk, with Δλ = λ_{MD}—λ_{inj}. For negative and small values of positive detuning, the Microdisk operates at its nonlinear regime, providing a non-linear TF suitable for NLAF in NN applications. As the injected laser moves towards higher positive detuning, the Microdisk operates as a bistable element with its transfer function following a hysteresis loop. Towards identifying the limits of the two operational regimes in our fabricated Microdisk prototype, a 10 kHz ramp optical signal was injected into the Microdisk with an optical power level and a wavelength detuning that were ranging between 0 to 120 μW and −25 pm to + 160 pm, respectively. Figure 2 (c) illustrates the two experimentally measured operational regimes, plotting the optical threshold versus Δλ for each case. More specifically, for the non-linear operational region, we plot the PNL that denotes the zeroing point of a fitted sigmoid-like function corresponding to approximately the half-power point of the resulting curve, as depicted in Fig. 2 (d) for Δλ = −10 pm. For the bi-stable case, we plot the PUP and PDOWN thresholds that correspond to the optical powers required for switching between the two states, with an illustrative example at Δλ = +140 pm depicted in Fig. 2 (e).

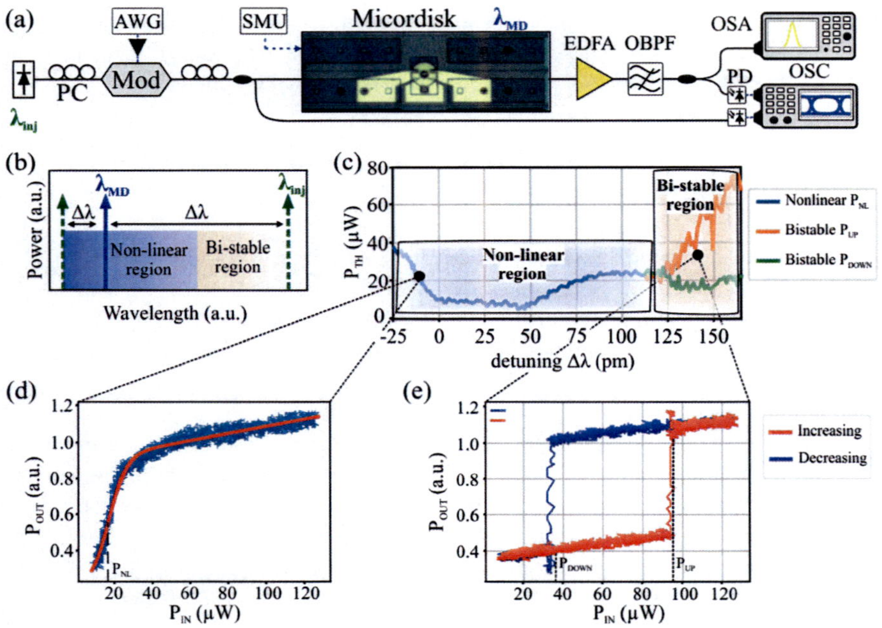

Fig. 2. (a) Experimental Setup for the characterization of the Microdisk diode as a NLAF (b) Qualitive operation mode diagram of the Microdisk nanolaser (c) Experimental definition of the operating regions of the fabricated devices. Transfer function of the Mircordisk at (d) its non-linear operational region and (e) the bi-stable operational region.

Having identified the two operational regimes of our fabricated sample, we focus on experimentally assessing the achieved TFs of the Microdisk LD for different detuning wavelengths in the non-linear optical regime. To extract the achieved optical TFs of the Microdisk, we generated a 32-level custom optical signal at 2 Gbaud and injected it into the laser. By monitoring the signal emerging at the Microdisk output and correlating

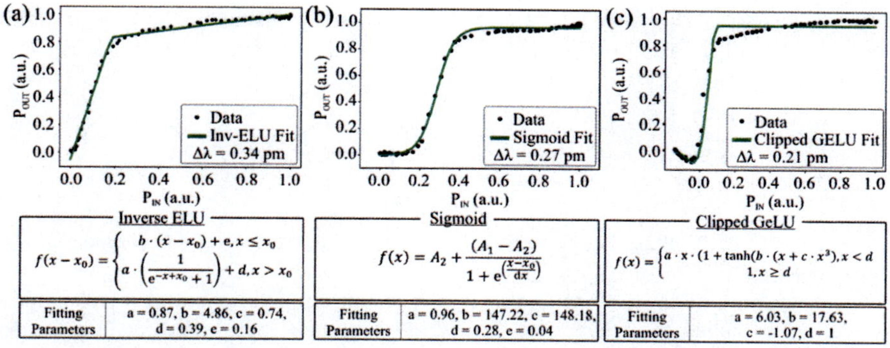

Fig. 3. (a), (b), (c) Experimentally acquired transfer functions for different detuning wavelengths $\Delta\lambda$, fitted to three popular NN activation functions i.e., Inverse ELU, Sigmoid and Clippeg GeLU, respectively. The related mathematical expressions are also depicted.

the output pulse peak powers with the respective input pulse peak powers, we can calculate the achieved optical TF of the Microdisk, following the same principles of [7, 8]. Figures 3 (a)–(c) illustrate three experimentally derived optical transfer functions of the Microdisk LD for a wavelength detuning of 34, 27 and 21 pm, respectively, when the Microdisk is electrically driven with 2.6 mA at 1.45 V. The optical input and output laser signals had an average power of 11.5 μW and 20 μW in all three cases. Fitting the experimentally acquired data points to three widely used NN AFs, i.e., the inverse-Exponential Linear Unit (ELU), Sigmoid and the Clipped Gaussian Error Linear Unit (GELU) revealed high-accuracy fitting, with the insets in Fig. 3 describing the mathematical expression of the targeted AFs.

Finally, in Table 1 we benchmark the performance of our prototype versus previously demonstrated all-optical AF. As can be observed, our implementation offers almost 4-orders of magnitude improvement in area efficiency compared to state-of-the-art implementations with a remarkable energy efficiency of only 1.89pJ/NOP.

Table 1. Comparison vs State-of-the-Art.

Reference	No of AFs	Freq. (GHz)	Footprint (mm^2)	Area Efficiency ($GSym/s/mm^2$)	Energy Efficiency (pJ/Sym)
[8]	1	10	9	1.11	164
[9]	1	10	1	10	15
[10]	4	1	0.105	9.5	25
[11]	3	0.25	0.2	1.25	0.016
[12]	3	0.01	8×10^{-4}	12.8	7.4×10^3
This work	**3**	**2**	$\mathbf{44.2 \times 10^{-6}}$	$\mathbf{44.2 \times 10^3}$	**1.89**

Acknowledgements. This work was supported by the EC via H2020 Projects PLASMONIAC (871391), SIPHO-G (101017194), ALLEGRO (101092766), and H2020 ERC HYPNOTIC (726420).

References

1. Shastri, B.J., et al.: Photonics for artificial intelligence and neuromorphic computing. Nat. Photon. **15** (2021)
2. Giamougiannis, G., et al.: Analog nanophotonic computing going practical: silicon photonic deep learning engines for tiled optical matrix multiplication with dynamic precision. Nanophotonics (2023)
3. Shen, Y., et al.: Deep learning with coherent nanophotonic circuits. Nat. Photon. **11**(7), 441–446 (2017)
4. Nahmias, M.A., et al.: Photonic multiply-accumulate operations for neural networks. IEEE JSTQE **26**(1), 1–18 (2020)
5. Totovic, A., et al.: WDM equipped universal linear optics for programmable neuromorphic photonic processors. NCE (2022)
6. Williamson, I.A.D., et al.: Reprogrammable electro-optic nonlinear activation functions for optical neural networks. IEEE JSTQE (2020)
7. Pappas, C., et al.: Programmable Tanh-, ELU-, Sigmoid-, and Sin-based nonlinear activation functions for neuromorphic photonics. IEEE J. Sel. Top. Quant. Electron. **29**(6), 1–10 (2023). Photon. Sig. Process.
8. Kovaios, S., et al.: Programmable Tanh- and ELU-based photonic neurons in optics-informed neural networks. J. Lightwave Technol. (2024) https://doi.org/10.1109/JLT.2024.3366711
9. Mourgias-Alexandris, G., et al.: An all-optical neuron with sigmoid activation function. Opt. Express **27** (2019)
10. Shi, B., et al.: InP photonic integrated multi-layer neural networks: architecture and performance analysis. APL Photon. **7**(1) (2022)
11. Jha, A., et al.: Programmable, high-speed all-optical nonlinear activation functions for neuromorphic photonics. OFC (2021)
12. Gordon, H.Y. L., et al.: All-optical ultrafast ReLU function for energy-efficient nanophotonic deep learning. Nanophotonics (2022)
13. Wu, B., et al.: Low-threshold all-optical nonlinear activation function based on a Ge/Si hybrid structure in a microring resonator. OMex (2022)
14. Demarchi, A., et al.: Low threshold hybrid InP-on-SOI microdisk laser diode with thin PIN junction. CSW (2023)

Waveguide Spectral Lens and Solar Spectrum Measurement in the Visible-Near-Infrared Region

Shijie Ke and Ziyang Zhang(✉)

Laboratory of Photonic Integration, School of Engineering, Westlake University, 18 Shilongshan Road, Hangzhou 310024, Zhejiang, China
zhangziyang@westlake.edu.cn

Abstract. We present a compact spectrometer based on integrated waveguide spectral lenses (WSL). The WSL directly focuses the on-chip dispersed light onto a camera at the designed distance without any free-space optics. Commercial polymer materials are chosen for wafer-scale fabrication with low propagation loss near visible-near-infrared region. Solar spectrum around 600–850 nm is obtained by a broadband WSL. The detailed absorption lines of O_2 (A band) are observed by a WSL of fine resolution. The results agree well with the reference from a commercial optical spectral analyzer and the literature.

Keywords: Astrophotonics · integrated photonic spectrometer · polymer waveguide

1 Introduction

The demand for next-generation astronomical spectroscopic instrumentation necessitates a compact and cost-effective integrated photonic spectrometer (IPS) that offers miniaturization and scalability [1]. Arrayed waveguide grating (AWG) functions as mature wavelength division multiplexers for optical communication systems [2, 3], rendering it a promising solution for spectroscopic observation. The feasibility of AWG-based IPS for astronomical observation has been previously demonstrated using commercial silica-on-silicon devices to disperse light gathered form starlight [4]. However, separated output channels of AWG need to be cleaved along the curved focal plane to meet the spectroscopic continuity requirement for astronomical observation. And additional lenses are needed to reimage the on-chip spectrum onto free-space camera with micrometer-scale accuracy. Above post-processing procedures may cause defocus aberrations and increase the complexity and volume of a IPS system.

To overcome these challenges, we present a novel IPS design called waveguide spectral lens (WSL) which simultaneously achieves the spectral dispersion and free-space focusing function on chip by tailoring the length difference in waveguide array [5]. It directly focuses the on-chip dispersed light onto camera at the designed distance without any free-space optics. Moreover, a millimeter-scale focal depth is achieved to relax the

J. Witzens et al. (Eds.): ECIO 2024, SPPHY 402, pp. 413–417, 2024.
https://doi.org/10.1007/978-3-031-63378-2_67

alignment accuracy between chip and camera which allows the fiber-chip assembly to be conveniently packaged in a tube mountable to a camera [6]. To further demonstrate its feasibility, we apply WSL to measure the solar spectrum in the visible-near-infrared region (VNIR) around 600–850 nm based on a polymer photonic integration platform [7]. We experimentally observed the absorption lines of O_2 (A and B bands) and H_2O through a broadband WSL (FSR = 405 nm) and the two-dip details of O_2 (A band) with a fine-resolution WSL (R~1000). The captured spectrum is in good agreement with the results from a commercial optical spectral analyzer (OSA) and the literature.

2 Design and Fabrication

Figure 1 shows the spectrometer system using a WSL chip. Initially, light is gathered through an aperture and filtered to prevent spectral line overlap beyond the chip's free spectral range (FSR). A commercial collimator then directs the light into a single-mode fiber connected to the WSL chip. The chip comprises a single-mode input waveguide, a beam broadening area (BBA), and a waveguide array. Light from the fiber is coupled to the waveguide and reaches the BBA, where it spreads horizontally while remaining vertically confined. Exiting the BBA, light enters equally spaced waveguide array around the slab's periphery, all equidistant from the input, ensuring all the output waveguides with equal phase. The waveguide array lengths then vary from each other to make phase difference for grating and focusing effects. Arranged evenly at the output as a straight line, the waveguide array emit light to the free space with phased modulation, producing a horizontally dispersed yet focused image on the camera. Finally, the original spectral picture is analyzed to get the spectrum by pre-calibration and summing pixel intensity in the vertical direction.

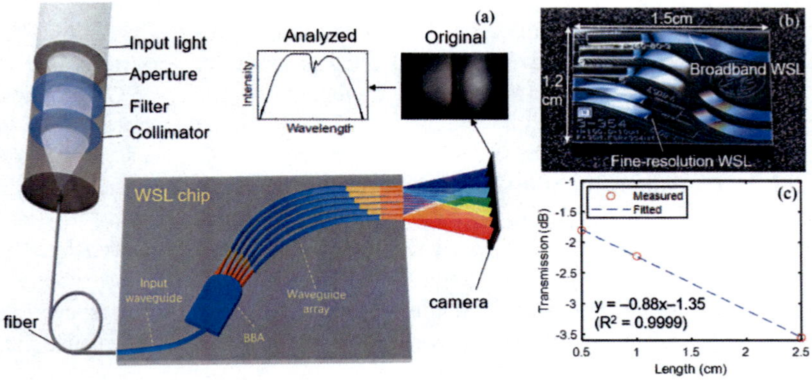

Fig. 1. (a) Schematic of a spectrometer system based on a WSL chip. (b) Photo of the fabricated WSL chip. (c) Cutback measurement of the polymer waveguides at 824.6 nm.

We designed and fabricated WSLs with different spectral resolutions (R~100–1000 at 600–850 nm) as shown in Fig. 1(b). The overall chip size is 1.2 cm × 1.5 cm with five different WSLs. For further demonstration, we selected a broadband WSL (top) and a fine-resolution WSL (bottom), featuring a designed free spectral range (FSR) of 409.6 nm and a resolution of 1000 at 824.6 nm, respectively. To work in the VNIR, commercial polymer materials (WIR30-RI series, ChemOptics) with low propagation loss at this region are adopted. The core and cladding materials have refractive indices of 1.466 and 1.45, respectively. The waveguide core's dimensions are set at 2.5 μm × 2.5 μm. Standard cutback measurement result of the waveguides at 824.6 nm is shown in Fig. 1(c). Low propagation loss (~0.88 dB/cm) and fiber coupling loss (~0.68 dB/facet) are achieved. The propagation losses at 784.2 nm and 636.1 nm are measured as ~1.20 dB/cm and ~1.83 dB/ cm, respectively. The propagation loss increases as the wavelength decreases.

3 Experimental Results

To evaluate WSL performance, we conducted a wavelength calibration using laser diodes with known spectral distributions to map camera pixel positions to wavelengths, including an 824.6 nm laser with a narrow spectral width (<0.02 nm) for assessing WSL's spectral resolution. The spectral measurements for the broadband WSL, displayed in Figs. 2(a) and (b), yielded a FSR of 404 nm and a Full Width at Half Maximum (FWHM) of 4.4 nm at 824.6 nm, aligning closely with the design specifications.

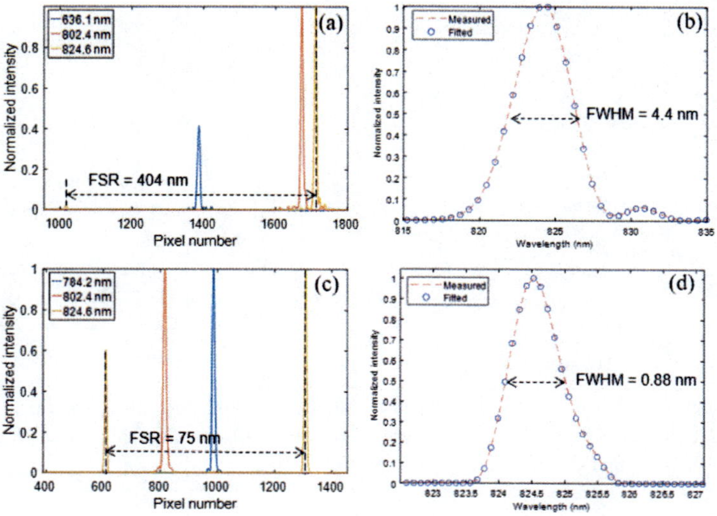

Fig. 2. (a) Normalized spectral distributions for wavelength calibration of the broadband WSL. (b) FWHM characterization of the broadband WSL at 824.6 nm. (c) Normalized spectral distributions for wavelength calibration of the fine-resolution WSL. (d) FWHM characterization of the fine-resolution WSL at 824.6 nm.

(FSR = 409.6 nm, FWHM = 4.2 nm). For the fine-resolution WSL, shown in Figs. 2(c) and (d), the findings were an FSR of 75 nm and FWHM of 0.88 nm at 824.6 nm, indicating a resolution of 937. The results also match the design (FSR = 74.5 nm, FWHM = 0.8 nm). Intensity calibration is then conducted to obtain the actual spectrum of an object by eliminating the spectrometer's intensity response from the measured spectrum [8]. This is realized using broadband light sources, i.e., LEDs with pre-known spectral intensity distribution.

After the calibration, we measure the solar spectrum by using these two WSL chips for broadband and fine-resolution analysis, respectively. Sunlight is collected, filtered, coupled into the fiber, and transmitted to the lab using the same system as shown in Fig. 1. Reference spectrum is simultaneously captured by a commercial OSA for comparison. The broadband solar spectrum is firstly analyzed to identify specific atmospheric absorption lines of H_2O and O_2 as a proof of concept. The OSA's spectral resolution is adjusted to 2 nm, providing finer detail compared to the WSL's resolution of 4.4 nm. Figure 3(a) displays the normalized sunlight intensity spectrum, where a significant intensity reduction below 700 nm of WSL reflects the polymer materials' higher absorption losses. The inset of Fig. 3(a) shows that the wavelengths and intensity of the absorption lines of O_2 (A and B bands) and H_2O are in good agreement with the spectrum measured by the OSA and the literature [9]. The absorption line O_2 (A band) is further resolved by the fine-resolution WSL. Figure 3(b) shows the normalized intensity spectrum. The spectral resolution of the OSA is set as 0.5 nm, which is close to this WSL (0.88 nm). The two dips of O_2 (A band) are well resolved. Both of the solar spectrum measured by WSL math well with the OSA, demonstrating its potential in providing integrated and robust photonic spectrometers for astronomical applications.

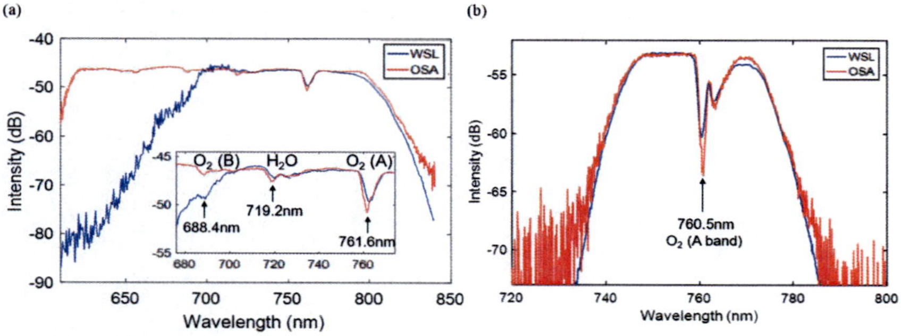

Fig. 3. (a) Calibrated Solar spectra measured by broadband WSL and OSA. The inserted picture shows the detailed absorption lines of O_2 and H_2O. (b) Calibrated Solar spectra measured by fine-resolution WSL and OSA.

4 Conclusion

We successfully measured the solar spectrum in the VNIR by integrated waveguide spectral lens (WSL). This compact chip disperses and focuses light onto a camera without any free-space optics. Commercial polymer materials with low propagation loss in this spectrum are used for wafer-scale manufacturing. Solar radiation is analyzed by two types of WSL chips for broadband and detailed spectral assessments, respectively. The detected absorption lines of O_2 and H_2O align closely with results from commercial OSA and literature. To the best of our knowledge, it is the first observation in this spectral region using integrated spectrometers. This demonstrates the potential of WSL chips as affordable, versatile tool for developing compact, portable IPS devices.

References

1. Allington-Smith, J., Bland-Hawthorn, J.: Astrophotonic spectroscopy: defining the potential advantage. Mon. Not. R. Astron. Soc. **404**(1), 232–238 (2010)
2. Smit, M.K.: New focusing and dispersive planar component based on an optical phased array. Electron. Lett. **24**(7), 385–386 (1988)
3. Takahashi, H., Suzuki, S., Kato, K., Nishi, L.: Arrayed-waveguide grating for wavelength division multi/demultiplexer with nanometer resolution. Electron. Lett. **26**, 87–88 (1990)
4. Cvetojevic, N., Jovanovic, N., Betters, C., et al.: First starlight spectrum captured using an integrated photonic micro-spectrometer. Astron. Astrophys. **544**(8), 453–478 (2012)
5. Jiang, X., Zhang, Z.: Planar waveguide-based fiber spectrum analyzer mountable to commercial camera. Photonics **9**, 456 (2022)
6. Jiang, X., et al.: Fiber spectrum analyzer based on planar waveguide array aligned to a camera without lens. Opt. Lasers Eng. **159**, 107226 (2022)
7. Ding, Z., Dang, Z., Jiang, X., Zhang, Z.: Solar spectrum measured by waveguide spectral lens. Opt. Laser Technol. **169**, 110110 (2024)
8. Thuillier, G., et al.: SOLAR/SOLSPEC: scientific objectives, instrument performance and its absolute calibration using a blackbody as primary standard source. Sol. Phys. **257**, 185–213 (2009)
9. Reference air mass 1.5 spectra. https://www.nrel.gov/grid/solarresource/spectra-am1.5.html

Fast Prototyping of Facet-Attached Microlenses Using 2PP Printing

Gandolf Feigl[1,2](✉), Matthias Jannach[1], Samuel M. Hörmann[1,3], Jakob W. Hinum-Wagner[1,3], and Alexander Bergmann[1,2]

[1] Institute of Electrical Measurement and Sensor Systems, Graz University of Technology, Inffeldgasse 33, 8010 Graz, Austria
gandolf.feigl@tugraz.at

[2] Christian Doppler Laboratory for Structured Matter Based Sensing, 8010 Graz, Austria

[3] ams OSRAM Group, Tobelbader Strasse 30, 8141 Premstaetten, Austria

Abstract. Photonic integrated circuits represent a promising approach for mass-produced biosensors that can be used for rapid medical testing in point-of-care diagnostics. An auspicious strategy to achieve more alignment-tolerant coupling of laser light is posed by using printed microoptics. In this work, printed microlenses are used to enhance the alignment tolerance of edge-coupled laser light into waveguides. A quick prototyping approach achieved a threefold improvement of the horizontal alignment tolerance. With the used printer, the fabrication time of a single lens is reduced to five minutes, enabling fast design iteration and potentially paving the way for cheap and user-friendly PIC sensors.

Keywords: Photonic integrated circuits · printed microoptics · microlenses · integrated photonics · silicon photonics · silicon nitride · biosensing

1 Introduction

In the past decade, much effort has been invested in designing and fabricating photonic integrated circuits (PICs). While they are already widely used in optical communication systems, PICs also play a crucial role in optical sensing applications, such as biomedical and environmental sensing [1]. They can be used to construct compact and highly sensitive sensors for detecting changes in light intensity, wavelength, or phase [2]. A commonly used platform for PICs is silicon nitride (SiN), which shows favorable optical properties and is established as the most common waveguide material used in biosensing [3]. Optical transmission of SiN shows a peak in the near-infrared, coinciding with a minimum in optical absorption of water, enabling SiN to be used for sensing analytes in aqueous solutions. The high refractive index contrast between SiN and cladding materials like silicon oxide or between SiN and aqueous environments renders SiN an excellent material for achieving high sensitivities in detecting various

J. Witzens et al. (Eds.): ECIO 2024, SPPHY 402, pp. 418–423, 2024.
https://doi.org/10.1007/978-3-031-63378-2_68

biomarkers [4]. A significant bottleneck hindering the widespread use of PICs in diagnostics is the efficient coupling of laser light. A promising solution to achieve higher alignment tolerance is the use of printed microlenses that can significantly enhance the efficiency of laser light coupling by focusing the incident light onto the input facet of edge-coupled (EC) waveguides [5]. With sub-micron resolution, two-photon polymerization (2PP) printers show great potential for the manufacturing of micro-optics. 2PP printed optics have been demonstrated for beam shaping, mode conversion, fiber-to-chip coupling, usually at telecom wavelengths, especially at 1550 nm [6]. In this work, we present a straightforward design and manufacturing approach for fast prototyping of 2PP microlenses that enhance alignment-tolerant fiber-to-chip coupling. We use PICs and lenses designed for visible laser light at 660 nm. With the presented approach, the application can easily be extended to wavelengths in the near-infrared used for chemical and biosensing. For the fabrication of the microlenses, we use a NanoOne 2PP printer system manufactured by UpNano, enabling possibilities for rapid prototyping.

2 Setup

2.1 Lens Design

Zemax OpticStudio was used to design the microlenses. The sequential ray-tracing method offers a computationally inexpensive technique to optimize the surface of the lenses in order to maximize the amount of light coupled into the waveguide and relax the alignment tolerance. For simplicity, the incident laser light is assumed to be a bundle of parallel beams. The lens surface is designed to be aspherical, focussing the light onto a single point. The lens is designed for the use with a 660 nm laser and the refractive index of the 2PP resin is $n = 1.505$. Two lens sizes were designed with focal lengths(diameters) of 80(40) μ and 160(50) μm, respectively. The curvature of the lens surface can be described with Eq. 1:

$$f(x) = \frac{\frac{x^2}{r}}{1 + \sqrt{1 - (1 + \kappa) * (\frac{x}{r})^2}} \tag{1}$$

The surface of the The curvature radii r of the small and the large lens are simulated to be 26.9 μm and 53.7 μm respectively. With a conic constant $\kappa = -0.44$, the surface is a prolate ellipsoid.

2.2 2PP Printing of Microlenses on Waveguide Facets

For the measurements, a 2 cm × 2 cm photonic chip was used containing SiN single-mode strip waveguides with dimensions of 450 nm × 160 nm width × height. The chips are glued to a glass substrate that was subsequently placed into the printer vat. Using the built-in substrate camera of the printer, the focal point of the lenses were aligned with the waveguide facets. The printing time of the lenses amounts to 5:21 min and 14:46 min for the small and large lenses, respectively. Figure 1 shows optical microscope images of the lenses.

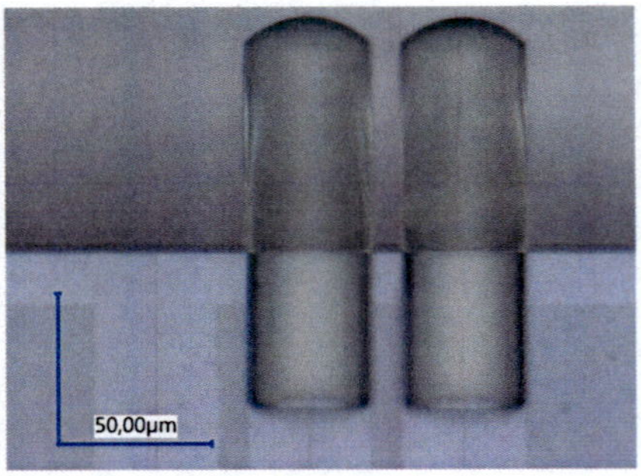

(a) Small lens printed onto the chip. The length(diameter) of the lenses are 80(40) µm length(diameter)

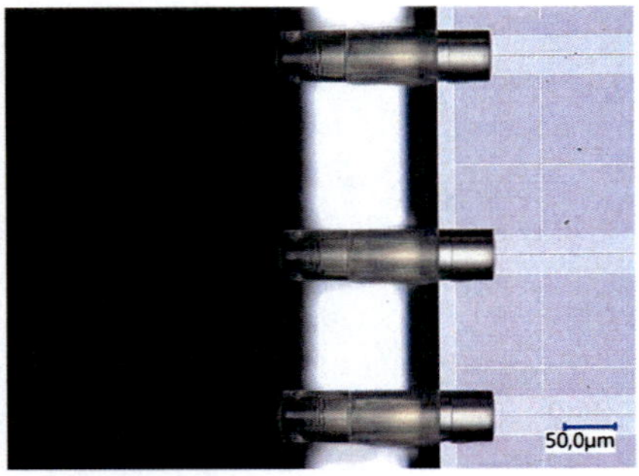

(b) Large lens printed onto the chip. The length(diameter) of the lenses is 160(50) µm

Fig. 1. Microscope images of the printed lenses

2.3 Edge Coupling Stage

In order to measure the alignment tolerance of the coupled light, a custom-built stage was used, see Fig. 2. The laser fiber was mounted in the vicinity of the waveguide facet, and piezoelectric stages can tune the horizontal and vertical shifts. For a precise measurement, the tuning characteristics of the piezo stages were measured with a Michelson interferometer yielding approximately 280 nm/V in the linear regime of the stages. The laser and the detector were manually aligned to optimize the detector signal. Using a function generator, the piezo stages were actuated to scan a two-dimensional space of approximately 35 µm × 35 µm, with a constant distance between the fiber tip and chip. A detector connected to an oscilloscope collects the light transmitted through the waveguides. As a laser source, a fiber-coupled laser with a center wavelength of 660 nm was used. Reference measurements of three waveguides without lenses were conducted. Identical waveguides were used for the lenses printed on the chip.

3 Results

The detector signal was used to create a two-dimensional representation of the alignment tolerance. Figure 3 show representative contourplots of the misalignment loss of the coupling process with a reference waveguide and a waveguide with a lens printed on it.

The innermost contour line represents the 3 dB loss border. With the reference waveguide, the loss exceeds 3 dB at a misalignment of <5 µm, exhibiting an almost symmetric behavior. The lensed waveguide shows a significantly increased horizontal alignment tolerance with an alignment tolerance of >10 µm, but no substantial improvement in vertical direction. This behavior is seen for all lensed

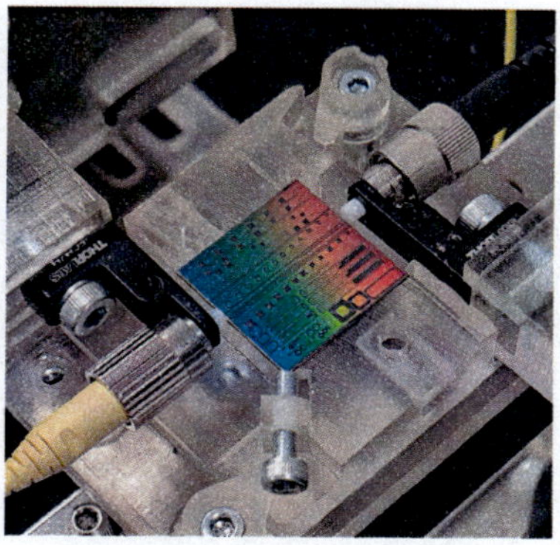

(a) Stage for edge coupling of the laser light. The laser fiber (yellow strain relief boot) is mounted in the vicinity of the waveguide facet.

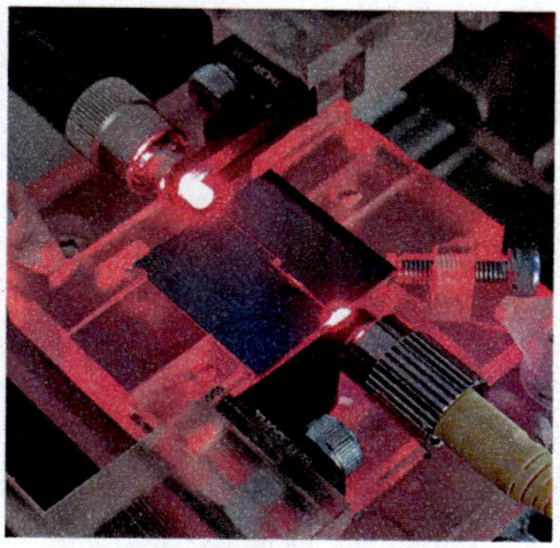

(b) Laser light coupled into the waveguide. Scattering on the waveguide surface makes the light clearly visible. In the final setup, a stray light blocker is used to suppress unwanted coupling into the detector fiber.

Fig. 2. Custom-built edge coupling stages

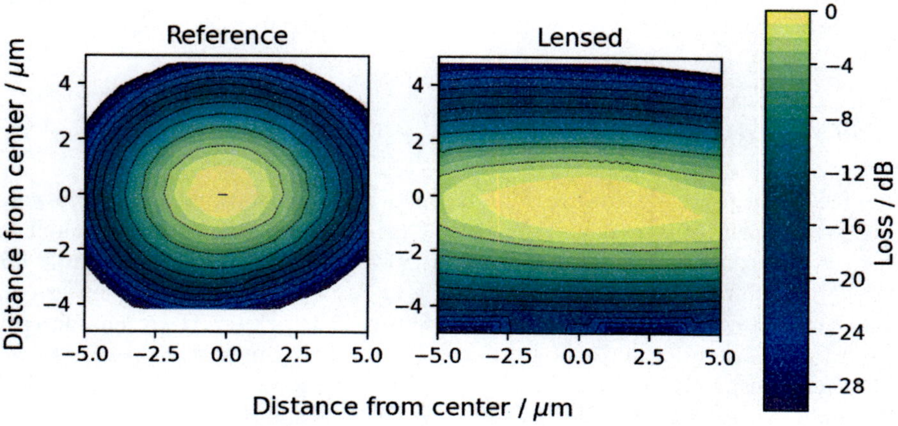

Fig. 3. EC stage with lens

waveguides and is attributed to the waveguide geometry, whose height is significantly smaller than its width. This leads to a delicate alignment of the lenses' focal point to the waveguide facet in the printing process. The results were very reproducible for all waveguides, whether with or without lens. Figure 4 shows the misalignment loss (a) and mean 3 dB alignment tolerance (b) for the reference, as well as the lensed chips. Figure 4b shows an almost two-fold improvement of the horizontal alignment tolerance when using chips with the small lens, while a three-fold improvement can be achieved with the bigger lens. The smaller radius of curvature of the bigger lens adds to an improved alignment tolerance as it more reliably couples light that exits the laser fiber at an angle. No significant change in the vertical alignment tolerance was achieved, regardless of the lens size.

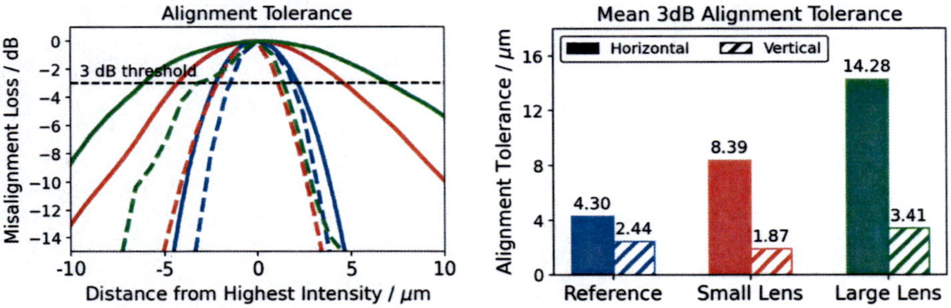

(a) Misalignment loss versus distance from the highest coupling efficiency

(b) Mean 3 dB alignment tolerances

Fig. 4. Misalignment loss (a) and alignment tolerance (b) of the chips when measuring the reference waveguide (blue) and the waveguides with small (red) and big (green) lens. The solid(dashed) lines represent the horizontal(vertical) tolerances.

4 Summary and Discussion

Microlenses for printing on waveguide facets of photonic integrated circuits were designed and printed onto chips using a 2PP printer with sub-μm resolution. The design is realized in a simple and straightforward manner, easily adaptable for laser light of different wavelengths, printing resins, or lens sizes. The lenses can quickly and easily be prototyped with the used 2PP printer. Two lens designs were printed: a smaller and a larger one, with printing times of approximately 5 and 15 min, respectively. With the smaller lens, the horizontal alignment tolerance could already be enhanced by a factor of two, while the bigger lens enabled a three-fold improvement. No substantial change in the vertical alignment tolerance could be observed. This is attributed to the delicate alignment between the lens's focal point and the waveguide facet, which is only as high as 160 nm. More sophisticated designs are expected to even further enhance the coupling efficiency. Lenses could be used for mode-field matching of the respective modes in laser fiber and waveguide. Future work is aimed to extend the experiments to other wavelengths, like 850 nm and 1550 nm, as prominently used in for sensing and telecommunication. The significant improvement of the alignment tolerance in combination with the straight-forward design and comparatively short prototyping times, 2PP printed microlenses show great potential in future solutions for the packaging of photonic integrated circuits.

Acknowledgements. The financial support by the Austrian Federal Ministry of Labour and Economy, the National Foundation for Research, Technology and Development and the Christian Doppler Research Association is gratefully acknowledged.

References

1. Torrijos-Moráin, L., Lisboa, B.D., Soler, M., Lechuga, L.M., García-Rupérez, J.: Integrated optical bimodal waveguide biosensors: principles and applications. Results Opt. **9**, 100285 (2022). https://doi.org/10.1016/j.rio.2022.100285. https://www.sciencedirect.com/science/article/pii/S266695012200075X
2. Milvich, J., Kohler, D., Freude, W., Koos, C.: Integrated phase-sensitive photonic sensors: A system design tutorial. Adv. Opt. Photon. **13**(3), 584 (2021). https://doi.org/10.1364/AOP.413399, https://opg.optica.org/aop/abstract.cfm?URI=aop-13-3-584
3. Muñoz, P., et al.: Silicon nitride photonic integration platforms for visible, near-infrared and mid-infrared applications. Sensors **17**(9) (2017). https://doi.org/10.3390/s17092088, https://www.mdpi.com/1424-8220/17/9/2088
4. Luan, E., Shoman, H., Ratner, D.M., Cheung, K.C., Chrostowski, L.: Silicon photonic biosensors using label-free detection. Sensors **18**(10) (2018). https://doi.org/10.3390/s18103519, https://www.mdpi.com/1424-8220/18/10/3519
5. Xu, Y., et al.: 3D-printed facet-attached microlenses for advanced photonic system assembly. Light: Adv. Manuf. **4**(2), 77 (2023). https://doi.org/10.37188/lam.2023.003. 43.32.03; LK 01
6. Gonzalez-Hernandez, D., Varapnickas, S., Bertoncini, A., Liberale, C., Malinauskas, M.: Micro-optics 3D printed via multi-photon laser lithography. Adv. Opt. Mater. **11**(1), 2201701 (2023). https://doi.org/10.1002/adom.202201701, https://onlinelibrary.wiley.com/doi/abs/10.1002/adom.202201701

Control of Programmable Optical Processors by Using a Neural-Network Digital Twin

Gabriele Cavicchioli(✉), Francesco Maria Sances, Andrea Melloni, and Francesco Morichetti

Dipartimento di Elettronica Informazione e Bioingegneria, Politecnico di Milano, Via Ponzio 34/4, 20133 Milan, Italy
gabriele.cavicchioli@polimi.it

Abstract. We propose a neural-network (NN) model to automatically configure programmable optical processors made of meshes of Mach-Zehnder interferometers (MZIs). The NN model predicts the phase shifts of the MZIs such that a MZI mesh implements a given matrix, even in presence of fabrication imperfections and thermal-cross talk. Using a simulated dataset, we validate the NN model and its scalability. Moreover, we show that our approach applies to arbitrary mesh topologies

Keywords: Programmable Optical Processors · Artificial Neural Network · Data-driven control systems · thermal cross-talk

1 Introduction

Programmable optical processors (POPs) provide a fast and efficient implementation of linear algebra operations, such as matrix-vector multiplication (MVM), directly in the optical domain and are finding many applications in analog computing, hardware accelerators [1] and machine learning [2]. POPs can be realized through networks, or meshes, of Mach-Zehnder interferometers (MZIs), which can be reconfigured by setting the values of phase shifters embedded inside each MZI. These architectures can implement arbitrary transformations between input and output [3], but their configuration is a quite complex task, especially in case of fabrication tolerances, which are responsible for unknown biasing of the MZI phases, and when thermal phase shifters are used, which suffer from mutual thermal cross-talk. Such issues are currently limiting the scalability of these architectures to large port counts, although many techniques have been proposed to implement control and calibration procedures [4].

In this work we propose a calibration technique based on the use of an artificial neural network (ANN) that allows us to configure a MZI-based POP to a desired transmission matrix. Such an ANN digital twin is trained to predict the working point of each phase shifter of the MZIs even in presence of fabrication imperfections and strong thermal cross-talk. Numerical simulations show the

J. Witzens et al. (Eds.): ECIO 2024, SPPHY 402, pp. 424–429, 2024.
https://doi.org/10.1007/978-3-031-63378-2_69

effectiveness of the approach with different MZI mesh topologies, including triangular [5] and rectangular meshes [6]. Moreover, we investigate how the dataset properties influence the performance of the network with respect to the size of the MZI mesh.

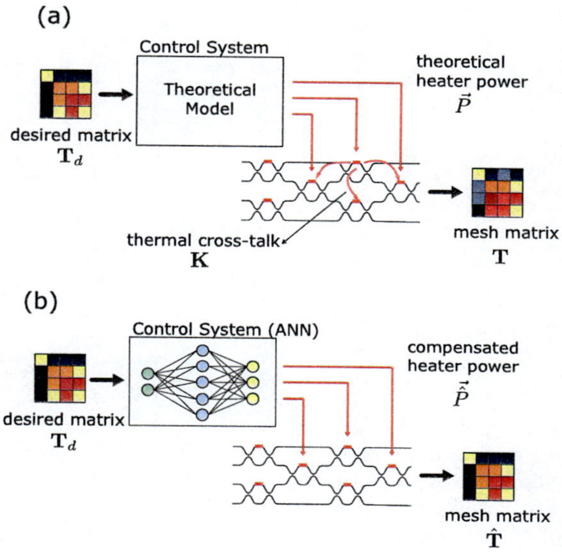

Fig. 1. (a) Control systems exploiting theoretical model fail in estimating non-idealities, including thermal cross-talk, resulting in inaccurate mesh matrix setting. (b) A neural network opportunely trained can take the matrix to be program and find the set of phase shiftings necessary to obtain it in the real device.

2 Neural-Network-Based Digital Twin of a Programmable Optical Processor

A MZI mesh is characterized by a transmission matrix T whose coefficients depend on the phase shifts $\boldsymbol{\theta}$ applied to each MZI (Fig. 1a). When thermal tuners are used, the phase shift θ_i in the i-th MZI is affected by the heat generated by surrounding heaters; a thermal cross-talk matrix K can be defined describing the relationship between the phase shift vector and the heater power vector $\boldsymbol{P}$, that is $\boldsymbol{\theta} = K\boldsymbol{P}$. The thermal cross-talk matrix strongly depends on the topology of the POP and on the position of the thermal tuners, and its analytical or experimental estimation is rather difficult, especially for large size MZI meshes. In order to bypass the estimation of K, in this work we adopted a neural network model to understand how the transmission matrix T of the POP and the heater power vector $\boldsymbol{P}$ are correlated.

The NN used to predict the mesh actuator states is a 5-layers feed-forward NN with the structure shown in Fig. 2a. The dataset used to train the network

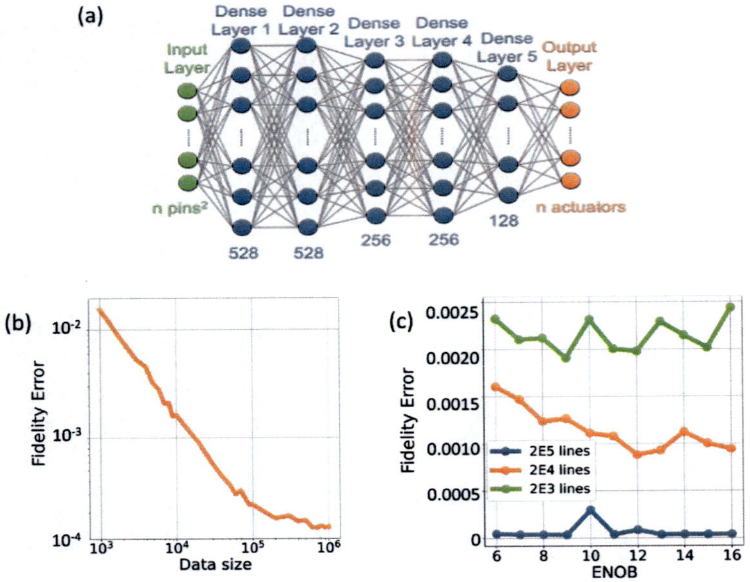

Fig. 2. (a) Structure of the ANN used to model the MZI-based programmable processor. (b) Model prediction performance (fidelity error) evaluated on a 3×3 triangular MZI mesh varying the dataset dimension D_{rand}, without D_{sweep} and quantization noise (ENOB = 12). (c) Fidelity error of a 3×3 triangular mesh with varying noise for fixed dataset size.

consists of the elements of T labelled by the values of $\boldsymbol{P}$ required to obtain that specific transmission matrix. Moreover the dataset is composed of two parts, namely D_{rand} and D_{sweep} . D_{rand} consists of entries obtained by first setting the MZI phase shifters in the $[0, \pi]$ interval, then analytically computing the matrix T. D_{sweep} is instead a dataset obtained by first setting all actuators but the sweeping one to a fixed value $\theta \in \{i * \frac{\pi}{8}\}, i = 1, 2, 3, ...8$; then, the T matrix is computed for n distinct values of the sweeping actuators sweeping range. This ensures that all actuators' effect is well represented in the dataset and corrects possible small random under-representation biases. The datasets used for training the model are numerically generated by using a MZI mesh simulator based on the transmission matrix method (TMM), which also emulates thermal cross-talk induced by neighbouring thermal tuners. The use of simulated datasets allows a more flexible exploration of the dataset size and composition. Training is performed using *Adam optimizer* with decaying learning rate and *Mean Squared Error* (MSE) between the predicted actuator phase shifts $\hat{\boldsymbol{\theta}}$ and the corresponding label in the dataset $\boldsymbol{\theta}_d$ as the loss function. The trained model is evaluated *a posteriori* according to the actuators mean squared error (MSE) but also to the fidelity

$$F = \frac{Tr(\hat{T}^T T_d)}{N}, \tag{1}$$

between the transmission matrix T_d used as input to the neural network and the estimated transmission matrix $\hat{T}$ obtained by setting the phase shifts to the predicted ones.

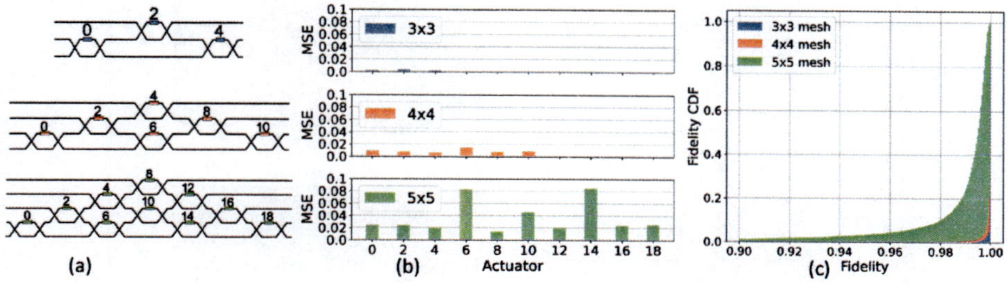

Fig. 3. Mesh performance scalability. (a) Triangular (Reck) mesh configuration schematics for 3×3, 4×4, 5×5 meshes. ANN performance comparison between 3×3, 4×4 and 5×5 triangular meshes, using datasets of 10^6 lines with 12 effective number of bits (ENOB), in terms of single actuators MSE (b) and fidelity (c).

3 Numerical Results

Figure 2b shows the *fidelity error* $\varepsilon_F = 1-F$ versus the dimension of the random dataset D_{rand}. An error $\varepsilon_F < 1\%$ is achieved with a dataset of 10^3 lines, while ε_F saturates to about 10^{-4} for $D_{rand} > 10^5$. It should be noted that, since the thermal phase shifters can be driven up to some kHz speed, such a large dataset can be collected in a time scale of a few seconds. Figure 2c shows the impact of a quantization noise in the dataset, which emulates for effective number of bits (ENOB) of the digital electronics controlling the thermal phase shifters. Simulations show that quantization noise does not affect the performance of the NN prediction, at least down to an ENOB of 6 bits. These results enable us to set the requirements in terms of dataset size and equipment performance for acquiring the training dataset in a real experimental scenario.

Figure 3 shows the application of the approach to increasingly larger MZI meshes. The same ANN structure and the same dataset size of 10^6 lines and ENOB = 12 is used to model 3×3, 4×4 and 5×5 MZI meshes. Figure 3b shows the prediction error of each phase shifter for different mesh dimensions. The actuators with the largest MSE are the one which are more difficult to observe because of their inner position, while the actuators situated in the first diagonal and anti-diagonal are more easily detected. The prediction accuracy reduces as the dimension of the mesh increases, and this behaviour is shown also in Fig. 3c, which represents the cumulative distribution function of the fidelity F. Nonetheless, in all the cases taken into account, the predicted fidelity reaches very high levels: the probability of getting a fidelity accuracy of 0.9 or less is 1% even for the 5×5 mesh.

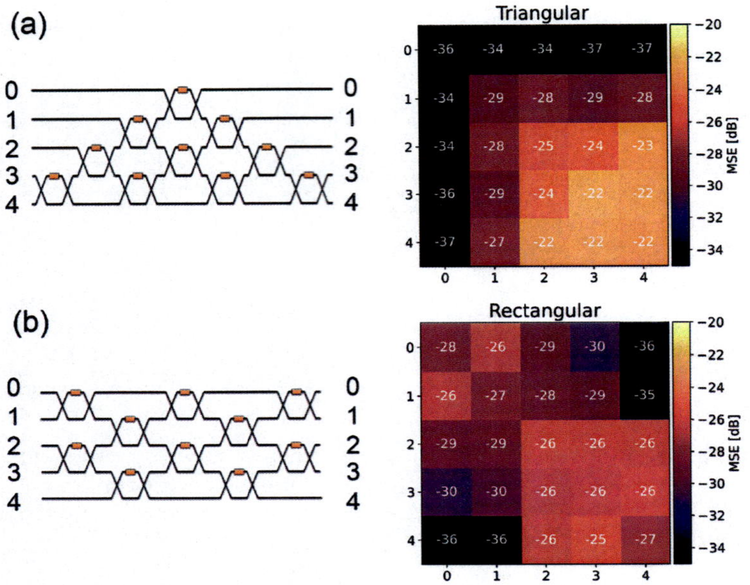

Fig. 4. (a) Comparison between MZI architectures with different topologies. MSE of the elements of the estimated transmission matrix $\hat{T}$ of a 5×5 MZI mesh with (b) triangular and (c) rectangular configuration.

To provide a more in-depth insight into the performance of the method, we also evaluated the MSE of each element of the transmission matrix T. Here, to show also the wide applicability of the NN-based approach, which is independent of the specific topology of the photonic circuit, we applied the model to different mesh architectures. For example, Fig. 4 shows the comparison between the prediction error of a 5×5 triangular (Reck) mesh (a) and that of a 5×5 rectangular (Clements) mesh (a). In both cases, a very good accuracy is achieved, the mean MSE (averaged on the 25 matrix elements) being -26.7 dB and -28.2 dB, respectively. As expected from the topological properties of the two architectures, the triangular mesh has a lower MSE in the elements associated to the "shortest" paths, that is those paths where the light travels through the minimum number of MZIs, and a progressively higher MSE in the "longer" paths, while the rectangular mesh has the advantage of more balanced paths, thus resulting in a reduced channel cross-talk.

4 Conclusions

In this work, we proposed and validated through numerical simulations an ANN-based model for a digital twin of programmable optical processors made of meshes of MZIs. The method can be applied to MZI meshes with arbitrary topologies, also in presence of non-idealities and thermal cross talk. Reported results for meshes with 5×5 optical I/Os show an average MSE of less than -25 dB on the estimation of the transfer matrix and a fidelity of 0.9 or more with 99% probability. The required size of the dataset is compatible with the tuning speed of conventional phase shifters enabling the acquisition of the training data in a time scale of a few seconds.

Acknowledgements. This work was supported by the Italian National Recovery and Resilience Plan (NRRP) of NextGenerationEU, "Telecommunications of the Future" (PE00000001 - program "RESTART", Structural Project "Rigoletto" and Focused Project "HePIC") and on MUSA: Multilayered Urban Sustainability Action.

References

1. Zhou, H., et al.: Photonic matrix multiplication lights up photonic accelerator and beyond. Light Sci. Appl. **11**(1), (2022). https://doi.org/10.1038/s41377-022-00717-8
2. Shen, Y., et al.: Deep learning with coherent nanophotonic circuits. Nat. Photonics **11**(7), 441–446 (2017). https://doi.org/10.1038/nphoton.2017.93
3. Miller, D.A.B.: Self-configuring universal linear optical component [Invited]. Photonics Res. **1**(1), 1–15 (2013). https://doi.org/10.1364/PRJ.1.000001
4. Alexiev, C., Mak, J.C.C., Sacher, W.D., Poon, J.K.S.: Calibrating rectangular interferometer meshes with external photodetectors. OSA Continuum **4**(11), 2892–2904 (2021). https://doi.org/10.1364/OSAC.437918
5. Reck, M., Zeilinger, A., Bernstein, H.J., Bertani, P.: Experimental realization of any discrete unitary operator. Phys. Rev. Lett. **73**(1), 58–61 (1994). https://doi.org/10.1103/PhysRevLett.73.58
6. Clements, W.R., Humphreys, P.C., Metcalf, B.J., Kolthammer, W.S., Walmsley, I.A.: Optimal design for universal multiport interferometers. Optica **3**(12), 1460–1465 (2016). https://doi.org/10.1364/OPTICA.3.001460

Combining Photonic Integrated Circuits and Detectors with Absolute Responsivity as an Absolute Power Source

Lars Kristian Skaar(✉), Johanne Heitmann Solheim, and Jarle Gran

Justervesenet Fetveien 99, Kjeller, Norway
lks@justervesenet.no
https://www.justervesenet.no/

Abstract. This paper presents for the first time a method of combining a Predictable Quantum Efficient Detector (PQED) and a cascaded multimode interference coupler (MMI) to generate a predictable laser source used to characterize detectors at low optical power levels in the sub nW range. The results show an agreement between predicted and measured power levels at the low output channel below 80 fW at a wide range of power levels. At higher power levels, the agreement between measured and predicted values are 0.2%.

Keywords: Integrated quantum optics · Multimode interference couplers · Optical power · Predictable Quantum Efficient Detector · Photonic integrated circuit

1 Introduction

Predictable Quantum Efficient Detectors (PQED)[3–5] have in recent years become a primary standard for the realization of the candela as its responsivity is given by its intrinsic reference to fundamental constants [1]. The PQEDs have a responsivity, R, directly proportional to the wavelength of the incoming light and fundamental constants by the following relation [6]:

$$R = \frac{e\lambda}{hc}(1 - \delta) \tag{1}$$

where λ is the vacuum wavelength of the incoming light, e is the elementary charge, h is Planck's constant and c is the speed of light in vacuum. The δ factor, estimated to 100 ppm, is included to account for reflectance and internal losses in the photodiode. The high accuracy at room-temperature and ease of use have made PQEDs a preferred method in characterizing sources and detectors.

In this paper we present a method of exploiting the PQED with an optical splitter to make an absolute power source which can be used for characterizing detectors at low optical power levels.

J. Witzens et al. (Eds.): ECIO 2024, SPPHY 402, pp. 430–436, 2024.
https://doi.org/10.1007/978-3-031-63378-2_70

The optical splitter consists of six cascaded 1:10 multimode interference couplers (MMI), integrated into a photonic integrated circuit (PIC) produced on the TriPleX platform [7]. The output from each MMI is collected and directed into optical fibres providing six output channels with a decadal power difference between them, as illustrated in Fig. 1. The output channels, labeled A1 to A6, have respectively a nominal power reduction ranging from 1:10^{-1} to 1:10^{-6}. The MMI splitter properties have been modelled in reference [2] and the physical parameters have been optimized for this application, with low propagation losses at wavelengths close to 850 nm.

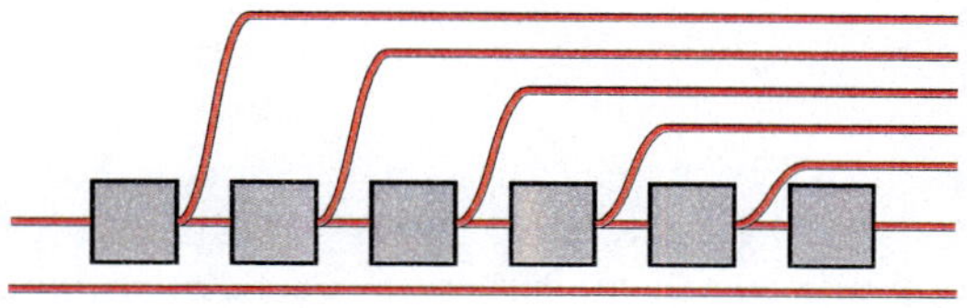

Fig. 1. Conceptual schematics of the cascaded MMI splitter chip. Each grey box represents the individual MMI, with one of the outputs being collected as output and the other output being fed into the next MMI.

The method entails attaching a PQED on one of the high power channels (e.g. A1) and a device under testing (DUT) on one of the lower power channels. The output power on the low power channel is predicted by monitoring the high power channel with the PQED and using the characterised split ratio of the PIC splitter. This ensures that the PQED receives sufficient power to operate effectively, and that the DUT is exposed to power levels matching its operational range.

2 Experimental Setup

The experimental setup is shown in Fig. 2. The incoming radiation is produced by a Thorlabs FPV852P laser with a measured wavelength of 851.8 nm, guided with single mode optical fibres into an attenuator (Agilent 8157A) and paddle polarization controller before entering the MMI splitter chip. The two output signals from the splitter are collected with Thorlabs FGA01FC photodiodes (PD) or PQEDs. The photocurrents from each detector are amplified by Femto DDPCA-300 transimpedance amplifiers (TIA) and measured with a 7 1/2 digit multimeter (Keithley 2002 and Keithley DMM7510). The setup was placed in a climate-controlled room at 21 °C to ensure thermal stability within 0.1 °C.

Prior to the measurement sequence, optimising the throughput of the low power channel was made by varying the polarisation. This procedure was followed in both the ratio characterisation and the prediction measurements. The measurements were conducted by first sampling the dark current (attenuator

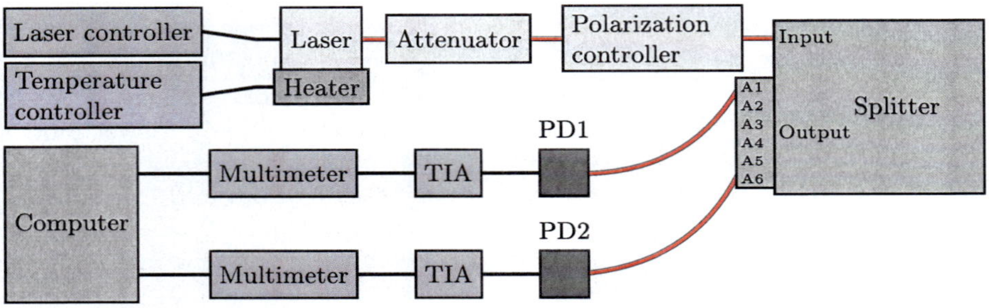

Fig. 2. Schematics of the experimental setup. The black (orange) lines represents electrical (optical) connections.

closed) before sampling the photocurrent (attenuator open). For both conditions, 10 consecutive readings were averaged, taken 1 s apart with a number of power line cycles set to 10 on each multimeter. A settling time of 20 s were included after each dark-and photocurrent measurement. This procedure was repeated for different attenuation levels of the optical power, specifically at 0, 5, 10, 15 and 20 dB.

3 Results and Discussion

In the following section we present the results achieved with the measurement procedure introduced above. Similar procedures were followed for both the ratio characterization and the prediction measurements.

3.1 Characterization of Ratio

Accurate characterization of the PIC split ratios and its stability is crucial to predict the output of other channels from a reference channel. The ratio characterization was performed with two identical detectors on the output channels, according to the measurement procedure given in Sect. 2. Using two identical detectors amplified with a gain G_1 (high power channel) and G_2 (low power channel), the photocurrent ratio, Γ, can be calculated as:

$$\Gamma = \frac{(V_{0,1} - V_{d,1})G_2}{(V_{0,2} - V_{d,2})G_1} \tag{2}$$

with $V_{d,i}$ and $V_{0,i}$ being the voltage signal from the dark-and photocurrent measurement for the high ($i = 1$) and low ($i = 2$) output channels. During the ratio calculation, only the 0 dB attenuation was used in order to optimize the signal to noise ratio received by the detectors in this characterisation. This results in a more accurate representation of the splitter ratio by reducing the relative contribution from the dark currents during the photocurrent measurements. By performing this calculation on the A1/A2 and A1/A5 channels, we get a ratio of 11.69 ± 0.17 and 10308 ± 177, respectively. As seen from the ratios above, and confirmed by measuring the ratios of other channel pairs, the values do not accurately follow the nominal decadal splitting, necessitating characterization of each channel pair individually.

3.2 Prediction of Output

By using the previously characterized ratio, we can predict the output of one of the low power channels by measuring the output from a high power channel, from the relation:

$$\Phi_p = \frac{(V_0 - V_d)}{GR\Gamma} \tag{3}$$

where G is the amplifier gain, R is the responsivity of the detector and V_0 and V_d is the light and dark voltage measurements, respectively. Such a measurement for A1/A2 is shown in Fig. 3a, illustrating the predicted and measured photon flux at the low power channel (A2) as a function of the measured photon flux at the high power channel (A1). Figure 3b shows the percent difference between the measured and predicted photon flux. From these figures, it is evident that the difference between the predicted and measured values fit well within 0.3%. The measurement series over the different attenuations was performed twice, leading to the two points on each power level.

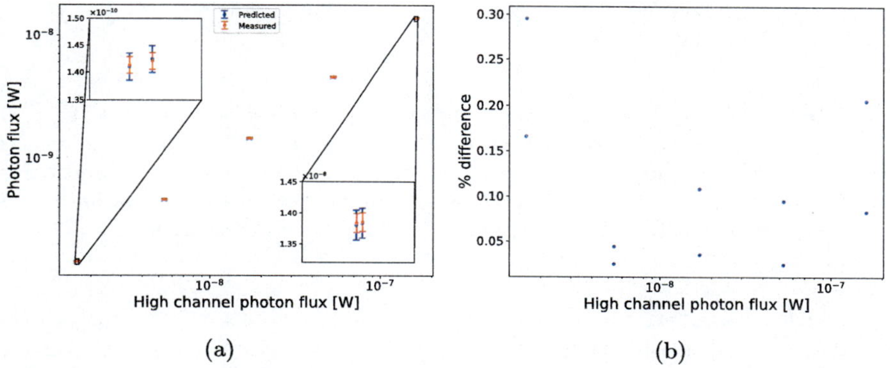

Fig. 3. Results of the (a) optical power of the predicted and measured output of the A2 channel as a function of the power from the A1 channel and (b) the difference between the measured and predicted values in (a). The ratio have been characterized with two PQEDs and the measurements have been performed with a PQED on the A1 channel and PD on the A2 channel.

As the optical power on the low power channels decreases, validating that the ratio value is maintained gets increasingly challenging as high gain is required and the photocurrent is approaching the dark current reading. Prediction measurements of the A1/A5 channels are shown in Fig. 4, where the ratio was characterized with two PDs as these have a lower surface area compared to the

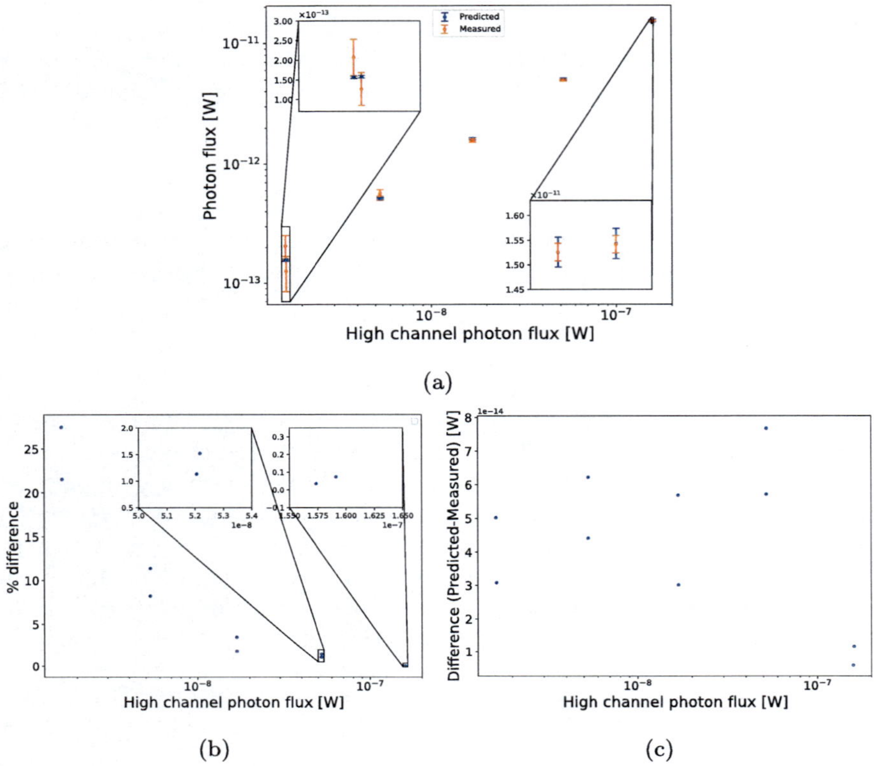

Fig. 4. Results of the (a) optical power of the predicted and measured output of the A5 channel as a function of the power from the A1 channel, (b) and (c) shows the relative and absolute difference between the predicted and measured values in (a), respectively. The ratio have been characterized with two PDs and the measurements have been performed with a PQED on the A1 channel and PD on the A5 channel.

PQEDs used, resulting in a lower dark current contribution and thus a more accurate characterization of the ratio. The prediction agrees with the measurements within 0.2% for 0 dB attenuation, at the signal level where the ratio have been characterized. However, the relative deviation increases rapidly with decreased optical power. Figure 4c shows the absolute difference between the predicted and measured value, which remains between 5 to 80 fW with no visible trend as a function of the power from the high output channel. When approaching the detection limit, the absolute uncertainty in fW is more consistent to use than relative uncertainty. The dominant factor for the difference between the predicted and measured values arise from a high dark current contribution in relation to the photocurrent, which can be mitigated by cooling the detectors.

3.3 Uncertainty Evaluation

In the following section, we present an evaluation of the uncertainties associated with the setup related to the ratio and optical power measurements. The uncertainties for the photocurrent measurements are determined by the standard deviation of the readings constituting the averaged value. The uncertainty in the amplifier is 1% of the gain, according to the instrument's manual. However, this can be significantly reduced through calibration of the amplifiers. The uncertainty in the responsivity of the PQEDs are calculated according to Eq. 1.

Tables 1 and 2 presents the uncertainty budget for the A1/A2 ratio and optical power calculation, respectively. The tables include the measurement value and its uncertainty, the sensitivity factor and the uncertainty factor in the ratio and power for each component included in Eq. (2) and (3). From the tables it is clear that the uncertainty from the gain accuracy is the primary contributor. However, the agreement between the predicted and measured values arises from the fact that the same amplifiers were used in both experiments. This does not reduce the uncertainty in the absolute value but suggests that the gain accuracy may be improved by calibrating the amplifiers. Analysing each component's contribution to the total uncertainty is a valuable method in order to examine what parts of the setup needs to be improved, in addition to stating the measurement capability.

Table 1. Uncertainty budget of the ratio calculation for A1 vs A2

Quantity	Value	Uncertainty	Sensitivity	Uncertainty in ratio	Relative uncertainty
$V_{0,1}$	−1.0096 V	1.1 mV	$-11.59\,V^{-1}$	−0.013	
$V_{d,1}$	−0.679 mV	15 µV	$11.59\,V^{-1}$	1.74×10^{-4}	
$V_{0,2}$	−0.8610 V	0.9 mV	$13.56\,V^{-1}$	0.012	
$V_{d,2}$	−7.271 mV	38 µV	$-13.56\,V^{-1}$	-5.14×10^{-4}	
G_1	1×10^7 V/A	0.01×10^7 V/A	-1.169×10^{-6} A/V	−0.12	
G_2	1×10^8 V/A	0.01×10^8 V/A	1.169×10^{-7} A/V	0.12	
Γ	11.69			0.17	1.45%

Table 2. Uncertainty budget of the predicted flux for A1 vs A2 with 0 dB attenuation.

Quantity	Value	Uncertainty	Sensitivity	Uncertainty in power[nW]	Relative uncertainty
V_0	−1.110 V	3 mV	1.245×10^{-8} W/V	0.04	
V_d	−0.902 mV	19 µV	-1.245×10^{-8} W/V	-2.31×10^{-4}	
G	1×10^7 V/A	0.01×10^7 V/A	1.380×10^{-15} WA/V	0.14	
R	0.68702 A/W	0.00011 A/W	2.009×10^{-8} W^2/A	2.13×10^{-3}	
Γ	11.69	0.17	1.180×10^{-9} W	0.20	
Φ	−13.80 nW			0.24	1.74%

4 Conclusion

By combining a well characterised cascaded MMI splitter and an absolute detector, such as the PQED, we are able to predict the optical power of one of the splitter channels with an agreement below 80 fW over a wide range of power levels. At higher power levels, the agreement between measured and predicted values are 0.2%. This initial work presents a proof of concept method, and better results are expected when the system components are calibrated and the setup is optimised.

Acknowledgements. This project (20FUN05 SEQUME) has received funding from the EMPIR programme co-financed by the Participating States and from the European Union's Horizon 2020 research and innovation programme.

References

1. BIPM: Mise en pratique for the definition of the candela and associated derived units for photometric and radiometric quantities in the SI. SI brochure, 9th edn., Appendix 2 (2019). https://www.bipm.org/documents/20126/41489685/SI-App2-candela.pdf. Accessed 14 February 2024
2. Gounaridis, L., et al.: A novel photonic integrated optical divider based on cascaded multimode interference (MMI) couplers, achieving a splitting ratio of 1:10^{-9} (2024). [Manuscript submitted for publication]
3. Koybasi, O., et al.: High performance predictable quantum efficient detector based on induced-junction photodiodes passivated with SiO2/SiNx. Sensors **21**(23) (2021). https://doi.org/10.3390/s21237807, https://www.mdpi.com/1424-8220/21/23/7807
4. Müller, I., et al.: Predictable quantum efficient detector: II. Characterization and confirmed responsivity. Metrologia **50**(4), 395 (2013). https://doi.org/10.1088/0026-1394/50/4/395.
5. Sildoja, M., et al.: Predictable quantum efficient detector: I. Photodiodes and predicted responsivity. Metrologia **50**(4), 385 (2013). https://doi.org/10.1088/0026-1394/50/4/385
6. Tran, T., Porrovecchio, G., Smid, M., Ikonen, E., Dönsberg, T., Gran, J.: Determination of the responsivity of a predictable quantum efficient detector over a wide spectral range based on a 3D model of charge carrier recombination losses. Metrologia **59**(4), 045012 (2022). https://doi.org/10.1088/1681-7575/ac604b.
7. Wörhoff, K., Heideman, R.G., Leinse, A., Hoekman, M.: TriPleX: a versatile dielectric photonic platform. Adv. Opt. Technol. **4**(2), 189–207 (2015). https://doi.org/10.1515/aot-2015-0016

Thermo-optical Robustness of Crystalline Sb_2Se_3/Si Waveguides at 1310 nm

Iñigo Lopez-Mulet, Jorge Parra, Miroslavna Kovylina, and Pablo Sanchis(✉)

Nanophotonics Technology Center, Universitat Politècnica de València, Camino de Vera s/n, 46022 Valencia, Spain
pabsanki@ntc.upv.es

Abstract. In recent years, there has been a rising effort in developing chalcogenides for integrated photonics due to their reversible and non-volatile material transition, which entails a large refractive index change, thereby enabling ultra-compact devices. Antimony selenide (Sb_2Se_3) has arisen for applications requiring high-performance phase shifters due to its negligible absorption loss in both states at datacom and telecom wavelengths. This work experimentally demonstrates the thermo-optical robustness of crystalline Sb_2Se_3/Si waveguides in the range from 20 °C to 65 °C. We show that our hybrid waveguide benchmarks similar performance against a standard silicon waveguide, thereby maintaining fully compatibility with silicon photonic temperature control systems.

Keywords: chalcogenide · phase change material · silicon photonics

1 Introduction

The field of silicon photonics has garnered special interest during the last two decades due to the growing need for integrated devices with high-speed and low-power operation. One of the main advantages of silicon photonics is that the fabrication processes are compatible with the already well-established complementary metal-oxide semiconductor (CMOS) processing [1]. However, the active properties of silicon are either not suitable or ideal for many applications, so a lot of efforts are being developed for research of CMOS-compatible materials with new and improved properties. Phase change materials (PCMs) stand out for their reversible and non-volatile transition between their crystalline and amorphous phase, accompanied by a large change in the refractive index [2]. Some PCMs, such as $Ge_2Sb_2Te_5$ (GST), are ideal for amplitude modulation due to their high contrast of absorption between states so they have been used in applications like neuromorphic computing [3], optical memories or reconfigurable PICs [4]. Antimony selenide (Sb_2Se_3), is an alternative PCM that can achieve a non-volatile refractive index change with almost negligible losses, allowing to perform optical phase-shifting in a power-efficient way [5, 6]. Its non-volatility and low losses at datacom wavelength (1310 nm) could allow Sb_2Se_3-based switches that drastically reduce the power consumption at data centers, which are experiencing a continuously growing energy demand. As PICs undergo temperature variations with regular use, it is necessary to be aware of how that

J. Witzens et al. (Eds.): ECIO 2024, SPPHY 402, pp. 437–440, 2024.
https://doi.org/10.1007/978-3-031-63378-2_71

will affect the optical performance of the different materials and devices involved in the circuitry, and Sb_2Se_3 is yet lacking that analysis. In this work, the thermo-optical robustness of crystalline Sb_2Se_3/Si waveguides is characterized to analyze the influence of temperate in the optical performance of the hybrid waveguide.

2 Results

The characterization of the temperature dependence of the optical properties of Sb_2Se_3 was performed at 1310 nm on 2 × 2 asymmetric Mach-Zehnder interferometers (MZIs), with an arm length difference of 120 µm. Optical switching was produced by a Sb_2Se_3/Si waveguide placed in the MZI short arm. The hybrid waveguide comprised a 450 nm × 220 nm silicon (Si) waveguide with a 40 nm thick and 14 µm long Sb_2Se_3 patch on top, as shown in Fig. 1(a). The Sb_2Se_3 patch was deposited by e-beam evaporation with a subsequent lift-off process. Crystallization of the Sb_2Se_3 was achieved by heating the sample at 250 °C for 10 min in Ar. Figure 1(b) shows a SEM (Scanning Electron Microscope) image of a crystalline Sb_2Se_3/Si waveguide with larger width. On the other hand, fiber-to-chip coupling was accomplished using transverse electric (TE) grating couplers. The temperature of the sample was controlled by using a Peltier device in the range of 20 °C to 65 °C using 5 °C/step with temperature resolution of 0.1 °C.

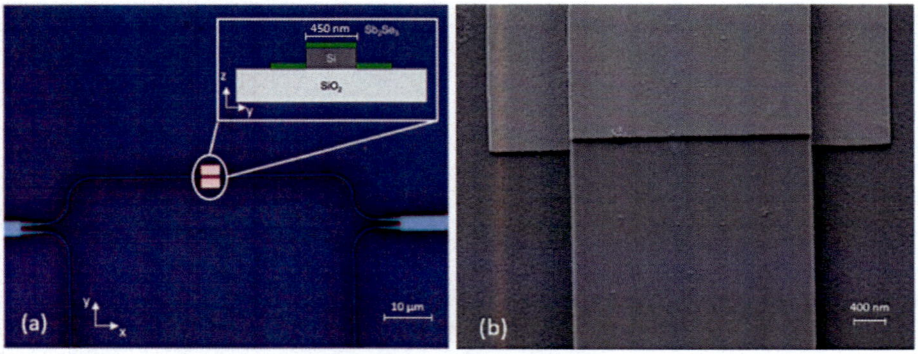

Fig. 1. (a) Optical microscope image of a fabricated Mach-Zehnder interferometer with a Sb_2Se_3/Si waveguide in one of its arms. The cross-section of the hybrid waveguide is shown in the inset. (b) SEM image of a 3 µm wide Sb_2Se_3/Si waveguide.

The thermo-optical response was first assessed for a Si-MZI as a reference and then for the MZI switch with the Sb_2Se_3/Si waveguide and Sb_2Se_3 in the crystalline state. Fig. 2 shows the experimental transmission response of the MZIs in the wavelength range of 1306–1316 nm. The evaluation of extinction ratio and phase shift dependence with temperature was extracted from the transmission spectra measured for both devices in the whole range of temperatures. Figure 3(a) shows the extinction ratio for the Si and the Sb_2Se_3/Si devices at different temperatures, along with the linear fits of the experimental data. The slope of the fitted lines is almost zero for both cases, and the experimental values show no clear tendency, so the extinction ratio can be considered invariant with temperature changes. This means that the temperature changes do not cause an increase in the absorption of the material or the coupling losses. On the other hand, it can be observed that the Si device exhibits a larger extinction ratio than the hybrid device, a difference that could be caused by small deviations in the fabrication process that produce an unbalance of optical losses of around 1 dB between the arms of the MZI.

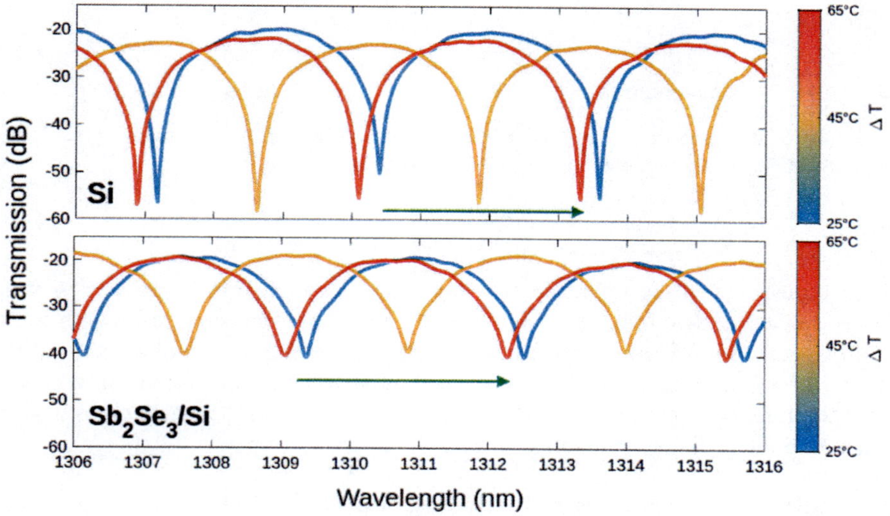

Fig. 2. Transmittance spectra of a Si MZI and a Sb_2Se_3/Si MZI at 25 °C, 45 °C and 65 °C. The arrows indicate the resonance peaks displacement with temperature variations.

Comparing the resonance peaks obtained in the MZI transmission spectra measurements, we calculated the deviation of the peak wavelength produced by the temperature changes for each device. The phase shift values are shown in Figure 2(b), where it is evident that the evolution with temperature is identical in both samples. The extracted values are 0.0458 π/°C for the Si waveguide and 0.0459 π/°C for the Sb_2Se_3/Si waveguide. Considering the thermo-optic coefficient of $1.89 \cdot 10^{-4}$/°C for Si at 1310 nm [7, 8], through simulations, we obtain a value of phase shift of 0.0398π/°C, which is in good agreement with the results obtained experimentally.

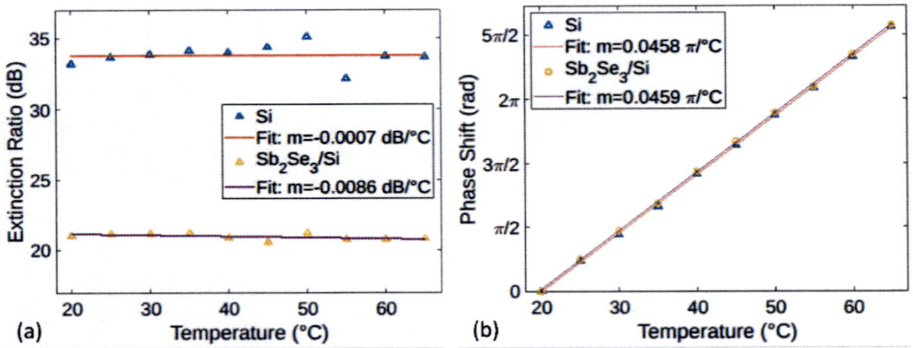

Fig. 3. (a) Extinction ratio and (b) optical phase shift variation with temperature for the Si and Sb_2Se_3/Si MZI devices with the linear fit of the values and its slope.

3 Conclusions

We have investigated the influence of temperature variations on the properties of an Sb_2Se_3/Si waveguide through experimental measurements and comparison with a regular Si waveguide. We have obtained that the optical absorption and phase shift of the hybrid waveguide with Sb_2Se_3 in the crystalline state is not dependent on temperature. Therefore, our results demonstrate the robustness of Sb_2Se_3/Si technology against temperature fluctuations in the range from 20 °C to 65 °C.

Acknowledgements. Agencia Estatal de Investigación (TED2021-132211B-I00, PID2022-137787OB-I00); Generalitat Valenciana (CIPROM/2022/14); Ministerio de Universidades (MS21-037); Universitat Politècncia de València (Grant PAID-01-22, PAID-10-23).

References

1. Lipson, M.: The revolution of silicon photonics. Nat. Mater. **21**, 974–975 (2022)
2. Zhang, Y., et al.: Myths and truths about optical phase change materials: a perspective. Appl. Phys. Let. **118**, 210501 (2021)
3. Shastri, B.J., et al.: Photonics for artificial intelligence and neuromorphic computing. Nat. Photon. **15**, 102–114 (2021)
4. Parra, J., Olivares, I., Brimont, A., Sanchis, P.: Toward non-volatile switching in silicon photonic devices. Laser Photon. Rev. **15**, 2000501 (2021)
5. Delaney, M., Zeimpekis, I., Lawson, D., Hewak, D.W., Muskens, O.L.: A new family of ultralow loss reversible phase-change materials for photonic integrated circuits: Sb2S3 and Sb2Se3. Adv. Funct. Mater. **30**, 2002447 (2020)
6. Ríos, C., et al.: Ultra-compact non-volatile phase shifter based on electrically reprogrammable transparent phase change materials. PhotoniX **3**, 26 (2022)
7. McCaulley, J.A., Donnelly, V.M., Vernon, M., Taha, I.: Temperature dependence of the near-infrared refractive index of silicon, gallium arsenide, and indium phosphide. Phys. Rev. B **49**, 7408–7417 (1994)
8. Kameda, A., Mizukawa, Y., Hanafusa, H., Higashi, S.: Precise measurement of the temperature of a silicon wafer by an optical-interference contactless thermometer during rapid plasma processing. J. Appl. Phys. **127**, 203302 (2020)

Towards Large-Language Model Assisted Layout of Silicon Photonic Integrated Circuits

Jason Liu[1], Ankita Sharma[1,2](✉), Cheick Doumbia[1], and Joyce K. S. Poon[1,2]

[1] Department of Electrical and Computer Engineering, University of Toronto, 10 King's College Road, Toronto, ON M5S 3G4, Canada
ank.sharma@mail.utoronto.ca, joyce.poon@mpi-halle.mpg.de
[2] Max Planck Institute of Microstructure Physics, Weinberg 2, 06120 Halle, Germany

Abstract. Recent breakthroughs in artificial intelligence (AI), especially large language models (LLMs) are revolutionizing the automation of complex language-based tasks. Here we report a software tool under development that uses the Llama2-70B language model for automating the layout of silicon photonic devices. Preliminary tests demonstrate the tool's effectiveness in accurately generating code to layout a single device from a natural language prompt.

Keywords: large language models · silicon photonics · chip design and layout

1 Introduction

Recent progress in both commercial (i.e. Open AI's GPT-4, Google's Gemini) and open-source (i.e. Meta's Llama2, xAI's Grok) large language models (LLMs) have been recognized as significant advancements in the field of artificial intelligence (AI), making it easier to automate complex tasks that require natural or programming languages [1–4]. LLMs are built on deep learning transformer architectures, trained on massive amounts of data to efficiently process and generate text with exceptional efficiency, scale and complexity [2, 4]. Numerous research efforts have already begun to apply LLMs to the design of electronic integrated circuits, leveraging their potential to enhance the existing capabilities of and accelerate electronic design automation (EDA) [5–7]. This includes deploying LLMs to develop autonomous agents for chip design that manage task planning, code generation for EDA tools, and oversee task execution [6]. Furthermore, LLMs have also been used to power engineering assistant chatbots or co-pilots that have the potential to help with identifying hardware constraints, summarizing and analyzing bugs, as well as clarifying design choices and theoretical concepts [5, 7].

Compared to EDA, the development of photonic design automation (PDA) tools that support automated design pipelines for circuit simulation and layout are still at an early stage [8]. The design and layout of photonic circuits and components typically requires significant human input, needing deep knowledge in semiconductor physics

Jason Liu and Ankita Sharma—These authors contributed equally to this work.

J. Witzens et al. (Eds.): ECIO 2024, SPPHY 402, pp. 441–447, 2024.
https://doi.org/10.1007/978-3-031-63378-2_72

and electromagnetic theory, along with skills in physics-based simulation and modelling. To reduce human intervention in the simulation and design process, photonics researchers and designers are increasingly turning towards deep learning approaches and optimization algorithms aiming to enhance efficiency in both forward and inverse design problems [8–10]. Nevertheless, a comprehensive set of tools enabling the full automation of photonic circuit design and layout does not yet exist. Advancements in LLMs offer an unprecedented opportunity to contribute to the advancement of PDA.

Previous work in [8] explored the use of ChatGPT as a co-pilot to automate the generation of electro-magnetic simulation and deep learning code for the inverse design of photonic crystal surface emitting lasers (PCSELs). However, in [8] a human designer was necessary to debug errors (hallucinations) in the LLM-generated code, which could only run successfully after several rounds of debugging. Here, we present our initial results in using the Llama2-70B model to automate the generation of code for the layout of conventional silicon photonic components and circuits, such as those that would be found in a foundry's process design kit (PDK). Our developed software tool automatically translates a natural language prompt into python code that generates a Graphic Data system Version II (GDSII) file for a device from a library of measured components in the C-Band/O-Band. We evaluated our layout automation tool using 7 distinct natural language prompts, each tested 100 times. Our preliminary results from these tests indicate that our tool generates an accurate layout in at least 99% of the test cases.

2 Software Overview

Due to concerns regarding data privacy and confidentiality in photonic design, we built our software using the Llama2-70B language model instead of GPT-4. Llama 2 is open-source and can be run locally on a computer. To eliminate hallucinations in LLM-generated code, our software breaks down the initial prompt (the "User Prompt") into several system prompts with clear instructions and in-context examples to instruct the LLM. We also provide the LLM with a vector index of layout functions from the open-source Python library GDSfactory to enhance the LLM's context and improve its output via retrieval-augmented generation (RAG) [5]. GDSfactory is an open-source package for chip design and validation, and it already supports the PDKs of numerous photonic foundries, including Advanced Micro Foundry, Smart Photonics, Tower Semiconductor and GlobalFoundries [11]. We utilized the Python package LlamaIndex to link our databases with the LLM and oversee data retrieval and querying. Finally, to reach a high level of correctness and to eliminate errors, we also implemented code verification steps, which prompt the LLM to revise its output if it is incorrect before delivering the final code to the user.

Figure 1 illustrates a block diagram of the overall approach. The first "System Prompt" in our code instructs the LLM to parse the "User Prompt" to identify the requested photonic device from an established list of devices as well as its operating wavelength. The LLM is then tasked with extracting all necessary physical parameters and the specific GDSfactory parametric cell needed for layout, retrieving this data from a verified design library stored in a Structure Query Language (SQL) database. Once the LLM has fetched the design parameters, the hard coded "SQL Response Check" assesses the accuracy of the data retrieved by the LLM by making sure it does not equal the default parameters of the GDSfactory parametric cell. After ensuring the correctness of the data, a second "System Prompt" asks the LLM to reference a vector index database containing all GDSfactory layout functions as well as in-context examples to produce the requisite code. Another external Python function then checks the parameters in the code against the parameters extracted from the SQL database to confirm that there are no errors in the generated code. Finally the LLM determines if the user has specified any specific values for physical parameters in the original "User Prompt". If so, it will replace the value extracted from the database with the user's request and perform a final comparison between the "User Prompt" and the generated code.

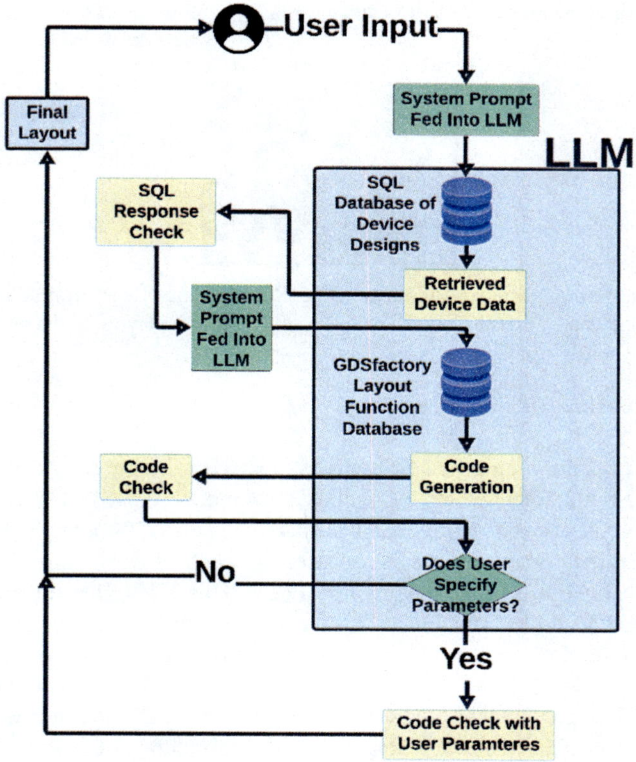

Fig. 1. Overview of software implemented for automated layout generation tool. The LLM used is the Llama2-70B model. To link the LLM with the SQL database of device designs and vector index database we used the python package LlamaIndex.

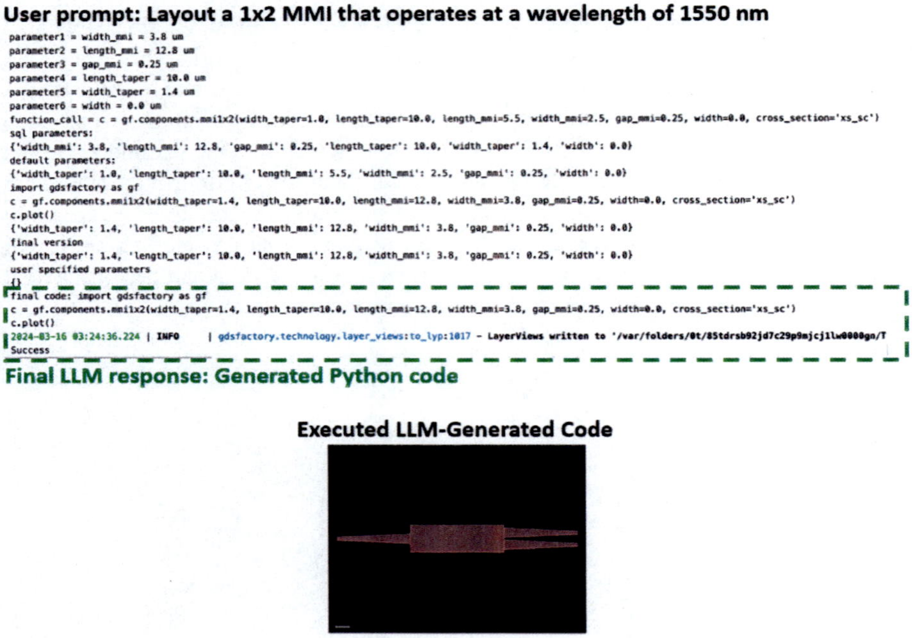

Fig. 2. Example output from our automation tool for a given "User Prompt". Our program generates code that can be executed by the designer to generate the layout of the requested device.

3 Characterization

First, to benchmark our program, we directly compared the output of our automation tool (Fig. 2) with the output of ChatGPT using the GPT-4 model (Fig. 3). To prompt ChatGPT we asked it to "Generate code to layout a 1x2MMI using the GDSfactory python library". ChatGPT generated-code contained several errors such as false function parameters for the python function gf.components.mmi1x2(). This highlights the importance of providing the LLM with sufficient context and also implementing code verification steps. In our first implementation of our tool with sufficient context but without the "SQL Response Check" and "Code Check", the LLM was only able to generate the correct layout for a C-Band 1x2 MMI 84 times out of 100 trials. By adding these additional verification steps the LLM was able to generate the correct layout for all 100 of our 100 tests.

In total, we tested our software with 7 distinct user prompts outlined in Table 1, with each prompt undergoing 100 trials. Both Prompts 5 and 6 encountered a single failure because the initial data fetched from the SQL database was incomplete, specifically missing one of the design parameters in the response. Nonetheless the generated code successfully met all of our code verification tests. An example of a failed generated layout, for example for Prompt 6 in Table 1, is code that generates a Mach Zehnder Interferometer (MZI) with 1x2 MMIs defined as the default parametric cells, which are not optimized for a wavelength of 1310 nm. We are currently improving the "SQL Response Check", which evaluates the accuracy of the retrieved data to include an assessment of the data's completeness.

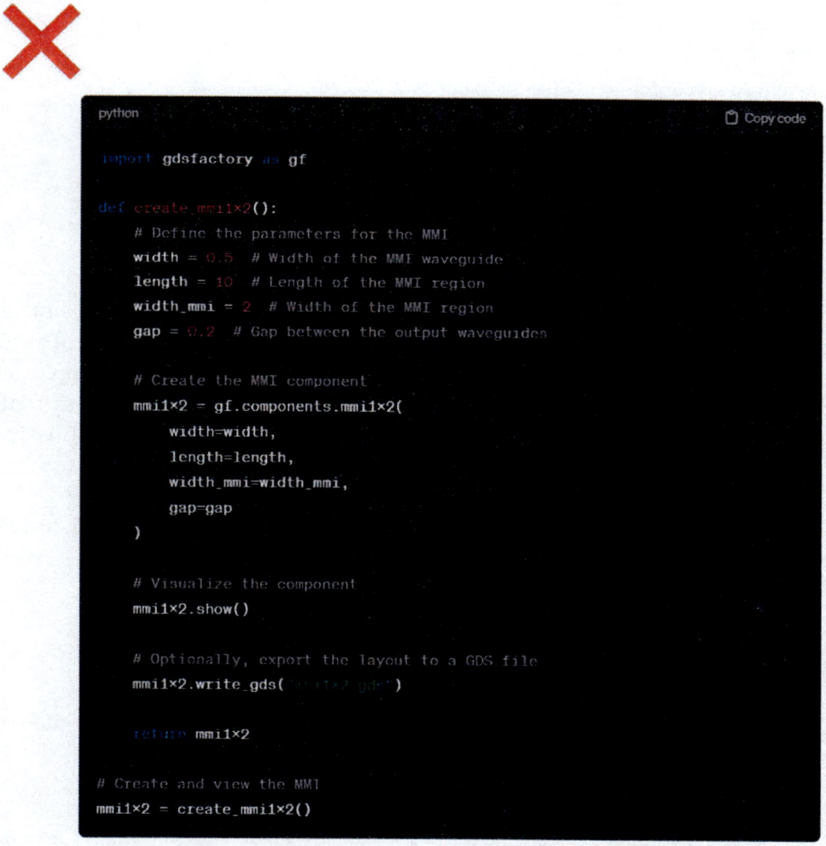

Fig. 3. . Code Generated by ChatGPT-4 when the user prompt is "Generate code to layout a 1x2MMI using the GDSfactory python library". The generated code contains errors and does not compile. The errors are the incorrect function parameters for the function of gf.components.mmi1x2().

Table 1. List of "User Prompts" test cases and the number of times our code generated the correct layout out of 100 trials for each prompt.

	Test User Prompts	Successful layout (out of 100 Trials)
1	Layout a 1x2 MMI that operates at a wavelength of 1550 nm	100
2	Give me a design for a 1x2 MMI for 1310 nm	100
3	Find an edge coupler that operates at a wavelength of 1535 nm	100
4	I need a grating coupler design for TE polarized light operating at 1550 nm	100
5	Give me the design of a 50–50 directional coupler optimized for 1550 nm	99
6	Develop a layout for an MZI at 1310 nm	99
7	Layout an edge coupler at 1550 nm where the taper length = 200.0 um	100

4 Conclusion

In summary we are developing a tool that uses the Llama2-70B large language model to automatically create layouts of photonic devices. Our tests with seven different prompts showed our tool is accurate 99% of the time. Further development of the tool is currently in progress, and we are investigating ways to handle complex tasks such as routing, component placement and optimization, necessary for the fully automated design of photonic integrated circuits.

Acknowledgements. We thank Prof. Dr. Englund, O. Akdeniz, G. Zu and Dr. V. Ansari for helpful discussions.

References

1. Brown, T., et al.: Language models are few-shot learners. In: Advances in neural information processing systems 2020. vol 33, pp. 1877–1901. Curran Associates Inc (2020)
2. Touvron, H., et al.: Llama 2: Open Foundation and Fine-Tuned Chat Models. Arxiv, 2307.09288 (2023)
3. Li, C., et al.: Common 7B Language Models Already Possess Strong Math Capabilities. Arxiv, 2403.04706 (2024)
4. Google. Gemini: A Family of Highly Capable Multimodal Models, Arxiv, 2312.11805 (2023)
5. Liu, M., et al.: ChipNeMo: Domain-Adapted LLMs for Chip Design.. Arxiv, 2311.00176 (2023)
6. He, Z., et al.: ChatEDA: A large language model powered autonomous agent for EDA. Arxiv, 2308.10204 (2023)

7. Blocklove, J., et al.: Chip-chat: challenges and opportunities in conversational hardware design. In 2023 ACM/IEEE 5th workshop on machine learning for CAD (MLCAD). IEEE (2023)
8. Li, R., et al.: From English to PCSEL: LLM helps design and optimize photonic crystal surface emitting lasers. HAL Open Science, 04175312 (2023)
9. Li, R., et al.: Deep reinforcement learning empowers automated inverse design and optimization of photonic crystals for nanoscale laser cavities. Nanophotonics **12**(2), 319–334 (2023)
10. Witt, D., et al.: Reinforcement learning for photonic component design. APL Photon. **8**, 106101 (2023)
11. GDS factory Homepage. https://gdsfactory.github.io/gdsfactory/index.html. Accessed 17 March 2023

Functional Programable Waveguide Engine and Arbitrary Beam Splitter

Zeyu Deng and Ziyang Zhang(✉)

School of Engineering, Westlake University, Hangzhou 310024, China
zhangziyang@westlake.edu.cn

Abstract. We demonstrated a waveguide engine which can update the thermal electrodes automatically and monitor the change of the interference in a synchronized and rapid way until the desired function is reached, all experimentally. We name this system "function programmable waveguide engine" (FPWE). Based on this technology, an arbitrary beam splitter is developed on a multimode waveguide by effectively manipulating the multimode interference through 4 microheaters. The device is designed by direct experiment data collection, model training with deep learning technology, and inverse calculation from the established mode. Result shows that any ratio can be achieved, including both lossless and lossy scenarios.

Keywords: Programmable waveguide · Multimode interference · Beam splitter

1 Introduction

Photonic integrated circuits (PICs) have tremendous developments in the past few decades due to its compact footprint, high efficiency and diversity properties, which can be applied not only in optical communication [1], but also to enable novel applications such as photonic computing [2]. Recently, the programmable PICs has attracted much attention as they can realize diverse function by adjusting the electronic control circuits on one chip [3]. Such PICs can be constructed by the basic elements such as Mach-Zehnder interferometers (MZIs) [4], multimode interference devices (MMIs) [5] and directional couplers (DCs) [6], into large switching network. Within the photonic network, the ability to create optical outputs with arbitrary power splitting ratio has advantages for several applications, such as signal quality monitor [7] and quantization resolution enhancement [8]. However, the MZI and DC-based solutions all work under the theoretically lossless condition. To implement extra attenuation, each output needs to connect to a variable optical attenuator (VOA), which increasing the device footprint and cost.

In this work, we developed an optical/switching engine based on a multimode waveguide and a series of thermal electrodes to alter the light interference in the multimode region experimentally, actively, and automatically, until the desired pattern is reached at the output plane, so as to define the target function [9]. We name it "function programmable waveguide engine" (FPWE). With this technology, we present a 3-port beam

J. Witzens et al. (Eds.): ECIO 2024, SPPHY 402, pp. 448–452, 2024.
https://doi.org/10.1007/978-3-031-63378-2_73

splitter based on a thermally tuned multimode waveguide, capable of achieving arbitrary power ratios [10]. The device is designed by direct experimental data collection, subsequent model training with neural network, and inverse optimization from the established model. Experiments result shows that both the lossless and lossy conditions can be achieved and match well with the target. We believe this work may inspire new ideas in developing compact, efficient, and low-cost photonic integrated circuits for optical switching and computing applications.

2 FPWE System

The diagram of the proposed FPWE system is shown in Fig. 1(a). The CW laser at 1550nm is adopted and the light is coupled into the chip by a standard single-mode fiber (SSMF). The polarization controller (PC) is added to manage the polarization state. An image system is built to capture the light from the open chip facet by an infrared (IR) camera. A central computer is used to capture the image and save the data on the local hard disk. The custom-made microcontroller unit (MCU) from the ARM-based 32-bit STM32F730XX series provides maximal 16 current source and drives the electrodes. The current (DC) can be varied from 0 to 20 mA with a minimum step of 1 μA. Its communication with central computer is established through the Modbus protocol via a USB serial cable. To complete the FPWE system, the MATLAB script is added to link the camera to the central computer via an Ethernet port by network cable through GigE vision protocol, and the MTALAB script is used to specify and update the current configuration. The Modbus module is used to send the control commands to the MCU. The actual photo of FPWE system is shown in Fig. 1(b). Figure 1(c) shows the detailed view of the optoelectronic assembly on the FPWE. A PCB adapter is developed to carry the chip and connect the microheaters via bond wires to the pins. The PCB adapter and the MCU are connected via plug-in bus cables. A microscopic photo of the integrated chip is shown in Fig. 1(d).

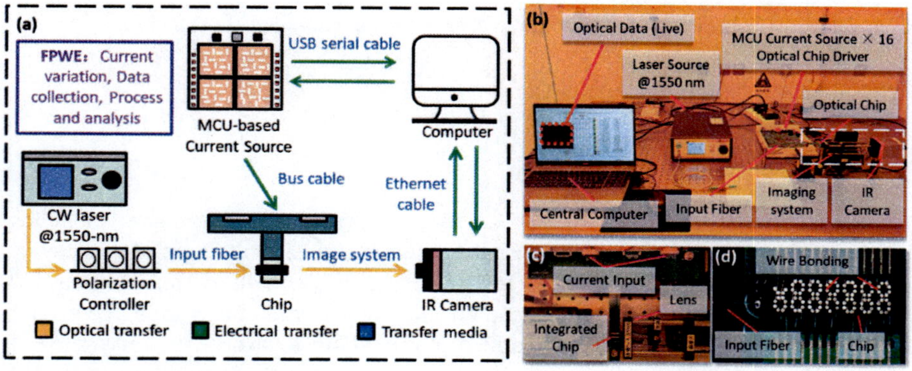

Fig.1. (a) The diagram and (b) the photo of the FPWE system. (c) Subsystem integrated with an input fiber and two electric bus cables to the MCU. The output of the chip is collected by the IR camera via an imaging system. (d) Microscopic photo of the optical chip attached and wired-bonded to the PCB pins.

3 Device Structure and Fabrication

Based on the FPWE system, a 3-port beam splitter with arbitrary power ratio is demonstrated. The structure of the device is shown in Fig. 2(a). Commercial polymer materials (ZPU-12 series from ChemOptics) are used for the core (ZPU-470) and cladding (ZPU-450). The refractive index of the core and cladding at 1550 nm wavelength is 1.47 and 1.45, respectively. The multimode region is determined as 70 μm in width and 2500 μm in length. The input/output waveguides have a core size of $3.5 \times 3.5\ \mu m^2$. The heaters are made of gold and each have a width of 8 μm, thickness of 100 nm and the same length as the multimode waveguide. The distance between the electrodes is set as 25 μm. The taper structures are applied to improve the coupling efficiency between the multimode waveguide and input/output single-mode waveguides. Figure 2(b) shows the cross-section of the multimode waveguide. The chip is fabricated using a standard process as described in our previous work [11]. After fabrication, the MMI chip is wire-bonded on the pins of the interconnect chip as shown in Fig. 2(c).

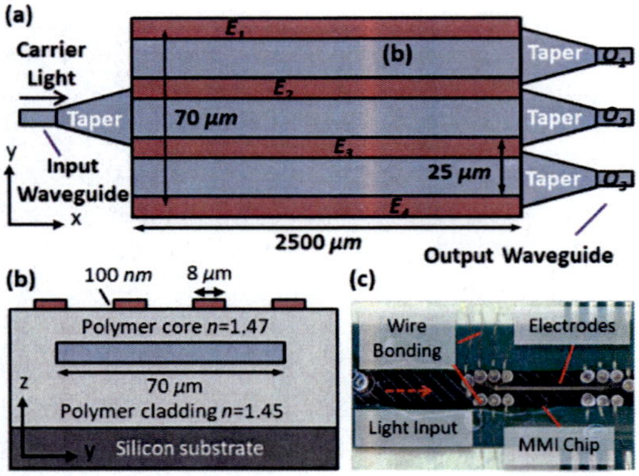

Fig. 2. (a) Scheme of the 3-port MMI beam splitter chip. (b) Cross section of the multimode waveguide with 4 electrodes. (c) Photo of the integrated and wire-bonded MMI chip.

4 Experiment Results

The FPWE system is controlled by MATLAB scripts on the central computer, including drive current updates (input) and synchronized image acquisition (output). In this work, we set the range of the current from 0 to 12mA and the step size to 1mA. In total 13^4 = 28,561 experiments are carried out and the computer would obtain an image in each sweeping process. Each image is normalized and converted to an array of three numbers $[O_1, O_2, O_3]$, including the respective applied currents $[I_1, I_2, I_3, I_4]$ on four electrodes. It is worth mentioning that each experiments takes about 1 s on the FPWE platform and the total duration of the data collection is 7.9h. To compare with, the thermal and optical simulation for a single setting of the device would take 30 min and the entire dataset would be completed over 595 days. After data collection, we then try to develop an equivalent neural network (ENN) to map the electronic input (current) and optical output (power) with minimal errors. The number of the layers is set to 4 and the neuron number in each layer is 8, 24, 48 and 3, respectively. The sigmoid activation function is implemented as it provides the generally adopted normalization function (output values ranging from 0 to 1).

Table 1. The actual shot photos for different power ratios in different experiments

Power Ratio			Applied Current (mA)				Live Spots
O_1	O_2	O_3	I_1	I_2	I_3	I_4	
1	0	0	10.7	0.8	6.3	12.0	
0.6	0	0	10.1	3.3	7.8	1.9	
0.2	0	0	10.4	10.7	11.2	12.0	
0	0	0	5.9	5.0	6.1	12.0	
0	1	0	4.7	2.5	10.8	11.9	
0	0.5	0	7.9	12.0	12.0	11.9	
0	0.1	0	8.1	2.8	7.2	12.0	
0.3	0.3	0	4.1	9.5	10.2	8.8	
0.4	0.2	0	7.5	2.1	3.3	10.4	
0	0.7	0.1	2.1	9.9	12.0	9.8	
0.2	0.2	0.2	11.0	10.8	7.1	8.0	
0.8	0.2	0	12.0	1.1	8.2	12.0	
0.1	0.2	0.4	10.2	4.2	9.8	0.0	

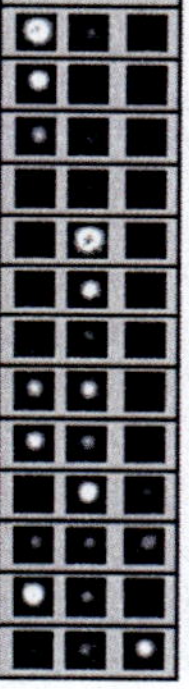

As the ENN is established, a reverse calculation process can be implemented from the ENN and the input [I_1, I_2, I_3, I_4] can be obtained for a given target output [O_1, O_2, O_3]. The function of the arbitrary beam splitter can be described as $O_1 + O_2 + O_3 \leq 1$. The algorithm follows a classic approach for the nonlinear optimization problems and would convergence in a very short time (less than 10 s). Table 1 summarizes some experimental splitter configurations with their light spots and applied currents. Both the lossless ($O_1 + O_2 + O_3 = 1$) and lossy ($O_1 + O_2 + O_3 < 1$) scenarios are achieved. We have hence demonstrated the function of reconfigurable power splitter on a thermally controlled MMI by establishing an ENN and thereupon the subsequent calculation by inverse optimization on our automatically FPWE system.

5 Discussion

To summarized, we have proposed the FPWE concept based on multimode interference device and a set of controlling electrodes. The FPWE system is simple, straightforward, and can define device functions in situ without time-consuming thermal and optical simulations. Based on the system, a 3-port arbitrary beam splitter is demonstrated. Different from the single-mode solutions such as MZIs and DCs, our device does not need extra VOAs to achieve attenuation in lossy conditions. We have also constructed an ENN of a thermally controlled MMI based on the experimental data. Furthermore, an inverse optimization method has been implemented to find the input that leads to the target power splitting output. It is worth noting that nonlinear neural network reveals the nonlinear relation on the MMI system between the input as current in the electronic domain and the output as light power variations in the optical domain. This finding hints that the multimode waveguide may be implemented as a compact optical computation unit, replacing one or more layers in a DNN, and has potential to solve AI problems in future work.

References

1. Nagarajan, R., et al.: Large-scale photonic integrated circuits. IEEE J. Sel. Top. Quant. Electron. **11**(1), 50–65 (2005)
2. Li, X.-K., et al.: High-efficiency reinforcement learning with hybrid architecture photonic integrated circuit. Nat. Commun. **15**(1), 1044 (2024)
3. Bogaerts, W., et al.: Programmable photonic circuits. Nature **586**(7828), 207–216 (2020)
4. Duan, F., et al.: Low-power and high-speed 2 × 2 thermo-optic MMI-MZI switch with suspended phase arms and heater-on-slab structure. Opt. Lett. **46**(2), 234–237 (2021)
5. Rosa, A., et al.: High performance silicon 2 × 2 optical switch based on a thermo-optically tunable multimode interference coupler and efficient electrodes. Opt. Exp. **24**(1), 191–198 (2016)
6. Wang, X., et al.: Polarization-insensitive mode-independent thermo-optic switch based on symmetric waveguide directional coupler. Opt. Exp. **27**(24), 35385–35393 (2019)
7. Qiu, J., et al.: Integrated In-band OSNR monitor based on asymmetrical parallel-MZIs for WDM signals. IEEE J. Sel. Top. Quant. Electron. **22**(6), 467–472 (2016)
8. Kang, Z., et al.: Resolution-enhanced all-optical analog-to-digital converter employing cascade optical quantization operation. Opt. Exp. **22**(18), 21441–21453 (2014)
9. Chen, T., et al.: Multibit NOT logic gate enabled by a function programmable optical waveguide. Opt. Lett. **47**(14), 3519–3522 (2022)
10. Deng, Z., et al.: 3-Port beam splitter of arbitrary power ratio enabled by deep learning on a multimode waveguide. Opt. Laser Technol. **169**, 109950 (2024)
11. Dang, Z., et al.: C/L-band 2-port broadband wavelength multiplexing switch using polymer waveguides. J. Lightwave Technol. **41**(8), 2451–2457 (2023)

Design of Microrings with Complex Waveguide Cross-Sections to Reduce Non-linear Effects of Silicon

Stefania Cucco(✉), Marco Novarese, and Mariangela Gioannini

Department of Electronic and Telecommunications, Politecnico di Torino, Torino, Italy
{stefania.cucco,marco.novarese,mariangela.gioannini}@polito.it

Abstract. This study presents a model for designing silicon and polysilicon (poly-Si) micro-ring resonators (MRRs) within the semiconductor-insulator-semiconductor capacitor platform (SISCAP). It employs generic waveguide cross-sections, including complex rib structures, to address nonlinear effects and self-heating. We highlight the crucial role of free carrier diffusion in rib waveguides and Shockley-Read-Hall recombination in mitigating nonlinearities within the ring resonator. In particular, we introduce a novel method for 2D simulations of MRRs, incorporating self-consistent generation of free carriers via TPA, their transport, non-radiative recombination, heat generation, and dissipation. These factors alter the MRR transmission coefficient, significantly diminishing the quality factor (Q) as injected power increases. The amount of input power injected into the ring is limited when high Q values ($1.3 \cdot 10^5$) are considered. Our objective is to propose new designs capable of increasing the input power by an order of magnitude while limiting the Q degradation to less than 10%, which is crucial for enhancing MRR performance in various silicon photonics applications.

Keywords: Microring resonator · silicon · nonlinear effects

1 Introduction

Microring resonators (MRRs) serve as passive optical components extensively utilized within silicon photonic integrated circuits, finding application across various photonics fields such as sensors, optical modulators, beam-steering, wavelength filtering, switching, and light modulation [1]. However, the nonlinear behavior of silicon, even under moderate input powers, presents challenges to achieving high Q-factor MRR performance. In this investigation, we present a rigorous simulation method to account for nonlinear and thermal effects in silicon MRRs featuring complex waveguide cross-sections. We apply this method specifically to optimize MRR designs within the SISCAP platform. This platform offers advantages in reducing both the cost and energy consumption of optical transceivers for optical interconnects, all while maintaining high data rates and volume across various applications [2].

Our study aims to improve the comprehension and performance of MRRs, which are crucial for advancing silicon photonics technologies.

J. Witzens et al. (Eds.): ECIO 2024, SPPHY 402, pp. 453–459, 2024.
https://doi.org/10.1007/978-3-031-63378-2_74

2 Model

Silicon introduces notable nonlinear effects, in particular two-photon absorption (TPA) and free carrier absorption (FCA). TPA results from the absorption of two photons, generating an electron-hole pair, while FCA arises from free carriers absorbing additional photons, elevating electrons and holes to higher energy states in the conduction and valence bands. The presence of free carriers modifies the refractive index, causing a blue shift in the resonant wavelength of the MRR transmission coefficient, and contributes to FC loss, degrading the Q-factor. Additionally, thermalization and recombination via Shockley-Read-Hall (SRH) mechanisms release energy as heat (self-heating), inducing a red shift in the transmission coefficient. To simulate these nonlinear effects and self-heating in MRRs, we utilize 2D simulations incorporating FC generation through TPA, FC drift-diffusion transport, SRH recombination, and thermal dynamics. By evaluating the impact of nonlinear effects on MRRs for various input powers and generic waveguide cross-sections, as those shown in Fig. 1(a), we improve our understanding of silicon photonics performance. Through a self-consistent model, integrating optical field distribution and solving drift-diffusion equations coupled with a thermal model, we obtain the distribution of free carriers and temperature variations, making it possible to compute the modal loss and refractive index changes due to free carriers and temperature fluctuations.

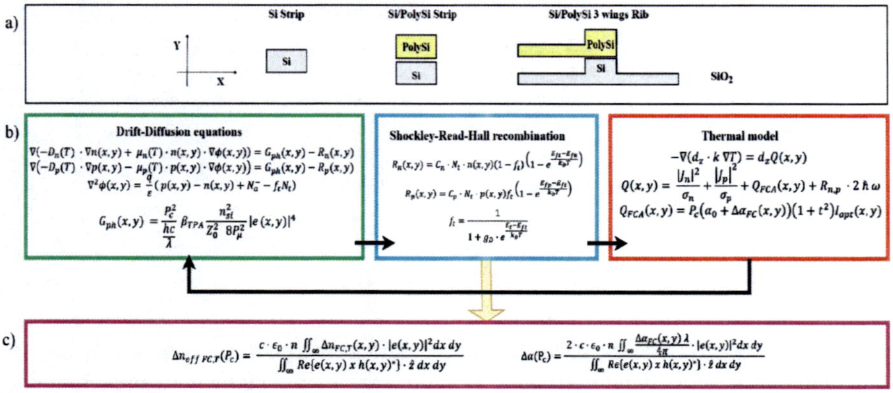

Fig. 1. (a) Cross sections of the waveguides. (b) Model summary. (c) Equations of effective refractive index variation and optical modal loss.

We summarize the self-consistent model, as shown in Fig. 1(b). we import in COMSOL Multiphysics (Semiconductor module with Thermal module) the spatial distribution of the fundamental guided mode optical field (e (x, y) and h (x, y)) in the ring waveguide and we solve self-consistently the drift-diffusion equations coupled with the thermal model. The solution provides the spatial distribution of free carriers, denoted as n(x, y) and p(x, y), and the temperature variation, represented as T(x, y), in relation to the circulating power across the waveguide cross-section. Having then the local variation of FC loss and refractive index due to free carriers and temperature [2], we can compute the optical modal loss ($\Delta\alpha$) and effective refractive index variation due to FCD or temperature ($\Delta n_{effFC,T}$) as shown in Fig. 1(c).

3 Results

We examine three MMRs with different radius, and different coupling coefficient (k^2) as reported in Table 1 to obtain a Q-factor fixed to $1.3 \cdot 10^5$, using waveguide cross-sections depicted in Fig. 1(a): Si Strip, Si/PolySi Strip and Si/PolySi 3 wings Rib. The silicon rib structure is omitted due to its bend loss exceeding 1 dB/cm.

In Fig. 2(a) and (b), we report an example of our model solutions, namely the spatial distribution of free carriers in the Si/PolySi 3 wings Rib. We observe that the silicon cross-section exhibits a higher free carrier density, leading to increased carrier diffusion compared to the poly-Si structure. The reduced carrier density in poly-Si can be attributed to a higher SRH recombination rate, resulting from a greater number of defects in poly-Si, which also shortens the diffusion length relative to silicon. Considering that Free Carrier Absorption (FCA) is directly proportional to the accumulated carrier density within the waveguide, Fig. 2(c) compares the total number of accumulated carriers across the three different waveguides under identical circulating power P_c. The integrated carriers are determined over an equivalent area, defined as 90% of the optical mode area, by integrating n(x,y) and p(x,y). From this graph, we observe that the Si/PolySi 3 wings Rib has fewer free carriers compared to the other waveguides. This advantage will lead to thermal benefits.

The third model solution is the temperature variation of the same example; our observations reveal that the Si/PolySi 3 wings Rib waveguide exhibits reduced heating in comparison to other strip waveguides operating at equivalent circulating powers. This reduced heating is primarily assigned to the presence of lateral wings, which enhance heat dissipation and prevent the concentration of carriers at the center of the waveguide. In Fig. 3(a, b) and (c), we present a comparative analysis of the temperature distribution.

In Fig. 3(d), we compare the integrated $Q_{i,FCA}$ (heat source due to FCA) across various waveguide cross-sections. For powers below 80mW, the Si/PolySi Strip experiences more heating compared to the Si Strip, which is the result of an higher local modal losses in polysilicon relative to silicon. In additional, the diminished accumulation of carriers in the Si/PolySi 3 wings Rib results in a reduced FCA, consequently minimizing carrier thermalization.

The Si/PolySi 3 wings Rib exhibits lower heat generation compared to others, as such this structure also demonstrates the lowest nonlinear loss for the same circulating power, which is a consequence of the electrons and holes diffusion in the lateral wings of the rib and the faster SRH carrier recombination in poly-Si due to higher trap density [3]. Consequently, this reduces free carrier density where the optical field is confined, leading to smaller modal loss and refractive index variation. The resulting changes in

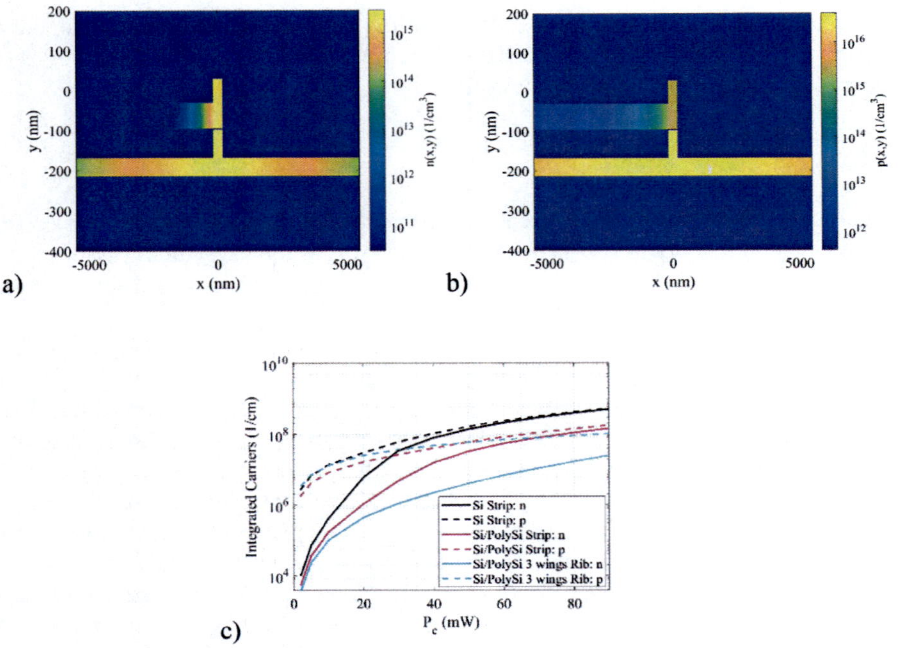

Fig. 2. Example of spatial carriers distribution: (a) electron distribution (n(x, y)) and (b) hole distribution (p(x, y)) in the Si/PolySi 3 wings Rib waveguide due to photon-generation and in for a circulation power equal to 50 mW and radius 10 μm. (c) Integrated total carrier density as a function of the circulating power for each waveguide analysed with radius 10 μm.

effective refractive index and modal loss are fitted with third-degree polynomials against circulating power, serving as input parameters for calculating the MRR transmission coefficient [4], as shown in Fig. 4. In this graph, we can observe that for low bus powers the 3 MRRs have a similar transmission coefficient; but as the power increases, the curve trend changes: for the Si/PolySi 3 wings rib, a smaller NL shift of the resonant wavelength is observed compared to Si Strip and Si/PolySi strip. From the transmission coefficient we can derive the degradation of the quality factor due to nonlinear effects, which varies with the bus power [4]. Consequently, the Si/PolySi 3 wings Rib presents less degradation in transmission coefficient compared to others. In Fig. 5 we report for each configuration the maximum input bus power to have a degradation of the Q of less than 10%; the Si/PolySi 3 wings Rib enables reaching a maximum power about 10 dB higher than the standard Si Strip waveguide available in this platform. In Table 2, we report the corresponding circulating power and the temperature increase in the core at the maximum input bus power. We observe that for the same waveguide design the circulating power remains almost constant with increasing radius because the increase in bus power is counterbalanced by an increase in radius in all structures. In addition, in the Si/PolySi 3 wings Rib the circulating power is greater than in the other two MRRs and consequently heats up more. However, between the Si/PolySi Strip and the Si/PolySi 3 wings Rib, the circulating power increased 7 times, whereas the temperature increased

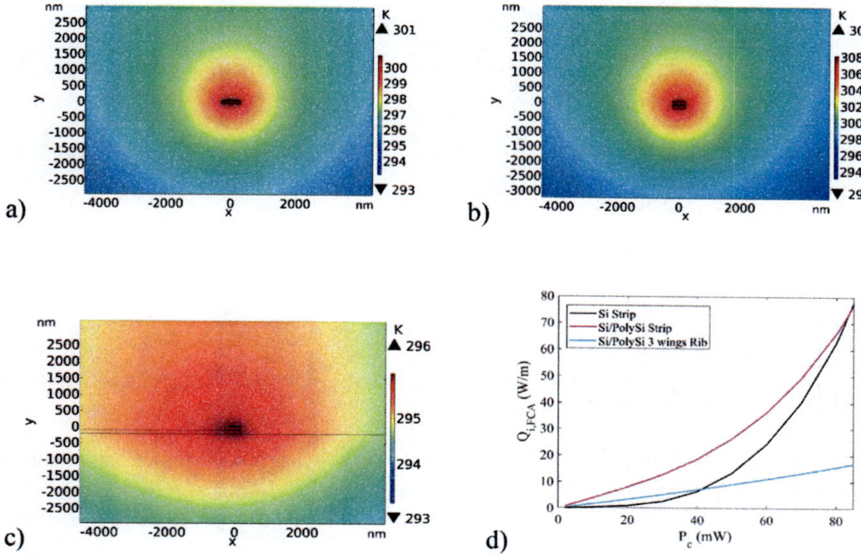

Fig. 3. Example of temperature fluctuation resulting from self-heating for $P_c = 50$ mW and radius 10 μm in (a) Si Strip, (b) Si/PolySi Strip and (c) Si/PolySi 3 wings Rib. (d) Integrated heat source due to FCA over the waveguide cross-section as a function of circulating power with R = 10 μm.

only 3 times. This difference is attributed to the greater heat dissipation in the latter waveguide design as demonstrated in Fig. 3(d). We also observed that the temperature variation for the same waveguide structure is constant for the 3 radii because there is a small variation in the circulating power in the three cases.

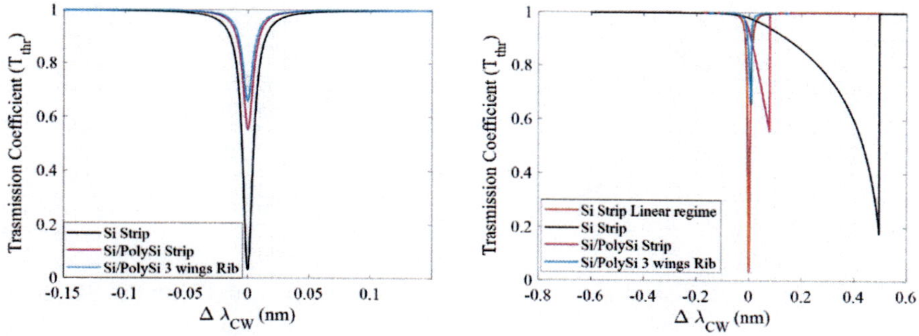

Fig. 4. Example of transmission coefficient of the 3 MRRs as a function of different bus input power: (a) linear regime, (b) $P_{bus} = -10$ dBm. The radius of MRRs is equal to 10 μm and $Q = 1.3 \cdot 10^5$.

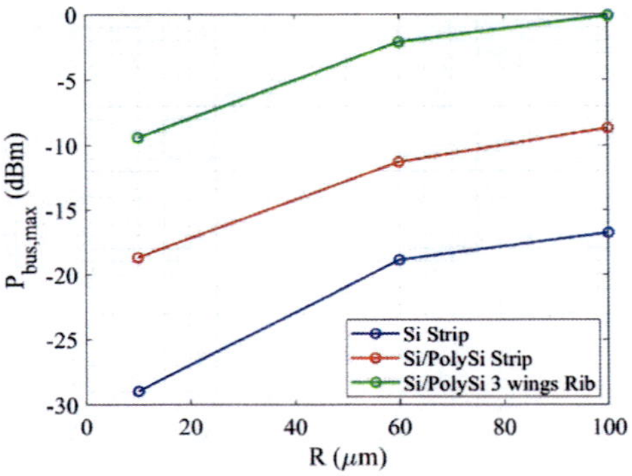

Fig. 5. Maximum bus power, with 10% Q, as a function of radius for each structure analyzed.

Table 1. The values of coupling coefficient for different radius and waveguide cross-sections.

	Si Strip	Si/PolySi Strip	Si/PolySi 3 wings Rib
R = 10 μm	0.3%	0.1%	0.08%
R = 60 μm	2%	0.7%	0.5%
R = 100 μm	3%	1%	0.7%

Table. 2. Summary of maximum circulating power with a 10% Q-factor dependence on radius for each structure, along with temperature variation in MRRs at resonance where the maximum circulating power is observed ($P_{c,max}$).

	Si Strip	Si/PolySi Strip	Si/PolySi 3 wings Rib
$P_{c,max}$ (mW) @R = 10 μm	0.256	0.798	4.6
$P_{c,max}$ (mW) @R = 60 μm	0.412	0.725	4.6
$P_{c,max}$ (mW) @R = 100 μm	0.423	0.705	4.5
ΔT (K)	0.03	0.15	0.4

4 Conclusions

In summary, our study has implemented a model capable of computing the 2D distribution of free carriers and temperature within the various waveguide cross-sections of MMRs on the SISCAP platform. Through our analysis, we have successfully identified a design solution that effectively mitigates nonlinear effects and self-heating. The Si/PolySi 3 wings Rib waveguide emerges as the most advantageous among the options examined. This structure benefits from the free carrier diffusion in the rib waveguides and heat dissipation, complemented by the addition of the poly-Si waveguide, which diminishes bend loss and reduces free-carrier lifetime through increased SRH recombination within poly-Si traps.

This work is supported by Italian National Recovery and Resilience Plan (NRRP) of NextGenerationEU: partnership on Telecommunications of the Future (PE00000001 – program "RESTART").

References

1. Atoum, I., et al.: Challenges of software requirements quality assurance and validation: a systematic literature review. IEEE Access **9**, 137613–137634 (2021). https://doi.org/10.1109/ACCESS.2021.3117989
2. Zhang, W., et al.: Towards subvolt and half-mm scale silicon MOS-capacitor MZI modulators. In: 2023 IEEE silicon photonics conference (SiPhotonics), pp. 1–2. Washington, DC, USA (2023). https://doi.org/10.1109/SiPhotonics55903.2023.10141906. (Author, F.: Contribution title. In: 9th International Proceedings on Proceedings, pp. 1–2. Publisher, Location (2010))
3. Novarese, M., Romero-Garcia,S., Bovington, J., Gioannini, M.: Dynamics of free carrier absorption and refractive index dispersion in Si and Si/PolySi Microrings. EEE Photon. Technol. Let. **35**(8), 450–453 (2023)
4. Novarese, M., Romero-Garcia,S., Bovington, J., Gioannini, M.: Study of nonlinear effects and self-heating in a silicon microring resonator including a Shockley-Read-Hall model for carrier recombination, vol. 30, no. 9, 25 Apr 2022. Optics Express 14341

Si_3N_4 Microring-Resonator-Based Integrated Photonic Sensor for Enhanced Label-Free Biofluid Analysis in the 850 nm Optical Band

Jakob Reck[1](✉), Klara Mihov[1], Martin Kresse[1], David de Felipe[1], Tianwen Qian[1], Madeleine Weigel[1], Csongor Keuer[1], Philipp Winklhofer[1], Crispin Zawadzki[1], Moritz Kleinert[1], Norbert Keil[1], and Martin Schell[1,2]

[1] Fraunhofer Institute for Telecommunications, Heinrich Hertz Institute, Einsteinufer 37, 10587 Berlin, Germany
jakob.reck@hhi.fraunhofer.de

[2] Technische Universität Berlin, Straße des 17. Juni 135, 10623 Berlin, Germany

Abstract. We present an innovative integrated photonic sensor for point-of-care applications using silicon nitride (Si_3N_4) tailored for biofluids analysis in the 850 nm optical band. With microring resonators (MRR) we detect smallest changes in the effective refractive index of the waveguide, achieving a sensitivity of 46 nm/RIU (refractive index unit) and a system limit of detection (LOD) of $5.8 \cdot 10^{-5}$ RIU. Furthermore, the successful integration of photodiodes directly into the Si_3N_4 photonic integrated circuit (PIC) output was demonstrated. Our preliminary work offers promising prospects for future use in lab-independent environments where compact, low-cost and fast sensors are needed that also support multi-channel applications.

Keywords: photonic sensor platform · microring resonators · lab-on-a-chip

1 Introduction

While PICs based on silicon nitride (Si_3N_4) are already extensively used in optical communication and quantum applications [1, 2], their potential in highly sensitive sensing applications, especially in fluid analysis, is becoming increasingly important [3]. We specifically focus on biosensing for point-of-care analysis, where there is a need for disposable and miniaturized sensors with compatible readout units. Our presented sensor utilizes racetrack microring resonators (MRRs) to detect local refractive index changes on the MRR surface that lead to a change in the effective refractive index (n_{eff}) within the waveguide. In the presented notch filter configuration, the resonant wavelength λ_{res} shifts according to Eq. (1), as well as with respect to the resonator radius r [4]. This shift can be recorded using optoelectronic sensors, which enables monitoring almost in real-time.

$$\lambda_{res} = (2\pi r \cdot n_{eff})/m; \;\; m = 1, 2, 3, \ldots \tag{1}$$

J. Witzens et al. (Eds.): ECIO 2024, SPPHY 402, pp. 460–466, 2024.
https://doi.org/10.1007/978-3-031-63378-2_75

When using MRR in a sensor system, the decisive performance indicator for high resolution and precision is the limit of detection (LOD). According to Eq. (2), the LOD is defined as the quotient of the resolution R and sensitivity S.

$$LOD = \frac{R}{S} \tag{2}$$

This paper focuses on the development and characterization of a sensor PIC tailored for the analysis of biofluids in the 850 nm optical band. Operating under this condition offers significant advantages compared to the use of the optical C-band around 1550 nm. On the one hand, the reduced absorption coefficient of water in this wavelength range is utilized, which leads to an improved intrinsic LOD of the PIC [5, 6]. On the other hand, it enables the realization of smaller and more compact PICs and improves the accessibility of VCSEL chips (Vertical-Cavity Surface-Emitting Laser) [7], also for the integration of tunable laser sources on the PIC.

2 Design and Fabrication

The primary objective of this research is a label-free photonic sensor that is user-friendly and flexible enough to integrate optoelectronic elements such as chip-based lasers or photodiodes as required. To ensure efficient light interaction between the single-mode waveguide core and the analyte, the waveguides have been designed as low-confinement TE-mode guiding structures. This ensures that the evanescent field radiates as far as possible into the measurement environment while maintaining acceptable losses for the relative measurement of the resonance wavelength shift.

The presented MRRs were specifically designed to achieve a free spectral range (FSR) of 301 pm, which matches with the wavelength tuning range of the VCSEL. The ring radius was set to $r = 320\ \mu m$ to ensure that at least two resonances fit within the tuning spectrum of the light source while exhibiting negligible bending losses. The CAD (computer-aided design) layout shown in Fig. 1 illustrates the waveguides, the multimode interference (MMI) coupler with 50% coupling ratio for light distribution and the MRRs that serve as functional sensing elements. In addition, the design focused on the integration of grating couplers to enable efficient light coupling in the single-mode waveguide plane. This design choice aims to simplify the handling for the user due to the increased coupling tolerance compared to etch coupling methods [8]. Fourth-order couplers with a coupling angle θ of 48° are used.

The initial fabrication step involved generating a lithographic mask using CAD. The PICs were subsequently manufactured on 4-inch wafers using stepper photolithography, dry etching and chemical vapor deposition. The single-mode waveguides designed for the base wavelength of 850 nm features a cross-section of 600 nm width by 200 nm height.

The functional elements were fabricated on a 525 μm silicon substrate as described in our previous work in [9, 10]. The layer stack, arranged from bottom to top, consisted of 3.4 μm of thermal silicon dioxide (SiO_2), 200 nm LPCVD (low pressure chemical vapor deposition) Si_3N_4 and 400 nm PECVD (plasma-enhanced chemical vapor deposition) SiO_2. To enable contact between the sensor waveguide and the environment, the

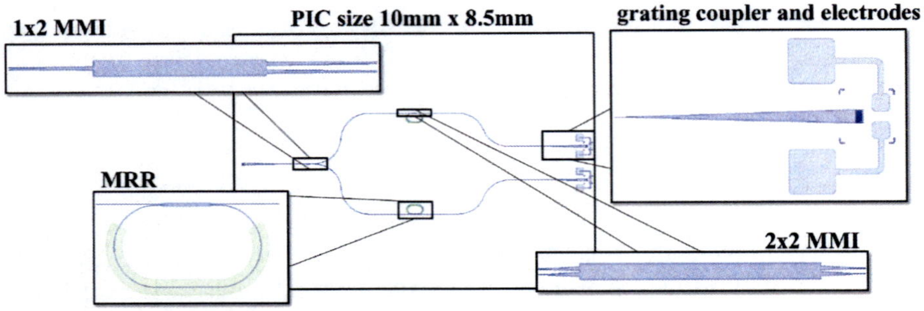

Fig. 1. Photonic building blocks within the PIC architecture include the grating couplers for efficient light coupling ($\theta = 48°$), MMIs as power dividers, MRR sensors and electrodes for the integration of electro-optical components on chip, if required.

waveguide was locally exposed and not buried with cladding in the region around the microring resonator using a lift-off process. In the final stage of fabrication, electrodes were deposited onto the PIC by vapor deposition of titanium and gold. The resulting chip exhibited dimensions of 10 mm by 8.5 mm (Fig. 2).

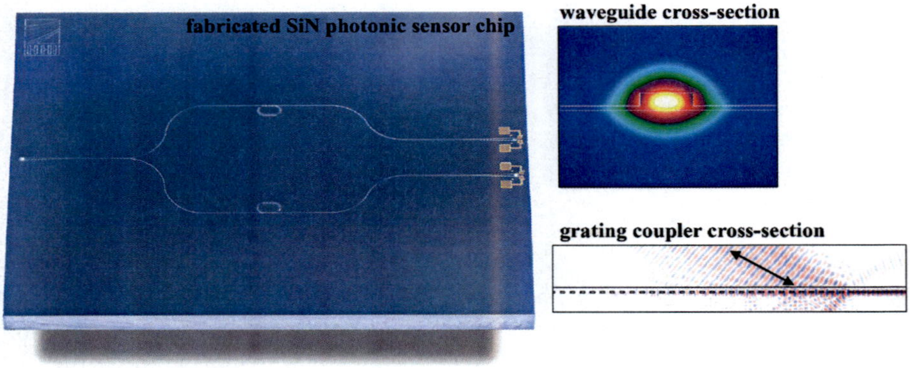

Fig. 2. Fabricated PIC (size 10 mm × 8.5 mm) with two channels for fluid analysis (left). Schematic view of the cross-section of the waveguide and the grating coupler structure with mode array (right).

3 Evaluation and Results

3.1 Optical Characterization

The optical characterization of the sensor was conducted with a broadband supercontinuum laser optimized for a wavelength range from 840 nm to 860 nm. The output signal was recorded with an optical spectrum analyzer. Analytical evaluation of the obtained spectrum resulted in a free spectral range (FSR) of 0.304 nm, aligning closely with the theoretically predicted value of 0.301 nm. The LOD of the sensor was determined by monitoring the shift of the resonance wavelength, as explained in the introduction. We used aqueous sodium chloride (NaCl) solution [11] as the analysis liquid at room temperature. The NaCl concentration was gradually doubled, starting at 0.1125% NaCl. To control the sensor, the VCSEL source with a tuning range of 1.9 nm was clocked every 100 ms (10 samples/s). Tuning was performed by varying the applied drive current. This fast cycle rate ensures real-time monitoring of the bulk shift responses. For the controlled flow of the calibration liquid, a microfluidic system was fabricated using a 3D printing process. The evaluation of the data obtained (see Fig. 3) shows a pronounced linear correlation between the change in the refractive index of the fluid and the shift in resonance wavelength λ_{res} with a variance (R^2) of 0.996.

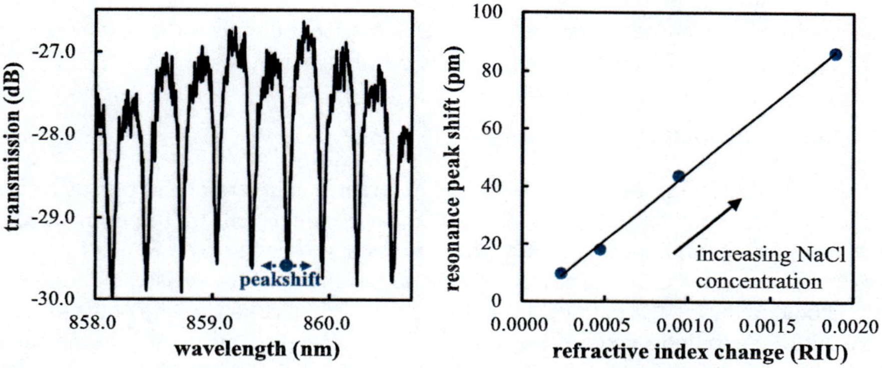

Fig. 3. Typical transmission spectrum of an optical microring resonator showing modulation dips (left). Various NaCl concentrations were used to illustrate the linear relationship between refractive index change [11] and resonance peak shift, indicating an R^2 value of 0.996 (right).

The data analysis yields a bulk sensitivity of 46 nm/RIU with a standard deviation (SD) of 2.7 pm. Consequently, the LOD of the overall measurement system is calculated to be $5.8 \cdot 10^{-5}$ RIU according to Eq. (2). Other microring resonator work within this optical window is described in the literature. SiON (silicon oxynitride) platforms have shown LODs of 10^{-6} RIU [5, 12, 13], while the only other Si_3N_4 platform called TriPleX from the company LioniX has also achieved LODs of 10^{-6} RIU [14], with the best reported value being $5.8 \cdot 10^{-7}$ RIU [15].

It is important to note that the comparability of the results may be limited depending on the measurement setup and methodologies used. When determining the LOD, factors beyond the PIC inherent LOD play a role, such as the sampling density, the noise of the optoelectronic components and the digital evaluation algorithms, among others [15–17]. When selecting the light source, for example, we considered the trade-off between cost and performance, opting for a low-cost and easy-to-integrate VCSEL with driver circuit. In this work the decision was driven by the intended use in a point-of-care device, where factors such as affordability and simplicity play an important role. Using a more complex and therefore more expensive measurement system would result in a better LOD with the same sensor chip. To make this impact quantitatively clearer, the random noise level of the MRR measurement platform from [15] can be used, which is given as SD = 0.04 pm (best value). Substituting this SD value and the sensitivity of our sensor into Eq. (2) also results in a LOD in the range of 10^{-7} RIU. However, the most significant source of systemic uncertainty in the field of biosensing is likely due to the biochemical surface functionalization [18] of the sensors with capture molecules. The impact of functionalization on receptor-ligand binding assays, which is applicable for the sensor being presented, is also apparent in the benchmarks for bioanalytical method validation published by the U.S. Food and Drug Administration (FDA) [19]. These benchmarks show a significant coefficient of variation of ±20%.

All of the platforms described above use edge coupling to connect the PIC and peripherals, often with a fixed fiber array unit. Our approach with grating couplers improves the viability of the sensor as a disposable element. Since we enable tolerant free space coupling, only the chip itself rather than the entire unit including the fiber array needs to be disposed after a single use. This concept minimizes the cost per sensor chip component and is therefore economically advantageous.

3.2 On-chip Photodiode Integration

The correct positioning of photodiodes on the grating couplers at the output side is crucial for their efficient integration on the Si_3N_4 photonic sensor. To ensure precise alignment, a semi-automatic assembly process was developed using a FINEPLACER® machine from finetech. The alignment marks on the PIC enable precise positioning of the photodiode during the passive flip-chip process. When capturing the optical output signal with a photodiode, consideration of the coupling angle is mostly negligible due to the anti-reflective photodiode surface in contrast to the coupling of a fiber in free space. This simplifies the integration process.

Commercially available GaAs PIN photodiodes from TRUMPF were used as photosensors for our sensor platform. In order to create robust electrical connections, a gold paste was used for the contact between the photodiode pads and the sensor electrode. The top view shown in Fig. 4 illustrates the positioning above the grating coupler, ensuring

Fig. 4. Microscope images of the electrodes and the output grating coupler after setting the photodiode on the Si_3N_4 PIC (left, center) and sampled spectrum using tunable VCSEL and integrated photodiodes (right).

efficient light detection. A corresponding photocurrent can be measured and resolved when operating the PIC using the VCSEL (Fig. 4). Fluctuations in the amplitude of the resonance depths are due to the inherently fluctuating output power of the VCSEL at different wavelengths, which, as described above, is tuned via its drive current.

4 Conclusion

Our work demonstrates the feasibility and potential of a miniaturized biosensor platform operating in the 850 nm optical band for enhanced fluid analysis. We have successfully established the basic functionality of the integrated photonic sensor system based on a Si_3N_4 microring resonator with grating couplers for easy light coupling, demonstrating an overall sensitivity of 46 nm/RIU and a LOD of $5.8 \cdot 10^{-5}$ RIU when measuring in sodium chloride (NaCl) solutions. In the future, optimization of the photonic elements, such as the efficiency of the grating coupler and minimizing of losses in the microrings, will be important to further improve the sensor performance. In addition, expanding the number of channels for multi-parameter analysis will enable a more comprehensive and detailed characterization of fluid. Furthermore, the sensor can be functionalized, enabling not only quantitative but also qualitative analysis of bioindicators or pathogens in point-of-care applications. This capability opens possibilities for real-time monitoring and diagnostics and could revolutionize healthcare and biotechnological research.

Our successful demonstration of on-chip integration of photodiodes in Si_3N_4 sensors highlights the potential for further integration of components such as VCSEL chips into the sensor architecture.

Acknowledgment. This work was partially funded by the German Federal Ministry of Education and Research (BMBF) through the projects optION (Project ID: 13GW0243B) and PolyChrome (Project ID: 200.024). Special thanks to leto digital Leontopoulos GbR for providing the chip photographs used in this paper.

References

1. Shekhar, S., et al.: Roadmapping the next generation of silicon photonics. Nat. Commun. **15**(1), 751 (2024). https://doi.org/10.1038/s41467-024-44750-0
2. Luo, W., et al.: Recent progress in quantum photonic chips for quantum communication and internet. Light Sci. Appl. **12**(1), 175 (2023). https://doi.org/10.1038/s41377-023-01173-8
3. Kazanskiy, N.L., Khonina, S.N., Butt, M.A., Kaźmierczak, A., Piramidowicz, R.: State-of-the-art optical devices for biomedical sensing applications—a review. Electronics **10**(8), 973 (2021). https://doi.org/10.3390/electronics10080973
4. Bogaerts, W., et al.: Silicon microring resonators. Laser Photon. Rev. **6**(1), 47–73 (2012). https://doi.org/10.1002/lpor.201100017
5. Samusenko, A., et al.: A SiON microring resonator-based platform for biosensing at 850 nm. J. Light. Technol. **34**(3), 969–977 (2016). https://doi.org/10.1109/JLT.2016.2516758
6. Steglich, P., et al.: Silicon photonic micro-ring resonators for chemical and biological sensing: a tutorial preprint (2021). https://doi.org/10.36227/techrxiv.14909901.v1

7. Marketsandmarkets Research Private Ltd. VCSEL Market by Type (Single-mode and Multi-mode), Material (GaAs, InP, GaN), Wavelength, Application (Sensing, Data Communication, Industrial Health & Printing, Emerging), Data Rate, Industry and Region - Global Forecast to 2028, SE 3835. Accessed 20 Feb 2024. https://www.marketsandmarkets.com/Market-Reports/vcsel-market-206588268.html
8. Cheng, L., Mao, S., Li, Z., Han, Y., Fu, H.: Grating couplers on silicon photonics: design principles, emerging trends and practical issues. Micromachines **11**(7), 666 (2020). https://doi.org/10.3390/mi11070666
9. Schulte-Osseili, C., Kleinert, M., Keil, N., Rosencrantz, R.R.: Rapid drop-test for lectin binding with glycopolymer-coated optical ring resonators. Biosensors **9**(1), 24 (2019). https://doi.org/10.3390/bios9010024
10. Reck, J., et al.: Eight-channel SiNx microring-resonator based photonic biosensor for label-free fluid analysis in the optical C-band *. In: 2023 45th annual international conference of the IEEE engineering in medicine & biology society (EMBC), pp. 1–4. IEEE (2023). https://doi.org/10.1109/EMBC40787.2023.10340677
11. Polyanskiy, M.N.: Refractiveindex.info database of optical constants. Sci. Data **11**(1) (2024). https://doi.org/10.1038/s41597-023-02898-2
12. Chalyan, T., et al.: Characterization of SION microring resonators for biosensing applications. In: 2015 International conference on BioPhotonics (BioPhotonics), pp. 1–4. IEEE, Florence (2015). https://doi.org/10.1109/BioPhotonics.2015.7304018
13. Guider, R., et al.: Sensitivity and Limit of Detection of biosensors based on ring resonators. Sens. Bio-Sens. Res. **6**, 99–102 (2015). https://doi.org/10.1016/j.sbsr.2015.08.002
14. Gaj, B., Rg, H.: Performance of arrayed microring resonator sensors with the TriPleX platform. J. Biosens. Bioelectron. **7**(2) (2016). https://doi.org/10.4172/2155-6210.1000209
15. Gounaridis, L., et al.: High performance refractive index sensor based on low Q-factor ring resonators and FFT processing of wavelength scanning data. Opt. Express **25**(7), 7483 (2017). https://doi.org/10.1364/OE.25.007483
16. Ghasemi, F., Chamanzar, M., Eftekhar, A.A., Adibi, A.: An efficient technique for the reduction of wavelength noise in resonance-based integrated photonic sensors. Analyst **139**(22), 5901–5910 (2014). https://doi.org/10.1039/C4AN01292E
17. Ghasemi, F., et al.: Multiplexed detection of lectins using integrated glycan-coated microring resonators. Biosens. Bioelectron. **80**, 682–690 (2016). https://doi.org/10.1016/j.bios.2016.01.051
18. Puumala, L.S., et al.: Biofunctionalization of multiplexed silicon photonic biosensors. Biosensors **13**(1), 53 (2022). https://doi.org/10.3390/bios13010053
19. Bioanalytical Method Validation Guidance for Industry. Food and Drug Administration (2018)

Computer-Generated MMI-Type Visible Multi-wavelength Combiner with Monitor Branch Using Zr-Doped PLC

Junji Sakamoto(✉), Yuji Fujiwara, Satomi Katayose, and Toshikazu Hashimoto

NTT Device Technology Labs, 3-1 Morinosato Wakamiya, Atsugi 243-0198, Kanagawa, Japan
Junji.sakamoto@ntt.com

Abstract. We fabricated a multimode interference (MMI)-type visible multi-wavelength combiner with a monitor branch by using a Zr-doped planar lightwave circuit (PLC), which is a more highly blue-light resistant waveguide compared with a conventional Ge-doped PLC, for a compact and high-power light source. The combiner was designed with a computer optimization method called the wavefront matching method. The fabricated combiner was 7.5 μm wide and 900 μm long and had a main port transmittance of approximately -1.0 dB and a monitor branch ratio of approximately 2% at R, G and B wavelengths.

Keywords: RGB combiner · photonic integrated circuit · AR/VR · visible wavelength · high power · computer generated · optimization · optical circuit · PLC · planar lightwave circuit · wavefront matching method

1 Introduction

Optical waveguide technology has been developed for various applications in the visible light range, such as smart glasses, analytical instruments, and quantum information processing [1–3]. In these applications, the wavelength combining circuit is an important component because the systems often require a coaxial output beam with multiple wavelengths. In order not to limit the application, the waveguide should have stable characteristics for high optical input powers and be able to combine multiple wavelengths. Here, the refractive index of the core of silica-based waveguides for optical telecommunications, called planar lightwave circuits (PLCs) [4], changes when blue light with powers of several mW is passed through them for a long time. This problem can be solved by changing the dopant from the conventional Ge to Zr, which has been reported to show stable characteristics even for blue light with input powers in excess of 50 mW [5]. On the other hand, PLCs with Zr as a dopant are difficult to vertically etch and should be used with circuit elements that do not have a narrow gap structure such as multimode interference (MMI) coupler. It is in principle difficult to use MMI coupler to combine light of different wavelengths, because the self-image length is determined by the width of the MMI coupler and the wavelength of the input light.

J. Witzens et al. (Eds.): ECIO 2024, SPPHY 402, pp. 467–470, 2024.
https://doi.org/10.1007/978-3-031-63378-2_76

To combine light of different wavelengths by using MMI coupler, we report the first demonstration of an RGB combiner with a monitor branch that incorporates a computer-designed MMI coupler using a Zr-doped PLC. The optical circuit was designed using an optimization method called the wavefront matching method (WFM), which is suitable for generating relatively large optical circuits such as silica PLCs because the calculation converges very quickly thanks to a locally defined updating rule and the use of the beam propagation method [6].

2 Design and Experimental Results

Figure 1(a) shows the functional diagram, (b) waveguide pattern designed by the WFM, and (c) light propagation calculated by a 3D beam propagation method (3D-BPM) at the R(635 nm), G(520 nm), and B(455 nm) inputs of the combiner. The combiner has R, G, and B input ports and two output ports; one of the two output ports is the main output and the other is the monitor output. The design values in this case were 2% branching for G and B and 1% branching for R. The total length of the combiner before optimization was set to 900 μm, consisting of 700 μm for the MMI part and 100 μm for the input and output ports. This is sufficiently compact compared with the about 1.5 mm length for a conventional RGB combiner with mode couplers[1]. The initial input/output port width and the gap between the input ports were set to 1.5 μm. In the WFM, the position of the waveguide sidewalls was optimized, and the inner sidewalls were filled with cores to reduce losses. The refractive index difference Δ was set to 1%. The sidewall position was changed to bring it closer to the target field by repeating the calculation loop. In the nth calculation loop, the target field was set to

$$\varphi^{n}_{target} = \sum\nolimits_{m=1}^{2} e^{i\theta^{n-1}_{m}} \varphi_{m} \tag{1}$$

Here, φ_m is the basic mode with its intensity matched to the branching ratio at output port m. $\theta^{n-1}_{m} = arg(\varphi^{*}_{m}\psi^{n-1}_{m})$ is the phase differences among the output ports when the input light propagates through the waveguide structure the $(n-1)$-th time. With this setting, the optimization calculation could be performed in such a way that only the branching ratio was matched to the target field, without the need for matching the phase between the two output ports. The propagation of light in the designed RGB combiner shows that the R, G, and B are combined into main port, and some of the light is split into the monitor port.

The transmission spectra of the R, G, and B input ports to the main and monitor ports of the fabricated RGB combiner are shown in Fig. 2. The solid line shows the experimental results and the dotted line shows the results of a 3DBPM simulation. Although the R peak position was slightly shifted due to manufacturing errors, the measured spectrum was in good agreement with the simulation results. Moreover, the combiner had a low loss of about −1.0 dB in transmittance at the main port.

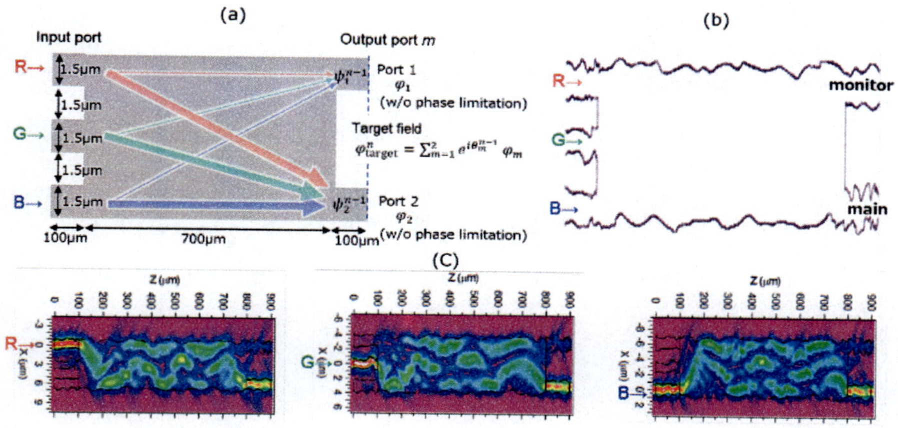

Fig. 1. (a) Functional diagram of the RGB combiner with monitor branch and (b) waveguide pattern designed by WFM and (c) propagation of light calculated by a 3D-BPM simulation at R,G,B wavelengths.

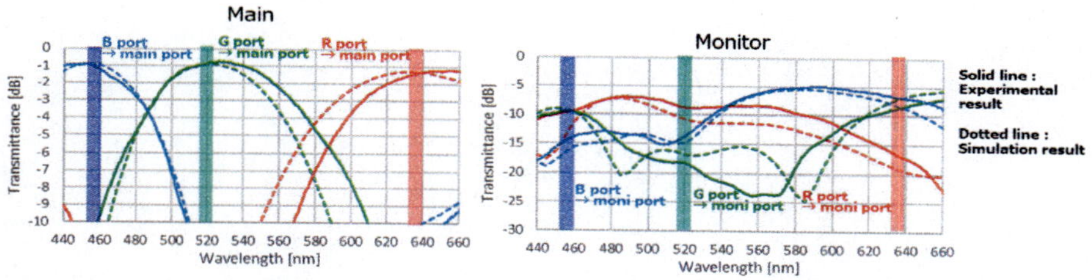

Fig. 2. Transmission spectra of signals sent through the R, G, and B input ports and measured at the main and monitor ports of the fabricated RGB combiner

3 Conclusion

We demonstrated a method to realize a wavelength combining circuit by changing the waveguide width of the MMI part. As a specific example, an RGB combiner with monitor branching was designed using a computer optimization method called the wavefront matching method. The fabricated combiner was compact, with a length of 900 μm and a width of 7.5 μm and low loss. The transmittance of each input port measured at the main port was approximately −1.0 dB. This technology can be applied to Zr-doped PLCs, which are highly blue-light tolerant but for which it is difficult to etch narrow-gap structures, for wavelength multiplexed circuits. We believe that this technology will greatly expand the range of visible-light applications of waveguide technology.

References

1. Sakamoto, J., et al.: Compact and low-loss RGB coupler using mode-conversion waveguides. Opt. Commun. **420**, 46–51 (2018)

2. Geuzebroek, D., et al.: Ultra-wide band (400–1700 nm) integrated spectrometer based on arrayed waveguide gratings for spectral tissue sensing. In: Proceedings of the IEEE 14th International Conference Group IV Photon, pp. 83–84 (2017)
3. Bogaerts, W., et al.: Programmable photonic circuits. Nature **586**, 207–216 (2020)
4. Kawachi, M., et al.: Recent progress in silica-based planar lightwave circuits on silicon. IEEE Proc.- Optoelectron. **143**(5), 257–262 (1996)
5. Fujiwara, Y., et al.: Experimental demonstration of stability of silica-based planar lightwave circuit for medium power RGB laser sources. In: Proceedings of the IDW 2023 FMC10/LCT10-3 (2023)
6. Hashimoto, T., et al.: Optical circuit design based on a wavefront-matching method. Opt. Lett. **30**, 2620 (2005)

All-Optical XOR Gate Based on a Saturable Cavity

Isaac Luntadila Lufungula(✉) and Dries Van Thourhout

Photonics Research Group, INTEC, Ghent University - Imec, Ghent, Belgium
isaac.luntadilalufungula@ugent.be

Abstract. We investigate the theoretical properties of a novel all-optical XOR gate design consisting of a saturable absorber in an all pass ring resonator. The two resonant inputs are combined in a 3 dB splitter which is then connected to the ring, where the pass port provides the XOR output. Using analytical methods we optimize the coupling constant of the ring and determine the power range over which the XOR operation is retained.

1 Introduction

All-optical signal processing could accelerate real-time tasks with optical inputs, such as inline denoising and photonic machine learning, by staying in the optical domain and avoiding slow analog-to-digital converters. All-optical gates are basic building blocks to realize these systems but these gates often require complex devices[1,2]. One of the basic logic gates is the XOR gate which provides a logical 1 as output only if one but not both of the inputs are 1. There have been all-optical demonstrations of an XOR gate but they are usually based on active elements [3]. We analyze an all-pass ring resonator augmented with a saturable absorber (SA) and show that by combining two inputs in a 3 dB splitter and coupling the output to the ring with SA , it can act as a passive all-optical XOR gate.

2 Methods and Results

For an all-pass ring resonator with an input field at the resonance frequency, we can express the field in the ring at steady state as [4]:

$$E_r = a_0 t_1 E_r + k_1 E_{in} \Rightarrow E_r = \frac{k_1}{1 - a_0 t_1} E_{in} \tag{1}$$

With E_r the field amplitude in the ring, a_0 the round trip loss, t the field transmission at the ring coupler, $k = \sqrt{1 - t^2}$ the field coupling constant and E_{in} the input field amplitude. From this we can get the field at the pass port E_p as:

$$E_p = tE_{in} - kE_r \Rightarrow E_p = \frac{t - a_0}{1 - a_0 t} E_{in} \tag{2}$$

J. Witzens et al. (Eds.): ECIO 2024, SPPHY 402, pp. 471–476, 2024.
https://doi.org/10.1007/978-3-031-63378-2_77

The relative power transmission to the drop port, $T_{pass} = (\frac{E_p}{E_{in}})^2$, only depends on the single round trip field transmission of the ring, a_0, and the field transmission coefficient t. By choosing the coupling gap so $t = a_0$ we get critical coupling, meaning $T_{pass} = 0$. In our design however, we add a saturable absorber (SA) in the cavity, resulting in a power-dependent loss in the ring:

$$\alpha_{SA} = \frac{\alpha_0}{1 + \frac{P_{ring}}{P_{sat}}} \tag{3}$$

$$a = a_0 e^{-\alpha_{SA}} \tag{4}$$

with α_{SA} the loss coefficient of the SA dependent on the power in the ring P_{ring}, the saturation power P_{sat} and the loss constant of the SA α_0. a is the power dependent round trip transmission coefficient taking into account the losses in the SA and the round trip loss of the ring itself a_0.
The steady state equations in the ring now becomes:

$$E_r = E_r a_0 e^{-\frac{\alpha_0}{1+\frac{P_{ring}}{P_{sat}}}} t + kE_{in} \tag{5}$$

By normalizing all powers and fields w.r.t. the saturation power $A^2 = \frac{P}{P_{sat}} = \frac{E^2}{E_{sat}^2}$ we can simplify this to:

$$A_r = A_r a_0 e^{-\frac{\alpha_0}{1+A_r^2}} t + kA_{in} \tag{6}$$

This is a transcendental equation and can not be solved analytically. Solving for A_{in} however results in a simple analytical equation:

$$A_{in} = \frac{A_{ring}}{k}(1 - a_0 e^{-\frac{\alpha_0}{1+(A_{ring})^2}} t) \tag{7}$$

Which we can use to investigate the behaviour of this system.

The mechanism to use this system as a logic gate is that the ring loss depends on the input power which in turn means that for a fixed coupling constant critical coupling will only be achieved at a specific power where: $t = a = a_0 e^{-\frac{\alpha_0}{1+A_r^2}}$. For our purposes we wan the limit of high ring powers to have critical coupling. Looking at the limit at infinity $\lim_{A_r \to \infty} a = a_0 e^{-\frac{\alpha_0}{1+A_r^2}}$ this reduces again to $t = a_0$ as the SA losses are completely saturated. As these high ring powers correspond to high input powers, this means at high input power the transmission will tend to $T_{pass} = 0$.

Using the parameters listed in Table 1 , where $t = a_0$, we show that the system has 3 regimes as shown in Fig. 1:

- for low input powers the SA has high losses, all light coupled in the ring gets absorbed so almost all the power is transmitted to the pass port. The transmission is high but the absolute output power will be low.
- for high input powers the SA is saturated, the light will resonate in the ring and the light will interfere destructively with the input light resulting in a low transmission to the pass port with a horizontal asymptote at $P_{pass} = 0$.

– for intermediate powers the SA gets somewhat saturated and the power starts building up in the ring so the pass transmission goes down, but since the absolute input power is going up a maximum will be reached at some power P_{in}^{max}.

Table 1. Parameter values

a_0	t	saturable loss [dB]/α_0
0.98	0.98	2.5/0.29

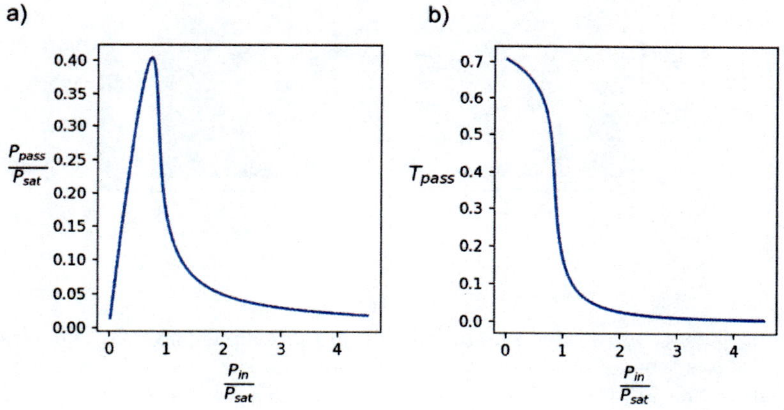

Fig. 1. Power response of the all pass ring with a saturable absorber (parameters from Table 1). (a) Pass power and (b) Pass transmission.

As an input to the saturable ring we will now consider an in-phase combination of 2 input bitstreams by using e.g. a 2×1 MMI:

$$E_{in} = \frac{E_1}{\sqrt{2}} + \frac{E_2}{\sqrt{2}} \tag{8}$$

$$\frac{P_{in}}{P_{sat}} = \frac{(A_1 + A_2)^2}{2} \tag{9}$$

Figure 2 shows the output the ring with SA as a function of the two input powers. By selecting the power corresponding to a 1-bit, equal to twice the input power with the maximum output $P_{on} = 2P_{in}^{max}$ this circuit can be used as an XOR gate. If both inputs are "off" there is no power at the output, if only one input is "on" the input to the saturable ring will be P_{in}^{max} and the output will be maximal since the 3-dB splitter will halve the power if only one port is "on" and if both the inputs are "on" the maximum is overshot , the ring input will be $4P_{in}^{max}$ and the transmission will be low. Figure 2 also shows that small deviations from $P_{on} = 2P_{in}^{max}$ will already result in a bit flip for the 01 input.

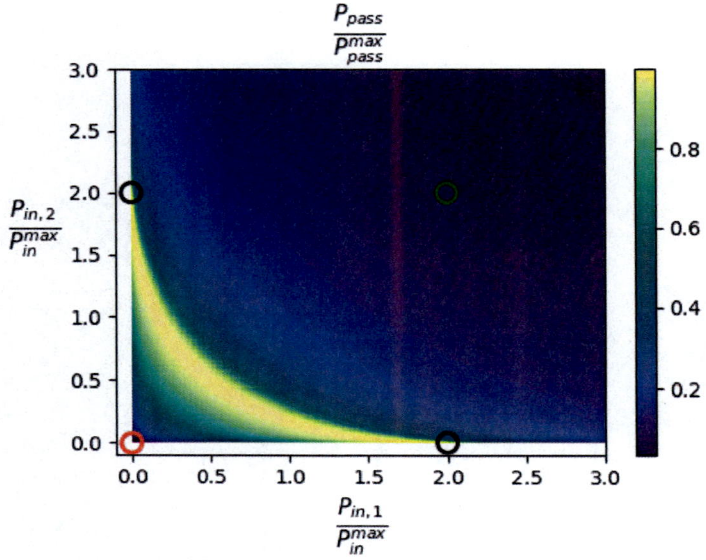

Fig. 2. output of the saturable ring dependent on the 2 input powers, the circles show the input states when $P_{on} = 2P_{in}^{max}$, black represents the 01 or 10 state with high output, green the 11 state and red the 00 state which both have low output power. This is exactly the expected response for the XOR operation.

To study this sensitivity and investigate if there are more optimal values for P_{on} we plot the output for each logical input state as a function of P_{on} corresponding to the logical one in Fig. 3. We also study errors on the logical zero by looking at different values for $P_{off} \neq 0$ proportional to the on-power $P_{off} = aP_{on}$. For the 00 state we see that for $P_{off} = \frac{P_{in}}{10}$ we get a high output power while the logical output should be 0. $P_{off} \leq \frac{P_{in}}{100}$ results in low outputs for the 00 input state and an output power > 80% of the maximum for the 10 and 01 inputs, if also $1.3P_{in}^{max} < P_{on} < 1.9P_{in}^{max}$. Lastly the 11 state results in a low output as long as $P_{on} > P_{in}^{max}$. The union of these result in an operating range of $1.3P_{in}^{max} < P_{on} < 1.9P_{in}^{max}$ and $P_{off} \leq \frac{P_{in}}{100}$. This is quite a narrow range, future analysis will focus on extending this range. However the narrow range might prove useful in pulsed neuron applications. Figure 2 also shows that different values for P_{on} can result in different gates e.g. the AND gate with $P_{on} = \frac{P_{in}^{max}}{2}$. Additionally, tuning the phase between the 2 inputs will result again in different gates and this will be investigated in future work.

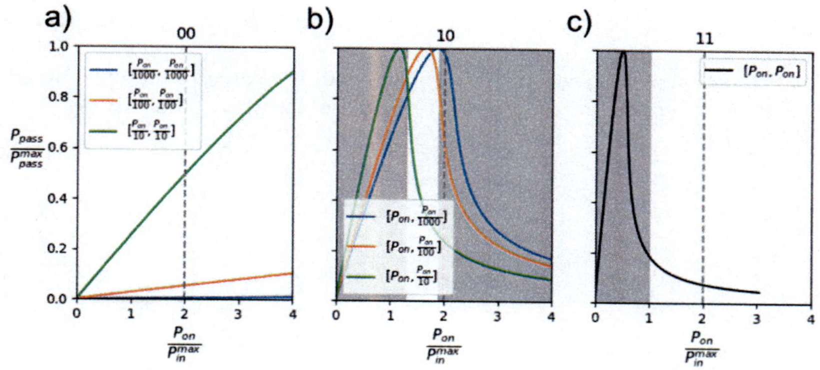

Fig. 3. Output of the XOR system for input powers deviating from the ideal powers, for a logical input of a) 00 b) 10 c) 11

3 Conclusion

We show that an SA ring resonator combined with a 3 dB splitter can be used as an all-optical XOR gate. The XOR behaviour is however very sensitive to power deviations making for an impractical device. Future work will focus on increasing the operation range and investigating if the XOR behaviour is retained for more realistic values of the SA insertion loss. However for use in pulsed neuron systems the narrow power range might be desirable[5]. To this end we will continue developing the analytic framework to characterize the gate properties as well as employ numerical techniques. Eventually a possible implementation of this system could be fabricated using quantum dots as a passive saturable absorber because they require no external power source and have comparably low saturation powers.

Acknowledgment. This project was funded within the QuantERA II Program that has received funding from the European Union's Horizon 2020 research and innovation program under Grant Agreement No 101017733.

References

1. Singh, P., Tripathi, D.K., Jaiswal, S., Dixit, H.: All-optical logic gates: designs, classification, and comparison. In: Advances in Optical Technologies, vol. 2014 (2014)
2. Kim, J.H., Byun, Y.T., Jhon, Y.M., Lee, S., Woo, D.H., Kim, S.H.: All-optical half adder using semiconductor optical amplifier based devices. Opt. Commun. **218**(4–6), 345–349 (2003)
3. Zhang, X., Thapa, S., Dutta, N.: All-optical XOR gates based on dual semiconductor optical amplifiers. Cogent Phys. **6**(1), 1660495 (2019)

4. Bogaerts, W., et al.: Silicon microring resonators. Laser & Photonics Rev. **6**(1), 47–73 (2012)
5. Zhang, Y., et al.: Evolution of neuron-like spiking response and spike-based all-optical XOR operation in a DFB with saturable absorber. J. Lightwave Technol. **42**(6) (2023)

Reconfigurable Photonic Integrated Circuit for All-Optical Matrix Inversion

Gabriele Cavicchioli[1(✉)], David A. B. Miller[2], Nader Engheta[3], Andrea Melloni[1], and Francesco Morichetti[1]

[1] Dipartimento di Elettronica Informazione E Bioingegneria, Politecnico di Milano, Via Ponzio 34/4, 20133 Milano, Italy
gabriele.cavicchioli@polimi.it

[2] Ginzton Laboratory, Spilker Building, Stanford University, Stanford, CA 94305, USA

[3] Department of Electrical and Systems Engineering, University of Pennsylvania, Philadelphia, PA 19104, USA

Abstract. We report on the design and experimental validation of a programmable silicon photonic architecture solving recursive mathematical problems directly in the optical domain. The reconfigurable kernel of the circuit is realized through a mesh of thermally tuneable Mach-Zehnder Interferometers, which is embedded into optical feedback loops and is used to implement arbitrary unitary matrices. The proposed architecture is employed to experimentally demonstrate matrix inversion without any optical-to-electrical conversion.

Keywords: Photonic integrated circuits · analog computing · matrix inversion · hardware accelerators

1 Introduction

Photonic computing is emerging as a viable alternative to digital electronic circuits, particularly in domains requiring solving mathematical problems with high throughput and low power consumption [1]. Programmable optical processors (POPs) already offer an efficient implementation of matrix-vector-multiplication (MVM) operations [2], successfully applied in several machine learning applications [3]. POPs inherently operates in the analogue domain [4], where inputs and outputs are arrays of optical signals requiring conversion to and from the digital domain. This makes the use of electronic circuits for digitization necessary, potentially introducing bottlenecks in the processor's performance. For this reason, numerous mathematical problems fail to fully leverage the advantages offered by POPs. Consequently, it becomes essential to broaden the number of mathematical operations which can be performed in optics without electro-optical conversions.

In this work we present the design of a programmable photonic solver based on a recursive topology for computing matrix inversion. In particular, we explain

J. Witzens et al. (Eds.): ECIO 2024, SPPHY 402, pp. 477–483, 2024.
https://doi.org/10.1007/978-3-031-63378-2_78

the proposed architecture and how it can perform matrix inversion, we show a silicon photonic realization of the proposed circuit, and we provide an experimental demonstration of all-optical matrix inversion.

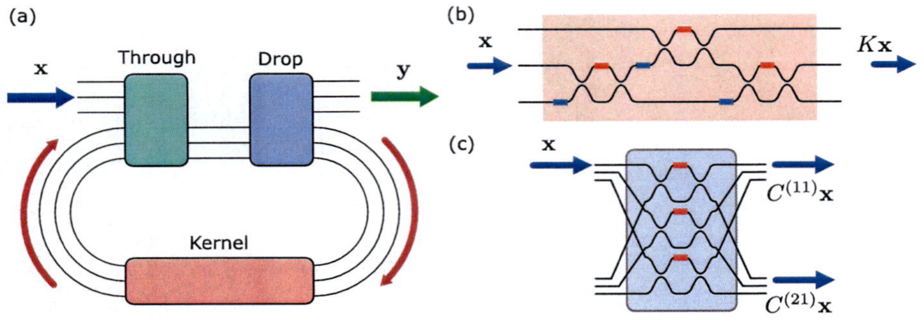

Fig. 1. Schematic of the recursive programmable optical processor and of its main components: (a) recursion is obtained through optical feedback loops, coupled together by a kernel K; (b) the kernel is implemented with a mesh of Mach-Zehnder interferometers (MZIs), and (c) each feedback loop is accessed through optical couplers, grouped into two distinct arrays (through and drop), which can be realized with programmable MZIs.

2 Programmable Optical Processors for Recursive Problems

Hardware analogue implementation of matrix inversion requires a recursive architecture [5–7]. We propose a photonic integrated circuit (PIC) implementing such a scheme consisting of N optical feedback loops which are coupled by a reconfigurable $N \times N$ mesh of Mach-Zehnder interferometers (MZIs), as shown in Fig. 1a, 1b. Each loop has the same round trip time (path-length equalization is not shown in Fig. 1 for simplicity) and can be accessed individually with optical couplers either to inject or collect light; in particular, each loop has two couplers, one used as input and one as output. Following the typical nomenclature of micro-ring resonators these couplers are called *through* and *drop* respectively. The circuit then accepts an input N-dimensional vector $\mathbf{x}$ at the through ports and returns an output N-dimensional vector $\mathbf{y}$ at the drop ports. Applying the transfer matrix method (TMM) to the schematic in Fig. 1 we can derive the transfer matrix T connecting the complex amplitudes of the elements of $\mathbf{x}$ with those of $\mathbf{y}$,

$$T = D^{(12)} e^{-j\phi} \frac{1}{I - e^{-j\Omega} T^{(22)} K D^{(22)}} T^{(21)}, \tag{1}$$

where $T^{(ij)}$ and $D^{(ij)}$ are the transfer matrices of the Through and Drop couplers, respectively (see Fig. 1c). ϕ_d is the phase delay accumulated in the propagation between the Through and the Drop couplers, and $\Omega = \omega/FSR$ is the frequency normalized to the free spectral range (FSR) of the circuit, which takes into account the round trip phase of each loop. Eq. (1) shows that, when $\Omega = 2M\pi$, the transfer matrix T is proportional to the inverse matrix of $I - K_{\text{eff}}$, where I is the identity matrix and $K_{\text{eff}} = T^{(22)}KD^{(22)}$ is the *effective kernel* of the circuit, that is the kernel scaled by the extrinsic loss due to the input $T^{(22)}$ and output $D^{(22)}$ coupling with the external waveguides [8]. Therefore, the computation of an inverse matrix A^{-1} requires setting the kernel such that

$$K_{\text{eff}} = I - A \tag{2}$$

Since, in principle, every general linear operator can be implemented optically using MZI meshes [9], the proposed circuit can invert arbitrary matrices. In practice, the implementation of arbitrary kernels depends on the technological platform used for realizing the POP. As example, in Silicon Photonics (SiPh), due to the lack of optical amplifiers, we are limited to kernels without internal gain, or equivalently with singular values within the unit circle. Moreover, the presence of $T^{(22)}$ and $D^{(22)}$ in (2) reduces further the space of reachable K_{eff}, making the quality factor of the optical loops a relevant parameter. In particular, as the round-trip losses of the cavities decrease, $T^{(22)}$ and $D^{(22)}$ tend to the identity matrix, and so K_{eff} approaches the actual kernel K implemented by the MZI mesh.

Exciting the circuit with the input vectors of the standard basis $\mathbf{e}_1 = (1,0,0)^T$, $\mathbf{e}_2 = (0,1,0)^T$, $\mathbf{e}_3 = (0,0,1)^T$, the drop outputs are proportional to the columns of the transfer matrix T, hence proportional to the inverse matrix of $1 - K_{\text{eff}}$. Each of these experiments requires a time which is roughly proportional to the inverse of the bandwidth of the optical circuit. Since typical bandwidths of such optical structures are in the order of some GHz, once the kernel is programmed, the time to compute the inverse matrix is in the order of ns, which represents a great improvement compared to digital electronic circuits.

3 Experimental Results

The experimental validation has been carried out on a prototype POP realized in a commercial 220-nm Silicon Photonics (SiPh) platform. In this platform, programmability is achieved through the thermo-optic effect using thermal phase shifters placed on top of the silicon waveguides. A photograph of the prototype is shown in Fig. 2. This device has 3 feedback loops, so it can process 3D vectors and invert 3×3 matrices. Through couplers are realized by means of tuneable MZIs and have variable coupling while drop couplers are made from tap directional couplers with fixed 10% power splitting ratio. All the feedback loops share the same length of 5.7 mm (including the kernel length), resulting in a FSR of 12.5 GHz. Finally, the kernel is a 3×3 triangular MZI mesh, implementing arbitrary unitary matrices.

To program the kernel each MZI of the circuit was characterized and its effect on the output ports isolated, providing information on the voltage levels

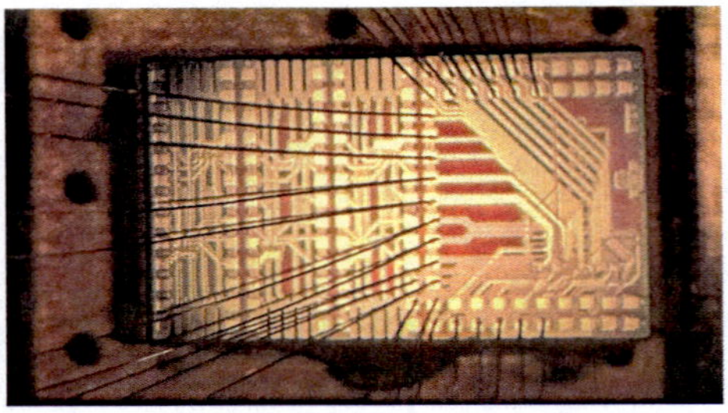

Fig. 2. Microscope photograph of the prototype chip wire-bonded to an electronic PCB.

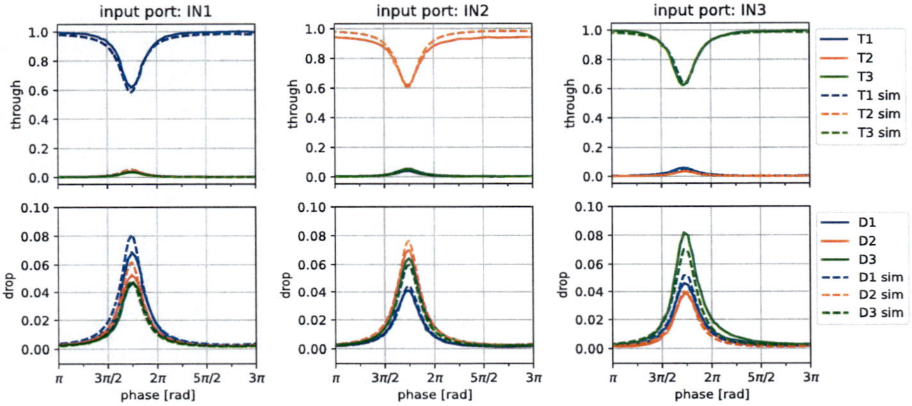

Fig. 3. Comparison of the power transfer matrices at the through ports (T1, T2, T3) and at the drop ports (D1, D2, D3) acquired experimentally (solid lines) and obtained through simulation (dashed lines) varying the round-trip phase of the optical loops.

required to drive each phase shifter in the circuit. Thanks to this calibration, the MZI mesh and the through coupler could be programmed to implement the desired kernel and coupler matrices K and $T^{(ij)}$.

The matrix inversion capabilities of the circuit are assessed by comparing the experimental behaviour of the prototype with the behaviour predicted by numerical simulations. Experimentally, the voltages of the phase shifters realizing the desired K and $T^{(ij)}$ are set and kept constant for the whole duration of the measurements, while the phase shifters controlling the resonance points of the feedback loops are fed with a voltage ramp. For each point in the ramp, we sample the optical power at the output of both the through and the drop port, obtaining a scan of the transfer matrix (1) for different values of the round trip phase contained in the term $e^{-j\Omega}$ in (1). The same measurement with a certain kernel is repeated three times, cycling through the input vectors $\mathbf{e}_1$, $\mathbf{e}_2$ and $\mathbf{e}_3$. As explained in Sect. 2, this gives at the drop output the columns of the transfer function and from the optical power we can obtain the square amplitude of its elements $|T_{nm}|^2$. For convenience, we define a *power transfer matrix* $\hat{T}$ whose elements are

$$\hat{T}_{nm} = |T_{nm}|^2. \tag{3}$$

$\hat{T}$ represents the amplitude response of the circuit and does not contain information on the relative phase between the elements of the transfer matrix. An example of the results of such a scan is shown in Fig. 3. In this experiment all the through couplers have a 10% coupling factor and the kernel implements the permutation matrix

$$K = \begin{pmatrix} 0 & 1 & 0 \\ 0 & 0 & 1 \\ 1 & 0 & 0 \end{pmatrix}; \tag{4}$$

the optical power at each output port is normalized to extract the experimental power transfer matrix $\hat{T}_{\text{exp}}$ and compared to the same matrix obtained through simulation $\hat{T}_{\text{sim}}$. Figure 3 shows a good agreement between the two and this validates the model expressed by (1). Each point of the scan corresponds to a different value of the round-trip phase $e^{-j\Omega}$ and so to a different matrix inverted by the system. When the system reaches a resonance point ($\Omega = 2n\pi$), the optical power at the drop port reaches its maximum and the transfer matrix of the system is proportional to the inverse of $A = I - K_{\text{eff}}$.

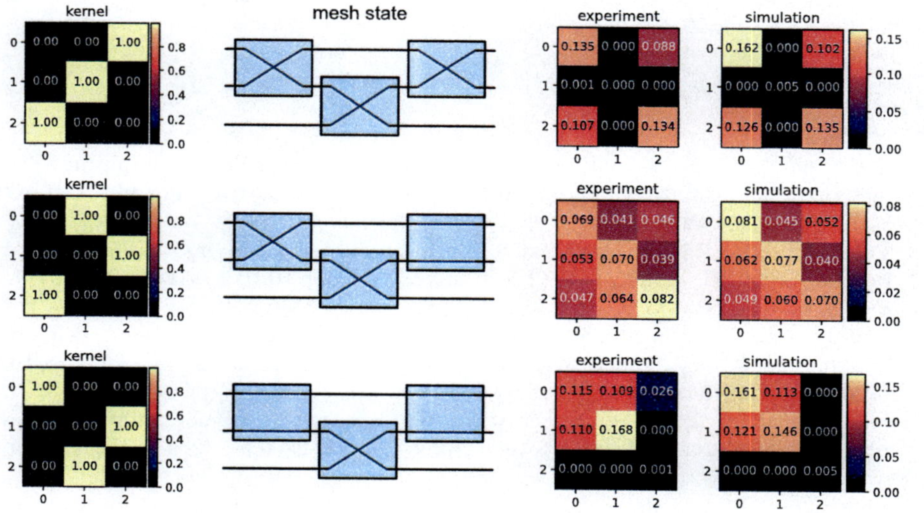

Fig. 4. Comparison of the experimental power transfer matrices obtained with three different kernels with the corresponding matrices obtained through simulation.

In Fig. 4 we report the power transfer function obtained using three different permutation matrices as kernels, which correspond to specific bar-cross configurations of the mesh MZIs; the matrices are all taken at resonance points.

Comparing $\hat{T}_{\mathrm{exp}}$ and $\hat{T}_{\mathrm{sim}}$ we see that the chip is indeed returning the expected results, except for some inaccuracies in the values of the optical power. The maximum deviation from simulation is 0.057, with an average l_1 error of 0.011. These are likely due to non idealities of the MZIs and programming errors because of which the effective kernel of the circuit is different from the nominal one.

4 Conclusions

In this study, we have conceived a programmable optical processor dedicated to matrix inversion. Embedding a programmable MZI mesh in a recursive topology, the resulting optical processor outputs are proportional to the columns of the inverse of a matrix, selected by programming the MZI mesh. This inversion process occurs without the need for optical-to-electrical conversions within the processor. A prototype managing 3×3 matrices has been implemented on Silicon Photonics (SiPh) platforms and has been tested experimentally. The device model has been validated by the experimental measurements and the capabilities of this new architecture have been demonstrated.

Acknowledgments. *This work was supported by the Italian National Recovery and Resilience Plan (NRRP) of NextGenerationEU, "Telecommunications of the Future" (PE00000001 - program "RESTART", Structural Project "Rigoletto" and Focused Project "HePIC") and "MUSA: Multilayered Urban Sustainability Action", and by AFOSR grant FA9550-21-1-0312.*

References

1. Zangeneh-Nejad, F., Sounas, D.L., Alù, A., Fleury, R.: Analogue computing with metamaterials. Nat. Rev. Mater. **6**(3), 207–225 (2021). https://doi.org/10.1038/s41578-020-00243-2
2. Zhou, H., et al.: Photonic matrix multiplication lights up photonic accelerator and beyond. Light Sci. Appl. **11**(1) (2022). https://doi.org/10.1038/s41377-022-00717-8
3. Pai, S., et al.: Experimentally realized in situ backpropagation for deep learning in photonic neural networks. Science **380**(6643), 398–404 (2023). https://doi.org/10.1126/science.ade8450
4. Macho-Ortiz, A., Pérez-López, D., Azaña, J., Capmany, J.: Analog programmable-photonic computation. Laser Photonics Rev. **17**(10), 2200360 (2023). https://doi.org/10.1002/lpor.202200360
5. Tzarouchis, D.C., Mencagli, M.J., Edwards, B., Engheta, N.: Mathematical operations and equation solving with reconfigurable metadevices. Light Sci. Appl. **11**(1), 263 (2022). https://doi.org/10.1038/s41377-022-00950-1
6. Mohammadi Estakhri, N., Edwards, B., Engheta, N.: Inverse-designed metastructures that solve equations. Science **363**(6433), 1333–1338 (2019). https://doi.org/10.1126/science.aaw2498
7. Tzarouchis, D.C., Edwards, B., Engheta, N.: Programmable wave-based analog computing machine: a metastructure that designs metastructures. arXiv (2022). https://doi.org/10.48550/arXiv.2301.02850

8. Camacho, M., Edwards, B., Engheta, N.: A single inverse-designed photonic structure that performs parallel computing. Nat. Commun. **12**(1) (2021). https://doi.org/10.1038/s41467-021-21664-9
9. Miller, D.A.B.: Self-configuring universal linear optical component. Photon. Res. **1**(1), 1–15 (2013). https://doi.org/10.1364/PRJ.1.000001

First Polymer-Based Passive Optical Waveguide for the Visible Range from 633 nm Down to 488 nm

Tianwen Qian[1(✉)], Robin Kraft[1], Thomas Wiglenda[3], Crispin Zawadzki[1], Klara Mihov[1], Martin Kresse[1], Madeleine Weigel[1], Jakob Reck[1], Csongor Keuer[1], Philipp Winklhofer[1], Moritz Kleinert[1], David de Felipe[1], Arne Schleunitz[3], Norbert Keil[1], and Martin Schell[1,2]

[1] Fraunhofer Institute for Telecommunications, Heinrich Hertz Institute, Einsteinufer 37, 10587 Berlin, Germany
tianwen.qian@hhi.fraunhofer.de
[2] Technische Universität Berlin, Straße des 17. Juni 135, 10623 Berlin, Germany
[3] Micro Resist Technology GmbH, Köpenicker Str. 325, 12555 Berlin, Germany

Abstract. We investigated the transmission properties of optical waveguide based on fluorinated acrylate polymer and Ormocer® based polymer in the visible range (VIS). Laser transmission-induced transparency (LTIT) and fluorescence were observed in the acrylate-type polymer, while Ormocer® based polymer seem not to show these issues.

Keywords: visible · polymer · hybrid photonic integration

1 Introduction

The polymer-based photonic platform has shown promising capabilities for hybrid integration with other active platforms such as InP [1, 2] and GaAs [3] in the near-infrared (NIR) wavelength range. Taking advantage of the low attenuation of polymers in the visible (VIS) wavelength range (see Fig. 1), hybrid integration holds great potential for augmented/virtual reality and spectroscopy applications. In this work, the transmission properties of the optical waveguide, which consists of the commonly used fluorinated acrylate polymer material (ZPU-12 series from ChemOptics Inc.), were investigated in the visible range, especially at the wavelengths red (633 nm), green (515 nm) and blue (488 nm). A novel organic-inorganic hybrid polymer material (Ormocer® series from micro resist technology GmbH) was also examined. The first result of the investigation indicates that the Ormocer® based polymer material as a potential candidate for applications of polymer-based waveguide for the VIS wavelength range.

J. Witzens et al. (Eds.): ECIO 2024, SPPHY 402, pp. 484–489, 2024.
https://doi.org/10.1007/978-3-031-63378-2_79

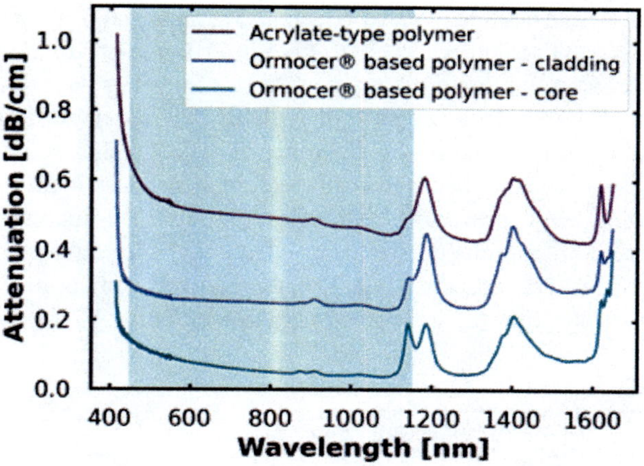

Fig. 1. Measured material loss of the bulk polymer materials, including the fluorinated acrylate polymer and the Ormocer® based polymer.

2 Acrylate-Type Polymer Waveguide in VIS Range

The typical waveguide loss obtained from the acrylate-type polymer is around 0.8 dB/cm at 1550 nm [4]. To investigate the waveguide loss in the VIS region, the single mode polymer waveguide with a geometry of 1000 nm × 800 nm (width × height) was selected from the effective refractive index simulation, and the schematic layout of the buried-channel acrylate-type polymer waveguide is also shown in Fig. 2(a). Stepper lithography in combination with dry etching is used to fabricate such a waveguide due to its small feature size. The optical transmission property is characterized by a 5-mm-long straight waveguide structure.

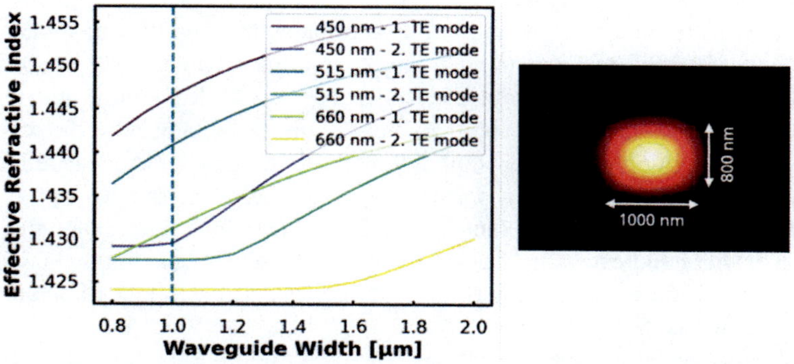

Fig. 2. (a) Effective refractive index simulation by varying the waveguide width, a 1000 nm × 800 nm (width × height) geometry is chosen to fulfil the single mode condition above 450 nm wavelength.

Three typical visible wavelengths of 633 nm, 515 nm, and 488 nm were used for characterization. At 633 nm, fiber to chip to fiber insertion losses of 24.4 dB was measured directly after coupling, with the waveguide loss reducing with time, as shown in Fig. 3(a). Within 60 min, a loss decrease of 20.1 dB was observed during characterization, which saturates at a level of 4.3 dB insertion loss and shows around 0.6 dB/cm waveguide loss. This effect agrees well with the so-called Laser Transmission Induced Transparency (LTIT) effect previously reported in [5]. After the first injection and measurement, two further measurements were carried out on the following days, each with a time span of 30 min. As can be seen in Fig. 3(b), the transmission improves in a saturated manner, which indicates the stability and storage capacity of this effect.

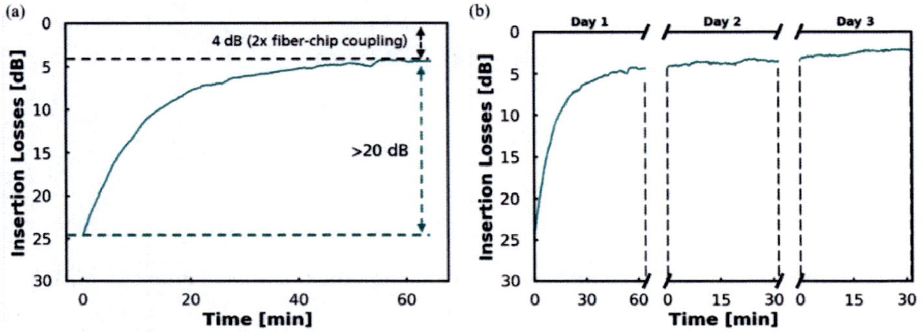

Fig. 3. (a) Optical transmission characteristics by injecting the 633 nm laser into the acrylate-type polymer waveguide and recording the insertion loss over time. A performance improvement of more than 20 dB was observed within 60 min (b). The tests conducted in the following two days show that the performance improvement persists and continues in a saturated manner.

A similar effect was also observed at 515 nm, where 36.5 dB fiber to chip to fiber insertion losses was measured directly after coupling, followed with 13.5 dB loss reduction. Assuming the fiber to chip coupling loss remains the same for this wavelength, waveguide loss of 38 dB/cm is extracted. In addition, fluorescence was observed in the waveguide (see Fig. 4c) [6]. The waveguide gleams orange while coupling with the green laser. The sudden power drops at about 15 and 23 min were damage to the waveguide facet due to the high power density (about 9.3×10^{11} mW·m^{-2}). A clear reflection at the coupling point can be detected by the microscope, indicating the damage, as shown in Fig. 4(d). However, during the long-term characterization, a loss reduction can still be observed, which indicates the LTIT effect hasn't reached its maximum, but since the waveguide facet is damaged, the transmitted power has a stronger power fluctuation and the loss reduction is limited to about 9.2 dB.

At 488 nm wavelength, the waveguide shows yellow color directly after coupling as in Fig. 5(a), which also indicates the fluorescence generated in the waveguide [6]. The spectral characterization is performed with a broadband white light source; the result is shown in Fig. 5(b). The characterization was first performed without the LTIT effect, and then the same waveguide was measured again after introducing the LTIT effect by injecting the 633-nm laser. Loss reduction can be observed from 525 nm to 700 nm,

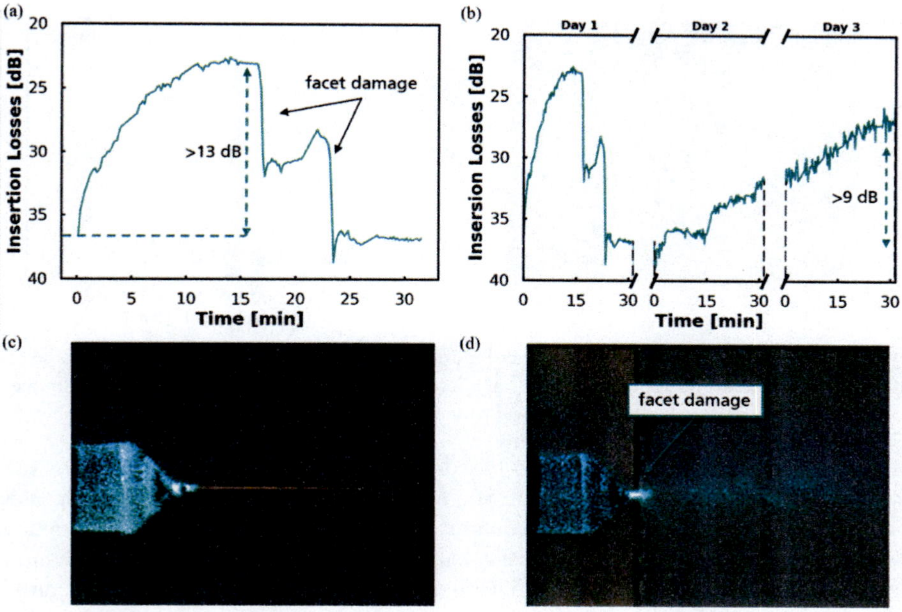

Fig. 4. (a) Optical transmission characteristics by injecting the 515 nm laser into the acrylate-type polymer waveguide and insertion loss evolution over time. A loss reduction of more than 13 dB was observed within 15 min, followed by a sudden drop in power, presumably due to damage to the facet caused by the high power density at the waveguide coupling point. (b) Consecutive tests show the loss reduction continues for 9.2 dB. The fluctuation indicates the damaged facet. (c) Waveguide gleams with orange color after coupling and injecting with 515 nm laser, which indicates fluorescence. (d) Clear reflection at the coupling point observed after the sudden drop in monitored power and indicates a facet damage.

which can be defined as the effective range of the LTIT effect, with a maximum of 43 dB at 613 nm. At shorter wavelengths, almost no power is detectable, which also indicates that this type of polymer exhibits broadband fluorescence in the blue to green wavelength range.

3 Ormocer® Based Polymer Waveguide: UV Nanoimprint Lithography (UV-NIL) and First Result

While investigating the reasons for LTIT and broadband fluorescence [5], we have focused on Ormocer® based polymers that connect organic and inorganic structural components. Specific polymers with a reduced C-H bond content and an optimized three-dimensional molecular structure are required in order to explore the possibility of using these polymers as waveguides even for the VIS range.

One way to further decrease the waveguide loss in the VIS rang is to decrease the scattering loss ($\alpha \propto \lambda^{-4}$) due to the sidewall roughness introduced by the etching process. The technique of UV nanoimprint lithography (UV-NIL) [7] is being investigated

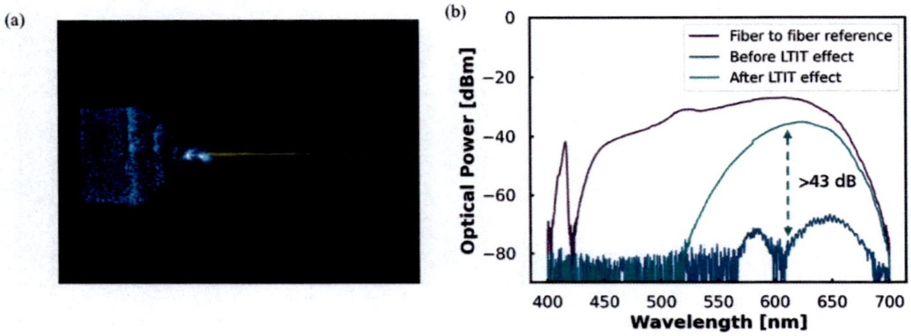

Fig. 5. (a) The waveguide shows a yellow color after coupling and injecting with a 488 nm laser, indicating the fluorescence effect. (b) Spectral characterization with a tunable white light source.

and applied for fabricating Ormocer® based polymer waveguide. Figure 6 shows the fabrication method and the SEM image of the first UV-imprinted waveguide with such polymer. In the first step, a high-precision master is produced using a standard contact lithography process using Si or photoresist. In the second step, a soft, transparent stamp is produced from the polymer polydimethylsiloxane (PDMS) on the basis of the structured master. Finally, the structured PDMS stamp is applied to the spin-coated Ormocer® based polymer material. The liquid polymer film is then cured through the transparent stamp with UV light, the target waveguide structure is formed and can be easily separated from the soft PDMS stamp.

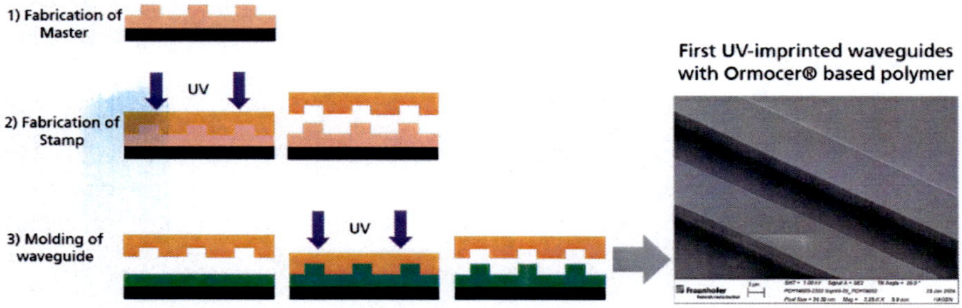

Fig. 6. Illustration of the UV-NIL technique and the UV-imprinted waveguide with Ormocer® based polymer.

Compared to the lithography method used by the acrylate polymer material, the UV-NIL technique does not require an additional etching process to form the waveguide, which in turn improves the sidewall roughness.

The latest measurement show that the waveguide formed with the Ormocer® based polymer has no significant fluorescence effect at 633 nm and 488 nm, as shown in Fig. 7. More detailed results on the propagation losses are currently being investigated and will be presented at the conference.

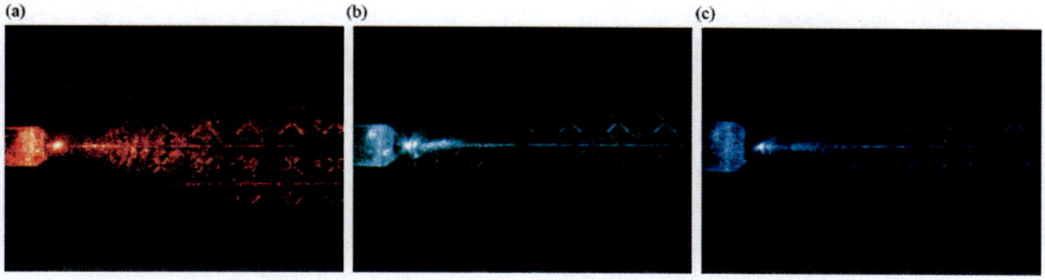

Fig. 7. Waveguide made of Ormocer® based polymer material produced using UV-imprint technique and coupled by (a) 633 nm, (b) 515 nm and (c) 488 nm laser signals. No fluorescence effect is observed.

4 Conclusion

We have investigated waveguides formed with the fluorinated acrylate polymer material and Ormocer® based polymer material in the VIS range. The result shows that for the acrylate-type polymer, although the bulk material has relative low attenuation, the additional fluorescence activated by the shorter wavelength may lead to additional loss in the target wavelength and thus limit the performance of the waveguide. On the other hand, the recent result of the waveguide formed from the Ormocer® based polymer shows no fluorescence effect and could be a good candidate for the fabrication of PICs in the VIS range.

References

1. Smit, M., Williams, K., van der Tol, J.: Past, present, and future of InP-based photonic integration. APL Photon. **4**, 50901 (2019). https://doi.org/10.1063/1.5087862
2. Soares, F.M., Baier, M., Gaertner, T., et al.: High-Performance InP PIC technology development based on a generic photonic integration foundry. In: Optical fiber communication conference:M3F.3 (2018). https://doi.org/10.1364/OFC.2018.M3F.3
3. Kleinert, M., de Felipe, D., Conradi, H., et al.: Hybrid polymer integration for communications, sensing and quantum technologies from the visible to the infrared. In: 2021 European Conference on Optical Communication (ECOC). IEEE (2021)
4. Kleinert, M., Nuck, M., Conradi, H., et al.: A platform approach towards hybrid photonic integration and assembly for communications, sensing, and quantum technologies based on a polymer waveguide technology. In: 2019 IEEE CPMT symposium Japan (ICSJ) (2019)
5. Liu, J., Wang, C., Dang, Z., et al.: Thermally resettable laser transmission induced transparency in polymer waveguides at 635 nm. Opt. Express **30**, 17529–17540 (2022). https://doi.org/10.1364/OE.456628
6. Liu, J., Wang, C., Zhang, Z.: Laser-transmission-induced Raman emission masked by progressive transparency in polymer waveguides. Opt Lett. OL **47**, 6117 (2022). https://doi.org/10.1364/OL.470832
7. Modaresialam, M., Chehadi, Z., Bottein, T., et al.: Nanoimprint lithography processing of inorganic-based materials. Chem. Mater. **33**, 5464–5482 (2021). https://doi.org/10.1021/acs.chemmater.1c00693

Wednesday Poster Session

On-Chip Calibration of an Optical Phased Array Through Chip Facet Reflections

Marco Gagino[1(✉)], Alonso Millan-Mejia[2], Erwin Bente[1], and Victor Dolores-Calzadilla[1]

[1] Eindhoven University of Technology, Eindhoven Hendrik Casimir Institute (EHCI), 5612 AP Eindhoven, The Netherlands
m.gagino@tue.nl

[2] SMART Photonics, 5656 AE Eindhoven, The Netherlands

Abstract. We present an InP optical phased array with on-chip calibration capabilities. The on-chip calibration relies on reflecting some of the light at the optical antennas back into the OPA input waveguide and measuring the phase difference between consecutive arms through an mREV calibration algorithm. We employ a spurious low-level (<0.1%) reflection at the OPA facet, and we demonstrate that it is enough to calibrate the PIC using this method. Moreover, we calibrate following both a free-space and on-chip method and measure negligible variations of sidelobe-suppression ratio and beam width respectively, showing comparable calibration accuracies for the two methods.

Keywords: Optical Phased Array · Calibration · Integrated Photonics

1 Introduction

Optical Phased Arrays (OPAs) are photonic integrated circuits (PICs) with beam forming and beam steering capabilities that find applications in LiDAR sensing (light detection and ranging) and optical wireless communication. Their functionality relies on the principle of interference between optical antennas, and on the control of the phase difference between the waveguide arms carrying light to the emitting elements. Due to the variations in the fabrication of photonic integrated circuits, non-uniformities in the OPA waveguides are introduced. These variations lead to phase differences between the arms of the OPA which, if not corrected for, deteriorate the free-space beam profile at the required steering angle [1]. For this reason, an initial calibration is often required, where a phase offset is provided by the phase tuning elements in the arms of the OPA.

Oftentimes the calibration routines are executed by monitoring the free-space beam using a far-field imaging system. The light intensity in a specific steering direction is collected as a figure of merit to be optimized through algorithms such as gradient descent, particle swarm optimization [2], genetic optimization, hill-climbing [3], modified rotating element electric field vector (mREV) [4] and neural network-based [5] methods. Although these free-space methods are well established in literature, calibration errors may arise due to potentially sub-optimal alignment of the imaging system with respect

J. Witzens et al. (Eds.): ECIO 2024, SPPHY 402, pp. 493–499, 2024.
https://doi.org/10.1007/978-3-031-63378-2_80

to the PIC under test, as well as the presence of chromatic and spherical aberrations. Moreover, these calibration systems are not optimal in real-case applications of OPAs due to the use of bulky and expensive equipment. Finally, repeated re-calibrations in the lifetime of the sensor may be necessary due to ageing of the electrical components on the chip, and due to external or internal environmental changes such as thermal fluctuations that may introduce a phase shift in the OPA arms [6].

To overcome these limitations, on-chip calibration methods have been researched in literature where the forward-propagating light is tapped and detected through photodetectors on the PIC. To the best of our knowledge, most of PIC architectures for on-chip monitoring and calibration of phase are divided into two categories. The first relies on tapping power from the waveguides before the light is emitted in free-space [7, 8] and combining it to measure phase variations between consecutive arms through photodetectors. This method is not scalable due to the use of at least 2N electrical contacts for an N-arms OPA (N photodetectors, and N phase tuners), and up to 3N electrical contacts when balanced photodetectors are employed. Other methods rely on tapping the power after the optical antennas [2, 9–12]. In this case, the detection is done either with waveguide-coupled photodetectors, or through an imaging element (e.g., star coupler) into one or multiple photodetectors. Although this method can be more scalable, it relies on the use of surface emitting antennas which can be limited in bandwidth and lossy, and it requires additional optical paths which may introduce further phase errors.

We propose an on-chip calibration PIC method that is scalable and is based on tapping power from edge-emitting optical antennas through a low-level on-chip spurious reflection at the chip's facet. In this way, the light follows the same optical path traveling back into the input power splitter, which also serves as a combiner. Since the reflected light is collected by a single waveguide and it is sensed off-chip, this method is more scalable than the ones relying on tapping power from consecutive arms as no additional electrical driver is needed. Moreover, a single power tapping element is required instead of multiple MMI splitters or directional couplers. We validated the method by calibrating the OPA through the on-chip reflection and free-space imaging, and by comparing the two results. The demonstration was performed on an InP generic platform operating in C-band, but the concept can be adapted to any photonic platform and wavelength range. The concept was patented prior this publication [13].

In the following Sections we show the PIC design, we explain the reflection-based calibration principle and we show the obtained measurement results.

2 On-Chip Calibration Method

The on-chip calibration method proposed here relies on coupling the light of an external laser source into the OPA through a waveguide-coupled fiber, reflecting some of the light at the OPA output facet back into the input waveguide and consequently into the fiber, and measuring the reflected light with an external powermeter. The amount of reflected light depends on the amount of phase shift in the OPA arm, and the PIC is calibrated by maximizing the reflection power level.

The light source and powermeter are isolated through a fiber-based circulator. Light is coupled out of the OPA in free-space, and it is imaged through an infrared camera placed at 14 cm distance. The setup is shown in Fig. 1.

2.1 Optical Phased Array Design

We designed and fabricated an OPA having 8 optically equivalent arms, and 1.2 mm long electro-optical phase modulators (EOPM) that are driven with reverse bias voltages in the 0–13 V range. The power splitting element is a 1×8 star coupler giving a Gaussian power distribution in the arms, with 6dB power ratio between the inner and outer arms. The optical antennas are 3 μm-wide edge-emitting deep-etched waveguides, uniformly separated by 10 μm gaps (13 μm pitch). A schematic representation of the designed PIC and a microscope picture of the fabricated device are shown in Fig. 1.

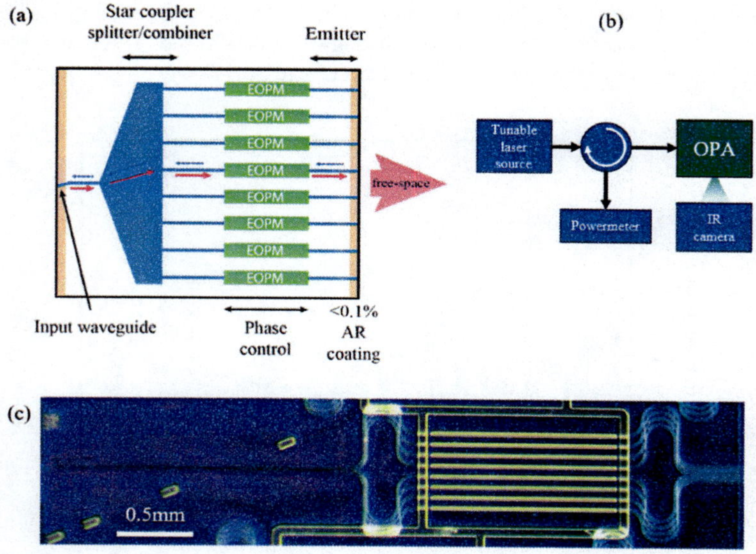

Fig. 1. (a) OPA circuit design. (b) Setup used to input light into the PIC, measure the reflected power, and image the OPA beam in free-space. (c) Microscope picture of the fabricated OPA. The metal contacts were wire-bonded to a printed circuit board (not shown here).

We chose to use the reflection from the facet interface of the edge-emitting output waveguides, which otherwise would be spurious optical power in the circuit. An anti-reflection coating is applied to the facet, whose design guarantees less than 0.1% reflection for the wavelength in use (1528 nm). With an input power of 10 dBm from the laser, we measured −4 dBm at the chip's facet, resulting in 14 dB losses through the chip, accounting for the fiber-to-chip coupling losses. We thus expect to measure a reflection level of 0.02 μW from the facet.

The star coupler is designed to distribute light from its input waveguide to the arrayed waveguides with zero phase difference between them. For a perfectly calibrated system and optically balanced path lengths in the OPA arms, the phase difference between arms at the emitter's edge will be zero as well, and the free-space beam will be oriented at the 0° direction. Since the reflection point is at the output edge of the OPA, the back-propagating light experiences an identical amount of phase shift compared to the forward-propagating light. Due reciprocity, the reflected light will also have zero phase difference at the star coupler's arrayed waveguide edge, and light will be focused back into the input waveguide of the star coupler. For this condition the reflected power is maximum, and the OPA is calibrated.

2.2 On-Chip Reflections

To analyze the origin of reflections on the OPA, we measured the reflections inside the chip through the OFDR method [14]. Two peaks are recognized at the cavity lengths (with respect to the input waveguide at the left facet of the PIC) of 2 mm and 5.5 mm corresponding to reflections at the star coupler and at the chip's facet (Fig. 2). Since the star coupler reflections occur before the splitting of power in the OPA arms, they only contribute to an offset in the measured reflected power.

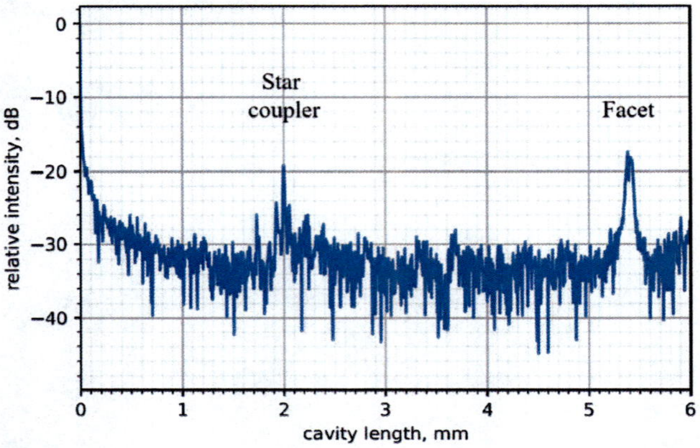

Fig. 2. On-chip reflections measured through the OFDR method.

The reflection peak from the PIC's facet is broader, showing that a distributed reflection is taking place. This reflection range (∼ 150 µm) matches an area close to the facet where the passivation layer is removed from the layer stack to facilitate facet cleaving, which may be responsible for additional reflections.

2.3 Validation of the On-Chip Calibration Concept

To validate the calibration concept, we employed a mREV algorithm to calibrate the OPA in free-space in a direction near 0°, and through the on-chip reflection using the setup shown previously. To execute the mREV algorithm in free-space, the reverse bias in each EOPM is swept over the 2π range, and the variation of the intensity profile in the central pixel of the camera is recorded. Since such profile corresponds to the interference between the arm under test and the rest of the OPA, the voltage for which the highest intensity is reached (constructive interference) is selected to calibrate the arm. In the case of the method relying on reflection, the reflected power recorded by the powermeter is observed.

Figure 3 shows a comparison between the interference curves recorded by the camera and the ones measured by the powermeter (i.e. by exploiting the facet reflectivity). We measured an average reflected power of 6.7 μW with up to 1 μm variations. These values are higher than expected and may be due to the presence of other reflections in the circuit (e.g., star coupler, and fiber tip).

The curves measured in reflection show a period that is half that of the forward direction. This is due to the fact that the phase in the arm under test is being shifted at twice the rate due to its double pass through the arm. Moreover, the two curves are offset, which indicates that the camera pixel that was chosen as a reference is not perfectly aligned to the 0° position, where constructive interference should be happening simultaneously at the input waveguide.

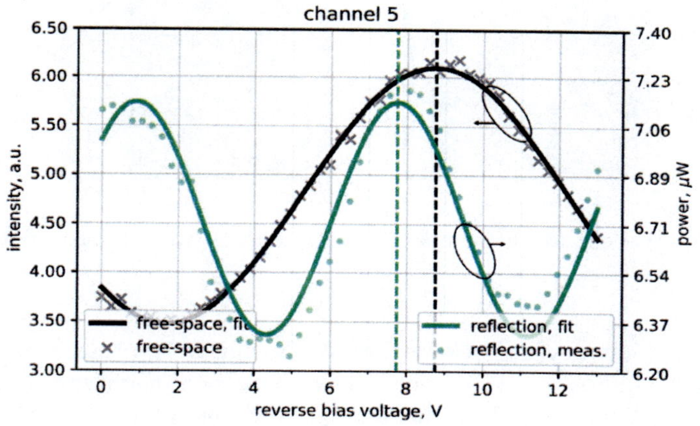

Fig. 3. Interference plots of one arm with the rest of the OPA in free-space (black crosses) and measured from the reflection powermeter (green, dots).

To support this claim, Fig. 4 shows the free-space beam recorded by the camera after the calibration in free-space and through the on-chip reflection. The two are shifted by a 0.25° angle. Moreover, the beam width (measured at the full-width at half maximum FWHM) and sidelobe suppression ratio (SLSR) measure 0.72° and −7.9 dB for the free-space calibration case and 0.74° and −7.6 dB for the on-chip calibration case, showing a good match between the two. The beam widths are measured by fitting the main beam to a Gaussian function, whereas each pixel in the camera corresponds to 0.06° of angular resolution. Since the variation in FWHM is below this resolution, and the resolution on the calculation of the SLSR is limited due the presence of interference fringes on the camera image, we can consider the two beams as identical, thus showing that the method employed for the on-chip calibration of the OPA has at least the same precision as the free-space-based calibration method.

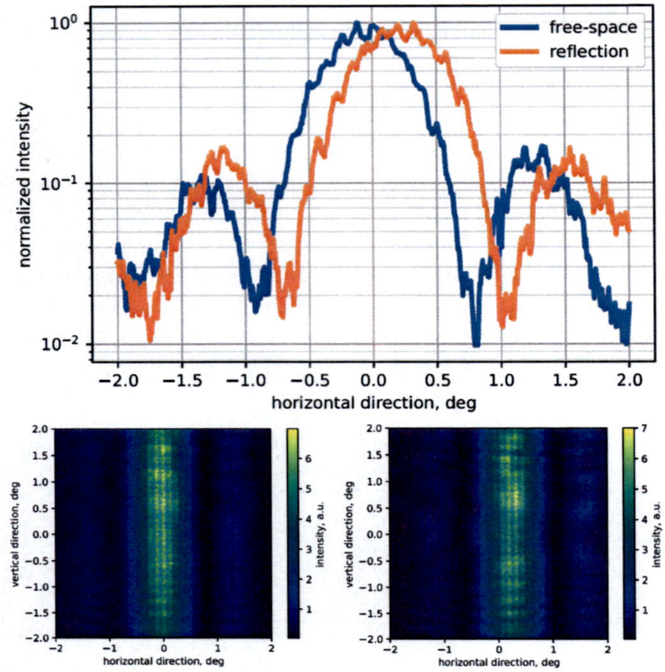

Fig. 4. Free-space (elongated) beam after calibration in free-space (left) and through on-chip reflection method (right).

The second sine peak of the backward light corresponds to a π phase shift (destructive interference) in free-space and a 2π phase shift in reflection. This causes a degeneracy of solutions which ultimately limits the method, which has to rely on the information obtained from the free-space calibration. This information, however, can be obtained once and recorded on a look-up-table. Any future re-calibration can be done on-chip though this method as, for example, gradual ageing effects and temperature variations are expected to cause phase shifts lower than $\pi/2$ in the OPA arms.

3 Conclusions

We demonstrated a method to calibrate the arms of an OPA using reflections generated at the edge of the emitting antennas. We show that the on-chip calibration gives an identical beam shape compared to a free-space based algorithm, with a small offset in the angular position given by the accuracy of the camera alignment in free-space.

This feature comes at no extra price on the design of the OPA since we make use of a weak (<0.1%) reflection given by the facet of our InP PIC. Although we used the spurious reflection of the waveguide with anti-reflection coating, a specific facet reflectivity could be realized to provide robustness to an OPA circuit using this concept.

References

1. Yang, Y., et al.: Phase coherence length in silicon photonic platform. Opt. Express **23**, 16890 (2015)
2. Komljenovic, T., Pintus, P.: On-chip calibration and control of optical phased arrays. Opt. Express **26**, 3199 (2018)
3. Guerber, S., et al.: Wafer-level calibration of large-scale integrated optical phased arrays. Op-tics Express **30**, 35246 (2022)
4. Lian, D., et al.: Grating-lobe-free optical phased array with 2-D circular sparse array aperture and high-efficiency phase calibration. Nanophotonics **13**, 29–37 (2024)
5. Leng, L., et al.: Phase calibration for integrated optical phased arrays using artificial neural network with resolved phase ambiguity. Photon. Res. **10**, 347 (2022)
6. Yepez, P.A.K., Scholz, U., Zimmermann, A.: Temperature dependence of the steering angles of a silicon photonic optical phased array. IEEE Photon. J. **12**, 1–13 (2020)
7. Ribeiro, A., Miura, K., Spuesens, T., Bogaerts, W. On-chip differential phase monitoring with balanced photodiodes. In: 2016 IEEE 13th international conference on group IV photonics (GFP) (2016)
8. Prost, M., et al.: Optical phased array with on-chip phase calibration. Opt. Lett. **47**, 6081 (2022)
9. Zhou, Z., et al.: An on-chip calibration method for optical phased array with no blind spot in the steering range. IEEE Photon. Technol. Lett. **35**, 585–588 (2023)
10. Hulme, J.C., et al.: Fully integrated hybrid silicon two dimensional beam scanner. Opt. Exp. **23**, 5861 (2015)
11. Shim, J., et al.: On-chip monitoring of far-field patterns using a planar diffractor in a silicon-based optical phased array. Opt. Lett. **45**, 6058 (2020)
12. Guo, W., et al.: Two-dimensional optical beam steering with InP-based photonic integrated circuits. IEEE J. Sel. Top. Quantum Electron. **19**, 6100212 (2013)
13. An optical phased array having integrated calibration and monitoring means, as well as a corresponding method and Integrated Circuit, IC., Patent Number: NL 2036375
14. Kamata, M., Baba, T.: OFDR analysis of Si photonics FMCW LiDAR chip. Opt. Exp. **31**, 25245 (2023)

Check for updates

Polymer-Ge Hybrid Waveguide for Flexible Photonic Integration

Jinyuan Liu, Zhenming Ding, and Ziyang Zhang(✉)

Laboratory of Photonic Integration, School of Engineering, Westlake University, Hangzhou 310024, China
zhangziyang@westlake.edu.cn

Abstract. Amorphous germanium nanostrip is embedded in polymer to form a waveguide. Such waveguides only transmit TM-polarized light while imposing a strong absorption on the TE-polarized light. Furthermore, this absorption is shown to saturate upon femtosecond pulse excitation. This property is used to construct on-chip saturable absorber by depositing a Ge strip onto a polymer waveguide. Such hybrid waveguide is sandwiched between an InP reflective gain chip and a fiber Bragg grating to construct a mode-locked laser. Experiments have shown a pulse train with a 147-ps pulse width at 50.58-MHz repetition rate. The electronic signal-to-noise ratio is larger than 50 dB.

Keywords: integrated optical waveguide · nanometer-scale thin film · heterogeneous integration · mode-locked laser

1 Introduction

Germanium (Ge) has become a key material for the development of high-speed photodetectors [1] at near infrared (NIR) wavelengths thanks to its narrow bandgap and high electron mobility upon doping [2]. Recently, Ge thin films have also been coated on fiber facets in fiber-based mode-locked lasers and shown good saturable absorption capability [3]. To realize compact photonic devices exploiting such features of Ge, developing new types of Ge-based optical waveguides becomes an inevitable step. The reported Ge micro/nano waveguides usually adopt channel or ridge structures based on crystalline Ge films with thickness on micrometre scale. However, as the main carrier of such thick Ge layers, the Ge-on-Insulator (GeOI) wafers require the critical and challenging wafer bonding technology for fabrication and the cost remains high [4]. Therefore, a simple and low-cost way to fabricate Ge-based waveguides for flexible photonic integration is highly desired.

Over the past decade, polymer-based waveguide technology has been developed for advanced photonic components but also as a hybrid integration platform [5]. This flexible feature allows various materials such as metal, 2D material and semiconductor thin films to be embedded in polymer as the key functional optical layer [6]. The fabrication technology for polymer waveguide devices is relatively simple and the optical loss can be suppressed at customized wavelength windows. Taking these advantages, tunable

J. Witzens et al. (Eds.): ECIO 2024, SPPHY 402, pp. 500–505, 2024.
https://doi.org/10.1007/978-3-031-63378-2_81

continuous-wave (CW) lasers have been demonstrated with an InP reflective gain chip and the thermally tunable polymer waveguide Bragg grating as external cavity. Such lasers have shown large tunable range, high reliability, and polarization diversity [7]. However, on the polymer platform, the feasibility for building a passively mode-locked laser that calls for a heterogeneous saturable absorber (SA) has not been well studied.

In this work, a thin amorphous Ge nanostrip is embedded in polymer as an optical waveguide using a simple deposition technology. The propagation loss of the waveguide is characterized rigorously at well-controlled TE and TM state to find that the waveguide only allows the transmission of the TM-polarized light. The TE-polarized light, on the other hand, is strongly absorbed by the Ge nanostrip. Furthermore, the Ge strip is deposited on top of a polymer channel waveguide to form a hybrid waveguide. The waveguide is further verified by femtosecond laser transmission experiment to show pulse width compression effect, which proves the saturable absorption effect of Ge. The hybrid waveguide is then integrated as a passive SA between a reflective gain chip and a fiber Bragg grating (FBG) to modulate the longitudinal modes in the cavity for pulsed output. This work demonstrates the feasibility of applying α-Ge films as practical saturable absorbing elements in photonics devices using a relaxed integration technology.

2 Polymer-Embedded Ge Nanostrip Waveguide

The schematic view of the polymer-embedded Ge nanostrip waveguide is shown in Fig. 1(a). An amorphous Ge (α-Ge) nanostrip of 4 μm in width and ~40 nm in thickness is embedded in a commercial perfluorinated polymer (Exguide ZPU12-450, ChemOptics) cladding ($n = 1.45$ at 1550 nm wavelength). The waveguide structure is defined by one standard contact photolithography, and transferred through a simple lift-off process [8]. Figure 1(b, c) illustrate the normalized profiles of the TE/TM fundamental waveguide mode at 1550 nm wavelength. For the TE mode, the dominating field component ($\mathbf{E_y}$) concentrates in the Ge core and results in strong absorption in the NIR. On the other hand, for the TM mode, the dominating field component is $\mathbf{E_z}$ and it expands more in the polymer cladding, resulting in a much lower propagation loss. The vastly different effective indices and mode field areas (MFAs) of each mode further confirm this highly birefringent eigen mode property.

A standard cut-back method measurement is performed using three sets of identical Ge nanostrip waveguides with different lengths. The overall transmission incorporating the system loss is measured through the fiber-waveguide-fiber coupling. The fiber-to-fiber transmission without the waveguide chip is also obtained to rule out the system influence. Figure 1(d) shows the waveguide transmission within the 1500–1600 nm range but only for the TM mode. The TE polarized light is supposed to undergo a strong absorption and the transmitted power dropped below the detection limit of the photodetector (−70 dBm). The propagation loss for the TM mode decreases from 20.56 dB/cm at 1500 nm to 11.75 dB/cm at 1600 nm as the gray dotted curve indicates. As plotted in Fig. 3(e), the propagation loss for the TM mode is measured and fitted as 4.89 dB/cm at 2004 nm wavelength, while the power of the TE polarized light still stays below the detection limit (−30 dBm) of another mid infrared power meter. These results show that the α-Ge thin film can be readily integrated in polymer and works as a TM-only waveguide.

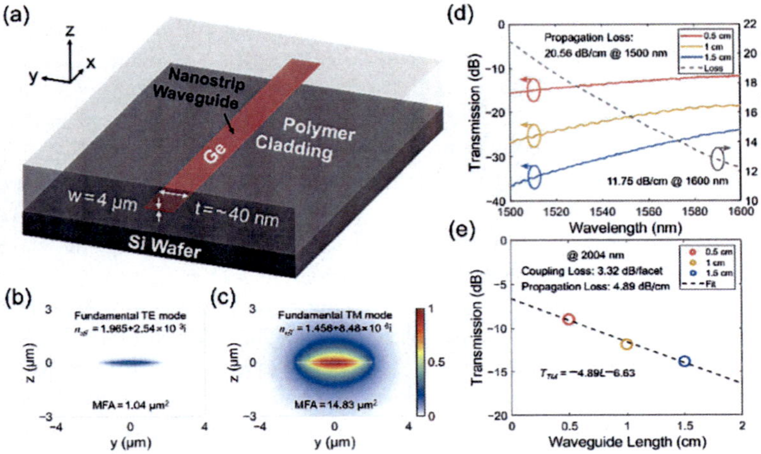

Fig. 1. (a) 3D schematic view of the polymer-embedded Ge nanostrip waveguide. (b, c) Normalized mode profiles of the TE/TM fundamental mode. (MFA: mode field area) (d, e) Measured transmission (TM polarization) of three Ge nanostrip waveguides at different length at 1500–1600 nm and 2004 nm wavelength.

3 Ge-Loaded Hybrid Waveguide

A similar α-Ge strip is further embedded on top of a single-mode polymer core to form the Ge-loaded hybrid waveguide. The cross-sectional design of the hybrid waveguide is shown in Fig. 2(a). The polymer core features a rectangular shape of 3.3 μm $\times$ 3 μm. The fabrication follows a standard process as described in our previous work [9]. As shown in Fig. 2(b), α-Ge strip of 4.5 μm in width and 50 nm in thickness is deposited by e-beam evaporation and defined through a lift-off process. The gap between the Ge strip and polymer core is 0.8 μm. The core and cladding are made of the same type of ZPU polymer but with a refractive index contrast of 1.47–1.45.

The characteristic of the hybrid waveguide under (CW) laser transmission is studied. As plotted in Fig. 2(c), hybrid waveguides with loaded Ge strips at all measured lengths exhibit flat transmission spectra in the 1500–1600 nm range with a wavelength dependent <0.5 dB. The transmission is attenuated with the increasing strip length, showing that the Ge strip indeed interacts actively with the TE-polarized light in the polymer core. In this work, we focus on the TE polarization as the InP gain chip used in the mode-locked laser integration is only sensitive to the TE-polarized light [7].

In addition, femtosecond laser transmission experiments are carried out on a series of hybrid waveguides. As plotted in Fig. 2(d), the laser pulses feature an average width of 455 fs after propagating through a polymer waveguide without Ge strip. On the other hand, the pulses are first compressed and then broadened with the increasing Ge strip length after the transmission through hybrid waveguides. The monotonic compression effect on femtosecond pulses for strip lengths below 200 μm is analyzed as the continuous depletion of low-energy pulse edges by the saturable absorption of Ge. The investigation on the α-Ge film further confirms this saturable absorption effect by a modulation depth of ~1.5% [10]. On the other hand, hybrid waveguides with longer Ge strips exhibit

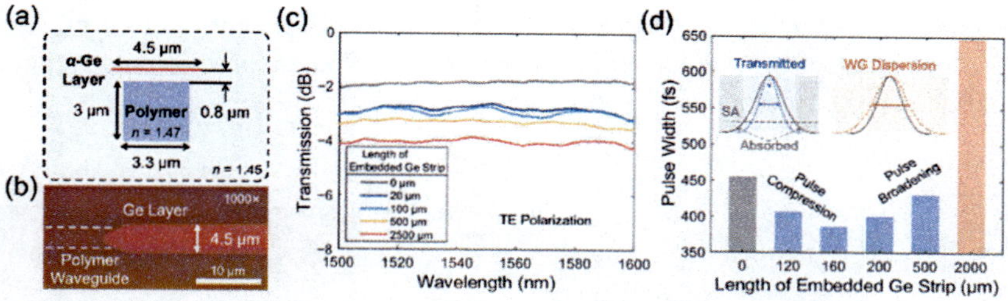

Fig. 2. (a) Cross-sectional design of the Ge-loaded hybrid waveguide. (b) Microscopic images of the fabricated waveguide. (c) Spectra under CW laser transmission of hybrid waveguides with different Ge strip lengths at the TE polarized state. (d) Pulse width characterization when the femtosecond laser transmits through hybrid waveguides with different Ge strip lengths.

stronger dispersion than the bare polymer waveguide, manifesting as the broadening of pulse widths for cases with the strip length >200 μm [9].

4 Picosecond Mode-Locked Laser Integration

The saturable absorption capability of the proposed polymer-Ge hybrid waveguide is used to construct the polymer-bridged mode-locked laser (PB-MLL). The schematic view of the PB-MLL is shown in Fig. 3(a). The gain chip incorporates a InP/InGaAsP multi quantum-well structure. The high-reflecting film against air is coated on the left facet, whereas the anti-reflecting film against $n = 1.46$ is coated on the right facet to match the effective index of the polymer waveguide [7]. The polymer bridge waveguide adopts the same cross-sectional design as shown in Fig. 2(a) for the waveguide mode field size to well match the optical mode of the gain chip. At the input side, a 20° tilting is also adopted on the polymer waveguide to receive the 9° tilting of the InP waveguide according to the Fresnel's law of refraction. The length of the loaded Ge strip is design at 160 μm to offer good saturable absorption and moderate intra-cavity dispersion, as the lowest pulse width of the transmitted femtosecond pulses is achieved at this length (see Fig. 2d). At the output side, the waveguide is inversely tapered to 1.1 μm in width, to improve the coupling efficiency with the fiber. A customized FBG is coupled to the output facet of the waveguide as the external cavity. The assembly of the PB-MLL is shown in Fig. 3(b).

The gain chip and FBG is first butt-joint coupled to a polymer bridge waveguide without Ge strip to verify the laser oscillation at CW state. As plotted in Fig. 3(c), the laser emits a quasi-CW output and exhibits a low threshold current of 5.4 mA. Next, the gain chip and FBG are moved to couple to another polymer bridge waveguide loaded with 160-μm-long Ge strip. The quasi-CW output is observed beyond the threshold current of 10 mA and the mode-locking begin to appear at the current > 14 mA. As depicts in Fig. 3(d), pulse trains are stabilized between 16–23 mA driving current, and the minimal pulse width of 147 ps is achieved at 20 mA as shown in Fig. 3(e). Figure 3(f) illustrates the profile of the laser pulse train with an interval of 19.8 ns, which is corresponding to a pulse repetition rate of ~50.5 MHz. The optical spectrum of the mode-locked laser output is plotted in Fig. 3(g). The inset of the detailed spectrum exhibits a central wavelength at

1549.95 nm. As shown in Fig. 3(h), the radio-frequency (RF) spectrum of the pulse train demonstrates an electronic signal-to-noise ratio of 52 dB and a precise pulse repetition rate at 50.58 MHz.

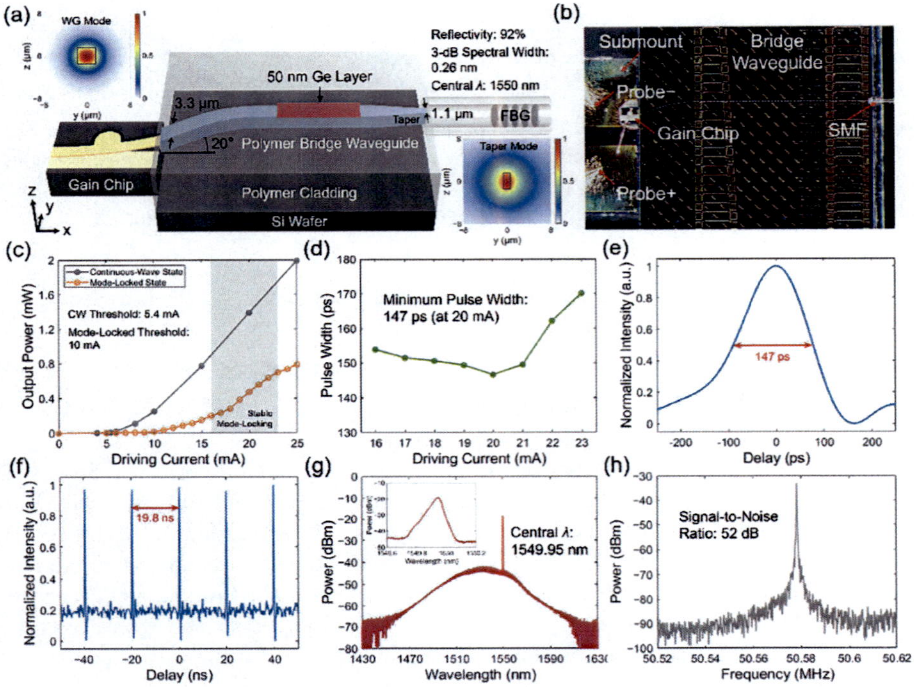

Fig. 3. (a) Schematic view of the polymer-bridged mode-locked laser (PB-MLL). (b) Assembly of the chips and the fiber. (c) P-I curve of the laser at CW state and mode-locked state. (d) Average pulse width of the PB-MLL at different driving currents. (e) Single pulse profile at 20-mA driving current. (f) Captured pulse train. (g) Optical spectrum of the laser pulses at 20-mA driving current. Inset: detailed spectrum. (h) RF spectrum of the pulse train.

5 Discussion

To summarize, we have proposed a polymer-Ge hybrid waveguide and demonstrated its integration as a saturable absorber but also as a bridge in a laser system. An amorphous Ge nanostrip is first embedded in the polymer cladding to form a waveguide. Measurements reveal that this high-aspect-ratio Ge strip can only transmit the TM mode while the TE-polarized light is strongly absorbed. The waveguide loss for the TM mode decreases gradually from 1500 nm toward mid-infrared region. Exploiting such enhanced light-mater interaction along the horizontal direction, the Ge strip is deposited on top of a buried core to modulate the TE-polarized light. The hybrid waveguide is further verified under femtosecond laser transmission to show saturable absorption effect, and then integrated inside the laser cavity as a passive saturable absorber for mode-locked output. The laser system is completed by discrete photonic components incorporating a reflective InP gain chip as the amplifier and a FBG as the low-loss external cavity. The mode-locked laser can emit pulses with 147-ps pulse width at a stable 50.58-MHz repetition rate. The electronic signal-to-noise ratio is 52 dB. We believe this work will prove useful in the flexible integration of light absorbing thin films as key functional layers on the polymer photonic platform.

References

1. Chen, G., et al.: High-speed photodetectors on silicon photonics platform for optical interconnect. Laser Photon. Rev. **16**(12), 2200117 (2022)
2. Liu, J., et al.: Ge-on-Si optoelectronics. Thin Solid Films **520**(8), 3354–3360 (2012)
3. Yang, C., et al.: Ultrathin amorphous Ge film enabling stabilized femtosecond fiber laser pulsation. Opt. Laser Technol. **136**, 106761 (2021)
4. Zhao, Z., et al.: Low-loss Ge waveguide at the 2-μm band on an n-type Ge-on-insulator wafer. Opt. Mater. Express **11**(12), 4097–4106 (2021)
5. Zhang, Z., et al.: Hybrid photonic integration on a polymer platform. Photonics **2**(3), 1005–1026 (2015)
6. Chu, Y., et al.: Birefringent and complex optical properties of monolayer graphene investigated by ellipsometry and waveguide integration. J. Phys. Chem. C **125**(3), 2124–2131 (2021)
7. De Felipe, D., et al.: Polymer-based external cavity lasers: Tuning efficiency, reliability, and polarization diversity. IEEE Photon. Technol. Lett. **26**(14), 1391–1394 (2014)
8. Liu, J., et al.: Polymer-embedding germanium nanostrip waveguide of high polarization extinction. Polymers **15**(20), 4093 (2023)
9. Wang, C., et al.: Transmission characteristics of femtosecond laser pulses in a polymer waveguide. Opt. Exp. **30**(17), 31396–31406 (2022)
10. Liu, J., et al.: Ge-polymer bridge waveguide for mode-locked laser pulse generation. Opt. Lett. **49**(3), 582–585 (2024)

Analyzing Frequency Combs in Multi-section Hybrid Silicon Quantum Dot Laser Using Hanbury Brown-Twiss Measurements

Mingzhao Shi[1](✉), Shihao Ding[1], Heming Huang[1], Geza Kurczveil[2], Raymond G. Beausoleil[2], and Frédéric Grillot[1,3]

[1] LTCI, Télécom Paris, Institut Polytechnique de Paris, 91120 Palaiseau, France
mingzhao.shi@telecom-paris.fr

[2] Large-Scale Integrated Photonics Lab, Hewlett Packard Labs, Hewlett Packard Enterprise, Milpitas, CA 95035, USA

[3] Center for High Technology Materials, University of New-Mexico, Albuquerque, NM 87106, USA

Abstract. We propose a novel method to characterize mode-locking regimes using a hybrid InAs/GaAs quantum dot frequency comb laser developed on silicon. By implementing Hanbury Brown-Twiss setup, measurements of zero-delay second-order correlation function $g^{(2)}(0)$ indicates that, in contrast to the frequency-modulated comb regime, the laser operating in amplitude-modulated comb regime exhibits higher $g^{(2)}(0)$ value, implying that photon bunching is enhanced when optical pulse starts to form. This work not only suggests a hardware-wise simple method to monitor frequency comb behaviors, but also unveil that there are more to understand for the comb generation from the perception of quantum physics.

Keywords: optical frequency comb · quantum dot · silicon photonics · photon bunching

1 Introduction

Over the past years, global data traffic saw steady exponential increases as the number of connected end users has been multiplied [1]. Moreover, as the society stepping into the artificial intelligence era, computing-power-craving and energy-hunger algorithms are incorporated into the existing network infrastructure, diverting even more available budgets for transmission volume and energy consumption. Surging demands are on the rise everywhere, requesting high bandwidth density, high data throughput and high energy efficiency optical interconnects and novel functionalities to help dealing with network congestion. To this end, photonics integration has emerged as one of the most preferred candidates by both academic and industrial worlds. Particularly, much devotion has been dedicated to photonic integrated circuits (PICs) on silicon, thanks to their compatibility with current CMOS technologies that are available for large scale production. Under

J. Witzens et al. (Eds.): ECIO 2024, SPPHY 402, pp. 506–510, 2024.
https://doi.org/10.1007/978-3-031-63378-2_82

such context, optical frequency comb (OFC) has been extensively studied recent years for their potential not only in complex applications such as optical metrology and optical sensing, but also in new-generation high-volume dense wavelength division multiplexing (DWDM) optical interconnects [2]. More specifically, OFC semiconductor lasers are gaining tractions lately for their future role as on-chip OFC generation, that will enable novel high performance DWDM optical transceiver, as well as optical intra-chip, inter-chip and inter-board datacom pipeline that can largely improve data exchange efficiency in datacenter and reduce latency in high-performance computing. Conventionally, OFC generation is linked to passive mode-locking effect, leading to amplitude-modulated (AM) comb that exhibits intense optical pulse train due to the modulation of photon density in the time domain [3]. Since the AM comb often offers narrow bandwidths, and the intrinsic optical pulses are often undesirable because of the risk in damaging other components on the transmission line, it is not so ideal for DWDM applications. On the other hand, frequency-modulated (FM) comb is prone to explore, thanks to their steady behavior at the output while preserving an equally fitting spectral profile as compared to AM comb. Nevertheless, it is challenging to fabricate semiconductor laser devices exclusively operating on FM mode, due to the nonlinear nature of the semiconductor gain medium. Commonly, to characterize the operating mode of the OFC are often conducted by measuring optical autocorrelation. Alternatively, we are proposing another approach to monitor comb behaviors by employing Hanbury Brown and Twiss (HBT) measurement [4] with the extraction of zero-delay second-order correlation function $g^{(2)}(0)$ [5]. By using a hybrid silicon InAs/GaAs quantum dot comb laser, experimental results indicate that as compared to FM comb, AM comb light are more bunched as its $g^{(2)}(0)$ measurements yield higher values. Thus, by assembling integrable HBT photonic circuits with semiconductor comb lasers, it is possible to design self-monitoring OFC modules that provides steady FM comb output. Moreover, this work also brings more insights into comb generation, the underlying mechanisms leading to more photon bunching effect in AM comb worth indeed more investigations.

2 Quantum Dot Laser Structure

The studied laser has a multi-section cavity design, which is schematically represented in Fig. 1(a). It is 2.6-mm-long, formed by two 50% reflectivity mirrors, both based on multi-mode interferometer (MMI). The optical gain is provided by a 1.4-mm-long semiconductor optical amplifier (SOA) section with a 176-μm-long saturable absorber (SA) placed in the center, in a colliding pulse configuration. Such a structure allows to take advantage of the second harmonic of the beatnote at around 32 GHz. The epitaxial structure consists of 5 layers of InAs QDs on GaAs substrate. The optical mode transfer between the III-V and Si waveguides is realized by modes converters at the edge of the active region. At the output, the light is guided out through grating couplers (GC). The fabrication process involves wafer-bonding of a commercial unpatterned GaAs-based QD epi wafer to a silicon-on-insulator (SOI) wafer and mesa formation. Standard III-V processing techniques are applied to perform for mesa etch, p- and n-contact metallization steps. The details of the design and fabrication of the epitaxial processes and laser structure can be found elsewhere [6].

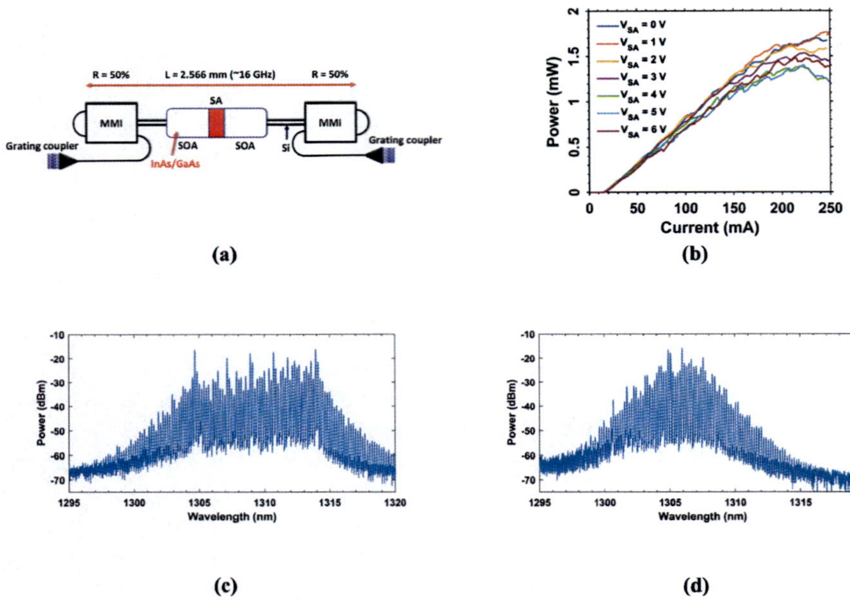

Fig. 1. (a) Schematics of the laser design. SOA: semiconductor optical amplifier; SA: saturable absorber; GC: Grating Coupler; MMI: Multimode Interferometer. (b) Light-current (L-I) curves of the QD laser under different reverse biases on the saturable absorber (SA) at 20 °C. (c) Optical spectrum of a FM comb. (d) Optical spectrum of an AM comb.

Figure 1(b) shows the variation of the output power with the bias current, at different applied voltages V_{SA} to saturable absorber. The threshold current I_{th} increases from 20 mA to 25 mA with the SA voltage going from 0 V to −6 V due to the increases of absorption. Exploiting the saturation of absorption in SA, which leads to synchronized emission of photons over a short duration, the laser modes is locked. In this laser, we observed the conventional AM comb spectrum where the comb lines are in phase, and the FM comb spectrum with a periodic modulation of the output frequency. Figure 1(c) and 1(d) present the optical spectrums of FM and AM combs. The FM combs exhibits broader spectral bandwidth and flattened top in the optical profile compared to the AM combs.

3 Results and Discussion

The $g^{(2)}(0)$ measurements are performed with standard HBT setup, i.e. two single photon detectors (ID Quantique ID200) are plugged for balanced detection, and the ensuing photocurrent signals, generated by the detectors, are subsequently directed into a time-to-digital converter (ID Quantique ID800) to ascertain the coherence interlinking the signals. Figure 2(a) and 2(b) shows the results of measuring the second-order correlation function $g^{(2)}(0)$ and the pulse extinction ratio under different bias currents and saturable absorber (SA) voltages. The pulse extinction ratio can be expressed as $\eta_p = 1 - V_{CW}/V_P$, where V_{CW} is the CW power level and V_p is the peak pulse power level in the time domain [7]. In both figures, a dashed line delineates the mapping into two distinct regions the AM region and the FM region, with the distribution of these regions being largely consistent across two graphs. For the pulse extinction ratio, the boundary between AM and FM is conventionally set at 0.5, while for $g^{(2)}(0)$, it approximates 1.19 for this laser. Overall, the higher $g^{(2)}(0)$ values of AM regime suggest that photon bunching effects are more pronounced than FM regime, this can be attributed to the formation of the optical pulse train that compresses photons into short period in the time domain.

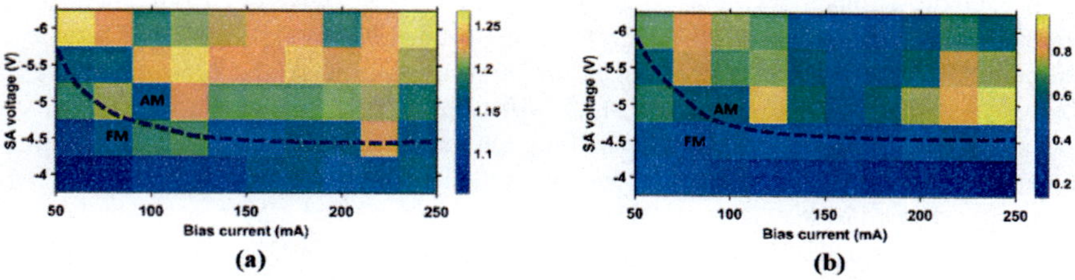

Fig. 2. (a) Mapping of the $g^{(2)}(0)$ under different biases on the gain and the SA. (b) Mapping of the pulse extinction ratio under different biases on the gain and the SA. The dashed line shows the boundary of AM and FM regions.

The consistency of these two regions can be attributed to the time-domain properties of the FM comb and AM comb. In the time domain, the emission of FM comb comprises a sequence of quasi-continuous-wave periodic short pulses. Nevertheless, for the AM comb, the intensity of pulses is much larger than the CW power level. Consequently, there exists large contrast of the intensity fluctuations of the comb laser under different operation conditions, which makes it possible to determine whether a state is AM or FM operation by $g^{(2)}(0)$.

4 Conclusion

This work proposes a novel approach based on HBT measurements for characterizing the AM and FM operations by analyzing the zero-delay second-order correlation function of the output light from a multi-section hybrid silicon InAs/GaAs QD comb laser. Our experimental findings reveal a distinct separation between the AM and FM regions, which corresponds to distinction of the extinction ratio measured by autocorrelation. This method offers multiple advantages such as ultra-lower input power requirement and integration-friendly setup. In the perspective of conceptualize steady FM comb based optical transceiver module, this work offers guidelines in designing the monitoring circuits capable of maintaining the comb stability at the output. Future work will focus on developing comprehensive models for QD lasers that incorporate parameters such as the second-order correlation function. This will enable us to gain deeper insights into the underlying physics driving this phenomenon, thereby advancing the understanding and further optimizing the performance of silicon-based QD comb lasers.

Acknowledgement. This work is financially supported by Hewlett Packard Enterprise (HPE).

References

1. Verified Market Research. Global Mega Data Center Market Size By Solution (Storage, Networking), By End-User (BFSI, Telecom and IT), Geographic Scope And Forecast (2024)
2. Fathololoumi, S., Nguyen, K., Mahalingam, H., et al.: 1.6 Tbps silicon photonics integrated circuit for co-packaged optical-IO switch applications. In: Optical Fiber Communication Conference. Optica Publishing Group (2020). T3H. 1. Brown R H, Twiss R Q. Correlation between photons in two coherent beams of light[J]. Nature, 1956, 177(4497): 27–29
3. Hillbrand, J., Auth, D., Piccardo, M., et al.: In-phase and anti-phase synchronization in a laser frequency comb. Phys. Rev. Lett. **124**(2), 023901 (2020)
4. Aspect, A.: Hanbury Brown and Twiss, Hong Ou and Mandel effects and other landmarks in quantum optics: from photons to atoms. In: Antoine Browaeys, and others (eds.), Current Trends in Atomic Physics, Oxford (2019)
5. McKeever, J., Boca, A., Boozer, A.D., et al.: Experimental realization of a one-atom laser in the regime of strong coupling. Nature **425**(6955), 268–271 (2003)
6. Kurczveil, G., Liang, D., Fiorentino, M., et al.: Robust hybrid quantum dot laser for integrated silicon photonics. Opt. Express **24**(14), 16167–16174 (2016)
7. Dong, B., Dumont, M., Terra, O., et al.: Broadband quantum-dot frequency-modulated comb laser. Light Sci. Appl. **12**(1), 182 (2023)

Development of PECVD SiN Thin Films for Integrated Photonic Applications on 300 mm Wafers

Margarita Lapteva[1](✉), Vinya Vibhuti[1], Mircea-Traian Catuneanu[2], Jens Knobbe[3], Peter Reinig[1], Sascha Bönhardt[1], and Kambiz Jamshidi[2]

[1] Fraunhofer Institute for Photonic Microsystems IPMS, Center Nanoelectronic Technologies CNT, An der Bartlake 5, 01099 Dresden, Germany
margarita.lapteva@ipms.fraunhofer.de

[2] Integrated Photonic Devices Laboratory, Technische Universität Dresden, Helmholtzstrasse 16, 01069 Dresden, Germany

[3] Fraunhofer Institute for Photonic Microsystems IPMS, Maria-Reiche-Str. 2, 01109 Dresden, Germany

Abstract. This work focusses on the development of CMOS compatible SiN processes for visible (VIS) and infrared (IR) applications in a 300 mm industry standard clean room. By utilizing PECVD for SiN deposition and RTA for annealing, hydrogen-related bonds were eliminated to reduce IR absorption. Thickness non-uniformity and process parameters have been optimized to reach improved optical properties across a broad wavelength range from 400 nm to 1600 nm. A versatile and uniform SiN process has been developed, which can act as a foundation for waveguides, optical couplers, resonators, and various applications targeting a universal photonic platform on 300 mm wafers.

Keywords: PECVD · Silicon Nitride · 300 mm platform

1 Introduction

The field of photonics has experienced steady growth in the last few decades, driven by a plethora of materials available for light emission, guiding, modulation and absorption in integrated circuits. Silicon nitride (SiN) platforms have shown remarkable properties and, aided by hybrid and heterogeneous integration techniques of III-V materials, can streamline the development of compact applications in telecommunication, sensing and metrology. A few of the most promising applications can profit from the availability of low loss waveguides [1]. Among these, high-quality resonators (Q-factors of 4.7 million) have been recently demonstrated using foundry compatible LPCVD processes [2]. Following this, integrating III-V gain sections can lead to chip-based lasers that approach the performance of external cavity diode lasers. Finally, SiN nonlinearities can be employed in the studies of 'microcombs', enabling threshold powers in the sub-mW regimes.

J. Witzens et al. (Eds.): ECIO 2024, SPPHY 402, pp. 511–515, 2024.
https://doi.org/10.1007/978-3-031-63378-2_83

Transferring standard PIC (photonic integrated circuit) processes from smaller wafer sizes to 300 mm wafers is crucial for future high-volume production. However, it also comes with challenges involving, in the case of SiN, decreasing thickness non-uniformity (NU) and handling the high mechanical stress [3]. Applications within the infrared (IR) range at a wavelength of 1550 nm require the utilization of thicker SiN waveguide layers, typically around 700 nm, in order to enable dispersion engineering and utilize nonlinear optical effects [1]. Additionally, post-process annealing at a temperature above 1000 °C is necessary to eliminate hydrogen-related bonds that absorb in the IR range. Especially, the first overtone of the N-H bonds causes significant absorption losses in the telecommunication band ranging from 1510 nm to 1565 nm [4, 5].

This study focuses on developing and optimizing CMOS compatible SiN processes in a 300 mm industry standard clean room for visible and IR applications while overcoming the scaling issues and at the same time meeting the requirements of integrated waveguides. This involves reducing the high stress in SiN thin films, lowering the absorption coefficient in the visible (VIS) range, and reducing the absorption in the IR range by removing residual hydrogen bonds through post-process annealing while maintaining low production times and high thickness uniformity across 300 mm wafers. For this sake, PECVD was used as a deposition technique and RTA (rapid thermal annealing) as a post process annealing technique. The usage of PECVD SiN was driven by the advantages of high deposition rate, and comparatively lower stress than LPCVD SiN [6]. The use of RTA provides better controllability of wafer temperature and wafer-to-wafer temperature uniformity, in addition to higher processing speed due to its single wafer processing mechanism compared to furnace annealing [7, 8].

2 Engineering Approach

The SiN films were deposited using silane (SiH_4), ammonia (NH_3), and nitrogen (N_2) as process gases at a deposition temperature of 400 °C. The RTA processes were carried out in a nitrogen gas atmosphere at different temperatures and anneal times.

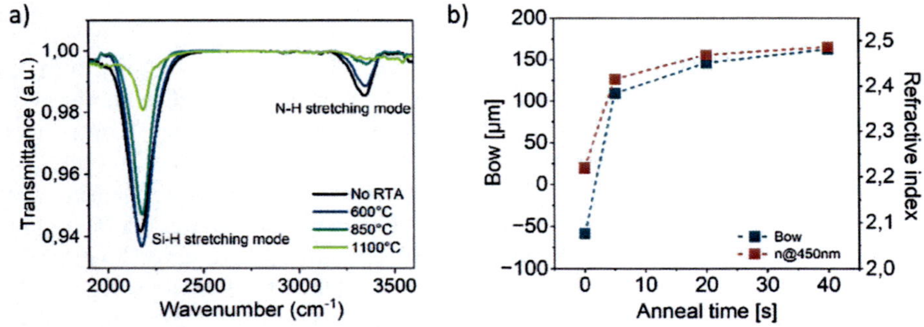

Fig. 1. a) Transmittance FTIR spectra monitoring the region of Si-H and N-H stretching modes at 2180 cm^{-1} and 3350 cm^{-1} for 300 nm SiN, annealed at different RTA temperatures for 20 s; b) Wafer bow and refractive index at a wavelength of 450 nm of SiN thin films (300 nm) at different RTA durations, where 0 s corresponds to films without an RTA step.

The first challenge of controlling the high mechanical stress of the SiN films was addressed by tuning the high and low radio frequency power percentages of the plasma [9]. As a next step, this work focused on achieving low optical absorption in the VIS and IR range, as well as reducing thickness non-uniformity (NU). This involved monitoring the complex feedback loop between the process iterations and the interlink between important factors like wafer bow originating from mechanical stress, optical dispersion, and IR absorption. Thickness and dispersion were determined by spectroscopic ellipsometry using the harmonic oscillator model with 2 oscillators. The NU was calculated from 49-point thickness measurements of the wafer surface and corresponds to the standard deviation (1σ). The IR absorption of the hydrogen bonds was monitored by FTIR.

3 Results

First, the IR absorption was minimized by testing different RTA temperatures. Figure 1. a) shows FTIR spectra of the SiN thin films without and with RTA processing at 600 °C, 850 °C, and 1100 °C. A significant decrease of hydrogen related bonds, both Si-H and N-H bonds, can be observed at 1100 °C, which agrees with literature [4].

The impact of RTA duration on wafer bow and refractive index ranging from 5 s to 40 s was tested. The trend in Fig. 1. b) shows that RTA had a considerable influence on refractive index and bow. At 5 s RTA time, bow is changing from ~ −59 μm to ~ 110 μm and the real part n of the refractive index is rising from 2.21 to 2.41, which indicates a densification of the film. At RTA durations above 5 s, both parameters show a saturated behavior. As a result, an annealing time below 40 s is sufficient. As a next step, the optical absorption in the VIS range was optimized. This required a PECVD process matrix focused on the pressure and the gas flows illustrated in Table 1, which shows the impact of these parameters on wafer bow, refractive index, NU, and VIS absorption. A subsequent process optimization led to a shift of the optical absorption to lower wavelengths.

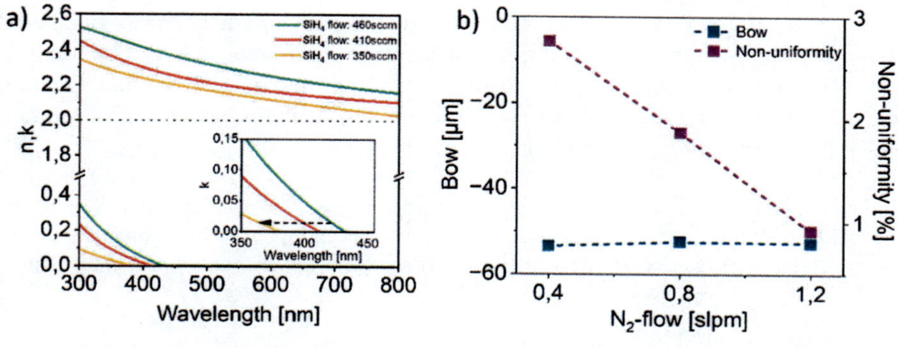

Fig. 2. a) Dispersion relations of SiN thin films at different silane gas flow rates. Lower gas flow rates correlate with an absorption shift to lower wavelengths; b) NU and wafer bow as a function of N_2 process gas variation. NU deacreases linearly while the bow remains constant.

As an example, Fig. 2. a) shows the final optimization step, where the silane flow was varied from 460 sccm to 350 sccm, which led to a shift of the optical absorption edge from 431 nm to 376 nm. It also reduced the refractive index at 450 nm from 2.32 to 2.20. The reduction in refractive index can be attributed to a lower Si/N ratio as the silane flow decreased. As a last step, the NU was optimized by increasing the N_2 flow. The N_2 flow was varied from 0.4 slpm to 1.2 slpm in steps of 0.4 slpm and as a result, a NU under 1% was achieved (see Fig. 2. b) while the bow remained unaffected at around −53 µm.

Table 1. Influence of different PECVD process parameters within an optimized process window on mechanical stress, refractive index in the visible spectral range, the VIS absorption quantified by the extinction coefficient as seen in Fig. 2. a), and the SiN thickness NU.

Process parameters	Bow (Stress)	Refractive Index	VIS Absorption	Non-uniformity
Pressure ↑	↓	—	↑	↑
Silane (SiH_4) flow ↑	—	↑	↑	↓
Ammonia (NH_3) flow ↑	↓	↓	↓	↓
Nitrogen (N_2) flow ↑	—	—	↓	↓

4 Outlook

This study successfully developed and optimized CMOS compatible SiN processes for visible and IR applications by using PECVD for SiN deposition and RTA for post-process annealing. Results demonstrated the removal of hydrogen bonds in the IR range, reduced optical absorption in the VIS range, and a NU under 1%. These advancements in SiN photonics open new possibilities for compact applications in telecommunication, sensing, and metrology, as well as the integration of III-V materials and the development of chip-based lasers [10, 11]. Patterned test structures such as unbalanced integrated Mach-Zehnder-Interferometers and ring resonators with different radii can be used for further characterization of waveguide losses in the visible and infrared regimes. Overall, this study contributes to the progress of the SiN platform and paves the way for future developments in photonics on 300 mm wafers.

Acknowledgement. We want to express our gratitude to Adel Kalache and Fred Schöne for supporting the PECVD and RTA processes. This work has been supported by Germany's Federal Ministry of Education and Research (BMBF) within the project VE-Silhouette under grant number 16ME0313K.

References

1. Xiang, C., Jin, W., Bowers, J.E.: Silicon nitride passive and active photonic integrated circuits: trends and prospects. Photon. Res. **10**(6), A82 (2022)

2. Ji, X., et al.: Ultra-low-loss silicon nitride photonics based on deposited films compatible with foundries. Laser Photonics Rev. **17**(3), 2200544 (2023)
3. Novak, A.V., Novak, V.R., Dedkova, A.A., Gusev, E.E.: Dependence of mechanical stresses in silicon nitride films on the mode of plasma-enhanced chemical vapor deposition. Semiconductors **52**(15), 1953 (2018)
4. Ay, F., Aydinli, A.: Comparative investigation of hydrogen bonding in silicon based PECVD grown dielectrics for optical waveguides. Opt. Mater. **26**(1), 33 (2004)
5. Mao, S.C., et al.: Low propagation loss SiN optical waveguide prepared by optimal low-hydrogen module. Opt. Express **16**(25), 20809 (2008)
6. Wang, L., Xie, W., van Thourhout, D., Zhang, Y., Yu, H., Wang, S.: Nonlinear silicon nitride waveguides based on a PECVD deposition platform. Opt. Express **26**(8), 9645 (2018)
7. Bae, M.-S., Kim, S.-H., Baek, J.-S., Koh, J.-H.: Comparative study of high-temperature annealed and RTA process β-Ga2O3 thin film by Sol-Gel process. Coatings **11**(10), 1220 (2021)
8. Xiao, H.: Introduction to Semiconductor Manufacturing Technology, 2nd edn. SPIE Press (2012)
9. Weyers, D., et al.: Electro-optical co-integration platform for high-density hybrid systems – SILHOUETTE (in press), Mikrosystemtechnik Kongress, Dresden (2023)
10. Sabouri, S., Mendoza Velasco, L.A., Jamshidi, K.: Efficiency comparison for frequency comb formations in a silicon nitride microring within anomalous dispersion regime. Opt. Continuum **1**(10), 2100 (2022)
11. He, M., Jamshidi, K.: Conditions for dual-pumped optical parametric oscillation in Kerr microresonators. Phys. Rev. Appl. **20**(5), 54036 (2023)

Discretely Tunable $GaSb/Si_3N_4$ Hybrid Laser Emitting at 2594, 2629, and 2670 nm

Samu-Pekka Ojanen(✉), Nouman Zia, Jukka Viheriälä, Eero Koivusalo, Joonas Hilska, Heidi Tuorila, and Mircea Guina

Optoelectronics Research Centre, Physics Unit, Tampere University, 33720 Tampere, Finland
samu-pekka.ojanen@tuni.fi

Abstract. Details on the development and characterization of a discretely tunable $GaSb/Si_3N_4$ hybrid multiwavelength laser are presented. The integrated laser combines a GaSb-based reflective semiconductor amplifier with a novel Si_3N_4 circuit. The Si_3N_4 is comprised of two Mach-Zehnder interferometers interconnected with three thermally tunable distributed Bragg reflectors to form a discretely tunable hybrid laser. It showcases discrete emission at 2670.42 nm, 2629.12 nm, and 2594.27 nm wavelengths, with maximum average output powers of 2.8 mA at 480 mA, 5 mW at 410 mA, and 3.3 mW at 495 mA, respectively. The unique characteristics of the hybrid laser are appealing for multiwavelength absorption spectroscopy of H_2S, CO_2 and H_2O.

Keywords: hybrid lasers · GaSb · silicon nitride photonics

1 Introduction

Many important gases like H_2S, C_2H_4, CH_4, CO, CO_2, and N_2O [1], as well as biomarkers such as lactate, glucose, and urea [2, 3], exhibit absorption in the 2–4 µm wavelength region. This makes this range particularly attractive for designing compact and efficient multiwavelength sensors. Mid-infrared hybrid integrated lasers, which combine GaSb-based semiconductor optical amplifiers (SOAs) with photonic integrated circuits (PICs), offer a promising avenue for achieving this goal.

Sensing applications typically require access to multiple absorption features to allow the detection of different analytes, and to increase the selectivity and accuracy of the measurement. In general, this can be achieved either with a tunable laser, such as a double ring resonator Vernier laser [4, 5], that covers the whole gain bandwidth of the SOA, or with a multiwavelength laser (MWL), which emits and can be tuned around specific wavelengths [6, 7].

Here, we present design and fabrication details of a tunable $GaSb/Si_3N_4$ hybrid MWL platform [9], employing a novel silicon nitride-on-insulator (SNOI) circuit. Using two thermally tunable Mach-Zehnder interferometers (MZIs) coupled to three thermally tunable narrow-band distributed Bragg reflectors (DBRs), we target lasing corresponding to absorption lines of H_2O, H_2S, and CO_2 of 2605.6, 2639.6, and 2682.0 nm, respectively. By tuning the MZI arms, the emission wavelength of the laser can be switched between

J. Witzens et al. (Eds.): ECIO 2024, SPPHY 402, pp. 516–520, 2024.
https://doi.org/10.1007/978-3-031-63378-2_84

the DBRs, while the DBRs themselves have integrated heaters, allowing to fine-tune the emission wavelength. A schematic illustration of the laser circuit is provided in Fig. 1.

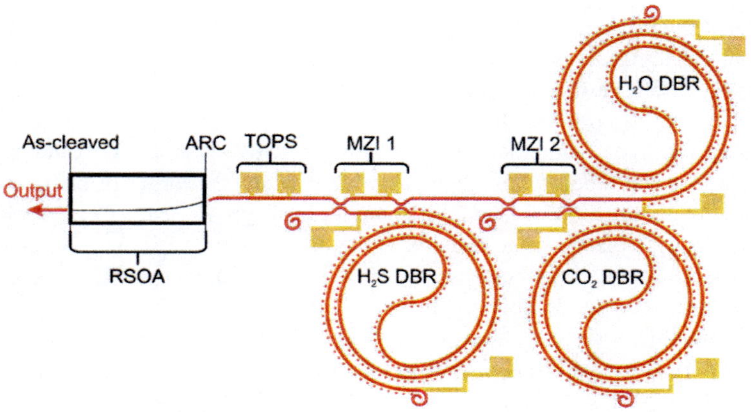

Fig. 1. Schematic of the GaSb/Si_3N_4 tunable hybrid MWL circuit.

The advantage of such a hybrid MWL design is that tuning the wavelength requires the simultaneous operation of only two heaters, the DBR and the TOPS. This is in contrasts with a Vernier hybrid laser, where wavelength tuning requires the simultaneous tuning of three heaters, the two ring resonators and the TOPS [5]. Moreover, compared to previous demonstrations of MWLs [6, 7], our design has the advantage of achieving multiwavelength emission with a single SOA component, decreasing the footprint, assembly effort, yield, and complexity of the laser compared to an assembly combining multiple SOAs. The hybrid MWL emits at wavelengths of 2670.42 nm, 2629.12 nm, and 2594.27 nm, showcasing its multiwavelength operation capability.

2 Measurements

The SNOI chip was manufactured utilizing an open-access foundry service (LIGENTEC). The laser includes a GaSb-based reflective SOA (RSOA) with a ridge waveguide (RWG) architecture. The RWG utilizes a J-shaped geometry with a tilted (7°) front facet with an anti-reflectance coating (ARC) to minimize back reflections and maximize transmission, and a straight as-cleaved rear facet to provide ~ 30% reflection. The light from the RSOA is coupled to an 800-nm-thick SNOI waveguide. An inverse taper tapering from 1.65 μm to 0.4 μm towards the facet was utilized to maximize coupling; the coupling loss was measured to be ~3 dB at 2.55 μm. The 1.65 μm SNOI waveguide width was chosen to prevent the excitation of higher order lateral modes. The waveguide propagation loss of 0.67 dB/cm was measured at 2.68 μm utilizing a Norcada distributed feedback (DFB) laser. A 700-μm-long thermo-optic phase shifter (TOPS), consisting of a resistive heater layer placed above the waveguide was utilized in both lasers to allow phase-matching the RSOA with the SNOI circuit to maximize power and spectral purity. The MZIs were formed with two 50/50 broadband phase-controlled directional couplers (PC-DC), exhibiting a low measured insertion loss of ~0.2 dB at 2.68 μm. A 700-μm-long TOPS was placed in both MZI arms to allow switching between the two output channels.

The DBRs were formed by placing periodic circular posts next to a spiral waveguide to achieve high-reflectance and narrow-bandwidth DBRs [8]. To take into account variation in the DBR resonances caused by geometric and index variations in the multi-project wafer (MPW) process, the resonances were designed 2.5 nm shorter than the target absorption lines, while also implementing resistive heaters above the DBRs to allow the laser to be tuned to the absorption lines. Thus, the target resonances for the DBRs were 2603.1 nm, 2637.1, and 2679.5 nm for the H_2O, H_2S, and CO_2, respectively.

The RSOA was placed on a heatsink that was stabilized to room temperature (23 °C), and the SNOI PIC was placed on a closed-loop piezo stage. The output signal was coupled to a multimode fiber connected to a photodiode, and the waveguides were actively aligned by maximizing the signal. The hybrid MWL was then switched to each of the three DBRs by applying current to the MZIs, and the IL characteristic of each DBR was measured by applying pulsed current (100 kHz frequency, 50% duty cycle) to the RSOA and measuring the average output power from 0 mA to 500 mA. The IL curves are shown in Fig. 2. When operating with the H_2O DBR, the laser has a threshold of 210 mA, and reaches a maximum average output power of 2.8 mW at 480 mA. Operation with the H_2S DBR led to a threshold of 185 mA with a maximum average output power of 5 mW at 410 mA. Finally, when operating at the CO_2 DBR, the laser had a threshold of 200 mA, and a maximum output power of 3.3 mW at 495 mA.

Fig. 2. IL curves of the hybrid MWL when operating at the different DBRs.

The spectrum of the hybrid MWL was measured for each DBR at 400 mA injection current is shown in Fig. 3 (left). These measurements reveal that the hybrid MWL achieves emission at 2670.42 nm, 2629.12 nm, and 2594.27 nm, showing successful switching between the DBR wavelengths. The wavelengths show on average ~8.6 nm blue shift comparted to the design wavelengths. By measuring the emission wavelength using several PICs from the same manufacturing process, we observed a ~1.0 nm standard variation in the wavelength, which is much smaller than the blue shift observed in the measurements. This implies that the shift can be attributed to inaccuracies in the Si_3N_4 refractive index data that was used in the simulations.

To measure the tunability of the hybrid MWL, the laser was switched to the CO_2 DBR, and the spectrum was measured without tuning, and with 140 mW power applied to the heater on top of the DBR. The results are shown in Fig. 3 (right), which shows that ~0.30 nm tuning is achieved at these power levels. The tuning was limited by the high 10-kΩ-level resistance of the DBR heater due to its long length. With an optimized heater design, multi-nm tuning is expected to be achievable with the hybrid MWL.

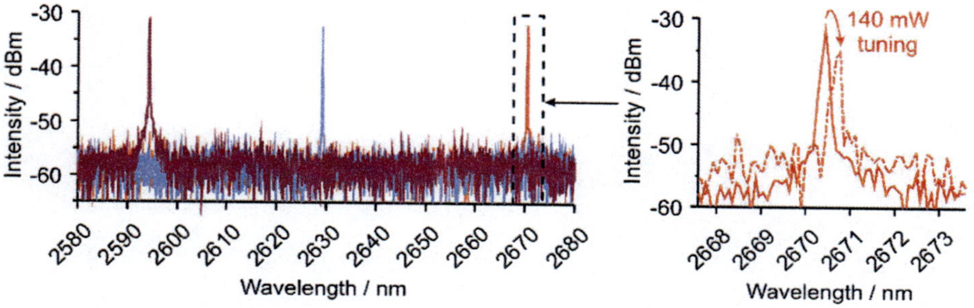

Fig. 3. Spectrum of the laser when the laser is switched to the different DBRs (left), and the CO_2 DBR emission spectrum without tuning, and with 140 mW applied to the heater on top of the DBR (right).

In conclusion, we have demonstrated a novel GaSb/Si_3N_4 hybrid MWL, which can be switched and tuned at 2670.42, 2629.12, and 2594.27 nm wavelengths. An output power of mW-level was achieved at each wavelength. The 2670.42 nm emission wavelength represents the longest wavelength emission achieved with a GaSb-based hybrid laser. Future research will focus on optimizing the DBR heater design, as well as integrating the RSOA on the PIC, and extending the wavelength towards and beyond 3 μm.

References

1. Gordon, I.E., et al.: The HITRAN2016 molecular spectroscopic database. J. Quant. Spectrosc. Radiat. Transf. **203**, 3–69 (2017)
2. Amerov, A.K., Chen, J., Arnold, M.A.: Molar absorptivities of glucose and other biological molecules in aqueous solutions over the first overtone and combination regions of the near-infrared spectrum. Appl. Spectrosc. **58**, 1195–1204 (2004)
3. Lin, P.T., et al.: Label-free glucose sensing using chip-scale mid-infrared integrated photonics. Adv. Opt. Mater. **4**, 1755–1759 (2016)
4. Guo, Y., et al.: Hybrid integrated external cavity laser with a 172-nm tuning range. APL Photonics **7**, 066101 (2022)
5. Ojanen, S., et al.: Widely Tunable (2.47–2.64 μm) Hybrid Laser Based on GaSb/GaInAsSb Quantum-Wells and a Low-Loss Si_3N_4 Photonic Integrated Circuit. Laser Photonics Rev. **17**(7), 2201028 (2023)
6. Kurczveil, G., Heck, M.J., Peters, J.D., Garcia, J.M., Spencer, D., Bowers, J.E.: An integrated hybrid silicon multiwavelength AWG laser. IEEE J. Sel. Top. Quantum Electron. **17**, 1521–1527 (2011)

7. Wang, R., Sprengel, S., Boehm, G., Baets, R., Amann, M.-C., Roelkens, G.: Broad wavelength coverage 23 μm III-V-on-silicon DFB laser array. Optica **4**, 972 (2017)
8. Xiang, C., et al.: Narrow-linewidth III-V/Si/Si_3N_4 laser using multilayer heterogeneous integration. Optica **7**, 20 (2020)
9. Ojanen, S., et al.: Discretely tunable (2594, 2629, 2670 nm) GaSb/Si3N4 hybrid laser for multiwavelength spectroscopy. Laser Photonics Rev. **17**(12), 2300492 (2023). https://doi.org/10.1002/lpor.202300492

Low-Loss, Buried InGaAs/InP Integrated Waveguides Operating in a Wide Mid-Infrared Range

Miguel Montesinos-Ballester[1(✉)], Lucius Miller[1], Victor Turpaud[2], Elsa Jöchl[1], Mathieu Bertrand[1], Delphine Marris-Morini[2], Emilio Gini[1], and Jérôme Faist[1]

[1] Institute for Quantum Electronics, ETH Zürich, CH-8093 Zürich, Switzerland
mmontesinos@ethz.ch

[2] Centre de Nanosciences Et de Nanotechnologies, Université Paris-Saclay, Univ. Paris-Sud, CNRS, 91120 Palaiseau, France

Abstract. This work introduces a photonic integrated platform based on buried InGaAs waveguides with InP cladding designed to operate in a wide mid-infrared (mid-IR) spectral range. Through the utilization of wet-etch fabrication patterning combined with Fe doping, the platform achieves remarkably low propagation losses. Experimental characterization reports losses below 1.2 dB/cm (equivalent to a 0.3 cm^{-1} loss coefficient) within the wavelength range of 4.6 to 11.2 μm (or 890 to 1960 cm^{-1} wavenumber), and for both TE and TM polarization modes.

Keywords: mid-infrared · long-wave infrared · photonic integrated circuits · integrated waveguides

1 Introduction

The mid-infrared (mid-IR) spectral range, spanning from 2 to 20 μm in wavelength, holds significant importance across various industrial and medical applications due to the pronounced absorption characteristics of numerous essential molecules within this range. In this regards, the development of broadband photonic integrated circuits (PICs) operating in the long-wave infrared (LWIR) spectrum (8–14 μm wavelength) is highly significant, as these circuits would offer compact and robust systems, while leveraging large-scale technologies to enable cost-effective solutions.

To date, various mid-IR platforms have been developed in different materials. However, those are typically limited to wavelengths shorter than 8 μm [1–3], as the long-wave infrared (LWIR) range poses challenges due to material transparency limitations and increasing free-carrier absorption [4]. Recent works have explored Ge and SiGe-alloy materials, achieving losses below 4.6 dB/cm between 5 and 11 μm [5]. Alternatively, III-V compounds emerge as compelling candidates for mid-IR PIC development due to their broad transparency window and the current capability to achieve exceptionally low free-carrier concentrations, reaching levels in the order or even below 10^{15} cm^{-3} [6]. Among the different possible compounds, InGaAs and InAlAs waveguides with

J. Witzens et al. (Eds.): ECIO 2024, SPPHY 402, pp. 521–525, 2024.
https://doi.org/10.1007/978-3-031-63378-2_85

InP cladding are highly interesting due to the technological maturity of existing semiconductor quantum cascade lasers (QCLs) based on these materials [7, 8]. For example, InGaAs membrane waveguides have been reported with propagation losses of 4.1 dB/cm at 6.1 μm [9]. Moreover, the use of lattice-matched III-V materials enables the monolithic QCL integration, typically emitting in TM polarization. For instance, recent works have demonstrate evanescent coupling between QCLs and InGaAs/InP waveguides in a homogeneous platform [10], where optical losses below 5 dB/cm were demonstrated in similar passive waveguides up to 11 μm wavelength in TE polarization [11]. Also, a butt-coupling configuration between a QCL and a passive InGaAs waveguide buried on InP was demonstrated, with estimated propagation losses of the passive section of 1.2 dB/cm near 8 μm wavelength [12].

The work we present here investigates a photonic integrated platform based on buried InGaAs core waveguides with InP cladding, targeting the major sources of propagation losses. Scattering losses are minimized through a wet-etch fabrication step and large bending radii, while free-carrier absorption is mitigated via Fe doping during material growth. Experimental results demonstrate propagation losses below 1.2 dB/cm between 4.6 and 11 μm wavelength, in both TE and TM polarizations. Interestingly, the optical waveguides, fully buried in InP material, enable butt-coupling to QCLs, and potentially detectors, with minimal coupling losses [12].

2 Results

2.1 Fabrication

In order to study the influence of the Fe-doing in the free-carrier concentration, three samples with a 2 μm thickness InGaAs layer were grown with different Fe concentrations by adjusting the gas flow in the MOPVE growth process. The overall free-carrier concentration of the three samples for increasing Fe-doping was estimated by Hall effect measurements to be 3.4×10^{15} cm^{-3}, 2.8×10^{15} cm^{-3} and 1.2×10^{15} cm^{-3}. The observed decrease of the averaged free-carrier concentration with increasing Fe-doping clearly proves the benefit of Fe-doping semiconductor materials to achieve low propagation losses in the long-wave infrared in integrated optical waveguides.

To better understand the remaining source of free-carriers, capacitance-voltage (CV) profiling characterization was performed along the InGaAs growth (vertical) axis of the sample with no Fe incorporated. An increase of the free-carrier concentration was observed near the interface between the InGaAs layer and the InP semi-insulating substrate. This is a clear indication that impurities (free-carriers) are introduced at the beginning of the MOPVE growth process, which is attributed to the initial precursor gas flow. To reduce the influence of those growth impurities, the addition of two 1 μm thickness InP:Fe layers on top and bottom of the InGaAs layer is proposed. If doping do, the effective free-carrier concentration that interacts with the optical mode is estimated to be approximately 1×10^{15} cm^{-3}. Deduced from the Drude model, this free carrier concentration introduces propagation losses of 0.1–0.8 dB/cm at 5–11 μm wavelength. These values are in good agreement with later experimental results of the optical propagation losses.

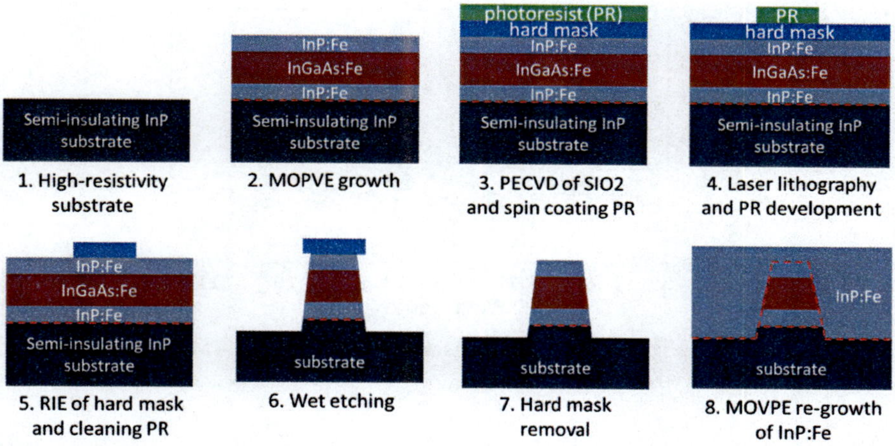

Fig. 1. Schematic of the passive waveguides fabrication process flow. The growth impurities caused by the precursor gas flow at the beginning of the MOPVE growth step is indicated in red color.

After those studies, we decided to fabricate a set of optical waveguides with increasing length and characterize its optical propagation losses. In addition to intrinsic material absorption, propagation losses can be introduced by scattering at waveguide sidewalls and absorption caused by oxidation or organic molecules on the surface. To mitigate these losses while maintaining compatibility with buried-heterostructure QCL fabrication processes, wet-etch patterning and buried passive waveguides are employed. The fabrication process flow is depicted Fig. 1, which begins with the growth of a 1 μm thick InP:Fe layer on a high-resistivity InP substrate using metal-organic vapor phase epitaxy (MOVPE), followed by 2 μm thick InGaAs:Fe, and another 1 μm thick top InP:Fe layer. A SiO_2 hard-mask is then patterned through laser lithography and reactive ion etching (RIE). Subsequently, waveguides are wet-etched using an isotropic HBr:Br:H2O solution (17:1:10 concentration in volume). Finally, the SiO_2 hard-mask is removed using a hydrofluoric acid (HF) solution, and another 3 μm thick top and lateral InP:Fe cladding is grown using MOVPE. For achieving a planarized surface profile, an additional lateral InP:Fe cladding growth can be performed prior to hard-mask removal.

2.2 Experimental Results

A set of integrated waveguides fabricated with high Fe-doping are characterized from 5.1 to 11.2 μm wavelength in a free-space experimental setup, where a bolometric mid-IR beam-profiler is used to ensure a correct coupling. Measurements in TE and TM polarization are performed thanks to a free-space polarization rotator. The experimental transmission spectra are smoothed with a Savitzky-Golay filter to mitigate atmospheric absorption peaks, followed by linear fitting for each wavelength data point. Loss values between 0.5 and 1.2 dB/cm are obtained in the 5.1–11.2 μm wavelength range, and are shown in Fig. 2.

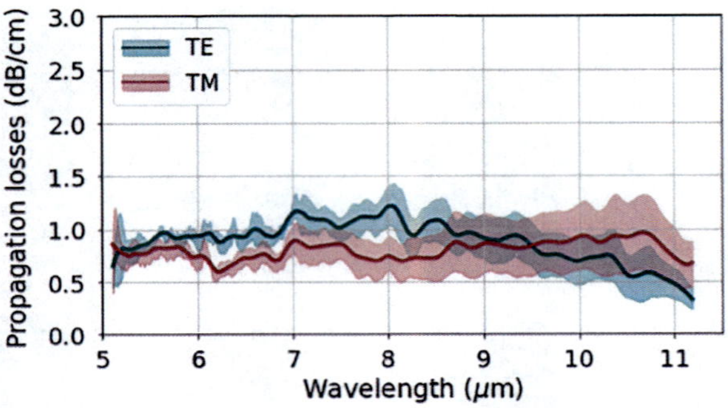

Fig. 2. Experimental results of propagation losses in TE (green) and TM (red) polarizations. Shaded color: estimated error from the linear fit of the cut-back method.

3 Conclusion

In conclusion, we have addressed the main sources of propagation losses to demonstrate high-performance mid-IR integrated platform based on Fe-doped, buried InGaAs/InP waveguides. Propagations losses as low as 0.5–1.2 dB/cm are here reported from near 5 to 11 μm wavelength. Moreover, the possibility to monolithically integrate such low-loss waveguides with mid-IR sources and detectors is highly interesting for the development of broadband and complex integrated photonic systems operating in a wide mid-IR range that includes the long-wave infrared.

References

1. Soref, R.: Mid-infrared photonics in silicon and germanium. Nat. Photonics **4**(8), 495–497 (2010)
2. Lin, H., et al.: Mid-infrared integrated photonics on silicon: a perspective. Nanophotonics **7**(2), 393–420 (2017)
3. Marris-Morini, D., et al.: Germanium-based integrated photonics from near- to mid-infrared applications. Nanophotonics **7**(11), 1781–1793 (2018)
4. Tsai, C.-Y., et al.: Theoretical model for intravalley and intervalley free-carrier absorption in semiconductor lasers: beyond the classical Drude model. IEEE J. Quantum Electron. **34**(3), 552–559 (1998)
5. Montesinos-Ballester, M., et al.: Ge-rich graded SiGe waveguides and interferometers from 5 to 11 μm wavelength range. Opt. Express **28**, 12771–12779 (2020)
6. Mattera, D., et al.: High-speed InP/Ga0.47In0.53As superlattice avalanche photodiodes with very low background doping grown by continuous trichloride vapor-phase epitaxy. J. Appl. Phys **60**, 2609–2612 (1986)
7. Vitiello, M.S., Scalari, G., Williams, B., De Natale, P.: Quantum cascade lasers: 20 years of challenges. Opt. Express **23**(4), 5167–5182 (2015)
8. Razeghi, M., et al.: Quantum cascade lasers: from tool to product. Opt. Express **23**, 8462–8475 (2015)
9. Yoo, K.M., Midkiff, J., Rostamian, A., Chung, C., Dalirand, H., Chen, R.T.: InGaAs membrane waveguide: a promising platform for monolithic integrated mid-infrared optical gas sensor. ACS Sens. **5**(3), 861–869 (2020)

10. Jung, S., et al.: Homogeneous photonic integration of mid-infrared quantum cascade lasers with low-loss passive waveguides on an InP platform. Optica **6**(8), 1023–1030 (2019)
11. Zhang, K., Böhm, G., Belkin, M.A.: Mid-infrared microring resonators and optical waveguides on an InP platform. Appl. Phys. Lett. **120**, 061106 (2022)
12. Wang, R., Täschler, P., Wang, Z., Gini, E., Beck, M., Faist, J.: Monolithic integration of mid-infrared quantum cascade lasers and frequency combs with passive waveguides. ACS Photonics **9**, 426–431 (2022)

Study of a SiN Microresonator with a Variable Coupler at 1550 nm

Sylvain Boust[1](✉), Jonathan Faugier-Tovar[2], Sylvain Guerber[2], Quentin Wilmart[2], François Duport[1], and Frédéric Van Dijk[1]

[1] III-V Lab, A Joint Laboratory Between Thales, CEA and Nokia, 1 Av. Augustin Fresnel, 91120 Palaiseau, France
sylvain.boust@3-5lab.fr
[2] University of Grenoble Alpes, CEA, LETI, 38000 Grenoble, France

Abstract. We propose a silicon nitride microresonator with an electrically controlled coupler, paving the way for a type of microresonator whose behavior can be perfectly tailored to the desired application. We have measured an intrinsic quality factor of 3.9 million for a 10 cm-long microresonator.

Keywords: Integrated Photonics · Silicon Nitride · Microresonator

1 Introduction

The microresonator (MR) in integrated photonics is a building block for low-loss materials or high non-linear susceptibility materials. For example, it can be used linearly for optical filtering, or non-linearly to concentrate the energy of an optical wave. It can thus be used in widely tunable extended-cavity lasers [1], or for the generation Kerr microcombs [2]. However, the design of a MR is complex, as it is subject to many degrees of freedom: its length, its coupler, the width of the guide used, or even the minimum bending radius used,... In this work, we focus on the most critical element: the coupler. To this end, we have designed a MR whose coupler is a Mach Zehnder modulator (MZM) (Fig. 1), enabling us to control the coupling rate of light injected into the microresonator. This allows us to adapt the profile of the MR resonance to the desired application.

2 Design and Simulation

Our component is fabricated on the low-loss silicon nitride (SiN) platform developed by CEA-Leti [3]. The waveguide used in the MR has a 800 nm-thick × 1700 nm-wide SiN core. Our design is a 2 × 2 MZM, one input port of which is connected to a 10 cm delay line then looped back to one of the MZM's output ports, forming an MR with a free spectral range (FSR) of 1.4 GHz, which we study at 1550 nm. Assuming that this design can be modeled by a MR with a variable coupling rate κ, we can simulate [4] its response and observe typical resonance profile (Fig. 2a). We thus find the different coupling regimes (under-coupled, over-coupled, critical coupling) distinguished by the extinction ratio (ER) and the full width at half-maximum (FWHM) of the notches.

J. Witzens et al. (Eds.): ECIO 2024, SPPHY 402, pp. 526–529, 2024.
https://doi.org/10.1007/978-3-031-63378-2_86

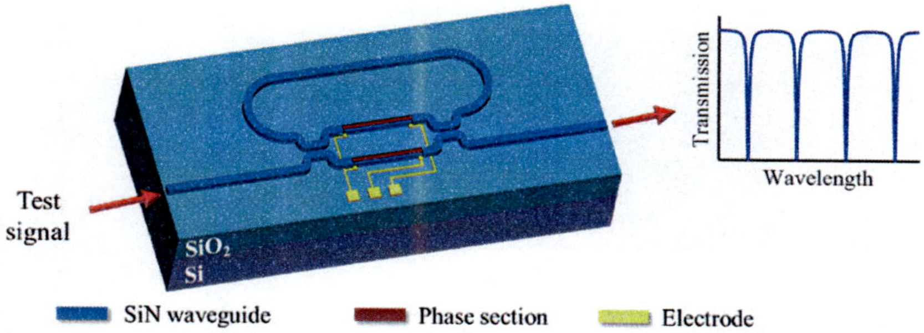

Fig. 1. Principle of the silicon nitride (SiN) variable-coupling microresonator (MR)

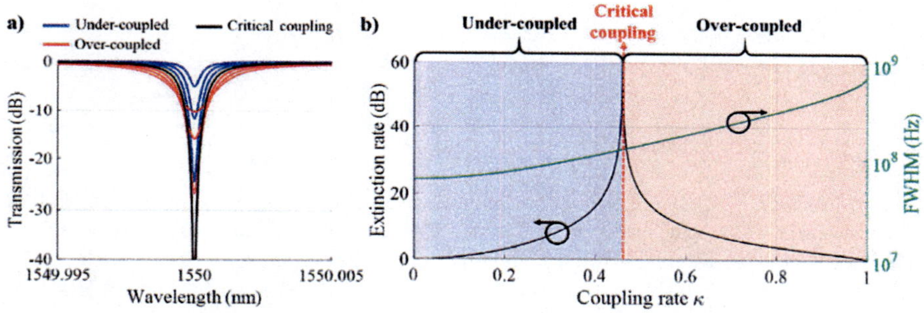

Fig. 2. **a)** Simulation of the transmission of a 1.4 GHz FSR MR resonance for different coupling rates. **b)** Extraction of the extinction ratio (ER) and the full width half maximum (FWHM) of the simulated transmissions.

In Fig. 2b, we plot these two parameters for a resonance of our MR, enabling us to clearly define the MR's two coupling zones, separated by the critical coupling ($\kappa_c = \sqrt{1-a^2}$ with a the losses in the MR). The value of κ_c depends only on the size of the MR (10 cm here), and on the linear losses, which we assume here to be 10 dB/m.

Furthermore, we observe as usual that configurations close to the critical coupling ($\kappa \approx \kappa_c$) offer the most contrasting resonances (high ER), while the highest loaded quality factors (inversely proportional to FWHM) are obtained with the least coupled MRs ($\kappa \ll 1$).

3 Characterization

In this work, we propose preliminary characterizations (Fig. 3a) where we measure the transmission spectral response of our MR for different settings of the variable coupler. These are performed by injecting a current into a 21 kΩ heater above one of the MZM arms.

Figure 3b shows the evolution of the spectral profile of a resonance when the current is increased from 2.9 mA (blue) to 3.6 mA (red), inducing localized heating proportional to the square of the current.

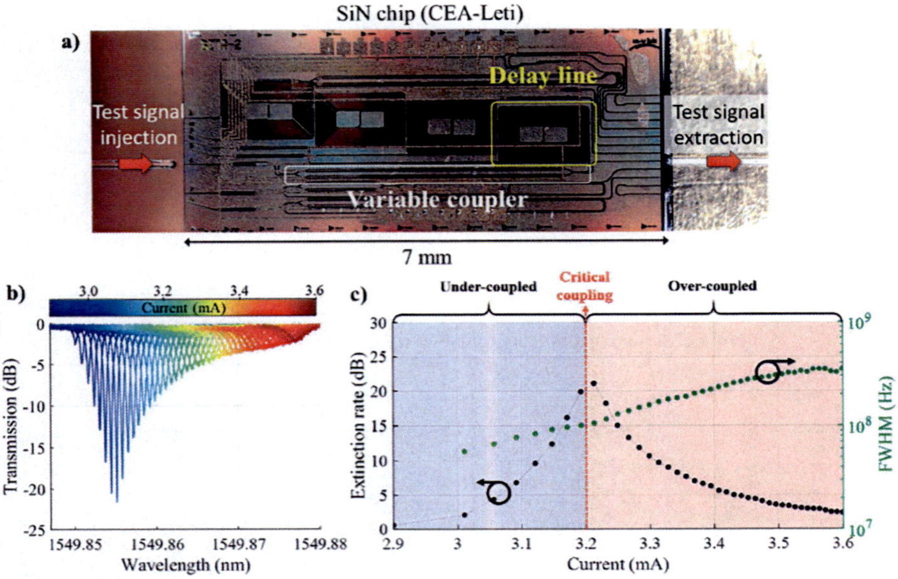

Fig. 3. **a)** Image of a 10 cm delay line connected to a variable coupler creating a 1.4 GHz-FSR MR. The signal is injected and extracted with lensed fibers. **b)** Spectral transmission measurements of a MR resonance for different variable coupler currents. The current source is a Keithley 2461, and the optical spectrum analyzer is an APEX AP2043B. **c)** Extraction of ER and FWHM from measured transmissions.

The temperature change has two thermo-optical consequences: the resonance is distorted as expected, and its wavelength is shifted. The first phenomenon is indeed linked to a phase change in the MZM modifying the MR coupling rate. The second is due to a change in the optical length of the MR, since the MZM is included in the MR: the FSR of the MR is therefore slightly modified. Note that thermal crosstalk certainly accentuates this phenomenon. In subsequent work, we'll cancel out this parasitic wavelength shift date by using both arms of the MZM in a push-pull configuration.

After performing Lorentzian regressions of these curves, we plot ER and FWHM in Fig. 3c. With the maximum of the ER, we identify the critical coupling, while with the evolution of the FWHM, we determine the zones of under- and over-coupling. At a later stage, we'll compare these curves more precisely with those simulated in Fig. 2b, enabling us to validate the model and establish a current-coupling rate relationship to control the MR perfectly for future applications.

Nevertheless, we can already note that the measurement-simulation agreement seems satisfactory for the overall shape of the curves. As the critical coupling is clearly identifiable in both measurement and simulation, we can now make a comparison: we find respectively a 20 dB-ER in measurement (vs. $+\infty$ in simulation), and a 100 MHz-FWHM (vs. 125 MHz in simulation). We believe that our ER is limited for different reasons: (i) the directional couplers forming the MZM have residual losses, (ii) our polarization of the test signal is not perfectly optimized, (iii) the measurement floor and resolution of our spectrum analyzer could be too limited to detect the sharpest resonances.

The difference in FWHM, on the other hand, is minimal. Note that the measured FWHM means a 3.9 million intrinsic quality factor, and by making the simplistic assumption that we neglect coupler losses, propagation losses are estimated to 9.5 dB/m.

4 Conclusion

We presented a microresonator with a variable coupler at 1550 nm, which is a promising building block for many applications. Previously, to obtain a resonator with a desired coupling, it was necessary to multiply the designs around a simulated geometry, taking up surface area on the photonic chip, and time to test. With our work, we hope to demonstrate that it will be possible to replace all these tests with a single design featuring a variable coupler.

We will then use this variable-coupling microresonator by injecting it with a DFB laser via edge-coupling. We will observe the efficiency of laser locking on the microresonator by optical self-injection, as a function of the resonance profile.

Acknowledgements. The authors would like to thank the Agence Nationale de la Recherche (ANR) for its support in the context of the ORCHID project ANR-21-CE24–0018.

References

1. Tran, M.A., et al.: Ring-resonator based widely-tunable narrow-linewidth Si/InP integrated lasers. IEEE J. Sel. Top. Quant. Electron. **26**(2), 1–14 (2019)
2. Shen, B., et al.: Integrated turnkey soliton microcombs. Nature **582**(7812), 365–369 (2020)
3. Wilmart, Q., et al.: A device library for the ultra-low loss Si3N4 platform. In: SPIE, Silicon Photonics XVII, vol. 12006, pp. 80–87 (2022)
4. Rabus, D.G.: Integrated Ring Resonators. Springer-Series in Optical Sciences (2007). https://doi.org/10.1007/978-3-540-68788-7

Electro-Optic Modulator in Thin-Film Lithium Niobate Foundry Process

Alberto Della Torre(✉), Homa Zarebidaki, Jacopo Leo, Arno Mettraux, Gregory Choong, Mattia Orvietani, Yves Petremand, Ivan Prieto, Olivier Dubochet, Michel Despont, and Hamed Sattari

Swiss Center for Electronics and Microtechnology (CSEM), Neuchatel, Switzerland
alberto.dellatorre@csem.ch

Abstract. The Characterization of thin-film lithium niobate modulators with electro-optical bandwidth higher than 50 GHz is presented. The devices, operating in the telecom C-band, are manufactured in a standard foundry process following design rules. The 7.95 mm long modulator has a half-wave voltage of 3.2 V and an extinction ratio of 15.4 dB. We also show that modulators, with the same design but fabricated in different wafers, present very similar performances, corroborating the reliability of the process design kit (PDK) component and the manufacturing process.

Keywords: Thin-Film Lithium Niobate · Electro-Optic Modulator · Foundry Process · Process Design Kit

1 Introduction

In the last decade, electro-optical (EO) modulation in photonics integrated circuits (PICs) has witnessed remarkable advancements, mainly driven by the increasing bandwidth demand of quickly evolving telecom applications. In this respect, among the different PIC platforms, thin-film lithium niobate on insulator (TFLN) technology excels, offering unique properties like wide transparency, minimal optical loss, and a high electro-optic coefficient, making it an excellent choice for low-power, high speed optical modulators [1].

Different approaches have been adopted for optical modulation in various PIC platforms. In silicon, for instance, due to the lack of second-order nonlinearity, optical modulation schemes based on plasma dispersion have been proposed [2]. However, the trade-off between voltage, bandwidth, and optical losses poses fundamental limitations to the overall performances achievable by electronic doping. Thermo-optical [3] and micro-electro-mechanic systems (MEMS) [4] modulators have also been widely explored, but the slow thermal response of most materials and the mechanical movement of microstructures make them inherently slower than their EO counterpart. Plasmonics-based approaches, while fast and compact, suffer from high losses and need further tech development to achieve high scalability [5]. On the contrary, the superior EO of TFLN

J. Witzens et al. (Eds.): ECIO 2024, SPPHY 402, pp. 530–535, 2024.
https://doi.org/10.1007/978-3-031-63378-2_87

offer the opportunity to simultaneously achieve low-voltage and high-speed modulation [6]. In particular, voltage-length product of 1.2 V.cm [7] and bandwidth beyond 100 GHz [8] have been achieved in TFLN modulators. Yet, as compared to more mature PIC platforms like silicon and InP, TFLN manufacturing technology still needs further development to meet PIC industry's demands. The development of an open-access, standardized TFLN PIC platform is a crucial step towards the widespread use of this technology.

Here, we present the performance metrics of TFLN Mach-Zender EO modulators (MZM) manufactured in our standard foundry process. We also investigate the reliability of the manufacturing process by measuring devices from different wafers, and we show that the devices present similar performances.

2 Design and Fabrication

The layout of our standard MZI modulator demonstrating the technology layers is presented in Fig. 1(a). The circuit encompasses a series of optimized building blocks from the process design kit (PDK) library, such as MMI beam splitters, waveguide crossing, and edge couplers [9, 10]. We adopted a push-pull configuration for the electrodes with a pair of ground-signal-ground pads to minimize the half-wave voltage and facilitate the high-speed probing. For designing the MZM, simulations have been conducted to achieve a 50-Ω match condition for the modulator, considering the specific dimensions of the layers provided by the platform criteria. A signal electrode width of 13.7 μm and a signal gap of 4.5 μm were chosen to balance optical and RF losses while also ensuring optical and RF phase matching and an efficient electric field distribution across the electrodes. This approach guarantees the modulator's high-speed functionality with low power consumption.

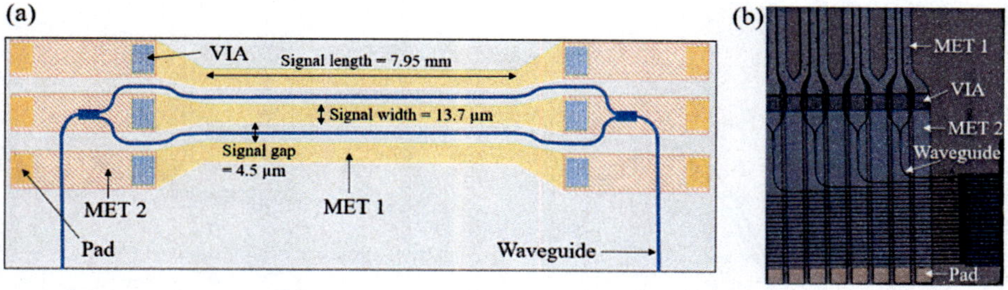

Fig. 1. (a) Schematic of the standard electro-optical modulator. (b) Optical image of the area of a fabricated TFLN with a series of EO modulators.

The devices are manufactured through a standard foundry process at CSEM on a 150 mm wafer scale. Our fabrication technique utilizes readily available thin-film lithium niobate on insulator (LNOI) wafers. These wafers consist of a layer of 600 nm thick mono-crystal x-cut $LiNbO^3$ on top of a 4.7 μm buried thermal oxide (BOX) layer. This platform offers three different types of waveguides, providing designers with flexibility to create circuits optimized for various wavelengths. The PIC platform incorporates a dual-layer metallization structure that improves the routing process for radio frequencies, enabling more efficient signal transmission. All layers in the platform are shielded by a silicon oxide cladding. Additionally, a specialized layer known as "clad open" is included, giving designers the option to selectively remove the cladding. This feature allows for specific applications, such as creating access points to the metal pads or directly interfacing with the waveguides. The final chip release process results in smooth facets, a critical factor for achieving effective light coupling at both the input and output interfaces of the device. We have fabricated two identical modulators from two different wafers. In the following section we will compare their performances.

3 Experiment and Discussion

The complete performance evaluation of the modulator focuses on two principal domains: DC (Direct Current) for assessing modulation efficiency, and RF (Radio Frequency) measurements for investigating the overall modulator performances. The latter comprises Electrical-Electrical (EE) measurements, to evaluate the electrical losses and the design quality in terms of impedance matching and microwave-optical index matching, and Electro-Optical (EO) measurements, to evaluate the performance of the modulator at high frequency.

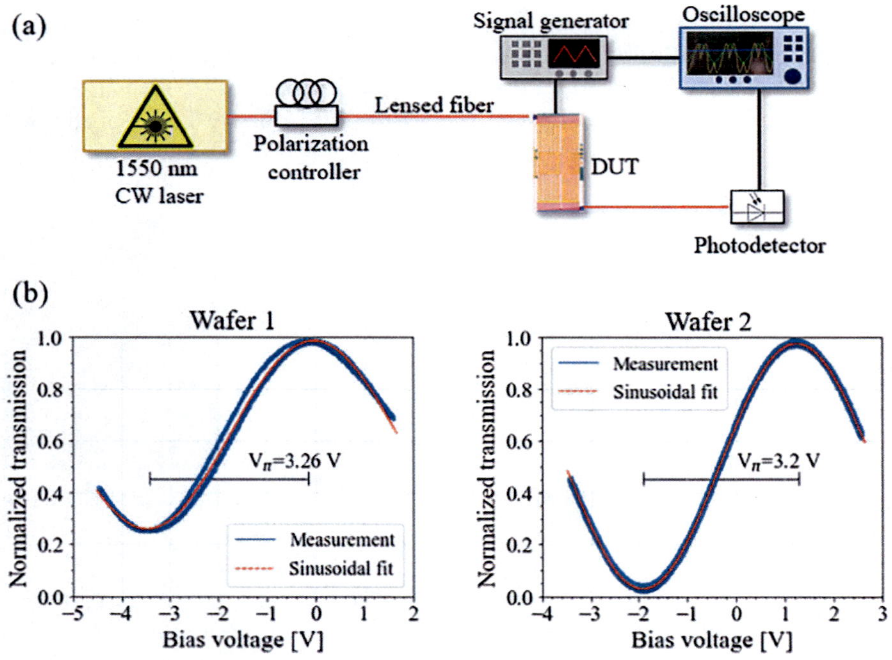

Fig. 2. Schematic of the experimental setups used for DC measurements (a) and DC electro-optical response of the modulators from wafer 1 and 2 (b).

Figure 2(a) depicts the characterization setups used for DC measurements. To measure the half-wave voltage V_{π} of the modulator, a low frequency characterization has been performed; a triangular electrical signal with a frequency of 5 kHz was directly applied to the electrodes, and the modulated optical output was recorded by a photodetector, whose response was acquired using a 500 MHz bandwidth oscilloscope. Modulators from two different wafers show very similar performances, with a half-wave voltage of 3.26 V and 3.2 V in wafer 1 and 2, respectively (Fig. 2(b)). The devices show a voltage-length products of 2.59 V.cm and 2.54 V.cm in Wafer 1 and Wafer 2, respectively. The difference in ER between Wafer 1 and Wafer 2 could be attributed to a degraded beam splitters in the Wafer 1 due to fabrication defects. We also observe a hysteresis profile in Wafer 1 that is attributed to the DC drift of the modulator due to a non-optimized manufacturing steps parameters in Wafer 1.

Figure 3(a) shows the setup used for RF measurements. Light from a mode hop-free laser emitting at 1550 nm was sent to the chip through a polarization controller and a lensed fiber. A DC bias was applied to drive the modulator at the quadrature point. A microwave signal was generated by a VNA (extending up to 50 GHz) and applied to the electrodes through ground-signal-ground electrical probes. The optical output was recorded by a fast photodetector and sent to the output port of the VNA. The relatively low electrical signal attenuation (Fig. 3(b)) allows to achieve high EE bandwidth (Fig. 3(d)), as the transmitted signal power decreases of about 6 dB at 50 GHz as compared to its value at the 1 GHz reference frequency. The microwave effective index extracted from measurements is around 2.85 at high frequency. This value is close to the optical group index of 2.28 retrieved from simulations (red line in Fig. 3(c)), yet an incomplete matching between the microwave and optical waves speed is a limiting factor for the modulator's EO performances at high frequency. This aspect will be addressed in future design to further improve the performances of the device. A very low electrical signal reflection of below 20 dB (see Fig. 3(d)) indicates an excellent impedance matched circuitry. Figure 3(e) shows the EO frequency response. The bandwidth exceeds 50 GHz, given that the output power hasn't decreased by 3 dB at this frequency (black dashed line in Fig. 3(e), calculated relatively to 1 GHz).

These findings highlight the device's significant potential to operate beyond 50 GHz bandwidth, a crucial milestone for required modulation schemes. Further refinements could focus on improving the metals quality to reduce the electrical losses, and to adjusting the modulator dimensions for a better microwave-optical index matching. These improvements could increase the bandwidth even further, potentially beyond 100 GHz, while pushing down the operation voltage towards CMOS levels.

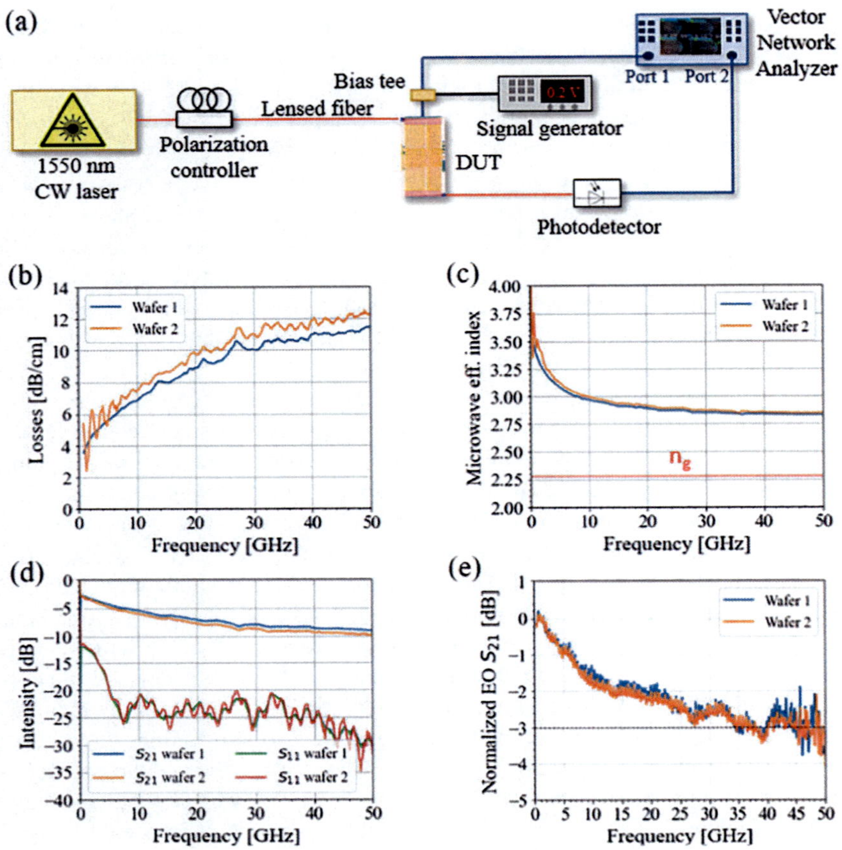

Fig. 3. (a) Schematic of the experimental setups used for DC measurements. (b) Measured electrical signal attenuation as a function of microwave frequency. (c) Measured microwave effective index (blue) and simulated optical group index (red). (d) Measured EE response. (e) Measured EO response normalized to 1 GHz.

4 Conclusion

We presented and compared a full characterization of two TFLN EO modulators, with an identical design but from two different wafers fabricated through our open-access foundry process. The modulators, operating in the telecom C-band show very similar performances, with an EO bandwidth surpassing 50 GHz and a half-wave voltage of around 3.2 V. This is an important starting point, evaluating the reliability of our foundry process, a crucial step towards the development of a standardized platform with a reliable PDK building blocks. Future work will focus on comparing the performances of modulators from a higher number of wafers, different locations on the wafer, and from different fabrication RUNs. According to the increasing demand for low power, high bandwidth photonic devices, developing of TFLN PICs in a standardized platform paves the way for fast, large-scale, and energy-efficient PIC systems, elusive via other PIC platforms.

Acknowledgments. This project has received funding from the European Union's Horizon 2020, GA No. 101016138; European Union Horizon under GA Nos. 101070506, 101070441 and 101070581.

References

1. Zhu, D., et al.: Integrated photonics on thin-film lithium niobate. Adv. Opt. Photon. **13**(2), 242–352 (2021)
2. Liu, A., et al.: High-speed optical modulation based on carrier depletion in a silicon waveguide. Opt. Express **15**(2), 660–668 (2007)
3. Harris, N.C., et al.: Efficient, compact and low loss thermo-optic phase shifter in silicon. Opt. Express **22**(9), 10487–10493 (2014)
4. Errando-Herranz, C., et al.: MEMS for photonic integrated circuits. IEEE J. Sel. Top. Quantum Electron. **26**(2), 1–16 (2020)
5. Melikyan, A., et al.: High-speed plasmonic modulators. Nat. Photonics **8**, 229–233 (2014)
6. Zhang, M., et al.: Integrated lithium niobate electro-optic modulators: when performance meets scalability. Optica **8**(5), 652–667 (2021)
7. Jin, M., et al.: Efficient electro-optical modulation on thin-film lithium niobate. Opt. Letters **46**, 1884–1887 (2021)
8. Weigel, P.O., et al.: Bonded thin film lithium niobate modulator on a silicon photonics platform exceeding 100 GHz 3-dB electrical modulation bandwidth. Opt. Express **26**(18), 23728–23739 (2018)
9. Leo, J., et al.: Wafer-scale fabrication of low-loss waveguides in lithium niobate on insulator (LNOI) integrated photonics platform. In: European Conference on Optical Communication, Basel (2022)
10. Monney, A., et al.: Statistical characterization of MMI beam splitters on thin film lithium niobate on insulator (LNOI) platform at telecom wavelength. In: Conference on Lasers and Electro-Optics/Europe and European Quantum Electronics Conference, Munich (2023)

Monolithically Integrated Active/Passive GaAs Laser Platform Including High-Q Ring Resonators

Jan-Philipp Koester(✉), Hans Wenzel, Jörg Fricke, Poojitha Sammeta, Olaf Brox, Matthias Reggentin, Pietro Della Casa, Markus Weyers, and Andrea Knigge

Ferdinand-Braun-Institut (FBH), Gustav-Kirchhoff-Str. 4, 12489 Berlin, Germany
Jan-philipp.koester@fbh-berlin.de

Abstract. Integrated ring resonators with high quality factors are of interest for a number of applications such as biosensing and spectroscopy. In the context of integrated laser systems, ring resonators can be utilized to achieve narrow linewidth and widely tunable lasers. In this contribution we present our efforts to realize passive high-Q ring resonators ($Q > 10^5$) together with gain regions emitting around 1064 nm on the same GaAs-based chip. The enabling technology utilizes spatially selective quantum well removal and two-step epitaxy, as well as shallow and deep etched waveguides. The resulting platform will enable the realization of GaAs-based ring-resonator-coupled lasers.

Keywords: GaAs · diode laser · photonic integrated circuit · ring resonator

1 Introduction

Gallium arsenide lasers can cover a wavelength range from 600 nm to 1200 nm, making them suitable for various applications, including classical and quantum metrology, spectroscopy, and biosensing/imaging [1, 2]. Most of these applications require laser sources with narrow linewidth and/or wavelength tunability. To achieve that, wavelength-selective elements must be integrated into the laser cavity. In the context of GaAs-based lasers, this is often done by integrating Bragg-type reflectors [3, 4]. In recent years, ring-resonator-coupled lasers have emerged as an alternative way to achieve narrow linewidth and widely tunable emission [5, 6]. However, the high-Q passive ring resonators required for these devices have not yet been realized on a GaAs laser platform.

This contribution presents our efforts to integrate passive ring resonators monolithically with active waveguides that provide optical gain around 1064 nm. The corresponding technology is based on spatially selective quantum well removal and two-step epitaxy, allowing the fabrication of active/passive and shallow/deep waveguides. Shallow-etched active waveguides are used in areas where optical gain is supplied, while deep-etched passive waveguides are used for tight waveguide bends needed to form ring resonators.

J. Witzens et al. (Eds.): ECIO 2024, SPPHY 402, pp. 536–539, 2024.
https://doi.org/10.1007/978-3-031-63378-2_88

2 Ring Resonator Design and Fabrication

The schematic design of the realized racetrack-type ring resonator is shown in Fig. 1 (a). This design is similar to the one presented in references [7] and [8]. Two Euler U-bends ($R_{eff} = 75$ µm), based on deeply etched waveguides, are connected to shallowly etched straight waveguides used as one side of a directional coupler connecting the ring to the bus waveguide. Spot-size converters are used to connect the waveguides of the two etch depths [9]. This ring design was chosen for two main reasons:

First, the design is compatible with i-line lithography as the critical dimensions are larger than 0.5 µm. This is possible since the directional coupler is realized using shallowly etched (low confinement) ridge waveguides.

Second, the waveguide width of the deeply-etched waveguide is chosen to be 2 µm wide to limit lateral scattering losses. These waveguides, however, support three transverse modes. Therefore, an adiabatic transition (negligible coupling between modes) from straight to curved waveguides is required, which is realized by Euler U-bends [10], see SEM-image in Fig. 1 (b).

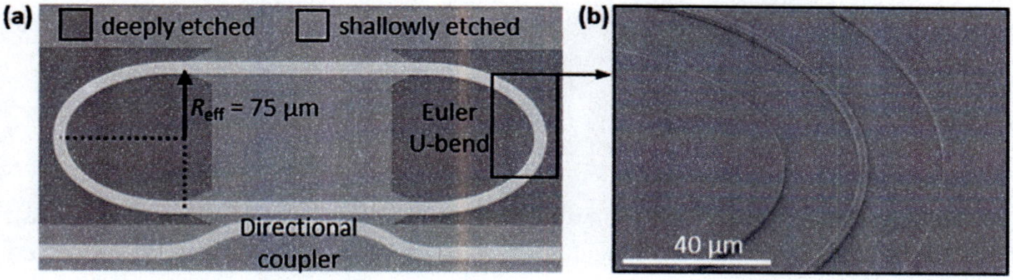

Fig. 1. (a) Schematic representation of the the all-pass racetrack ring resonator design. (b) Scanning electron microscope image of the Euler U-bend.

The vertical waveguide structure is grown using metalorganic vapor phase epitaxy (MOVPE) in two steps [4]. The first step ends with the double InGaAs quantum well and a GaAs top layer. In passive device areas these layers are removed in a selective wet chemical etch step. After the i-line resist is removed, the GaInP stop layer is etched using selective wet chemistry. The second growth process, which takes place in the same MOVPE reactor, involves in-situ CBr_4 surface cleaning and completion of the full vertical structure.

After the epitaxial growth, the waveguides were defined using i-line lithography and formed by dry etching trenches into the p-side. This process was repeated twice to form the shallow and deep-etched regions separately, similar as in reference [9]. Next, an insulating layer of silicon nitride (SiN) was deposited. The electrical p-contacts were created by selectively removing the SiN from the top of the active ridge regions, followed by the deposition of a thick Ti-Pt-Au layer. The n-contact was formed by depositing a Ni-Au-Ge alloy on the backside of the thinned wafer. Finally, the wafers were separated into smaller chips (laser bars) and the facets were coated with appropriate thin-film coatings.

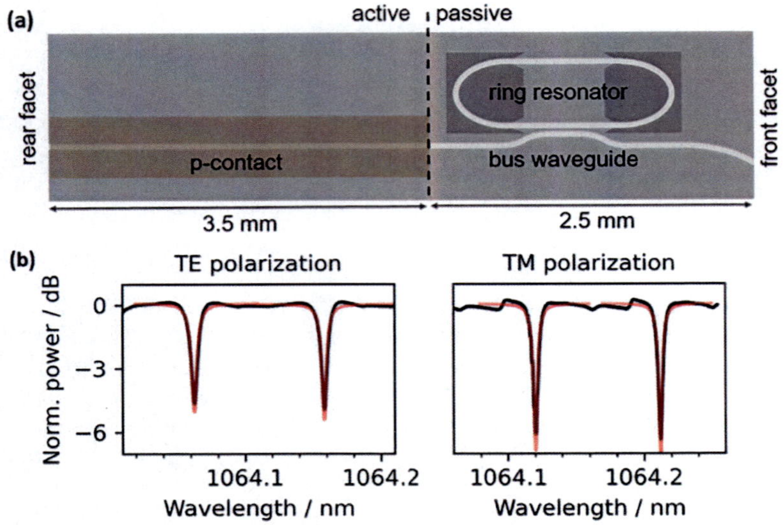

Fig. 2. (a) Schematic representation of the ring resonator test device. (b) Normalized and baseline corrected transmitted and ring resonator filtered power. The black and red lines correspond to measured data and Lorentzian fits, respectively.

3 Ring Resonator Characterization

The characterization of the ring resonators is based on the test device shown in Fig. 2 (a). The device comprises a 3.4 mm long active section located at the rear facet, connected to a 2.5 mm long passive section that incorporates an all-pass ring resonator following the design presented in Sect. 2. All test devices were measured unmounted and had a facet reflectivity of 30% and 0% at the rear and front facets, respectively.

First, the ring resonators were measured using the active region as a direct source of TE-polarized amplified spontaneous emission. The optical spectra at the front facet were measured using an optical spectrum analyzer (OSA). However, the resulting spectral width (at full width half maximum) of 23 pm was suspected to be limited by the spectral resolution of the OSA used (Yokogawa AQ6373B).

To obtain more accurate results, the ring resonances around 1064 nm were measured by coupling polarization-controlled, narrow-linewidth probe light emitted by a DFB laser (EYP-DFB-1064-00025-1500-BFY12-0002) into the bent waveguide at the front facet. The wavelength was tuned from 1064.01 nm to 1064.25 nm by adjusting the injection current from 50 mA to 150 mA. The transmitted and ring resonator filtered light was detected using the active region at the pass port of the bus waveguide as a monitor diode where the photocurrent was measured. The resulting data was baseline-corrected and is shown in Fig. 2 (b). The TE and TM resonances were fitted using Lorentzian fits (red lines), revealing a spectral width of 5.8 pm and 4.6 pm, resulting in loaded quality factors of $1.9 \cdot 10^5$ and $2.3 \cdot 10^5$ for TE and TM polarized light, respectively.

4 Conclusion

We presented a GaAs-based active/passive laser platform capable of incorporating ring resonators with loaded quality factors higher than 10^5 ($Q_{TE} = 1.9 \cdot 10^5$ and $Q_{TM} = 2.3 \cdot 10^5$). These results will enable the development of fully integrated GaAs-based narrow linewidth and widely tunable ring-coupled lasers.

References

1. Chauhan, N., et al.: Ultra-low loss visible light waveguides for integrated atomic, molecular, and quantum photonics. Opt. Express, OE **30**(5), 6960–6969 (2022). https://doi.org/10.1364/OE.448938
2. Kazanskiy, N.L., Khonina, S.N., Butt, M.A.: A review of photonic sensors based on ring resonator structures: three widely used platforms and implications of sensing applications. Micromachines **14**(5), 5 (2023). https://doi.org/10.3390/mi14051080
3. Brox, O., et al.: Realisation of a widely tuneable sampled grating DBR laser emitting around 970 nm. Electron. Lett. **53**(11), 744–746 (2017). https://doi.org/10.1049/el.2017.0521
4. Wenzel, S., et al.: Monolithically integrated extended cavity diode laser with 32 kHz 3 dB linewidth emitting at 1064 nm. Laser Photonics Rev. **16**, 2200442 (2022). https://doi.org/10.1002/lpor.202200442
5. Kumar, R., et al.: Fully integrated tunable III-V/Si laser with on-chip SOA. J. Lightwave Technol. **42**(9), 3314–3319 (2024). https://doi.org/10.1109/JLT.2024.3361080
6. Tran, M.A., Huang, D., Guo, J., Komljenovic, T., Morton, P.A., Bowers, J.E.: Ring-resonator based widely-tunable narrow-linewidth Si/InP integrated lasers. IEEE J. Sel. Top. Quantum Electron. **26**(2), 1–14 (2020). https://doi.org/10.1109/JSTQE.2019.2935274
7. Zhang, B., et al.: Compact multi-million Q resonators and 100 MHz passband filter bank in a thick-SOI photonics platform. Opt. Lett. **45**(11), 3005 (2020). https://doi.org/10.1364/OL.395203
8. Marin, Y.E., Bera, A., Cherchi, M., Aalto, T.: Ultra-High-q racetrack resonators on thick SOI platform through hydrogen annealing smoothing. J. Lightwave Technol. **41**(11), 3642–3648 (2023). https://doi.org/10.1109/JLT.2023.3262413
9. Koester, J.-P., et al.: Comparative study of monolithic integrated MMI-coupler-based dual-wavelength lasers. IEEE J. Sel. Top. Quantum Electron. **28**(1), 1–9 (2022). https://doi.org/10.1109/JSTQE.2021.3084344
10. Cherchi, M., Ylinen, S., Harjanne, M., Kapulainen, M., Aalto, T.: Dramatic size reduction of waveguide bends on a micron-scale silicon photonic plat-form. Opt. Express OE **21**(15), 17814–17823 (2013). https://doi.org/10.1364/OE.21.017814

Integrated Photonic In-Plane Beam Converter for Bessel-Gaussian Beam Generation

Jaewhan Lee[1], Jae-hoon Lee[1,2], and Sangsik Kim[1,3](✉)

[1] Graduate School of Quantum Science and Technology, Korea Advanced Institute of Science and Technology, Daejeon 34141, Republic of Korea
[2] Korea Research Institute of Standards and Science, Daejeon 34113, Republic of Korea
[3] School of Electrical Engineering, Korea Advanced Institute of Science and Technology, Daejeon 34141, Republic of Korea
sangsik.kim@kaist.ac.kr

Abstract. We present an integrated photonic in-plane beam converter generating a Bessel-Gaussian (BG) beam with a strong localized intensity along its propagation axis. The device is designed in two stages: the first stage transforms the waveguide mode into the Gaussian-distributed slab mode; the second stage produces the BG beam via an in-plane axicon. This axicon consists of apodized subwavelength gratings, expanding the vertical beam waist to minimize divergence. Our simulation generates a BG beam, achieving a diffraction-free zone over 2 mm that can be used for atomic-photonic hybrid integration.

Keywords: Mode converter · Axicon · Subwavelength grating

1 Introduction

Atom-photon interaction using atomic vapor in laser fields is regarded as foundational to various quantum sensing platforms, ranging from optical atomic clocks [1] to magneto-optic experiments [2]. Alongside these experiments, there has been an attempt to integrate the atomic vapor cell within a nanophotonic platform, utilizing a Gaussian beam from a large grating to facilitate atom-photon interactions within an atomic vapor cell [3, 4]. Despite these advancements, the integration faces significant challenges, notably in beam alignment and conversion efficiency. These issues stem from the out-of-plane emission of the beam by gratings, which radiates in a direction almost perpendicular to the chip surface, albeit at a slight angle. This configuration necessitates precise spatial alignment for coupling between multiple elements. To address these challenges, this work introduces a design for an in-plane beam converter that transforms a waveguide mode into a Bessel-Gaussian (BG) beam. This BG beam, characterized by diffraction-free and highly focused local intensity properties, offers a solution to the alignment and efficiency hurdles inherent in previous out-of-plane approaches.

Figure 1 illustrates the proposed chip, which consists of two stages: a waveguide-to-slab expander and an apodized subwavelength grating (SWG) axicon. The generated beam exhibits a well-defined BG beam profile horizontally while maintaining a Gaussian

J. Witzens et al. (Eds.): ECIO 2024, SPPHY 402, pp. 540–544, 2024.
https://doi.org/10.1007/978-3-031-63378-2_89

distribution vertically. The axicon slab is responsible for shaping the wavefront of the BG beam horizontally, whereas the apodized SWG is designed to expands the beam waist vertically. The BG beam, also known as a diffraction-free beam, strongly localizes field intensity over extended propagation distances [5–7]. Moreover, the BG beam's ring-shaped k-vector introduces a peculiar self-healing characteristic [8]. These attributes of BG beam, i.e., strong field localization over long propagation distances and self-healing capability, make it suitable for applications in chip-scale atomic-photon interactions.

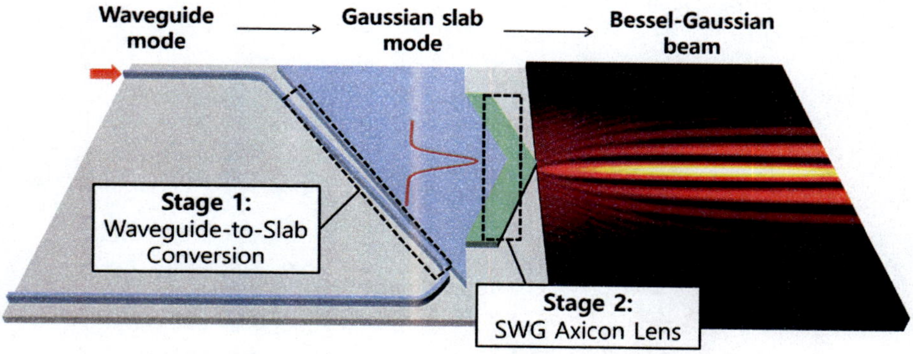

Fig. 1. Illustration of the device, which consists of two stages: stage one is a waveguide-to-slab mode converter, and stage two is an axicon for creating a Bessel-Gaussian beam with a subwavelength (SWG) grating. The green area represents the SWG region.

2 Designs and Results

Among other approaches, axicon is a simple method for generating a BG beam through superposition of multiple Gaussian beams, where each beam's k-vector forms a cone along the propagation axis at an angle θ [9]. For this, we first utilize a beam expander technique to create a large Gaussian slab mode [4]. This first stage involves engineering the gap profile between the input waveguide and the slab. Since the waveguide mode is evanescently coupled to the slab, the coupled supermode is leaky and is represented by a complex effective index. This index plays a crucial role in defining the slab mode. The generated slab mode propagates at an angle relative to the waveguide, due to the phase difference represented by the real part of the effective index contrast. For this study, the thickness of the silicon nitride layer is set to 150 nm, and the slab and waveguide (width is 350 nm) modes effective indices are approximately 1.69 and 1.51, respectively. This gives the tilt angle $\theta_{\text{tilt}} = \tan^{-1}(1.51/1.69) \approx 42°$. The imaginary part of the waveguide mode determines the power radiation to the slab, quantifying the loss coefficient α according to the gap size. Following the process detailed in Ref. [4], we designed the gap profile $g(z')$ along the 500 μm long expander. The target beam waist is 55.8 μm. Figure 2(a) shows the designed gap profile, with an inset illustrating the top view of the expander. This gap profile converts the waveguide mode to the slab mode, having a flat wavefront with $\theta_{\text{tilt}} \approx 42°$ and Gaussian power distribution. To check this design, we ran 2.5D finite-difference time-domain (FDTD) simulation. Figure 2(b) shows the simulated $|E|^2$ from the top view. The waveguide is located at $y = 0$ μm, and the wave propagates along the $+z'$ direction. The upper region $y > 2$ μm shows the field distribution in the slab. Note that the optical power in the waveguide decreases as wave propagates (from left to right), while field intensity distribution in the slab exhibits a well-defined Gaussian power distribution. To examine this slab mode profile, we set the power monitor at $y = 30$ μm as indicated with black dashed line. The blue line in Fig. 2(c) shows the normalized $|E|^2$ at this line. We conducted the Gaussian fitting (orange dashed line), quantifying the beam waist as 55.1 μm that closely matches with our target design.

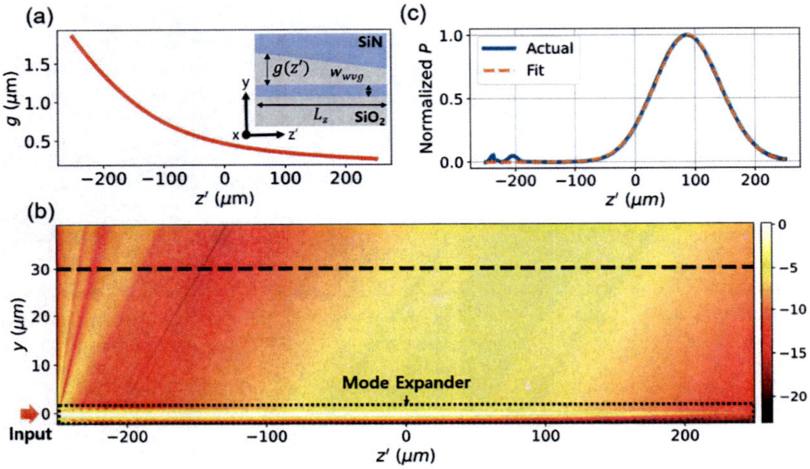

Fig. 2. Stage 1: waveguide-to-slab expander. (a) Designed gap profile along the propagation direction z'. Inset shows the top view of the stage 1 region with the gap profile $g(z')$ between the waveguide and the slab. The Si_3N_4 film thickness is 150 nm, and the waveguide width is $w_{wvg} = 350$ nm. The total expander length is $L_z = 500$ μm. (b) The simulated field intensity profile $|E|^2$ in the slab (top view). The arrow indicates the location of input waveguide at $y = 0$ μm. The optical power in the waveguide decreases as the wave propagates and evanescently couples to the adjacent slab with Gaussian intensity distribution. (c) The obtained power distribution (normalized P, blue) from the simulation monitor located at $y = 30$ μm [along the black dashed line indicated on (b)]. The power distribution follows the Gaussian profile, and the fitting result (orange dashed line) shows the beam waist approximately 55.1 μm. The wavelength is at 780 nm.

In the second stage, the Gaussian slab mode is converted to a BG beam in free space, facilitated by the in-plane axicon. This axicon ensures the generation of BG beam profile along the horizontal plane, while maintaining a Gaussian distribution vertically. Since a beam divergence angle is inversely proportional to its beam waist, it is essential to expand the beam vertically. To achieve this, we employed a linearly apodised SWG. Figure 3(a) illustrates the cross-sectional yz-plane of the apodized SWG, with a fixed periodicity $\Lambda = 200$ nm and duty cycles D_i decreasing linearly. The SWGs can be effectively represented by Rytov's formula [10]. Figure 3(b) shows the simulated input (orange) and output (blue) beams' vertical profiles and confirms a two-fold vertical expansion via apodized SWGs. Figure 3(c) shows the top view of Fig. 3(a), and Fig. 3(d) shows the simulated beam propagation after the linearly apodized SWG when a 100 μm wide Gaussian slab mode is excited. The Rayleigh range of the beam is ≈ 10.1 mm, thus exhibiting well-collimated beam propagation. However, a wide spread of beam gives a low intensity beam profile. Figure 3(e) shows a schematic top view, akin to Fig. 3(c), but with an axicon angle θ introduced at the center of the Gaussian slab mode. Figure 3(f) shows the beam's intensity profile generated by the axicon design in Fig. 3(e). A distinct BG beam pattern emerges, exhibiting strong localization that is approximately five times more intense than what is observed in Fig. 3(d). We expect a stronger atom-photon interaction utilizing this BG beam.

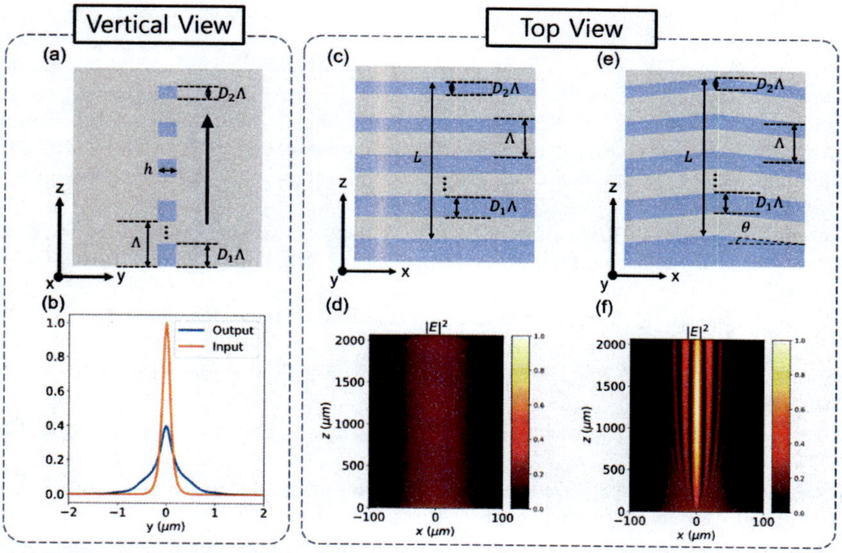

Fig. 3. Stage 2: In-plane SWG axicon. (a) Schematic *yz*-plane of the apodized subwavelength gratings (SWG) along the beam propagation direction z. The film height h and the SWG periodicity Λ are fixed to 150 nm and 200 nm respectively. The duty cycle D of the silicon nitride grating is linearly apodized from $D_1 = 0.5$ to $D_2 = 0.2$, along the length $L = 50$ μm. The wavelength is set to 780 nm. (b) Simulated intensity profiles ($|E|^2$, normalized to the input peak) at input (orange) and output (blue). The full width at half maximums (FWHM) at input and output are approximately 210.5 nm and 421 nm respectively. (c) Schematic xz-plane (top view) of the apodized SWGs. Geometric parameters are the same as in (a). (d) Simulated Gaussian beam propagation ($|E|^2$) along the z direction up to 2 mm after the in-plane SWG beam converter. The $z = 0$ is the chip edge. (d) Similar to (c), schematic xz-plane (top view) of the apodized SWGs but with the axicon angle $\theta = 5°$. (f) Simulated Bessel-Gaussian (BG) beam propagation ($|E|^2$). The result shows the strong localization with the BG beam pattern.

3 Summary

We have presented an in-plane SWG axicon that converts the waveguide mode into a BG beam. This component eases alignment between atomic vapor cell and photonic chip and provides a localized intensity over millimeter-scale propagation distances.

Acknowledgements. This work was supported by Institute for Information & Communications Technology Planning & Evaluation (IITP) grant No. RS-2023-00223497 and the National Research Foundation of Korea (NRF) grant No. RS-2023-00210997 funded by the Korea government (MSIT).

References

1. Poli, N., Oates, C.W., Gill, P., Tino, G.M.: Optical atomic clocks. La Rivista del Nuovo Cimento **36**(12), 555–624 (2013). https://doi.org/10.1393/ncr/i2013-10095-x
2. Budker, D., Kimball, D.F., Rochester, S.M., Yashchuk, V.V., Zolotorev, M.: Sensitive magnetometry based on nonlinear magneto-optical rotation. Phys. Rev. A **62**, 043403 (2000)
3. Hummon, M.T., et al.: Photonic chip for laser stabilization to an atomic vapor with 10–11 instability. Optica **5**, 443–449 (2018)
4. Kim, S., Westly, D.A., Roxworthy, B.J., et al.: Photonic waveguide to free-space Gaussian beam extreme mode converter. Light Sci. Appl. **7** 72 (2018). https://doi.org/10.1038/s41377-018-0073-2
5. Durnin, J., Miceli, J., Eberly, J.: Diffraction-free beams. Phys. Rev. Lett. **58**, 1499–1501 (1987)
6. Bickel, G., Hausler, G., Maul, M.: Triangulation with expanded range of depth. Opt. Eng. **24**(6) (1985)
7. Gori, F., Guattari, G., Padovani, C.: Bessel-Gauss beams. Opt. Commun. **64**(6), 491–495 (1987)
8. Aiello, A., Agarwal, G.S.: Wave-optics description of self-healing mechanism in Bessel beams. Opt. Lett. **39**, 6819–6822 (2014)
9. Indebetouw, G.: Nondiffracting optical fields: some remarks on their analysis and synthesis. J. Opt. Soc. Am. A **6**, 150–152 (1989)
10. Rytov, S.: Electromagnetic properties of a finely stratified medium. Sov. Phys. JEPT **2**, 466–475 (1956)

Additive Manufacturing of Strip-Loaded Thin-Film Lithium Niobate Waveguides by Means of Two-Photon Polymerization

Alexandra Rittmeier[1,2](✉), Elisavet Chatzizyrli[1,2], Angeliki Afentaki[1,2], Jörg Neumann[1,2], Andreas Wienke[1,2], Dietmar Kracht[1,2], Michael Kues[1,2,3], and Moritz Hinkelmann[1,2]

[1] Laser Zentrum Hannover e.V., Hollerithallee 8, 30419 Hannover, Germany
a.rittmeier@lzh.de
[2] Exzellenzcluster PhoenixD, Leibniz Universität Hannover, Welfengarten 1A, 30167 Hannover, Germany
[3] Institute of Photonics, Leibniz Universität Hannover, Nienburger Str. 17, 30167 Hannover, Germany

Abstract. Rapid manufacturing of optical waveguides in highly nonlinear lithium niobate paves the way for enhanced photonic integrated circuit (PIC) development applied in e.g. quantum photonics. Conventional fabrication of PICs requires usually multiple production steps while lacking rapid prototyping and reusability of substrates. Here, we present the fabrication of strip-loaded waveguides on 900 nm thin-film lithium niobate on insulator (LNOI) with a single-step production approach of two-photon polymerization (2PP) that addresses these shortcomings. We report propagation losses between 1.2 to 5.1 dB/cm for these waveguides and support the experimental work with simulations.

Keywords: Two-photon polymerization · strip-loaded waveguides · lithium niobate · additive manufacturing · photonic integrated circuits

1 Introduction

The increasing demand for high-speed communication and data-intensive applications necessitates photonic components with high bandwidth capabilities and energy efficiency. Photonic integrated circuits (PICs) enable the integration of multiple optical functions on a single chip, leading to smaller form factors, reduced power consumption, and cost-effective manufacturing techniques [1]. PICs have been developed within several material platforms like silicon on insulator [2], silicon nitride [3], indium phosphide [4] and lithium niobate on insulator (LNOI) [5]. Hereby, LNOI is a promising platform due to its large second-order nonlinearity and high refractive index [6, 7] paving the way for miniaturized quantum photonic devices such as single-photon sources and frequency converters [8, 9].

Reusability of PIC-related substrates can enhance efficiency in chip development, reduce waste, and facilitate more sustainable practices in the field. However, PICs

J. Witzens et al. (Eds.): ECIO 2024, SPPHY 402, pp. 545–551, 2024.
https://doi.org/10.1007/978-3-031-63378-2_90

often involve intricate photonic structures and materials, such as optical waveguides [7, 10], modulators [11], and micro ring resonators [12]. Conventional high-resolution fabrication processes like lithography and etching [8, 9] or fs laser direct writing [13] result in irreversible changes to the substrate and the chip itself, making straightforward reusability challenging.

In contrast, additive manufacturing (AM) of optical components has gained major interest in the last years to a more automated single-step production with great potential for rapid manufacturing and, simultaneously, higher design degree of freedom [14–16]. Among the 3D printing methods, two-photon polymerization (2PP) exhibits the highest spatial resolution [17, 18]. Compared to conventional optical lithography, the nonlinear two-photon absorption process enables a higher resolution with feature sizes below the diffraction limit [14, 19]. In addition, 2PP is inherently mask-free resulting in less complex manufacturing processes. Different 2PP-printed optical elements such as microlenses [14, 20, 21], modulation elements [14, 22] and waveguides [14, 23] have been reported.

In this paper, we present a new approach of combining the advantages of the highly nonlinear LNOI platform with the rapid production of 2PP-written strip-loaded waveguides. We show first experimental results of propagation loss characterization at a wavelength of 780 nm, which are supported by simulations. Erasing the 2PP-written strip does not damage the LNOI and allows for printing new waveguide designs onto the same substrate. Thus, the presented production enables reusability of the LNOI substrate, which paves the way for rapid prototyping while maintaining sustainable practices and, in general, enhanced PIC development.

2 Simulations of 2PP Strip-Loaded LNOI Waveguides

We use commercially available x-cut LNOI substrates (NanoLN) and the photopolymer IP-Dip2 (Nanoscribe GmbH) as the strip material. Details on the waveguide production follow in Sect. 2.2. Our LNOI substrates are composed of a silicon substrate with layer thickness of 500 μm followed by a SiO_2 insulator layer with thickness of 2 μm and a lithium niobate (LN) layer with 900 nm thickness, see Fig. 1(a). We chose this LN thickness to allow for a larger and more circular optical waveguide mode, which simplifies the transmission optimization of the coupling to standard single-mode fiber. The strips consist of photopolymer material with a height of 3 μm. To study the influence of the geometric waveguide design on the optical properties we chose the fabrication of strips with widths between 1 and 5 μm.

To support the experimental work, the finite-difference eigenmode method (Ansys Lumerical) was used to simulate the waveguide's optical properties. Figure 1(b) shows exemplarily a mode profile of a strip-loaded waveguide featuring a strip cross-section of 3×3 μm. The mode is guided in the LN layer, confined in y-direction due to the layer thickness and in x-direction due to the effective refractive index contrast caused by the printed strip. The waveguide fulfills the single-mode condition for the TM mode. Due to the large LN layer thickness, the TE modes do not fulfill the single-mode condition. Higher-order modes can be present and cannot be suppressed by different strip geometries. Nevertheless, in the experimental setup we use the TE mode because of a higher nonlinear efficiency needed for possible future applications.

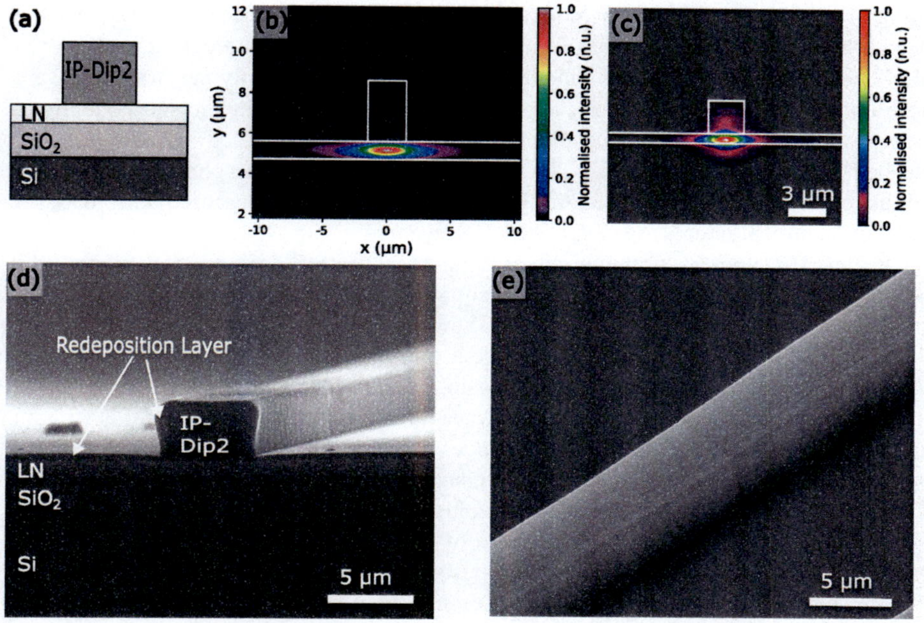

Fig. 1. (a) Cross-sectional sketch of LNOI-based strip-loaded waveguide. (b) Simulated and (c) measured mode profiles of a strip-loaded waveguide with strip size of 3×3 µm. (d) SEM image of the end facet after polishing. The redeposition layer can be mitigated with protective PVA layer, which needs to be selectively removed after polishing process. (e) SEM image of a waveguide with a stitching position visible in the center.

Figure 2(b) shows propagation loss values for the TE_{00} mode depending on the strip width. The decrease with higher strip width can be explained by the confinement of the TE_{00} mode. In the simulations we see that for smaller strip width the mode tends to leak more into the substrate which explains the propagation loss trend in Fig. 2(b). For the chosen wavelength of 780 nm the intrinsic material losses are not significant (in the range of 10^{-2} dB/cm).

3 Fabrication of 2PP Strip-Loaded LNOI Waveguides

The waveguides were produced according to the following procedure: We first polished the substrate edges of the chip with an argon ion beam milling system (Leica Microsystems EM TIC 3X) to ensure good optical quality of the end facets. We protected the LN surface during this procedure with a PVA (polyvinyl alcohol) coating from redeposition of the sputtered material. Afterwards we removed the PVA with distilled water. The substrate was rinsed with acetone, distilled water and isopropanol for cleaning and was subsequently dried with nitrogen. Before applying the photopolymer (IP-Dip2) we verified the cleanness with an optical microscope (Leica DM4000 M). The waveguides were written with the commercial 2PP printer (Photonic Professional GT2, Nanoscribe GmbH). The process parameters are a laser average power of 17.5 mW and a writing speed of 10000 μm/s. The waveguides are written directly from one substrate edge to the other while using coordinate transformations to ensure a perpendicular angle between the waveguide's propagation axis and the end facet. The printed polymer structures were developed in PGMEA (1-methoxy-2-propanol acetate) for 15 min and subsequently in isopropanol for 2 min. After developing the substrates were dried and if necessary, cleaned again with isopropanol. The printing parameters of the photopolymer were optimized to have a small surface roughness S_a below 20 nm measured with 3D laser scanning confocal microscope (VK-X1100, Keyence).

In Fig. 1(d) and (e), scanning electron microscope (SEM) images taken from the printed structures are shown. Figure 1(d) shows the substrate edge including the strip-loaded waveguide. The polished end facet showed no defects and a smooth surface with low roughness. On this specific waveguide a redeposition layer (mainly consisting of silicon) from the polishing process was visible due to a different order of production steps. In this sample, we first printed the waveguide and polished it afterwards without any protection layer. Note that the waveguides used for the optical measurements were fabricated in the production sequence of polishing first and later printing the waveguide. These samples do not show this redeposition layer.

Figure 1(e) shows a stitching position of a waveguide. These stitching positions were required since the waveguides are designed to feature a propagation length in the mm-range and a stitching-free structure can only be realized for up to 140 μm, depending on the range of the galvo scanner. We optimized the stitching parameters so that the geometric deviation at these points was minimized. Nevertheless, an influence of the stitching points on the optical properties was not evident in the measurements as expected since such small deviations in the strip have a barely measurable impact on the local effective refractive index and the mode profile.

4 Experimental Results

We coupled the emission of a laser diode with wavelength of 780 nm by means of lensed fibers with focal spot size diameter of 1 μm into the strip-loaded waveguide and collected the outcoupled light with an 60x magnification objective to characterize the mode profile or transmitted optical power. A typical mode profile image is shown in Fig. 1(c). When comparing to the simulations the same elliptical shape could be measured. In addition, we observed a small amount of light scattered into the polymer strip itself, which might be caused by a slight mode mismatch between waveguide and fiber mode.

To investigate the propagation losses of the strip-loaded waveguides, we used the cut-back method established in literature [24, 25]. In this method, usually for a single sample, the in- and output powers are measured for different propagation lengths. The power along the waveguide length z follows an exponential decay according to $\sim e^{-az}$, where a is the intrinsic propagation loss (including the material and scattering losses) [25]. The y-axis intercept is related to the coupling losses at both input and output facets. Advantages of this method are that it is feasible with low effort and the coupling losses can be extracted as additional information. However, the end facets and hence the coupling efficiency can vary strongly over one sample and induce an additional source of error in the cut-back method. Therefore, waveguides with different lengths were printed

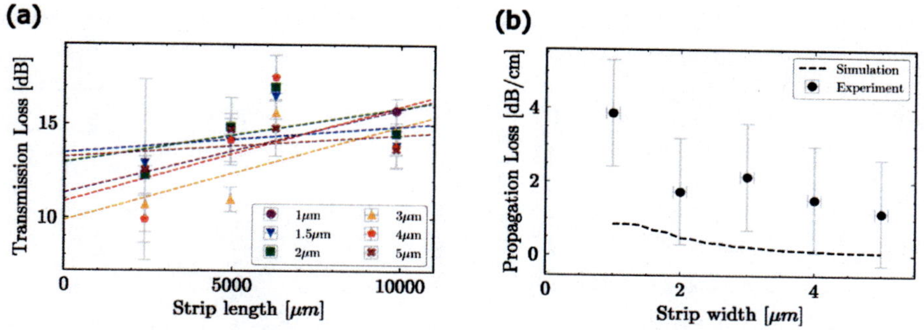

Fig. 2. (a) Propagation losses for six different strip widths, the strip height is 3 μm for all cases. Dotted lines are linear regressions. (b) Propagation loss, excluding coupling losses as a function of the strip width. Dotted line shows simulation results for the TE_{00} mode. Experimental results were gained from one sample and the mean coupling loss is subtracted.

on different chips. To overcome the problem that the coupling or the waveguide itself might vary, the propagation losses of many waveguides with the same parameters and geometries were measured to calculate a statistic mean. The results are shown in Fig. 2(a). The results were fitted and the propagation loss of the waveguide as well as the total coupling loss were extracted, see Table 1. We measured propagation losses between 1.2 and 5.1 dB/cm, which is comparable to other fabrication techniques reported in literature [1, 26, 27], e.g. Li et al. reports propagation losses of 5.8 dB/cm for TiO_2 strip-loaded waveguides on LN thin films [28]. However, we benefit from the rapid manufacturing of our production method.

During this investigation we analyzed the noise and the reproducibility of our new production platform. The propagation losses between waveguides with the same geometries differ by 1.1 ± 0.4 dB including different coupling losses and misalignment errors which is acceptable for e.g. single-photon applications.

In addition, we found that removing the waveguide structures from a chip with an ultrasonic bath of acetone or PGMEA and reprinting the structure will not influence the chip itself and does not affect the efficiency of the waveguides. This can be a major benefit when compared to other fabrication methods, especially in the context of multifunctional, integrated chips with non-polymer materials. With our production method the waveguide design could be changed fast without requiring a completely new chip.

Comparing experimental and simulation results for the propagation loss dependence on the strip width (Fig. 2(b)), the same trend is visible. The experimental results were gained from only one sample to reduce influences of different substrate and coupling quality. The mean coupling loss was subtracted. For larger strip widths the loss drops due to the confinement of the mode. The offset of the experimental data to the simulation results can be explained by additional loss due to imperfection of the waveguide or higher order mode loss since for TE polarization the waveguide is not single-mode.

Table 1. Propagation loss results, calculated with linear regression. Coupling loss combined of both end facets.

Strip width [μm]	Propagation loss [dB/cm]	Coupling loss [dB]
1	4.49 ± 0.49	11.31 ± 0.86
1.5	1.44 ± 1.26	13.45 ± 1.18
2	2.94 ± 1.47	12.92 ± 0.93
3	5.04 ± 1.55	9.83 ± 1.49
4	5.10 ± 2.32	10.83 ± 1.10
5	1.20 ± 0.85	13.21 ± 1.08

5 Summary

In this contribution, we showed a new approach for fabricating LN strip-loaded waveguides by means of 2PP. We report propagation losses between 1.2 to 5.1 dB/cm, which is comparable to LN strip-loaded waveguides fabricated by e.g. electron beam lithography procedures [1]. Furthermore, we investigated the reproducibility of the waveguides where the polymer structures enable the possibility of reusing chips for rapid manufacturing which is a major benefit against standard lithographic production. In future, this platform could be used for nonlinear applications e.g. frequency mixing or photon-pair generation. We are convinced that this new production platform enables the possibility for enhanced PIC development and more design freedom in waveguides devices.

Acknowledgements. This work is funded by the German Federal Ministry of Education and Research (BMBF) under the funding program "Quantum Technologies – from basic research to market" within the project "QPhotLab" (FKZ 13N15984) and by the Deutsche Forschungsgemeinschaft (DFG, German Research Foundation) under Germany's Excellence Strategy within the Cluster of Excellence PhoenixD (EXC 2122, Project ID 390833453).

References

1. Liu, Y., Huang, X., Li, Z., Guan, H., Fan, Z., Han, W., Li, Z.: Silicon rich nitride-lithium niobate on insulator platform for photonics integrated circuits. arXiv:2105.07880. (2021)
2. Wang, J., Long, Y.: On-chip silicon photonic signaling and processing: a review. Sci. Bull. **63**(19), 1267–1310 (2018)
3. Blumenthal, D.J., Heideman, R., Geuzebroek, D., Leinse, A., Roeloffzen, C.: Silicon nitride in silicon photonics. Proc. IEEE **106**(12), 2209–2231 (2018)
4. Stabile, R.: Towards large-scale fast reprogrammable SOA-based photonic integrated switch circuits. Appl. Sci. **7**(9), 920 (2017)
5. Sun, D., Zhang, Y., Wang, D., et al.: Microstructure and domain engineering of lithium niobate crystal films for integrated photonic applications. Light Sci. Appl. **9**, 197 (2020)
6. Chen, P.K., Briggs, I., Cui, C., et al.: Adapted poling to break the nonlinear efficiency limit in nanophotonic lithium niobate waveguides. Nat. Nanotechnol. **19**, 44–50 (2024)

7. Boes, A., Corcoran, B., Chang, L., Bowers, J., Mitchell, A.: Status and potential of Lithium Niobate on Insulator (LNOI) for photonic integrated circuits. Laser Photonics Rev. **12**, 1700256 (2018)
8. Lomonte, E., Wolff, M.A., Beutel, F., et al.: Single-photon detection and cryogenic reconfigurability in lithium niobate nanophotonic circuits. Nat. Commun. **12**, 6847 (2021)
9. Saravi, S., Pertsch, T., Setzpfandt, F.: Lithium niobate on insulator: an emerging platform for integrated quantum photonics. Adv. Optical Mater. **9**(22), 2100789 (2021)
10. Volk, M.F., Suntsov, S., Rüter, C.E., Kip, D.: Low loss ridge waveguides in lithium niobate thin films by optical grade diamond blade dicing. Opt. Express **24**(2), 1386–1391 (2016)
11. Cao, L., Aboketaf, A., Wang, Z., Preble, S.: Hybrid amorphous silicon (a-Si:H)–LiNbO3 electro-optic modulator. Opt. Commun. **330**, 40–44 (2014). https://doi.org/10.1016/j.optcom.2014.05.021
12. Witmer, J.D., Valery, J.A., Arrangoiz-Arriola, P., et al.: High-Q photonic resonators and electro-optic coupling using silicon-on-lithium-niobate. Sci. Rep. **7**, 46313 (2017)
13. Lin, J., Xu, Y., Fang, Z., et al.: Fabrication of high-Q lithium niobate microresonators using femtosecond laser micromachining. Sci. Rep. **5**, 8072 (2015)
14. Berglund, G., Wisniowiecki, A., Gawedzinski, J., Applegate, B., Tkaczyk, T.S.: Additive manufacturing for the development of optical/photonic systems and components. Optica **9**(6), 623–638 (2022)
15. Zolfaghari, A., Chen, T., Yi, A.Y.: Additive manufacturing of precision optics at micro and nanoscale. Int. J. Extrem. Manuf. **1**(1), 012005 (2019)
16. Camposeo, A., Persano, L., Farsari, M., Pisignano, D.: Additive manufacturing: applications and directions in photonics and optoelectronics. Adv. Optical Mater. **7**(1), 1800419 (2019)
17. Gissibl, T., Thiele, S., Herkommer, A., et al.: Sub-micrometre accurate free-form optics by three-dimensional printing on single-mode fibres. Nat. Commun. **7**, 11763 (2016)
18. Schmid, M., Thiele, S., Herkommer, A., Giessen, H.: Three-dimensional direct laser written achromatic axicons and multi-component microlenses. Opt. Lett. **43**(23), 5837–5840 (2018)
19. Burmeister, F., Steenhusen, S., Houbertz, R., Zeitner, U.D., Nolte, S., Tünnermann, A.: Materials and technologies for fabrication of three-dimensional microstructures with sub-100 nm feature sizes by two-photon polymerization. J. Laser Appl. **24**(4), 042014 (2012)
20. Kiekens, K.C., Barton, J.K.: 3D printed lens for depth of field imaging. OSA Continuum **2**(11), 3019–3025 (2019)
21. Dietrich, P.I., et al.: Printed freeform lens arrays on multi-core fibers for highly efficient coupling in astrophotonic systems. Opt. Express **25**(15), 18288–18295 (2017)
22. Lightman, S., Gvishi, R., Hurvitz, G., Arie, A.: Comparative analysis of direct laser writing and nanoimprint lithography for fabrication of optical phase elements. Appl. Opt. **55**(34), 9724–9730 (2016)
23. Adão, R.M.R., Alves, T.L., Maibohm, C., Romeira, B., Nieder, J.B.: Two-photon polymerization simulation and fabrication of 3D microprinted suspended waveguides for on-chip optical interconnects. Opt. Express **30**(6), 9623–9642 (2022)
24. Panusa, G., Pu, Y., Wang, J., Moser, C., Psaltis, D.: Fabrication of sub-micron polymer waveguides through two-photon polymerization in polydimethylsiloxane. Polymers **12**(11), 2485 (2020)
25. Tong, X.C.: Advanced Materials for Integrated Optical Waveguides. Springer, Switzerland (2014). https://doi.org/10.1007/978-3-319-01550-7
26. Roussey, M., et al.: Strip-loaded horizontal slot waveguide. Opt. Lett. **42**(2), 211–214 (2017)
27. Zhu, H., et al.: Hybrid mono-crystalline silicon and lithium niobate thin films. Chin. Opt. Lett. **19**(6), 060017 (2021)
28. Li, S., Cai, L., Wang, Y., Jiang, Y., Hu, H.: Waveguides consisting of single-crystal lithium niobate thin film and oxidized titanium stripe. Opt. Express **23**(19), 24212–24219 (2015)

Integrated GeSn Heterojunction Phototransistor on Si

Michael Oehme(✉), Christian Spieth, Sören Schäfer, Maurice Wanitzek, Lukas Seidel, Michael Hack, Erich Kasper, and Daniel Schwarz

Institute of Semiconductor Engineering, University of Stuttgart, 70569 Stuttgart, Germany
oehme@iht.uni-stuttgart.de

Abstract. This work investigates on the behavior of p-n-p GeSn heterojunction phototransistors integrated on Si substrates. We found that the GeSn heterojunction phototransistor performs significantly better in inverse operation mode than in normal operation mode. The investigations show that wavelengths up to 2000 nm can be detected with the GeSn heterojunction phototransistor. At the wavelength of 1550 nm, an emitter-collector-voltage of −1 V, and an optical power of 0.1 W, an optical responsivity of 2 A/W was measured. Compared to the Ge reference phototransistor, this is an increase by a factor of 50.

Keywords: Silicon photonics · GeSn alloys · Heterojunction phototransistors

1 Introduction

Modern photonic technologies have a wide range of applications in various areas such as information transfer, communication, quantum computers, automotive applications, medicine, and many others. Over 90% of microelectronics is dominated by the material Si. For this reason, Si photonics or Group IV photonics has been developed to integrate photonic functions into Si chips. This enables photonic integrated circuits to be built in which photonic and electronic components work perfectly together.

Due to its band gap of 1.1 eV, Si can only be used to a very limited extent for the fabrication of active optical components in the near and short-wave infrared (wavelength 1 μm to 2 μm). Ge and GeSn compounds exhibit better physical properties in this wavelength range. These materials can be monolithically integrated on Si. In recent years, proof of the functionality of integrated optoelectronic Ge or GeSn components on Si substrates has been repeatedly provided [1–3].

Photodetectors (PD) rely on different functions. For PD with a pin structure, the Signal-to-Noise Ratio (SNR) is limited due to the lack of internal amplification. An alternative device is the Avalanche photodiode (APD), since it exhibits an internal amplification due to the avalanche multiplication. However, the APD suffers from high excessive noise due to the multiplication process. Furthermore, the need for a high operating voltage required to trigger impact ionization, also hinders practical use in low

J. Witzens et al. (Eds.): ECIO 2024, SPPHY 402, pp. 552–557, 2024.
https://doi.org/10.1007/978-3-031-63378-2_91

power applications. Another device with internal amplification is a photo transistor (PT), which is designed like a bipolar transistor with a floating base. Thus, the base of the PT is not connected, which its design as a 2-terminal (2T) device. The base current is formed by the absorption of photons and therefore the generation of electron-hole-pairs. This current is internally amplified by appropriately designing the structure of the PT. In recent decades, many standalone and top-coupled Ge bipolar transistors on Si have been demonstrated, and more recently several waveguide-coupled PTs have been reported for use in fiber optic communications [4–6]. However, the optical responsivity decreases at wavelengths larger than 1550 nm due to the direct absorption edge of Ge at this value. By adding small amounts of Sn to the Ge matrix, the absorption edge can be shifted significantly towards the infrared. If this Sn addition is only carried out in the base, a hetero bipolar transistor or a heterojunction phototransistor (HPT) is created. Simulations of such GeSn HPT have been intensively carried out by Kumar's group [7]. Yeo's group demonstrated an n-p-i-n GeSn HPT on a Ge substrate with a free-floating base with a Sn concentration of 6.5% [8]. The group of Chang reported p-n-i-p GeSn HPT on a Si substrate, however the Sn is incorporated in the collector region [9].

In this paper, we demonstrate a p-n-p GeSn HPT in which the base consists of pseudomorphic GeSn. The PT is fabricated as a 2T device. A Ge reference PT is fabricated for comparison.

2 Device Fabrication

The GeSn HPT are grown on p^+-Si (100) substrates with a resistivity between 10 mΩ cm and 20 mΩ cm. For this, the molecular beam epitaxy (MBE) method is used [10]. The growth begins with the fabrication of the Ge collector. In order to overcome the lattice mismatch between Si and Ge, a virtual substrate (VS) technology is used (see [11]). Subsequently, intrinsic Ge is deposited with a thickness of 400 nm. The background doping of the i-Ge is p-type with a concentration of 1×10^{16} cm^{-3}. In the next step, the n-type doped $Ge_{0.94}Sn_{0.06}$ base is grown with a doping concentration of 1×10^{18} cm^{-3}. Here it is important that the GeSn grows pseudomorphically to avoid the formation of new defects which could be detrimental to the device performance. The emitter consists of a 100 nm thick Ge layer and a 111 nm thick Si top layer. Both layers are doped with B (1×10^{20} cm^{-3}). In addition, a Ge reference PT is grown. The complete layer stack is shown in Fig. 1. The layer structure on a p^+-substrate offers the possibility of monolithic integration. To do this, the p^+-substrate must be replaced by an n-type substrate with corresponding highly doped p^+-implants as a buried layer. The implants can be made by ion implantation as in [12].

The PT fabrication is carried out using a single mesa process with an ultra-low thermal budget. The highest process temperature is 250 °C. In this process, circular PT with different diameters are fabricated. The light is coupled into the devices from the top via an optical window. A detailed description of the technology can be found in [13].

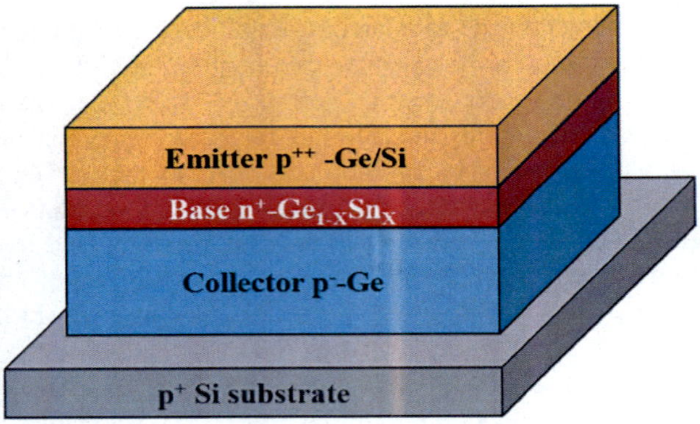

Fig. 1. Schematic layer stack of the investigated GeSn HPT (X = 0.06) and the Ge PT (X = 0).

3 Results

The pseudomorphic growth of the GeSn base was demonstrated using high-resolution X-ray diffractometry. For this purpose, the asymmetrical $(\overline{22}4)$) reflection was measured as a reciprocal space map and given in a space map presentation. From the measured in-plane and out-off-plane lattice constants, the Sn content was calculated to be 6% with a compressive stress of −0.8.

The electro-optical characterization of both phototransistors is carried out using on-wafer measurement systems. In all electro-optical measurements, the collector is connected to the ground and the voltage is changed at the emitter. This means that the transistor is in normal operation (NO) when the emitter-collector voltage is positive. If there is a negative voltage is applied at the emitter, the transistor works in inverse operation (IO), respectively. A super continuum laser with an acoustic-optical tunable wavelength filter is used as a light source. The light is coupled into the phototransistor

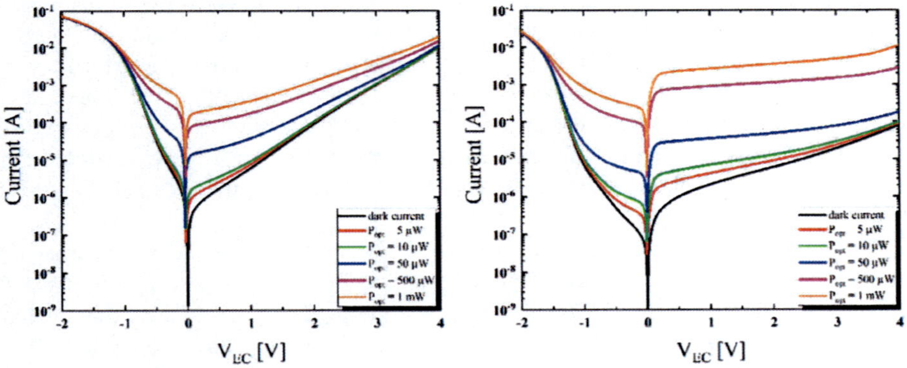

Fig. 2. Collector current voltage characteristics of the GeSn HPT (left) and the Ge PT (right) under illumination (λ = 1310 nm) with different optical powers. The black curve shows dark current as a function of the emitter-collector voltage

via a Single-Mode fiber. The incident optical power is measured with an absolutely calibrated InGaAs detector.

Figure 2 shows the collector current measurements as a function of the emitter-collector voltage for different illuminated optical powers at a wavelength of 1310 nm (left –GeSn HPT and right–Ge PT). The black curves represent the respective dark characteristic curve fields. The dark measurements show that both types of transistors work differently. The Ge PT works as a normal bipolar transistor, i.e. the collector current is significantly higher in NO than in IO. This behavior does not change even under illumination. The collector current of the GeSn HPT behaves exactly the opposite (see left illustration in Fig. 2), as it is significantly higher in IO than in NO. This shows the influence of the GeSn/Ge heterostructure.

Due to the amplification properties of the transistor, the generated photocurrent does not show a linear behavior with the illuminated optical power P_{opt}. As a result, the optical responsivity changes with P_{opt}. In order to compare both transistors, the optical responsivity is calculated at a constant optical power of 0.1 mW. Figure 3 shows this calculation for the IO at $V_{EC} = -1$ V. The GeSn HPT has a significantly higher optical responsivity across the entire wavelength range compared to the reference Ge PT. The influence of GeSn can be observed at wavelengths larger than the direct band gap of Ge ($\lambda = 1550$ nm). With the GeSn HPT, light can be detected up to the measurement limit of $\lambda = 2000$ nm. The inset in Fig. 3 shows an enlarged view around the wavelength of the direct band gap of Ge. At $P_{opt} = 0.1$ mW the optical responsivity of the GeSn HPT increased by a factor of 50 in comparison to the Ge PT.

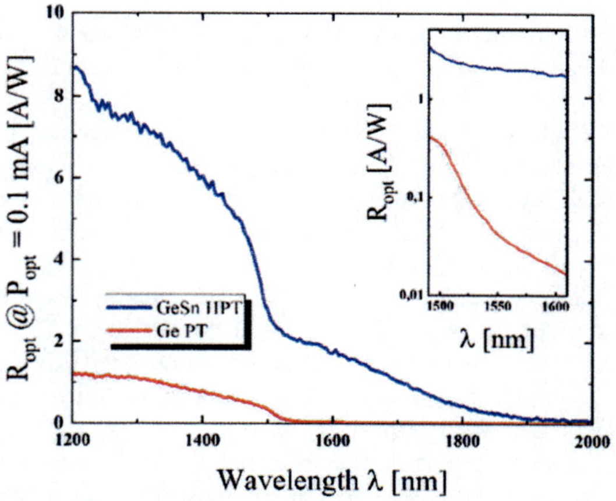

Fig. 3. Optical responsivity as a function of wavelength at a fixed optical power of $P_{opt} = 0.1$ mW and $V_{EC} = -1$ V for both transistors. Inset: Enlarged view around the optical telecommunications wave of $\lambda = 1550$ nm.

4 Conclusion

In conclusion, we have investigated the growth of GeSn p-n-p HPT with a Sn content of 6% and a Ge reference PT. The transistors were grown on Si substrates using a MBE process. An ultrathin Ge-VS was used to accommodate the lattice mismatch between Si and Ge. Dark current analysis shows that the Ge PT performs similarly to a typical bipolar junction transistor, with a much larger current in NO mode than in IO mode. However, GeSn HPT shows the opposite behavior. It performs significantly better in IO than in NO mode.

The photocurrent of our fabricated device does not have a linear relationship with the incident optical power. At lower optical powers, there is a significant amplification of the photocurrent. For this reason, the responsivity measurements were carried out with a fixed optical power across the entire wavelength range. The GeSn HPT detects light up to wavelengths of around 2000 nm. At a wavelength of 1550 nm, an emitter-collector bias of -1 V, and an optical power of 0.1 W an optical responsivity of 2 A/W is determined. Compared to the Ge PT, the optical responsivity increases here by a factor of 50 with the GeSn HPT.

Acknowledgments. This work was supported by the Deutsche Forschungsgemeinschaft (431314977/GRK2642).

References

1. Liu, J., et al.: Ge-on-Si optoelectronics. Thin Solid Films **520**, 3354–3360 (2012)
2. Asghari, M., Krishnamoorthy, A.V.: Energy-efficient communication. Nat. Photonics **5**, 268–270 (2011)
3. Chen, G., et al.: High-speed photodetectors on silicon photonics platform for optical interconnect. Laser Photonics Rev. **16**(12), 2200117 (2022)
4. Ang, K.W., Yu, M.B., Lo, G.Q., Kwong, D.L.: Low-voltage and high-responsivity germanium bipolar phototransistor for optical detections in the near-infrared regime. IEEE Electron Device Lett. **29**(10), 1124–1127 (2008)
5. Going, R., Seok, T.J., Loo, J., Hsu, K., Wu, M.C.: Germanium wraparound photodetectors on silicon photonics. Opt. Express **23**(9), 11975–11984 (2015)
6. Frounchi, M., Tzintzarov, G.N., Ildefonso, A., Cressler, J.D.: High responsivity GE phototransistor in commercial CMOS Si-photonics platform for monolithic optoelectronic receivers. IEEE Electron Device Lett. **42**(2), 196–199 (2021)
7. Kumar, H., Basu, R.: Noise analysis of group IV material-based heterojunction phototransistor for fiber-optic telecommunication networks. IEEE Sens. J. **18**(22), 9180–9187 (2018)
8. Wang, W., et al.: Floating-base germanium-tin heterojunction phototransistor for high-efficiency photodetection in short-wave infrared range. Opt. Express **25**(16), 18502–18507 (2017)
9. Hung, W.T., Barshilia, D., Basu, R., Cheng, H.H., Chang, G.E.: Silicon-based high-responsivity GeSn short-wave infrared heterojunction phototransistors with a floating base. Opt. Lett. **45**(5), 1088–1091 (2020)
10. Schwarz, D., Oehme, M., Kasper,E.: Molecular beam epitaxy of Si, Ge, and Sn and Their compounds. In: Thin Films - Growth, Characterization and Electrochemical Applications, F. Sarf, E. Yakar, und I. Karaduman Er, Hrsg., IntechOpen (2024)

11. Oehme, M., Werner, J., Kaschel, M., Kirfel, O., Kasper, E.: Germanium waveguide photodetectors integrated on silicon with MBE. Thin Solid Films **517**(1), 137–139 (2008)
12. Oehme, M., et al.: Backside illuminated Ge-on-Si NIR camera. IEEE Sens. J. **21**(17), 18696–18705 (2021)
13. Werner, J., et al.: Germanium-tin p-i-n photodetectors integrated on silicon grown by molecular beam epitaxy. Appl. Phys. Lett. **98**(6), 061108 (2011)

Check for updates

Self-aligned Fibre-to-Chip Edge Coupling Structure with Suspended Taper

Imene Si Hadj Mohand(✉), Paul Baroux, Etienne Herth, Abdelmounaim Harouri, Jean René Coudevylle, and Xavier Checoury(✉)

Université Paris-Saclay, CNRS, C2N, Palaiseau, France
{imene.si-hadj-mohand,xavier.checoury}@c2n.upsaclay.fr

Abstract. We propose a self-aligned fiber-to-chip coupling structure to address optical fiber alignment and mode-mismatch issues between the modes of cleaved single-mode optical fibers and integrated Silicon on Insulator (SOI) waveguides. The structure consists of a suspended linear taper and an innovatively shaped on-chip groove fabricated with a two-step inductively coupled plasma deep reactive ion etching (ICP-DRIE) independent of the silicon crystal lattice orientation. We achieved coupling losses of the order of 2.5 dB per facet, with the optical fiber automatically aligning with the integrated waveguide.

Keywords: Integrated silicon photonic circuits · coupling techniques · fiber-to-chip alignment

1 Introduction

Coupling integrated photonic circuits with optical fibers is challenging because the optical fiber alignment can be very time-consuming when done manually. The optical mode mismatch between the optical fiber and the integrated waveguide is also problematic: the Mode-Field Diameter (MFD) is 10.4 ± 0.5 µm for a cleaved SMF28e fiber at 1550 nm, and the integrated waveguide typical size is 450×220 nm. In this paper, we propose to solve these two issues. The structure we developed falls into the category of edge couplers [10]. We have opted for a passive approach to fiber alignment that uses an on-chip groove to make fiber alignment more straightforward and fast than usual manual fiber alignment using micrometric translation stages. The groove design enables a three-dimensional (3D) self-positioning of the optical fiber. The groove is etched using inductively coupled plasma deep reactive ion etching (ICP-DRIE) [8], which avoids the constraints associated with potassium hydroxide (KOH) or tetramethyl ammonium hydroxide (TMAH) wet etchings commonly used in the literature. Indeed, the ICP-DRIE process allows the etching of the groove in any direction, regardless of the silicon crystal lattice orientation. In addition, the dry etching does not require the protection of the sample's backside. Moreover, ICP-DRIE is expected to be more reproducible than wet etching. To demonstrate the efficiency of our

J. Witzens et al. (Eds.): ECIO 2024, SPPHY 402, pp. 558–564, 2024.
https://doi.org/10.1007/978-3-031-63378-2_92

groove design, we use a suspended linear taper to achieve the coupling between the fiber and the waveguide [9]. In this work, we have fabricated structures with 125-µm long linear tapers and suspended them to ensure that the expanded mode does not leak into the silicon substrate. Thanks to the groove design, we achieved successful self-alignment of the optical fiber and coupling losses of the order of 2.5 dB per facet. Other tapering shapes can be used (exponential, quadratic, constant-loss) [3,9] and are expected to reduce the losses further. This coupling structure, which does not require a matching cladding (e.g., silicon dioxide, polymer), exhibits many potential use cases. For example, it can be used in applications involving suspended membranes, such as photonic crystals [4] and optomechanical devices [2]. The proposed groove design can be utilized in other applications with encapsulated tapers.

2 Design and Fabrication

The coupling structure includes a groove for fiber self-alignment and a silicon taper for the mode conversion from the optical fiber to the integrated waveguide. Figure 1 shows all design parameters of the groove. SOI platform, which consists of a 220 nm thick silicon layer on a 2-µm thick oxide layer, is used here as it is common in silicon photonics. Two types of ICP-DRIE etching of the silicon bulk can be performed: an isotropic and an anisotropic one [7]. The characteristic groove cross-section shown in Fig. 1 is achieved when these two types of etching are done successively. In this groove, The fiber is supported vertically and horizontally by two points formed by the intersection of the isotropic and anisotropic silicon etching profiles. Longitudinally, as we will see later in the fabrication section, the fiber is blocked by a stop with the same height as the support points and was created by design.

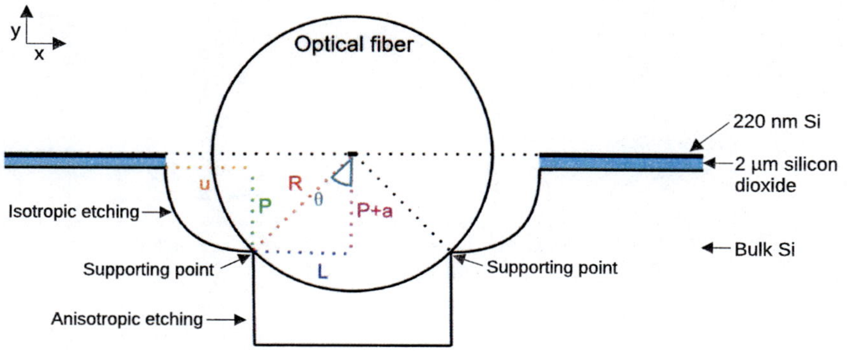

Fig. 1. Cross-sectional view of the groove. **2R** represents the diameter of the optical fiber (125.0 ± 0.7 µm), **2L** is the opening of the lithographic mask for the groove etching (108.26 µm); **P** is the depth of the isotropic etch (29.14 µm); **u** is the underetching; **a** the distance between the center of the Si waveguide and the Si bulk surface (2.11 µm); 2θ, the opening angle of the fiber supporting points.

Knowing the diameter (125.0 ± 0.7 μm) of the optical fiber and choosing an opening angle θ of 60°, we have set the width of the groove etching aperture **2L** (108.26 μm) and the depth of isotropic etching of the silicon bulk **P** (29.14 μm). Note that our design allows flexibility in the choice of θ angle from 0 to 90°, the angle at which a U-groove (found in the literature) is achieved [12]. Such a U-groove allows for less precise lateral alignment than our design. The depth **P** determines the vertical alignment of the fiber. With these parameters, the optical fiber automatically aligns with the integrated waveguide when inserted in the groove.

We illustrate the efficiency of this groove design for self-alignment of the optical fiber using a suspended linear silicon taper that couples light from an optical fiber into an integrated waveguide. To determine the optimal dimensions of the silicon taper tip, we performed frequency-domain 2D simulations to obtain the overlap of the optical fiber and the taper modes. For a given sufficiently long taper, the coupling efficiency between the optical fiber and the taper is given by [3,9,11]:

$$\eta = \frac{\left|\iint E_{fb} \times H^*_{wvg} dxdy\right|^2}{\left|\iint E_{fb} \times H^*_{fb} dxdy\right| \cdot \left|\iint E_{wvg} \times H^*_{wvg} dxdy\right|} \tag{1}$$

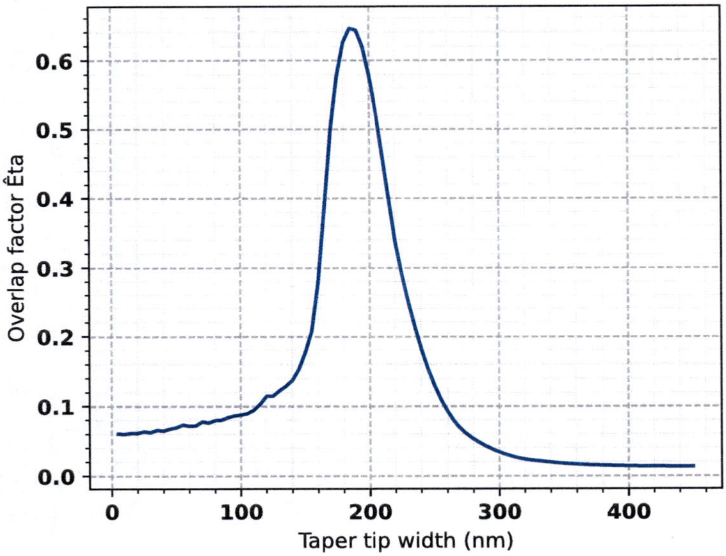

Fig. 2. Calculated overlap integral η between the mode of an SMF28e optical fiber (MFD 10.4 μm) and the TE mode of a silicon taper tip whose width ranges from 10 to 450 nm.

where E_{fb}, H_{fb} are respectively the electric field and the magnetic field of the optical fiber mode and E_{wvg}, H_{wvg} the electric and magnetic field of the taper tip mode. Figure 2 shows the overlap integral (1) between the electromagnetic fields of the cleaved SMF28e optical fiber and the mode of a silicon waveguide whose tip width varies from 10 nm to 450 nm. The maximum overlap value, 0.64, was achieved at a tip width of 185 nm. To achieve an efficient conversion between the taper tip mode and a standard 450-nm wide waveguide, we make a smooth transition between the taper tip and the waveguide by linearly varying the taper width over 125 μm.

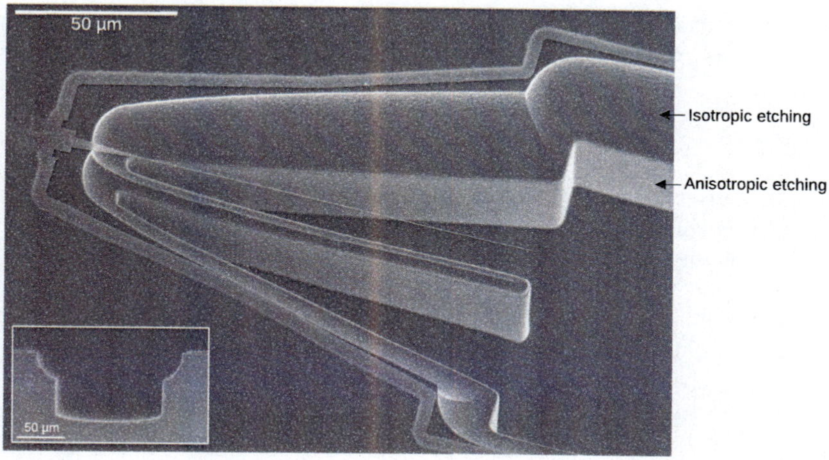

Fig. 3. SEM image showing the groove for optical fiber self-alignment and a 125-μm long suspended taper. We observed the curved part of the groove obtained by isotropic silicon etching and the straight part obtained by anisotropic etching. Also visible below the taper tip and on both sides, we can see the stops that prevent the optical fiber from hitting the taper. **Inset**: Cross-section view of the groove at the edge of the SOI chip.

The silicon tapers and waveguides are fabricated with two electron lithographies employing Raith Electron Beam Pattern Generator (EBPG) 5200 at 100 keV, followed by ICP-DRIE etching of the top silicon layer (220 nm) using a $SF_6 : C_4F_8$ gas mixture. We perform an optical lithography aligned with the previously fabricated integrated circuit to fabricate the grooves. After that, we carry out the groove ICP-DRIE etching in three successive steps using three different etching recipes. An initial anisotropic etching of the 2-μm thick silicon dioxide layer is performed; then an isotropic etching of the silicon bulk ($SF_6 : O_2$) followed by anisotropic silicon etching ($SF_6 : C_4F_8$) to obtain the particular shape of the groove that we can see in Fig. 3. Finally, we remove the photoresist and etch the silica under the taper using vapor hydrofluoric (HF) acid to suspend it. We have calibrated the etching process time to obtain structures corresponding to the design presented in Sect. 2. The horizontal alignment of the fiber depends on the optical lithography alignment. The vertical alignment depends on the etch depth of the silicon. These alignments can be achieved with less than 1-μm error for the first and less than 200 nm for the second.

Figure 3 shows a scanning electron microscope (SEM) image of the fabricated coupling structure. On the right, at the edge of the taper, is seen the groove that allows the optical fiber self-alignment. A cross-section view of the groove is shown in the inset. We can see the two supporting points on the sides that support the fiber vertically and horizontally. In the center of Fig. 3, we can see the 125-μm long suspended taper supported at its base by a weakly etched silicon triangle transitioning between a fully etched waveguide and a shallowly etched one (in the very left of the figure) [5]. A stop point under the taper tip, separated by a height of 29 μm, prevents the fiber from advancing too far forward.

3 Results

To measure our structure's coupling and propagation losses, we have fabricated a series of waveguides of different lengths (1210, 1710, and 5556 μm) with the same taper shape (185 nm tip width and 125 um length). We insert roughly SMF28e cleaved fibers into the grooves, which self-align with the waveguides. Each waveguide has two tapers at its ends and two grooves to receive the fibers used to inject light into the waveguide on one side and collect it on the other. We then sweep the laser wavelength from 1480 to 1640 nm and record the transmitted power.

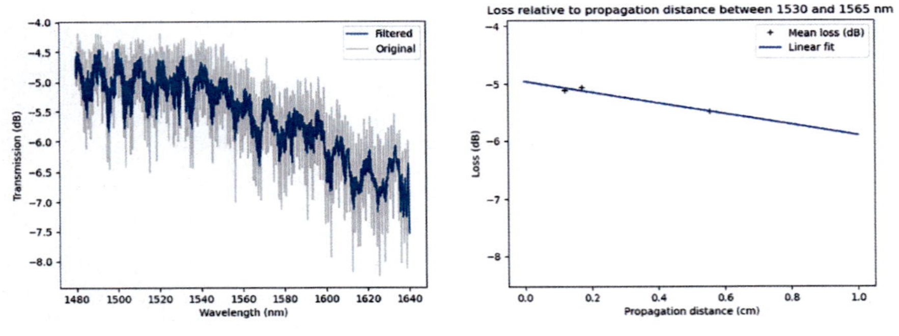

Fig. 4. (a) Waveguide transmission normalized to fiber-to-fiber transmission as a function of wavelength. (b) Insertion loss in dB as a function of propagation distance between 1530 and 1565 nm wavelengths.

Figure 4 (a) shows the transmission normalized to a fiber-to-fiber transmission (grey curve) and then filtered to eliminate setup-induced fluctuations (blue curve). The transmission is relatively flat between 1480 and 1550 nm and decreases slightly beyond that. The transmission is broadband, as there is no cut-off between 1480 and 1640. Figure 4 (b) shows the average losses as a function of the propagation distance between 1530 and 1565 nm wavelengths. The black cross points in the figure indicate the mean losses of three waveguides with different lengths. We fit these three points to obtain the blue curve representing the mean losses relative to the propagation distance. When we set the propagation distance to 0 and divide it by 2, we obtain the coupling losses per facet. The coupling losses are of the order of 2.5 dB per facet, and the propagation losses are 0.95 dB per cm. These coupling losses are relatively satisfactory considering what has been demonstrated in the literature lately: T. Barwicz et al. achieved 0.7 dB losses using silicon tapers with subwavelength periodic features [1] and J. He et al. obtained 2 dB coupling losses with SiN linear tapers [6]. These losses are expected to decrease further with a longer or better-designed taper.

4 Conclusions

We have designed and fabricated a self-aligned coupling structure between a cleaved optical fiber and an integrated silicon photonic circuit. This coupling structure contains a groove with an original shape formed by a two-step ICP-DRIE etching (isotropic and anisotropic) independent of the direction of the silicon crystal axes. The intersection of the isotropic and anisotropic etching profiles forms two supporting lines that hold the optical fiber. Once deposited in the groove, the optical fiber self-aligns with the taper. As proof of concept, we used a suspended linear taper to measure the coupling losses, achieving around 2.5 dB per facet. The advantages of our proposed structure lie in the design and the manufacturing process, which is compatible with the industrial microfabrication process and avoids difficulties related to wet etching with KOH or TMAH. Looking ahead, we plan to improve the coupling losses of our structure through a better taper design. The taper can also be encapsulated in silicon dioxide or polymer for more robustness.

Acknowledgments. This work has been partly funded by the ANR OFELIA (ANR-21-CE24-007). This work was done within the C2N micro nanotechnologies platforms and partly supported by the RENATECH network and the General Council of Essonne. Imene SI HADJ MOHAND thanks the EOBE doctoral school for her scholarship.

References

1. Barwicz, T., et al.: integrated metamaterial interfaces for self-aligned fiber-to-chip coupling in volume manufacturing. IEEE J. Selected Topics Quant. Electron. **25**(3), 1–13 (2019), May-June Art no. 4700313 (2019)
2. Cohen, J.D., et al.: Optical coupling to nanoscale optomechanical cavities for near quantum-limited motion transduction. Opt. Express **21**, 11227–11236 (2013)
3. Guanghui, R., et al.: Study on inverse taper based mode transformer for low loss coupling between silicon wire waveguide and lensed fiber. Opt. Commun. **284**(19), 4782–4788 (2011)
4. Zheng, H., et al.: Optimized design for 2×106 ultra-high Q silicon photonic crystal cavities. Opt. Commun. **283**(21), pp. 4387–4391 (2010). ISSN 0030-4018

5. Han, Z., et al.: High-performance and power-efficient 2×2 optical switch on Silicon-on-Insulator. Opt. Express **23**, 24163–24170 (2015)
6. He, J., et al.: V-Groove assisted passive assembly of single-mode fibers to ultra-broadband polarization-insensitive edge couplers for silicon photonics. In: 45th European Conference on Optical Communication (ECOC 2019), Dublin, Ireland, pp. 1–4 (2019)
7. Herth, E., et al.: Screening of spherical moulds manufactured isotropically in plasma etching conditions. Contrib. Plasma Phys. **61**(7), 1–8 (2021)
8. Herth, E., et al.: Fast ultra-deep silicon cavities: toward isotropically etched spherical silicon molds using an ICP-DRIE. J. Vac. Sci. Technol., B **37**(2), 021206 (2019)
9. Horth, A., et al.: Ideal, constant-loss nanophotonic mode converter using a Lagrangian approach. Opt. Express **24**, 6680–6688 (2016)
10. Marchetti, R., et al.: Coupling strategies for silicon photonics integrated chips [Invited]. Photon. Res. **7**, 201–239 (2019)
11. Neumann, EG., The fundamental fiber mode. In: Single-Mode Fibers. Springer Series in Optical Sciences, vol. 57. Springer, Berlin, Heidelberg (1988). https://doi.org/10.1007/978-3-540-48173-7
12. Wang, P.-H., et al.: CMOS-compatible silicon etched U-grooves with groove-first fabrication for nanophotonic applications. IEEE Photonics Technol. Lett. **34**(22), 1230–1233 (2022)

Silicon Photonic Wafer-Scale Yield of Single Mode Resonator with Broadband DBR Mirrors

Arnab Goswami[1(✉)], Pratyasha Priyadarshini[1], Gan Yih Loong[2], Ng Chew Yan[2], Deleep Nair[1], Anjan Chakravorty[1], and Bijoy Krishna Das[1]

[1] Department of Electrical Engineering, Centre for Programmable Photonic Circuit and Systems (CPPICS), IIT Madras, Chennai 600036, India
ee17d011@smail.iitm.ac.in

[2] SilTerra Malaysia Sdn. Bhd, 09000 Kulim, Kedah, Malaysia

Abstract. We demonstrated compact DBR cavities with broadband DBR mirrors using SilTerra's 200 mm silicon photonics foundry process line. The wafer-scale yield has been analyzed in terms of resonance wavelength and quality factor. The standard deviation of the resonant wavelength and quality factor are found to be 3.735 nm and 2650 respectively.

Keywords: Silicon Photonics · Wafer Scale Analysis · Broadband DBR Mirrors

1 Introduction

Resonator devices like microring resonator (MRR), microdisk resonator (MDR), photonic crystal (PhC) and distributed Bragg reflector (DBR) based Fabry-Pérot resonator are useful for various photonic integrated circuit (PIC) applications. MRR based devices have been studied extensively for PIC design and fabrication in commercial silicon photonics foundries [1]. However, in our recent studies, we found that Fabry-Pérot resonator with broadband DBRs can be a suitable alternative for mitigating certain application specific limitations of an MRR [2]. A singly resonant compact DBR cavity with ultrabroad sideband suppressions can be designed for CMOS compatible foundry fabrication. Such devices can be potentially used for on-chip noise filtering from a pump laser source, especially for quantum photonic applications. Moreover, these type of devices can also be used for electro-optic modulators, switches, sensors and many other integrated photonic signal processing applications [1]. Though narrowband DBR structures have been studied earlier for wafer-scale fabrication using CMOS silicon photonic foundries [3], DBR-based resonators have not been investigated for wafer-scale manufacturing to the best of our knowledge. In our previously reported work [2], the apodized DBR resonators were patterned with e-beam lithography, hence the resolution of the fabricated structure could be obtained with high accuracy, especially in terms of obtaining the smallest corrugation width. However, the commercial foundry processes predominantly use DUV lithography for

J. Witzens et al. (Eds.): ECIO 2024, SPPHY 402, pp. 565–570, 2024.
https://doi.org/10.1007/978-3-031-63378-2_93

patterning large scale circuits aimed at high volume production. Hence, it is essential to analyze the fabrication reproducibility as well as the performance of such complex structures (apodized DBR) in a commercial foundry process. In this work, we present the fabrication yield of a singly resonant DBR cavity design with broad sideband suppression using SilTerra's 200 mm EMO1 silicon photonics process platform in 220 nm top silicon layer thickness (H) on 2.5 µm buried-oxide (BOX) thickness (t_{BOX}). The detailed device architecture along with design parameters of the proposed DBR cavity as well as fabrication yield in terms of resonance wavelength, Q value and insertion loss at the resonance have been discussed based on 25 dies from each of the 2 wafers.

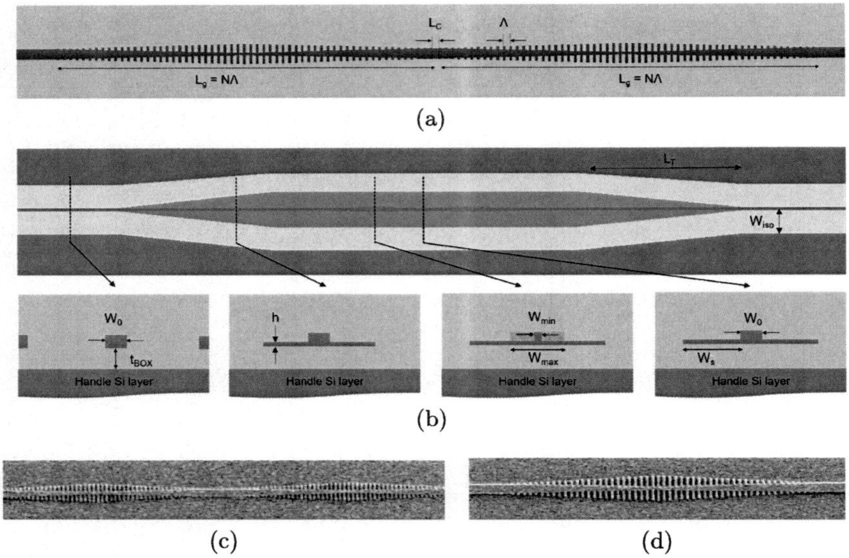

Fig. 1. (a) Schematic top view of the proposed device, (b) top view layout of the device and cross-sectional views at four important locations, (c) SEM image of the fabricated DBR cavity and (d) zoomed SEM image of one of the DBRs of the cavity. The important design parameters have been annotated.

2 Device Architecture and Design Parameters

In designing the device, we adhered to the approach outlined in [4], making adjustments to the design parameters in accordance with the fabrication process at SilTerra's silicon photonics foundry (193 nm Deep UV for waveguide fabrication). For device fabrication, we opted for a 70 nm slab thickness (h) out of two available options (70 nm and 150 nm). This choice was made to facilitate future integration of a p-n junction for active device control. The devices are integrated in between strip input/output waveguides (h = 0 nm), with a taper length (L_T) of 30 µm. Figure 1(a) depicts a top-view schematic of the devices, while Fig. 1(b) illustrates the layout of the devices for fabrication, providing cross-sectional views at different sections of the resonator. Critical design parameters are emphasized in Table 1.

Table 1. The values of design parameters annotated in Fig. 1, which are used for device fabrication

t_{BOX}	W_0	W_{min}	W_{max}	h	Λ	N	L_C	L_T	W_{iso}
2.5 µm	500 nm	100 nm	1 µm, 1.5 µm	70 nm	312 nm	60,70	312 nm	30 µm	4 µm

The device parameters were carefully selected to achieve a broadband rejection spectrum of over 40 nm centered around 1540 nm. As has been discussed in [4], the maximum grating width (W_{max}) is one of the important parameters in deciding the maximum coupling co-efficient of the apodized DBR which, in turn, determines the DBR reflectivity, rejection bandwidth and cavity Q factor. Similarly, increasing the mirror length (L_g) or the number of gratings (N) at a fixed W_{max} and W_{min} enhances the Q factor due to an increase in the DBR mirror reflectivity, but increases the insertion loss of resonance transmission at the same time. The overall objective is to achieve high broadband rejection with minimum resonance loss with a reasonable Q factor at the cavity. To investigate how the number of gratings and maximum width impact the device response, four devices corresponding to two sets of N (60 & 70) and W_{max} (1 µm & 1.5 µm) values have been fabricated. Figure 1(c) and 1(d) show the SEM images of the device (W_{max} = 1 µm and N = 60). As expected, the rectangular edges of the gratings have undergone slight rounding during the fabrication process. Nonetheless, the SEM images confirm the effective fabrication of the apodized DBRs.

3 Characterization Results

Grating couplers have been integrated at both ends of the devices to facilitate fiber-optic interfacing. Additionally, reference straight waveguides have been included to analyze the wavelength dependency of the devices without grating couplers. The devices were characterized through an optical source spectrum analyzer (OSSA) with a resolution bandwidth of 0.8 pm (APEX AP2683A). Typical transmission spectra of the devices are plotted in Fig. 2 from the center of the wafer. Figures 2(a) and 2(b) show the response of the devices for W_{max} of 1 µm with N = 60 and N = 70, respectively. Due to the limited wavelength range of OSSA, the left stopband edge of the resonator is not visible. Both devices exhibit negligible insertion loss (<1 dB) at the cavity passband and a broad stopband (>40 nm) response. The responses of W_{max} of 1.5 µm devices are depicted in Fig. 2(c) and 2(d). We observe two resonances for N = 70 device. When W_{max} is increased from 1 µm to 1.5 µm, the insertion loss at the cavity passband increases along with the quality factor, as expected due to higher coupling strength from increased perturbations. As a result, the devices' stopbands also improve (>50 nm). Therefore, careful selection of design parameters is necessary to achieve specific figure of merit.

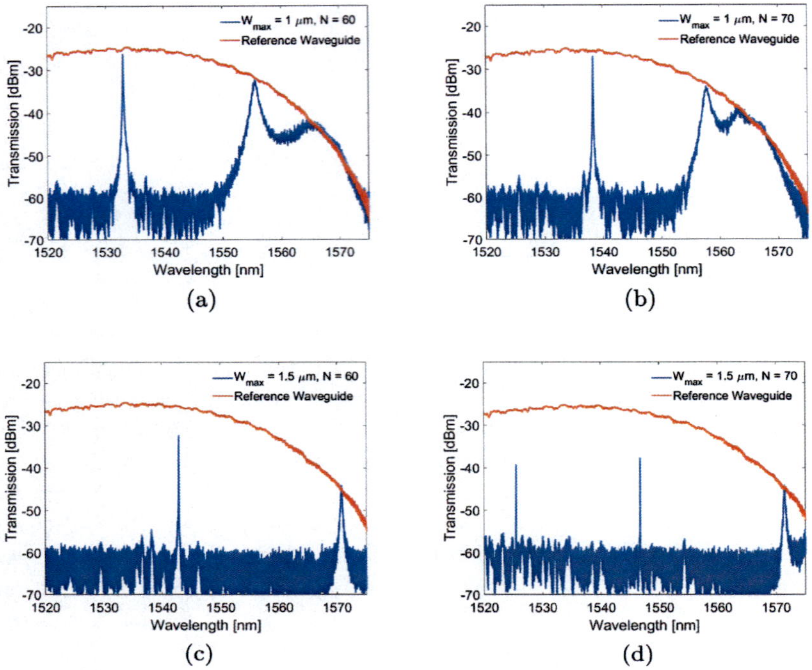

Fig. 2. Typical transmission characteristics of the fabricated devices along with the reference waveguide; (a) $W_{max} = 1\,\mu m$ and N = 60, (b) $W_{max} = 1\ \mu m$ and N = 70, (c) $W_{max} = 1.5\ \mu m$ and N = 60 and (d) $W_{max} = 1.5\ \mu m$ and N = 70.

4 Yield Analysis

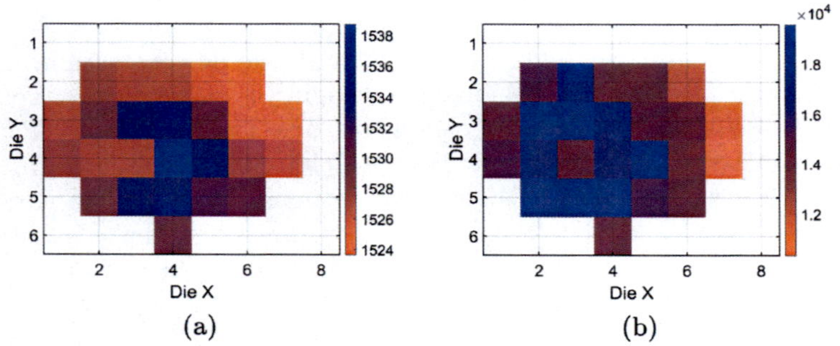

Fig. 3. (a) Resonant wavelength [nm] map across the Wafer#1 and (b) corresponding quality factor map across the Wafer#1.

We have presented the wafer scale yield analysis of the device with W_{max} of $1\,\mu m$ and N = 60. The position of the resonant wavelength varies across the wafer. Despite this variation in resonant wavelength across different dies, all devices exhibit negligible insertion loss (<1 dB) and a broad stopband (>40 nm), with out-of-band rejection exceeding 30 dB. The spatial distribution of the resonant wavelength and the quality factor are shown in Fig. 3(a) and Fig. 3(b) respectively. The standard deviation of the resonant wavelength is approximately 3.735 nm while that for the quality factor is about 2650.

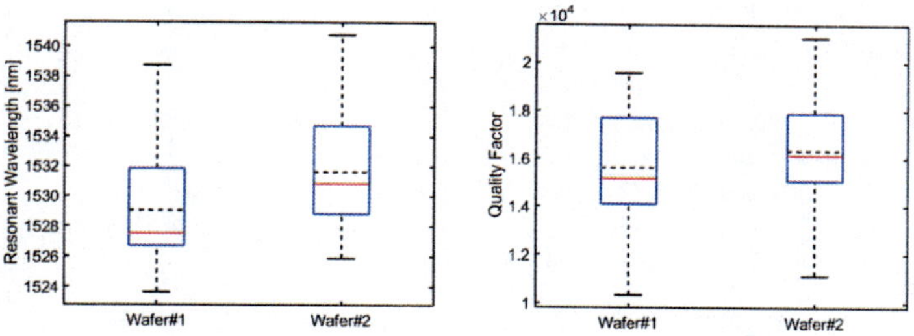

(a) (b)

Fig. 4. BOX plot of (a) resonant wavelength and (b) quality factor for two different wafers. Solid red lines and dashed black lines within BOX signify the median and mean respectively.

We present the distribution of resonant wavelengths for two separate wafers in Fig. 4(a). Wafer#1 exhibits a mean wavelength of 1529.056 nm, with a median of 1527.498 nm and an interquartile range of 4.712 nm. In contrast, Wafer#2 shows a mean wavelength of 1531.62 nm, a median of 1530.811 nm, and an interquartile range of 5.827 nm. In Fig. 4(b), we explore the quality factors of the same wafers. Wafer#1 demonstrates a mean quality factor of 15610, with a median of 15130 and an interquartile range of 3325. Wafer#2, on the other hand, showcases a mean quality factor of 16330, a median of 16112 and an interquartile range of 2615. These results underline the similarity in the performance of devices across wafers, indicating minimal variation within each wafer and good process yield.

5 Conclusion

We have studied wafer-scale fabrication yield of a novel compact resonator device design (operating in C band) using SilTerra's CMOS compatible silicon photonic fabrication process lines. The devices were analyzed in terms of resonant wavelength and quality factor across the wafer divided into 25 dies. The fabrication yield is reasonably good but there is a scope for improvement by fine tuning the design and process parameters. Additionally, thermo-optic tuning of the resonant wavelength can be enabled via the integration of microheaters. Nevertheless, these results are promising for the future inclusion of a DBR-based Fabry-Pérot resonator in the device libraries of commercial silicon photonics PDK, as various active and passive components for large-scale integration.

Acknowledgments. The authors gratefully acknowledge the sponsorship of MeitY, Govt. of India to establish silicon photonics CoE-CPPICS at IIT Madras.

References

1. Sun, P., et al.: Statistical behavioral models of silicon ring resonators at a commercial CMOS foundry. IEEE J. Sel. Top. Quant. Electron. **56**(2), 1–10 (2019)
2. Priyadarshini, P., et al.: Thermo-optically tunable DBR resonator with ultra-broad rejection band for silicon photonic applications. In: European Conference on Integrated Optics (ECIO 2023), 19–21 April 2023
3. Wang, X., Shi, W., Yun, H., Grist, S., Jaeger, N.A.F., Chrostowski, L.: Narrow-band waveguide Bragg gratings on SOI wafers with CMOS compatible fabrication process. Opt. Express **20**(14), 15547–15558 (2012)
4. Priyadarshini, P., et al.: Compact tunable resonance filters with ultra-broad rejection for silicon photonics. Authorea Preprints (2023)

Silicon Nitride C-Band Grating Coupler with Reduced Waveguide Back-Reflection Using Adaptively Corrected Elliptical Grates

Ibrahim Ghannam(✉), Florian Merget, and Jeremy Witzens

Institute of Integrated Photonics, RWTH Aachen University, 52074 Aachen, Germany
ighannam@iph.rwth-aachen.de

Abstract. We present experimental results for a fully etched C-band grating coupler with reduced back reflection fabricated in an 800 nm silicon nitride platform. Back-reflections are reduced by symmetrically interrupting the first few grates around the center axis of the propagating light. The span of the etched grates is gradually increased until they cover the full width. By interrupting the grates, light is reflected back obliquely, which leads to the excitation of higher-order modes that are scattered out of the structure. While this approach has been previously shown in silicon, it comes with a significant penalty in coupling efficiency of around 2.4 dB of extra loss in the layer stack investigated here. In this work, we present the design and measurement results of a grating coupler in which waveguide-to-waveguide back-reflections are suppressed by ~10 dB with this technique, while at the same time mitigating excess insertion losses by reshaping the grates as ellipses of varying eccentricity. This helps to compensate the phase front error induced by the interruption of the grates. This correction does not affect the level by which the back-reflection is suppressed, but reduces the insertion loss penalty from 2.4 dB to 1 dB.

Keywords: Light coupling · grating coupler · back-reflection suppression · silicon nitride

1 Introduction

Grating couplers (GCs) enable dense photonic integrated circuits (PICs) by allowing coupling from a single-mode fiber to anywhere on the chip's surface. This has led to the wide adoption of GCs as a preferred method to couple light into the silicon-on-insulator (SOI) and silicon nitride (SiN) platforms [1, 2], particularly as it comes to individual device characterization or accessing taps inside a system. More recently, there has been a focus on optimizing GCs to reduce waveguide-to-waveguide back-reflections [3–6], since these can cause parasitic laser feedback that can otherwise limit the usability of GCs in systems such as external cavity lasers (ECLs) and transceivers [7–9].

In this work, we focus on reducing the back-reflection that occurs at the boundary between the input taper of a focusing GC and the first few grating trenches. This type of back-reflection is particularly high in fully-etched GCs, which is a requirement in

J. Witzens et al. (Eds.): ECIO 2024, SPPHY 402, pp. 571–578, 2024.
https://doi.org/10.1007/978-3-031-63378-2_94

some open access foundry processes. While partial etches and apodization allow for a substantial reduction in back-reflection, increased requirements as for example required to obtain ECL linewidths in the sub-kHz range [10, 11] warrant here too the introduction of advanced design concepts.

Here, we present a continuation of our work in [6], showing the experimental realization of a fully etched C-band SiN GC featuring both suppressed back-reflection and a very moderate insertion loss penalty that was proposed and modeled there.

The reduction in waveguide-to-waveguide back-reflection is achieved by altering the field profile of the back-reflected light and exciting higher-order modes that are not supported by the single-mode input waveguide. This is achieved by gradually modifying the span of the first few grates, as also previously shown in the SOI platform [4, 5]. However, this also distorts the phase front of the field propagating through the grating coupler, after emission from the single mode waveguide, so that significant excess losses are induced [6]. To mitigate this, the shape of the grates is adapted on an individual basis to compensate for the incurred phase error. This modification does not have a negative effect on the back-reflection reduction.

The devices are implemented in the 800 nm nitride thickness Ligentec platform. The reference GC, prior to any of these modifications, has an insertion loss of 7.8 dB. Modification of the grate span suppresses back-reflections by ~10 dB, but adds an additional 2.4 dB of losses. After modification of the grates into elliptical shapes with grate-dependent eccentricity, these excess losses drop to 1 dB.

2 Grating Coupler Design

The design of the reference GC is shown in Fig. 1(a). The device has a pitch of 1.17 μm with a fill factor of 44.87%, defined as the ratio of the width of the unetched section to the total pitch. The length of the input taper is 30 μm, spanning from the input waveguide to the onset of the grates. The taper's full angle is 25.6 degrees. The input waveguide only supports a single transverse electric (TE) ground mode and has a width of 600 nm. The light is coupled in and out through a fiber array with an 11° fiber polish and the GC is operated as a forward-coupling device. The reference device provides a baseline against which to compare the performance of the modified devices.

Figure 1b, 1c shows the designs modified for reducing the back-reflection. In both devices, the span of the trenches in the first three grates, that are interrupted in the center of the beam propagating inside the slab, is increased gradually. The first trench covers an angle ϕ on both sides of the beam, the second trench an angle 2ϕ, and so on until the trenches fill the entire span. This essentially forms a segmented boundary that reflects light at an odd angle that cannot reenter the single mode waveguide.

Device B has an identical pitch and fill factor to device A, so that the out-coupling angle is essentially maintained. However, the missing grates in the center of the device result in a perturbation of the phase front, that is inherited by the emitted beam and increases the insertion losses. A second excess loss mechanism results from the scattered field simply missing in the center region of the truncated grates. While the latter cannot be helped without compromising the mechanism leading to the back-reflection reduction, the phase front distortion can be accommodated by modifying the grate shape accordingly.

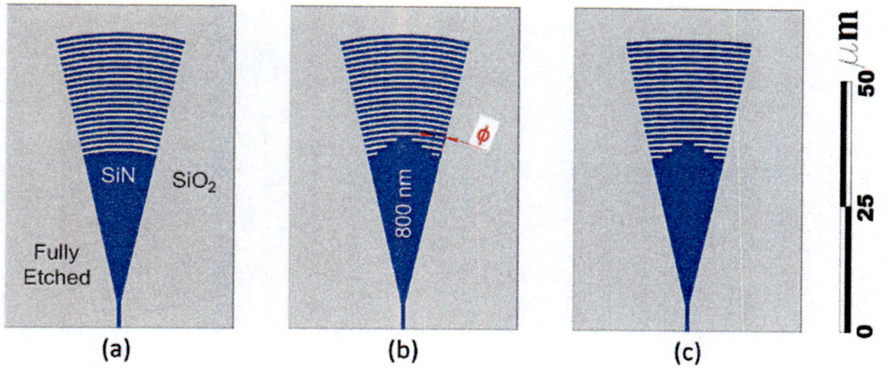

Fig. 1. Layouts of (a) the reference GC with circular grates ("Device A"), (b) the modified GC with three interrupted grates but otherwise identical geometry ("Device B"), and (c) the modified GC in which the grates are further converted into ellipses of varying eccentricity ("Device C"). In b and c, the span of the trenches in the first three grates is controlled by the angle ϕ and the grate index. The trench spans, from the edge of the taper, over an angle given by the grate index multiplied by ϕ (on both sides). ϕ is set to 3.65 degrees in both b and c.

Figure 2 illustrates the design flow by which the modified grate eccentricity is obtained. The correction is performed in two steps. First, the grates are segmented into seven angular sectors (2 times the number of interrupted grates + 1) as shown in Fig. 2(a). In each sector, grates are moved towards the input beam by an increment *d* for each left-out grate. This displacement *d* is calculated using the equation

$$\frac{2\pi n_{sl}}{\lambda_0}d = \frac{2\pi(n_{sl} - n_{ox})}{\lambda_0}w + \frac{2\pi n_{ox}}{\lambda_0}sin(\theta)d \tag{1}$$

with n_{sl} the effective index of the SiN slab (1.88), λ_0 the free-space wavelength (1550 nm), n_{ox} the refractive index of SiO_2 (1.45), w the width of the trenches (645 nm), and θ the angle of the scattered beam in SiO_2 (11°). After algebraic manipulation, the equation can be rewritten as

$$d = \frac{n_{sl} - n_{ox}}{n_{sl} - n_{ox}sin(\theta)}w \tag{2}$$

This modification is targeted to scatter the field with the correct phase for obtaining a beam emission with a flat free-space wave-front. However, the abrupt transitions at the interfaces between the angular sectors leads to scattering of the field, so that smoothing of the resulting geometry is required.

In a second step, the segmented grates are replaced by elliptical grates that best match the segments. This is done by modifying the eccentricity of the ellipses, which vary between 0.85 and 0.64 for the first 20 grates. This also allows the device to be in compliance with fabrication restrictions. The final geometry is shown in Fig. 2(b).

A series of simulations were performed to illustrate the devices' functionality further. Figure 3 shows the intensity of the back reflected field as it reaches the interface between the input waveguide and the taper. The data is generated using 3D finite difference time

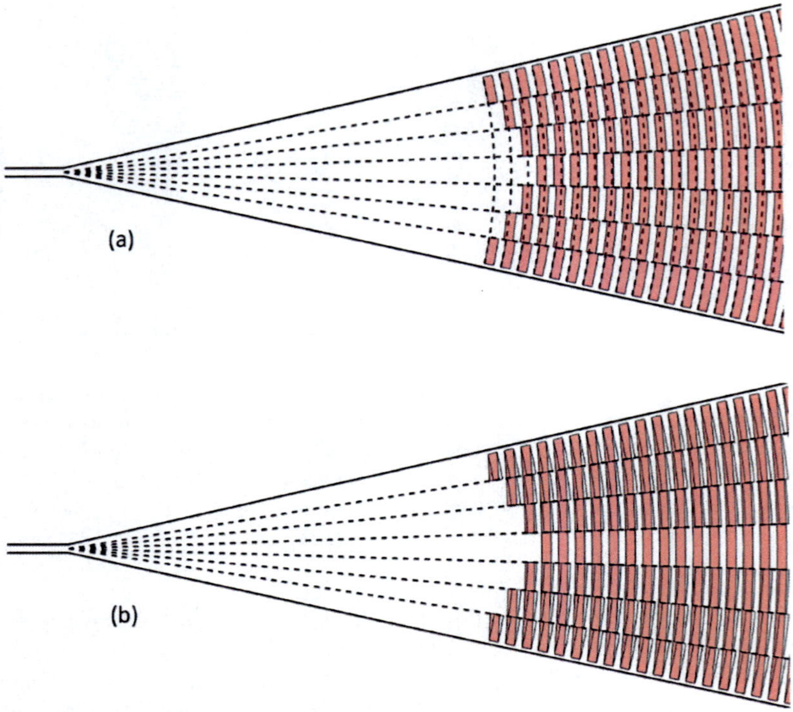

Fig. 2. (a) First approximation of the modification required for accommodating the phase front distortion. Grates in each sector are shifted towards the direction of the input waveguide to compensate for the increased phase accretion in the angular sectors with unetched regions. The dashed curves correspond to the original grates (identical in device B and the reference GC) and are plotted to visualize the displacement. (b) Grates after smoothing by fitting to an ellipse. Here, the outline of the segmented grates is overlaid for comparison.

domain (FDTD) simulations. A packet of light centered around 1550 nm is launched from a mode source into the waveguide on the left side of Fig. 3. The field is recorded with a monitor that takes a snapshot of the intensity along the center plane of the SiN film at a time at which the reflected field reaches again the input waveguide. For the reference device, most of the light is reflected back into the ground mode of the input waveguide. For devices B and C, light is reflected obliquely due to the modification of the first three grates. This leads to the excitation of higher-order modes that are not supported by the input waveguide and couple out, reducing the intensity of the waveguide back-reflection.

3 Measurement Results

The devices were characterized and evaluated in a similar way to [6]. Test structures consist of two GCs connected by a waveguide loop with a 184 μm length. One GC was used as an input, and the other as output. A tunable laser source was used to obtain the wavelength dependent transfer function. Besides the insertion losses, the waveguide-to-waveguide

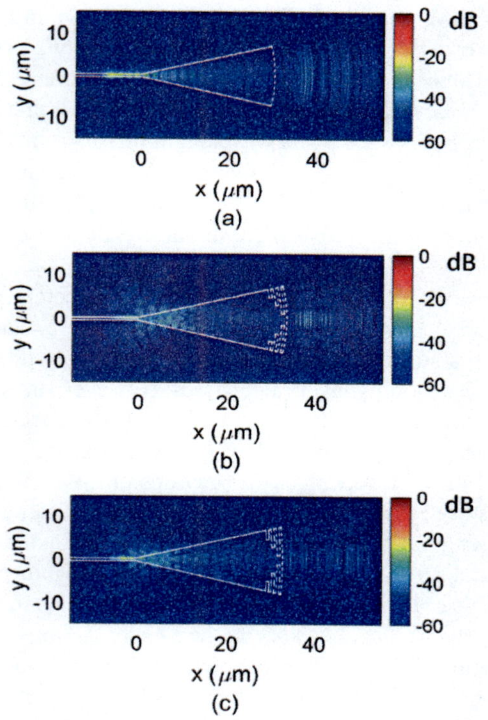

Fig. 3. Intensity of back-reflected light for (a) the reference device, (b) device B, and (c) device C, recorded at the center of the SiN film from an FDTD simulation at a time step at which the back-reflected light has again reached the input waveguide (identical time step for all panels).

back-reflection can be extracted by analyzing the ripples in the device response, wherein Fourier filtering is used to reject ripples arising from reflections occurring elsewhere in the setup.

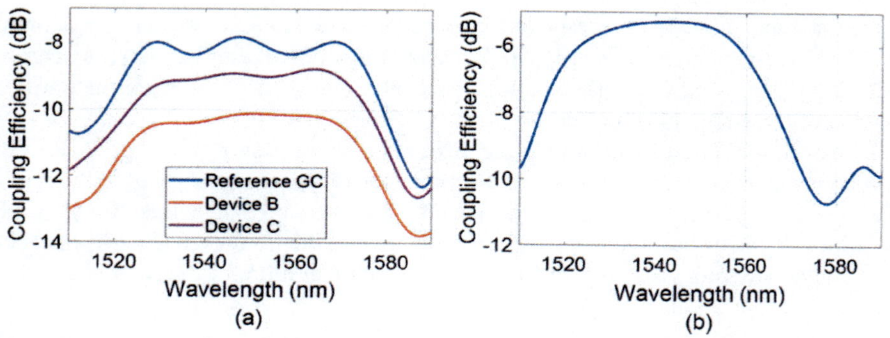

Fig. 4. (a) Smoothed coupling efficiency of the three devices. (b) Smoothed coupling efficiency of the reference device from an earlier run.

The smoothed coupling efficiency of a single GC is shown in Fig. 4(a) for all three devices. The losses recorded from the loops were divided by two to extract the insertion loss for a single GC. The best coupling efficiency of the reference device and device B are −7.8 dB and −10.2 dB, respectively, obtained at a wavelength of 1550 nm. Device C has a coupling efficiency of −8.8 dB at a wavelength of 1564 nm. The 3-dB bandwidth is around 65 nm for all three devices. The large ripples that can be seen in Fig. 4(a) correspond to a length that matches that of the GC input taper (30 μm), which indicates multiple reflections between its interfaces. These can already be seen to be less prominent in the modified GCs.

In an earlier run, the reference device and device B had better insertion losses of −5.4 dB and −8.1 dB, respectively, 2.1–2.4 dB better than here. The cause of this drop in coupling efficiency is still under investigation, but is likely due to a change in the under-cladding and over-cladding layer thicknesses. However, the relative improvement in the performance of device C relative to device B matches expectations very well. The data from previous runs gives reason to believe that the insertion losses of device C can be improved to ~−6.4 dB by reverting to the previous process conditions.

Fourier filtering is used to extract the ripples with the wavelength periodicity corresponding to the loop length. Considering a path length of 248 μm including the waveguide path and tapers and a group index of 2.1, we expect the ripples' free spectral range (FSR) to be 2.3 nm. The optical spectra are taken with a 1 pm resolution (80,000 points between 1510 and 1590 nm). For such spectra, an FSR of 2.3 nm corresponds to a spectral peak at the data point of index 35 after applying a discrete Fourier transform (DFT), which can be observed in Fig. 5(b). It can be seen that this peak is very significantly attenuated for devices B and C. The second peak is caused by the measurement setup, as it was present in different measurements at the same data point index regardless of the length of the loop on the SiN chip. The peak around index 35 is filtered by removing the background spectrum and transformed back with an inverse Fourier transform. This results in the ripples shown in Fig. 5(c). To convert these into a power back-reflection coefficient *R*, the peak-to-peak ripple values are used in the following equation [6, 12]

$$R = \frac{\sqrt{ER} - 1}{\sqrt{ER} + 1} \tag{3}$$

where *ER* stands for extinction ratio and corresponds to the ratio between the maximum and minimum transmission recorded in the transfer function of the GC loop (i.e., before dividing the losses by two). The ripples represented in Fig. 5(c) are from the transfer function of the whole loop and can be directly plugged into Eq. (3).

The back-reflection spectrum for the different devices is shown in Fig. 5(d). Both Devices B and C achieve significant reduction in back reflection compared to the reference device, in the range of ~10 dB over a bandwidth of 60 nm that includes the C-band. In addition, it can be observed that device C outperforms device B even in terms of back-reflection suppression, which is attributed to the fact that more of the light is coupled out.

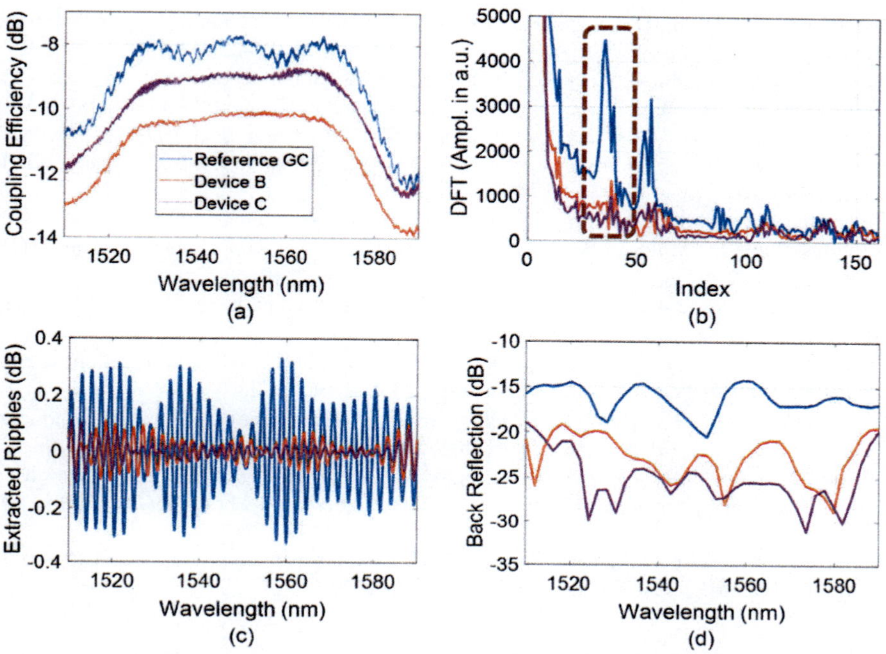

Fig. 5. (a) Raw coupling efficiency of a single GC, (b) DFT applied to the GC transfer function in dB, (c) extracted ripples after filtering and the inverse DFT, with results normalized for the whole loop (i.e., multiplied back by two relative to panel (a)), and (d) extracted back-reflection (color coding identical in all panels).

4 Conclusion

We present a fully etched SiN grating coupler with a reduced back-reflection based on truncated grates. The reduction in back-reflection usually comes at the expense of a reduced coupling efficiency. To counter this and reduce the insertion loss caused by the truncation of the grates, we adapt the grate geometry in a way that compensates for phase errors. The new grating coupler has elliptical grates of varying eccentricity. ~10 dB of back-reflection suppression is obtained across the 60 nm 3-dB bandwidth while reducing excess insertion losses from 2.4 dB to 1 dB.

References

1. Marchetti, R., Lacava, C., Carroll, L., Gradkowski, K., Minzioni, P.: Coupling strategies for silicon photonics integrated chips. Photon. Res. **7**, 201–239 (2019)
2. Romero-García, S., Merget, F., Zhong, F., Finkelstein, H., Witzens, J.: Silicon nitride CMOS-compatible platforms for integrated photonics applications at visible wavelengths. Opt. Expr. **21**(12), 14036–14046 (2013)
3. Benedikovic, D., et al.: L-shaped fiber-chip grating couplers with high directionality and low reflectivity fabricated with deep-UV lithography. Opt. Lett. **42**, 3439–3442 (2017)

4. Song, J.H., Snyder, B., Lodewijks, K., Jansen, R., Rottenberg, X.: Grating coupler design for reduced back reflections. IEEE Photon. Technol. Lett. **30**(2), 217–220 (2018)
5. Zou, J., Zhang, Y., Hu, J., Wang, C., Zhang, M., Le, Z.: Grating coupler with reduced back reflection using $\lambda/4$ offset at its grating sub-teeth. J. Lightwave Technol. **37**, 1195–1199 (2019)
6. Ghannam, I., Merget, F., Witzens, J.: Silicon nitride C-band grating couplers with reduced waveguide back-reflection. In: Proceedings of the SPIE 12007, Optical Interconnects XXII, 120070N (2022)
7. Ghannam, I., Shen, B., Merget, F., Witzens, J.: Silicon nitride external cavity laser with alignment tolerant Multi-Mode RSOA-to-PIC interface. J. Sel. Top. Quant. Electron. **28**(1), 1501710 (2022)
8. Tran, M.A., et al.: Extending the spectrum of fully integrated photonics. arXiv:2112.02923 (2021)
9. Shang, C., et al.: Perspectives on advances in quantum dot lasers and integration with Si photonic integrated circuits. ACS Photon. **8**(9), 2555–2566 (2021)
10. Xiang, C., Morton, P.A., Bowers, J.E.: Ultra-narrow linewidth laser based on a semiconductor gain chip and extended Si_3N_4 Bragg grating. Opt. Lett. **44**(15), 3825–3828 (2019)
11. Maier, P., et al.: Sub-kHz-linewidth External-Cavity Laser (ECL) with Si_3N_4 resonator used as a tunable pump for a Kerr frequency comb. J. Lightwave Technol. **41**(11), 3479–3490 (2023)
12. Shen, B., et al.: Reconfigurable frequency-selective resonance splitting in chalcogenide microring resonators. ACS Photon. **7**(2), 499–511 (2020)

Spot Size Converters for Enhanced Coupling Efficiency Between Single Mode Fibers and High-Confinement Si_3N_4 Waveguides

Klara Mihov[1(✉)], Aron Elias Lutz[2], Moritz Kleinert[1], Martin Kresse[1], Daniel Preuß[1], Jakob Reck[1], David de Felipe[1], Tianwen Qian[1], Madeleine Weigel[1], Csongor Keuer[1], Philipp Winklhofer[1], Crispin Zawadzki[1], Norbert Keil[1], and Martin Schell[1,2]

[1] Fraunhofer Institute for Telecommunications, Heinrich Hertz Institute, Einsteinufer 37, 10587 Berlin, Germany
klara.mihov@hhi.fraunhofer.de
[2] Technische Universität Berlin, Straße des 17. Juni 135, 10623 Berlin, Germany

Abstract. For highly loss-sensitive applications of photonic integrated circuits (PICs), efficient coupling of high-confinement silicon nitride (Si_3N_4) passive waveguides with active optoelectronic components and standard single mode fibers (SMF) is important. To achieve this, width and height of the waveguide are varied to enable polarization-independent low-loss coupling. A mode-field converter is realized by focused ion beam etching, and a proof of concept is provided with three different prototypes that differ in their final spot size and are fabricated by focused ion beam milling. For these prototypes, a broadened mode field diameter and a more efficient coupling to a SMF are demonstrated.

Keywords: edge coupling · silicon nitride · spot size converter

1 Introduction

Over the last two decades, silicon nitride (Si_3N_4) has emerged as an important platform for photonic integration alongside III-V and silicon on insulator (SOI) technologies [1, 2]. Its performance outperforms others, especially in areas such as ultra-high Q resonators, high coherent laser resonators, nonlinear applications such as frequency comb generators and on-chip biophotonics [1–5]. Si_3N_4 also enables the development of low-loss waveguides over a broad spectral range from about 400 nm to 2350 nm [1]. However, since Si_3N_4 is a passive material, hybrid or heterogeneous integration with active materials or components is required for many applications [1]. Therefore, effective coupling of Si_3N_4 waveguides with high confinement to other waveguiding technologies is crucial. In addition, efficient coupling with single-mode fiber (SMF) is essential. Two primary coupling methods are commonly used: vertical coupling utilizing grating structures, and edge coupling. In general, edge coupling with a broadened spot size provides a wider spectral bandwidth and enables lower losses [6]. However, the large differences in mode sizes between high-confinement Si_3N_4 waveguides and SMF pose a challenge for chip-to-fiber coupling.

J. Witzens et al. (Eds.): ECIO 2024, SPPHY 402, pp. 579–584, 2024.
https://doi.org/10.1007/978-3-031-63378-2_95

LIGENTEC and LioniX are current providers of spot-size converters (SSC) for coupling Si_3N_4 to an SMF. They achieve coupling losses of <1 dB per facet and <0.5 dB per facet for TE polarization, respectively, whereby the LioniX platform is mainly designed for TE polarization [7, 8]. For some applications, however, it is necessary to use waveguides with polarization-independent low losses and thus also polarization-independent SSCs. Therefore, a polarization-independent taper is needed.

In this work, a polarization-independent SSC for Si_3N_4 waveguides with an initial cross section of 800 nm by 800 nm is investigated. In the following, a design for such a taper is shown, which can be used for low-loss coupling to a SMF, but also for other materials with a mode field diameter (MFD) different from that of Si_3N_4 waveguides.

2 Design and Fabrication

Efficient coupling relies on the similarity of the modes of the two components to be coupled. In the following, a method for effective coupling of an 800 nm x 800 nm Si_3N_4 waveguide with a SMF is presented. The initial mode field diameter (MFD) of the Si_3N_4 waveguides is ~ 1 μm in both axes, while the MFD of a single-mode fiber is ~ 10 μm. For efficient coupling, the modal overlap η between waveguide and SMF given in Eq. 1 must be maximized.

$$\eta = \frac{|\int_C E_{WG} E_{SMF} dA|^2}{\int_C |E_{WG}|^2 dA \int_C |E_{SMF}|^2 dA} \tag{1}$$

where E_{WG} and E_{SMF} stand for the complex electrical field amplitudes of the waveguide and the SMF, and the integration is carried out using the cross-sectional area C. Lumerical MODE from Ansys is used to determine the design parameters for maximum overlap. The modal analysis and overlap calculations are performed using a finite difference eigenmode solver (FDE). The simulation model includes the Si_3N_4 waveguide, the silica cladding, the Si substrate and the surrounding air. All simulations are carried out at a wavelength of $\lambda = 1550$ nm.

The simplest SSC design involves only a change in the waveguide width, known as a lateral taper. This adjustment can be easily integrated into the standard wafer fabrication process using either stepper lithography or e-beam lithography, depending on the target width. The overlap of the fundamental TE modes of the waveguide and SMF exhibits two peaks at waveguide widths of about 104 nm and 13.7 μm. However, above a width of 13.7 μm, the waveguides become multi-modal, and a long taper is required for an adiabatic mode transition. At a width of 104 nm, the waveguide exhibits low confinement enabling large mode field diameters MFDs both laterally and vertically, resulting in minimal coupling losses of 0.62 dB for TE polarization. Nevertheless, this approach leads to a significant asymmetry in the width and height of the waveguide and the MFDs, so that the overlap is strongly polarization-dependent and polarization-dependent losses of around 1.5 dB are to be expected.

Hence, in this work we adjust not only the width but also the height of the waveguide. Figure 1.a) illustrates the minimum overlap, as defined in Eq. 1, obtained for all simulated widths and heights in TE and TM polarization with the SMF. When the widths and heights are below 100 nm, the waveguide mode becomes larger than the SMF mode, reducing the overlap. The dashed line represents the 1 dB limit. It is observed that the overlap is greatest for square waveguide cross-sections. With a square waveguide cross-section, the minimal coupling losses of 0.61 dB is achieved with a waveguide width and height of 242 nm. A sketch of this simulated taper, including a pre-taper, is shown in Fig. 1.b).

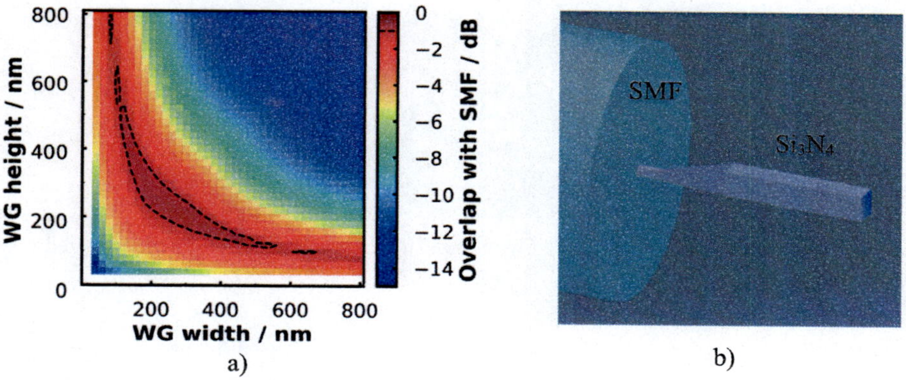

Fig. 1. a) The minimum overlaps for TE and TM polarization as a function of waveguide width and height. The dashed line shows the 1 dB limit. In b) a sketch of the Si_3N_4 taper is shown.

A combination of different lithography processes, such as stepper, e-beam and contact lithography, is required to produce the simulated SSC. The height and width variations of the SSC are achieved with a shifting mask and with e-beam lithography, respectively. The remaining structures are produced with the stepper lithography.

3 Evaluation and Results

For an initial validation of the simulations, various tapers were fabricated using FIB (Focused Ion Beam) milling. Waveguides with a height of 400 nm and a width of 800 nm were used as initial structures. An example of a prototype is shown in Fig. 2. The waveguide widths can be tapered down to around 60 nm.

Due to the limitations in FIB milling, the initial tapering process cannot precisely control the vertical and lateral dimensions during the process, resulting in a large variation in the cross-sections obtained. The prototypes differ in their length, limited to 125 μm, and estimated cross-sections at the facet. A summary is given in Table 1.

The functionality of the tapers is evaluated for wavelengths in the range from 1300 to 1600 nm (O-, E-, S-, C- and L-bands). The insertion loss measurements for the different prototypes are shown in Fig. 3, where measured transmission is averaged over a range of ±5 nm for each data point. The waveguides comprise a spot-size converter at one facet, where a SMF is coupled, and no spot size converter at the other facet, where a lensed-fiber with 2 μm MFD is coupled. A waveguide without SSC has been also included and characterized in order to serve as a reference. It can be seen that the high insertion losses observed around 1390 nm and 1520 nm are due to absorption by O-H and N-H bonds,

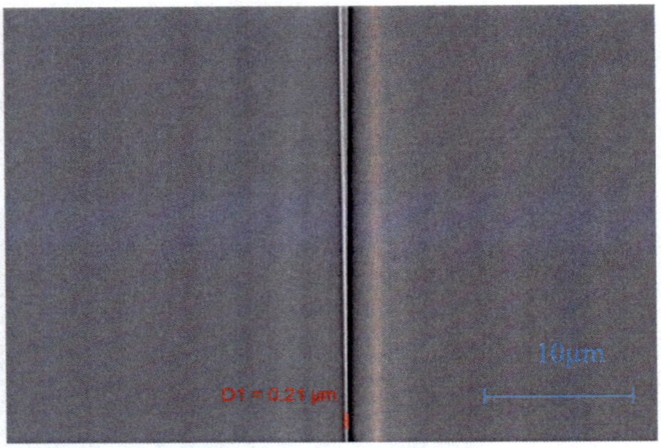

Fig. 2. An example of a taper produced with the FIB is shown. The waveguide is tapered to a width of 210 nm.

Table 1. Details for the final size of the waveguide for the three different prototypes.

Prototype	Taper length [μm]	Width [nm]	Height [nm]
1	25	340–360	190–210
2	75	120–140	90–110
3	125	50–70	60–80

respectively. In order to break the N-H bonds and thus minimize the insertion losses, an annealing will be used in our upcoming generations of Si_3N_4 chips.

It is evident that using any of the three taper types for coupling with a cleaved SMF reduces the insertion loss by several dB compared to using no taper. This improvement is consistent across the entire recorded spectrum. The expected mode sizes increase from prototype 1 to 3, with the restriction that the mode size of prototype 1 is smaller than the SMF mode size, while prototype 3 exceeds it. For all prototypes, the coupling losses per facet are measured for TE mode. Without any tapering, coupling losses to the SMF are found to be 10.1 dB per facet. When using prototype 1, the coupling losses can be reduced to 6.1 dB, with prototype 2 even to 4.2 dB. With prototype 3 coupling losses of 5.2 dB per facet were measured. The deviation from the simulations is caused by the fact that the tapers produced with the FIB are clearly too short for efficient coupling to the SMF. In addition, it should be noted that the cross-sections achieved do not correspond to the simulated dimensions. However, it was possible to investigate the taper at the facets with regard to mode expansion using a short-wave infrared camera to obtain information about the size and shape of the modes. A mode broadening can be observed for all prototypes. Due to the asymmetrical cross-sections at the end of the taper, the MFDs differ depending on the direction. Differences in MFD can be observed in the wafer plane and perpendicular to it. With prototype 2 we achieve the best results, i.e., a simulated and measured MFD in wafer plane of ~ 12 μm. For prototypes 1 and 3,

MFD in the wafer plane of ~ 5 μm and ~ 20 μm, resp., are obtained. From this we can conclude that the prototypes produced with the FIB are promising.

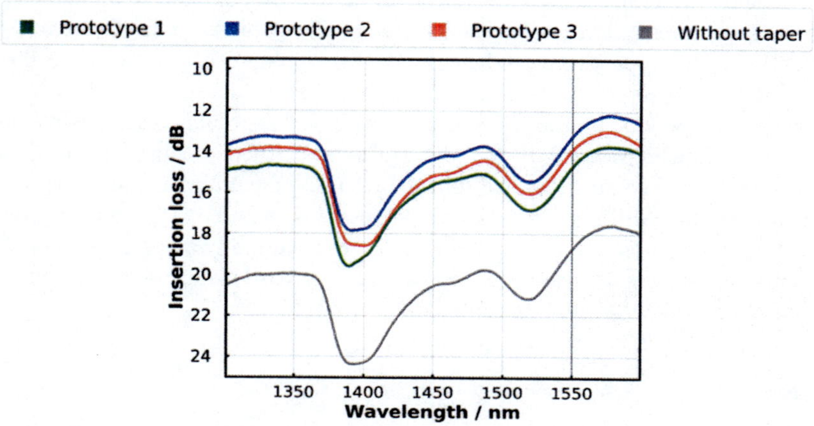

Fig. 3. The insertion loss of 3 cm long spirals is measured using either a different prototype or w/o a taper in combination with a SMF. All measurements are performed in TE polarization. The different dimensions of the prototypes are listed in Table 1.

As already mentioned, three different lithography processes are required for tapers with well-defined dimensions of around 240 nm in height and width and longer taper lengths. To realize structures around 240 nm, e-beam lithography is required instead of stepper lithography. For the fabrication of a vertical taper, contact lithography is used in combination with a shifting or displacement mask. All three steps were tested separately on Si_3N_4. So far, e-beam and stepper lithography have been tested as a combination and are used as standard for grating couplers in the visible range, where gratings with periods smaller than 500 nm are required. The combination of contact lithography and stepper lithography has also been successfully tested.

Due to the promising results of the preliminary tests, a combination of these three types of lithography is currently being worked on and the process development for Si_3N_4 waveguides with the dimensions 800 nm by 800 nm is being carried out as well.

The fabrication of a polarization-independent taper is an important building block for the Si_3N_4 platform. Other suppliers offer polarization-independent SSCs with losses <1 dB per facet [7]. According to simulations, the reduction of the coupling losses is possible. Therefore, controlling the final dimensions of the Si_3N_4 taper and reducing the roughness due to etching are crucial. Further process development and the combination of lithography types will lead to an efficient edge coupling. Moreover, this type of taper is customizable, making it suitable for materials other than SMF simply by adjusting the final dimensions of the taper.

4 Conclusion and Outlook

To take advantage of the many benefits of Si_3N_4 PICs, the passive Si_3N_4 structures must be able to be coupled with optical fibers or active optoelectronic components. Our simulation shows that it is possible to fabricate an SSC with low coupling losses of 0.6 dB per facet. First prototypes were produced with FIB etching. They have shown that the widening the Si_3N_4 mode is possible and can reach more than 20 μm. The coupling losses with the tapers being too short and having not the simulated dimensions at the facets, can be reduced from 10 dB to 4 dB. For a more efficient coupling it is essential to control the fabrication process and thus the resulting taper dimensions. Therefore, a combination of three different lithography processes is required. The combination and optimization of the combination of all of three types is currently being developed. Once developed, this tapering process for Si_3N_4 SSCs can be easily adapted to different mode field diameters or materials to be coupled to.

References

1. Blumenthal, D.J., et al.: Silicon Nitride in Silicon Photonics. Proc. IEEE **106**(12), 2209–2231 (2018)
2. Shekhar, S., et al.: Roadmapping the next generation of silicon photonics. Nat. Commun. **15**(1), 751 (2024). https://doi.org/10.1038/s41467-024-44750-0
3. Luo, W., et al.: Recent progress in quantum photonic chips for quantum communication and internet. Light Sci. Appl. **12**(1), 175 (2023). https://doi.org/10.1038/s41377-023-01173-8
4. Pfeiffer, M.H.P., et al.: Photonic Damascene process for integrated high-Q microresonator based nonlinear photonics. Optica **3**(1), 20 (2016). https://doi.org/10.1364/OPTICA.3.000020
5. Kazanskiy, N.L., et al.: State-of-the-art optical devices for biomedical sensing applications—a review. Electronics **10**(8), 973 (2021). https://doi.org/10.3390/electronics10080973
6. Zhang, Y., et al.: High-efficiency and compact polarization-insensitive multi-segment linear silicon nitride edge coupler. Photonics **10**(5), 510 (2023). https://doi.org/10.3390/photonics10050510
7. LIGENTEC. https://www.ligentec.com/technology/. Accessed 26 Feb 2024
8. Roeloffzen, C.G.H., et al.: Low-loss Si_3N_4 TriPleX optical waveguides: technology and applications overview. IEEE J. Sel. Top. Quantum Electron. **24**(4), 4400321 (2018)

Highly Flexible Dielectric Platform for Post-CMOS Photonics

Marcus Westhues[1(✉)], Thomas Geruschke[1], Julia Hauser[1], Roman Burkard[2], Aleksandar Nesic[1], and Anna Lena Schall-Giesecke[1,2,3]

[1] Fraunhofer Institute for Microelectronic Circuits and Systems (IMS), Finkenstr. 61, 47057 Duisburg, Germany
marcus.westhues@ims.fraunhofer.de

[2] Department of Electronic Components and Circuits (EBS), University of Duisburg-Essen (UDE), Bismarckstr. 81, 47057 Duisburg, Germany

[3] Center for Nanointegration Duisburg-Essen (CENIDE), Carl-Benz-Strasse 199, 47057 Duisburg, Germany

Abstract. We present a photonic technological development service that includes an 8″ wafer-scale optical-lithography photonic platform customizable to application-specific needs. Our platform offers a choice of waveguide materials and deposition techniques. Low-temperature deposition techniques, in particular, allow for CMOS-compatibility and electronic-photonic co-integration (post-CMOS). We show results of experimental characterization of photonic waveguides and ring resonators fabricated in initial fabrication runs in silicon nitride (Si_3N_4 and Si_xN_y) and tantalum pentoxide (Ta_2O_5) layers. The results of transmission measurements in the C-band and at visible wavelengths indicate that performance of our devices is on par with those of established players in the field.

Keywords: integrated optics · post-CMOS · waveguides · visible wavelengths · optical lithography

1 Introduction

Photonic integrated circuits (PICs) have emerged as a promising solution for developing massively parallel architectures. While silicon-on-insulator (SOI) waveguides initially laid the foundation for PICs, the further development of silicon nitride photonics with a higher bandgap from the visible to the mid-infrared spectrum has added additional advantages [1]. For instance, silicon nitride offers a better ability to exploit Kerr nonlinearity, e.g., in generating frequency combs [2, 3]. An overview of recent developments in silicon nitride photonics can be found in Ref. [4].

These advanced technologies are available for research and commercial use through various fabs, e.g. AMF [5, 6], imec [7] or CORNERSTONE [8, 9]. Nevertheless, customization options for processes and devices are somewhat limited. To address this limitation, Fraunhofer IMS provides a photonic technological development service equipped

J. Witzens et al. (Eds.): ECIO 2024, SPPHY 402, pp. 585–592, 2024.
https://doi.org/10.1007/978-3-031-63378-2_96

with expertise from design and simulation of photonic devices and circuits, to prototyping in our in-house cleanroom, allowing for the customization and fabrication of photonic devices according to application-specific requirements.

Within this manuscript, we demonstrate the viability of our technology by fabricating and experimentally characterizing waveguide-based cutback structures and ring resonator filters in silicon nitride (Si_3N_4 and Si_xN_y) and tantalum pentoxide (Ta_2O_5) by using optical lithography and different deposition techniques. Measured waveguide attenuation of structures fabricated in the initial fabrication runs show a good match to the state-of-the-art.

2 Technology

The dielectric photonic platform presented in this work utilizes the infrastructure of the Fraunhofer Institute for Microelectronic Circuits and Systems (IMS) clean room facility, which has been extensively used for microelectronic purposes. Automatically handled cassette-to-cassette qualified processes on 8″ wafers are available in 2400 m^2 of cleanroom facilities, where standard CMOS technology has been developed for almost 40 years. The last research highlights were silicon single-photon detectors (SPADs) with world record lowest dark count rate [10], which find their use in quantum photonic applications.

The fabrication steps of our flexible photonic platform start at the deposition of the bottom oxide (BOX) SiO_2 on silicon wafers. This can be done by thermal growth or via plasma-enhanced chemical vapor deposition (PECVD) processes. Then the waveguide layer is deposited, which can be done by low-pressure chemical vapor deposition (LPCVD), PECVD, or by atomic layer deposition (ALD). To meet different application requirements, to date, we have three different types of silicon nitride waveguide materials with controllable thicknesses of up to 800 nm that can be applied on top of the bottom oxide. In addition, we can also use tantalum pentoxide (Ta_2O_5) as waveguide material. The waveguide structures are defined using deep ultraviolet lithography (DUV) at a wavelength of 248 nm, with a minimum feature size of approximately 250 nm. After the lithographic definition of the waveguides and developing, the structure transfer is realized via reactive ion etching (RIE) processes. The etching processes are optimized in order to have smooth sidewalls and a rectangular waveguide profile. On top of the waveguide structures a SiO_2-cladding is deposited via a PECVD process. Thermal-optical heaters can be defined on top of the cladding layer.

The proposed photonic platform at IMS also enables the use of electronic circuitry underneath the photonic layers (post-CMOS), see Fig. 1. Starting with a standard CMOS wafer, a SiO_2 BOX layer is deposited and planarized. Vias allow for contacting of the electronic circuits or for interconnects between different electronic layers, e.g., the underlying electronic chip and heater structures implemented above. Post-CMOS integration however, is only possible, if solely CMOS-compatible low temperature deposition techniques are used. In Fig. 1(b) a waveguide layer with ring resonators on top of a CMOS circuit is depicted. Both layers stay functional after the waveguide integration processes.

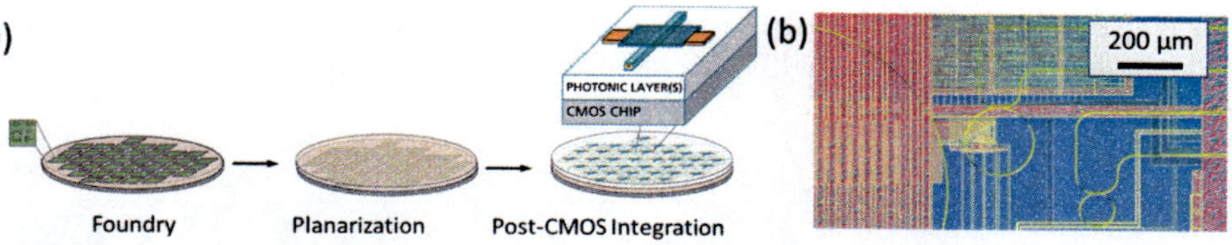

Fig. 1. Post-CMOS integration of photonic waveguides. **(a)** A CMOS (foundry) wafer is planarized and covered by bottom oxide. On top of this layer, the photonic waveguides can be integrated. **(b)** Optical microscope image (taken with a polarization filter) of CMOS circuits (rectangular lines) with photonic ring resonators on top.

3 Fabrication and Experimental Characterization

To demonstrate the feasibility of our technology, we fabricate photonic waveguides by depositing four different waveguide materials using three different deposition techniques: stoichiometric Si_3N_4 via LPCVD, Si_xN_y via PECVD, optimized for visible wavelengths Si_xN_y via PECVD (PE-VIS), and Ta_2O_5 via ALD – schematic cross-sections of these waveguides are shown in Fig. 2. To quantify propagation losses of the waveguides, we fabricate cutback structures with waveguide lengths varying from 2 cm to 17 cm. In addition, we fabricate ring-resonator filters, see Fig. 4a.

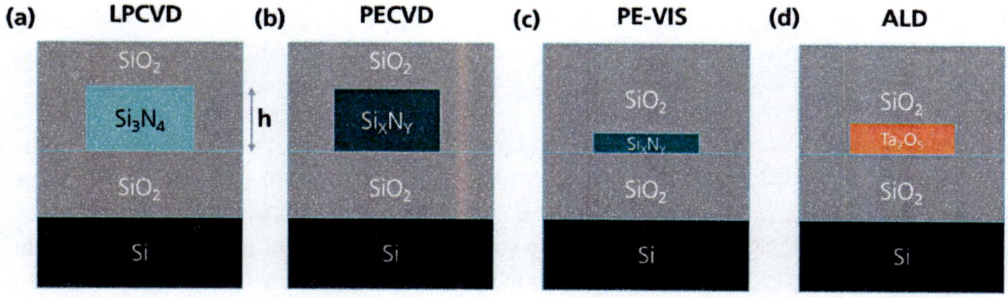

Fig. 2. Schematic cross-sections of the waveguides studied in this paper. On top of a Si substrate, a SiO_2 bottom oxide (BOX) of 2.7 μm thickness is deposited. Four different waveguide material layers with a layer thickness h were deposited on top: **(a)** $h = 360$ nm Si_3N_4 via LPCVD, **(b)** $h = 360$ nm Si_xN_y via PECVD, **(c)** $h = 100$ nm of optimized for visible wavelengths Si_xN_y via PECVD (PE-VIS) and **(d)** $h = 200$ nm of Ta_2O_5 via ALD. The waveguides are then lithographically patterned and subsequently covered with 1 μm of SiO_2-cladding.

We experimentally characterize the fabricated devices in the C-band, as well as at three individual wavelengths in the visible spectrum. To characterize the devices in the C-band, we use a tunable laser source, while in the visible spectrum we use three laser diodes emitting at vacuum wavelengths of 783 nm, 633 nm, and 405 nm. In all instances, the laser sources are used to generate continuous-wave (CW) light, that is coupled to an input single-mode fiber (SMF) and launched to the fabricated devices under test (DUT) via edge-couplers or grating couplers – after propagating through the DUT, the light is probed by an output fiber, see Fig. 3. For the output fiber, we used a multi-mode fiber (MMF) for measurements in the C-band and a SMF for measurements at visible wavelengths. Finally, the output fiber is connected to an optical spectrum analyzer (OSA) or to an optical power meter.

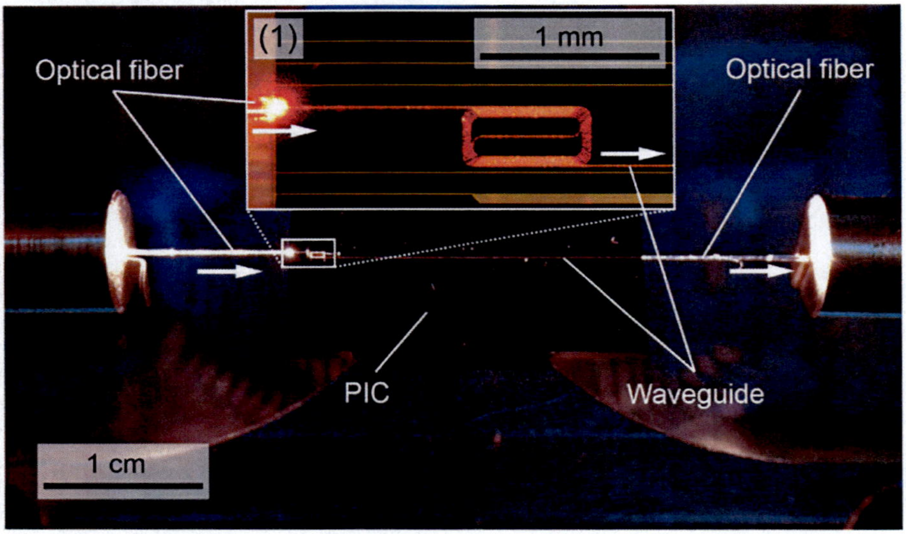

Fig. 3. Photograph of edge coupling into and out of a cutback structure with light of a wavelength of 633 nm. **Inset (1):** Microscope image of the cutback structure and coupling region.

The waveguide attenuation is obtained by measuring the transmission through cutback structures of different waveguide lengths – the results are presented on top of Table 1. For the LPCVD Si_3N_4 and the PECVD Si_xN_y we provide attenuation values at 1550 nm and at 1580 nm. We observe a minimum attenuation near 1580 nm, while at wavelengths near 1550 nm the attenuation is elevated – we attribute this behavior to an increased absorption on N-H and Si-H bonds near 1550 nm [11]. Attenuation for LPCVD stoichiometric Si_3N_4 is smaller than for PECVD Si_xN_y. For waveguides patterned in the PE-VIS Si_xN_y, which are optimized for applications in the visible spectrum, with a cross-section $w \times h = 800 \times 100\ \text{nm}^2$, the attenuation is approximately 1 dB/cm at 405 nm and 633 nm, and <3 dB/cm at 783 nm. If the cross-section area is increased to $w \times h = 800 \times 360\ \text{nm}^2$, the waveguide is multi-mode at 783 nm, but due to reduced scattering losses on waveguide side walls, the attenuation drops to 2.1 dB/cm. If LPCVD is used with this cross-section, the attenuation at 783 nm additionally drops to 1.9 dB/cm. Finally, we characterize Ta_2O_5 waveguides deposited by ALD – we fabricated waveguides with two different widths, single-mode with a 1000 nm width and multi-mode with a 1600 nm width. Both waveguide types show an attenuation of approx. 1 dB throughout the C-band, with multimode waveguides having slightly smaller attenuation due to reduced scattering losses on the side walls.

To benchmark our photonic platform, in Table 1 we include measured attenuation of silicon nitride waveguides reported in the literature by other foundries. The results

Table 1. Comparison of Fraunhofer IMS photonic platform to state-of-the-art silicon-nitride foundries.

Platform	Waveguide material*	Thickness h [nm]	Width w [nm]	Wavelength [nm]	Attenuation [dB/cm]
Fraunhofer IMS (this work)	LP PE	360 360	800 800	1550 / 1580 1550 / 1580	0.62 / 0.46 2.64 / 1.19
	PE-VIS PE-VIS LP	100 360 360	800 800 800	405 / 633 / 783 783 783	1.03 / 1.2 / 2.9 2.1 1.9
	ALD Ta2O5 ALD Ta_2O_5	200 200	1000 1600	1520–1620 1520–1620	<1.15 <1.0
AMF [5, 6]	LP PE PE	400 400 135	N/A N/A 520	1550 / 1580 1550 / 1580 430–648	<0.42/ <0.22 <3.01/ <1.25 <4
imec [7]	PE	180	700	532 / 900	<1
CORNERSTONE [8, 9]	LP PE	300 300	N/A 1200	C-Band 1550	<0.6 1.5
CEA-LETI [12]	LP	750	750	1550	1.5
Tower Semiconductor [13]	LP	24	800	405	0.93
Ligentec [14]	LP LP	150 150	N/A N/A	450/520 630/750	6.85/4.39 0.98/0.87
AIM Photonics [15]	PE	150	N/A	500 – 1000	<3.5

* LP – Silicon nitride deposited by LPCVD; PE – Silicon nitride deposited by PECVD; PE-VIS – Silicon nitride optimized for visible wavelengths deposited by PECVD; ALD Ta_2O_5 – tantalum pentoxide deposited by ALD

measured in the C-band are on par with results reported by AMF [5], CORNERSTONE [8, 9], and CEA-LETI [12] for similar waveguide geometries and for both LPCVD and PECVD deposited silicon nitride. At visible wavelengths, our platform exhibits superior performance compared to results reported by AMF [6] and AIM Photonics [15], on par with Tower Semiconductors [13] and imec [7], and mixed performance compared to Ligentec platform [14], which exhibits higher losses at lower wavelengths and lower losses at higher wavelengths of the visible spectrum. It should be noted that there are other silicon-nitride platforms that use more complex technological processes such as multiple etching, damascene, or planarization for obtaining advanced waveguide geometries with ultra-low waveguide attenuation at infrared and visible wavelengths [16–19], which we do not include in Table 1. Finally, the waveguides made in our first ALD Ta_2O_5 run show attenuation values in the C-band between those obtained by LPCVD and PECVD deposited silicon nitride waveguides, thereby demonstrating the viability of using ALD for expanding the platform by introducing novel materials for photonic applications.

In addition to characterizing simple waveguides, we also characterize ring-resonator filters (RRF) consisting of a ring-resonator coupled to a single bus waveguide by directional coupling. Figure 4(a) shows a scanning-electron-microscope (SEM) image of a RRF with a diameter of 100 μm, taken before applying the SiO_2 cladding layer. Figure 4(b) shows a resonance dip in the vicinity of 1550 nm. By applying a Lorentzian fit, we found an extinction ratio of 14.6 dB and the full width at half maximum (FWHM) of 24.5 pm, which determines the Q-factor of the RRF, Q = 63 300.

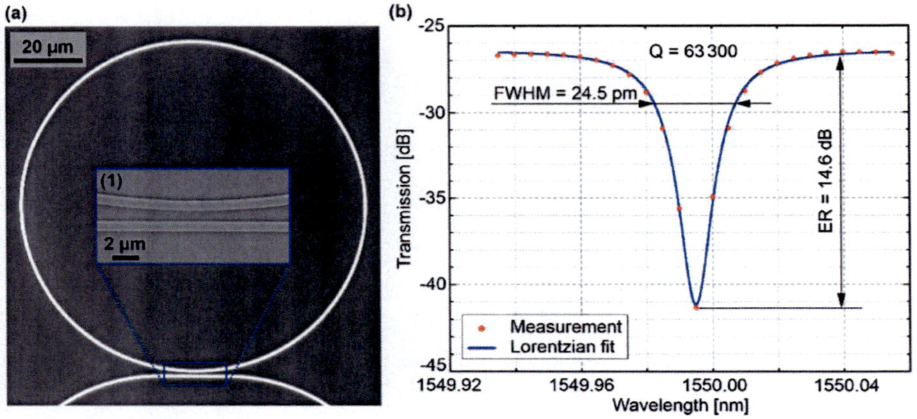

Fig. 4. Ring resonator filters realized in 360 nm thick PECVD silicon nitride. **(a)** SEM image of a resonator filter with a diameter of 100 μm. **Inset (1)**: Close-up of the coupling region. **(b)** Measured transmission near a resonance frequency of another ring resonator filter with a circumference of 13.14 mm – the transmission has been fitted with a Lorentzian fit to determine the extinction ratio (ER), the full width at half maximum (FWHM), and the Q-factor (Q).

4 Summary and Outlook

We demonstrated the viability of our photonic technology development service aimed at application-specific requirements. Experimental characterization in the C-band and at the visible wavelengths of silicon nitride waveguide-based photonic devices fabricated in initial fabrication runs show results comparable to state-of-the-art, with a potential for further optimization. In addition, we demonstrated Ta_2O_5 photonic waveguides, and therefore possibility of integrating more photonic materials. Furthermore, different deposition techniques can be chosen according to requirements, e.g., LPCVD that provides ultra-low-loss silicon nitride waveguides or post-CMOS compatible PECVD. Our platform leaves room for further expansion by, e.g., introducing additional novel photonic materials, or post-CMOS integration with SPADs [10] for highly sensitive sensing applications or for quantum applications. The platform is ready to use for electronic-photonic co-integration.

Acknowledgement. This work was supported by "terahertz.NRW". The project "terahertz.NRW" is receiving funding from the programme "Netzwerke 2021", an initiative of the Ministry of Culture and Science of the State of Northrhine Westphalia. The sole responsibility for the content of this publication lies with the authors.

References

1. Romero-García, S., Merget, F., Zhong, F., Finkelstein, H., Witzens, J.: Silicon nitride CMOS-compatible platform for integrated photonics applications at visible wavelengths. In: Optics Express, vol. 21, pp. 14036–14046 (2013)
2. Tan, D.T.H., Ooi, K.J.A., Ng, D.K.T.: Nonlinear optics on silicon-rich nitride—a high non-linear figure of merit CMOS platform. In: Photonics Research, vol. 6, no. 5, pp. B50-B66 (2018)
3. Kippenberg, T.J., Holzwarth, R., Diddams. S.A.: Microresonator-based optical frequency combs. In: science, vol. 332, no. 6029, pp. 555–559 (2011)
4. Poon, J.K., Govdeli, A., Sharma, A., Mu, X., Chen, F.D., Xue, T., Liu, T.: Silicon photonics for the visible and near-infrared spectrum. In: Advances in Optics and Photonics, vol. 16, no. 1, pp. 1–59 (2024)
5. Siew, S.Y., et al.: Review of silicon photonics technology and platform development. In: Journal of Lightwave Technology, vol. 39, no. 13, pp. 4374–4389 (2021)
6. Sacher, W.D., et al.: Visible-light silicon nitride waveguide devices and implantable neuro-photonic probes on thinned 200 mm silicon wafers. In: Optics express, vol. 27, no. 26, pp. 37400–37418 (2019)
7. Subramanian, A.Z., et al.: Low-loss singlemode PECVD silicon nitride photonic wire waveguides for 532–900 nm wavelength window fabricated within a CMOS pilot line. In: IEEE Photonics Journal, vol. 5, no. 6, p. 2202809 (2013)
8. Cornerstone Design Rules. https://www.cornerstone.sotonfab.co.uk/design-rules/. Accessed 21 Feb 2024
9. Bucio, T.D., et al.: Material and optical properties of low-temperature NH3-free PECVD SiNx layers for photonic applications. In: Journal of Physics D: Applied Physics, vol. 50, no. 2, p. 025106 (2016)
10. Grosse, S., Steuer, A., Stein, P.V., Zeidler, C., Haase, J.F.: A 64 × 48 BSI SPAD sensor based on 8" wafer 3D stacking technology. In: Sensor and Measurement Science International 2021 (2021)
11. de Ridder, R.M., Warthoff, K., Driessen, A., Lambeck, P.V., Albers, H.: Silicon oxynitride planar waveguiding structures for application in optical communication. In: IEEE Journal of Selected Topics Quantum Electronics, vol. 4, no. 6, pp. 930–937 (1998)
12. Wilmart, Q., et al.: A versatile silicon-silicon nitride photonics platform for enhanced functionalities and applications. In: Applied Sciences, vol. 9, no. 2, p. 255 (2019)
13. Morin, T.J., et al.: CMOS-foundry-based blue and violet photonics. In: Optica, vol. 8, no. 5, pp. 755–756 (2021)
14. Smith, J.A., Francis, H., Navickaite, G., Strain, M.J.: SiN foundry platform for high performance visible light integrated photonics. In: Optical Materials Express, vol. 13, no. 2, pp. 458–468 (2023)
15. Tyndall, N.F., Pruessner, M.W., Fahrenkopf, N.M., Antohe, A., Stievater, T.H.: A visible-light foundry platform from AIM photonics. In: 2023 Optical Fiber Communications Conference and Exhibition (OFC), pp. 1–2. IEEE (2023)
16. Wörhoff, K., Heideman, R.G., Leinse, A., Hoekman, M.: TriPleX: a versatile dielectric photonic platform. In: Advanced Optical Technologies, vol. 4, 189–207 (2015)

17. Pfeiffer, M.H.P., et al.: Photonic damascene process for low-loss, high-confinement silicon nitride waveguides. In: IEEE Journal of Selected Topics in Quantum Electronics, vol. 24, no. 4, pp. 1–11 (2018)
18. Chauhan, N., Isichenko, A., Wang, J., Blumenthal, D.J.: Sub-dB/m loss integrated 103 and 90 million Q resonators for laser stabilization at rubidium and strontium wavelengths. In: 2022 IEEE Photonics Conference (IPC), pp. 1–2. IEEE (2022)
19. Chauhan, N., Isichenko, A., Liu, K., Blumenthal, D.J.: Visible 780 nm SBS laser with mW level threshold in an ultra-high 145 million Q integrated waveguide resonator. In: 2023 Conference on Lasers and Electro-Optics (CLEO), pp. 1–2. IEEE (2023)

Author Index

J. Witzens et al. (Eds.): ECIO 2024, SPPHY 402, pp. 593–598, 2024.
https://doi.org/10.1007/978-3-031-63378-2